T0255746

THE ROUTLEDGE HANDBOOK OF GLOBAL DEVELOPMENT

This Handbook provides a comprehensive analysis of some of the world's most pressing global development challenges – including how they may be better understood and addressed through innovative practices and approaches to learning and teaching.

Featuring 61 contributions from leading and emerging academics and practitioners, this multidisciplinary volume is organized into five thematic parts exploring: changes in global development financing, ideologies, norms and partnerships; interrelationships between development, natural environments and inequality; shifts in critical development challenges, and; new possibilities for positive change. Collectively, the Handbook demonstrates that global development challenges are becoming increasingly complex and multi-faceted and are to be found in the Global 'North' as much as the 'South.' It draws attention to structural inequality and disadvantage alongside possibilities for positive change.

The Handbook will serve as a valuable resource for students and scholars across multiple disciplines including Development Studies, Anthropology, Geography, Global Studies, Indigenous and Postcolonial Studies, Political Science, and Urban Studies.

Kearrin Sims is a lecturer in Development Studies at James Cook University, Australia.

Nicola Banks is a senior lecturer in Global Urbanism and Urban Development at the Global Development Institute, University of Manchester.

Susan Engel is an associate professor in Politics and International Studies at the University of Wollongong, Australia.

Paul Hodge is a senior lecturer in Geography and Environmental Studies at The University of Newcastle, Australia.

Jonathan Makuwira is a professor in Development Studies and Deputy Vice Chancellor of Malawi University of Science and Technology.

Naohiro Nakamura is a senior lecturer in Geography at the University of the South Pacific, Fiji.

Jonathan Rigg is a professor in Geography at the University of Bristol, UK.

Albert Salamanca is a senior research fellow at the Stockholm Environment Institute's Asia Centre, Thailand.

Pichamon Yeophantong is a senior lecturer at the University of New South Wales, Canberra.

"This path-breaking Handbook moves thinking from its conventional 'international development' approach to a genuinely 'global development' framing. Drawing on contributions from a diverse and broad-based set of authors (not just the usual suspects) it examines today's big issues – sustainability and inequality – and explores the war of ideas that is needed if we are to reimagine and redirect human and planetary futures. The Handbook's chapters powerfully critique the retroliberalism that shapes contemporary policy and action and introduce the reader to emancipatory and transformative ways of understanding global problems and changing what individuals, communities, businesses and states can do. This is a 'must-have-on-my-bookshelf' publication."

David Hulme, Professor of Development Studies at the University of Manchester, Executive Director of the Global Development Institute; CEO of the Effective States and Inclusive Development Research Centre, UK

"Gradual changes in the political economy of the global order and the unprecedented increase in climate, health and biodiversity risk demands a collective rethink of the fundamentals of international development. In this watershed contribution, that not only distils problems of the current development machine but charts new ways ahead, the Handbook of Global Development is provocative and inspiring. Drawing from a new generation of development leadership and foregrounding fresh voices from the across the world, the book breaks new ground by setting out new modes of thinking supranationally, alternative ways of acting on transnational grand challenges and lays out innovative teaching approaches that, taken together, reshape the paradigm of global connections and challenges."

Susan Parnell, Global Challenges Research Professor in the School of Geography at the University of Bristol and Emeritus Professor at the African Centre for Cities (ACC) at the University of Cape Town

"This is a timely and invaluable handbook for anyone working in global development, or anyone wishing to. The contexts, actors, narratives, and challenges shaping development are constantly changing. It is incumbent on all of us – from students to the more experienced – to continually consider our own practice and positionality. Are we really 'doing good' for the world's poorest and most disadvantaged? How can we do better? By taking a wide-ranging, multi-disciplinary approach, and explicitly addressing critical cross-cutting issues such as climate change, inequality and population growth, the chapters in this volume provide a rich resource to guide ongoing reflection and learning on these difficult questions. I can't recommend it highly enough."

Praveena Gunaratnam, DrPH, Global Public Health Specialist and Human Rights Activist

"This book is a valuable guide through a range of pressing issues for policy experts and students alike, who are grappling with the future of development from within and beyond the sector. Its established and emerging authors explore big questions like what to make of deglobalisation, changing donor systems and aid chains, and whether neoliberalism is really dead or just evolving. The book is a must-read for practitioners and scholars aiming to keep ahead of global trends, like the future of development finance and sustainable development."

Dr Amrita Malhi, Senior Advisor Geoeconomics, Save the Children

"*The Routledge Handbook of Global Development* stands poignantly at the cutting-edge of new thinking on challenges, prospects, possibilities, and desired development futures. Contributors have done a splendid job in bringing to the fore of academia and public policy the most recent challenges of Anthropocene and existentialism, extractivism and violence of development, migration and Covid-19, decolonization and many other topical themes; opening up important epistemological questions in the field of global development. This is a must-read Handbook and resource for scholars and policymakers alike, which fundamentally refreshes and nourishes the mind of all those who care to know the state of the world we live-in."

Professor Dr. Sabelo J. Ndlovu-Gatsheni, Chair of Epistemologies of the Global South, Faculty of Humanities and Social Sciences, Africa Multiple Cluster of Excellence, University of Bayreuth

"This Handbook provides an invaluable resource for all those concerned with contemporary global challenges. It goes beyond the usual description of the world's problems to address head on the ways in which these can be addressed through pedagogy, policy and practice. Importantly, in making a critical intervention into a field that is currently in flux it reveals shifting geographies of power and global relations. This truly international and interdisciplinary volume includes contributions from leading scholars in their field that illuminate the multiple influences and dynamics of contemporary development thinking and practice. It moves beyond despondency, to provide innovative and more hopeful engagements with global concerns, ones that can work towards advancing more equitable and sustainable futures. This Handbook encourages us to reflect more deeply on the ideologies and practices that have for so long characterised international development and development studies."

Professor Uma Kothari, Professor of Migration and Postcolonial Studies, Global Development Institute, University of Manchester, UK

THE ROUTLEDGE HANDBOOK OF GLOBAL DEVELOPMENT

*Edited by Kearrin Sims, Nicola Banks, Susan Engel,
Paul Hodge, Jonathan Makuwira, Naohiro Nakamura,
Jonathan Rigg, Albert Salamanca, and
Pichamon Yeophantong*

LONDON AND NEW YORK

Cover image credit: Getty images

First published 2022
by Routledge
2 Park Square, Milton Park, Abingdon, Oxon OX14 4RN

and by Routledge
605 Third Avenue, New York, NY 10158

Routledge is an imprint of the Taylor & Francis Group, an informa business

© 2022 selection and editorial matter, Kearrin Sims, Nicola Banks, Susan Engel, Paul Hodge, Jonathan Makuwira, Naohiro Nakamura, Jonathan Rigg, Albert Salamanca and Pichamon Yeophantong; individual chapters, the contributors.

The right of Kearrin Sims, Nicola Banks, Susan Engel, Paul Hodge, Jonathan Makuwira, Naohiro Nakamura, Jonathan Rigg, Albert Salamanca and Pichamon Yeophantong to be identified as the authors of the editorial material, and of the authors for their individual chapters, has been asserted in accordance with sections 77 and 78 of the Copyright, Designs and Patents Act 1988.

With the exception of Introduction, no part of this book may be reprinted or reproduced or utilised in any form or by any electronic, mechanical, or other means, now known or hereafter invented, including photocopying and recording, or in any information storage or retrieval system, without permission in writing from the publishers.

Introduction of this book is available for free in PDF format as Open Access at www.taylorfrancis.com. It has been made available under a Creative Commons Attribution-Non Commercial-No Derivatives 4.0 International License.

Trademark notice: Product or corporate names may be trademarks or registered trademarks and are used only for identification and explanation without intent to infringe.

British Library Cataloguing-in-Publication Data
A catalogue record for this book is available from the British Library

Library of Congress Cataloging-in-Publication Data
A catalog record has been requested for this book

ISBN: 978–0–367–86202–2 (hbk)
ISBN: 978–1–032–15734–4 (pbk)
ISBN: 978–1–003–01765–3 (ebk)

DOI: 10.4324/9781003017653

Typeset in Bembo
by Newgen Publishing UK

The Open access version of Introduction was funded by University of Manchester.

CONTENTS

Contents

Contents

Contents

Contents

Contents

FIGURES

TABLES

CONTRIBUTORS

Ashraful Alam is a lecturer in planning and environmental management at the University of Otago. His research is at the intersection of development, planning and urban geography. He is particularly interested in migrant politics of housing, home, and homelessness in Bangladeshi and Australian cities.

Sara N. Amin is senior lecturer and coordinator of Sociology at the University of the South Pacific. Her research focuses on changing gender relations; identity politics, violence, and security; migration; and education. She has three ongoing projects: Innovations in Policing Domestic Violence in the Pacific, Religion and Policing in the Pacific and Changing Gender Relations in Families in South Asia. Her research has been funded by The Australian Research Council, the British Academy, the Global Religion Research Initiative, the Ford Foundation and the Social Sciences and Humanities Research Council of Canada. She recently coedited the volume *Mapping Security in the Pacific* (Routledge, 2020).

Edo Andriesse is an Associate Professor at the department of geography, Seoul National University. He teaches a wide range of human geography courses at undergraduate and graduate levels. His research interest is rural development in Southeast Asia through the lenses of economic geography and development geography (environmental pressures, poverty issues, and local politics), focusing on both fisherfolk and farmers. He has fieldwork experience in Thailand, Malaysia, Lao PDR, Indonesia, and the Philippines. He has also actively contributed to the internationalization of Seoul National University.

Hurriyet Babacan is the Research Director for Rural Economies Centre of Excellence at the Cairns Institute, James Cook University. She has a distinguished career over the last 25 years in senior roles in higher education and the public sector. Her work focuses on inclusion, economic and social development, gender, policy, and governance.

Nicola Banks is a senior lecturer in Global Urbanism and Urban Development at the Global Development Institute, University of Manchester. Her research explores issues of urban poverty, employment, and livelihoods, with a specific focus on the influence of these on young people's development. Alongside this, she has a keen research interest in development NGOs.

Her research here has critiqued development NGOs and mapped the UK's development NGO sector, and she is interested in how development NGOs can play a catalytic, transformative role in a more equitable system of international aid.

Tanja Bastia is a reader at the Global Development Institute at the University of Manchester. She is author of *Gender, Migration and Social Transformation: Intersectionality in Bolivian Itinerant Migrations* (Routledge, 2019), co-editor (with Ronald Skeldon) of the *Routledge Handbook of Migration and Development*, and sole editor of *Migration and Inequality* (2013, Routledge).

Joseph Besigye Bazirake is a postdoctoral fellow at the Research Chair for Critical Studies in Higher Education Transformation (CriSHET) at Nelson Mandela University in South Africa. His research interests revolve around institutional change, higher education and peacebuilding theories of change.

Walden Bello is an international adjunct professor of sociology at the State University of New York at Binghamton and co-chairperson of the Bangkok-based research and advocacy institute Focus on the Global South. He is the author or co-author of 25 books, with recent additions including *Counterrevolution: The Global Rise of the Far Right* (Nova Scotia: Fernwood, 2019), and *Paper Dragons: China and the Next Crash* (London: Bloomsbury/Zed, 2019).

Lucy Benge holds a Master of Arts in Development Studies from the University of Auckland. Her research interests include disaster-induced migration and displacement, climate change adaptation, disaster recovery and community-centred approaches to risk reduction and education.

Rebecca Bilous is a human geographer whose research opens up spaces for engaging with a multiplicity of knowledges and challenging both learners and educators to change the ways in which they tell, hear, teach and learn from stories. At Macquarie University she worked with PACE international partners to co-create curriculum resources that would better prepare students for their PACE activities with international community development organizations.

Michelle Bishop is a Gamilaroi woman from Western NSW and has been grown up on Dhawaral Country in south-west Sydney. She is an associate lecturer in the School of Education at Macquarie University, Australia, and is completing a PhD which envisions Indigenous education sovereignty.

Patrick Bond is a Professor at the University of Johannesburg Department of Sociology. He has also taught political economy and political ecology at Western Cape, Witwatersrand KwaZulu-Natal and Johns Hopkins universities.

Johanna Brugman Alvarez is planning practitioner and lecturer at the University of Technology Sydney. She holds a PhD from The University of Queensland and a Master of Urban Development Planning from the Development Planning Unit, UCL. Her practice and research aim to contribute towards the construction of more socially just cities through the inclusion of diverse knowledges and collective action in urban and regional planning processes and instruments. She has learnt from and supported urban poor communities in Cambodia and other countries in Southeast Asia to organise and develop strategies to access land and housing together with local partners and organisations.

Badru Bukenya is a Uganda-based academic, policy analyst and development practitioner. He is currently a senior lecturer in the Department of Social Work and Social Administration, Makerere University Kampala. He is also a research associate at ESID, the University of Manchester. His research interests include civil society; the politics of service delivery; state and citizenship-building; and governance of natural resources.

Feifei Cai is a PhD candidate in the School of Humanities and Social Sciences at the University of New South Wales (UNSW), Canberra. She was previously acting director and researcher at the Beijing-based NGO Social Resources Institute. Her research interests include the role of companies in sustainable community development, business-civil society conflicts, and Chinese investment policy.

Domenica Gisella Calabrò holds a PhD in Cultural Anthropology and is currently lecturer and coordinator of Gender Studies at the University of the South Pacific. Her research interests include gender constructions and relations, indigeneity, sport, migration, and violence in the Pacific region. She has recently coedited the volume *Sport, Migration, and Gender in the Neoliberal Age* (Routledge 2020).

Jenny Cameron is Conjoint Associate Professor in the discipline of Geography and Environmental Studies at the University of Newcastle, Australia. She is currently Deputy Chair of the Community Economies Institute. Throughout her academic career, she has conducted action research with communities in a range of contexts.

Aidan Craney is a development scholar, anthropologist, and social worker. His research looks at youth civic engagement and livelihoods in Oceania, and the practical and philosophical challenges for aid donors in supporting locally led development practices. Aidan has worked with development initiatives in the Pacific region on how to integrate local values systems into developmental reforms and advised youth activists in Australia and the Pacific on thinking and working politically. He holds a PhD in Anthropology from La Trobe University.

Mia Chung is senior research manager of the East-West Management Institute's Open Development Initiative. She is a social justice-oriented research and advocacy professional working in the Mekong Region. Mia holds a JD from Dalhousie University's Schulich School of Law and a BA in gender and family studies from the University of British Columbia.

Pyrou Chung is director of the East-West Management Institute's programs on natural resource, land and data initiatives in Southeast Asia under the Open Development Initiative. Her work focuses on the nexus between environmental governance, conservation and human rights, with an emphasis on implementing integrated technical solutions with indigenous, forest dependent communities.

Lisa Denney is a senior research fellow at the Institute for Human Security and Social Change at La Trobe University and a research associate with the Overseas Development Institute. Her work focuses on the politics of development and the need for external actors to operate in more politically astute, locally led ways that respond to and generate learning. Lisa has worked extensively with donors, non-government organizations and the private sector in Asia, the Pacific and sub-Saharan Africa on implementing politically-informed and adaptive development programmes. Lisa has a PhD in International Politics from Aberystwyth University.

Poonam Pritika Devi is a PhD student in geography at the University of the South Pacific, based on Laucala campus, Fiji. Her doctoral project is examining the effectiveness of squatter settlement upgrading plans in Fiji.

Kevin Gavi Duncan is an Awaba, Gomeroi and Mandandanji custodian who has over 40 years' experience working with students in schools, TAFE and universities to teach culture and connections to the land. Gavi is also a musician and dancer and has performed across Australia in numerous cultural festivals. Gavi is currently senior Tourism Education Officer with the Darkinjung Local Aboriginal Land Council (DLALC).

Lorraine Elliott is Professor Emerita in International Relations at the Coral Bell School, Australian National University. Her research focuses on global governance and human security, transnational environmental crime, regional environmental governance in Southeast Asia, environmental security, climate security and human security. She chaired the International Board of the Academic Council on the UN System (2015–2018).

Susan Engel is an associate professor in Politics and International Studies at the University the Wollongong, Australia. She has a book with AR Bazbauers on *The Global Architecture of Multilateral Development Banks: A System of Debt or Development?* (Routledge, 2021) and a 2010 book on the World Bank plus over 30 articles and book chapters. Her main research interests are: multilateral development finance, theories and practices of development, the role of emotions in development interventions, and teaching and learning.

Rachel Etter-Phoya is a senior researcher with the Tax Justice Network, based in Lilongwe, Malawi. Prior to this, Rachel worked in civil society and with government in Malawi in researching and improving the management of natural resources. She holds a Master of Law in Natural Resources Law and Policy (University of Dundee, UK), a Master of Science in International Rural Development (Royal Agricultural University, UK), and a Bachelor of Arts in Anthropology and English (University of Zurich, Switzerland).

Aqil Teguh Fathani is a research fellow at the Jusuf Kalla School of Government, Universitas Muhammadiyah Yogyakarta (UMY), Indonesia. His research focuses on sustainable policy, sustainable energy, and energy policy.

Bina Fernandez is an associate professor in Development Studies at the University of Melbourne, Australia. Her research focuses on gender, migration, and social policy, and illuminates how a feminist analysis of social reproduction is critical to the understanding of poor and ethnic minority women's precarious access to resources, and consequently to the construction of policy interventions to improve their access to and control over such resources. Key publications include the monographs *Ethiopian Migrant Domestic Workers: Migrant Agency and Social Change* (2020) and *Transformative Policy for Poor Women: a new feminist framework* (2012).

Tom Gillespie is a Hallsworth research fellow in the University of Manchester's Global Development Institute. His research sits at the intersection of urban geography, political economy, and development studies, and he is interested in the relationship between capitalism, urban inequalities, and social change in the Global North and South.

Christian Girard is a lecturer in Management and Public Administration at the University of the South Pacific. His research interests include the political economy of development, vulnerability, poverty, informality, housing, governance, public policy, and social entrepreneurship and he has conducted research in Asia, Africa, Latin America and the Pacific. He is coeditor of the book *Mapping Security in the Pacific: A Focus on Context, Gender and Organizational Culture* (Routledge, 2020).

Narayan Gopalkrishnan is the course coordinator of Social Work and fellow of the Cairns Institute at James Cook University. He has worked over three decades in Australia and overseas in universities, NGOs, and the private sector. He has a deep passion for social justice and sustainable development practice.

Moran Harari completed her LLB at the Hebrew University of Jerusalem, and her LLM in human rights law at University College London (UCL). Alongside her role at the Tax Justice Network as a lead researcher of indices, she is also the founder and director of Tax Justice Network Israel. Before joining the Tax Justice Network, she worked for several years as a tax lawyer in Tel- Aviv and volunteered in human rights organizations. She also worked in the Corporate Social Responsibility fields both in London and Israel.

Vandra Harris Agisilaou teaches courses in applied ethics, humanitarian practice and child-focused development practice in the postgraduate International Development programme at RMIT University in Melbourne, Australia. With a background in early childhood development and community development, her research and teaching are centrally concerned with the ways diverse groups interact in humanitarian and development spaces, especially militaries, police, NGOs, and local actors.

Pascale Hatcher is a senior lecturer at the Department of Political Science and International Relations, University of Canterbury, New Zealand. Her research interests are focused on the political economy of extractivism in Asia.

Laura Hammersley is a human geographer at the University of Wollongong Australian Centre for Culture, Environment, Society and Space (ACCESS). Her PhD research critically examined the postcolonial practice of international service-learning and the foundational concept of reciprocity by focusing on the perspectives and experiences of Indigenous community-based organizations in the Northern Territory Australia, and Sabah, Borneo. Her recent work focuses on building an ethics of care in the context of international service-learning relationships and in centring the knowledges, experiences, and expertise of community-based partners in service-learning research, education and practice.

Jessica Hawkins is a lecturer in Humanitarian Studies at the Humanitarian and Conflict Response Institute at the University of Manchester. She has researched a historical sociology of state formation and development in Uganda and continues to research histories of state humanitarianism and displacement in northern Uganda. More recently, she has worked on pedagogies of humanitarianism and development.

Anna Hayes is a senior lecturer in International Relations in the College of Arts, Society and Education at James Cook University, Cairns. She is also an honorary research fellow at the East

Asia Security Centre. Anna specializes in non-traditional threats to security, with a particular focus on China. She has presented numerous papers in Beijing, on topics ranging from the situation in Xinjiang, how the BRI has been viewed outside of China, as well as the Quad and the Indo-Pacific from the Australian perspective. Anna has published numerous articles, book chapters and edited books on these topics. She is currently co-editing a book examining the Chinese in Papua New Guinea. In 2016 Anna coedited *Inside Xinjiang: Space, place and power in China's Muslim Far Northwest* (Routledge, 2016) with associate professor Michael Clarke from the Australian National University.

Philip Hirsch is Emeritus Professor of Human Geography at the School of Geosciences, University of Sydney. Since the 1980s he has written extensively on agrarian change and the politics of land, environment, and natural resource governance in Southeast Asia.

Paul Hodge is a senior lecturer in the discipline of Geography and Environmental Studies at the University of Newcastle (UON), Australia. The Indigenous-led collaborations Paul is part of, seek to nourish, support, and learn from more-than-human, inter-generational, and multi-temporal ways of being, knowing and doing in highly colonized semi-urban contexts. This research on and with Gumbaynggirr and Darug Countries tries to centre people and place in ways that foreground more-than-human relationality.

David Hudson is a professor of Politics and Development in the International Development Department, University of Birmingham, director of the Developmental Leadership Program (DLP) and co-director of the Development Engagement Lab (DEL). He has written widely on the politics of development, in particular on the role of coalitions, leadership and power in reform processes and how development actors can think and work politically as part of the Developmental Leadership Program; migration decision-making; and how people in rich countries engage with global development issues.

Heiner Janus is a researcher at the German Development Institute / Deutsches Institutfür Entwicklungspolitik (DIE) and holds a PhD from the University of Manchester. His research focuses on aid and development effectiveness, the role of rising powers in development cooperation and the 2030 Agenda for Sustainable Development. Previously, he has worked as a consultant for the United Nations Secretariat in New York in the Department of Economic and Social Affairs. He is an editor of the 2021 open access *Palgrave Handbook of Development Cooperation for Achieving the 2030 Agenda*.

Susie Jolly is a honorary associate at the Institute of Development Studies (IDS). She is an activist, freelance consultant, researcher, communicator, facilitator and trainer on gender and sexuality. She previously directed the Ford Foundation sexuality education portfolio in China and founded and convened the IDS Sexuality and Development programme.

Bernard Kelly-Edwards is a Gumbaynggirr, Bundjalung and Dhunghutti custodian and practitioner of cultural connections to the land through his work with BKE consultancy. Bernard works closely with individuals and groups to develop and maintain connection to themselves and loved ones with blessings from Gumbaynggirr Country through visual art and cultural photography. Along with his consultancy work Bernard works part time as a lecturer at the University of Newcastle.

Ilan Kelman (www.ilankelman.org and Twitter/Instagram @ILANKELMAN) is a professor of Disasters and Health at University College London, England and a professor II at the University of Agder, Kristiansand, Norway. His overall research interest is linking disasters and health.

Sophie King is Development Manager for Community Led Action and Savings Support (CLASS) – the support agency for the UK-based Community Savers movement (www.communitysavers.net). Sophie is also honorary research fellow at the School of Environment, Education and Development, University of Manchester; and visiting practice fellow at the Urban Institute, University of Sheffield.

Leakhana Kol is a researcher based in Phnom Penh with experience conducting research and consultancies in the topics of community development, housing rights, land, natural resource management, disaster risk response, and child protection in Southeast Asia.

Krishna Kumar Kotra did his Masters' and PhD in Chemistry at Andhra University, India. He is currently working as lecturer at The University of the South Pacific's Emalus Campus, Vanuatu. His specializations include water quality, WASH and climate change, and currently involved in multiple collaborative research studies both as lead and co-investigator.

Ujjwal Krishna is a specialist doctoral research scholar with the Developmental Leadership Program (DLP) at the Institute for Human Security and Social Change, La Trobe University. His PhD research on the political economy of development policy involves working with the Australian Government's Department of Foreign Affairs and Trade (DFAT) on its aid investments in leadership and coalitions. He was previously with the Indian Council for research on International Economic Relations (ICRIER), where he supported the Government of India's digital development interests. Ujjwal holds an MA in Development Studies (with Distinction) from the Institute of Development Studies (IDS), University of Sussex.

Aarti Krishnan is a research fellow at the Global Development Institute, at the University of Manchester. She is an economist, working at the nexus between environmental, trade and development economics, with experience as a commodity derivate market analyst. Her areas of expertise include value chain analysis, green industrial policy, and sustainable digital development. Prior to Manchester, she worked as a researcher at the Overseas Development Institute and in investment banking. Her research is across a range of countries in Asia and Africa, as well a breadth of sectors including agriculture and light manufacturing.

Antonio G.M. La Viña is a human rights, sustainable development, and climate justice scholar and advocate. An LLM and JSD graduate of Yale Law School, he is currently senior fellow on climate change and Energy Collaboratory Director of Manila Observatory. He teaches philosophy, environmental law and policy, international negotiations, constitutional and other political law subjects, and public international law in universities of the Philippines, while also publishing books, journal articles, columns, and other journalistic pieces on these subjects.

Pranee Liamputtong is Professor of Behaviour Sciences at the College of Health Sciences, VinUniversity, Vietnam. She is a health–social science researcher with knowledge and skills in qualitative research methodologies and inclusive research methods. She has worked extensively with vulnerable/marginalized groups including trans women, refugees, and migrants in

Australia, and women in Asia in the areas of sexual and reproductive health, sexuality, breast cancer, HIV/AIDS.

Kate Lloyd is an associate professor in Geography and Planning at Macquarie University. Her research, learning and teaching experiences focus on Indigenous and Development Geographies, University-Community engagement, and practice-based learning. She takes an applied, action-oriented and collaborative approach to research and teaching characterized by community partnerships, co-construction of knowledges and an ethics of reciprocity. As a result, at the heart of her life-work are innovative co-authored and co-created projects with academic colleagues, community partners, students and Country that focus on learning, nurturing, healing and sensing.

Isaac Lyne has worked on various research projects and NGO programmes in Cambodia since 2008. He continues to work with Buddhism for Social Development Action (BSDA) in Cambodia on community-based economic initiatives and has taught international development course units at universities in Australia, Cambodia, and the United Kingdom. He is currently a postdoctoral research fellow at the Institute for Culture and Society, Western Sydney University.

Sango Mahanty is a human geographer and Associate Professor in the Department of Resources, Environment and Development at the Australian National University's Crawford School of Public Policy. She studies the politics of social and environmental change, having completed an Australian Research Council Future Fellowship on related themes along the Cambodia-Vietnam border.

Jack Makau is an administrator at SDI Kenya, heading up the NGO that provides professional and technical support to Muungano wa Wanavijiji, the Kenyan federation of slum dwellers. He has been a support professional for Muungano since 2000.

Jonathan Makuwira is a professor of Development Studies and Deputy Vice Chancellor of Malawi University of Science and Technology (MUST). He is a research associate in the Department of Development Studies at Nelson Mandela University (NMU), South Africa. He is also a visiting fellow at Airlangga University, Surabaya, Indonesia. His research is multidisciplinary as he researches the role of the NGO sector in development; the political economy of foreign aid; disability-inclusive development, and the role of education in development.

Danny Marks is an assistant Professor of Environmental Politics and Policy in the School of Law and Government of Dublin City University. He has spent a number of years conducting research and working in Southeast Asia, particularly in the field of environmental governance. His research interests are political ecology, environmental justice, climate governance, disaster risk reduction, with a focus on Southeast Asia.

Emma Mawdsley is a Professor in Human Geography and Fellow of Newnham College at the University of Cambridge. Her research is on the politics of global development, in particular South-South development cooperation, with a particular but not exclusive focus on India. A recent interest is how DAC donors are re-purposing their development narratives, tools, and agendas to a changing global landscape. This includes work on the domestic politics of aid in the UK, donor-middle income country development transitions, and re-theorizing the changing governance of development finance.

Linsey McGoey is a professor in the Department of Sociology at the University of Essex. Research interests include philanthropy (*No Such Thing as a Free Gift,* Verso, 2015) and was instrumental in developing the field of ignorance studies, where her key publications include *The Unknowers: How Strategic Ignorance Rules the World* (Zed Books, 2019) and *The Routledge International Handbook of Ignorance Studies* which she co-edited Matthias Gross (2015).

Andrew McGregor is an associate professor of Geography at Macquarie University. He is author of *Southeast Asian Development* (Routledge: 2008) and Lead Editor of *The Routledge Handbook of Southeast Asian Development* (2018). His research focuses on climate mitigation strategies within food and forestry sectors in Indonesia and Australia.

Thomas McNamara is a lecturer in development studies at La Trobe University, and the deputy director of the Master of International Development. He also holds an FNRS mandate (N. 34785392) at the University of Liège. His primary research interest is how Global South economies and aspirations are shaped by and guide affective relationships and moral norms, especially norms and narratives relating to 'development.'

Diana Mitlin is a professor of Global Urbanism in the Global Development Institute at the University of Manchester (www.sed.manchester.ac.uk/research/gurc). She is also research associate at the International Institute for Environment and Development, and editor of IIED's journal, Environment and Urbanization. Since 2020, Diana has been CEO of the FCDO-funded African Cities Research Consortium.

Markus Meinzer is the director of the Financial Secrecy and Governance workstream at the Tax Justice Network. He authored the book *Tax Haven Germany ('Deutschland')*, published in 2015 at C.H. Beck, and was the Tax Justice Network's principal investigator on the COFFERS EU research project under Horizon 2020 (Combating Fiscal Fraud and Empowering Regulators). He obtained his PhD in economics at Utrecht University ('Countering cross-border tax evasion and avoidance. An assessment of OECD policy design from 2008 to 2018').

Tolu Muliaina teaches courses in Human Geography at the University of the South Pacific's Laucala Campus, Suva, Fiji. He supervises local and external postgraduate students' research projects on education, development, migration, gender and change in the Pacific.

Warwick E. Murray is a professor of Human Geography in the School of Geography, Environment and Earth Sciences, Victoria University of Wellington, Aotearoa / New Zealand. He writes in the fields of development, rural, and economic geography, focusing especially on Chile and Latin America, as well as the Pacific Islands, the Asia Pacific, and New Zealand. His most recent book co-authored with colleagues is *Aid, Ownership and Development: The Inverse Sovereignty Effect in the Pacific Islands* (Routledge, 2018).

Naohiro Nakamura is a senior lecturer in geography at the University of the South Pacific, based on Laucala campus, Fiji. His research focuses on indigenous issues in Japan, Canada, and Fiji. Currently, he is striving for working with Pacific Island nation students.

Andreas Neef is a professor of Development Studies at the University of Auckland. His current research focuses on adaptation and resilience to climate change, climate-induced migration and displacement, post-disaster response and recovery, and land and resource grabbing, predominantly in Southeast Asia and the South Pacific.

Etienne Nel is a professor in the School of Geography, University of Otago, Dunedin, Aotearoa / New Zealand. He has written 12 books and over a hundred journal articles and chapters. Etienne is currently undertaking research in economic and urban history and development in South Africa, local development and urban agriculture in Zambia and regional, local and small-town development in New Zealand.

John Overton is a professor of Development Studies in the School of Geography, Environment and Earth Sciences, Victoria University of Wellington, Aotearoa / New Zealand. He has written a number of books and many chapters and journal articles. His two main research interests are, first aid and development with a focus on New Zealand and the Pacific Islands and second, the relations between wine, place, and capital. His most recent book is the co-authored *Aid and Development* (Routledge, 2021).

Miroslav Palanský is a postdoctoral researcher at the Institute of Economic Studies, Charles University, Prague, and a data scientist at the Tax Justice Network. Through his research, he aims to help in the ongoing fight against corruption, tax abuse and financial secrecy. He holds a Masters' and a PhD in Economics from Charles University and a Masters' in Econometrics from Aix-Marseille University.

Joseph Palis is an associate professor in Geography at the University of the Philippines-Diliman. He currently sits as director of the Third World Studies Center and project leader of the Geonarratives Mapping Project at UP Diliman. He co-manages and co-directs FilmGeographies.com and the Film Geography Specialty Group at the American Association of Geographers with Jessica Jacobs. Currently he co-edits the Palgrave Macmillan Pivot Series devoted to Geographies of Media with Torsten Wißmann. He actively serves as music DJ for WXYC-Chapel Hill whenever he is in North Carolina.

Eko Priyo Purnomo is a senior lecturer at Department of Government Affairs and Administration, Jusuf Kalla School of Government, Universitas Muhammadiyah Yogyakarta (UMY), Indonesia. He has two research projects, one focusing on political ecology, sustainability and green politics, and another on smart cities and social media. He has completed joint research and was a visiting professor at Korea University, University of Maryland College Park, Khonkaen University, and the University of Cambridge amongst others.

Jameela Joy M. Reyes is a lawyer and collaborator of Dean La Viña. Her practice is mostly in the fields of human rights and environmental law, and she currently works as the senior communicator at the Manila Observatory. She obtained her Juris Doctor degree from the University of the Philippines and her undergraduate degrees in Psychology and Political Science from the Ateneo de Manila University.

Zoe Sanipreeya Rice is an independent scholar currently working in London, UK. She graduated from Monash University and the University of Melbourne, in Australia. She has co-authored various book chapters on social inclusion, the social determinants of health, and research methods in the health and social sciences.

Jonathan Rigg is a professor of Human Geography based at the University of Bristol in the UK. He was formerly director of the National University of Singapore's Asia Research Institute. Jonathan's research focuses on agrarian change in the Asian region and he has undertaken fieldwork in Laos, Nepal, Sri Lanka, Thailand and Vietnam. He is the author of nine books, most recently *Rural development in Southeast Asia: dispossession, accumulation and persistence* (Cambridge University Press, 2020) and *More than rural: textures of Thailand's agrarian transformation* (Hawaii University Press, 2019).

Etienne Roy Grégoire is an assistant professor at Universit e du Qu e bec a Chicoutimi, Canada, and a member of the Centre for Indigenous Conservation and Development Alternative (CICADA). His research focuses on the global extractive sector as a vector of normative innovations and political transformations. He has conducted fi eldwork in Guatemala and Colombia. His current research focusses on internormative dynamics and their impact on Indigenous self- determination in Canada's extractive territories.

Kristian Saguin is an associate professor in the Department of Geography, University of the Philippines with teaching and research interests at the intersection of political ecology, urban studies, and agrarian change. He has published on resource geographies and politics surrounding aquaculture, fisheries, and food production in peri-urban environments in the Philippines.

Albert Salamanca is a senior research fellow at the Stockholm Environment Institute's Asia Centre where he leads its Climate Change, Disasters and Development cluster. He previously led SEI's global initiative on Transforming Development and Disaster Risk, the Regional Climate Change Adaptation Knowledge Platform (AKP) and the Partnership in Governance Transition: the Bali Cultural Landscape. He has a PhD in Geography from Durham University (UK). He is a co-editor of *Climate Change, Disasters, and Internal Displacement in Asia and the Pacific* (Routledge: 2020).

Kearrin Sims is a lecturer in Development Studies at James Cook University. He researches regional connectivity and South-South cooperation within Mainland Southeast Asia, with a focus on ethical development. His recent work examines the intersectional violence of large-scale infrastructures, political oppression, and development geopolitics. Kearrin is the author of numerous academic and media publications and a founding member of the Development Studies Association of Australia. More information about his work at: www.rethinkingdevelopment.net

Ronald Skeldon is an emeritus professor at the University of Sussex and an honorary professor at Maastricht University in The Netherlands. Trained at the University of Glasgow in Scotland and the University of Toronto in Canada, his career has spanned both academic and international civil service positions in Latin America, Asia/Pacific, and Europe.

Diana Suhardiman has over 15 years' experiences in natural resources governance with particular focus on water governance in Southeast Asia. She is currently research group lead governance and inclusion at the International Water Management Institute, based in Vientiane, Lao PDR. Her research puts power and politics central in natural resource governance across scales, emphasizing the need to better understand the scales interlinkages, from transboundary to local. Diana's most recent research looks at the scalar politics and power struggles in Myanmar's water governance and the politics of representation in hydropower decision-making in Laos.

Carolina Suransky is an associate professor of Social Change and Higher Education at the University of Humanistic Studies in the Netherlands and research associate at the Unit for Institutional Change and Social Justice at the University of the Free State in South Africa. In her teaching and research, she focuses on social change in a globalizing world, pluralism and decoloniality with a particular focus on the role of higher education in the pursuit of social – and ecological justice.

Youyenn Teo is an associate professor, Provost's Chair, and Head of Sociology at Nanyang Technological University, Singapore. Her research focuses on poverty and inequality, governance and state-society dynamics, gender, and class. She has published in journals such as *Economy and Society*, *Signs*, *Social Politics*, and *Development and Change*. She is the author of *Neoliberal Morality in Singapore: How family policies make state and society* (Routledge, 2011) and *This is What Inequality Looks Like* (Ethos Books, 2018). More information about her work at: https://teoyouyenn.sg

Stephanie M. Topp is an associate professor in the discipline of Public Health at James Cook University, Australia. Her research and teaching centres on explorations of health policy and health systems, and the way power, trust and accountability influence their form and function.

Éric Toussaint is a historian and political scientist who completed his PhD at the universities of Paris VIII and Liège. He is the spokesperson of CADTM International and sits on the Scientific Council of ATTAC France. He is the author of many books including *Debt System* (Haymarket books, 2019), *Bankocracy* (2015) and *Glance in the Rear View Mirror. Neoliberal Ideology From its Origins to the Present* (Haymarket books, 2012).

Laura Trajber Waisbich is a researcher at the South-South Cooperation Research and Policy Centre (Articulação SUL) and at the Brazilian Centre for Analysis and Planning (Cebrap) and PhD candidate in Geography at the University of Cambridge. She has worked at think-tanks and international non-governmental organizations based in Brazil, Belgium, and the United States. Her research interests lie at the crossroad of international development cooperation and South-South development cooperation, particularly Brazilian development cooperation and citizen participation and mobilization

Lauren Tynan is a trawlwulwuy woman from tebrakunna country in northeast Tasmania. She is a PhD candidate at Macquarie University in the Department of Geography and Planning. Her PhD focuses on relationality with Country, particularly through the teachings of fire and cultural burning.

Yvonne Underhill-Sem is a Cook Island New Zealander with close family ties to PNG. She is a Pacific feminist decolonial development geographer currently working at the University of Auckland. She has also taught at the University of Papua New Guinea, worked at the ACP Secretariat and been an independent scholar based in Samoa and Germany. Since 2002, she taught development studies before happily moving to Pacific Studies in 2020. The intricacies of mobilities, maternities and markets continue to catch her attention as she works to resist the erasure of diversity within colonial differences, while simultaneously finding epistemic strength in one's partial position.

Henry Veltmeyer is a research professor of Development Studies at the Autonomous University of Zacatecas, Mexico and annually engages in an extended programme of research and public lectures across Latin America. He conducts research, writes, and teaches about diverse issues related to the political economy and sociology of development, with a particular focus on issues of Latin American development, globalization processes, government policies, alternative models and approaches and social movements.

Penny Vera-Sanso is a senior lecturer at Birkbeck College, University of London. Her research into how age, class, gender, and caste shape livelihoods and inter-generational relations provides the basis for interrogating the ageism embedded in public, policy and disciplinary discourse.

Archana Preeti Voola is a social policy researcher and Adjunct Fellow with the School of Social Sciences at Western Sydney University, interested in interdisciplinary research on gender, marketplaces, poverty, and inequality. Her research is underpinned by her passion to be a voice for the voiceless, and is best encapsulated in the saying, 'Be the change that you want to see in the world.' Archana's research expertise is focused on ultra-poor women, financially excluded people, people experiencing food insecurity and domestic violence. Her interdisciplinary work is published in journals such as *Australian Journal of Social Issues*, *European Journal of Marketing*, *Australasian Marketing Journal*, *Third Sector Review*, *Social Business* and *The International Education Journal: Comparative Perspectives*.

Hilary Whitehouse is an educator, editor for the *International Journal of Environmental Education*, and a life member of the Australian Association for Environmental Education. She volunteers for the small, conservation NGO, the Bats and Trees Society of Cairns. A fellow of the Cairns Institute, James Cook University, Hilary has published on education for sustainability, climate change education, reef education, rainforest education and on ecofeminism and gender.

Sarah Wright is a professor and Future Fellow in geography and critical development studies at the University of Newcastle in Australia. Her research focus is on the geographies of food (particularly working with networks of small-scale farmers and Lumads in the Philippines) and in Indigenous geographies. She is part of the Yandaarra Collective, led by Gumbaynggirr Elder Aunty Shaa Smith, which aims to shift camp together towards Gumbaynggirr-led, decolonizing ways of caring for and as Country. She is also part of the Bawaka Collective and has had the privilege of working/living/loving with Yolngu grandmothers and other families for over 12 years.

Pichamon Yeophantong is an Australian Research Council fellow and senior lecturer at the University of New South Wales (UNSW), Canberra. A chief investigator on multiple research projects focused on strengthening environmental justice and governance in the Asia-Pacific, she is a recipient of the Australia Future Leader Prize (2018) from the Council for the Humanities, Arts and Social Sciences and was also recognized as a 'human rights fighter' by the Advance Awards.

Abitassha Az Zahra is a junior researcher at the Jusuf Kalla School of Government, Universitas Muhammadiyah Yogyakarta (UMY), Indonesia. His research interests include green politics, e-government, and social media.

INTRODUCTION

Kearrin Sims, Nicola Banks, Susan Engel, Paul Hodge,
Jonathan Makuwira, Naohiro Nakamura, Jonathan Rigg,
Albert Salamanca, and Pichamon Yeophantong

Following the emergence of 'development' as a global project some 75 years ago, many parts of the world have seen notable improvements in living standards. Poverty rates have declined, life expectancy has increased, formalized education has expanded, and new infrastructures have provided greater access to water, electricity and other social services. Yet alongside such 'good change' (Chambers 2004), the world has also witnessed new and enduring challenges, including rising socio-economic inequality, unsustainable resource depletion, new conflicts, democratic backsliding, and the rapid acceleration of anthropogenic climate change. Where parts of Asia have experienced exceptional economic growth and strong improvements in human development, in other regions such progress has been halting. Inequities and inequalities related to gender, race, ethnicity, place, religion, and other social categories remain pervasive, as do global health threats, more frequent and intense disasters, and a myriad of other challenges (UN 2015).

The above paragraph provides an unremarkable summary of where the globe is now in terms of 'development': the evidence is mixed, contested and geographically variegated. This, though, is also the nub of the matter. In 2019, Abhijit Banerjee and Esther Duflo won the Nobel Prize in Economics (along with Michael Kremer) for their 'experimental approach to alleviating global poverty.' In *Poor Economics: A Radical Rethinking of the Way to Fight Global Poverty*, they write:

> Economists (and other experts) seem to have very little useful to say about why some countries grow and others do not. Basket cases, such as Bangladesh or Cambodia, turn into small miracles. Poster children, such as Côte D'Ivoire, fall into the 'bottom billion.' In retrospect, it is always possible to construct a rationale for what happened in each place. But the truth is, we are largely incapable of predicting where growth will happen, and we don't understand very well why things fire up.
>
> (Banerjee and Duflo 2011, 267)

Depending where – and, importantly, how – we look, there is every reason to avoid being sanguine about the prospects for development. Across and within countries we see enormous disparities in wealth, opportunity, and power. Depending on the criteria used, billions of people continue to live in absolute poverty, with projections that up to 150 million additional people will be pushed into poverty during 2020–2021 as a result of the COVID-19 pandemic (World

DOI: 10.4324/9781003017653-1

This chapter has been made available under a CC-BY-NC-ND license.

Bank 2020). In addition to, and interrelated with poverty, 78 million people are affected by forced displacement globally (UNHCR 2020), and if the worst projections of global climate change are realized, the next few decades will bring widespread hunger, accelerating inequality and mass species extinction. Poverty persists, structural and systemic inequality are deepening in many places, and critical challenges are becoming more entrenched.

COVID-19: global development amidst a global pandemic

Global development challenges constantly evolve, such that as one is addressed, another seemingly emerges to take its place. Many global development challenges are also becoming increasingly complex. Since this Handbook was first conceived, the world experienced one of the most tumultuous periods since the end of World War II, with the COVID-19 pandemic generating social and economic turbulence around the world.

On 31 December 2019, the first signs of a new virus were detected in the Chinese city of Wuhan. One month later, on 30 January 2020, the World Health Organization (WHO) declared the outbreak of a novel coronavirus (later termed COVID-19) a public health emergency of international concern. By 11 March, two days after Italy entered a national lockdown to curb its exploding cases of the virus, WHO declared COVID-19 a global pandemic. Less than a month later there were more than a million reported cases of the virus. In another two weeks there were 2 million cases, and by 12 May there were more than 4 million reported cases (including 280,000 deaths) across 213 countries. At the time of writing, the Johns Hopkins Resource Centre database (n.d.) records over 162 million cases globally, and more than 3.3 million deaths. *The Economist's* (2021) model based on excess deaths provides a much higher figure suggesting that by May 2021 the pandemic had claimed between 7.1 million and 12.7 million lives, most in low and middle-income countries. While many countries within the Global South initially performed well in containing the spread of the virus, by early to mid-2021 second and third 'waves' were resulting in continued and widespread loss of life – most notably in India and Nepal, even as some countries began vaccination rollouts.

As local economies ground to a halt to limit the spread of the virus, and as countries around the world closed their international borders to prevent the spread of the virus, the global economy tumbled into recession. According to Asian Development Bank (ADB) estimates, in Asia alone COVID-19 may see as many as 167 million jobs lost, and more than 399 million people pushed into poverty (ADB 2019). In Bangladesh, for example, an additional 21 per cent of the population joined the pre-existing 20.5 per cent living under the poverty line, leading the government to announce a US$150million social protection programme for the 'new poor' (Rahman et al. 2021). Around the world, support packages to deal with the social, economic and health costs of the crisis have led to huge expenditures even as fiscal revenues have shrunk.

Through the pandemic, old patterns of inequality between countries of the Global North and South have been reproduced. Regarding increasing poverty, modelling suggests that South Asia will be the worst affected region, with up to 57 million people being pushed into extreme poverty (World Bank 2020, 5). Sub-Saharan Africa is expected to be the next most affected region, where up to 40 million additional people are predicted to fall into extreme poverty (ibid., 5). In both cases, these projections are based on the heavily criticized US$1.90/day poverty line, meaning poverty rates at higher thresholds will be much greater.

Existing patterns of inequality are also evident in the global vaccine rollout. As of 30 March 2021, 86 per cent of global vaccinations had been administered in high-income countries, with low-income countries accounting for just 0.1 per cent of vaccinations (Collins and Holder

2021). By May 2021, it seems likely that children in some high-income countries will receive a vaccine before high-risk population cohorts in low and middle-income countries. Kenya anticipates that just 30 per cent of its population will be vaccinated by 2023 (ibid.). As the virus mutates, vaccination delays in one country may result in new strains that are resistant to available vaccines. Attempts to waive patents in order to provide affordable vaccines to low-income countries have faced resistance from pharmaceutical companies and their shareholders. Under the leadership of President Donald Trump, the US showed disdain for a collective global response to the pandemic by withdrawing its funding from WHO and buying up (in early July 2019) the world's supply of the drug Remdesivir – which research suggests could speed the recovery of coronavirus patients (Martin 2020).

The scramble to respond domestically to the virus has been accompanied by some concerning retractions in global development and humanitarian work and reduced international cooperation across critical socio-economic sectors. As of 2021, Australian aid reached a record low as percentage of GDP, despite its economy being comparatively less affected than other donor countries. Cuts to aid in the most recent budget continued a long trend of reduced funding, which has seen Australia shift from contributing 4.3 per cent of total OECD aid in 2012 to only 1.6 per cent in 2020 (Pryke 2021). In 2015, the UK enshrined in law its commitment to the Monterrey Consensus to provide 0.7 per cent of GDP in aid funding; this was slashed to just 0.5 per cent of GDP in 2020, justified due to the pandemic's pressure on public finances. This occurred alongside the incorporation of the UK aid-administering body, the Department for International Development (DfID), into the Foreign and Commonwealth Office – a move that has resulted in reduced autonomy and seen significant staff losses (Worley 2020). Such funding cuts and other effects of the pandemic have also seen many International Non-Governmental Organizations (iNGOs) under financial and operational strain, with Oxfam closing its operations in 18 countries and cutting 1,500 staff (Beaumont 2020).

Politically, an important correlation requiring further analysis regarding the pandemic is that, as of May 2021, six of the top seven countries for total reported COVID-19 case numbers have/had conservative populist leaders: Boris Johnson in the UK; Donald Trump in the US; Narendra Modi in India; Jair Bolsonaro in Brazil; Recep Tayyip Erdoğan in Turkey; and Vladimir Putin in Russia. Of course, there are major differences between these countries that bring caution to any generalizations, yet across the world the pandemic has unfurled alongside – and been shaped by – democratic rollback and more assertive forms of nationalism. In the US claims of election fraud by the outgoing President Donald Trump saw his supporters storm the US Capitol building on 6 January 2021, in what many considered a coup attempt in the world's self-ascribed champion of global democracy. Less than two months later, a(nother) military coup occurred in Myanmar, a country long considered as one of the most authoritarian states in the world, but which had been making some notable progress on enhanced political freedoms. In Hong Kong, Beijing has imposed draconian new security laws after more than a year of ongoing protests, while in India President Narendra Modi's government has suppressed public reporting on COVID-19 case rates.

As tensions have mounted during the pandemic, public protests have occurred across multiple countries. Black Lives Matter protests against racial injustice in the US – spurred foremost by the murder of George Floyd Jr by an acting police officer – spilled over to other parts of the world, sparking a range of new social movements. In the UK, race-related protests were accompanied by monuments of former colonial figures being torn down. In Australia, attention was again drawn to the country's abhorrent statistics on Indigenous incarceration and deaths in custody. Protests against racial injustice have been accompanied by climate change movements across multiple countries, tax reform protests in Colombia, and protests against police violence in Nigeria, and women's rights protests in Pakistan, Paris and Turkey – to name just a few examples. Global crises generate politically charged environments and create the potential for the (re)emergence of conflict.

COVID-19 has both reinforced existing patterns of inequality and revealed the deeply problematic nature of binaries such as 'developed' and 'developing' countries. Policy failures across the 'Global North' and commendable successes within the 'Global South' have further exposed 'the falsity of assumptions that the Global North has all the expertise and solutions' (Oldekop et al. 2020, 2). Failures of governance and social policies have been criticized across a range of countries, including the US, UK, Italy and Spain, leading to late responses in containing the virus, and in turn, high (and unevenly spread) mortality rates. The causal factors behind the high case rates in each of the above-mentioned countries are highly complex, and shaped by social, cultural, political, economic, demographic, and geographic forces. But it is notable that the US and UK have the highest levels of inequality within the 'Global North,' and were also ranked first and second in 2019 for pandemic preparedness (NTI et al. 2019).

The great (un)equalizer

Initial commentary suggested that COVID-19 might be a 'great equalizer' in terms of its effects; viruses do not discriminate between rich and poor, men and women, or other social or cultural cohorts. These claims echo those sometimes made about climate change which is said to be globally indiscriminate in its effects. What is ignored in such commentary, however, is that the effects of the pandemic – and climate change – are shaped by the pre-existing structures of power and privilege as well as by the discriminatory politics and policies of governments. Across the world, people have been unevenly exposed to the pandemic and its social and economic effects, including uneven access to health services and job security. Not only has COVID-19 brought forth many new forms of inequality and disadvantage, but it has also exposed and deepened sociopolitical and ethnocultural cleavages. It has produced new challenges for global development while amplifying and reworking existing challenges.

Wealth inequality grew dramatically during the pandemic, with already marginalized and disadvantaged groups being the most seriously affected. Globally, the pandemic has led to the worst setback on poverty reduction in decades (World Bank 2020). At the same time, some of richest individuals saw their wealth grow. Billionaire Elon Musk's net worth grew by US$140 billion during the pandemic, just at the time when he was threatening to sue the government of California and relocate production to Texas in opposition to the state's coronavirus restrictions (Siddiqui and Romm 2020). Similarly, Amazon's CEO Jeff Bezos saw his wealth grow by US$70 billion at a time when Amazon employees were protesting unsafe working conditions exposing them to risks of COVID-19 (Sainato 2021). Surging share prices following the approval and production of COVID-19 vaccines has also created at least nine new billionaires (Ziady 2021).

Gender inequalities have also been shaped, and grown, during the pandemic. Globally, women have experienced greater exposure to COVID-19 due to their overrepresentation in frontline health sector professions and service industries, as well as their additional caring responsibilities in many households (World Bank 2020). Women have also experienced increased rates of domestic violence during pandemic lockdowns and have been more likely than men to step out of the labour force to cover additional caregiving or domestic work (ibid.). In the US, research has also demonstrated that transgender people have been disproportionately affected by COVID-19 due to their greater likelihood to be low-income, their higher rates of HIV and asthma, and the barriers that they experience in accessing healthcare (inequality.org n.d.).

In 2020-2021 COVID-19 dominated international news in such a way that other critical global challenges have, arguably, received insufficient attention – including other health threats. They have been pushed from news reports, out of world attention, and therefore off the globe. COVID-19 has also become a reason – and a justification, as noted above – for richer countries to scale back their Official Development Assistance commitments and to focus on the needs of their own populations rather than those in less prosperous places. The COVID-19 pandemic is unquestionably important in and of itself. But it also makes visible many long-standing and entrenched issues of global development, as well as revealing new ones. It provides a valuable antidote to the lazy notion that 'things are getting better.' For some, yes; but for many millions of others such generalizations are incorrect, if not insulting. To begin with, the pandemic shows the world to be truly – and deeply – interconnected – reflected in one of the aphorisms of the age, 'No one is safe, until everyone is safe.' But COVID-19 has also shown that poverty reduction can be thrown into reverse in a matter of months, that hundreds of millions are living just barely above the poverty line, wherever it is drawn, and that inequalities of all colours and stripes are enduring and sometimes deepening.

Global development

Collectively, the mounting interrelated challenges discussed above point to the need for new thinking and practices in global development. Readers with familiarity in the field of development will likely have already noted the usage of the term 'global' in this introduction (and Handbook title), as opposed to some of the other common prefixes that are used to define different forms of, or approaches to, development: sustainable development, economic development, community development, and so forth. Such prefixes matter – they are more than academic wordplay. Of foremost importance to the chosen nomenclature of this Handbook is the distinction between *global development* and *international development*.

'Development' remains a heavily contested concept. Without rehashing a conversation that has been extensively examined elsewhere (Cowen and Shenton 1996, Rist 1997, Kothari 2005, Pieterse 2010), it bears noting that the study and practice of development extends across multiple fields. Development, albeit contested, encompasses two key sets of concerns. These have come to be known by the 'Big-D/ Little-d' (D/development as practice versus progress) distinction. The first are attentive to processes of social, cultural, ecological, economic and political change – be it through attempts to pursue *progressive* change, or through critique of uneven, contradictory and negative consequences of 'd'evelopment. The second are oriented towards the landscape of actors and surrounding 'architectures' that have been established to pursue 'D'evelopment in its various forms. This landscape includes, to name just some examples, states, multilateral organizations, non-government organizations, multinational corporations, small-scale enterprises, community-based organizations, and local volunteers (DSAA n.d.). One of the broadest and most widely accepted definitions of development is Robert Chamber's (2004) framing of development as 'good change.' Yet even here there is the important question of what constitutes 'good change.' This remains highly subjective, contested, and embedded within power relations that privilege some perspectives and voices, while silencing others.

Until recently, international development has been the prevailing term used to describe efforts – typically by Western countries – to bring about 'good change' in countries around the world. Like any complex term, international development is interpreted in many different ways, but also has some commonly agreed-upon attributes. Examples include a focus on international aid and inter-state relations, attention to international development institutions (most notably multilateral banks and governance institutions and international non-governmental organizations), and – perhaps most significantly – the idea that development involves high-income 'Northern' countries providing assistance to address development challenges that are limited to countries in the Global South (Oldekop et al. 2020) that are still 'catching up.' Debates in international development reach far beyond, and heavily critique, each of these three topic areas, and the diversity of the field(s) of study and practice(s) that fall under the heading of international development should not be ignored as a result of efforts to identify common themes. Nonetheless, international development is a term that carries a lot of baggage, and this baggage often establishes entry-points for thinking that limit the potentialities for thinking about global justice and wellbeing in new ways.

Global development offers an alternative language to international development that seeks to move beyond some of the more antiquated ways of thinking about global efforts to pursue good change. It seeks to be more attentive to development challenges in *all* (Northern and Southern) countries, as well as to collective challenges that transcend nation-state boundaries. This includes a questioning of former, problematic and poorly defined categories such as 'developed' and 'developing' countries, which underpinned much thinking that has occurred under the conceptual umbrella of 'international development' (Horner and Hulme 2019).

Two notable advocates calling for a paradigm shift from international to global development have been Horner and Hulme (2017; 2019; 2020), who see the need for a global development paradigm as made most pertinent via: the global interconnectedness of contemporary capitalism; the global nature of sustainability and climate change challenges; and the increasingly diverse forms of global inequality which cut across North-South boundaries and present challenges in all countries. To these themes could also be added the global challenges of rising authoritarianism (witnessed in all world regions), the growing global interconnectedness of social movements, global debt, and the still-emerging global transformative effects of artificial intelligence and big data. If all of the above themes were not convincing enough evidence for the need to think about development through a global lens, COVID-19 has provided even further evidence of the global, collective, and connected nature of development: it is a development challenge that pervades all countries, and that will only be successfully addressed through a collective global response.

In providing a starting point for a global development paradigm, Horner and Hulme focus their attention on four key vectors. First, they note important geographic shifts from 'poor people, poor countries and the global South' to interconnected and shared issues across North and South, as well as tackling development challenges wherever they exist (Horner and Hulme 2017, 26). As Sumner (2012) noted a decade ago, geographies of impoverishment have shifted such that the majority of the world's poor now live in middle-income countries. This shift to recognize that development is also a 'northern' issue is captured in the United Nations Sustainable Development Goals (SDGs), which call on all countries to report of their actions and progress towards the 17 goals. The turn to the global represents an important shift from the SDGs' predecessor of the Millennium Development Goals (MDGs), which were focused much more firmly on development as a 'Southern' problem.

Spatially, and as discussed above, Horner and Hulme see the need for a shift in nomenclature of North-South and developed/developing countries to global convergences and national and sub-national differences. In both Northern and Southern countries, we see highly affluent neighbourhoods and low socio-economic neighbourhoods, and differential life outcomes that are shaped by gender, ethnicity and other common themes of disadvantage and privilege. Furthermore, rising middle classes across the Global South have expanded alongside the stalling, and backsliding, of many middle-class incomes in the Global North (Horner 2020).

Conceptually, Horner and Hulme call for a shift from the technocratic and econocentric thinking on development that was so firmly entrenched within 'international development' to a greater focus on sustainability and social justice, as well as efforts to bring about good change that extend beyond poverty alleviation. As Horner notes, 'poor countries' and 'poor people' remain important to global development, but global development is about much more than poverty alleviation (Horner 2020). It is perhaps important to emphasize here that the call for a global development paradigm is not a call to shift focus away from the Global South, or from the needs of the world's most impoverished, marginalized or persecuted peoples. Rather, it is a call for greater attentiveness to global interconnections. In March 2021, for example, the need to think more globally about development – as well as the interdependencies and fragilities of global trade – was exemplified by the grounding of the *Ever Given* container ship, which came to block the Suez Canal. Approximately 12 per cent of global trade passes through the canal, and consequently the grounding of a single ship caused daily trade disruptions of around US$10 billion and prevented the movement of goods between countries around the world in the midst of a global pandemic.

Finally, regarding actors, Horner and Hulme note a shift from both understandings of development as requiring charity and development aid from Northern to Southern states to more complex thinking about South-South cooperation, and a more diversified development practice landscape where South-South cooperation is increasing – and becoming increasingly influential in shaping global development norms and modalities. South-South cooperation has long been a – underexamined – feature of global development cooperation, but its increasing scale and influence over the past decade contributes to the need to think about global development in new ways.

Contribution of this Handbook

When development first emerged as a post-World War II and post-colonial project, there was much enthusiasm for the opportunities that lay ahead. Economic growth, coupled with technological modernization, was seen as a means to alleviate poverty and to bring widespread progress and prosperity for all. Since then, scathing critiques have challenged the idea that development can achieve widespread prosperity. Today, continued faith in business-as-usual (economic) development persists alongside pervasive scepticism, as well as expanding and increasingly complex development challenges. Calls for more and better development sit alongside demands for radical change.

This Handbook seeks to contribute to the global development turn through a contemporary analysis of emergent challenges and crises around the world, the relationships between them, and the persistent structural inequalities that continue to (re)produce forms of marginalization and disadvantage – as well as through the ways that we might approach these myriad issues in our teaching, research, and practice. A handbook on Global Development is overdue. Climate change and COVID-19 make the timing of this volume particularly important, but in an increasingly interconnected and interdependent world, we can expect future global crises.

Part 1, 'Changing development configurations,' brings attention to shifting geographies, spatialities, actors and modalities of development. It sets some of the key context for contemporary development debates about structures and configurations that shape the sector and provides multiple entry-points for rethinking development as global. This includes a focus on the harmful effects of globalization and possibilities for more progressive alternatives, the expansion of retroliberal modes of development (Ch3), global debt relations (Ch5), and regional development challenges within Northern countries (Ch4). Further chapters consider changes within the 'traditional' donor landscape (Ch6), the ways in which South-South cooperation (and associated research) have shifted in recent years (Ch7), the growth of new development financing and activities through philanthropy and social enterprise (Ch10, Ch11), and the enduring tensions that exist in efforts to 'localize' Northern funding within Southern contexts (Ch9).

In Part 2, 'Sustainability and the environment,' chapters examine the many tensions around the relationship between environmental sustainability and development. It looks at the notion of planetary boundaries (Ch13), the Anthropocene (Ch14) and the collective global challenges of natural resource management (Ch18, Ch20, Ch21, Ch23), extractivism (Ch17), climate change/crisis, food systems (Ch22), the plastic crisis (Ch24) and mass extinction (Ch19). The pandemic has seen a partial – but temporary – slowing of global emissions, but global emission reduction targets remain far off-track. In June 2020, parts of Siberia reached astonishing world record temperatures of 38 degrees Celsius (Gardner 2020), and by 2030 the World Bank (2020, 1) projects that up to 132 million people may fall into poverty due to the manifold effects of climate change. Thus, 'taking climate change into account makes the case for a global development approach inescapable' (Horner and Hulme 2019, 497). More-than-human approaches (Ch15), caring economies (Ch16), and creative ways of teaching sustainability (Ch25) are needed to resolve the sustainability dilemmas that the current development paradigm has imposed on the planet.

In Part 3, 'Inequality and inequitable development,' attention is given to the multi-scalar and multi-sectoral challenges that inequality presents for global development. Inequalities of global development exist at the local, national, and global scales, and across and within different genders, classes, ethnicities, religions, age groups, levels of education, and other demographic cohorts. In considering these wide-ranging and often intersecting challenges, the contributions to Part 3 consider different approaches to understanding and measuring poverty (Ch27), structural inequalities within the global financial system (Ch28), and the widespread injustices of global extractive industries (Ch29). Further chapters explore spatial distributions of inequalities within countries (Ch30), global land grabbing (Ch31), forced displacement and resettlement (Ch32; Ch33), and inequalities in gender and education (Ch34; Ch35; Ch36). A final case study chapter considers how new ambitions to remake global development via China's belt and road initiative (BRI) is bringing new forms of 'violent' development (Ch37).

Part 4, 'Game changers,' considers some highly significant global shifts that are taking place across the world, and the implications of these shifts in presenting opportunities and challenges for the global population, national and internationally. Chapters explore the game-changing nature of the pandemic and how COVID-19 (Ch39), disability and other health threats intersect with social determinants of wellbeing (Ch40; Ch41), as well as the different development needs of young and ageing populations (Ch48; Ch49). Urbanization and housing accessibility are also explored, through case studies that draw out commonalities and shared lessons across Northern and Southern contexts (Ch42; Ch43). Regarding game-changing dynamics in global development and mobility, contributors examine questions of equity in global value chain governance (Ch44), complexities of forced, voluntary and involuntary migration (Ch45; Ch46), and the need to better understand intersections between conflict and development (Ch47).

Finally, Part 5, 'Reimagining futures,' asks what the re-imaging of global development futures might look like, and how development students, educators, researchers and practitioners are contributing to such reimagining. Contributors offer a range of practices, orientations and methodologies that current and future people working in the vast and changing field of development might do well to consider and take on as part of reimagining development futures beyond what we have come to know. A strong thread working through all of the chapters is the importance of attending more deeply to the people, knowledges, and non-human kin relations that have for far too long been relegated to development's margins. Each chapter makes a case for why development, in the diverse contexts within which the authors are writing, needs to change and what this change might encompass, leading to more equitable, creative, and nourishing human/more-than-human futures. Contributors explore more-than-human kin relations (Ch51), curriculum and activism (Ch52), Indigenous-led pedagogy (Ch53), decolonialism and gender (Ch54), service learning (Ch55) capacity building in development education (Ch56) and poetry as a form of decolonial practice (Ch61). In addition, attention is also given to adaptive programming (Ch57), southern research methodologies (Ch58), community economies (Ch59) and geonarratives and countermapping (Ch59).

Collectively, the contributions to this Handbook demonstrate the need for heavily contextualized studies of, and responses to, global development challenges. It provides multi-scalar analyses that are attentive to local, national, and global development challenges. Case studies focused on Singapore, Australia, the UK, and others treat the minority world as a subject of development, while chapters examining Cambodia, Sri Lanka, China, Nairobi, Chile, the Philippines and elsewhere ensure a wider, global focus. Across all parts of the Handbook, there has been an attempt to capture development challenges that transcend 'North-South' or 'Majority-Minority' boundaries. Where it is necessary for the purpose of analysis to construct a binary between country groupings our preference is for Majority-Minority world, which seeks to overturn previous framings of development that place a small-number of high-income countries as a global norm for development to aim for and aspire to. However, in recognition of the multi-disciplinary nature of development, contributors have each used the terminology that speaks best to their disciplinary audience.

There has also been an attempt to capture new thinking, and to read development through lenses of sustainability and social justice. While notions of 'progress' sit uneasily with many development scholars, development continues to be about the pursuit of varying forms of good change. What has changed from earlier years is the widening of voices and interpretations regarding what progress means, its different forms, as well as how it is best pursued. Development has become more critical. The Handbook emphasizes that processes of development have influenced people's lives in both positive and negative ways, drawing attention to structural inequality and disadvantage alongside possibilities for positive change.

The call for a global development paradigm is, in our view, also strongly connected with efforts to decolonize development studies and development practice. While it calls for more attention to development challenges within high-income countries, this should not result in a recentring of the West (Horner 2020). It is not a call for universalism, but for the recognition of a greater plurality of voices and ideas. Accordingly, in this Handbook we have sought to include a broad representation of authorial voices from across the globe, with particular effort to seek contributions from Majority world countries. Producing the Handbook in the context of COVID-19 created some challenges for this aspiration, but we have seen some success. An editorial team situated across five countries similarly reflects the global reach of development as an area of study.

In addition to Majority world authorship, the Handbook has sought contributions from highly established and emergent scholars, and from development researchers (and educators) and development practitioners. We hope it will serve as a resource for mutual learning and associated collaboration across North and South, and between academics and practitioners.

Finally, a critical aim of this Handbook is to strengthen the nexus between development research, theorizing, practice, and pedagogy. It is not sufficient for discussions of development to exist only within an academic vacuum, and as such, there is a need for critical reflection on how development research and theorizing can inform and is also informed by development pedagogy and practice. To this end, all of the authors who have contributed to this volume have, in various different ways, thought through the pedagogical implications of theories and debates in their fields of study. Critical engagement across disciplinary boundaries here is a reoccurring message, alongside the fact that a process of 'unlearning' is as important as learning, for students to look beyond the traditional (predominantly Western) theories and models that have dominated thinking and practice in the field of international – and now global – development.

Of course, and despite its breadth, this book is anything but exhaustive. Some notable omissions that we would have liked to include are chapters on Artificial Intelligence and Big Data, further contributions on activism and social movements, more attention to race and racism, and more on China's expanding presence within global development – most notably via its Belt and Road Initiative (BRI). What these omissions reflect is the exigent need for constant interdisciplinary dialogue between 'development scholars' and experts in other fields. This Handbook represents a collective attempt at encouraging and broadening such dialogue, as its contributors hail from a variety of disciplines, but clearly more still needs to be done in this space.

There is one matter which the shift from international to global development can sometimes hide, and which remains central to everything that this book surveys: deep and enduring unevenness in living standards and prospects for the future, within and across countries. While the debate is becoming increasingly 'global,' this unevenness has a spatial signature that is yet to be erased. The poorest people on the planet live in low-income countries, which remain as poor today as they were 75 years ago at the start of the age of 'international development.' But while the popularity of Development as an idea, a project or a set of interventions may have waxed and waned, the need for development as the pursuit of positive change and social justice has not disappeared.

References

Asian Development Bank (ADB) (2019). An Updated Assessment of the Economic Impact of COVID-19, [online] www.adb.org/sites/default/files/publication/604206/adb-brief-133-updated-economic-impact-covid-19.pdf

Banerjee, A. V., and E. Duflo (2011). *Poor Economics: A radical rethinking of the way to fight global poverty*, New York: Public Affairs.

Beaumont, P. (2020). *Oxfam to close in 18 countries and cut 1,500 staff amid coronavirus pressures,* The Guardian [online] www.theguardian.com/global-development/2020/may/20/oxfam-to-close-in-18-countries-and-cut-1500-staff-amid-coronavirus-pressures

Chambers, R. (2004). *Ideas for development: reflecting forwards*, IDS Working Paper 238, Institute of Development Studies, Brighton, England. See: https://citeseerx.ist.psu.edu/viewdoc/download?doi=10.1.1.487.8068&rep=rep1&type=pdf

Collins, K. and Holder, J. (2021). *See How Rich Countries Got to the Front of the Vaccine Line*, New York Times, [online] www.nytimes.com/interactive/2021/03/31/world/global-vaccine-supply-inequity.html

Cowen, M. and Shenton, R.W. (1996). *Doctrines of Development*, New York: Routledge.

DSAA (n.d). What is development studies [online] www.developmentstudies.asn.au/about/

Economist, The (2021). 'Briefing: The covid-19 pandemic', *The Economist*, 15 May, pp. 16–20.

Gardner, D. (2020). *Siberia temperature hits record high amid Arctic heatwave*, The Guardian [online] www.theguardian.com/world/2020/jun/24/siberia-temperature-hits-record-high-amid-arctic-heatwave

Horner, R. (2020). Towards a new paradigm of global development? Beyond the limits of international development. *Progress in Human Geography*, 44(3), 415–436.

Horner, R. and Hulme, D. (2017). From international to global development: new geographies of 21st century development, *Development and Change*, 50(2), 347–378.

Horner, R. and Hulme, D. (2019). Global Development, Converging Divergence and Development Studies: A Rejoinder, *Development and Change*, 50(2), 495–510.

Inequality.org (n.d.). [online] https://inequality.org/facts/inequality-and-covid-19/#gender-inequality-covid

Johns Hopkins Coronavirus Resource Center, (n.d.). *COVID-19 Dashboard*, Johns Hopkins Coronavirus Resource Center, [online] https://coronavirus.jhu.edu/map.html

Kothari, U. (ed.) (2005). *A Radical History of Development Studies: Individuals, Institutions, and Ideologies*, London: Zed Books.

Martin, I. (2020). *U.S. Buys The World's Supply Of Breakthrough Coronavirus Drug Remdesivir*, Forbes, [online] www.forbes.com/sites/iainmartin/2020/07/01/us-buys-the-world-supply-of-breakthrough-coronavirus-drug-remdesivir/#6ded20715472

Nuclear Threat Initiative (NTI), Johns Hopkins Center for Health Security, The Economist Intelligence Unit (2019). Global Health Security Index: Building Collective Action and Accountability, [online] www.ghsindex.org/wp-content/uploads/2019/10/2019-Global-Health-Security-Index.pdf

Oldekop, J. A., R. Horner, D. Hulme, R. Adhikari, B. Agarwal, M. Alford, O. Bakewell, N. Banks, S. Barrientos, T. Bastia, A. J. Bebbington, U. Das, R. Dimova, R. Duncombe, C. Enns, D. Fielding, C. Foster, T. Foster, T. Frederiksen, P. Gao, T. Gillespie, R. Heeks, S. Hickey, M. Hess, N. Jepson, A. Karamchedu, U. Kothari, A. Krishnan, T. Lavers, A. Mamman, D. Mitlin, N. M, Tabrizi, T. R. Müller, K. Nadvi, G. Pasquali, R. Pritchard, K. Pruce, C. Rees, J. Renken, A. Savoia, S. Schindler, A. Surmeier, G. Tampubolon, M. Tyce, V. Unnikrishnan, Y.-F.Zhang, (2020). 'COVID-19 and the case for global development', *World Development*, 134, https://doi.org/10.1016/j.worlddev.2020.105044

Pieterse, J. N. (2010). *Development Theory: Deconstructions/Reconstructions,* London: Sage.

Pryke, J. (2021). *Australian Aid: How low can it go?* [online] www.lowyinstitute.org/the-interpreter/australian-aid-how-low-can-it-go?mc_cid=0d75d72d1d&mc_eid=61ee0c4ebe

Rahman, H. Z., I. Matin, N. Banks and D. Hulme, (2021). 'Finding out fast about the impact of Covid-19: The need for policy relevant methodological innovation' *World Development* 140, 105380.

Rist, G. (1997), *The history of development*, London: Zed Books.

Sainato, M. (2021). *Billionaires add $1tn to net worth during pandemic as their workers struggle*, The Guardian [online] www.theguardian.com/world/2021/jan/15/billionaires-net-worth-coronavirus-pandemic-jeff-bezos-elon-musk

Siddiqui, F. and Romm, T. (2020). *Tesla files suit in response to coronavirus restrictions after Musk threatens to relocate operations*, The Washington Post [online] www.washingtonpost.com/technology/2020/05/09/elon-musk-tesla-threat/

Sumner, A. (2012). 'Where do the poor live?' *World Development*, 40(5), 865–877.

United Nations (2015). TRANSFORMING OUR WORLD: THE 2030 AGENDA FOR SUSTAINABLE DEVELOPMENT (online) https://sdgs.un.org/sites/default/files/publications/21252030%20Agenda%20for%20Sustainable%20Development%20web.pdf

UNHCR (2020). *1 percent of humanity displaced: UNHCR Global Trends Report,* [online] (www.unrefugees.org/news/1-percent-of-humanity-displaced-unhcr-global-trends-report/)

World Bank (2020). *Poverty and Shared Prosperity 2020: Reversals of Fortune*, doi: 10.1596/978–1–4648–1602–4.

Worley, W. (2020). Exclusive: DFID seeks cuts of up to 30% on aid projects [online] www.devex.com/news/exclusive-dfid-seeks-cuts-of-up-to-30-on-aid-projects-97600

Ziady, H. (2021). *Covid vaccine profits mint 9 new pharma billionaires,* [online] https://edition.cnn.com/2021/05/21/business/covid-vaccine-billionaires/index.html

PART 1

Changing development configurations

1

INTRODUCTION

Changing development configurations

Susan Engel and Kearrin Sims

This section on 'Changing Development Configurations' sets some of the key context for development debates about the structures and configurations that shape the sector. The Handbook takes a broad, multidisciplinary approach to the difficult concept of development. Cowan and Shenton's (1996) idea of 'little D' and 'big D' development offers a useful guiding frame for this section, as it highlights the difference between 'little D' development as broad forces of societal change, and 'big D' development as specific practices or interventions to promote development. The global system that has built up around the latter is substantial, and some of its key ideological and structural aspects – as well as the inter-relationships between ideology and structure – are discussed in this section.

Walden Bello's chapter on deglobalization provides an important critical framing for this section as it both outlines the major forces shaping global development for countries all around the world and how we might think about a more just, sustainable global vision. Bello highlights how corporate globalization and neoliberalism became dominating global forces from the 1980s and deepened over time through liberalization, deregulation, and privatization to produce a profoundly inegalitarian planet. Even seemingly centre-left or 'third way' governments have promoted neoliberal agendas with the result that the far- and centre-right have now detached themselves from this agenda, and are promoting a xenophobic version of deglobalization that is out of step with the original agenda.

An emancipatory agenda, in contrast, is based on ethical cooperation, solidarity, community, and ecological stabilization. Despite the emphasis on community, deglobalization prioritizes locally specific responses and visions of sustainability along with local production and consumption (subsidiarity) and related industrial, trade, and financial policies that would facilitate this. It demands redistributive justice in both land and income. It demands precedence be given to the quality of life over economic growth, while creating a healthy balance between population and ecology. Deglobalization does not reject industry and manufacturing but demands they and agriculture are ecologically and socially sensitive, thus providing an important link to the food sovereignty movement. Some adjustments to growth and consumption are required and these need to fall disproportionality on the Minority World, as improvement in life quality in the Majority World are essential. Overall, Bello's chapter provides the case for constant rethinking of progressive ideas to ensure that key issues like gender, indigenous paradigms, and more are properly addressed.

DOI: 10.4324/9781003017653-3

Murray and Overton's chapter on retroliberalism encourages us to think about the state of development thinking today. The Global Financial Crisis, the rise of China and the centrality of the Millennium and Sustainable Development Goals (MDGs and SDGs) in development debates has led many scholars and practitioners to argue that neoliberalism is dead. Murray and Overton show that neoliberalism has in, fact, evolved and argue that retroliberalism is a better term to describe today's development paradigm. Given path dependence, some key elements of neoliberalism have persisted, in particular the primacy of the markets and limiting the size of states, but development thinking today is drawing on some older traditions too. It has echoes first, of the mercantilist use of state-supported, private companies used to expand operations; second, the age of imperialism's focus on nationalist self-interest; third, Modernization theory's prioritization of economic growth facilitated by the state through infrastructure investment; and fourth, the neostructuralist or Third Way agenda of the 2000s, which saw poverty reduction and increased grant aid budgets for a short time head the global consensus on aid. We think the focus on neostructuralism may be a little overstated because, as Murray and Overton note, the MDGs and SDGs were a globalization friendly project and were dominated by an income-based measure of absolute poverty set at an absurdly low level. Still, there was a sense for a time that that aid agenda had changed, but that optimism had, as they suggest, disappeared by 2010 with the beyond aid or retroliberal agenda. This is an important argument as there has been too little focus on development paradigms of late.

Etienne Nel's chapter reflects the Handbook's philosophy that thinking about big and little D development processes needs to be global. It is dichotomizing to talk of development in the North and South, although sometimes still useful as a political device and parsimony does facilitate thinking. Still, it is problematic and Nel's chapter effectively demonstrates how much thinking about development in the North has changed since the Second World War in parallel with changes in development economics. Examining the issue of regional development strategies in the North, Nel shows how they shifted fundamentally from the Bretton Woods era where state strategies to assist 'lagging regions' included state subsidies for economic activity, infrastructure support and indirect support through comprehensive public welfare. With the rise of neoliberalism, markets were meant to solve this problem of lagging regions but when they did not, governments could not ignore the spatial inequality, so focused on promoting strategies based on so-called regional comparative advantage and devolution of power. Regions were expected to promote local innovation, creativity, and leadership, and more. Critical human geography has challenged the premises and practices of such strategies, showing how resilience, history, and path dependence all shape regional paths but are influenced by both endogenous and exogenous factors.

Debt relations are another key issue in contemporary development configurations, with large increases in global debt in the 2010s leaving reduced space for states to borrow in the face of the fallout from the COVID-19 crisis. In his somewhat polemical but powerful chapter on debt, **Éric Toussaint** argues that the outcomes of debt crises are directed by the big banks and the governments that support them. Toussaint begins with a historical overview of debt, providing a number of examples of imperial debt burdens during the late 1800s (Tunisia to France, Egypt to Britain, etc.), before considering debt relations in the post-colonial period in more detail. He argues that the imposition of debt has been a central means by which imperial powers regained controlling influence over their former colonies following independence movements. Traversing the 1980s debt crisis and onwards through the 1990s–2000s to the 2008 Global Financial Crisis and the COVID-19 pandemic, he brings us into what he describes as a looming debt crisis for the Global South, which began in 2016–2017 and has accelerated since. The chapter also summarizes the debt profiles of Africa and Latin America and the Caribbean, to reveal some of the detrimental development effects of debt repayments. Toussaint introduces the concept of 'odious' and/or illegitimate debt and demonstrates historical precedents of countries opposing such debt.

The chapter by **Heiner Janus** provides an overview of the mainstream donor system framed by the emergence of providers of non-traditional development assistance or South-South Cooperation (SSC) as a challenge to that system. The 30 members of the Organisation for Economic Co-operation and Development's (OECD) Development Assistance Committee (DAC), while still significant donors, are being challenged to first, think about what counts as development assistance or Official Development Assistance (ODA). The increases in ODA that accompanied the MDGs and SDGs seem to have reached their peak, indeed a number of DAC donors are reducing aid. SSC also seems to have reach a nadir for the time being, nevertheless, its rapid emergence saw the DAC start to redefine development cooperation. Second, recipient, or partner countries are questioning donors' approaches to aid priorities, channels, and modalities; they are asking that donors actually address partner countries' needs and goals. As a result, we have seen donor programs that emphasize results-based management, thinking and working politically and doing development differently and, as Janus notes, it will be important to see which of these, or mix of them, will become the new standard for donors. Still, such ideas are competing with an increased emphasis on national interests in both donor and partner countries.

While Janus does not discuss this issue, we think that COVID-19 has brought into stark relief the limitations of the shift away from grant aid and toward debt-based development cooperation. The increase in ODA appears limited and SSC providers, in particular China, have reduced their footprint, yet many countries in the Majority World, even those regarded as middle-income, simply do not have the resources to adequately respond to the health and economic impacts of the pandemic.

A further vital piece of the development cooperation system is the multilateral development banks (MDBs). While not all of their loans count as ODA, they are counted in the DAC's revised broader approach to development finance discussed by Janus. As **Susan Engel and Patrick Bond** outline, the MDBs are not only very major providers of development finance, the largest banks, and in particular the World Bank, have had an outsized influence on development thinking and discourse. There are now 32 or so MDBs and they all have similar mandates and structures. Their mandate reflects the Keynesian, modernization, welfare thinking that predominated at their founding, but their structure as banks meant they always prioritized economic growth and markets over welfare and statist development. The two new MDBs, the China-backed Asian Infrastructure Investment Bank (established 2015) and the New Development Bank (2014) have duplicated the structures of the old MDBs, which in itself indicates they are not a major threat to the current neo- or retro-liberal order, though their founders are certainly interested in expanding their place in their global order and their extraction of capital to their own benefit. MDBs are a key source of finance for countries in the wake of the COVID-19 pandemic but the World Bank and others are again imposing neoliberal conditions on their lending. This is even clearer in the growing number of states that defaulted on debt in 2020 and early 2021, and have hence turned to the World Bank and International Monetary Fund for support.

Laura Trajber Waisbich and Emma Mawdsley give the collection a sophisticated engagement with SSC, which they define broadly as an array of transfer and exchange modalities, including resources, technologies, and knowledge, between 'developing' countries. Reviewing the 'critical turn' in South-South Cooperation (SSC) research, they discuss how SSC research has evolved in tandem with changes in SSC practices, leading to what they consider to be the current (post-2015) 'consolidation and politicization' phase of SSC. Regarding the latter, attention is given to the growing convergence between Southern and Northern development norms, modalities and practices, the so-called Southernisation of development through the rise prominence of SSC, growing political competition across SSC development cooperation, and increased contestation regarding development effectiveness. In weaving these different sub-domains together, Waisbich and Mawdsley propose that an important task for SSC research is continued questioning of the gaps between the transformative claims of Southern partners and their uneven materialization on the ground.

Yet another piece of the big D development architecture is non-governmental organizations (NGOs). **Nicola Banks and Badru Bukenya's** chapter evaluates the role of both those headquartered in the North that raise funds there to spend on projects in the South and local NGOs and civil society organizations based in the South. Southern NGOs are too often reduced to project implementers for Northern NGOs or other donors despite calls for 'shifting the power' from the North to the South, which Banks and Bukenya discuss in the chapter. The aim of the call is a transformation towards effective, locally-led and owned development approaches, that challenge the managerialism of the aid chain, outlined in the other chapters in this section. A Dutch-government initiative for partnership-based funding of political action demonstrates both potential for change and the extent of transformation that is required to get beyond the managerialist mindset.

Following on from NGOs, the chapter by **Linsey McGoey** provides both an overview and critical evaluation of the role of private philanthropy and foundations in development. Foundations are different to NGOs as they are not answerable to their members overall, rather their boards and their role has come into stark relief since COVID-19 because of the Bill and Melinda Gates Foundation's role as one of the three largest funders to the World Health Organisation (WHO). As Jonathan Glennie (2021, 28) wrote, this 'tells you all you need to know about the global community's commitment to the organization.' Still as McGoey points out, private philanthropy is only a fraction of the funds provided for health and development by donor governments as ODA. Nevertheless, it has often been able to have an outsized policy influence, for example in promoting market-led solutions to health care, which will in fact intensify the health and economic inequities that philanthropy is supposed to address. McGoey outlines some of the limitations to market-based approaches, including the costs of profit-extraction through microfinance and the way that the Gates Foundation are subsidizing private firms providing mobile funds transfer schemes and labelling this charity. These programs on even a kind reading have produced mixed results for the global poor; a more critical lens highlights how they extract profit from the poor in ways they cannot afford. This highlights the problems of philanthrocapitalism and social enterprise.

In somewhat of a contrast to McGoey's critique, **Hurriyet Babacan and Narayan Gopalkrishnan** provide a positive view of for-profit development initiatives in their chapter on 'inclusive economic development' and social enterprises. Beginning with the former, they argue that future efforts to drive economic growth must better incorporate notions of equity, rights and citizenship, as well as be led by inclusive institutions that allow the broad citizenship participation, uphold the rule of law, and support plurality. It is here where Babacan and Gopalkrishnan see social enterprises as useful. Social enterprises operate at the intersection of the profit and the not-for-profit sectors and are increasingly globally. As the authors summarize, social enterprises are led by a mission to bring positive change. They trade to fulfil this mission, deriving income to reinvest the majority of profits into fulfilment of their broader missions. By operating in this way, social enterprises present a strong and viable alternative to traditional for-profit/not-for-profit silos, and an important driver of inclusive economic development. This is not to suggest – as the authors point out – that social entrepreneurship represents a panacea for all social and ecological issues. Social enterprises have been critiqued in a number of ways, and Gopalkrishnan and Babacan are attentive to this. Nonetheless, they do see these enterprises as important stakeholders in working towards inclusive development in many countries across the world.

Overall then, this diverse section of the Handbook sets the context for development today.

References

Cowan, M.P., and Shenton, R.W. (1996). *Doctrines of Development*. London: Routledge.
Glennie, J. (2021). *The Future of Aid: Global Public Investment*. London and New York: Routledge.

2

DEGLOBALIZATION[1]

Walden Bello

Introduction

The paradigm of deglobalization was advanced by Focus on the Global South in 2000, at a time when corporate-driven globalization appeared to be irresistible. It first attracted attention and provoked discussion in progressive circles but only after the 2008 financial implosion it attracted attention from the mainstream, with *the Economist* (2009) writing that with the 'integration of the world economy in retreat on almost every front,' the economic melt-down 'has popularized a new term: deglobalization.'

As a program for organizing the economy, deglobalization was, interestingly, first proposed in France. It inspired the platform of Arnaud Montebourg, a socialist running for the presidency of France in 2012 (Haski 2011). Unfortunately, deglobalization was also embraced by the right in France, and today deglobalization and the critique of globalization more generally are most strongly identified with the right.

This situation makes it imperative to clarify what deglobalization is all about, and why it is something that should be associated with liberation, not xenophobic nationalism. Clarification is needed, not only to detach deglobalization from its embrace by the right but also to assess its relevance as a liberating paradigm for today's world. In other words, the world has moved forward – or backward, as the case may be – since the 2000s, when we first articulated deglobalization. Aside from the 2008 financial crisis, other developments have come to the fore, such as the acceleration of climate change, automation, inequality, and now the COVID-19 pandemic. In the light of these changes, it is worthwhile to revisit deglobalization to see if it still provides a way to cope with these developments and to compare it to other alternative paradigms that have drawn attention more recently.

How the right hijacked deglobalization

Before we embark on this broader task, it is important to briefly consider how the critique of globalization was hijacked by the right. The left's critique of neoliberalism and globalization took off in the mid-1980s. In the South, it unfolded as part and parcel of the opposition to neoliberal 'structural adjustment' in developing countries imposed by the International Monetary Fund (IMF) and the World Bank, the key aims of which were accelerated liberalization of trade,

DOI: 10.4324/9781003017653-4

deregulation, and privatization. In the North, it was triggered by the drive of transnational corporations to relocate their facilities to cheap labour locations, which accelerated with China's integration in the 1980s. It was also a response to neoliberalism which elected governments introduced in the US and Britain in the early 1980s. The establishment of the North American Free Trade Area (NAFTA) in 1994 and the World Trade Organization (WTO) in 1995 added fuel to the spread of what came to be known as the anti-globalization or alter-globalization movement. September 11 dented the momentum of the anti-globalization movement, but the World Social Forum provided a North-South avenue for the elaboration of anti-globalization strategies. With the outbreak of the global financial crisis in 2008, the anti-globalization movement re-emerged in force in the North in the Occupy Movement, the coming to power of Syriza in Greece and the rise of Podemos in Spain.

The left's ability to ride on the anti-globalization agenda, however, was severely compromised by the fact that, since the 1990s, the centre-left in the US and Europe had bought into and aggressively promoted the free trade, neoliberal agenda. In the US, under the leadership of Democratic Clinton administration, NAFTA, and the WTO came into being and the New Deal-era Glass Steagall Act was repealed. Obama's Democratic presidency promoted the Trans-Pacific Partnership (TPP) that, to the working class, meant a continuation of the export of their jobs to China. Similar processes occurred in the UK with New Labour and the 'Third Way,' in Germany, under the Social Democratic Party led by Gerard Schroeder and in France, with socialist figures being enthusiastic proponents of the Euro (Bello 2017). Established workers' parties became defenders of the pro-globalization agenda, leading not only to failure to expand their mass base but also to part of that base leaving their ranks. Liberal non-governmental organizations also played a role in promoting globalization as they bought into the neoliberal trade paradigm but sought to tweak it with 'reformist' measures. Their critics saw this as providing the corporate capture of global trade with a human face (see Bello 2002, Cleary 2002).

As this occurred, the extreme right was detaching itself from the free trade, neoliberal agenda that it had formerly supported along with the centre-right. In the US, Donald Trump broke with the Republican Party and big business when he opposed the TPP. Seeking to make inroads into the working class, right-wing parties in Europe gradually abandoned anti-tax, anti-big-government, and free-market concerns and opportunistically embraced an anti-neoliberal and anti-globalization agenda and the welfare state. The strategy paid off (Bremner 2014). It became the extreme right's passport to power or to the antechamber of power throughout Europe. But there were two ingredients that the right added that marked off its anti-globalization agenda from the left: racism and a reactionary nationalism aimed at migrants. This happened in Denmark, Norway and Austria (Judis 2016), in Greece (Varoufakis 2017) and in the US with welfare proving a fertile ground for debate (Lowery 2018). In the Global South, there were right wing figures that also engaged in cherry picking of left-wing themes. Foremost among them was Rodrigo Duterte, who was elected president of the Philippines in 2016. In summary, the subordination of trade to the social good, the expansion of social protection, and the re-embedding of the market in society are progressive themes that are appearing in the right's deglobalization discourse, but they have been articulated within a racist and nationalist framework, within an exclusionary political economy that marginalizes large numbers of people on account of their race, ethnicity, nationality, and culture. The challenge then, is how to bring deglobalization back to the left, how to regain its original appeal as an emancipatory paradigm for all rather than just for some. Undertaking this task means revisiting not only the strategic prescriptions of deglobalization but also its foundational concepts, as they were originally articulated by Focus on the Global South.

Revisiting the foundational concepts of deglobalization

Deglobalization is, at its core, an ethical perspective (Bello 2002). It prioritizes values above interests, cooperation above competition, and community above efficiency. It does not say that interests, competition, and efficiency are bad but that their pursuit must be subordinated to values, cooperation, and community. Translated into economics, the aim of the deglobalization paradigm is to move beyond the economics of narrow efficiency, in which the key criterion is the reduction of unit cost, never mind the social and ecological destabilization this process brings about. It is to move beyond a system of economic calculation that, in the words of John Maynard Keynes (1933), made 'the whole conduct of life… into a paradox of an accountant's nightmare.' It aims to promote effective economics, which strengthens social solidarity by subordinating the operations of the market to the values of equity, justice, and community and by enlarging the range of democratic decision-making in the economic sphere. To use the language of the great Hungarian thinker Karl Polanyi in his book *The Great Transformation* (1957), deglobalization is about 're-embedding' the economy and the market in society, instead of having society driven by the economy and the market.

The deglobalization paradigm asserts that a 'one size fits all' model like neoliberalism or centralized bureaucratic socialism is dysfunctional and destabilizing. Instead, diversity should be expected and encouraged, as it is in nature. As Pablo Solon (2017a) has pointed out, deglobalization 'does not seek to replace the homogenizing vision of globalization with another model that can be universally applied to all countries and communities.' Shared principles of alternative economics do exist, and they have already substantially emerged in the struggle against and critical reflection over the failure of centralized socialism and capitalism. However, how these principles – the most important of which have been sketched out above – are concretely articulated will depend on the values, rhythms, and strategic choices of each community.

What was missing in the original formulation was an elaboration of the idea of community, for it was this ambiguity that allowed the right to step in and hijack the paradigm. For the right, community is determined by race, ethnicity, and blood. It is narrow in terms of who is included in it rather than expansive. For the left, community is principally a matter of shared values that transcend differences in blood, gender, race, class, and culture. Community tends towards continual expansion and incorporation of people that share the same values, though of course this sharing may be imperfect, limited, and open to different interpretations. Central to this interpretation of community is the assumption that all people are entitled to the full range of political, civil, economic, social, and human rights, including the right to join a desired community. This does not mean that there are no procedural rules governing the acquisition of citizenship or migration. It does mean though that these rules and regulations are guided by a fundamental openness towards accepting those who wish to join a community.

Reassessing the deglobalization program

Moving beyond these contrasts between our paradigm of deglobalization and that of the right, we now look at the concrete program for deglobalization as it was initially formulated to see how relevant it is to current conditions. Fourteen key thrusts were proposed:

1. Production for the domestic market must again become the centre of gravity of the economy rather than production for export markets.

2. The principle of subsidiarity should be enshrined in economic life by encouraging production of goods at the level of the community and at the national level if this can be done at reasonable cost in order to preserve community.

3. Trade policy – that is, quotas and tariffs – should be used to protect the local economy from destruction by corporate-subsidized commodities with artificially low prices.

4. Industrial policy – including subsidies, tariffs, and trade – should be used to revitalize and strengthen the manufacturing sector.

5. Long-postponed measures of equitable income redistribution and land redistribution (including urban land reform) can create a vibrant internal market that would serve as the anchor of the economy and produce local financial resources for investment.

6. Deemphasizing growth, emphasizing upgrading the quality of life, and maximizing equity will reduce environmental disequilibrium.

7. The development and diffusion of environmentally congenial technology in both agriculture and industry should be encouraged.

8. The power and transportation systems must be transformed into decentralized systems based on renewable resources.

9. A healthy balance must be maintained between a society's population and ecology.

10. A gender lens must be applied in all areas of decision-making so as to ensure gender equity.

11. Strategic economic decisions cannot be left to the market or to technocrats. Instead, the scope of democratic decision-making in the economy should be expanded so that all vital questions – such as which industries to develop or phase out, what proportion of the government budget to devote to agriculture, etc. – become subject to democratic discussion and choice.

12. Civil society must constantly monitor and supervise the private sector and the state, a process that should be institutionalized.

13. The property complex should be transformed into a 'mixed economy' that includes community cooperatives, private enterprises, and state enterprises, and excludes transnational corporations.

14. Centralized global institutions like the IMF, World Bank, and the WTO should be replaced with regional institutions built not on free trade and capital mobility but on principles of cooperation.

While the foundational concepts of deglobalization were generally welcomed, the strategic program drew a number of useful questions, comments, and criticisms. The principal ones were:

- Does deglobalization propose delinking the local economy from the international economy?
- It was still articulated within a developmental framework whereas what is now needed is 'an alternative to development.'
- It did not appreciate the urgency of climate change.
- It did not give sufficient stress to food security and food sovereignty.
- It did not pay adequate attention to the gender issue.
- It was not sensitive to structural changes in the economy.

There is, in varying degrees, merit to all these criticisms, and one problem they point to is that we devoted most of our effort to elucidating the foundational concepts of deglobalization and left the economic program proper rather brief, like the items in a doctor's prescription. An extended explanation could have been given on some of the key points, one that could have addressed some of the issues or problems associated with their implementation in real life. But perhaps the more important reason for the paucity of substantive articulation of some of these proposals is that we ourselves were still grappling with their implications when we first drew up the deglobalization program. The following section briefly addresses each of these points in turn.

Does deglobalization propose delinking?

In her initial forays on international trade, Marine Le Pen came across as advocating 'an exit by France from the euro and erection of barriers at France's borders' (Haski 2011). Le Pen was superseded by Brexit, but both Le Pen's position and Brexit are caricatures of deglobalization. Contrary to the claims of some critics, deglobalization in our formulation was never about delinking from the international economy. It was always about achieving a healthy balance between the national economy and international economy, one largely presided over by a state that pragmatically employed tariffs, quotas, and other mechanisms to ensure the survival and health of local industries competing against highly subsidized Northern agriculture and industrial corporate giants with deep pockets. It is about modifying the rules of trade to protect the welfare of all sectors of the community from the predatory acts of corporations that are justified by appeal to free trade, and it was never proposed as a nationalist weapon in global economic and political rivalries.

Development alternative or alternative to development?

In recent years, development has become a controversial word, with its connotations that the non-Western countries had only one way to go, and that was to follow the stages of development pioneered by the West. Few of those proposing deglobalization envision development as following the trajectory of the West or even of the newly industrializing countries. In so far as we have used development, we have meant it as the building of a society's capacity to meet people's needs. That industry must eventually supplant agriculture as the largest economic sector in terms of Gross Domestic Product (GDP), and that the services sector will eventually be the biggest employer, are not views that we hold. In fact, a vibrant agriculture that not only keeps its workforce but expands it while raising incomes would be seen as positive. Deglobalization, however, is not anti-industry. Industry and manufacturing are not, however, seen as important because they add more value to GDP than agriculture, but owing to their building capacity, making an economy more self-reliant because it has a diversified production base. That a country's industrial structure must be environmentally congenial or sustainable goes without saying, which means extractive industries and ecologically and socially disruptive industrial processes that trigger land grabs, are wasteful in terms of their material inputs, or cannot minimize their industrial waste, pollution, or carbon output, would have to be banned or phased out.

This brings up the question of growth, which is more fully discussed in the next section. But in connection with the idea of development as the building of capacity, agricultural and industrial output do need to grow in order to raise the population's standard of living. However, growth rates would be much less than the 6–10 per cent per annum that have been characteristic of the newly industrialized countries or emergent economies owing to three factors. First, in the deglobalization paradigm, production would be accompanied by radical redistributive policies, which was not the case with these countries. Second, recognition of the reproductive rights of women, along with greater income equality and better health and social services, would lessen the pressure to have large families, which is a key force behind high population growth rates that growth apologists use as a justification for high economic growth rates. Third, a conscious effort by civil society would be made to discourage consumption along the lines of the unsustainable western model, and to delink the measurement of social worth from material consumption and link it with the growth of non-material sources of psychological satisfaction.

If sustainable development is seen as undertaking radical reform in order to build the capacity of a community to survive and flourish while enhancing the environment, then it would probably meet with little objection. But orthodox sustainable development discourse has depoliticized the process of transformative change and concealed the fact that unequal social structures have to be changed to bring about a better life for the world's billions who are currently marginalized, and that inevitably, there will have to be losers in this process which will undoubtedly involve much conflict (Kothari, Demaria and Acosta 2014). Thus, more socially imaginative alternatives are needed for deglobalization.

Deglobalization, decoupling, and degrowth?

In recent years, the most contentious debate in ecological economics has been that between decoupling and degrowth. This debate is central to defining the substance of deglobalization.

Decoupling means delinking the rise in GDP from the rise in carbon emissions. Relative decoupling refers to a GDP growth rate that rises faster than the carbon emissions rate. Absolute decoupling means the GDP growth rate continues to rise while the carbon emissions rate either flattens out or decreases. Absolute decoupling is what matters for climate activists. Two things make this possible. One is the 'dematerialization' of commodities, that is, the amount of materials that go into a product becomes greatly reduced. The second is more efficient use of energy throughout the economy. The first phenomenon leads to reduced draw on natural resources, the second to lower carbon emissions in the production process and consumption. Most of the debate has centred on the role of energy efficiency in cutting down greenhouse gas emissions, a position associated with both liberal advocates and progressive ones like Robert Pollin (2018, 9) who argues that 'absolute decoupling' of GDP growth from carbon emissions is possible, citing data that showed that in 21 countries expanded their GDP while reducing their carbon emissions.

The problem with the decoupling argument is twofold. First, as pointed out by Tim Jackson (2009) and conceded by Pollin (2018), there is no evidence of absolute decoupling of growth and carbon emissions either at a global level or among, respectively, low-, medium-, high-income country groupings between 1965 and 2015, although it may have occurred in some countries. Also, there is no evidence for absolute decoupling of growth and resource consumption. Indeed, 'Global resource intensities (the ratios of resource use to GDP), far from declining, have increased significantly across a range of non-fuel minerals. Resource efficiency is going in the wrong direction. Even relative decoupling isn't happening' (Jackson 2009, 51).

Second, what appears to be happening is the rebound effect or 'Jevons effect,' after the British economist William Jevons, who observed that by raising the productivity of coal, that is, making its use more efficient in the production of iron, the price of iron would drop, creating more demand for iron and consequently increasing the use of coal. Efficiency gains in one area translate into savings that fuel energy consumption in other areas, generally raising fossil fuel use overall and raising carbon emissions. Thus Jackson (2009, 8) says that 'simplistic assumptions that capitalism's propensity for efficiency will allow us to stabilize the climate and protect against resource scarcity are nothing short of delusional.' The problem is that this may not simply be delusional but politically dangerous. As Jorgen Norgaard writes, 'Unfortunately, the notion of decoupling has served as a peacemaker between environmentalists and growth-oriented politicians by conveniently exempting economic growth of any responsibility for environmental problems' (cited in Owen 2011, 26).

Once the decoupling myth is brushed aside, it becomes evident that the addiction to growth must be confronted squarely. Since the demands of addressing poverty and respecting global justice and equity demands that the countries of the South will have to experience some growth, then it is clear that the adjustment in terms of radically restraining growth and consumption must fall for the most part on the rich countries, though of course, in both the rate of growth and consumption the poorer countries must not follow the way of the West by putting the emphasis on equity-enhancing economic strategies.

The shift in the West must not be simply one based on cutting down on material consumption but psychological in nature too, embracing less material consumption as a precondition for a superior way of life. As Kothari et al. (2014, 369) articulate:

> the emphasis should not only be on 'less' but also on 'different.' Degrowth signifies a society with a different metabolism (the energy and material throughput of the economy), but more importantly, a society with a metabolism which has a different structure and serves new functions. In a degrowth society everything will be different from the current mainstream: activities, forms and uses of energy, relations, gender roles, allocations of time between paid and non-paid work, and relations with the non-human world.

While integrating bioeconomics and ecological macroeconomics, degrowth is a non-economic concept as it attempts to challenge 'the omnipresence of market-based relations in society (i.e., commodification) … replacing them with the idea of frugal abundance' (Kothari et al. 2014, 369). It also promotes deep democratic engagement, and 'a redistribution of wealth within and across the Global North and South, as well as between present and future generations' (Kothari et al. 2014, 369). A similar vision is offered by Tim Jackson, one of the most respected proponents of degrowth. He writes:

> The rewards from these changes are likely to be significant. A less materialistic society will be a happier one. A more equal society will be a less anxious one. Greater attention to community and to participation in the life of society will reduce the loneliness and anomie that has undermined wellbeing in the modern economy. Enhanced investment in public goods will provide lasting returns to the nation's prosperity.
>
> (Jackson 2009, 93)

As far as the relationship between decouplers and degrowthers is concerned, it is clear that deglobalizers would be on the side of the latter. So long as it is understood that the main adjustment in terms of 'degrowing' is understood to lie with the North and that the so-called emerging economies must also rein in their high growth rates, the degrowth paradigm is not only compatible with but must be assimilated into deglobalization.

Deglobalization and food sovereignty

Deglobalization, when originally formulated, had as a major concern the protection of small agricultural producers from being bankrupted by cheap imports by big agro-corporations facilitated by the WTO, leading to the loss of a country's food security. We soon realized that this was a rather limited formulation of larger concerns we had. Fortunately, the international peasant organization Via Campesina formulated ideas we were grappling with far better than we ever could under the rubric of 'food sovereignty.' As articulated by its representatives, the key propositions of food sovereignty include following: food self-sufficiency, a focus on quality, low input, small-scale, locally produced food, locally-controlled food systems, a focus on farmer and consumer welfare and a balance between urban and rural needs, delinking from global production chains and transnational corporations, an end to land grabs, democratic control over trade and more (Desmarais 2007). Just as proponents of degrowth clarified the principles of climate science and activism that needed to be incorporated into it, Via Campesina and its allies articulated the elements of food sovereignty that needed to be integrated into – and enrich – the deglobalization paradigm.

Deglobalization and feminist economics

The gender question barely figured in the original articulation of deglobalization but feminist economics advances and complements the paradigm. Of particular significance are two ideas. One is the demand that women's reproductive work, that is, creating the social and economic conditions for recreation of work traditionally regarded as 'productive' work, must be recognized as creating value and this must be recognized in the national accounts. With its focus on GDP and its bias towards productive work as the source of value, traditional economics as well as Marxist economics have disparaged and denigrated women's work despite the fact that research has shown that the total amount of unpaid work in a national economy is greater than the amount of paid work (Aguinaga et al. 2013).

The issue of the value of reproductive work has brought up the broader issue of the 'economics of care,' which 'identifies the need for the care of children, the sick, those with special abilities, and the elderly, as one of the most important human needs for living a full life with dignity' (Aguinaga et al. 2013, 52). Once we talk about the economics of care, we move away from the criterion of efficiency to a less tangible and more comprehensive measurement of the value of a product or service than the reduction of unit cost. This approach ties into our discussion in the original deglobalization paradigm of the transition from 'efficiency economics' to 'effective economics,' with the contribution a product or service makes to enhancing the welfare of a community – not reduction in unit cost as in efficiency economics – being the decisive measure of value or worth. Measuring value within this framework will be a challenge but a common measure would be a major advance towards the kind of 'absolute decoupling' – of non-material 'flourishing' and material consumption – that really matters.

Deglobalization and 'Emancipatory Marxism'

Another issue which deglobalization must address more fully are structural changes in the local and international economies. Swift technological change and neoliberal 'reform' within a global capitalist system has created a social upheaval the end to which is not yet in sight. The negative consequences of this fatal combination have become all too evident and alarming and include increasing wealth and income inequalities in most countries and among countries; concentration of power in a few corporations linked to the rise of information technology (McAfee and Erik Brynjolfsson 2017); and the vulnerability to workers created by advances in artificial intelligence. For the South this latter issue is particularly problematic, as Lowery (2018, 20–1) writes:

> *'Premature industrialization' might turn lower-income countries into service economies long before they have a middle class to buy those services … A common path to rapid economic growth, the one that aided South Korea, among other countries, might simply disappear … Mass unemployment would likely hit high-income countries first. But it could hit developing nations hardest.*

Unless there is radical structural change, the deadly combination of neoliberal policies, corporate and income concentration, and advances in artificial intelligence would not only give capital unparalleled power over labour but it would make a very large part of society redundant, marginal, and miserable.

If, on the other hand, technology was no longer controlled by the imperatives of capital, artificial intelligence could be liberating, allowing the vast masses of people to truly move from the realm of history to the realm of freedom. This is the intriguing view of what we might term 'Emanicipatory Marxism.' According to Paul Mason (2016, 138), in a piece of writing known as the Fragment on the Machine, Marx imagined that in a society where information technology was no longer bound by the laws of the market and profit, the main objective of the working class would be 'freedom from work.' In such a society, 'liberation would come through leisure time' (Mason 2016, 138).

Short of a radical transformation of the economic system, so that leisure becomes the condition for the collective creation of socially useful work, there are other strategies that can be undertaken to counteract job-displacing technology. First, strong controls over the diffusion of labour-saving technology, which can be done via trade and investment restrictions. Second, focus on creating or developing jobs that cannot be replaced by robots, such as occupations involving intensive emotional care (Mason 2016). Third, promote agrocecology-industry and industrial agriculture are extremely vulnerable to automation and its accompanying massive job displacement, but agroecology would be much less vulnerable to robotics and thus much more absorptive of human labour. Fourth, states can provide a basic income to everyone.

Buen Vivir, deglobalization, and the convergence of alternatives

A major contribution to the debate on alternatives to capitalist globalization in recent years has been the paradigm of 'Buen Vivir,' or 'Living Well,' distilled by progressive scholars from Latin America from the 'cosmovisions' of Andean indigenous communities (Solon 2017b; Prada 2013). This paradigm was largely articulated after deglobalization made its appearance as an alternative. However, it fits remarkably well with the spirit and goals of deglobalization.

Like deglobalization, degrowth, ecofeminism, and post-capitalism, Buen Vivir is critical of development as growth. It also shares deglobalization's stress on respecting diverse roads to the achievement of the 'good life,' as well as the latter's stress on the primacy of values and community. The different paradigms do have different emphases and some real differences – for instance, Buen Vivir's acceptance of inequality as a part of the cosmic condition while the other paradigms value equality (Solon 2017b), but the overriding thrust is convergence in their values and directions, in particular, the subordination of the market to social solidarity, the value placed on cooperation over competition, the move away from GDP as a measure of social well-being, the reintegration of the economy and the environment, and the respect for different roads to the 'good life.' Buen Vivir's key principles are outlined in the section on pedagogy below as well as exercise to get students to think about the similarities and differences with deglobalization.

Evolving ideas of deglobalization: some pedagogical considerations

The chapter offers a range of issues that can be used to engage people in debates about the structure of the global economy and the meaning of a good life. A few examples follow, all based on having this chapter as background reading.

a. *Unsustainable development*: show a video of a Jeffrey Sachs discussion on sustainable development. Discussion: in his book, *The Age of Sustainable Development,* Sachs (2015) posits that the four pillars of sustainable development are: economic prosperity, social inclusion, environmental sustainability, and good governance. He says these can be achieved if only elites can see that it is in their interest that the poor are lifted up from their current condition, further they will be willing to support sustainable development if they see that it will cost them nothing or little. There is no mention of economic exploitation, of capitalism and capitalism's drive to create poverty and inequality. In so far as corporations contribute to poverty, inequality, and environmental destruction, it is owing to misguided policies which are remediable because there really is no reason deep social conflicts should persist. Sachs says change will be brought about by the power of ideas and exhortation to achieve the sustainable development goals plus foreign aid. Discuss what apolitical means? Is this approach apolitical, why or why not? How does this differ from the chapter by Bello?

b. *Food sovereignty* – read the Declaration on Nyéléni on food sovereignty (www.viacampesina.org/en/declaration-of-nyi/). Write down by yourself which ideas you find the most and least appealing and why you think those ideas are/are not appealing. In small groups share your thoughts and discuss what these principles add to those of deglobalization? Are there any differences in emphasis or ideas with deglobalization?

c. *Buen Vivir/Living Well*: the four key principles underlying this approach based on indigenous cosmovisions are:
    1. It is not centred on man and society but on the whole, the universe, the Pacha, of which human society and the earth are parts.
    2. There is a unity of past, present, and the future. 'In the Aymara way of thinking,' writes Solon (2017b, 17), 'there is no death as understood in the West, in which the body disappears into a hell or a heaven. Here, death is just another moment in life, because one lives anew in the mountains or the depths of the lakes or rivers.'
    3. The thrust of Buen Vivir is towards equilibrium, not progress or growth.
    4. Buen Vivir does not aim towards homogeneity but towards diversity and not only respects difference but sees it as a source of strength.

What are the similarities or complementarities with deglobalization (and/or other paradigms) and differences? Where there are differences ask the students which they prefer and why.

d. *Overall discussion*: does one paradigm need to subsume the others under one overarching framework, with a unified narrative? Or should we simply acknowledge the convergence and leave the different paradigms the space to borrow from each other's conceptual wealth? To conclude, note that in Bello's view deglobalization is enriched by perspectives from the other narratives even as it expresses them in its distinctive discourse.

Conclusion

Deglobalization was first articulated by the left, but it was taken over by the right, which invoked its progressive propositions on trade and welfare but placed them in an exclusionary political economy based on racial and ethnic supremacy. While the foundational concepts of deglobalization met with approval in progressive circles, the economic program was regarded by some as needing more substantive articulation. Key issues that needed to be addressed more fully were whether deglobalization favoured delinking from the international economy, whether it was a 'development alternative' or an 'alternative to development,' whether it favoured decoupling or degrowth as the way forward in addressing climate change, what its relationship was to the food sovereignty paradigm, what its stand was on feminist economics, how it related to structural changes in the economy like advances in artificial intelligence, and what its relationship was to Buen Vivir and other paradigms influenced by indigenous perspectives.

The chapter clarified that deglobalization did not mean delinking from the international economy; was more of an alternative to development than a development alternative; favoured degrowth over decoupling; was enriched by the perspectives of food sovereignty and agroecology; and sought to integrate the insights of feminist economics, in particular, the value of reproductive work and the centrality of work related to care in the post-growth economy. Also, like emancipatory Marxism, deglobalization recognized both the massive threat posed to workers by advances in artificial intelligence and the liberating potential of the latter in terms of releasing people from the burden of work to concentrate on fulfilling their potential as creative beings.

Deglobalization has much in common with the perspectives and values undergirding Buen Vivir. Indeed, the shared perspectives among all the alternatives to development are striking. However, rather than subsuming all the discourses under one overarching discourse, it is probably more productive for each of these discourses to be articulated and developed separately, with their complementarities being pointed out along with their differences. The insights each paradigm now delivers might be lost under a homogenizing conceptual framework.

Key sources

1. Bello, Walden (2002). *Deglobalization: Ideas for a New Word Economy*, London: Bloomsbury/Zed.

This book highlights the failings of the current global economy, in particular the negative role of transnational corporations and footloose capital. It provides a comprehensive introduction and key concepts of deglobalization. This includes how to shift to a more decentralized, plural system of global economic governance. It emphasizes how states around the world need to be able to follow development strategies that are sensitive to their own context and needs.

2. Kothari, A., Demaria, F. and Acosta, A. (2014). 'Buen Vivir, Degrowth, and Ecological Swaraj: Alternatives to Sustainable Development and the Green Economy,' *Development* 57(3–4): 362–375.

This article argues that we need to go beyond ideas like the 'Green Economy' and develop alternative socio-economy systems that radically disrupt contemporary inegalitarian global systems.

3. Pollin, R. (2018). 'Degrowth Versus a Green New Deal,' *New Left Review* 112 (July–August): 5–25.

This article explores two divergent progressive approaches to the ecological crisis: the green new deal and degrowth. Pollin is largely a supporter of a green new deal, which argues for a global program of investing 1.5–2 per cent of GDP in energy efficiency and renewable energy. The article addresses a range of challenges around such a program including land use, jobs, industrial policy, and ownership forms.

Note

1 An earlier, longer version of this chapter was published as: Bello W. (2019). Revisiting & Reclaiming, Focus on the Global South, URL: www.focusweb.org/wp-content/uploads/2019/05/Revisiting-Reclaiming-Deglobalization-web.pdf access date 20 October 2020.

References

Aguinaga M., Lang, M., Mokrani, D. and Santillana, A. (2013). Critiques and Alternatives to Development: A Feminist Perspective. In M. Lang and D. Mokrani (eds.) *Beyond Development: Alternative Visions form Latin America*. Quito and Amsterdam: Rosa Luxemburg Foundation and Transnational Institute, 41–60.

Bello, W. (2017). Social Democracy's Faustian Bargain with Global Finance Unravels, Transnational Institute, 18 May. Available: www.tni.org/en/publication/europe-social-democracys-faustian-pact-with-global-finance-unravels [Accessed 4 January 2019].

Bello, W. (2019). Revisiting & Reclaiming, Focus on the Global South, URL: www.focusweb.org/wp-content/uploads/2019/05/Revisiting-Reclaiming-Deglobalization-web.pdf access date 20 October 2020.

Bello, W. (2002). The Oxfam Trade Debate: From Controversy to Common Strategy. *Focus on Trade*, No. 78 (May). Available: www.focusweb.org/number-78-may-2002/ [Accessed 4 January 2019].

Bremner, C. (2014). At the Gates of Power: How Marine Le Pen is Unnerving the French Establishment. *The New Statesman*, 4 December. Available: www.newstatesman.com/politics/2014/12/gates-powerhow-marine-le-pen-unnerving-french-establishment [Accessed 4 January 2019].

Cleary, A. (2002). Oxfam's Response to Walden Bello. *Focus on Trade*, No. 78 (May). Available: www.focusweb.org/number-78-may-2002/ [Accessed 4 January 2019].

Desmarais, A. (2007). *La Via Campesina: Globalization and the Power of Peasants*. London: Pluto Press.

Haski, P. (2011). Is France on Course to Bid Adieu to Globalization? *YaleGlobal Online*, July 21. Available: www.yaleglobal.yale.edu/content/france-coursebid-adieu-globalization [Accessed 4 January 2019].

Jackson, T. (2009). *Prosperity Without Growth? The Transition to a Sustainable Economy*. UK: Sustainable Development Commission. http://www.sd-commission.org.uk/data/files/publications/prosperity_without_growth_report.pdf

Judis, J. (2016). *The Populist Explosion: How the Great Recession Transformed American and European Politics*. New York: Columbia Global Reports.

Keynes, J. M. (1933). National Self Sufficiency. *Yale Review*, 22(4) (June 1933). Available: www.panarchy.org/keynes/national.1933.html [Accessed 6 January 2019].

Kothari, A., Demaria, F. and Acosta, A. (2014). Buen Vivir, Degrowth, and Ecological Swaraj: Alternatives to Sustainable Development and the Green Economy. *Development* 57(3–4): 362–375.

Lowery, A. (2018). *Give People Money*. New York: Crown.

Mason, P. (2016). *Post-Capitalism: A Guide to the Future*. London: Penguin.

McAfee, A. and Brynjolfsson, E. (2017). *Machine, Crowd Platform*. New York: Norton.

Owen, D. (2011). *The Conundrum*. New York: Riverhead Books.

Polanyi, K. (1957). *The Great Transformation*. Boston: Beacon.

Pollin, R. (2018). Degrowth Versus a Green New Deal. *New Left Review* 112 (July–August): 5–25.

Prada, R. (2013). 'Buen Vivir' as a Model for the State and Economy. In M. Lang and D. Mokrani (eds.). *Beyond Development: Alternative Visions form Latin America.* Quito and Amsterdam: Rosa Luxemburg Foundation and Transnational Institute, 145–158.

Sachs, J. (2015). *The Age of Sustainable Development.* New York: Columbia University Press.

Solon, P. (2017a). Deglobalization. *Focus on the Global South, Systemic Alternatives.* Available: www.systemicalternatives.org/2017/05/10/deglobalisation/ [Accessed 27 October 2020].

Solon, P. (2017b). Vivir Bien. *Focus on the Global South, Systemic Alternatives.* Available: www.systemicalternatives.org/2017/05/10/deglobalisation/ [Accessed 27 October 2020].

The Economist (2009). Turning Their Backs on the World: Globalization. February 19. Available: www.economist.com/international/2009/02/19/turning-theirbacks-on-the-world [Accessed 4 January, 2019].

Varoufakis, Y. (2017). *Adults in the Room: My Battle with Europe's Deep Establishment.* London: Bodley Head.

3

RETROLIBERALISM AND DEVELOPMENT

Warwick E. Murray and John Overton

Introduction

Retroliberalism is a paradigm in development thinking that has increasingly influenced policy-making, particularly in the aid sector, since the Global Financial Crisis (GFC) of 2007–2008. The approach highlights that the state exists to protect and facilitate private capital, which is seen as the main engine of growth. Critically, the state supports the way capital extends its reach beyond the domestic economy. Retroliberalism links to the past as it draws on some of the early ideas of classic liberalism and mercantilism, as well as the policies and attitudes associated with both modernization and neoliberalism. As with all paradigms in development, its roots can be traced to a point earlier than the major shock, in this case the GFC, which brought it to the forefront.

In this chapter we explore the relationship between retroliberalism and development, its roots, outcomes, and prospects from both empirical and theoretical viewpoints. First, we briefly outline the theoretical argument for retroliberalism and compare it directly with other influential paradigms in development in rough chronological order from modernization, through structuralism, neoliberalism, and neostructuralism. In the second section, we trace the history of the concept and address the question: what makes the paradigm *retro*? We focus here on the legacy of the modernization and mercantilist periods. In the third section, we discuss some of the policies associated with retroliberalism in the aid sector in particular and trace out the rise of the 'shared prosperity' and 'exporting stimulus' discourses. This is illustrated through aid policies and patterns across the world and shifts in policies in a number of case study countries. The fourth section raises a number of controversies about retroliberalism and suggests a number of areas for future research. The chapter concludes with a discussion of the future of retroliberalism in the context of the rise of autarkic nationalism and the multipolar geopolitical world that characterizes the early 2020s.

Crisis – development paradigms and the rise of retroliberalism

The GFC that struck the global economy in 2007–08 profoundly threatened the prosperity of economies the world over, but especially in the West. Faced with the imminent collapse of banks and very large corporations, many governments resorted to Keynesian-inspired stimulus

DOI: 10.4324/9781003017653-5

packages. Huge bailouts allowed firms such as General Motors to survive and not contribute to growing unemployment. While there were cuts in public expenditure in the face of reduced state revenue, many governments in the West sought ways to use fiscal policy to bolster demand and encourage economic activity. This marked a departure from the dominant neoliberal prescriptions for economic management: balanced budgets, structural adjustment, and supply-side economics. Along with the buoyant Chinese economy (which survived the GFC relatively unscathed through assertive state policies to support economic activity), the partial reversion to Keynesian economics seemed to help the global economy recover from the financial crisis and not descend into prolonged and deep economic depression.

Unlike the recovery from the Great Depression of the 1930s, where governments had attempted to stimulate economic activity through inward-focused public works schemes and welfare payments, post-GFC policies added a new element: support for the private sector in externally-directed activities. Whilst many donor governments cut real expenditure sharply in many sectors, such as health and education and even defence, they maintained and even, in the case of the UK, slightly increased their aid budgets (Banks et al. 2012; Mawdsley et al. 2018). Elsewhere, such as Canada, Australia, and New Zealand, aid budgets were maintained at current levels (even if experiencing a slight fall in real terms) before some later cuts. But within such budgets, we saw a major shift from aid spending on established poverty-related programmes ostensibly led by recipient states, to incentives for domestic (donor) companies to conduct business overseas in developing economies. Whilst centre-right governments might have baulked at explicit and direct subsidy for companies operating domestically, it was possible to justify support for these same companies if they were ostensibly contributing to some grander global development mission.

This key shift in aid spending – what we have termed 'exporting stimulus' (Mawdsley et al. 2018) – was encompassed within a new mantra for development assistance. Gone was the former neostructural central concern for poverty alleviation (as articulated in the Millennium Development Goals (MDGs)), and instead focus was on 'sustainable economic development' and the principle of 'shared prosperity' (see DfID 2014). In practice, 'shared prosperity' justified the use of aid budgets and the expansion of development finance programs to support directly or indirectly the international operations of domestic companies if they contributed to economic growth in some way. This was an extension of the old 'trickle-down' principle to one that could be seen as 'trickle-out.' It shared the neoliberal belief that aggregate economic growth, led by dynamic firms, was the best means to ensure growth – and ultimately development for all – but it allowed, and encouraged, direct state stimulus for development. Furthermore, in a break with neostructural strategies, it put emphasis on national institutions and interests, rather than global agencies and agreements. National self-interest of donors – primarily economic but also strategic and diplomatic – largely replaced supposed commitment to global goals and cooperation.

As well as the shift to the 'shared prosperity' motto, these significant post-GFC changes to Western aid budgets and operations were accompanied by several common and telling institutional transitions. In a range of countries (Australia, Canada, New Zealand, and most recently the UK), former independent government aid agencies (AusAID, CIDA, NZAID and DfID) were disestablished and re-incorporated within wider foreign affairs and trade ministries. Aid agencies were rebranded with national self-interest, which was explicit in agencies' logos (the Union Jack, the kangaroo, the maple leaf, and silver fern for example). New partnerships were also forged, especially with donor companies, to cooperate in development activities overseas. In several cases, government aid agencies began to act as a development broker: providing funding for feasibility studies etc.; supporting civil society organizations and the private sector to work together (the former providing invaluable local contacts and knowledge for new business operations); and negotiating with recipient government agencies to help facilitate operations.

Subsequently, aid agencies have moved further to explore new forms of subsidy through, for example, notions of 'viability gap funding' (covering possible risk and non-viable profit margins so that businesses are encouraged to invest) or promoting new forms of 'blended' financing for development (such as development impact bonds) (Overton and Murray 2021). One of the most prominent, and controversial, examples of these new approaches has been the activities of the former Commonwealth Development Corporation (now the CDC) in the UK. In 2015, the British government announced that CDC would receive an additional £735 million in funding over three years through the aid budget. At the time, DfID justified the funding by stating: 'This will ensure countries can grow and trade their way out of poverty while building future markets that British businesses can compete in' (Anderson 2015).

Thus, we have seen a profound change in the nature of development policies since 2008. There is still a strong emphasis on the role of the private sector in promoting economic growth but practices of bailouts in times of crisis, direct state subsidies for businesses, and increased aid budgets represent a marked shift from the neoliberalism of the 1980s and 1990s. Yet it can also be seen as a significant shift away from the first decade of the new millennium and the neostructural concerns for internationalism, poverty-focused aid, and the strengthening of state institutions. For recipient states, these presented major challenges in terms of adaptation. Neostructural use of direct funding for government activities (such as Sector Wide Approaches, or SWAps), had been largely effective, predictable, and welcome sources of support. They used recipient systems and institutions in ways which conformed to the Aid Effectiveness program principle of recipient ownership. However, retroliberalism tended to shift back to project aid modalities that often bypassed the local state in favour of business partnerships, and which were typically smaller-scale and narrower in scope. Local ownership faded as donors sought to promote their own aid brands. In many ways, this was a partial return to the aid landscape of the neoliberalism of the early 1990s: projects rather than programs particularly focusing on infrastructure and often tied to use of donor-state firms, private sector rather than the state, and economic growth rather than poverty alleviation.

This new approach has been noted and discussed by a number of authors, who see it largely as a new manifestation of neoliberalism (Hendrikse and Sidaway 2010; Peck et al., 2010; Peck and Theodore 2015). We, however, regard it as a significant shift in development – a new paradigm that is largely superseding neostructuralism. We term it *retro*liberalism because, we argue, it has reintegrated several elements from older paradigms, including neoliberalism, modernization, and mercantilism.

Table 3.1 depicts and summarizes the main development paradigms of the past 70 years and how these have interacted with aid regimes. These are neither discreet nor exhaustive, as such approaches overlapped in time and space and some utilized an eclectic variety of ideologies and strategies – modernization, for example spanned neoclassical and Keynesian economics, and incorporated elements of Latin American structuralism. We also suggest that there is a degree of path dependence in these paradigms, in that each grew out of what came before, often with global crises punctuating and prompting a break from the status quo and a deep reform of institutions and practices.

History

Retroliberalism incorporates elements from a range of older paradigms. At first glance, retroliberalism has much in common with modernization theory of the 1950s and 1960s. The

Table 3.1 Development paradigms and aid regimes 1950–present: selected events, principles, goals and policies

	Modernization 1950–1980	Neoliberalism 1980–2000	Neostructuralism 2000–2010	Retroliberalism c. 2010 to present
Global events	Allied War victory, evolving Cold War, Truman's four point programme	Debt crisis; fall of the USSR	MDGs, 9/11 and the hollowing-out of the state	GFC, the rise of China and other Southern 'emerging powers'
Domestic political context in West	Cold War politics, Kennedy Alliance for Progress	Thatcherism, Reaganomics	The rise of Tony Blair's New Labour (UK) and Clinton's democrats	Swing back to the right – Cameron, Republican control of Senate in US culminating in Trump and Johnson in the UK and US.
Principles	Modernist and structuralist ideas concerning role of industrialization and backwardness of rural development. Geopolitical imperative of preventing domino effect across the Third World; socialist modernities, dependency theorists	Neoliberal theories. The state crowds out the private sector and leads to inefficiency and corruption. The market will arrive at Pareto Optimum. Benefits of export growth will trickle down to poor through employment growth	The state tackles social justice and poverty but in the context of a globalized economy that remains open. Delivering the benefits of globalization and ensuring its trickle-down and pro-poor growth	The state exists to facilitate economic growth, the private sector should not be crowded out by the state, the state sponsors and facilitates the private sector. Ricardian comparative advantage
Development goals	Grow industrial sector, promote regional alliances, promote urbanization and reduce rural inefficiencies	Reduce size and scope of government, raise productivity, stimulate exports	Poverty alleviation, equality promotion, aid effectiveness through mixed market and state mechanisms	Economic growth, infrastructure development, stimulate trade and investment, increased financialization
Aid policies and modalities	Import-substitution. Industrialization, land reform, state planning, mix of programme and project modalities	SAPs, export-orientation, deregulation, privatization, hollowing-out of the state, reduction in social expenditure, 'good governance' and re-regulation. Focus on project-based activities	MDGs, national interest and development agenda (formally) separate, poverty reduction-based projects, SWAps, general budget support, reconstruction of the state for security	Infrastructure, semi-tied aid projects, new (returnable) forms of development financing, development for diplomacy and the rolling together of national interest and developmentalism, partial return to project modalities

Source: Adapted from Overton et al. 2019: 31–32.

emphasis on economic growth as the driving factor in development (sometimes packaged as 'sustainable economic development'), and the central role of infrastructure in order to open up markets almost seem lifted from the textbooks of the 1960s written by W.W. Rostow and others associated with this ideology. The role of the state at the centre of this and its role in facilitating the private sector is also important. The implicit and sometimes explicit tying of aid to the domestic agendas of donor counties harks back to the role of aid in the post-WWII order – particularly in the burgeoning of the Cold War. As such the retroliberal model has also seen the re-entanglement of diplomacy, aid, and trade, though they were never truly separated.

In some ways then, retroliberalism can be cast as the new modernization – it is centred on infrastructure and state-building and directly followed, as did modernization, a major economic and political restructuring crisis in the global economy (the Great Depression and WWII and the GFC respectively). This version of modernization is different however, as it is more explicitly outward-oriented or globalized than the earlier approach, and the renewed focus of comparative advantage means that the range of commodities and products for export includes tertiary and quaternary goods, as well as traditional manufacturing and primary product exports. In a nutshell it is 'post-industrial globalized modernization.' As with the earlier version of modernization it retains a significant urban bias.

What makes retroliberalism 'retro'?

Retroliberalism is more than just post-industrial globalist modernization. There are areas where the new regime diverges from the old modernist project – some of which are more reminiscent of neoliberalism and some of which are completely new. The focus on private sector accumulation and outward orientation as the model of development in the era of globalization has been clearly perpetuated. On the other hand, the rise of bailing out the private sector and state support is neo-Keynesian.

Yet the new regime not only echoes the modernist past of development thinking but the deeper colonial history, particularly colonial companies of the expansionist period of late mercantile and early industrial colonialism. Of course, the companies of today are larger and more globalized, but the representation that their expansion is of mutual benefit to the core and periphery is reminiscent of that time. There is also a darker underlying element to the new discourse implicit in retroliberalism, namely that opening new markets and opportunities somehow 'civilizes' ('develops') the periphery.

In tracing the historical antecedents of retroliberalism, we point to several epochs where we see institutions, strategies, and policy discourses that resonate with the present. In this sense the model is deeply regressive.

Mercantilism

Prior to the extension of formal colonial control and the scramble for overseas territories, Western economies employed a more subtle – but often no less violent and disruptive – strategy to advance their economic interests globally. Mercantilism involved trade as the key engine of growth for European economies. Here states, otherwise loathe to intervene in domestic economies, supported companies to operate overseas, often by granting monopoly powers through state charters. This fulfilled the dual purpose of promoting economic growth and the national interest, but in this case, without direct intervention through the seizing and governing of overseas territories. There was also an implicit recognition that the extension of trade and markets amounted to a sort of 'civilizing mission' through the spread of capitalism. Retroliberalism echoes this approach, with state-supported private companies extending their operations overseas in the name of 'development.' And they do so, often employing modern-day consortia combining aid budgets, foreign investment, local government agencies, and even development NGOs.

Imperialism

The nineteenth century age of imperialism, when European powers grabbed overseas territories, founded settlements, appropriated key resources, and violently suppressed indigenous economies and polities, seems a far cry from the aid world of the present day. However, there are some important similarities. Unlike the brief era of international cooperation in the neostructuralist decade of the early 2000s, retroliberalism has been associated with some old-style nationalism and superpower rivalry in the form of 'spheres of influence' and the promotion of national brands, combining private enterprises with highly visible bilateral aid programmes and projects. These are seen in concepts such as 'co-prosperity' and 'shared prosperity.' China, for example, is often accused of blatant promotion of national self-interest through its foreign policy and aid policies – and often questionable use of loans and debt as forms of soft power – yet Western donors increasingly operate in very similar ways, with national self-interest now an explicit goal of aid programmes.

Modernization

Modernization strategies of the 1950s and 1960s were founded on Keynesian economics and the role of the state as the key agent to stimulate and manage economic development and growth. Within this, infrastructure and market development were posited as critical, linking people with markets and aligning domestic market growth with, and modelling their development on, Western economies. We see many shadows of this modernization approach in contemporary development discourses. The roads and bridges approach to infrastructure of the 1960s is matched by concern for mobile phone networks and 'financial inclusion' strategies today. Similarly, the new iconography of aid matches the optimistic promise of modernization: images of starving babies and squalid settlements are replaced by visions of thriving businesses and bustling markets. Furthermore, while Cold War geopolitics provided the backdrop for the utopian modernization imaginaries of the 1960s, we see in increased donor competitiveness in the past decade (particularly between China and the West) new constructions of modernity centred on business vitality, trade, and consumerism.

Neoliberalism

Neoliberalism extols the primacy of market as the driver of growth and prosperity. There should be a minimal role for state in economic sectors and state regulation should facilitate not restrict the operations of markets. The 1980s and early 1990s were characterized by neoliberalism in the aid sector, where aid, and development more generally were guided by policies of structural adjustment. The latter had come about in response to the debt crises of the 1980s. Based on the ideas of influential economists from the USA (Milton Friedman in particular), there was an earlier rolling out of neoliberal ideas, including most notoriously Chile, following the coup of 1973 where Pinochet was convinced by economists trained at the University of

Chicago to adopt almost pure free-market measures (Barton and Murray 2002). Aid quickly became contingent upon free-market reforms across the Global South. Following the end of the Cold War, the more explicitly pro-capitalist policies of the World Bank were toned-down as the threat of socialism was removed. Retroliberalism certainly re-emphasizes the primacy of the market and economic growth – and promotes globalization – though it involves a more pragmatic and assertive – hence neo-Keynesian – approach by the state to intervene and stimulate economic activity.

Neostructuralism

Retroliberalism emerged from a period in the first decade of the 2000s which was characterized by mixed state-market 'Third Way' strategies. This has been termed neostructuralism by a number of authors (Leiva 2008; Murray and Overton 2011; Overton et al. 2013) and echoes those who talk of post-neoliberalism, particularly as it emerged in Latin America (Grugel and Riggirozzi 2012; Laing 2012; Ruckert et al. 2017; Springer 2015; Yates and Bakker 2013). This had a marked imprint on the global aid regime, being associated with a focus on poverty alleviation and the MDGs, a marked increase in aid volumes and high degrees of international cooperation. It was still strongly associated with globalization – its export-orientation contrasted with the import-substitution focus of structuralism of the 1950s. Yet neostructural reforms did not have long to become firmly embedded before they were disrupted by the GFC. Goals such as the shift to SWAps and General Budget Support (GBS), let alone many of the ambitious targets of the MDGs, were not realized. However, retroliberalism drew on this foundation of large aid budgets, use of poverty and sustainability as aid justifications, and global-orientation. Yet it has also significantly shifted and remoulded these approaches to development, although the practices of development, on the whole, did not advance as far as the neostructralist and linked MDGs agenda might like to suggest.

Thus, while on one hand retroliberalism can be seen as a new development paradigm, emerging in response to the GFC, we argue that, in doing so, it incorporated and expressed many of the past discourses, strategies, agencies, and policies utilized by Western capitalism as it spread globally over the preceding three centuries (Murray and Overton 2016). Six key features summarize the aspects of retroliberalism and their links to past development paradigms in roughly chronological order:

1. The state is co-opted and subservient to the market, which is reminiscent of classical liberalism.
2. The role of state is to be largely subservient to capital – unlike neostructuralism which sees capital and the state mutually reinforce each other with the objective of improvements in broader wellbeing.
3. Sees the purpose of international interaction – especially in the developing world – as a means of stimulating and supporting domestic capital including industry and services (in this there are echoes as far back as the chartered East India Company).
4. Places private capital and the market first – as does neoliberalism.
5. Puts economic development at the fore – and investment in infrastructure – which harks back to modernization.
6. Is often neo-colonial in its aspirations to maintain the economic and political dominance of the Global North.

Places/policies

What are the policies associated with retroliberalism and how have they been rolled out in specific sectoral and country cases? In this section, we focus on the evolution of the paradigm in the aid sector and its influence on the aid regime. As previously noted, the retroliberal regime in aid came about following the GFC and involved a significant turn away from the more inclusive and holistic policies of the neostructural 2000s. Gone were the aid agencies separated from broader diplomatic missions intended to facilitate objectivity and the non-politicization of the aid sector. The focus on development efforts returned in various forms in various parts of the world to economic development to be stimulated by infrastructural investment. Finally, and most tellingly, there was a shift away from the rhetoric of altruistic aid to the new 'shared prosperity.' In the face of the austerity of the GFC, governments in places as diverse as New Zealand, the Netherlands, the US and beyond, re-defined their aid objectives placing national interest largely first. In some way this mirrored what China was already doing – through its soft-loan system of cooperation across the world (McEwan and Mawdsley 2012; Sears 2019).

Elsewhere we have referred to this evolution as 'exporting stimulus' (Mawdsley et al. 2018) and it helps explain why immediately after the GFC aid per capita from most donors did not decline. What in fact happened is that the nature of aid shifted and became aligned much more closely with the interests of the donors. This, to us, at the time, echoed the relationship between the colonies and the various European empires of the late 1800s. Retroliberal policy in the aid sector was observed in New Zealand, Australia, Canada, the United Kingdom and the Netherlands as is discussed directly in Mawdsley et al. (2018, see Table 3.2). Although the evolutions of these sets of policies differed in detail they were unified by a diluting of the previous poverty-focus, a weakening and re-branding of the institutions responsible for aid programmes, an increase in the role of the private sector in terms of aid modalities, and increased emphasis on infrastructural aid projects.

Controversies

In this section we focus upon three controversies posed as questions regarding the evolution, nature, implications, and future of retroliberalism with regard to development which could form the basis for discussions. We finish the section with a set of suggested empirical studies that would help illuminate some of the posed controversies.

Is retroliberalism a new aid regime?

We argue that retroliberalism should be perceived as a fundamental shift in the way that aid relations are undertaken and evolve. There can be little doubt that the core objective of aid has shifted from poverty reduction to economic growth. In many cases donors undertook significant political readjustments as ministries were reshuffled to reflect the new objectives. There has been a new geography of aid relations as a consequence too, as former colonial powers often reset their focus to a narrower set of countries with whom they had political and often continued trading relations. This also meant that aid was not explicitly directed to those most in need (as defined by poverty indices – though this was rarely the case previously). Rather it seeks countries with a high potential for economic growth and trade, where a nascent business sector can be worked with.

There are signs in some places, in New Zealand for example, that some aspects of the retroliberal turn are being reversed and the focus has shifted once again back to more humanistic

Table 3.2 Retroliberal aid regimes in selected cases

	Regime change	Central mission	Institutional change	Private sector	Total budget
New Zealand	National Govt (Key/ McCully) 2008	'Poverty alleviation' changed to 'sustainable development in developing countries in order to reduce poverty and contribute to a more secure, equitable and prosperous world'	NZAID (semi-autonomous) reintegrated into MFAT	Direct involvement of NZ companies (Fonterra, Meridian) tying of aid (e.g. tertiary scholarships increase) Infrastructure projects (airports, energy)	Aid budget increased but at lower rate of increase
Australia	Liberal Govt (Abbot/ Bishop) 2013	Poverty focus diluted: 'promoting prosperity, reducing poverty, enhancing stability'. 'Aid for trade' 'Australia's national interest'	AusAID (standalone) folded into DFAT and disestablished 2013	Move to infrastructure projects	Cuts to aid budget (12% in 2013, more in 2014) Capped at $5 bill for 5 years
Canada	Conservative Govt (Harper/ Fantino) 2011	Poverty reduction enshrined in law but … 'sustainable economic growth' given prominence 'economic diplomacy'	CIDA amalgamated with FATDC (alongside trade and foreign affairs) 2013	Involvement of Canada's private sector Interest in countries with mineral resources	Aid budget cuts then stabilization beyond 2015 at $4.62 bill (0.3% GNI)
UK	Conservative/ LibDem Govt (Cameron) 2010	long-term programmes to help tackle the underlying causes of poverty … 'economic development for shared prosperity'	DfID retained but rebranded (UKAid)	Accusations that Africa funding is used to support land grabs by MNCs	Aid budget increased (30% in 2013- to 0.7% GNI)
Netherlands	Conservative/Labour coalition (Rutte) 2012	'sustainable economic growth in developing countries … global stability and security and to foster human rights' shift from aid to trade	Part of Ministry of Foreign Affairs Major review in 2010	'new markets to explore' 'an enabling environment for economic activity'	Cuts in 2012 – achieved then abandoned 0.7% target

Source: Adapted from Mawdsley et al. 2018.

41

objectives under the guidance of the current centre-left coalition government (Labour, Green and New Zealand First parties) led by Prime Minister Jacinda Ardern. However, this case seems to buck the international trend and may not last, particularly as the effects of the current COVID-19 pandemic hit the global economy. In addition, even in New Zealand, a number of the retroliberal reforms remain untouched including, crucially, the reintegration of the formerly independent New Zealand aid ministry into the Ministry of Foreign Affairs and Trade. In short, there is nothing radically new in retroliberalism – it remains a collection of well-tried (and often discredited) strategies and policies. We need to look more closely at different countries and how retroliberalism has evolved and been shaped in different political environments and in various multilateral agencies.

Can retroliberalism be considered a development paradigm?

Although retroliberalism has found its first and expression in the aid sector, it is our contention that retroliberalism is broader than that – the shift to nationalistic self-interest, the rise of South-South cooperation and the evolution of a multipolar world in the post-GFC era has implications across the global political economy. The naivety with which the market and the invisible hand are viewed have declined. The political acceptability of open borders has faded. In this sense, retroliberalism can be seen as a broader shift in political economy.

We are heading for a post-aid world (Mawdsley et al. 2014). The idea of non-reciprocal altruistic aid (if it was ever so) is rapidly disappearing. This aligns and is spurred by the example of China – which prefers the term 'development cooperation' (like 'shared prosperity') rather than 'aid.' It is about investment, doing business, and returning a dividend to donor companies – as well as furthering national diplomatic and strategic interests. So retroliberalism is not just about aid. In fact, it may be about the dismantling of aid as a concept – and replacing it with a new and wider form of global economy: nationally competitive, state-supported, and with more explicit (shame-faced) donor power being exercised. Fundamentally, retroliberalism as a paradigm for global development is about doing good through doing deals – for business and for national interests.

How does retroliberalism articulate with contemporary autarkic nationalism?

Retroliberalism has links to nationalistic new-right domestic politics: harder borders, the retreat from multilateralism, anti-immigration, pro-business, and less concern for domestic welfare, poverty, and inequality – the nationalistic new-right. It started with a centrist/centre-right response to the GFC (Cameron, Obama, Key, Merkel) but appears to be moving to a stance that is farther to the right (Trump, Johnson, Bolsonaro, Piñera), being more overtly nationalist and breaking with internationalism. At the present time, retroliberalism sits apparently comfortably alongside forms of autarkic nationalism. Trump has cut aid and has more explicitly tied it to national interest (and to narrow political interests – such as cuts to Planned Parenthood). The USA already had a rather retroliberal aid structure – especially through use of private contractors and long-standing explicit self-interest. Perhaps this – and Johnson's post-Brexit UK – are signposts: retroliberalism is becoming even more about national interest and less engaged with global discourses of good practice such as untied aid – and reduced aid budgets are now following. Will the aid world in retroliberal times becomes more competitive and removed even further from global agreements and projects such as the Sustainable Development Goals (SDGs)?

Proposed studies of retroliberalism

There are a range of studies that could throw light on the questions posed above. First, it is important that the aid studies that inform this work are extended where possible – especially beyond the Pacific. There is no doubt, as we briefly described above, that evidence exists in the case of aid relations in the broader Asia-Pacific, Africa, and Latin America that policies are increasingly retroliberal. The rise of China as a 'development partner' in these regions – where heavy investment infrastructure, based on soft loans and often utilizing Chinese companies, parts and labour in many ways mirrors the retroliberal turn that has taken place among Western aid powers elsewhere. In fact, retroliberalism can be seen in part as a reaction to these changes in the way China engages with the Global South.

A second set of possible studies regarding the rise of retroliberalism relates to the impacts of Chinese aid and growth. As – or if – Chinese growth slows and its thirst for resources diminishes at least in a relative sense, this will alter the context within which retroliberalism takes place. It may be that this demand is replaced by increases in economic dynamism elsewhere (such as India and possibly even Latin America – especially Brazil – in the longer term). The geopolitics of aid is an important consideration – if China reduces its imprint on the Global South and withdraws some of its soft-loan programs, how will this impact the nature of the aid regime?

Thirdly, a set of studies relating to studies of retroliberal tendencies beyond aid is required. Much of the work to date has focused on this area as it is where retroliberalism has been most apparent. Thinking moves quickly in the aid world, regimes shift and modalities are adjusted. As such movement in aid policy is often a good barometer regarding the general political economic conditions – this is probably because it does not figure heavily in the public debate and is not an election issue for the majority. Governments are relatively free from public demands in setting policy. However, there can be little doubt that we have seen the incorporation of retroliberal ideas in other development sectors, such as trade policy, investment and even peacekeeping, and work in these areas is sorely required.

Finally, we would like to encourage work that deals with the theoretical questions regarding retroliberalism. Can we consider it a new regime of accumulation? How do we articulate the relationship between the state and the market? How do we measure the influence of private capital on states and their co-option in the process? Retroliberalism requires more detailed theoretical elucidation.

Futures

The nature of development policy emanating from powerful countries and applied to the Global South has changed. We have witnessed this very clearly in the aid world, where there was a shift from the fundamental forms of neoliberalism in the 1990s to neostructuralism in the 2000s, embodied in the MDGs and later the SDGs. The GFC created a counter-turn, sending aid policy back towards the self-interest and econo-centrism of the past. We coined the term retroliberalism to reflect this new regime, which incorporates elements of modernization and liberalism, while at the same time seeing a strong facilitating and protective role for the state in the Global North especially, over which private capital has gained increasing influence. The retroliberal model has seen the reintegration of diplomacy, aid, and trade. Although it was only ever very partially and imperfectly separated under the neostructural aid regime, at least it constituted a formal ideal and to some extent acknowledged norm. There are areas where the new model diverges from modernization – some of which are more reminiscent of neoliberalism and some of which are completely new. Private sector accumulation and export-orientation have never been absent, of course, but they are now being rejuvenated.

All aid regimes and development paradigms more generally are of their time – history makes theory, theory makes history. Modernization arose out the ashes of World War II and embodied an attempt ostensibly to rebuild the world in the image of the prosperous West and also to prevent the spread of communism. Structuralism and, even more so, dependency, was a reaction to the Western-centric nature of development thinking and arose in the periphery as a challenge to it. The debt crisis facilitated the spread of neoliberalism as the expensive follies of modernist ambitions and resultant obligations precipitated a major crisis. Just so, retroliberalism has come about in response to a set of historic factors that are, by definition, unique. The GFC and the resultant turning inwards of Western economic and political interests, coming at the same time as the rise of China as a major economic force have been the main factors in the rise of retroliberalism. Notwithstanding this point, the term retroliberalism reminds us that while historic configurations are always unique, they are also inextricably linked to processes of the past. There is path-dependency in the evolution of development thinking and theory. New paradigms pick up on ideas of the past just as they react to conditions that echo the past. There has been a shift in the approach to development that is mirrored across the world in various guises under banners such as 'shared prosperity,' as aid and development relations shift in the context of an increasingly nationalistic and multipolar geopolitical moment. In this multipolar reality, the lines between the Global North and Global South are increasingly blurred and the concept of aid is called into question.

There is an intricate geography of development paradigms that we often forget in our rush to explain everything through one set of propositions. They do not evolve neatly across space. They are not uniform in their application. This is most certainly the case with retroliberalism. It is not equally as applicable everywhere; it has not manifested itself in the same ways and its consequences are far from uniform. As the world appears to be shifting into a multipolar geopolitical moment, we would do well to remember the importance of geography. The myth of a borderless world, a level playing field, and mobile labour has been laid bare. Retroliberalism is a paradigm for the age of self-interested, inward-looking, non-sustainable, capitalism. It will be reworked and remoulded by all of the powerful blocs in this multipolar world – the USA, the EU, China, India, and Russia – as the rush for resources from the Global North and South continues unabated. The scramble for Africa of the late 1800s, which reflected competition among European powers and was a major factor in precipitating the First World War and resonates today in terms of conflict and poverty, may well pale into insignificance when we consider the power of and the unsustainable demands from the new set of great powers.

Retroliberal development policy is a sign of self-interested, neo-colonialism, econo-centric thinking, and sits well, and in some ways has underpinned, the rise in autarkic nationalism sweeping the globe currently from Brazil to the USA onto the UK and beyond. These trends are taking place exactly at the point when they are perhaps potentially most damaging given the climate emergency, the pandemic crisis, and catastrophic development inequality. The potential for global conflict sits in a tinderbox, rather like it did between the Great Wars – only this time on a much larger scale and with the potential to be deeper and more rapidly devastating than ever. In this sense retroliberalism should be researched, understood, resisted, and demolished if development is to ever approximate anything nearing human and environmental progress.

Key sources

1. Hendrikse, R. P., and Sidaway, J. D. (2010). Neoliberalism 3.0. *Environment and Planning A*, 42(9), 2037–2042.

This article reflects on the evolution of neoliberalism through three stages. It argues that while change can be observed there is an essential continuity in the model across the world.

2. Leiva, F. I. (2008). *Latin American neostructuralism: The contradictions of post-neoliberal development,* University of Minnesota Press.

This book reflects on the rise of a new and adapted form of structuralist political economy in Latin America. The neostructuralist approach has much in common with neoliberalism according to Leiva.

3. Murray, W. E. and Overton, J. (2016). Retroliberalism and the New Aid Regime of the 2010s. *Progress in Development Studies*, 16(3) 1–17.

This article coins the term retroliberalism, defines it and offers case studies of its unfolding in the international aid sector.

4. Overton, J. and Murray, W.E. (2021). *Aid and Development*, London: Routledge.

This upper-level textbook discusses the nature, history, patterns, and futures of aid and pays particular attention to the influence of retroliberalism and other development paradigms.

5. Springer, S. (2015). Postneoliberalism? *Review of Radical Political Economics,* 47(1): 5–17.

This is a radical and critical discussion of the concept of postneoliberalism which is argued to be a rhetorical device.

References

Anderson, M. (2015). DfID to pump £735m into investment arm for private sector projects. *The Guardian.* www.theguardian.com/global-development/2015/jul/17/department-for-international-development-cdc-group-735m-uk-aid-private-sector (accessed 6 April 2018).

Banks, G., Murray, W.E., Overton, J., and Scheyvens, R. (2012). Paddling On One Side of the Canoe? The changing nature of New Zealands' development assistance programme. *Development Policy Review* 30(2): 169–186.

Barton, J. R., and Murray, W. E. (2002). The end of transition? Chile 1990–2000. *Bulletin of Latin American Research* 21(3): 329–338.

DfID (2014). Economic Development for Shared Prosperity and Poverty Reduction: A Strategic Framework. Available at www.gov.uk/government/publications/economic-development-for-shared-prosperity-and-poverty-reduction-a-strategic-framework (accessed 6 March 2015).

Grugel, J., and Riggirozzi P. (2012). Post-Neoliberalism in Latin America: Rebuilding and Reclaiming the State after Crisis. *Development and Change* 43(1): 1–21.

Hendrikse, R. P., and Sidaway, J. D. (2010). Neoliberalism 3.0. *Environment and Planning A* 42(9): 2037–2042.

Laing, A. F. (2012). Beyond the Zeitgeist of 'Post-Neoliberal' Theory in Latin America: The Politics of Anti-Colonial Struggles in Bolivia. *Antipode* 44(4): 1051–1054.

Leiva, F. I. (2008). *Latin American neostructuralism: The contradictions of post-neoliberal development.* University of Minnesota Press.

Mawdsley, E., Murray, W.E., Overton, J., Scheyvens, R. and Banks, G.A. (2018). Exporting stimulus and 'shared prosperity': Re-inventing aid for a retroliberal era. *Development Policy Review* 36: O25–O43. DOI: 10.1111/dpr.12282

Mawdsley, E., Savage, L. and Kim, S-M (2014). A 'Post-Aid World'? Paradigm shift in foreign aid and development cooperation at the 2011 Busan High level Forum. *Geographical Journal* 180(1): 272–38.

McEwan, C. and Mawdsley, E. (2012). Trilateral development cooperation: Power and politics in emerging aid relationships. *Development and Change* 43(6): 1185–1209.

Murray, W.E. and Overton, J. (2016). Retroliberalism and the New Aid Regime of the 2010s. *Progress in Development Studies* 16(3): 1–17.

Murray, W.E. and Overton, J. (2011). Neoliberalism is Dead, Long Live Neoliberalism. Neostructuralism and the new international aid regime of the 2000s. *Progress in Development Studies* 11(4): 307–19.

Overton, J. and Murray, W.E. (2021). *Aid and Development*. Routledge: London (in press).

Overton, J., Murray W.E. and McGregor A. (2013). Geographies of aid: A critical research agenda. *Geography Compass* 7(2): 116–127.

Overton, J., Murray, W.E., Prinsen, G., Ulu, J.A. and Wrighton, N. (2019). *Aid, ownership and Development: The Inverse Sovereignty Effect in the Pacific Islands*. Routledge: London.

Peck, J., and Theodore, N. (2015) *Fast policy: Experimental statecraft at the thresholds of neoliberalism*. University of Minnesota Press.

Peck, J., Theodore, N., and Brenner, N. (2010). Postneoliberalism and its malcontents. *Antipode* 41(s1): 94–116.

Ruckert, A., Macdonald, L. and Proulx, K.R. (2017). Post-neoliberalism in Latin America: a conceptual review. *Third World Quarterly* 38(7): 1583–1602.

Sears, C. (2019). What counts as foreign aid: Dilemmas and ways forward in measuring China's overseas development flows. *The Professional Geographer*, 71(1): 135–144.

Springer, S. (2015). Postneoliberalism? *Review of Radical Political Economics* 47(1): 5–17.

Yates, J. S., and Bakker K. (2013). Debating the 'Post-Neoliberal Turn' in Latin America. *Progress in Human Geography* 38(1): 1–29.

4

DEVELOPMENT IN THE GLOBAL NORTH

Etienne Nel

Introduction

The period after World War II has been a tumultuous one for the economies and regions of the Global North. The initial post-war stability, partially anchored on Keynesian economic and spatial management, which sought to reduce regional economic disparities, stabilize economies and extend welfare support, gave way to the uncertainties of the neoliberal era from the late 1970s (Hudson 2005). Parallel processes of globalization, on-going global crises, including: the Cold War, the economic and debt crises of the 1970s and 1980s, the Global Financial crises of 2008 and the recent 2020 COVID-19 pandemic, have all severely tested national and regional resilience and led to fundamental shifts in economic theory and practice, accelerating social and economic inequality (Pike et al. 2017a). Parallel processes of deindustrialization, globalization and the rise to prominence of the service sector have taken place in the context of reduced state economic and regional intervention, which have often led to worrying increases in regional economic and social inequality (Sheppard 2012). Weak state action in this space has meant that growing social inequalities within and between societies have emerged as one of the key global challenges of the contemporary world.

This led, in the 1990s, to the recognition that there are regions of both 'recession' and 'resurgence' in the North (Chisholm 1990). In spatial terms underperforming or 'lagging regions' have become a stark reality not only between different parts of the world, with the most obvious being between the Global North and the South, but also within countries (Pallares-Barbera et al. 2012). Later writings have drawn attention to the processes and realities of 'uneven geographical development,' which argue that macro-economic forces and the crisis-prone nature of capitalism can significantly advantage or disadvantage regions (Harvey 2005, 2014). Efforts to understand, theorize and respond to such differences form the core focus of this chapter.

National development policy has clearly aligned with market principles in the North, however, approaches to spatial development have often only partially shifted to a reliance on market forces. There is a continued recognition that regional disadvantage is an issue which requires intervention and, in response, policy has generally evolved away from state-centric interventions to correct regional disadvantage to trying to put in place mechanisms to promote regional competitiveness, albeit often with only limited evidence of success (van Staden and Haslam McKenzie 2019). In parallel, a rich academic discourse has emerged which seeks to

DOI: 10.4324/9781003017653-6

better understand the development challenges and experiences of regions in the North. This has a particular focus on uneven development and structural change within regions, as well as the various forms of resilience and adaptation which may evolve, conditioned by past geographies and historical dependencies (Pike et al. 2017a).

This chapter overviews changes in the theories, practices, and understandings of development in the North with a specific focus on regional and local spaces. Key themes covered in the first section are: shifts in state economic policy, uneven geographical development, new regionalism and localism, place-based development, the importance of regional resilience, and the conceptual contributions made by thinking on evolutionary economic geography. After overviewing these themes, attention then shifts to examine applied policy and practice, with a specific focus on the approaches adopted in the European Union (EU), before concluding with an examination of how these topics could be incorporated within teaching and learning practice.

Key themes in regional development

Shifts in dominant political-economy discourse

Post-World War II the countries of the North, in broad terms, entered a 'golden age of prosperity' marked by widespread welfare gains and state economic management anchored on Keynesian-style economic policy. At the regional and local levels, this era was characterized by state subsidies for economic activity, nationalization of industries and state economic management. The crisis of capitalism in the 1970s following the oil price shocks, the Vietnam War and the American repudiation of the Bretton Woods system including the gold standard, plunged the world into a period of sustained economic uncertainty. By the end of the decade, there was a resurgence of pro-market and minimalist state approaches, as propounded by neoliberal thinking and the writings of the monetarists (Pike et al. 2017a,b). The rise of Thatcherism and Reaganomics from the late 1970s was associated with the embedding of neoliberalism globally and the interventions of the IMF and World Bank following the 1980s debt crisis fundamentally transformed state economic management (Tomaney et al. 2010). The so-called 'hollowing-out' of the state advocated by neoliberalism led to privatization, de-regulation, the scaling back of the welfare state, a reliance on market forces and a new devolution mandate (Hudson 2005). Together with the deindustrialization wave sweeping most countries of the North, areas such as the US Rustbelt, Germany's Ruhr and the 'North' in the UK experienced significant and protracted economic decline as production globalized and state support reduced, forcing regions to seek local responses, often with limited success (Pike et al. 2017b).

The net result has been an increase in the number of 'lagging regions.' At a broader level this reflects the inevitable emergence of uneven development and uneven geographical development as cogently articulated by Neil Smith (1984) and David Harvey (2005, 2014). It is the outcome of capitalism's spatial fixes and periodic crises, which have exacerbated, not reduced, regional inequalities.

In the prevailing economic and policy context, in which states have significantly reduced their influence, the local, regional, private and community sectors have often been ill-prepared to assume the mantle of responsibility for welfare and development. As a direct result, there has been experimentation with a range of endogenous responses. Evolving theory – such as new understandings of what the 'region' is, place-based development, regional resilience and evolutionary economic geography have sought to offer an explanation as to what the catalysts or barriers to change are. Attention now turns to look at evolving thinking about the embedded geographical unevenness that shapes prospects for development and change.

Uneven geographical development

Implicit within studies on globalization, economic change and development is the recognition that spatial factors and regional differences are key to the operation of the global economy, with globalization variously privileging or marginalizing different areas. To understand processes of global and national change, development opportunities and responses to inequality, adopting a regional lens has gained in popularity despite growing interest in globalization (Tomaney et al. 2010). As Scott and Storper (2003, 579) argue, 'the theory of development must incorporate the role of cities and regions as active and casual elements in the economic growth process.' New thinking on the region recognizes the role played by power dynamics, human agency, social structures, the intersection of endogenous, exogenous forces and global production chains.

The era of spatial Keynesianism, from the 1930s to the 1980s, had sought, often with limited success, to promote regional equality through state infrastructural support and economic subsidies. The subsequent switch to the promotion of regional comparative advantage has, inevitably, led to the often dramatic growth of leading and/or 'sunbelt' areas characterized by the knowledge economy and what Florida (2005) refers to as the presence of creative industries, creative cities and the creative class. In parallel, the 'triumph of the city,' or rather certain privileged and well-located cities, has been celebrated in certain parts of the world (Glaeser, 2011). However, as Florida (2017) acknowledges, success for some places and social groups has come at a very real cost to less advantaged social groups and places. These stark and growing realities have led to a resurgence of interest in both the reality and theory of 'uneven geographical development.' As a net result of this concentration of economic activity in certain locations, what are termed 'lagging regions' have emerged as places which 'occupy an inferior position in the landscape of economic unevenness … [they] struggle to gain in rank and to receive a … share of the wealth created in the global economy' (Pallares-Barbera et al. 2012, 4). This situation was exacerbated by the 2008 Global Financial Crisis which significantly impacted on uneven development in the wake of this crisis.

In applied terms, responses to persistent geographical unevenness have variously led to a reliance on the assumption that market forces will provide the long-term solution, or alternatively to more proactive responses, such as by the EU (European Commission, 2017), which is examined in the policy section below, or finally to efforts to boost 'localism' as has characterized recent responses to persistent spatial inequalities in the UK (Pike et al. 2017b), which is touched on in the following sections.

Understanding regions: new regionalism and localism

A key first step in a regional focus is to better understand the place-based dynamics operating within regions, how they are constituted and how they respond to change. How regions are understood theoretically, has evolved from seeing them as passive spatial entities and as recipients of external support to a more dynamic understanding of the relationality of regions, and the degree to which they are socially constructed and dynamic, in terms of the role which they play within broader political-economic contexts (Turok et al. 2017). Recent research on regional innovation and on co-existing processes of regional/local innovation and external engagement have encouraged a more sophisticated understating of regions and their functions based on notions of 'flow,' interaction and agency (Jonas 2012).

This in turn has led to a new understanding of regions through the lens of 'new regionalism' with its focus on roles of multiple level governance, political processes and social movements (Zimmerbauer and Paasi 2013). Earlier writings on new regionalism focused on the critical components of regional competitive advantage in the globalized economy and the role played by localized concentrations of knowledge, relationships and motivation. Such approaches were criticized for their focus on internal factors and the unique attributes of successful regions and for down-playing the role of territory and politics (Jonas 2012). In partial response to this critique, there is a revival of interest in the concept, with recent work focusing on the inter-action of regional-external process in defining regional identity, action and outcomes (Vodden et al. 2019). This new focus also considers resilience to change mediated through multi-level governance, innovation and learning, integrated development, rural-urban interactions, and place-based development, making the concept of new regionalism, flexible, adaptive and con-text appropriate.

A parallel line of reasoning, but one focused on lower-level spatial entities, has been the focus on 'localities' and 'localism' and the unique attributes of place, local action and leadership. This approach has links with the devolution of power to local places, communities, and institutions, which has been seen as a way to strengthen local democratic engagement, and as a pragmatic way of dealing with local challenges. Interest in this area has links to the work of Massey (1984) on the spatial division of labour that recognized the importance of relationality, contingency, the impermanence of localities and processes of locality making. This approach has, however, been challenged for being apolitical, down-playing external factors and empowering places ill-equipped to respond to global challenges. Recently, this line of thinking has been given impetus by political shifts in the UK where a new focus on 'new localism' as a government policy initia-tive emerged (Jones 2019). Yet this shift has raised queries about the capacity of different places to respond to the new devolution mandate and the capacity of local institutions to champion development (Pike et al. 2016).

Place-based development

The limited success of traditional forms of state support and subsidies for lagging regions has catalysed new thinking about the role of place in development. This approach partially draws from the arguments presented in the last section, but more specifically from the quest by the EU and OECD to address regional challenges by having a focus on 'place-based development' (PBD). It has evolved to become mainstream practice in many parts of the world (Thissen and van Oort 2010). The report of Barca to the European Commission on the topic and subse-quent writings shaped the approach, it argued that 'the relationships between geography and institutions, and also how these vary across space ... really matter' and are critical to develop-ment outcomes (Barca et al. 2012, 135). This has sparked renewed interest in the 'region,' par-ticularly in the EU, about how to achieve regional development though a focus on supporting local competitive advantage, local participation and multi-level governance, and through the provision of place-tailored public goods and services. According to Pike et al. (2016, 8), PBD focuses on the 'promotion of locally rooted human and knowledge-based assets through fine-grained locally conceived and executed development strategies that provide public goods aimed at improving the local business environment.' PBD thinking influenced EU Cohesion Policy for the period 2014–2020, leading to the promotion of community-led economic development and integrated territorial investments, based on the perceived need to revive regional identity and to rely on cross-boundary and cross-sectoral initiatives, open governance and leadership (European Commission 2014). In turn, this has helped galvanize OECD thinking around the potential to promote development in all regions – anchored on bottom-up initiatives, self-help, self-directed and self-financed development (Barca et al. 2012; Pike et al. 2016).

One of the more interesting outcomes of the rise to prominence of PBD has been its association with long-established thinking about the role which leadership – both by individuals and institutions – can play in catalysing community buy-in and successful development outcomes. Writing about 'place-based leadership' argues that effective leadership helps to explain why certain places adapt and exploit opportunities in a rapidly changing world and how deliberate action can alter economic trajectories (Beer et al. 2019). This has led to growing interest in the role played by local actors, both formal and informal, networks of key actors and institutions active in mobilizing resources and driving regional and local level change (Beer et al. 2019). These approaches accord with principles of democratic governance and decentralized control, however success is not guaranteed for all places because of variations in local comparative advantages and the risk of over-relying on the potential of market forces and key individuals (van Staden and Haslam McKenzie 2019). It is apparent that not all regions successfully adapt and transform through exogenous or endogenous interventions and this, in turn, has spawned interest in the latent and embedded characteristics of regions that may determine their potential to adapt positively. Factors such as levels of resilience, and historically contingent and path dependent processes, to which attention now turns, may predispose regions to particular development paths.

Regional resilience and evolutionary economic geography

Persistent uneven development has spawned an extensive research literature that seeks to understand and interrogate how regions respond to crises in terms of their inherent 'regional resilience.' The ability of regions to respond to crises is often determined by the social and economic attributes embedded in particular places which variously encourage efforts to return to past economic paths – 'path dependency' or promote adaptability and the pursuit of new 'path creation' or alternatively 'path diversification.' Authors such as Simmie and Martin (2010), Dawley et al. (2010) and Pike et al. (2010) have written extensively on these ideas, the concept of regional resilience and the characteristics of resilient regions. They have analysed the degree to which differing regions are able to reinvent themselves or struggle to transition to new economic realities following economic change or shocks. Drawing on regional case studies, Simmie and Martin (2010) have shown how the different components of regional economies, be they firms, organizations or institutions, have enabled certain regions to adapt to new economic activities, guided by the decisions of local actors drawing on endogenous knowledge. A key focus in recent writing has been on 'how and why local communities respond to major disturbances, including the balance between continuity and change' (Bailey and Turok 2016, 557) and the degree to which different regions manifest different levels of resilience to recession and shocks. This led to an analysis of the differential roles played by: industrial and business structures, finance, governance, agency, and labour market considerations in shaping regional reactions. Martin (2012) also identified a four-fold process of how regions react to and recover from recessionary shocks, namely through: resistance, recovery, renewal, and re-orientation. Successful adaptation to change relies on having high levels of regional innovation, modern infrastructure, proactive governance, entrepreneurship, local skills, supportive financial systems, proactive communities and a diversified economy.

The degree to which regions are 'locked-in' to historic development paths, which they may or may not be able to break-out from, has encouraged another rich strand of recent research, namely that of 'evolutionary economic geography' (EEG). It proposes a new lens to understand the factors determining regional path dependency or new path formation that considers: the actors and firm within regions, the influence of embedded historical factors and processes, the role of local institutions and the level of innovation (Henning 2019). Boschma and Frenken (2011, 296) explain that their EEG starts 'from the definition of economic geography as dealing with the uneven distribution of economic activity across space. An evolutionary approach specifically focuses on the historical processes that produce these patterns. The current distribution of economic activity across space is thus understood as an outcome of largely contingent, yet path dependent, historical processes. EEG thus focuses on: evolutionary economics, self-transformation within systems, uneven geographical development, adaptation and 'creative destruction' (Boschma and Martin 2007, 539).

This growing field of research has spawned investigation into historical processes of change within regional economies, geographical clusters, firm behaviour, and the role of local institutions. New thinking around 'path dependence' has, in parallel with EEG, sought to understand the opportunities and constraints which impact on the capacity of regions to diversify or alter their economic trajectory, dependent on historical factors, the role of local agency and institutions and external opportunities (Henning 2019). Particular strengths of EEG are that it helps to periodize change, it considers the impact of economic shocks at the regional level and how historical factors impact on the ability of regions to respond, while also seeking to account for firm behaviour, institutional response and the role of human agency. Pike et al. (2010) and Boschma and Frenken (2011) have usefully hypothesized the conceptual bridge between resilience thinking and EEG. They have done this by drawing attention to the historical factors and the inherent capacities within regions that help to promote resilience and adaptation to either existing or new pathways. Having examined how thinking about 'regions' has evolved, attention now shifts to applied policy and practice, which has, partially been informed by thinking around place-based development, neoliberalism and resilience.

Regional development policy and practice in Europe

The transition from Keynesian economic management to neoliberal orthodoxy at the state level, has encouraged clear transitions in the policy and practice of regional and local development interventions and support. Top-down state led interventions, financial subsidies paid to firms operating in lagging regions and redistributive interventions in much of the OECD gave way to the reduced role of the state from the 1970s. Devolution passed development responsibilities to lower levels of authority, paralleling an increased reliance on market-based interventions and the promotion of regional comparative advantage, rather than responding to regional disadvantage and pursuing equity. Here, it is noteworthy that these new approaches, while enhancing the success of leading regions, have generally failed to address regional inequalities, which have often grown in many countries (van Staden and Haslam McKenzie 2019).

From the 1980s, applied regional development saw a transition from the provision of regional subsidies, in many countries, to locally focused interventions designed to catalyse local growth based on perceived local comparative advantages. This was undertaken through support for a range of spatial interventions, such as 'enterprise zones,' 'export processing zones,' 'special economic zones,' 'empowerment zones,' and the encouragement of urban and local development partnerships. These interventions sought to draw in external investment, catalyse local initiative and tap into local, regional, national, and global production networks (Pike et al. 2017b). Individuals and localities, in response to both recurring crises and new-found powers have variously pursued place-promotion activities through investment attraction, 'imagineering' and the promotion of the creative sector and the creative city (Florida 2005; Glaeser 2011).

Regional development policy within many countries has shifted from top-down to bottom-up approaches but tensions have emerged over the appropriateness of sectoral versus territorial approaches and over support for infrastructure versus industry. There has been a growing focus on human capital development, support for young people, innovation, support for partnerships, participation, sustainable development and what are regarded as 'place-neutral' interventions (Pike et al. 2017b).

Arguably the most interesting interventions in terms of regional development in the North have been those pursued by the EU since its formation in the 1970s. In spite of the neoliberal transition, the EU has remained firmly committed to the principle of cohesion designed to improve the well-being and potential of 'lagging regions' with below averages scores in terms of key development indicators (European Commission 2017). This has led to significant investment over the decades to promote growth and address the structural challenges faced by the least developed areas in the Union. In recent years, policy has shifted from a redistribution to a growth focus, based on the limited success achieved, and the difficulties experienced in trying to transform intractable structural challenges. In parallel, there has been the recognition of the need to tailor development interventions to the unique needs and opportunities of each region and place i.e. PBD (Barca et al. 2012). According to the European Commission (EC 2014), regional policy now focuses on all regions and cities to boost their economic growth and improve quality of life, albeit that special support does remain for the most disadvantaged regions. The EC claims that between 2007 and 2012 an estimated 594,000 jobs were created through regional interventions and over 275,000 businesses were supported (EC 2014). Yet despite this apparent success and notable gains in terms of parallel support for human capital development, commentators regard the outcomes as mixed and making little impact on regional disparities (Pike et al. 2017b).

Since 2014, the concept of 'smart specialization' has come to underlie EU support in an effort to respond to the often-limited success of past interventions. Smart specialization is linked to the aforementioned concept of PBD and focuses on the use of technology and productivity increases, particularly in the knowledge economy sectors. Between 2014 and 2020, the EU allocated €351.8 bn to promote 'smart, sustainable and inclusive development' (Kristensen et al. 2018). In practice, this meant a focus on promoting growth enhancing sectors, territorial cooperation, training and retraining throughout the EU with additional targeted investment in regions with the lowest per capita GDP. Supported projects need to promote smart specialization (i.e. regional specialization in the sectors providing them with the highest potential for growth and competitiveness), reduce youth unemployment, promote reforms, and comply with environmental laws. Supported projects tend to focus on information and communication technology, research and development and low carbon initiatives (EC 2014). These approaches allow a focus on regional strengths and advantages, but have been criticized as potentially favouring winning regions, while being difficult to apply and adopt in the weakest regions (Kristensen et al. 2018).

In general terms, regional development interventions have encouraged the success of 'winning regions' but have often struggled to improve the fortunes of 'lagging regions,' while uneven development persists within and between countries. The switch from addressing regional disadvantage to promoting regional competitiveness through targeted spatial strategies, the promotion of regional competitive advantage and smart specialization has led to selective benefits that seldom address the development backlog experienced by the most disadvantaged regions. The persistence of 'lagging region' and regional inequalities, despite the investment over decades of billions of Euros in the case of the EU, indicates that key challenges remain and refining both theoretical insights and applied policy and practice is a necessary process to improve applied outcomes. To encourage critical thinking on the part of educators and students, the next section explores ideas and concepts that students can be encouraged to explore, to better understand the regional and development challenges and opportunities, which exist around them.

Pedagogy

There are fruitful opportunities for lecturers and their students to explore concepts of development by looking at how the world and the regions in which we live have transformed over time as a result of recurring crises, changes in state economic policies and applied development practice.

Given the significance of these processes in the life-worlds of students, there is ample opportunity for educators to encourage students to become critically aware of how the features and processes discussed in this chapter are apparent in our life-worlds and which shape our choices and opportunities. In this final section of the chapter suggestions are provided to educators about the types of processes and trends students can develop an awareness of, followed by a set of applied mini-exercises that students can undertake.

Key processes to be aware of

Key economic trends and policy processes that students could be encouraged to be aware of are:

- The constancy of economic and global changes and the impact of recent crises – economic and natural – on countries.
- Globalization processes and changes and associated crises in the operation of capitalism.
- The degree to which neoliberalism is both pursued and contested.
- Where state intervention in economies does occur.
- How 'uneven geographical development' and the existence of 'lagging regions' are distinct features in national and global space economies.
- Whether there are tensions between policies to encourage growth and policies to address the needs of lagging regions.
- How the role of nations, regions and places is understood and is contested and respond to endogenous and exogenous processes.
- The degree to which regions and localities, following crises return to path dependency or develop new development paths, influenced by local levels of resilience and adaptation.
- Within this context, is there PBD and does place-based leadership significantly influence development outcomes?

Applied exercises

Based on the preceding reflections, students could undertake a series of mini-exercises to identify the degree to which places and regions they are familiar with have been shaped by processes identified in this chapter. Students can also consider the degree to which there are similarities and differences between international examples of regional growth and decline – such as Silicon Valley and the Rustbelt – and regions in their own country or locality.

Below are a set of questions and tasks for students to guide lesson planning:

-   Identify what the role and place of their nation/region/place is in the global economy. How it is linked to other places and what are the prevailing opportunities and threats? Can they identify any lagging regions?
-   Comment on how the broader development of capitalism, globalization and free trade has impacted on their nation/region/place both positively and negatively in the past and the present.
-   Chart how their local economy has evolved over the last 50 years. Have economic activities changed and if so why and how? Have past changes been related to local or global crises and has the local economy successfully transitioned to a new growth path?
-   Identify the characteristics of the political economy of their nation and/or region and/or local place. What form does such policy take and how is it practised?
-   Does their country have national or regional (or EU) development policy and support? If so, what form does it take and what has the local impact been?
-   Based on the preceding points, determine the degree to which development outcomes in the place or region being studied are shaped by inherited economic systems, embedded social processes, and 'lock-in.'
-   To what degree have new development paths been identified, actively pursued and have they succeeded or not?
    -   Who identified or undertook the intervention – a regional council, the national government or a supra-national body like the EU?
        -   Local spatial interventions might include support for: local partnerships, cluster formation, enterprise zones, city development strategies and creative cities.
        -   New development priorities or trends emerging might include sustainable development, climate change responses, green or alternate economic activity and social initiatives.
    -   If they succeeded, what catalysed these new activities?
    -   Were such changes driven internally or externally and who led them?
    -   What role has endogenous development in the form of community action, local governance, local and regional state initiatives, business support and partnership development played?
    -   Is local resilience a factor in local / regional transformation or economic retention, and how did it influence development outcomes?
    -   How have local, national, and international market forces impacted?
-   Can they identify and contrast a local growth focused development project and targeted support for a lagging area? Do these exist independently, or do they have links with other areas and institutions?
-   Are there examples of 'new regionalism,' localism and place-based development and leadership which they can identify?

To investigate, students may be able to use locally available census and statistical data. This allows them to trace how, over time, employment, economic output, well-being, and inequality levels have narrowed or grown within and between different regions. They can also consider using alternate sources of information such as oral histories, media, blogs, social media etc.

Undertaking exercises such as these can help students to actualize an understanding of the processes described in this chapter, through examples which have fundamentally shaped their life-worlds. Awareness of the role of both internal and external forces in shaping regional and local outcomes, prevailing levels of inequality and the processes which facilitate or restrict new development opportunities can help students to be more critically aware as they navigate the world.

Conclusion

Key events and crises in the post-World War II era in the Global North have led to fundamental shifts in political economies, prevailing economic theory, and international linkages. In turn, these processes have shaped national development policy in the Global North as much as in the South, and more specifically the degree to which regions and localities are directly supported or encouraged to develop their own unique comparative advantages. Failure of many regions to prosper and growing social and regional inequality have encouraged critical research into what regions are, how they develop and the degree to which they are able to change their economic trajectory. The roles which resilience, leadership, governance, place-based development, community and state policy plays in shaping such outcomes have become critical. While theoretical advances have clearly developed in sophistication, this is generally not paralleled by tangible socio-economic improvements in many regions and places, leading to the persistence of disadvantage and the continual quest for new applied instruments to promote growth and address ingrained structural disadvantages.

Key sources

1. Barca, F., McCann, P., and Rodríguez-Pose, A. (2012). The case for regional development intervention: place-based versus place-neutral approaches. *Journal of Regional Science*, 52(1), 134–152.

This article captures a key evolving debate within regional studies with respect to the role of place relative to external interventions. 'Place-neutral' approaches which do not factor in unique place specific needs are giving way to 'place-based' approaches which encourage a focus on local uniqueness in terms of both needs and capacities. Such changes reflect broader debates within the EU about the effectiveness of regional intervention and the role of market forces.

2. Dawley, S., Pike, A., and Tomaney, J. (2010). Towards the resilient region? *Local Economy*, 25(8), 650–667.

This was one of the first analyses to consider the role of 'resilience' thinking in understanding and explaining regional development. It challenges the static notion of 'equilibrium' with respect to how resilience helps regions respond to crisis. Instead of seeking a return to past economic activities, regions can also show dynamic responses which evolve into new forms of economic activity.

3. Harvey, D. (2014). *Seventeen contradictions and the end of capitalism*. New York: Oxford University Press.

This book critically examines and critiques how capitalism has shaped the world and it details the challenges and contradictions which it has spawned. The chapter on uneven geographical development explains how capitalism takes advantages of economic differences between regions and of periodic economic crises to strengthen its position, albeit often at a cost to the least advantaged regions.

4.   Pike, A., Rodríguez-Pose, A., and Tomaney, J. (2017a). Shifting horizons in local and regional development. *Regional studies*, 51(1), 46–57.

This article explains how local and regional development has evolved over time. It details how theory and practice have changed, often in response to competing ideologies and differing understandings of the role of both government and governance. In addition, it provides insight into what has been achieved and what challenges have been experienced. How regional and local development may evolve in the future are also key elements in the argument.

5.   Pike, A., Rodríguez-Pose, A., and Tomaney, J. (2017b). *Local and regional development*. London: Routledge.

This book covers what local and regional development is and how it is understood. It details key theories relevant to this topic, it analyses the different strategic interventions that have been applied over time and it draws on case studies from around the world to illustrate the key arguments made. It is framed in the context of current global challenges and both endogenous development responses and externally driven interventions are detailed.

References

Bailey, D., and Turok, I. (2016). Resilience revisited. *Regional Studies,* 50(4), 557–560.

Beer, A., Ayres, S., Clower, T., Faller, F., Sancino, A., and Sotarauta, M. (2019). Place leadership and regional economic development: A framework for cross-regional analysis. *Regional Studies*, 53(2), 171–182.

Boschma, R., and Frenken, K. (2011). The emerging empirics of evolutionary economic geography. *Journal of Economic Geography*, 11(2), 295–307.

Boschma, R., and Martin, R. (2007). Constructing an evolutionary economic geography. *Journal of Economic Geography,* 7, 537–538.

Chisholm, M. (1990). *1990: Regions in recession and resurgence*. London: Unwin Hyman.

European Commission (2014) *Regional Policy*. Brussels: European Commission.

European Commission (2017). *Final report: Economic challenges of lagging regions*. Brussels: European Commission.

Florida, R. (2005). *Cities and the creative class*. London: Routledge.

Florida, R. (2017). *The new urban crisis: How our cities are increasing inequality, deepening segregation, and failing the middle class-and what we can do about it*. New York: Basic Books.

Glaeser, E. (2011). *Triumph of the city*. London: Pan Macmillan.

Harvey, D. (2005). *Spaces of neoliberalization: towards a theory of uneven geographical development* (Vol. 8). Stuttgart: Franz Steiner Verlag.

Henning, M. (2019). Time should tell (more): evolutionary economic geography and the challenge of history. *Regional Studies*, 53(4), 602–613.

Hudson, R. (2005). Region and place: devolved regional government and regional economic success? *Progress in Human Geography*, 29(5), 618–625.

Jonas, A.E. (2012). Region and place: Regionalism in question. *Progress in Human Geography*, 36(2), 263–272.

Jones, M. (2019). *Cities and regions in crisis*. Edward Elgar, Cheltenham.

Kristensen, I., Dubois, A., and Teräs, J. (Eds.). (2018). *Strategic approaches to regional development: Smart experimentation in less-favoured regions*. London: Routledge.

Massey, D. (1995). *Spatial divisions of labour: social structures and the geography of production*. London: Macmillan International Higher Education.

Martin, R. (2012). Regional economic resilience, hysteresis and recessionary shocks. *Journal of Economic Geography*, 12(1), 1–32.

Pallares-Barbera, M., Suau-Sanchez, P., Le Heron, R., and Fromhold-Eisebith, M. (2012). Globalising economic spaces, uneven development and regional challenges: Introduction to the special issue. *Urbani izziv*, 23, S2–S10.

Pike, A., Dawley, S., and Tomaney, J. (2010). Resilience, adaptation and adaptability. *Cambridge Journal of Regions, Economy and Society*, 3(1), 59–70.

Pike, A., MacKinnon, D., Cumbers, A., Dawley, S., and McMaster, R. (2016). Doing evolution in economic geography. *Economic Geography*, 92(2), 123–144.

Scott, A., and Storper, M. (2003). Regions, globalization, development. *Regional Studies*, 37(6–7), 579–593.

Sheppard, E. (2012). Trade, globalization and uneven development: Entanglements of geographical political economy. *Progress in Human Geography*, 36(1), 44–71.

Simmie, J., and Martin, R. (2010). The economic resilience of regions: towards an evolutionary approach. *Cambridge Journal of Regions, Economy and Society*, 3(1), 27–43.

Smith, N. (1984). *Uneven development, nature, capital and the production of space.* Oxford: Blackwell.

Thissen, M., and Van Oort, F. (2010). European place-based development policy and sustainable economic agglomeration. *Tijdschrift voor economische en sociale geografie*, 101(4), 473–480.

Tomaney, J., Pike, A., and Rodríguez-Pose, A. (2010). Local and regional development in times of crisis. *Environment and Planning A*, 42(4), 771–779.

Turok, I., Bailey, D., Clark, J., Du, J., Fratesi, U., Fritsch, M., and Mickiewicz, T. (2017). Global reversal, regional revival? *Regional Studies*, 51, 1–8.

van Staden, J. W., and Haslam McKenzie, F. (2019). Comparing contemporary regional development in Western Australia with international trends. *Regional Studies*, 53(10), 1470–1482.

Vodden, K., Douglas, D.J., Markey, S., Minnes, S., and Reimer, B. (Eds.). (2019). *The theory, practice and potential of regional development: The case of Canada.* London: Routledge.

Zimmerbauer, K., and Paasi, A. (2013). When old and new regionalism collide: Deinstitutionalization of regions and resistance identity in municipality amalgamations. *Journal of Rural Studies*, 30, 31–40.

5

DEBT

Éric Toussaint

Introduction

Over the last ten years Greece has been a prime example of how a country and a people can be deprived of their liberty through clearly illegitimate debt. Since the 19th century, from Latin America to China, Haiti, Greece, Tunisia, Egypt and the Ottoman Empire, public debt has been used as a coercive force to impose domination and pillage (Toussaint, 2017a). The combination of debt and free trade constitute the fundamental factors subordinating whole economies as from the 19th century. Through these two inter-related processes, local elites ally themselves with big financial powers in order to subject their own countries and peoples to methods of power that transfer wealth towards local and foreign creditors.

Contrary to commonplace ideas, it is generally not the indebted weaker countries that are the cause of sovereign debt crises. These crises break out first in the biggest capitalist countries or are the result of their unilateral decisions that produce effects of great magnitude in indebted countries. It is not so-called 'excessive' public spending that builds up unsustainable debt levels, but rather the conditions imposed by local and foreign creditors. Real interest rates are abusively high and so are bankers' commissions. The indebted countries unable to keep up with repayments have to continually find new loans to repay old loans. In the past, when that became impossible, the great powers had license to resort to military action to ensure they were repaid. Today, debt crises and their outcomes are directed by the big banks and the governments that support them.

Over the last two centuries, several countries have successfully repudiated or unilaterally restructured debts by arguing that they were either illegitimate or odious. Portugal (1837), Mexico (1861, 1867, 1883, 1914, 1943), the USA (1837, 1865, 1898), Russia (1917–1918), Costa Rica (1919), Brazil (1931, 1946), Cuba (1909, 1934, 1959), China (1949), Indonesia (1956), Iran (1979), Paraguay (2005), Ecuador (2007–2009), Iceland (2008–2009) have all done this (Toussaint, 2017a). Conflict involving debt non-payment has given birth to a judicial doctrine known as Odious Debt which is to this day pertinent (see box).

Odious Debt

According to the Odious Debt doctrine theorized by Alexander Sack in 1927 a debt may be considered odious if it fulfils two conditions:

1) **The population does not enjoy the benefits:** the debt was incurred not in the interests of the people or the State but against their interest and/or in the personal interest of the leaders or persons holding power; and
2) **Lenders' complicity:** the lenders had foreknowledge, or could have had foreknowledge, that the funds concerned would not benefit the population.

The democratic or despotic nature of a regime does not influence this general rule.

The father of the Odious Debt doctrine clearly states that '**regular governments** (may) incur debts that are incontestably odious.' Sack defines a regular government as follows:

By a regular government is to be understood the supreme power that effectively exists within the limits of a given territory. Whether that government be monarchical (absolute or limited) or republican; whether it functions by 'the grace of God' or 'the will of the people'; whether it express 'the will of the people' or not, of all the people or only of some; whether it be legally established or not, etc., **none of that is relevant to the problem we are concerned with.** (Sack 1927, emphasis added).

Sack says that a debt may be considered odious if:

a) that the purpose which the former government wanted to cover by the debt in question was odious and clearly against the interests of the population of the whole or part of the territory, and
b) that the creditors, at the moment of the issuance of the loan, were aware of its odious purpose.

He continues:

Once these two points are established, the burden of proof that the funds were used for the general or special needs of the state and were not of an odious character would be upon the creditors.

This doctrine has been invoked and applied several times in history (King 2016).

Historical examples

Creditors, whether powerful states, multilateral organizations that serve them, or banks, have become very adroit at imposing their will on debtors. From early in the 19th century Haiti, the first independent Black republic, was an early testing ground. The island gained freedom from the yoke of the French empire in 1804, but Paris did not abandon its claims on the country and obtained from Haiti payment of a royal indemnity granted to the former colonial slave owners. The 1825 agreements signed by the new Haitian leaders created a monumental debt of independence untenable from 1828, which took a full century to pay off, thus preventing any real development.

Debt was also used to subjugate Tunisia under France in 1881 (Toussaint 2016a) and Egypt to the British in 1882 (Toussaint 2016b). The lending powers used unpaid debt to impose their will on countries that had so far been independent. Greece too, was born in the 1830s with a burden of debt that held it in the sway of Russia, France and the British (Toussaint 2016c). Newfoundland, which had become the first autonomous dominion of the British Empire in 1855, well before Canada and Australia, had to renounce its independence in 1933 because of a grave economic crisis and, in order to face up to its debts, it was eventually incorporated into Canada in 1949. Canada agreed to take charge of 90 per cent of Newfoundland's debt (Reinhardt and Rogoff, 2010).

Debt during the 1960s and 1970s

Similar processes were again repeated after the Second World War, when Latin American countries sought capital to fund their development and Asian and African colonies gained independence. Debt became the principal instrument used to impose neocolonialist relations. It became frowned upon to use force against a debtor country, and new means of coercion had to be found.

The massive loans granted from the 1960s to an increasing number of peripheral countries (not least those in which the Western powers had a strategic interest such as Mobutu's Congo, Suharto's Indonesia, the military regimes in Brazil, Argentina, and Pakistan) oiled a powerful mechanism that took back the control of countries that had begun to adopt policies that were independent of former colonial powers and Washington.

Three big players have incited post-colonial countries into debt by promising relatively low interest rates:

1. leading global (Western) banks seeking to put massive amounts of liquidities to work;
2. industrialized countries seeking to stimulate their economies after the 1973 oil crisis and the world recession of 1973–1974; and
3. the World Bank seeking to increase US influence and to avoid being edged out by the increasing expansion of private banks.

Local elites in borrower countries also encouraged higher debt and made gains, contrary to the populations, who derived no benefit.

Theoretical ideas promoting high levels of foreign debt

In neoclassical theory, savings should precede investment and are insufficient in developing countries. This means that shortages of savings are seen as a fundamental factor explaining why development is blocked. An influx of external funding is required. Paul Samuelson (1980), in *Economics*, took the history of US indebtedness in the 19th and 20th centuries as a basis for determining four different stages of borrowing that would supposedly lead other states to prosperity:

1. young borrowing nation in debt (from the War of Independence in 1776 to the end of the Civil War in 1865)
2. mature indebted nation (from 1873 to 1914)
3. new lending nation (from the First to Second World Wars)
4. mature lending nation (1960s)

Samuelson and his emulators slapped the model of US economic development from the late 18th century until the Second World War onto one hundred or so countries which made up the Third World after 1945, as though it were possible for all those countries to quite simply imitate the experience of the US.

On the need to resort to foreign capital (in the form of loans and foreign investments) Paul Rosenstein-Rodan (1961: 107) identified the following formula:

> Foreign capital will be a pure addition to domestic capital formation, i.e. it will all be invested; the investment will be productive or 'businesslike' and result in increased production. The main function of foreign capital inflow is to increase the rate of domestic capital formation up to a level which could then be maintained without any further aid.

This statement contradicts the facts. It is not true that foreign capital enhances the formation of national capital and is all invested. Often, a large part of foreign capital rapidly leaves the country where it was temporarily directed, as capital flight and repatriation of profits.

Rosenstein-Rodan who was the assistant director of the Economics Department in the World Bank between 1946 and 1952, made another monumental error in predicting the dates when various countries would reach self-sustaining growth. He reckoned that Colombia would reach that stage by 1965, Yugoslavia by 1966, Argentina and Mexico between 1965 and 1975, India in the early 1970s, Pakistan three or four years after India, and the Philippines after 1975. What nonsense that has proved to be!

Development planning as envisaged by the World Bank and much Western academia amounts to pseudo-scientific deception based on mathematical equations. It is supposed to give legitimacy and credibility to the intention to make the developing countries dependent on obtaining external capital. There follows an example, advanced by Max Millikan and Walt W. Rostow (1957: 158):

> If the initial rate of domestic investment in a country is 5 per cent of national income, if foreign capital is supplied at a constant rate equal to one-third the initial level of domestic investment, if 25 per cent of all additions to income are saved and reinvested, if the capital–output ratio is 3 and if interest and dividend service on foreign loans and private investment are paid at the rate of 6 per cent per year, the country will be able to discontinue net foreign borrowing after fourteen years and sustain a 3 per cent rate of growth out of its own resources.

This theoretical assumption has never been confirmed by a single practical example.

In fact, these authors who favoured the capitalist system, dominated by the US, refused to envisage the deep reforms that would have allowed for forms of development that were not dependent on external funding.

The debt crisis of the 1980s

At the end of 1979, in what is generally called the Volker Shock, the US decided to increase its interest rates. This had a flow-on effect on the rates applied to indebted Southern countries whose borrowing rates were variable and had already been subject to sharp rises. Coupled with a downturn in export commodities prices (coffee, cacao, cotton, sugar, ores, oil, etc.,) which caused reduced revenues for the countries, the trap was sprung.

In August 1982, Mexico, among other countries announced that they were unable to assure debt repayments. So, the International Monetary Fund (IMF) was asked, by the creditor banks, to lend the countries the necessary funds at high interest rates, on the double condition that they continue debt repayments and apply the policies decided by the IMF 'experts.' These included: abandoning subsidies on goods and services of primary necessity; reducing public spending; devaluing the currency; introducing high interest rates in order to attract foreign capital; directing agricultural production towards exportable products; freeing access to interior markets for foreign investors; liberalizing the economy, including suppressing capital controls; introducing a taxation system that aggravates inequalities, including increases in consumption taxes; and privatizing profitable publicly owned industries. This list is not exhaustive.

These structural adjustment loans were aimed at the suppression of independent economic and financial policies in the peripheral countries and tying them to world markets. Also, they ensured access by the industrialized economies to the raw materials and fossil fuels they needed. By gradually putting the developing countries into competition with each other and encouraging the adoption of an economic model based on exports and the extraction of raw materials for foreign markets, the goal was to reduce the price of those exported commodities, which in turn favoured the developed economies in that it reduces their production costs and increases the profits of their companies.

A new form of colonialism sprang up. It was no longer necessary to maintain an administration and an army to put the local population to heel; debt did the job of creaming off the wealth produced and directing it to the creditors. Of course, the colonialists continued to interfere in local politics and economic policies whenever they considered that it suited them.

Developments in the 2000s

From 2003 to 2004, in a context of strong world demand, commodity prices started to increase. Exporting countries improved their foreign exchange incomes. Some developing countries increased their social spending but most preferred to buy US treasury bonds and thus put their increased means at the disposal of the principal economic powers. This increase in developing countries' incomes whittled down the power of the World Bank and the IMF.

Chinese economic expansion became a major factor in this period. China had become the world's principal sweatshop and was accumulating important financial reserves and using them to significantly increase funding to developing countries in competition with the offers of funding from the industrialized countries and the multilateral institutions. While some scholars underline the fact that this funding is provided with no structural adjustment programs attached, it must be noted that many of the loans provided by China seem to be collateralized by public assets such as infrastructure and national resources, allowing Chinese private or state-owned enterprises to take control of national assets of debtor countries unable to repay.

During the 2000s, the reduction of interest rates by the central banks in the industrialized countries in the North decreased the costs of debt in the South. Because of the 2007–8 financial crisis in North America and Western Europe, massive amounts of liquidity were injected into the financial system to save the big banks and corporations that were themselves too heavily indebted. A decrease in the costs of financing the debts of the developing countries followed naturally and the governments of developing countries gained a false sense of security.

A new debt crisis is looming in the Global South

The situation began to degrade in 2016–17 when the US Federal Reserve started to raise its interest rates, from 0.25 per cent in 2015 to 1.5 per cent in October 2019 and tax breaks were granted by the Trump administration to big business to attract US foreign investment back to the US. What's more, commodities prices fell, and exporter countries' revenues slipped with them, making debt repayments that are mostly owed in strong currencies more difficult. In consequence, since 2018–19 a new debt crisis is hitting countries such as Argentina, Venezuela, Turkey, Indonesia, Nigeria, and Mozambique. Repayments are mostly in dollars, so difficulties are further aggravated by any devaluations of the local currencies.

The COVID-19 crisis saw interest rates fall again but is accelerating and further aggravating the debt crisis in other ways. With governments implementing lockdowns all over the world and with the just-in-time production brought to a halt for several months – a halt which will not be completely reversed any time soon as countries take differentiated approaches in order to limit the spread of the virus – export revenues dropped. Domestic consumption obviously declined as well as households were forced to stay at home, millions of workers became unemployed, and general insecurity about the future set in for many of those who were not yet in a state of precarious living conditions. Still with limited capacity to raise revenue domestically many states are turning to the World Bank and other international financial institutions (IFIs) for more loans.

States are facing an economic crisis the size of which was not seen at least since the Great Depression of the 1930s, if not ever. They have used large amounts of money to bail out corporations and enable the economy to stay afloat. What is more, many transnational corporations and rich households accelerated the capital flight that had been going on for a few years with the rise of interest rates in the Global North. All this contributes to a depreciation of most of the developing countries' currencies and an important decrease in their domestic as well as foreign exchange reserves, to the point that some of them are on the brink of partially or fully defaulting on the payment of their public debt – this happened in Argentina and in Lebanon in 2020, and it is likely to happen to other countries in the near future.

Debt in the South

These last years have seen a significant increase in constant values of foreign debt; between 2000 and 2019 it has more than tripled (Table 5.1; Rivié 2021). The greater part is in the private sector.

Foreign public debt has also increased although less abruptly than in the private sector.

Table 5.1 Foreign debt by region (USD billions)

Country name	1980	1990	2000	2010	2012	2019
Latin America & Caribbean	227	413	723	1.064	1.360	1.927
Sub-Saharan Africa	60	175	215	296	374	625
Middle East & North Africa	64	137	144	191	200	340
South Asia	37	124	163	410	528	789
East Asia & Pacific	54	218	456	1.192	1.738	2.993
Europe & Central Asia	42	138	339	1.148	1.412	1.465
Total	**485**	**1,205**	**2,040**	**4,301**	**5,613**	**8,138**

Source: World Bank (2020a; 2020b)

Whatever the World Bank and the IMF may cheerfully repeat, the debt of developing countries is still a major obstacle to meeting basic needs and safeguarding human rights. Inequalities have sharply increased and progress in terms of human development has been very limited.

Africa

In sub-Saharan Africa, outgoing flow of capital via debt service and corporations garnering their profits are significant. In 2012, the profits repatriated from the poorest area on earth amounted to 5 per cent of its Gross Domestic Product (GDP) versus 1 per cent in public aid to development (Toussaint et al. 2015). In this context, it is legitimate to raise the question: who is helping who? If we take into account the plundering of Africa's natural resources by private corporations, the brain drain of African intellectuals, embezzlement of goods by the African ruling class, manipulations of transfer prices by private corporations and other misappropriations, we cannot but be aware that Africa has been drained dry.

European Union (EU) relations with Africa illustrate the continuation of neocolonial policies. These have developed beyond the framework of the African, Caribbean, and Pacific (ACP)-EU Partnership Agreements (2000), generally known as the Cotonou Agreements, which covered over 100 countries. Nowadays, the EU has enforced other frameworks that are more significant in its relation with Africa such as an EU partnership framework for migration (the Valletta Action Plan with the Khartoum and Rabat processes), to which we should add the bilateral frameworks and agreements that European countries have with African countries or regions. Not forgetting the two CFA franc currencies used in 15 African countries, which has been guaranteed by the French treasury. Though, in order to regain monetary and fiscal independence from France, the eight states of the West African CFA are aiming to introduce the Eco.

Many European citizens have no idea of the extent to which conditions and clauses imposed under such agreements are setting the ground for a new debt crisis in the developing countries. Some basic facts that are not known by most people are that, whereas the total volume of aid received annually by Africa from Europe stands at around $21 billion, African migrants in Europe remit around $30 billion to their families in their home countries, almost 50 per cent more than the amount of the European aid. The funds currently available from the European Investment Fund (the small and medium enterprise lending arm of the European multilateral development bank, the European Investment Bank) for the whole African continent stand at just $3.3 billion, which is equivalent to the cost of one mid-sized infrastructure project like a port. Furthermore, the EU proposed budget for 2021–2027 plans to allocate more than $34.9 billion to various mechanisms of migration control (Valero 2018). It will end up costing Europe more to patrol its borders than what is allocated to Africa as development aid or what Africa suffers from trade losses with Europe. Indeed, it seems that the new trade deals have worsened things in this latter regard. From 2003 to 2014, Africa always had a trade surplus with Europe, whereas since 2015, the trend has reversed amounting close to a $30 billion deficit.

Latin America and the Caribbean

Latin America has one of the highest negative external debt balances among developing continents for 1985–2019. As Table 5.2 demonstrates this debt stock grew significantly over the 2010s and the repayment burden is large.

Table 5.2 Debt and resources devoted to repayment: Latin America and the Caribbean

Debt and repayments	External debt USD (billion)	Public external debt USD (billion)
Debt stock in 1970	30	14
Debt stock in 2010	1064	479
Debt stock in 2012	1360	583
Debt stock in 2017	1789	878
Debt stock in 2019	1927	919
Repayments 1970 = 2010	2969	1612
Repayments 1970 = 2012	3332	1833
Repayments 1970 = 2017	4493	2222
Repayments 1970 = 2019	5059	2484

Source: World Bank (2020a; 2020b) DataBank using indicators for total external debt (DT.TDS.DECT.CD) and public external debt and guarantee (DT.TDS.DPPG.CD).

Note: repayments cover the total of depreciation and debt interests

Table 5.3 Net transfers on external debt 1985–2017: Latin America and the Caribbean

Net transfers on external debt	1985–2017 (USD billion)	1985–2019 (USD billion)
External debt	205	244
Public external debt	-90	-123

Sources: World Bank (2020a; 2020b), DataBank, indicator public external debt and guarantee.

Table 5.3 on net transfers on external debt indicates that Latin American and Caribbean countries paid back much more in debt service between 1985 and 2019 than they received in loans during the same period. The net transfer on debt is the difference between what a country or a region receives as loans and what it pays (capital and interest included, also called debt servicing). If the amount is negative, it means that for that year the country or the region paid more than it received.

Table 5.4 shows that public debt servicing consumes a greater percentage of both GDP and the budget than does expenditure public expenditure on education and health care in three of the four Latin American countries listed. Only in Ecuador in the case of education – but not health care – does public expenditure exceed debt payments.

Debt servicing's impact on government expenditure

If we examine the evolution of public expenditure of some 50 low-income countries from 2015 to 2017, we notice an increase of expenditure related to debt repayment, a decrease of health-related expenditure and a stagnation in terms of education (see Figure 5.1).

There was also an increase in public expenditure related to debt repayment in Africa, South Asia and in general for Least Developed Countries (LDCs). According to Milan Rivié (2019) using IMF information, in July 2019, among low-income countries, nine were over indebted and 24 were on the brink of being over indebted, that is 39 per cent of them (IMF 2019, UN 2019). As evidence of the inability (and the lack of determination) of IFIs to find an adequate and sustainable response to over indebtedness, half of those countries had strictly applied the adjustment policies of the Heavily Indebted Poor Country (HIPC) initiative launched by the G7, the World Bank and the IMF in 1996. And according to a German NGO, 122

Table 5.4 Distribution of expenditure in national budgets (as % of GDP and as % of the budget) in Select Latin American States in 2013

	% of the GDP			% of the budget		
	Public debt servicing	Public expenditure for education	Public expenditure for health care	Public debt servicing	Public expenditure for education	Public expenditure for health care
Argentina	9.6	1.8	1.0	38.4	7.3	4.0
Brazil	22.7	1.8	2.1	42.2	3.9	3.4
Columbia	6.3	3.5	1.6	24.3	13.4	6.2
Ecuador	3.7	7.1	3.1	8.3	15.9	6.8

Sources: Argentina Ministry of Economy and Public Finance (2013); Brazil Fattorelli (2014); Colombia Ministerio de Hacienda y Crédito Público (2013); and Government of the Republic of Ecuador (2012).

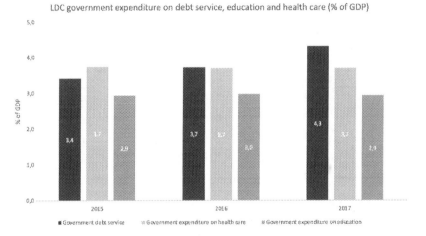

Figure 5.1 LDC government expenditure on debt service, education and health care (% of GDP)
Source: UNCTAD secretariat calculations based on World Bank, World Development Indicators data and IMF DSA LIC country reports published between 2015 and 2018. Based on 50 low-income countries.

were actually in a critical debt situation in 2019, before the COVID-19 crisis even erupted (Kaiser 2019).

It is possible not to repay an illegitimate debt

It is quite possible to resist creditors, as evidenced by Mexico under Benito Juárez, who in 1867 refused to repay loans contracted by emperor Maximilian from the Société Générale de Paris two years earlier in order to finance the occupation of Mexico by the French army (Toussaint 2017b). In 1914, at the height of the revolution, when Emiliano Zapata and Pancho Villa were victorious, Mexico completely suspended payment of its external debt, which was considered to be illegitimate; the Mexican government only repaid symbolic amounts from 1914 to 1942, just in order to pacify creditors. From 1934 to 1940, President Lázaro Cárdenas nationalized the railway and the oil industry without any compensation; he also expropriated over 18 million hectares of landed estates to give them over to Indigenous communities. His tenacity paid off: in 1942, creditors renounced about 90 per cent of the debt value and said they were satisfied with limited compensations for the companies they had been evicted from. Mexico was able to undergo major social and economic development from the 1930s to the 1960s. Other countries such as Brazil, Bolivia and Ecuador successfully suspended debt repayment from 1931. In the case of Brazil, selective suspension of repayment lasted until 1943, when an agreement made it possible to reduce debt by 30 per cent.

More recently, in July 2007, in Ecuador, then President Rafael Correa set up a committee to audit public debt. After 14 months of work, its findings gave evidence that a large part of the country's public debt was illegitimate and illegal. In November 2008, the government decided to unilaterally suspend repayment of debt securities sold on international financial markets and were maturing in 2012 and in 2030. Eventually, the government of this small country won its case opposing North American bankers who held those securities. It bought for US $900 million securities that had been worth $3.2 billion. Through this operation Ecuador's Treasury saved about $7 billion on the borrowed capital and the remaining interest. It freed resources to finance new social spending (as shown in Table 5.4). Ecuador has not been targeted by international reprisals (Toussaint 2021).

It is obvious that refusing to repay illegitimate debt is a necessary measure, but it is not enough to generate development. A consistent development programme must be implemented. Financial resources have to be generated through increasing the State's resources through taxes that respect social and environmental justice (Millet and Toussaint 2018).

Questions for consideration

1. Consider the concept of 'odious debt' (discussed on page 1–2). Can you think of any recent examples of development lending that might constitute odious debt? Working in small groups, identify a development project of your choice and demonstrate (1) how debt for this project was incurred 'not in the interests of the people or the state,' and (2) evidence that the lenders had foreknowledge, or could have had foreknowledge, that the funds concerned would not benefit the population.
2. Global debt levels have increased substantially following the 2019-onwards COVID-19 pandemic. Choose two countries of your choice and collect data on their debt profiles since the pandemic commenced.
3. This chapter argues that debt functions as a tool of neocolonial imperialism. Discuss this claim.
4. China has become a leading financier of global infrastructure projects, many of which are linked to its ambitious Belt and Road Initiative. Alongside China's increased lending, there has been much criticism that it is expanding unsustainable debt burdens, particularly within heavily indebted poor countries (HIPCs). However, while some scholars have accused China of engaging in 'debt diplomacy' others have rejected such suggestions. Interrogate these debates. Resources on the CADTM website will provide a useful starting point. www.cadtm.org/China?lang=en
5. As south-south development cooperation has increased, a number of new multilateral development banks (MDBs) have been established. Two examples include the Asian Infrastructure Investment Bank and the New Development Bank. How do these new MBDs differ from existing multilateral banks? Are they reproducing the same problems regarding debt expansion or do they present new challenges and opportunities?

6. A key historical event for global development lending was the 1980s debt crisis. What caused this crisis and where did it begin? How did lenders and borrowers respond?

7. The heavily indebted poor countries (HIPC) initiative was initiated by the International Monetary Fund and the World Bank in 1996 to provide debt relief to the most indebted countries. (1) Examine the criteria to be eligible to HIPC relief. (2) Examine critique of the HIPC initiative.

Key sources

1. Toussaint, É. (2017a). *The Debt System: A History of Sovereign Debts and their Repudiation.* (Chicago: Haymarket, 2019)

Detailed history of debt since the late 1700s and its relationship with the long cycles of growth and crisis in capitalism. Explains the concept of odious debt in detail.

2. CADTM (2020). Committee for the Abolition of Illegitimate Debt. Available: www. cadtm.org/English

A regularly updated site on debt with publications, news and analysis on key debt topics from around the world.

3. Reinhardt, C., and K. Rogoff (2009). *This Time Is Different: Eight Centuries of Financial Folly.* (Princeton, NJ: Princeton University Press)

An influential mainstream economic history of financial crises which generated much controversy thanks to its claim the debt above 90 per cent of GDP dramatically harms economic growth. The claim was used to support austerity programs but it was found to be methodologically flawed.

References

ACP-EU Partnership Agreement (2000). Cotonou Agreement 23 June. Available www.ec.europa.eu/international-partnerships/acp-eu-partnership_en [Accessed 7 October 2020].

Argentina Ministry of Economy and Public Finance (2013). Nation's General Budget for 2013 (Nation's Presidency, Presupuesto 2013 Resumen, Buenos Aires). Available: www.mecon.gov.ar/onp/html/presupresumen/resum13.pdf [data no longer available].

Columbia Ministerio de Hacienda y Crédito Público (2013). *Presupesto general de la Nación, 2013* (Nation's General Budget for 2013), República de Colombia. Available: www.minhacienda.gov.co/presupuesto/index.html [data no longer available]

Fattorelli, M.L. (2014). Dívida consumirá mais de um trilhão de reais em 2014 (Brazil's Citizens' Audit of the Debt), Auditoria Cidadã da Dívida. Available: www.auditoriacidada.org.br/wp-content/uploads/2013/09/Artigo-Orcamento-2014.pdf [Accessed 15 August 2019].

Government of the Republic of Ecuador (2012). *Presupuesto General del Estado, 2012* (Nation's General Budget for 2012), Ministry of Finance. Available: www.finanzas.gob.ec/el-presupuesto-general-del-estado [Accessed 2015].

International Monetary Fund (IMF) (2019). List of LIC DSAs for PRGT-Eligible Countries. As of July 31, 2019. Available: www.imf.org/external/Pubs/ft/dsa/DSAlist.pdf [Accessed 15 August 2019].

Kaiser, Jürgen (2019). *Global Sovereign Debt Monitor* (Erlassjahr and Misereor). Available: www.erlassjahr.de/en/news/global-sovereign-debt-monitor-2019/ [Accessed 7 October 2020].

King, Jeff (2016). The Doctrine of Odious Debt in International Law: A Restatement (Cambridge: Cambridge University Press).

King, Jeff (2007). Odious Debt: The Terms of Debate. *North Carolina Journal of International Law and Commercial Regulation*, 32(4): 605–668.

Millet, Damien and E. Toussaint (2018). Once upon a time there was a popular government that wanted to do away with the export-oriented extractivist model. Available: www.cadtm.org/once-upon-a-time-there-was-a-popular-government-that-wanted-to-do-away-with-the [Accessed 7 October 2020].

Millikan, M.F. and W.W. Rostow (1957). *A Proposal: Keys to An Effective Foreign Policy.* Harper: New York.

Rivié, Milan (2019). New Debt Crisis in the South. Available: www.cadtm.org/New-debt-crisis-in-the-South [Accessed 7 October 2020].

Rivié, M. and E. Toussaint (2021). Evolution of the external debt of developing countries between 2000 and 2019. Available: www.cadtm.org/Evolution-of-the-external-debt-of-developing-countries-between-2000-and-2019 [Accessed 16 April 2021].

Rosenstein-Rodan, P. (1961). International Aid for Underdeveloped Countries. *Review of Economics and Statistics* 43(2): 107–138.

Sack, A. N. (1927). *Les Effets des Transformations des États sur leurs Dettes Publiques et Autres Obligations financières (The effects of the transformation of States on their public debt and other financial obligations).* Paris, France: Sirey.

Samuelson, P. (1980). *Economics,* 11th edition. New York: McGraw Hill.

Toussaint, É. (2016a). Debt: how France appropriated Tunisia, cadtm.org, 13 June. Available: www.cadtm.org/Debt-how-France-appropriated [Accessed 7 October 2020].

Toussaint, É. (2016b). Debt as an instrument of the colonial conquest of Egypt cadtm.org, 6 June. Available: www.cadtm.org/Debt-as-an-instrument-of-the [Accessed 7 October 2020].

Toussaint, É. (2016c). Newly Independent Greece had an Odious Debt round her Neck, cadtm.org, 26 April. Available: www.cadtm.org/Newly-Independent-Greece-had-an [Accessed 7 October 2020].

Toussaint, É. (2017b). Mexico proved that debt can be repudiated 22 July. Available: www.cadtm.org/Mexico-proved-that-debt-can-be [Accessed 7 October 2020].

Toussaint, É., Munevar, D., Gottiniaux, P. and A. Sanabria (2015). World Debt Figures 2015, *Debt in the South,* Chapter 3 CADTM. Available: www.cadtm.org/Debt-in-the-South [Accessed 16 April 2021]

Toussaint, É. (2021). Ecuador: Resistance against the policies imposed by the World Bank, the IMF and other creditors between 2007 and 2011. Available: www.cadtm.org/Ecuador-Resistance-against-the-policies-imposed-by-the-World-Bank-the-IMF-and [Accessed 16 April 2021]

United Nations (UN) (2019). Financing for Sustainable Development Report 2019. Available: www.developmentfinance.un.org/sites/developmentfinance.un.org/files/FSDR2019.pdf [Accessed 7 October 2020].

Valero, J. (2018). EU will spend more on border and migration control than on Africa. Euractiv 1 August. Available: www.euractiv.com/section/africa/news/for-tomorrow-eu-will-spend-more-on-border-and-migration-control-than-on-africa/ [Accessed 7 October 2020].

World Bank (2020a). International Debt Statistics. Available www.datatopics.worldbank.org/debt/ids/region/lmy [Accessed 7 October 2020].

World Bank (2020b). DataBank. Available: www.databank.worldbank.org/home.aspx [Accessed 7 October 2020].

6

OECD DAC DEVELOPMENT COOPERATION

Heiner Janus

Introduction

Addressing the world's most pressing development challenges like poverty, inequality and climate change requires international cooperation. The COVID-19 pandemic and the associated economic recession has further exacerbated these challenges. Reversing previous trends of decreasing global poverty, the pandemic, armed conflicts and climate change have pushed some 100 million people into extreme poverty in 2020 (World Bank 2020). In addition, progress towards achieving the 2030 Agenda is uneven and there are major setbacks in realizing the Sustainable Development Goals (SDGs). Modern global crises are increasingly overlapping and are characterized by an unprecedented scale and complexity.

Over recent decades, development cooperation has been a fairly stable external resource for low- and middle-income countries to address these development challenges. Yet, development cooperation is becoming increasingly fragmented and the idea of development cooperation itself is contested among policymakers and scholars (Chaturvedi et al. 2021). Unresolved points of debate include the purpose, the effects, and the legitimacy of development cooperation. Addressing these debates requires unpacking the unhelpful aggregate of 'development cooperation' into more specific categories. A first differentiation lies in defining development cooperation narrowly or broadly.

Following a narrow understanding, development cooperation is defined as 'Official Development Assistance' (ODA) provided by members of the Organisation for Economic Co-operation and Development (OECD) Development Assistance Committee (DAC). The DAC is an intergovernmental forum of rich countries that have been in charge of defining, collecting, and publishing the statistics on ODA flows since 1969. Today, 30 DAC members and over 80 other providers of development cooperation, including other countries, multilateral organizations and private foundations, record their assistance based on the ODA reporting standard (OECD 2020c).

According to the OECD DAC definition (OECD 2020c), ODA covers (financial) flows to countries and territories on the DAC list of ODA recipients and to multilateral development institutions that are: i) provided by official agencies; ii) concessional (i.e. grants and soft loans); iii) administered with the promotion of the economic development and welfare of developing

DOI: 10.4324/9781003017653-8

countries as the main objective. Often, the ODA reporting standard is equated with the terms 'foreign aid' or 'development aid' and members of the OECD DAC are referred to as 'traditional donors,' although these terms can be problematic. The term 'aid' may indicate benevolence and generosity while obscuring more exploitative, hierarchical, and self-interested relations (Waisbich and Mawdsley 2022). The term 'traditional donors' might falsely suggest that countries outside the OECD, especially emerging economies, do not have a long track record of engaging in development cooperation.

A broader understanding of development cooperation addresses some of these concerns by expanding the scope of actors beyond the members of the OECD DAC and the type of cooperation beyond the ODA definition. Emerging economies or so-called providers of South-South Cooperation (SSC), as well as some non-governmental actors (including civil society, philanthropy, and businesses) typically take a wider perspective on development cooperation. Although there is no agreed methodology for a broad definition, Alonso and Glennie (2015) have suggested that development cooperation covers any 'activity that aims explicitly to support national or international development priorities, is not driven by profit, discriminates in favour of developing countries, and is based on cooperative relationships that seek to enhance developing country ownership.'

Currently, both narrow and broad understandings of development cooperation co-exist and different actors have vested interests in maintaining this status quo. On the one hand, OECD DAC members hold on to the definition of ODA and only propose gradual reforms to the existing system through discussions that they lead themselves. On the other hand, emerging economies like China and India put forward their own definitions of development cooperation, often under the umbrella of South-South cooperation. Overall, the tensions between competing definitions of development cooperation represent a major fault line in the overarching policy field of development cooperation.

The entrance of 'new' actors into the policy field has been the key trend driving changes in development cooperation over the recent decade as part of a broader transformation of the policy field, often summarized as 'beyond aid' (Janus et al. 2015). Emerging economies have offered more choice for recipient countries, created competition in the international aid architecture and brought their own ideas, principles and approaches to development cooperation (Purushothaman 2020). Researchers have varying assessments of whether emerging economies are 'socialized' into the existing development cooperation architecture and in how far they are driving a 'Southernization' of development cooperation (Waisbich and Mawdsley 2022). A detailed assessment of competing definitions of development cooperation is a research agenda in itself (Fejerskov et al. 2016) and lies beyond the scope of this chapter, but the described tensions form an important background for analysing bilateral donors who are members of the OECD DAC.

The focus of this chapter lies in describing key trends that affect bilateral OECD DAC donors, who provide the majority of all ODA. In 2019, about 70 per cent of the total ODA of USD 152.8 billion was bilateral and less than 30 per cent was multilateral ODA (channeled through a multilateral development agency) (OECD 2020a). Further, the members of the OECD DAC form an epistemic community around the only globally available financial resource that is explicitly targeted at promoting sustainable development (Tomlinson 2018).

The chapter summarizes key challenges and sketches possible ways forward for bilateral ODA. It does not provide a comprehensive review of the literature but rather highlights some key publications that enrich understanding of the overall transformation of development cooperation as a policy area. The key theme of this analysis is that even narrowly defined development cooperation in the form of ODA provided by bilateral donors in the OECD DAC remains an unhelpful aggregate that needs to be broken down further. To unpack the challenges faced by bilateral donors, the analysis is structured into three main parts: i) global trends; ii) domestic trends in donor countries; and iii) donor relations with low- and middle-income countries. Each of these areas addresses a fundamental component of development cooperation: the global landscape, the domestic politics of aid and the donor relationships with low- and middle-income countries. The following sections examine each of these areas in turn and the final section presents conclusions.

Global trends affecting bilateral OECD DAC donors

ODA figures in absolute terms have steadily increased over the last decade. In relative terms, ODA in 2019 was equivalent to 0.30 per cent of DAC countries combined gross national income (GNI), a slight decrease from 0.31 per cent in 2018 and still below the targeted 0.7 per cent of ODA to GNI (OECD 2020a). Decreased economic growth due to the COVID-19 pandemic and cuts to ODA budgets of bilateral donors will put further pressure on globally available ODA. In 2019, ODA to least developed countries (LDCs) increased in real terms, continuing a slight improvement in directing aid flows to the most under-aided countries in recent years.

The 2030 Agenda for Sustainable Development functions as the main formal normative framework for development cooperation, providing measurable goals and indicators for achieving sustainable development. Importantly, the 2030 Agenda is universal in its application to all countries and its content. The SDGs range from global public goods challenges like addressing climate change to local challenges across different regions and countries, for instance providing basic service delivery in fragile states.

In addressing the broad scope of development challenges contained in the SDGs, ODA has been a small, albeit important, source of financing compared to other financial flows. According to OECD estimates, the annual 'financing gap' for the SDGs is expected to increase from USD 2.5 trillion to USD 4.2 trillion due to the effects of the COVID-19 pandemic (OECD 2020b). In development finance, domestic resources raised through taxes and external financial flows like foreign direct investments or remittances have traditionally outweighed ODA as a source of development finance (OECD 2020b).

Yet, over recent years, the relative importance of ODA has decreased due to several factors. First, the entrance of new actors into the policy field has ushered in an 'age of choice' for countries to choose between different providers of development finance (Prizzon et al. 2017). Next, countries are graduating from low- to middle-income countries and from middle- to high-income countries, leading to an increased discussion on 'exit from ODA' and transitioning to increasingly diversified sources of financing (Calleja and Prizzon 2019). Although income per capita thresholds are arbitrary and impractical due to data quality reasons, international financial institutions and donors continue to use income thresholds in determining the eligibility of countries to receive internationally recognized forms of concessional finance. In 2014, the OECD estimated that by 2030 a total of 28 countries with a combined population of two billion people will exceed the thresholds of ODA eligibility (Sedemund 2014). Still, ODA allocation patterns reveal that since the 1980s (despite some fluctuations), only 25–30 per cent of total aid has gone to low-income countries and the majority of total global ODA has been allocated to middle-income countries. A large share of the world's poorest people still live in middle-income countries, but ODA has potentially higher developmental impact in low-income countries.

The thematic allocation patterns of ODA have diversified over the past couple of decades. The SDGs as a framework for reference have not incentivized stronger prioritization of particular sectors or challenges in ODA allocation but rather increased the complexity of goal setting. The SDG global indicators include 13 targets that explicitly mention ODA across 11 goals, including issues like food, health, education, water, climate, biodiversity, energy, infrastructure and inequality (Cohen and Shinwell 2020). Allocating ODA across these 13 targets in itself already creates a complex web of interdependencies with many synergies and trade-offs across sectoral policies to consider. Updates in the ODA reporting system by the OECD DAC added to the complexities of linking ODA to the SDGs. Donors can report that individual ODA financed activities (on a voluntary basis) contribute to up to ten targets, goals, or a combination of goals and targets (Gualberti and Benn 2020).

ODA allocation also addresses thematic issues on different levels. ODA can fund targeted poverty-related initiatives in one country, it can address development challenges with spill-over effects across multiple countries, or it can contribute to the provision of global public goods like a stable climate or safety from communicable diseases. Unfortunately, bilateral donors have not developed common strategies for addressing these complexities yet. Further, there are methodological challenges in counting ODA for global public goods. As a result, the contribution of ODA to the SDGs is recorded in rather piecemeal ways on a country-by-country basis or in global thematic aggregates, rendering it difficult to better coordinate ODA allocation based on thematic interconnections and cross-country spillovers.

ODA on the global level is allocated in a fragmented manner and remains thinly stretched across a range of countries and themes included in the 17 SDGs. The key global macro-economic trend underlying this phenomenon is the decreasing relative importance of ODA as a financial resource across different countries and thematic issues.

The OECD DAC has addressed this trend mainly through technical proposals on updating the current ODA definition and introducing an alternative indicator called Total Official Support for Sustainable Development (TOSSD) (OECD 2020c). These proposals address the macro-trend of the decreasing importance of aid through safeguarding the developmental aspects of the ODA definition on the one hand, while adding measures for capturing contributions to sustainable development beyond ODA.

Some of the challenges in this context include debt relief, global public goods accounting (especially concerning climate finance and refugee costs) and the transparency of reforms (Rogerson and Ritchie 2020). Further, both the ODA modernization and the piloting TOSSD have been harshly criticized as methodologically fraught and incomplete (Chaturvedi et al. 2016; Scott 2020). One key issue is that these definitions are not solely technical questions but rather reflect political decisions that are mostly made by the member states of the OECD DAC (although they conduct consultations with 'recipient' countries) (Rogerson and Ritchie 2020). Therefore, it remains open whether current ODA reporting reforms will be able to address the overall trend of decreasing relative importance of ODA.

Domestic trends in OECD DAC donor countries

From a donor country perspective, the key trend of recent years has been the so-called 'return of national interests' and the rise of populist movements that challenge the current practices of bilateral OECD DAC donors (Gulrajani 2017). Since the Second World War, development cooperation has been an instrument for higher-income countries to support lower-income countries in achieving development outcomes and addressing global development challenges. In addition to this developmental purpose, development cooperation has always been a tool for governments to pursue national interests, particularly political and economic interests. These underlying motivations – pursuing national interests and tackling development challenges – have traditionally co-existed, although in varied forms across donor countries and across time (Gulrajani and Calleja 2019).

Analysing donor country aid allocation over longer periods indicates that aid allocation patterns during the Cold War were strongly driven by security and economic ties, whereas more recent aid allocation patterns reflect a broader set of interconnections between countries (Bermeo 2017). Modern globalization challenges, such as violent conflicts, migration, and refugee crises, as well as climate change have created more shared interests in cooperation. The specific factors influencing donor motivations of a given bilateral donor vary and can include a mix of international and domestic drivers, including homogenizing aid practices among OECD DAC donors or domestic party politics in donor countries (Yanguas 2018).

In recent years, however, domestic discourses on development cooperation have increasingly turned away from development-oriented motives towards the strategic interests of development cooperation providers, such as expanding their own political and economic interests. Increasingly, OECD DAC donors adopt terms like 'mutual benefit' or 'mutual interest' to describe their development cooperation, to underline national agendas (Keijzer and Lundsgaarde 2018). This trend is more a rhetorical turn than a fundamental shift in practice, as DAC donors have long track records of tying their aid to domestic agencies and inflating ODA figures by reporting domestic spending on refugees for instance (Tomlinson 2018). But the open acknowledgement and communication of own interests mark a clear break from a previous dominant aid discourse that was focused on partner country needs and the principle of 'country ownership,' as put forward in the 2005 Paris Declaration on Aid Effectiveness.

From a global perspective, the reorientation towards national interests and mutual benefits can be read as a sign of convergence between OECD DAC donors and emerging economies whose South-South cooperation is based on the principle of mutual benefit. From a domestic perspective in individual donor countries, the rhetorical shift relates to justifying aid spending abroad to domestic taxpayers with a different narrative that emphasizes national benefits. The rise of populist movements, especially on the political right, has influenced government agendas across Europe towards framing development cooperation as a contribution to reducing migration to Europe for instance (Bergmann et al. 2021). Promoting national interests above international norms of needs-based cooperation and solidarity has therefore become the new norm for many bilateral donors (Gulrajani 2017).

Another important factor in aid effectiveness is the positioning of bilateral donor agencies. These agencies are accountable to domestic as well as foreign constituents. They face the challenge of mending the 'broken feedback loop' of aid relations (Martens et al. 2002), where constituents in low- and middle-income countries do not have a direct way of providing feedback to bilateral donors. To navigate a contested global field of development cooperation, domestic polarized politics and the needs of partner countries, development agencies need to maintain their autonomy. Gulrajani (2017) has demonstrated that such autonomy needs to be multi-dimensional, including financial, interventional, legal, policy, managerial and structural autonomy.

Structural autonomy, defined as the formal independence and freedom of a development agency from foreign affairs, is key but agencies need to be embedded in a broader governmental system that allows development agencies to balance national and developmental goals. Still, recent cases where standalone development agencies have been subsumed into foreign affairs departments indicate a risk of decreased developmental orientation. For instance, the previous Canadian International Development Agency (CIDA) was merged into the Global Affairs Canada department in 2013 and Brown (2016) describes how Canada has increasingly instrumentalized foreign aid for own benefits to the detriment of poverty reduction abroad. In 2013, a conservative government in Australia integrated the standalone agency AusAID into the Department of Foreign Affairs and Trade and in 2009 the New Zealand agency NZAID was reintegrated into the Ministry of Foreign Affairs and Trade. In the United Kingdom, the previous Department for International Development (DfID) was merged into the Foreign, Commonwealth and Development Office (FCDO) in 2020. Despite DfID having a reputation for being the leading donor agency and the existence of an aid law protecting the 0.7 per cent goal, the Johnson administration has moved towards using aid for its agenda of 'Global Britain' and cut aid to 0.5 per cent of GNI.

There are several attempts to redirect the renewed focus on national interests among OECD DAC donors towards a form of 'enlightened self-interest' that puts the national interest in the service of a greater common good like achieving the SDGs. Gulrajani and Calleja (2019) have developed the 'Principled Aid Index' that ranks OECD DAC donors according to whether their foreign aid allocations support a principled or parochial national interest. Keijzer and Lundsgaarde (2018) have proposed to broaden aid evaluations to include aspects of 'mutual benefit' to enable donors to be accountable to taxpayers for the full scope of development policy. At the OECD DAC, there are efforts to recast the narrative of development cooperation. The key motivation is to adopt a new communications strategy to 'begin a conversation with citizens about development' (OECD 2019), appealing to their sense of community and showing citizens how they can get involved personally. An index for ranking donors, broader evaluation approaches and a new communications strategy are all valuable attempts to address the challenge of reconciling national with developmental interests. Whether these attempts can be successful will shape the future of OECD DAC donors. The impact of COVID-19 crises and the potential decreases of ODA volumes in the post-crisis period, however, will exacerbate the challenge of reframing national interests among donors.

Trends affecting donor relations with low- and middle-income countries

From a low- and middle-income country perspective, the key trend of recent years is that long-standing demands for aid reforms remain unmet and that the gap between developing country needs and the current practices of ODA provision continue to diverge.

When analysing thematic priorities, partner country priorities seem to largely align with donor priorities. An AidData survey (Custer et al. 2018) of nearly 3,500 leaders, including government officials, civil society leaders, and private sector representatives from 126 low- and middle-income countries found that donors have largely allocated aid in areas that are prioritized by leaders and citizens, although they were considered to be underinvesting in jobs and schools. A smaller survey (Ingram and Lord 2019) of 93 leaders across the development industry notes that poverty reduction once provided a common goal for development actors, but no longer fulfils this function. Development priorities across low- and middle-income countries have diversified, and leaders increasingly view poverty as a challenge that is interrelated with other challenges, like inequality, and cannot be addressed in isolation. Still, the authors note that there is a 'growing bifurcation between countries trapped in a toxic blend of conflict, state fragility, and poverty and those that have escaped to middle-income status' (Ingram and Lord 2019). This divide is expected to grow further.

Turning to the specific channels and modalities of aid delivery, donors do not sufficiently address partner country needs and priorities. The latest monitoring round of the Global Partnership for Effective Development Co-operation (OECD/UNDP 2019) documents this trend. From 2016 to 2018, the alignment of donor strategies and projects to country-led development priorities and results has declined and the use of country systems when channelling ODA to the public countries' budgets has decreased. Donors continue to struggle to implement their commitments that date back to the 2005 Paris Declaration on Aid Effectiveness. A survey drawing on observations of more than 1000 public sector officials from 70 low- and middle-income countries, highlights the continued importance of channelling funds through the public financial management systems of partner countries (Masaki et al. 2021). It found that using country systems builds trust with partner country officials, and promotes an environment of mutual interdependence and accountability to ensure that external funds effectively support domestic reform efforts.

As low- and middle-income countries reduce their reliance on ODA, many donors are refocusing on engaging in policy and institutional reforms that they hope will promote development in the long-run (Masaki et al. 2021). For years, aid agencies have struggled with the necessary but challenging move towards a more political and locally adapted engagement with their partner countries (Carothers and De Gramont 2013). Yanguas (2018) summarizes the last two decades of this attempt as a struggle between three competing norms – results, politics, and adaptation – that all represent different solutions to the problem of improving donor and partner relations in a context of competing accountabilities.

Each of these schools of thought is 'politically smart' in slightly different ways, combining a specific narrative for domestic constituents in donor countries with a matching set of aid modalities for low- and middle-income countries. As a part of a stronger push towards results, development agencies have introduced results-based management tools that are intended to better demonstrate the effects of their aid, but which often come with heavy reporting requirements and potential adverse impacts for partner countries like a focus on short-term and easily measurable development indicators (Holzapfel 2016). Under the politics banner, donor agencies have piloted tools like political economy analysis and pilot programs for 'thinking and working politically' that combine a narrative of 'messy politics of change' with more tailor-made and context-sensitive support for partner countries (Dasandi et al. 2019). The related school of thought of 'adaptive development' has sought to introduce higher autonomy and flexibility into aid management, often with an emphasis on 'doing development differently' (Honig and Gulrajani 2018). None of these ideas is entirely new and each approach has different advantages and disadvantages (Vähämäki and Verger 2019). Still, either one of these ideas or a mix of them are likely to become the new paradigm for OECD DAC donors.

At the level of donor relationships with partner countries, several key dynamics mirror the larger trends at the global level. The role of ODA has diversified in terms of objectives (from poverty reduction to the SDGs) and evolved from financing service delivery directly to a stronger role in providing knowledge and engaging in policy reforms. Yet, the key underlying problem is that the gap between low- and middle-income country needs, and specific forms of ODA provision, continues. Reform attempts have evolved from the aid effectiveness agenda to updated narratives and aid modalities built on the ideas of results, politics, and adaptation.

Whether these updated norms will change the long-standing obstacles in improving donor relations with partner countries remains open.

Conclusion

The policy field of development is characterized by competing definitions of key concepts. A narrow definition of development cooperation revolves around the concept of ODA and a broader definition includes emerging economies who have their own approaches to development cooperation. This basic tension stems from a changing geopolitical and economic context, but it affects the current practices of bilateral OECD DAC donors across three main areas discussed in this chapter.

On the global level, the key trend affecting OECD DAC donors is the decreasing relative importance of ODA as a financial resource, which is partly driven by emerging economies and private actors who provide new sources of finance to low- and middle-income countries. On the domestic level in donor countries, the key trend is the return of national interests in discourse and practice, again partly driven by emerging economies who have traditionally highlighted 'mutual benefit' as a main principle of development cooperation. Finally, in donor-partner country relations, the critical trend is that demands for aid reforms remain unmet, despite new attempts to improve aid effectiveness through focusing on results, politics, and adaptation. In this context, the influence of emerging economies is more indirect as they mainly act as competition at the country level, thereby putting pressure on OECD DAC donors to improve the value-added of ODA.

Two main conclusions are drawn from the analysis in this chapter. First, the rise of emerging economies in the policy field of development cooperation is primarily a political challenge. However, the response of OECD DAC donors has been at the level of technical solutions, for instance updating the ODA definition or introducing a measure for total official flows (TOSSD). Yet, maintaining and improving the relevance of ODA also requires better outreach by OECD DAC countries at the political level, including through multilateral fora like the G-20 or the United Nations and in coordinating ODA at the country level. Being transparent, hosting open consultations and giving low- and middle-income countries real authority in norm and standard setting at the OECD DAC and in other multilateral fora are important steps in this context.

Second, there are ongoing reforms of ODA that need further efforts, regardless of the influence of emerging economies. The common theme across the three areas analysed is that OECD DAC donors face the challenge of bridging an increasing discrepancy between the discourse used to justify aid to domestic constituents and the reality of decreasing relative importance of aid at the country level. Reconciling the ODA discourse with practice will require better technical tools for accounting for the effects of ODA, improved political communication and aid modalities that meet partner country needs.

Overall, this chapter has provided an overview of key trends affecting OECD DAC donors in the 21st century and pointed out some potential directions of change. Looking beyond the gradual changes analysed here, there are also bolder visions for the future that see ODA being replaced by 'global public policies' or 'global public investment' (Glennie 2020). The aftermath of the COVID-19 pandemic might present a window of opportunity for accelerating this more radical shift of integrating ODA into global-level policy, potentially overcoming the current 'North-South' divide and abolishing the structures created by the OECD DAC. Even in this scenario, the challenges outlined in this chapter should offer useful criteria for assessing the potential impacts of such a shift.

Key sources

1. Chaturvedi, S., Janus, H., Klingebiel, S., Li, X., Mello e Souza, A. d., Sidiropoulos, E., and Wehrmann, D. (2021). Development Cooperation in the Context of Contested Global Governance. In S. Chaturvedi, H. Janus, S. Klingebiel, X. Li, A. d. Mello e Souza, E. Sidiropoulos, and D. Wehrmann (eds.), *The Palgrave Handbook of Development Cooperation for Achieving the 2030 Agenda: Contested Collaboration*. Cham: Springer International Publishing, 1–21.

This article provides an overview of the current landscape of development cooperation and competing approaches to development cooperation. It maps different narratives, norms and measurement frameworks for development cooperation and explains how they contribute to the SDGs in a context of 'contested collaboration.'

2. Prizzon, A., Greenhill, R., and Mustapha, S. (2017). An 'age of choice' for external development finance? Evidence from country case studies. *Development Policy Review*, 35 (S1), O29–O45.

In this article, the authors analyse the key trend of a decreasing relevance of aid as a financial resource across nine country case studies in sub-Saharan Africa and Asia. They map the expanding access of partner country governments to external development finance beyond ODA and analyse governments' priorities for the terms and conditions of development finance flows they would like to access.

3. Gulrajani, N. (2017). Bilateral donors and the age of the national interest: what prospects for challenge by development agencies? *World Development*, 96, 375–389.

Gulrajani analyses the trend of increasing national interests among donor countries and the tensions between national interests and the developmental orientation of foreign aid. She interrogates how development agencies manage this tension through the analytical lens of autonomy and identifies ways forward for development agencies to maintain different types of autonomy.

4. Yanguas, P. (2018). *Why we lie about aid: Development and the messy politics of change*. London: Zed Books Ltd.

Yanguas provides a comprehensive overview of the domestic politics of foreign aid in donor countries and the challenges that donors face when they want to engage in promoting political change in low- and middle-income countries.

5. Vähämäki, J., and Verger, C. (2019). Learning from Results-Based Management evaluations and reviews. *OECD Development Co-operation Working Paper 5*.

Vähämäki and Verger review current trends in the management systems of development agencies and highlight best practices and key challenges. They also analyse the potential of alternative approaches like adaptive development to introduce innovative change into the relations between donors and partner countries.

References

Alonso, J. A. and Glennie, J. (2015). What is development cooperation? *Development Cooperation Forum Policy Brief*. New York: United Nations. www.un.org/en/ecosoc/newfunct/pdf15/2016_dcf_policy_brief_no.1.pdf

Bergmann, J., Hackenesch, C. and Stockemer, D. (2021). Populist Radical Right Parties in Europe: What Impact Do they Have on Development Policy? *JCMS: Journal of Common Market Studies*, 59, 37–52, www.doi.org/10.1111/jcms.13143

Bermeo, S. B. (2017). Aid allocation and targeted development in an increasingly connected world. *International Organization*, 71, 735–766, www.doi.org/10.1017/S0020818317000315

Brown, S. (2016). The instrumentalization of foreign aid under the Harper government. *Studies in Political Economy*, 97, 18–36.

Calleja, R. and Prizzon, A. (2019). *Moving away from aid: Lessons from country studies*. London: Overseas Development Institute. www.cdn.odi.org/media/documents/191128_summary_report_final_v1.pdf

Carothers, T. and De Gramont, D. (2013). *Development aid confronts politics: The almost revolution*. Washington, DC: Carnegie Endowment for International Peace.

Chaturvedi, S., Chakrabarti, M. and Shiva, H. (2016). TOSSD: Southernisation of ODA *Forum for Indian Development Cooperation*. New Delhi: Research and Information System for Developing Countries (RIS). www.fidc.ris.org.in/sites/default/files/9.pdf

Chaturvedi, S., Janus, H., Klingebiel, S., Li, X., Mello e Souza, A. d., Sidiropoulos, E. and Wehrmann, D. (2021). Development Cooperation in the Context of Contested Global Governance. In: Chaturvedi, S., Janus, H., Klingebiel, S., Li, X., Mello E Souza, A. D., Sidiropoulos, E. and Wehrmann, D. (eds.) *The Palgrave Handbook of Development Cooperation for Achieving the 2030 Agenda: Contested Collaboration*. Cham: Springer International Publishing, pp. 1–21.

Cohen, G. and Shinwell, M. (2020). How to measure distance to SDG targets anywhere. Paris: Organisation for Economic Co-operation and Development. www.doi.org/10.1787/a0ac1413-en

Custer, S., DiLorenzo, M., Masaki, T., Sethi, T. and Harutyunyan, A. (2018). Listening to leaders 2018: Is development cooperation tuned-in or tone-deaf. Williamsburg, VA: AidData at the College of William & Mary. www.docs.aiddata.org/ad4/pdfs/Listening_To_Leaders_2018.pdf

Dasandi, N., Laws, E., Marquette, H. and Robinson, M. (2019). What does the evidence tell us about 'thinking and working politically in development assistance? *Politics and Governance*, 7, 155–168, www.dx.doi.org/10.17645/pag.v7i2.1904

Fejerskov, A. M., Lundsgaarde, E. and Cold-Ravnkilde, S. M. (2016). Uncovering the dynamics of interaction in development cooperation: A review of the 'new actors in development' research agenda. *DIIS Working Paper 2016:1* Copenhagen: Danish Institute for International Studies. www.econstor.eu/bitstream/10419/148178/1/85730450X.pdf

Glennie, J. (2020). *The Future of Aid: Global Public Investment*, New York: Routledge.

Gualberti, G. and Benn, J. (2020). DAC Working Party on Development Finance Statistics = [DRAFT] Handbook for reporting the SDG focus of development co-operation activities. Paris: Organisation for Economic Co-operation and Development. www.oecd.org/officialdocuments/publicdisplaydocumentpdf/?cote=DCD/DAC/STAT(2020)7&docLanguage=En

Gulrajani, N. (2017). Bilateral donors and the age of the national interest: what prospects for challenge by development agencies? *World Development*, 96, 375–389, www.doi.org/10.1016/j.worlddev.2017.03.021

Gulrajani, N. and Calleja, R. (2019). Understanding Donor Motivations: Developing the Principled Aid Index. London: Overseas Development Institute (ODI). www.apo.org.au/sites/default/files/resource-files/2019-03/apo-nid227586_1.pdf

Holzapfel, S. (2016). Boosting or hindering aid effectiveness? An assessment of systems for measuring donor agency results. *Public Administration and Development*, 36, 3–19, www.doi.org/10.1002/pad.1749.

Honig, D. and Gulrajani, N. (2018). Making good on donors' desire to Do Development Differently. *Third World Quarterly*, 39, 68–84, www.doi.org/10.1080/01436597.2017.1369030

Ingram, G. and Lord, K. M. (2019). *Global development disrupted: Findings from a survey of 93 leaders*. Washington, DC: Brookings Institution. www.brookings.edu/research/global-development-disrupted-findings-from-a-survey-of-93-leaders/.

Janus, H., Klingebiel, S. and Paulo, S. (2015). Beyond aid: A conceptual perspective on the transformation of development cooperation. *Journal of International Development*, 27, 155–169, www.doi.org/10.1002/jid.3045

Keijzer, N. and Lundsgaarde, E. (2018). When 'unintended effects' reveal hidden intentions: Implications of 'mutual benefit' discourses for evaluating development cooperation. *Evaluation and Program Planning,* 68, 210–217, www.doi.org/10.1016/j.evalprogplan.2017.09.003

Martens, B., Mummert, U., Murrell, P. and Seabright, P. (2002). *The institutional economics of foreign aid.* Cambridge: Cambridge University Press.

Masaki, T., Parks, B. C., Faust, J., Leiderer, S. and DiLorenzo, M. D. (2021). Aid management, trust, and development policy influence: New evidence from a survey of public sector officials in low-income and middle-income countries. *Studies in Comparative International Development,* 56, 1–20, www.doi.org/10.1007/s12116–020–09316–3

OECD (2019). Development Co-operation Report 2019. Paris: Organisation for Economic Co-operation and Development. www.doi.org/10.1787/9a58c83f-en

OECD (2020a). *Aid by DAC members increases in 2019 with more aid to the poorest countries.* Paris: Organisation for Economic Co-operation and Development. www.oecd.org/dac/financing-sustainable-development/development-finance-data/ODA-2019-detailed-summary.pdf

OECD (2020b). *Global Outlook on Financing for Sustainable Development 2021.* Paris: Organisation for Economic Co-operation and Development. www.doi.org/10.1787/e3c30a9a-en

OECD (2020c). *What is ODA? Paris: Organisation for Economic Co-operation and Development.* www.oecd.org/dac/financing-sustainable-development/development-finance-standards/What-is-ODA.pdf

OECD/UNDP (2019). *Making Development Co-operation More Effective: 2019 Progress Report.* Paris: OECD Publishing. www.oecd-ilibrary.org/development/making-development-co-operation-more-effective_26f2638f-en

Prizzon, A., Greenhill, R. and Mustapha, S. (2017). An 'age of choice' for external development finance? Evidence from country case studies. *Development Policy Review,* 35, O29–O45, www.doi.org/10.1111/dpr.12268

Purushothaman, C. (2020). *Emerging Powers, Development Cooperation and South-South Relations.* Cham: Springer.

Rogerson, A. and Ritchie, E. (2020). *ODA in Turmoil: Why Aid Definitions and Targets Will Come Under Pressure in the Pandemic Age, and What Might Be Done About It.* Washington, DC: Center for Global Development (CGD). www.cgdev.org/sites/default/files/PP198-Ritchie-Rogerson-ODA-Turmoil.pdf

Scott, S. (2020). Lessons from the Crisis in Foreign Aid Statistics. *Statistika: Statistics and Economy Journal,* 100 (4), 431–443, www.czso.cz/documents/10180/125507861/32019720q4_431–443_scott_consultation.pdf

Sedemund, J. (2014). *An outlook on ODA graduation in the post-2015 era.* Paris: Organisation for Economic Co-operation and Development (OECD). www.oecd.org/dac/financing-sustainable-development/ODA-graduation.pdf

Tomlinson, B. (2018). *Trends in the Reality of Aid 2018: Growing diversions of ODA and a diminished resource for the SDGs.* Quezon City, Philippines: Reality of Aid. www.realityofaid.org/wp-content/uploads/2018/12/Final-July-Global-Aid-Trends-Chapter.pdf

Vähämäki, J. and Verger, C. (2019). *Learning from Results-Based Management evaluations and reviews.* Paris: Organisation for Economic Co-operation and Development. www.oecd-ilibrary.org/development/learning-from-results-based-management-evaluations-and-reviews_3fda0081-en

Waisbich, L. T. and Mawdsley, E. (2022). South-South Cooperation. In: Sims, K., Banks, N. Engel, S., Hodge, P., Makuwira, J., Nakamura, N., Rigg, J., Salamanca, A., and Yeophantong, P. (eds.). *The Routledge Handbook of Global Development.* London: Routledge, 82–92.

World Bank (2020). *Poverty and Shared Prosperity 2020: Reversals of Fortune.* Washington, DC: The World Bank. www.openknowledge.worldbank.org/handle/10986/34496

Yanguas, P. (2018). *Why we lie about aid: Development and the messy politics of change.* London: Zed Books Ltd.

7

SOUTH-SOUTH COOPERATION

Laura Trajber Waisbich and Emma Mawdsley

Introduction

South-South (Development) Cooperation refers to an array of transfer and exchange modalities, including resources, technologies, and knowledge, between (so-called) 'developing' countries. Although not a new phenomenon, SSC has gained visibility and relevance in the last two decades, with the rise in activities of the (re)emerging development cooperation 'partners' or 'providers' from the South (Mawdsley 2012). Southern partners include large economies – like China, Brazil, India, South Africa, Mexico, Turkey and Indonesia – and some smaller, low income and even fragile Southern states. Their growing role in global development has spurred researchers and institutions to respond, leading to the consolidation and deepening of particular themes and approaches in what is a globally interdisciplinary field of studies. In this chapter we review the critical turn in South-South Cooperation (SSC) research, discussing first the different approaches to date. Next, we bring this analysis to the current 'consolidation and politicisation' phase of SSC occurring in many places, exploring some of the themes and perspectives for understanding SSC politics in the years ahead.

Critical theory and South-South Cooperation

Critical SSC studies emerged in the mid-2000s as a lens to investigate South-South exchanges beyond the grand narratives and over-simplistic scripts that were prevalent in some media, policy analysis, and in some of the early scholarship on 'the SSC boom.' In an explicit attempt to move beyond the classic foreign policy and/or geopolitical focus of much initial scholarship, critical scholars brought insights from disciplines like anthropology, geography, development studies and area studies, as well as critical political economy.

Attentive to power within and across the South – as well as between 'North' and 'South' – analysts turned to how SSC is formulated, enacted, and performed. Critical researchers deconstructed SSC principles and official narratives, and sought to understand the contested nature of SSC and its uneven social, political, economic and/or environmental effects. This was often a sympathetic debunking of some of its political and epistemic 'myths' (Bergamaschi, Moore, and Tickner 2017, 10), but also a response to overly hostile attacks. While recognizing the differential nature of South-South development exchanges, and its transformative potential

DOI: 10.4324/9781003017653-9

outside traditional North-South aid/assistance colonially-rooted discourses (Six 2009; Mosse 2011), this critical gaze emphasizes that SSC principles and axiomatic watchwords (such as 'horizontality' and 'solidarity') 'also need to be empirically tested rather than taken at face value' (Mohan 2014, 288).

Covering a large set of issues, sectors and partnerships, topics of inquiry included the conflicts and negotiations of SSC policymaking 'at home' and those resulting from South-South encounters 'on the ground.' Studies underscored the emergent, under expansion, and/or shifting institutional configurations of SSC (Gu, Shankland, and Chenoy 2016; Mawdsley 2019), as well as the dilemmas of its policymaking and stances, and even the anxieties of its practitioners (van der Westhuizen and Milani 2019; Doucette 2020) They were also attentive to the iteratively changing context within which very different Southern providers were consolidating, strategizing and 'feeling for stones' in their own journeys: operating within and influencing shifts among donors in the Development Assistance Committee from the Organization for Economic Co-operation and Development (OECD-DAC), and in and with 'recipient'/partner countries. Below, we provide a brief account of how SSC research has evolved in tandem with changes in SSC practices – in time and across geographies – while firmly establishing the need to place SSC as an always moving target. Inevitably this is a partial history – we have not attempted a comprehensive exercise here, but drawn out some of the more prominent threads of a changing field.

From understanding actors and flows to investigating practices and their effects

Like the phenomena itself, research on SSC goes back decades. But around the turn of the millennium, SSC started – unevenly and very diversely – to grow in scale, breadth, and importance. Global scholarship tracked this growth, initially tending to focus on defining and delineating flows and mapping emerging actors, discourses, and practices. Within this body of work scholars asked questions about the challenges and opportunities brought by the rise of SSC to an already increasingly fragmented and decentred global development field, and in particular to poor peoples, places and countries (e.g. Cheru and Obi 2010). As the field and its analysis deepened, there was a growing insistence on the importance of exploring SSC from recipient perspectives (e.g. Mohan and Lampert 2013), and, at the same time, on understanding the effects of the shifting geographies of knowledge and power on global development (e.g. Zoccal and Esteves 2018).

Uneven geographies of fieldwork-based research

Not infrequently early studies were limited by the paucity of empirically-backed and fieldwork-informed knowledge (Bräutigam 2009; Bergamaschi, Moore, and Tickner 2017). Some Northern analysis was based on uncritical and Eurocentric perspectives at best, and outright ignorance and/or hostility at worst. Concurrently, some Southern commentators were (too) close to policymakers, as is not uncommonly the case in both development studies and international relations scholarship. Furthermore, in several Southern contexts, structural lack of funding for social science research hindered researchers' ability to design research projects that include extensive components of international or multi-sited fieldwork – although this is now changing. As a consequence, knowledge around SSC is produced differently and unevenly across geographies (Mawdsley, Fourie and Nauta 2019), with the inequalities of global academia meaning that especially in this early phase, comparatively more research and more fieldwork-based research on SSC was done by scholars based in Northern institutions than by academics based in the South (Waisbich, Pomeroy, and Leite, 2021).

Both a focus on large 'emerging economies' (China mostly, but also Brazil, India and to a lesser extent South Africa) and the slow growth in the number and depth of 'field'-based studies, has undeniably imprinted SSC research and debates. Echoing much of the polarized discussion on rising powers in international affairs, several assessments of SSC – particularly in the early years of the millennium – relied on rather simplistic scripts that projected euphoria or scepticism. These tones have not fully disappeared in more recent years but have been significantly challenged by the emerging critical wave, to which we now turn.

Critically researching South-South Cooperation

In an effort to deepen and refine SSC research, critical scholars have investigated the topic through at least four overlapping critical lenses and/or research streams: political economy, postcolonial and subaltern studies, knowledge and power, and governance.

Political economy. In this stream, scholars examined the reconfiguration of transnational capital and the new forms of dependence in the 21st century. These studies, often focusing on 'emerging powers' engagement in their own regions and discuss ambiguities and contradictions in the South-South 'solidarity project.' In particular, they unpack how South-South engagements can turn into sub-imperialist or dispossessing enterprises, as 'rising powers' export their national capital-labour and capital-nature contradictions across boundaries. These studies also emphasize the emerging forms of contestation – particularly from below, from civil society actors – of SSC agricultural, extractive and/or infrastructure building initiatives, and the disconnect between official and range of societal perspectives on the purposes and outcomes of SSC (e.g. Bond and Garcia 2015; Taylor 2016; Shankland and Gonçalves 2016).

Postcolonial and subaltern studies. This stream applies postcolonial lenses to study what is framed as 'non-Western' development cooperation. Many within this paradigm also sought to re-inscribe gender, sexuality, and race in SSC, going beyond the state-centric dominant lenses of the 'West versus the rest dichotomy' and paying attention to social injustice issues cross-cutting South-South exchanges. As such, these studies have deconstructed solidarity, affinities, brotherhood/sisterhood narratives to reveal new forms of South-South power relations and inequalities, adding societal and individual lenses, beyond the traditional focus on states and their institutions (e.g. Six 2009; Mohan 2014; Mawdsley 2020).

Knowledge and power. This cluster deals with 'Southern development knowledges' and the ways they have been packaged and transferred to other contexts, through SSC, embedded in similarity, best-fit, and/or appropriateness discourses. Scholars within this cluster also explored how competing knowledges and policy ideas 'from the South' circulate through SSC and how traveling policy ideas, narratives, and programmes generate friction and disputes within and among Southern partners (e.g. Shankland and Gonçalves 2016; Milhorance and Bursztyn 2017).

Governance. This approach came into dialogue with studies of the 'domestic politics' of foreign aid policymaking, such as Lancaster's (2007) seminal book on ideas, interests and institutions shaping aid-making in five OECD-DAC donor countries. Studies within this cluster critically investigate SSC governance, moving beyond flat accounts of foreign policymaking and looking at competing interests and ideas on development at home and abroad, and on the relationship between them (e.g. Gu, Shankland and Chenoy 2016; Bergamaschi, Moore and Tickner 2017). Their contribution is to offer detailed pictures of the interplay and tensions between SSC diplomatic discourse, policy, and practice (and between planners and implementers), while recognizing the constantly shifting features and open-ended configurations of SSC across the different countries.

Many contributions within this last stream are found in the work of scholars studying Brazil, who have extensively dealt with the contradictions, disjunctions, and/or contestations of Brazilian development engagements, notably in the field of agricultural development. Brazilian SSC has been an important locus for critical scholarship across the four different streams mentioned above, contrasting with the more classic international relations approaches dominating much scholarship on other large Southern providers, India and China. One reason is that scholarship on Brazil has more explicitly and consistently examined the multiple dimensions of the domestic and transnational politics of Brazilian foreign policy and global development engagements. This is observable not only in studies around Brazilian bureaucratic politics and SSC policymaking, but also on studies bridging policy diffusion and SSC (van der Westhuizen and Milani 2019; Waisbich, Pomeroy and Leite, 2021). In India, alternatively, the debate has remained in many ways less diverse and less critical, when not openly nationalistic (as noted by Chenoy and Joshi 2016), with the exacerbation of this trend in the last years under Narendra Modi and Bharatiya Janata Party rule.

Current South-South Cooperation politics

A recurrent question for SSC scholars has been the relationship between this subfield, its actors, and practices, and the existing norms and practices governing so-called 'traditional aid,' as practiced by members of the OECD-DAC. Not uncommonly, questions about compliance and/or resistance to OECD-DAC norms and standards by the Southern providers have been embedded in broader reflections on the shifts in global governance under the more multipolar configurations of the 21st century. Critical SSC scholarship offers a fresh look into these broad geopolitical dynamics, asking how power disputes interact with the politics of foreign policy in SSC partners, and what are the implications for South-South encounters on the ground. By doing so, it has offered new responses to these ongoing questions in a complex and multi-layered way, with insights from particular relations, initiatives, and dynamics. These investigations have been conducted in a rapidly growing and evolving context, and the underlying evidence base, normative expectations, and social field have constantly changed. As examples, we explore three features of current SSC politics: convergence-divergence dynamics with OECD-DAC development norms and practices; SSC consolidation and politicization dynamics; and finally, the surge in results/outcomes and effectiveness concerns in the context of Southern-led development cooperation.

Co-option, convergence or Southernization?

Different and sometimes opposing evidence exists on how SSC and 'traditional aid' providers interact and on who is converging with whom, on what: something that is hardly surprising given the enormous heterogeneity of actors, contexts, and multi-layered interactions. Some scholars describe the current landscape as a progressive 'Southernization' of development. This is mostly based on the diagnosis of a 're-emergence' of SSC as a substantive, ideational and ontological challenge to the existing aid paradigm. For some, Southernization is also a synonym of a 'BRICS effect' (Zoccal and Esteves 2018), while for others its more about the attractiveness of a so-called 'Beijing consensus' on win-win economic cooperation and infrastructure building. This model is seen as appealing to other developing countries that seek new funding sources for building connectivity as well as for Northern donors increasingly concerned to leverage aid spending to 'national interests' (Gulrajani and Faure 2019). Another version of the 'Southernization' tale looks at social policy travel 'from the South' and at how poverty alleviation-related SSC has been influencing traditional donors and international organizations' thematic priorities and ways of working, notably within the United Nations' development system. Examples include the renewed global fight against hunger and the diffusion of certain social protection schemes and instruments, like Conditional Cash Transfers (Stone, Porto de Oliveira and Pal 2019). These studies highlight the historical conjuncture in which certain Southern partners, notably rising powers, acquired power, authority, and resources (political, symbolic, and material) to spread their 'models' and policy experiences to other developing countries and to international organizations (Milhorance and Soule-Kohndou 2017). In both accounts, traditional donors' ways of working and thematic priorities are seen as being increasingly influenced by their emulation of, or learning from the South (Constantine and Shankland 2017).

Alongside 'Southernization,' scholars have studied the opposite process, namely SSC convergence with existing aid norms and practices. Convergence is due, firstly, to direct and indirect socialization pressures or even co-optation attempts. This translates into dynamics ranging from 'naming-and-shaming' and stigmatization of what is seen as 'deviant behaviour' by Southern providers, as well as opening-up and outreach efforts by traditional donors and in international organizations (Eyben 2013). Outreach efforts include seeking to influence Southern providers,' and in particular rising powers, to shape their global behaviour: partnering with them to reform SSC management structures, and working together in the benefit of third countries through triangular cooperation (Abdenur and Da Fonseca 2013; Zhang 2017). Two examples of this kind of (attempted) influencing strategy include the 2009 creation of the China-DAC Study Group, and the funding of knowledge actors within Southern providers to produce academic and/or applied research on their countries' growing international development roles.

Concurrently, convergence with existing OECD-DAC standards on how to practice development cooperation has also been explained by a will to integrate with existing norms by Southern providers. This more conciliatory stance arguably reflects a waning of the global Bandung revisionist impetus and the adoption of increasingly more 'pragmatic,' result-oriented, SSC approaches (Esteves and Assunção 2014; Mawdsley 2019). Rather than a pure acceptance of existing norms and standards, Southern providers have shown a will to integrate *differently*, keeping certain 'Southern characteristics' and ways of doing development cooperation. Examples of Southern providers' critical engagement with existing 'donorship' norms and practices include crafting Southern alternative ways to count and report SSC flows rather than following OECD-DAC metrics and reporting standards, challenging DAC's 'soft norm' of untying aid, and selectively adopting socio-environmental norms when financing infrastructure projects rather than embracing the existing 'do no harm' consensus in international development finance. Pragmatic, yet still critical, these stances reflect SSC providers' responses to growing internal and external pressures on powerholders and SSC bureaucracies to justify the purposes and outcomes of their development cooperation to multiple constituencies, including domestic audiences (van der Westhuizen and Milani 2019). They also reflect the political and management strategies from those leading on SSC to deal with the political and diplomatic consequences of their SSC initiatives becoming more disputed, and in some cases eventually 'going wrong' on the ground (Yeophantong 2020).

While seemingly conflicting, these scripts – of co-option, convergence and Southernization – are not mutually exclusive. Rather they reveal unfolding 'mutual' or 'two-way socialization' dynamics (Milhorance and Soule-Kohndou 2017) happening in the field, with negotiations between 'traditional/Northern' and 'new/Southern' providers leading to the formation of new shared development norms, understandings and/or agreed expectations on 'appropriate' development cooperation provider behaviour and organizational structures for managing and doing development cooperation; as well as points of enduring tension.

SSC consolidation and politicization

Shifts and negotiations are not only taking place between old and new providers, but are also happening inside SSC providers, making the current phase one of both consolidation and politicization. As a moving target, spatial-temporal sensitivity is key to the study of SSC. Mawdsley (2019) suggested one periodization of SSC, with three phases (as a heuristic rather than any claim to a defining chronology). The first corresponds to the emergence of SSC during the Cold War and in particular under the Third World 'Bandung Spirit.' This first and rather long phase also includes the progressive adoption of the SSC agenda and normative/axiomatic principles under the label of 'technical cooperation between developing countries' by the United Nations, as well as its subsequent demobilization in the late 1980s, due, among other things, to the successive political-economic crises in the South, in the debt crisis and structural adjustment years.

The second (heuristic) phase corresponds to the re-emergence of SSC in the early 2000s, and its intense expansion and diversification in terms of actors and modalities, strongly championed by, although not limited to, the large 'emerging economies.' Finally, the third and current 'SSC consolidation phase' is located from around 2015 onwards. The year of 2015 was undeniably a global watershed, with the approval of important international agreements including the Sustainable Development Goals (SDGs), the Addis Ababa Action Agenda on financing for development, and the Paris Agreement on climate change. According to Mawdsley, the 'SSC 3.0 moment' is marked by Southern providers having to 'manage the success' from the previous expansionary phase in the early 2000s. As a consequence, important shifts in narratives, modalities, and institutions are taking place within key SSC providers and at the global level, including: (a) more pragmatic, outcome-oriented SSC narratives; (b) less strong non-interference modalities of engagement; and (c) less ideational and operational distinction from (so-called) 'established' donors. It remains to be seen if the recent COVID-19 pandemic will prompt the emergence of a new phase of SSC, possibly with the accelerated retreat of some previously active countries (something witnessed in Brazil before COVID-19), the consolidation of new forms of 'health diplomacies' and exchanges, and the rise of new forms of bargains between Southern partners and their domestic constituencies.

While tentative, and certainly not the only way available to assess the phenomenon, this proposed periodization attempts to understand shifts in SSC across time. It also captures the more recent changes unfolding in large and/or paradigmatic SSC providers, even if the exact dates and the nature of the consolidation dynamics varies across countries. Consolidation includes the institutionalization of development cooperation and the creation of new agencies, such as the Chinese development cooperation agency (CIDCA), set-up in 2018.[1] In other cases, it has meant the creation of more ambitious development cooperation programmes and funds, such as the Brazil-World Food Programme Centre of Excellence Against Hunger, set-up in 2011; China's South-South Cooperation Assistance Fund, established in 2015 and co-managed by China and a range of UN agencies; and the India-United Nations Development Partnership Fund, set up in 2017. In some contexts, like Brazil, South Africa or Mexico, debates around institutionalization started around 2010, but have unfolded under increasingly politically competitive and polarized contexts, leading to a mix of consolidation and stalled or even downsizing reforms (van der Westhuizen and Milani 2019).

The 'effectiveness agenda' and South-South Cooperation

Rather than purely technical or managerial issues, the unfolding shifts and institutional reforms we have touched on here illustrate the politicized nature of the current consolidation moment. This politicization is observed in the increased domestic debate over SSC policymaking, and over its results, and in the rise of 'SSC effectiveness' debates at the global and local-national levels. As Southern providers have expanded their global development engagements, investing more resources as well as symbolic and political capital in this agenda, questions around benefits and purposes of SSC, around trade-offs between promoting development at home and abroad, and around outcomes and impacts of SSC specific initiatives, have become more present in domestic politics. This is now materializing in subnational politics and in national media, electoral, and legislative debates both within large providers but also across several smaller Southern partners (Mohan and Lampert 2013; Waisbich 2019).

At the same time, as the role of Southern providers expands, concerns with the developmental outcomes and impact of Southern-led development cooperation has also emerged at the global level. The 'Aid Effectiveness Agenda' is an important component of traditional aid landscape since the 2000s. Under this Agenda, a wave of 'aid effectiveness reforms' were put in place seeking to improve aid management and maximize its results. The will to reform *Aidland* (at least half-heartedly and imperfectly), responded to a multitude of geopolitical and domestic changes in both donors and recipients resulting in increased pressure on aid agencies in DAC members and their counterparts in recipient countries to improve the management of aid resources and to show results (Eyben 2013). Such pressures grew during the 2010s under a more challenging landscape for resource mobilization after the 2008 financial crisis, while raising the global ambitions for concerted action under the SDGs agenda. In this context, many in the development community started to treat and frame SSC as an additional source (beyond 'traditional aid') of financing for global development or, in UN-language, as a means to implement the SDGs. Progressively, and not without controversies, Southern providers have joined development effectiveness debates on their own terms. By doing so, '*SSC-land*' has become a new site for questions and unresolved debates on how to re-negotiate global development roles and responsibilities in an increasingly complex, and for some even post/beyond aid, world.

Conclusion

This chapter offers a brief overview of the critical turn in South-South Cooperation studies and the ways in which a diverse array of scholars have tried try to unpack a highly heterogenous and rapidly changing field. It shows how critical scholars, across different and intertwined 'Southern' and 'Northern' geographies, have challenged the first wave of SSC studies bringing new research themes and questions to the forefront, and/or fresh lenses into longstanding issues, including approaching SSC from political economy, postcolonial and subaltern studies, the anthropology of international development, and the domestic politics of development cooperation policymaking.

Looking at the last decade or so, we have briefly unpacked the linkages between the current consolidation shifts and reforms, and the growing politicization of this agenda at different scales: at the global level, domestically in both large and small Southern countries involved in different ways, and roles in South-South development exchanges. In that context, and thinking in particular about the more active SSC protagonists, we observe that emerging internal and external calls for Southern providers to justify their policies and practices – what one could call

SSC accountability politics – constitutes an interesting and important lens into the politics and the (re)politicization of SSC.

SSC was born out of an explicitly politically radical discourse within the Third World in the midst of formal decolonization and during the Cold War, and was animated by demands for global economic and political justice. Its (re)emergence in the 2000s also reflected strong claims about its transformative potential. 'SSC 2.0' was presented as desired and desirable, embedded in proud official narratives from both partners in the South-South equation, celebrating the similarities and adaptability of Southern-grown knowledges and/or technologies, as well as the welcome expansion of room for manoeuvre vis-à-vis Northern aid paradigms and conditionalities. If Northern aid was permeated by (supposedly) sanitized technical evidence-based discourses about development (Li 2007), Southern claims to development were alternatively situated, and self-proclaimed as better and more appropriate exactly because of this political situatedness. Put another way: while Western donors denied and concealed history (above all, colonialism and ongoing structural inequality), the emerging South decided to build upon it (Six 2009).

Beyond the promises, as SSC grew materially and politically it has also, in many ways, changed its nature towards less radical, techno-scientific forms of development cooperation among Southern countries. Such de-politicization happened, for instance, through upholding conventional modernizing assumptions of how development works, or through practicing development cooperation through unidirectionally sharing 'best-practices' from middle-income countries in the South (or from emerging powers) to lower-income countries. The turn to, or deepening, of various forms of right-wing authoritarian nationalism, have also played a role in unfolding competitive and complementary 'development' agendas within and across the South, with different implications for different elites and citizens. South-South development exchanges in the post-2015 era have also led towards renewed forms of partnerships capable of strengthening political ties and expanding connectivity between Southern countries; and also of perpetuating, deepening and reinventing hierarchies and extraction. While officially portrayed as win–win partnerships, these ventures are also subject to tensions and fractures both at the geopolitical level as well as within countries, generating news forms of politics.

Growing contestation around the impacts of South-South initiatives on the ground, increased debates over Southern providers common but differentiated responsibilities in the global development realm, and over how to measure the flows and the results of SSC in the context of the Agenda 2030, or the growing political visibility of SSC in domestic policy arenas, are all examples of new forms of SSC politics that deserve further critical inquiry in the years ahead. One task for those interested in critically researching South-South (Development) Cooperation in the decade ahead resides therefore in continued questioning of the gaps between the transformative claims by Southern partners and their uneven materializations on the ground, while unpacking the fragmented and divisive state of contemporary global development, the disputes around power and responsibility in global development, and the political-institutional-citizenship dilemmas of consolidating and institutionalizing SSC as a policy field for Southern providers.

Key sources

1. Gu, Jing, Alex Shankland, and Anuradha M. Chenoy. (2016). *The BRICS in International Development*. Basingstoke: Palgrave Macmillan.

This edited volume includes a collection of insightful pieces on the actors, institutions, and ideas shaping the first years of the re-emergence of BRICS countries as 'rising powers in international development.' The authors are Northern and Southern-based scholars with sound understanding and empirical knowledge of national and global development dynamics shaping BRICS countries rising roles in the field.

2. Mawdsley, Emma, Elsje Fourie, and Wiebe Nauta, eds. (2019). *Researching South–South Development Cooperation: The Politics of Knowledge Production.* 1st ed. Routledge. www.doi.org/10.4324/9780429459146

This recent edited volume reflects on politics of knowledge production in the field of South-South Cooperation for development. It gathers a collection of conceptual, methodological, and reflexive first-hand accounts by both Northern and Southern, senior and junior scholars and offers honest accounts on the promises, opportunities and impossibilities in co-producing critical knowledge on SSC.

3. Mohan, G., and B. Lampert. (2013). Negotiating China: Reinserting African Agency into China-Africa Relations. *African Affairs* 112 (446): 92–110. www.doi.org/10.1093/afraf/ads065

One of the first and most accomplished papers on African agency in China-Africa relations, offering a complex account of how different actors within African countries negotiate the Chinese partner. The paper highlights the relationships between African and Chinese political and business elites and indicates emerging forms of contestation of China presence within African societies beyond simplistic 'anti-Chinese' narratives often portrayed in Western circles.

4. Shankland, Alex, and Euclides Gonçalves. (2016). Imagining Agricultural Development in South–South Cooperation: The Contestation and Transformation of ProSAVANA. *World Development* 81 (May): 35–46. www.doi.org/10.1016/j.worlddev.2016.01.002

An empirically rich contribution on one particular triangular cooperation initiative, the Brazil, Japan, Mozambique ProSAVANA agricultural development program, and its multiple negotiations on the ground. Based on a multi-sited ethnographic work, this paper offers a powerful illustration of how South-South affinities and imaginaries are build and contested in the course of one initiative.

5. Westhuizen, Janis van der, and Carlos R. S. Milani. (2019). Development Cooperation, the International–Domestic Nexus and the Graduation Dilemma: Comparing South Africa and Brazil. *Cambridge Review of International Affairs* 32 (1): 22–42. www.doi.org/10.1080/09557571.2018.1554622

An important contribution to the conceptualization of the domestic politics of South-South Cooperation in two emblematic Southern partners: Brazil and South Africa. Here the authors emphasize the multiple constituencies bargaining around SSC at home and abroad, and the tensions and dilemmas Southern providers face while promoting development beyond borders.

6. Zoccal, Geovana, and Paulo Esteves. (2018). The BRICS Effect: Impacts of South–South Cooperation in the Social Field of International Development Cooperation. *IDS Bulletin* 49 (3): 129–144 www.doi.org/10.19088/1968–2018.152

A good representative of 'second generation' South-South Cooperation studies, which aims to unpack the effects of the political and material rise of Southern development actors in the norms, governance, and practice of global development.

Note

1 In the last decade, for example, many other development cooperation agencies were set-up, including the Mexican Agency for International Development Cooperation and Colombian Presidential Agency of International Co-operation (in 2011), India's Development Partnership Administration (in 2012), the Palestinian International Cooperation Agency (in 2016), and the Indonesian Agency for International Development (in 2019).

References

Abdenur, Adriana Erthal, and João Moura Estevão Marques Da Fonseca. (2013). The North's Growing Role in South–South Cooperation: Keeping the Foothold. *Third World Quarterly* 34 (8): 1475–91. www.doi.org/10.1080/01436597.2013.831579

Bergamaschi, Isaline, Phoebe V. Moore, and Arlene B. Tickner, eds. (2017). *South-South Cooperation Beyond the Myths*. Basingstoke: Palgrave Macmillan UK.

Bond, Patrick, and Ana Garcia. (2015). *BRICS: An Anti-Capitalist Critique*. London: Pluto Press.

Bräutigam, Deborah. (2009). *The Dragon's Gift: The Real Story of China in Africa*. Oxford: Oxford University Press.

Cheru, Fantu, and Cyril I. Obi, eds. (2010). The Rise of China and Africa in India: Challenges, Opportunities and Critical Interventions. Africa Now. London: Zed Books (in Association with the Nordic Africa Institute, Uppsala).

Chenoy, Anuradha, and Anuradha Joshi. (2016). India: From Technical Cooperation to Trade and Investment. In *The BRICS in International Development*, edited by Jing Gu, Alex Shankland, and Anuradha Chenoy. Basingstoke: Palgrave Macmillan.

Constantine, Jennifer, and Alex Shankland. (2017). From Policy Transfer to Mutual Learning?: Political Recognition, Power and Process in the Emerging Landscape of International Development Cooperation. *Novos Estudos CEBRAP* 36 (1): 99–122.

Doucette, Jamie. (2020). Anxieties of an Emerging Donor: The Korean Development Experience and the Politics of International Development Cooperation. *Environment and Planning C: Politics and Space* 38 (4): 656–73. www.doi.org/10.1177/2399654420904082

Esteves, Paulo, and Manaíra Assunção. (2014). South–South Cooperation and the International Development Battlefield: Between the OECD and the UN. *Third World Quarterly* 35 (10): 1775–90. www.doi.org/10.1080/01436597.2014.971591

Eyben, Rosalind. (2013). Struggles in Paris: The DAC and the Purposes of Development Aid. *The European Journal of Development Research* 25 (1): 78–91. www.doi.org/10.1057/ejdr.2012.49

Gu, Jing, Alex Shankland, and Anuradha M. Chenoy. (2016). *The BRICS in International Development*. Basingstoke: Palgrave Macmillan.

Gulrajani, Nilima, and Raphaëlle Faure. (2019). Donors in Transition and the Future of Development Cooperation: What Do the Data from Brazil, India, China, and South Africa Reveal? *Public Administration and Development*, September. www.doi.org/10.1002/pad.1861

Lancaster, Carol. (2007). *Foreign Aid: Diplomacy, Development, Domestic Politics*. Chicago: University of Chicago Press.

Li, Tania Murray. (2007). *The Will to Improve: Governmentality, Development, and the Practice of Politics*. Durham, NC; London: Duke University Press.

Mawdsley, Emma. (2012). *From Recipients to Donors: Emerging Powers and the Changing Development Landscape*. London: Zed Books.

Mawdsley, Emma. (2019). South–South Cooperation 3.0? Managing the Consequences of Success in the Decade Ahead. *Oxford Development Studies*, March, 1–16. www.doi.org/10.1080/13600818.2019.1585792

Mawdsley, Emma. (2020). Queering Development? The Unsettling Geographies of South–South Cooperation. *Antipode* 52 (1): 227–45. www.doi.org/10.1111/anti.12574

Mawdsley, Emma, Elsje Fourie, and Wiebe Nauta, eds. (2019). *Researching South–South Development Cooperation: The Politics of Knowledge Production.* 1st ed. Routledge. www.doi.org/10.4324/9780429459146

Milhorance, Carolina, and Marcel Bursztyn. (2017). South-South Civil Society Partnerships: Renewed Ties of Political Contention and Policy Building. *Development Policy Review. Volume 35 (2017) Supplement 2.*

Milhorance, Carolina, and Folashade Soule-Kohndou. (2017). South-South Cooperation and Change in International Organizations. *Global Governance: A Review of Multilateralism and International Organizations* 23 (3): 461–81. www.doi.org/10.1163/19426720–02303008

Mohan, G., and B. Lampert. (2013). Negotiating China: Reinserting African Agency into China-Africa Relations. *African Affairs* 112 (446): 92–110. www.doi.org/10.1093/afraf/ads065

Mohan, Giles. (2014). China in Africa: Impacts and Prospects for Accountable Development. In *The Politics of Inclusive Development: Interrogating the Evidence*, edited by Sam Hickey, Kunal Sen, and Badru Bukenya, 279–304. Oxford: Oxford University Press.

Mosse, David. (2011). *Adventures in Aidland: The Anthropology of Professionals in International Development.* New York: Berghahn Books.

Shankland, Alex, and Euclides Gonçalves. (2016). Imagining Agricultural Development in South South Cooperation: The Contestation and Transformation of ProSAVANA. *World Development* 81 (May): 35–46. www.doi.org/10.1016/j.worlddev.2016.01.002

Six, Clemens. (2009). The Rise of Postcolonial States as Donors: A Challenge to the Development Paradigm? *Third World Quarterly* 30 (6): 1103–21. www.doi.org/10.1080/01436590903037366

Stone, Diane, Osmany Porto de Oliveira, and Leslie A. Pal. (2019). Transnational Policy Transfer: The Circulation of Ideas, Power and Development Models. *Policy and Society*, September, 1–18. www.doi.org/10.1080/14494035.2019.1619325

Taylor, I. (2016). Dependency redux: why Africa is not rising. *Review of African Political Economy* 43 (147): 8–25.

Waisbich, Laura Trajber. (2019). Democracy and South-South Cooperation in IBSA Countries: Emerging Legislative Debates. *Development Cooperation Review* 1 (9): 3–13.

Waisbich, Laura Trajber, Melissa Pomeroy, and Iara Costa Leite. (2021). Travelling across Developing Countries: Unpacking the Role of South-South Cooperation and Civil Society in Policy Transfer. In *Handbook of Policy Transfer, Diffusion and Circulation*, edited by Osmany Porto de Oliveira. Cheltenham and Camberley: Edward Elgar Publishing, p. 214–236.

Westhuizen, Janis van der, and Carlos R. S. Milani. (2019). Development Cooperation, the International–Domestic Nexus and the Graduation Dilemma: Comparing South Africa and Brazil. *Cambridge Review of International Affairs* 32 (1): 22–42. www.doi.org/10.1080/09557571.2018.1554622

Yeophantong, Pichamon. (2020). China and the Accountability Politics of Hydropower Development: How Effective Are Transnational Advocacy Networks in the Mekong Region? *Contemporary Southeast Asia* 42 (1): 85–117. www.doi.org/10.1355/cs42–1d

Zhang, Denghua. (2017). Why Cooperate with Others? Demystifying China's Trilateral Aid Cooperation. *The Pacific Review* 30 (5): 750–68. www.doi.org/10.1080/09512748.2017.1296886

Zoccal, Geovana, and Paulo Esteves. (2018). The BRICS Effect: Impacts of South–South Cooperation in the Social Field of International Development Cooperation. *IDS Bulletin* 49 (3), 129–144. www.doi.org/10.19088/1968–2018.152

8

MULTILATERAL DEVELOPMENT BANKS

Old and new

Susan Engel and Patrick Bond

Introduction

When the topic of multilateral development banks (MDBs) arises, most people think of the original project lender – the World Bank whose 1944 mandate was to lend for 'reconstruction and development.' It is one of two Bretton Woods institutions based in Washington, DC. The vision for its sister, the International Monetary Fund (IMF), was, in some ways, also 'developmental,' although most of the pathbreaking designs of British economist (and lead negotiator of the debtor states) John Maynard Keynes were rejected. Those visions included the IMF managing a world currency and imposing penalties for excessive trade surpluses, so as to limit 1920s-style uneven development. By the 1960s, both the Bank and IMF had become focused on orthodox financial values, reminiscent of US banks, with all the national, class, race, gender, and ecological biases those entail.

Subsequently, more than 30 MDBs were created, of which two thirds were between 1956 and 1979 (Bazbauers and Engel 2021, 2). From 1980 to 1997, six MDBs were created, and from 1998 to 2015, another five were established, including the New Development Bank (NDB, established 2014) of the Brazil, Russia, India, China, South Africa (BRICS) grouping and the China-backed Asian Infrastructure Investment Bank (AIIB, 2015). These latter two have garnered the most attention, because their backers are considered potential threats to the current economic hegemon, the US. Given that challenges to the slowly-reforming Bretton Woods development finance system have failed since the early 1980s, there were great expectations that the AIIB and NDB would chart a different – more equitable – path. Instead, the latest MDBs have facilitated, relegitimized and recapitalized global and regional financial maldevelopment. Once perceived as a threat to extant multilateral norms, they soon amplified some of the most dangerous development finance trends.

This chapter starts by summarizing central functions of multilateral development banking, and follows with comparisons to the two new MDBs. The third section briefly reflects on MDB responses to the COVID-19 pandemic, focusing on the World Bank and NDB, while the final section considers pedagogical approaches to the MDBs.

DOI: 10.4324/9781003017653-10

The multilateral development bank model

The creation of MDBs can be traced to number of precedents including growing international monetary cooperation and the rise of export-import banks, but most explicitly from a proposal by Mexico's delegation to 1939 negotiations over inter-American financial cooperation. The Party of the Mexican Revolution government led by Lázaro Cárdenas was still drawing inspiration from the 1910–20 struggle for sovereignty, and boldly proposed that an Inter-American Bank be tasked with channelling investment to promote economic development (Bazbauers and Engel 2021). Although the proposal failed, one of the Americans involved in the negotiations, Harry Dexter White, became a principal architect for the International Bank for Reconstruction and Development (IBRD) – or World Bank – negotiated at the Bretton Woods Hotel in New Hampshire in 1944 and in Savannah, Georgia in 1946. The World Bank was established at a time the previously-dominant global private banks were in retreat following the 1929–33 world financial crash that had resulted from their exploitative, crisis-prone character (Oliver 1975). This also led to new regulations on banks such as the US Glass-Steagall Act, which split stock market investments from lending to avoid conflicts of interest.

The IBRD established an International Finance Corporation in 1956 to manage for-profit lending to, and ownership-investments in, the private sector, as well as an International Development Association wing in 1960 to provide concessional low interest loans with a grace period, to very poor countries. That aside, the 'World Bank Group' increasingly reflected banking norms more so than developmental ones. With a fierce post-War ideological battle between 'modernizationists' based at the US State Department (led by W.W. Rostow) and '*dependistas*' at the UN Commission on Latin America (led by Raul Prebisch), the main Bank officers periodically shifted their orientation between lending for mega-projects and for basic needs and from the 1980s began imposing macroeconomic conditionality. Overall, though, its development model has prioritized growth over human wellbeing and ecological sustainability. When the Bank deems that austerity policies are required, then invariably social policy, civil service wages and parastatal ownership are usually on the chopping block. Other MDBs have largely followed the model established by the World Bank.

Seven key factors fundamentally shape the work of the World Bank and all the MDBs, including the two latest ones. First, *MDBs aim to promote business interests more than poverty reduction or sustainability*. In many countries, uneven development means that economic growth occurs at the expense of unsustainable resource extraction and does not result in re-investment of profit within those economies and societies. Indeed, if 'natural capital depletion' were recognized as a debit in national wealth accounts of most African countries, as it should be, the continent's actual net growth is negative: the profits from extracting non-renewable minerals are largely exported and the debits to the extracted wealth leave 88 per cent of African countries as net losers even when royalties, taxes, foreign exchange revenue, physical capital, jobs, backward/forward linkages, and infrastructure are factored in. Even the Bank concedes this (Bond 2016).

Second, *the ideology behind its commitment to economic growth is market-centric*. Most MDBs, the World Bank in particular, have traditionally been hostile to the kind of state-led development – including extensive parastatal corporations – that propelled much of East Asia to middle- or high-income status. For states where capitalist development is not a viable pathway (e.g. small island developing states), the banks still promote markets, rather than sustainability and wellbeing. The World Bank's leadership in development thinking over the decades has been questioned because so many Asian experiences contradict the neoliberal ideology (George and Sabelli 1994, Wade 2006). After one maverick chief economist, Joseph Stiglitz, encouraged the Bank to consider an augmented or 'post-Washington Consensus' view of development (Engel 2010), the Bank briefly abandoned heterodoxy. Next followed the mid-2010s fall in commodity prices and the re-emergence of debt crisis reinforcing the shift (Güven 2018). The 2012–19 World Bank president – once-leftist NGO leader and medical anthropologist, Jim Yong Kim – initially reinforced this shift. However, he was fairly quickly enraptured by markets, in particular the idea of public–private partnerships (especially with Wall Street investment banks) to dramatically expand Bank lending. At the same time, partly due to competition with Chinese state capital in poorer countries, infrastructure regained a predominance in Bank lending patterns during the 2010s.

Third, *MDBs are structured much like private banks* with the main difference being that their shareholders are members of the interstate system. Some MDBs also have other development bank representatives as board members and a few non-state members. The richest states have the largest shareholdings in most MDBs and hence the most say because votes are closely linked to shares. Despite rounds of reforms aiming to balance ownership quotas at the Bank (and IMF), most recently in 2015, the US (17.4 per cent), Japan (6.5 per cent) and Europe still dominate, although China is now the third largest shareholder (6.4 per cent).

The World Bank and IMF suffer a democratic deficit and lack of meritocracy because most borrowing states have little say in the two institutions. Although management personnel from poorer countries – albeit mainly with *status quo* perspectives – are found in the second and third tiers of power, the tradition that top leaders are respectively from the US and Europe has not yet been seriously challenged. Excessively slow reform of the Bretton Woods institutions was one reason China decided to establish a new bank, the AIIB, of which it is the largest shareholder. The BRICS alternative, the NDB is an exception to the weighted shareholding system as the five founding states having equal shareholdings, though this may change if other states join. Further, the BRICS also initiated a $100 billion Contingent Reserve Arrangement (CRA) in 2014 at the same time as the NDB, which has a wealth-biased shareholding. Revealingly, in the worst days of the COVID-19 crisis when South Africa's finance minister claimed to require a $4.3 billion IMF loan, the CRA was not mentioned, and it seems to have disappeared from once-celebratory accounts of BRICS alternatives.

Fourth, *MDBs raise most of their funds from private capital markets and on-lend them at a small margin*. This means their actions are profoundly shaped by the analysis and views of rating agencies and private investors (Bond and Brown 2020). But given the ratings agencies' exceptionally dubious track records – e.g. in their investment-grade ratings of Lehman Brothers and AIG just before both collapsed in 2008 – should they be trusted? To protect its AAA credit rating, in the midst of the COVID-19 crisis in 2020, the Bank refused its borrowers debt cancellation – preferring mere deferment of payments. Its middle-income borrowers, meanwhile, suffered an average 'Ba2' credit rating but rarely fell behind on World Bank payments, so Moody's Investors Service (2020) was impressed with 'only 0.2% of total outstanding development assets qualifying as non-performing over the past three fiscal years.' The Bank's callable capital was then a comfortable 1.14 times the amount of the Bank's total outstanding debt and it had just issued $54 billion in medium- and long-term securities in 2019, borrowing at the world's cheapest rates (Bond and Brown 2020).

Ratings agencies' analyses of MDBs focus on the creditworthiness of their major shareholders and the volume of their paid-in capital to the banks, because they would be responsible for paying out most of the callable capital should it be required (Ben-Artzi 2016, Humphrey 2014). MDBs without rich states as shareholders have found it difficult to access financial markets, which has constrained their growth. In some cases, rich states were admitted and even allowed to dominate, so as to facilitate asset growth, e.g. the African Development Bank. In such cases, MDBs place conservative banking priorities above other ideals to remain afloat – a point too often missed in analysis of their developmental ideologies and investment agendas (Engel and Bazbauers 2020). One of the five leading NDB officials, Leslie Maasdorp (2020) acknowledged this:

> *I think we recognized in 2015 when we started here, as we studied the business model of the multilateral banks in general, we recognized how central and critical a very high credit rating is to the effective functioning of the institution. So, we really sought to create a triple-A institution from scratch. And how do you do that? You mirror and create a balance sheet that looks like a triple-A institution.*

Fifth, *MDBs mostly still fund individual projects, and project quality is a problem* (George and Sabelli 1994). The model for project lending established by the World Bank's (2012) Articles of Agreement is to lend where the project cannot find support through the market. In that sense, MDBs fund second-best projects, or projects that require interest rates artificially lower than market rates to make them viable. Further, MDB projects are often large and complex, and infrastructure is prone to cost overruns, shortfalls in the promised benefits, or inadequate costing of social and ecological damage. This means many projects are more a case of 'survival of the unfittest' (Flyvbjerg 2009). Displacement of local residents, or ignoring their concerns in public participation processes, are typical problems. MDB financing of mega-projects amplifies existing power relations in a society and economy, no matter how skewed these are.

The main beneficiaries of MDB procurement have historically been core capitalist states and their corporations: the US, Germany, Japan, the UK, France, and Italy. By the mid-2000s, private and parastatal firms from China, India and Brazil emerged as major beneficiaries, especially in civil works contacts (Bracking 2009). Around this time, more attention was paid to corruption, with blacklisting of guilty firms increasingly common. Still, the process left much to be desired, as in 2010 the Bank's largest-ever loan – $3.75 billion to the South African electricity supplier Eskom for a coal-fired power plant – was bedevilled by a bribery-type relationship between the main beneficiary (Hitachi) and the country's ruling party, which led to successful US Foreign Corrupt Practices Act (FCPA) prosecution of the former in 2015. Civil society groups had warned the Bank president not to lend for this project, but Robert Zoellick was politically inclined to do so, and taxpayers and Eskom customers subsequently repaid the tainted loan, instead of the Bank taking lender liability.

Sixth, *MDB macroeconomic ideology remains suffused with neoliberal dogma.* As the 1980s Third World Debt crisis followed the Paul 'Volcker Shock' – the US Federal Reserve chair's tripling of interest rates, leading to foreign debt repayment problems that threatened the world's largest commercial lenders – the MDBs provided budget support to desperate borrowers. The new MDB funding bailed out many Northern commercial banks. Moreover, structural adjustment lending came with excessive Washington Consensus conditionality, following 'ten commandments' promoted by John Williamson (2004): fiscal discipline, public expenditure reorientation, more reliance upon value added taxes and lower marginal income tax rates, higher interest rates, a lower currency valuation, liberalized trade and foreign investment, privatization, deregulation and stronger property rights. An attempt to introduce a post-Washington Consensus by then World Bank Chief Economist Joseph Stiglitz (1998) failed to stick and he was fired shortly thereafter. Although neoliberal ideology was discredited by the 2007–09 world financial meltdown and then went into (another temporary) retreat in some Global North settings due to COVID-19, which required dramatic fiscal and monetary loosening, it prevails elsewhere. In most poorer and even middle-income countries, austerity was imposed from 2021 onwards, with well-worn Washington Consensus ideological conditions attached to World Bank and IMF pandemic lending (Glennie 2021, Bretton Woods Observer 2021).

Seventh, *MDB loans add to countries' debt burden,* and their loan terms are not particularly concessional once currency mismatch is considered. The MDB share of developing country debt grew significantly after the outbreak of the 1980s Debt Crisis, as public banks took on the debt collection role of private banks through structural adjustment loans. The highly-delimited Multilateral Debt Relief Initiative of 2005 (based on Washington Consensus prerequisites) saw only a partial Global North write-down of unrepayable liabilities (Bond 2006). Long-standing complaints from organizations like Jubilee South about 'odious debt' (see Chapter 5) taken out by prior dictators and still owed to MDBs and the IMF, were ignored. Debt relief was terminated in 2015 even though developing country debt was again on the increase (Culpeper and Kappagoda 2016).

Many MDBs expanded their capital base and loan portfolios from the mid-2000s, especially after the 2007–09 crisis. For example, IMF recapitalization, Special Drawing Right issuance and new borrowing powers raised nearly $1 trillion, reviving that institution's prior influence. The Bretton Woods twins acquiesced to the emergence of the two new MDBs, the AIIB and NDB, which further expanded the total capitalization of MDBs looked at as a group. At the World Bank, Kim's late-2010s 'Maximizing Finance for Development' strategy aimed to leverage billions of dollars of MDB lending into trillions via hedge funds, private-sector pension funds and the like, confirming that the ongoing Washington Consensus was also a Wall Street one. This expansion of debt relations took place in a context where there is still no orderly or humane mechanism for dealing with debt default or accountability for what should be lender liability (Soederberg 2006). The debt build up in emerging markets and developing economies from 2010 to 2018 was, according to World Bank researchers, the 'largest, fastest and most broad-based increase' ever (Kose et al. 2020, 111). The World Bank and IMF have been publicly warning about debt repayment capacity, especially in several African countries, yet they and other MDBs have expanded their lending capacity and promote external debt-based development as the key pathway for the Global South. All this suggests that the next debt crisis is just a matter of time and the MDBs will be a key player. The COVID-19 crisis, discussed a little later in the chapter, could well be a trigger for a crisis.

The new MDBs

The 2010s rise of the BRICS and the establishment of two new MDBs fuelled debate in international relations about potential reforms of global governance, including in financial markets. The Trump era (2017–20), Brexit and the rise of far-right leadership in several middle-tier countries (notably Brazil, Hungary, India, the Philippines, and Turkey) suggested that, in contrast to the long period of fused neoconservative and neoliberal international leadership by the US and its allies (1980s–2010s), power within and between states had begun to shift. However, the idea that the BRICS would disrupt structural injustices imposed from the Global North was soon disabused, as they mostly became ever greater beneficiaries of, and participants in, the global capitalist order (Thakur 2014). While the three primary-product exporters – Russia, Brazil, and South Africa – suffered from the peak and decline of the commodity super-cycle in 2015, and Russia was subject to sanctions for its annexation of Crimea, neither they, China nor India challenged global power. The AIIB and the NDB simply facilitate the further integration of global capitalism (Robinson 2014). Indeed, the BRICS can be understood as sub-imperialist powers that, as Ruy Mauro Marini (1965) first posited with respect to Brazil, facilitate imperialism's expansion through newly-relegitimized and better-financed multilateralism.

The BRICS have regularly articulated their dissatisfaction with the slow process of existing MDB reform, especially the informal provision that World Bank leadership is reserved for US citizens and IMF leadership for Europeans. Still, they only once offered alternative candidates – in 2012 when Zoellick was not reappointed. When Trump nominated David Malpass – a neoliberal, climate-denialist, renowned China-basher, and former Bear Stearns chief econo-mist who predicted in 2007 that financial markets would *not* suffer turbulence (shortly before Bear Stearns went bankrupt) (Malpass 2007) – to replace Kim in 2019, the BRICS acquiesced, failing to nominate an alternative.

Given the failure to reform either leadership or systems at the Bretton Woods institutions, there was initial hope that a new MDB, the Bank of the South would accompany the 'Pink Tide' of Latin American governments from 1998 when Hugo Chávez's successful Venezuelan presi-dential campaign launched the idea. Chávez and Argentine leader Néstor Kirchner promoted the Banco del Sur in 2006, gaining endorsements in 2007 from other regional presidents as well as Stiglitz. Former Ecuadorian Minister of Economic Planning Pedro Páez attempted to introduce into its provisional articles of agreement several important innovations in eco-social project assessment criteria as well as trade financing to support the Bolivarian Alliance for the Peoples of Our America (ALBA) Peoples' Trade Treaty. However, the bank's $20 billion capit-alization never appeared, and its champion Chávez died in 2013.

When the two new MDBs of the BRICS and China were established in 2014–15 it was along the lines of the old MDBs. They have similar goals and language to the World Bank's and Asian Development Bank's founding charters but focus on promoting infrastructure and sustainable development, with no mention of poverty reduction. They conform with the 'GDP-centred, Northern-development-model approach,' as 50 Years is Enough NGO advo-cate Sameer Dossani (2014) complained. Indeed, as Nicholas Stern explained in an unguarded revelation, he hoped that the NDB would tie down member governments to more responsible partnership arrangements with Western corporations:

> If you have a development bank that is part of a [major business] deal then it makes it more difficult for governments to be unreliable... What you had was the presence of the European Bank for Reconstruction and Development (EBRD) reducing the potential for government-induced policy risk, and the presence of the EBRD in the deal making the government of the host country more confident about accepting that investment. *And that is why Meles Zenawi, Joe Stiglitz and myself, nearly three years ago now, started the idea. And are there any press here, by the way? Ok, so this bit's off the record. We started to move the idea of a BRICS-led development bank* for those two reasons.
>
> *(Stern 2013, emphasis added)*

Reflecting such pressure, the new MDBs adopted similar structures to the old ones. For example, they have limited paid-in capital, so are reliant on capital markets to raise funds. Appointments of governors and staff to the new MDBs follow the old model insofar as they comprise largely of neoliberal economists, ex-bankers, or staff who have worked at other MDBs and are thus socialized into the epistemic community (Haas 1992). To illustrate, two initial lead South African appointments to the NDB were Vice President Maasdorp, who had been a key privat-izer of South African state assets and an employee of major Western banks; and Tito Mboweni as a Director, at a time when he was an international advisor to Goldman Sachs, and after having been celebrated as *Euromoney* Central Banker of the Year in 2001 and 2008 – in both years, the local currency collapsed by 30 per cent in spite of extremely high local interest rates. There are undoubtedly staff in MDBs concerned about poverty and sustainability, and who are opposed to mindless privatization, systemic corruption, and illicit financial flows – but there appear to be many more who are ideologically commitment to neoliberal models (Bond 2020).

Both new MDBs also have environmental and social safeguards or frameworks. The AIIB's Articles of Agreement refer to such policies, and the NDB postures that its Compliance Officer takes these into consideration, and has 'zero tolerance' for corruption, despite all evidence to the contrary. Constructivist analysis explains the expansion of such safeguards and account-ability mechanisms across the MDBs (Park 2014). But if done properly, with strong civil society watchdogging, such safeguards would threaten national sovereignty, which both the old and new MDBs need to respect, to ensure ongoing borrowing. So there has been very little real progress in ensuring better accountability and transparency. As Humphrey (2016) pointed out, typical MDB safeguards sidestep national frameworks and do not challenge local practices and power structures, which often feature elite arrangements damaging to local human and envir-onmental wellbeing. Both old and new MDB safeguards are therefore often merely tokenistic one-size-fits-all policies, with insufficient political power and financial resources to make any difference.

As pointed out by various NGO watchdogs – NGO Forum on the ADB, the Bank Information Centre, the Bretton Woods Project and Inclusive Development International – project-affected communities find MDB processes extremely frustrating. They typically lead to limited reparations for harm and unsatisfactory meagre reforms, rather than the whole-scale rethinking of mega-projects and recalibration of development philosophies that are required. If the safeguard systems were to mature, they would need to be expanded to projects funded through the MDBs' financial intermediaries (e.g. national banks that on-lend to small and medium enterprises or to a particular sector). In the case of the AIIB, such on-lending comprises a significant per cent of the bank's portfolio. Watchdogs have also criticized financial intermediary lending that hides fossil fuel investments.

Like the old MDBs, the AIIB and NDB suffer from challenges with project quality and corruption, and they push countries into a neoliberal export orientation, in part because they frequently lend in inappropriate currencies. For example, the first four loans of the NDB to South Africa were US dollar denominated. They were for: (1) linking privatized renewable energy producers to the South African power grid (which Maasdorp admitted to *Euromoney* in 2019 was delayed by corruption); (2) expanding the Durban port (also curtailed by corruption within a few months of its announcement in 2018); (3) a major loan to Eskom for the same corrupt coal-fired power plant for which Hitachi had bribed the ruling party; and (4) an untransparent loan to the Development Bank of Southern Africa to on-lend to municipalities. These loans were all capable of being financed with local currency and all were granted without a hint of community consultation. They demonstrate that South Africa's new democratic lead-ership follows in the footsteps of apartheid (World Bank-financed) institutions, insofar as the 'extractivist' priorities are identical: an overreliance on coal-fired electricity, which provides subsidized power to Western and BRICS corporations' operations (while households regularly suffer disconnections due to inability to pay). These economic structures lock South Africa further into a world economy which does not pay fair value for its non-renewable mineral resources (Bond 2018).

The new MDBs tie borrowers into transnational corporate networks and inappropriate hard-currency debt repayment no less than old MDBs do. The AIIB and NDB have memo-randa of understanding with the World Bank and other MDBs so as to share project co-finan-cing opportunities and more. Partly this serves to win the confidence of the all-important credit rating agencies (Bond and Brown 2020). Partly, though, the mindset of the new MDBs mirrors the old. There is little reason, then, to expect that the AIIB and NDB will differ in any substantive ways when it comes to lending or knowledge production and this is evident in the way old and new MDBs united to offer and implement highly inappropriate COVID-19 financing.

MDBs and COVID-19

The COVID-19 pandemic hit many of the world's poorest and middle-income countries just as hard as it did the US, Great Britain, Italy, and other rich countries that were unprepared for economic lockdown, social distancing requirements, and severe health system stresses. In many of those rich countries, governments dispensed with some neoliberal precepts during 2020 in order to cushion the blows of lockdowns. The fiscal space to do so in other countries was far less, however, so the World Bank and IMF roles were suddenly amplified, especially as a new round of debt crises caused a half-dozen sovereign defaults in 2020 alone, and enormous pressure to adopt austerity programs. The World Bank committed to lending $160 billion to help states address the pandemic (including $12 billion for vaccines) from June 2020 to June 2021 and other large MDBs promised a further $80 billion.

In health care, the key context is decades of MDB promoted under-investment public health care. From the 1980s, health care was among the social spending commitments poor countries had to cut in line with the Washington Consensus. Further, World Bank conditionality pushed private-sector health provision. By the early 2000s, with the public primary health care system irreparably damaged in many countries, the Bank identified a minimal package of public health services covering a fraction of the disease burden for those that could not afford private insurance. The results for most countries was a non-universal, fragmented health system comprised of: limited government-run health infrastructure with bilateral and multilateral donor-funded projects and programs picking up some of the slack for the poor; while wealth patients in urban areas could access often quite substantial private health care services (Schneider et al. 2006).

In a majority of Southern states, most hospital beds are now located in the private system. The pandemic underscored this system's profound weaknesses, such as the lack of coordination between the different parts of the system, the financial vulnerability of some private providers, and private providers which were unwilling to accept COVID-19 patients for treatment, without receiving price 'premiums.' Yet, according to Oxfam (2020), 'just 8 of the 71 World Bank COVID-19 health projects include any plans to remove financial barriers to accessing health services' (Oxfam 2020, 6). Oxfam found that two thirds have no plans 'to increase the number of health workers, and that the 25 projects which do, have substantial shortcomings' (Oxfam 2020, 7). The Kampala-based Initiative for Social and Economic Rights argued: 'countries like Uganda that already borrowed to mitigate the pandemic can't afford more loans. The World Bank in true solidarity should only provide grants to support COVID vaccination and strengthen public health systems, which are the first point of call for the poor' (Bretton Woods Observer 2021). Further, indications are that MDB funding to the health sector may underpin private health care profits (Engel et al. 2020). The poor will likely, as in the past, be left paying for debts that have not benefitted them.

The World Bank endorsed the World Health Organization's COVID-19 Vaccine Global Access (Covax) project in April 2020, which uses donor funds to purchase vaccines from private firms (especially AstraZeneca) on the open market and pass these to poor countries. But that left Covax subject to artificial market distortions and to outbidding by rich countries, which obtained far greater quantities of vaccines at high prices, e.g. Canada had, by early 2021, reserved vaccines sufficient for *five times its citizenry.* In a system described as vaccine apartheid, the Northern countries enforced Trade Related Intellectual Property System (TRIPS) rules to prevent local generic production of vaccines in poor countries, something achieved two decades earlier with AIDS medicines. A waiver request from South Africa, India and more than one hundred other countries was rejected, and the next step proposed is 'vaccine passports' to regulate international migration, again favouring rich countries' citizens.

In this context of extreme global-health injustice, World Bank President David Malpass prioritized working with private health care. A contrasting strategy would have been what a respected NGO watchdog, the Bretton Woods Project (Bretton Woods Observer 2021, 2) suggested: 'support for enhanced distributed local production, the TRIPS waiver and … vaccine pricing … [to] address the costs and other structural barriers associated with the vaccination efforts.' Bank lending for *status quo* politics could 'exacerbate the debt load of some countries' due to increased 'borrowing to fund vaccine purchases and cutting essential services or support for vulnerable communities, while pharmaceutical companies continue to reap the rewards'.

Addressing poor countries' foreign exchange shortages, the World Bank did make front-end grants through the COVID-19 Fast Track Facility (though these were typically followed by loans). It also helped develop the G20 Debt Service Suspension Initiative (DSSI) through which eligible countries can temporarily suspend (not repudiate or cancel) some bilateral – not MDB – debt repayments. But again, the conditionality of Bank lending included what the Bretton Woods Project (2021, 2) termed 'the financialisation of healthcare. The Bank's failed Pandemic Emergency Financing Facility bond is a clear case in point.'

Finally, the World Bank is also imposing what can only be described as neoliberal conditionality on other COVID-19 loans. Malpass (2020) called for structural reforms to end 'excessive regulations, subsidies, licensing regimes, trade protection or litigiousness as obstacles …' which exacerbates corporate domination of poor countries. As just one example, Ecuador (one of the countries that defaulted on debt in 2020) received three Development Policy Loans (formerly known as Structural Adjustment Loans) of $500 million each. While they do contain provisions for expanding social protection, they increase 'targeting of the poor,' which has a divisive impact compared to universal treatment plus cross-subsidization. But the main objectives of the loans are government fiscal consolidation, including cutting staffing and advancing private sector participation through reduced regulation. The loan documentation encourages public-private partnerships, removing 'distorting labour market regulation' (i.e. wages and conditions for workers) (World Bank 2020b, 6), and expresses concern about regulation that gave the government a higher share of oil revenue as it supposedly 'discouraged foreign direct investment' (World Bank 2020a, 4).

Pedagogical considerations for the MDBs as 'learning institutions' and MDB teaching

A trite phrase found regularly since the mid-1960s, after popularization by leading economist Kenneth Arrow, is 'learning by doing.' It stresses marginalist epistemology: each step of the way, adjusting the strategy in incremental mode, and as a result never challenging structural processes or power relations that may be the cause of underdevelopment. The MDBs as learning institutions reflect the strengths and weaknesses of this approach, especially for development-studies pedagogy, because as financiers they tend to react to deeper economic processes, with little opportunity to shift these.

Against this is learning by rupture, such as in the increasingly successful attempts of climate activists to shift MDBs away from fossil fuel lending, which has helped lower some countries' reliance on coal. Or the demand made to the Bretton Woods institutions to halt apartheid financing made from the mid-1960s – financial sanctions against Pretoria helped end apartheid in 1994. In both cases, activists demand that profits made in the process – a 'climate debt' the North owes the South, and 'apartheid reparations' – be paid. Admitting mistakes is something the World Bank has not had to do thanks to its historic enjoyment of full-fledged diplomatic immunity, but this changed somewhat in 2019, when the US Supreme Court stripped away that right in the case of a Bank-financed coal-fired power plant whose pollution damaged a fishing community in Mundra, India.

Key debates about MDBs can be interrogated from a pedagogical lens. For example, World Bank research demonstrates the intellectual damage of incrementalism as it has led to unreflexive, self-referential research that does not acknowledge critique. Wade (1996) explained the Bank's research as a case of 'paradigm maintenance' and even the Bank's own review of its research noted the tendency to 'proselytize on behalf of Bank policy' and excessive self-citation (Banerjee et al. 2006). Broad (2006, 387) concluded that World Bank research became unjustifiably 'skewed toward reinforcing the dominant neoliberal policy agenda.' There is a related constructivist literature on the Bank as a 'norm entrepreneur,' which highlights its role in spreading neoliberal development ideals. This constructivist literature also demonstrates a second form of MDB learning, socialization, for example, the World Bank Institute, the Bank's training arm, is important in norm internalization through socializing participants from MDBs and governments around the world (Bazbauers 2016).

In terms of teaching about MDBs, a central challenge is to provide both practical applications of development theories (e.g. modernization and neoliberalism), and explore the internal logics and contradictions of lending mechanisms in the context of power relations encompassing class, race, gender, generation (age), society–nature and North-South positionality, sometimes all in the same favoured project or policy. There is no literature to be found on teaching about development banks, so we hope the two ideas below might prompt sharing.

1. To explain the disbursement pressures that MDBs face: the instructor represents the MDB, divide the class into three groups in your head. Move to the middle of group one and explain you are from the World Bank and you'd like to offer them a loan. Involve students: 'Mika you're from the Indonesian government, what are your top priorities … Great, I can lend you $150 million for that … That was my 2020 lending targets met, now its 2021 and here's my next group of countries that I'm targeting.' Move to group 2, and so on to group 3 for 2022. 'Well now its 2023 and I still need to be lending – as a development bank it does not look good if I have more money coming back to me in repayments that I'm putting out the door. Also, my job has lending targets built in and I want my pay rise. So, Mika (or another student), you must need some more money, can I interest you in a new power plant?' To generate discussion, ask the students what the government official might be thinking and what the consequences of the lending imperative might be?

2. For a discussion group or seminar with at least 50 minutes, pick a loan from an MDB on a country, topic, or theme that your class has been discussing, for example community development or agriculture in East Timor or infrastructure in India. The World Bank is useful as it publishes quite a bit of information in the Projects and Operations section of its website, but other MDBs publish at least summary information too. Have the students read the summary information then divide them into small groups of two or three to research and debate the project. The topics for debate need to reflect the project, but a basic structure would have groups examining the developmental approach representing government, local business and different NGOs and

affected communities (and workforces) focused on the environment, the theme of the project, project-affected people, women, indigenous groups, future generations and so on. There could also be groups working looking at who won procurement contacts for the project, their links to government, multinational business and to corruption concerns.

Most obviously, in any such exercises, the tendencies towards friction between different perspectives are most valuable to surface. That in turn allows a development financing window into various kinds of conflicting narratives to then open a wider door, one that allows explorations of contrasting developmental principles, analyses, strategies, tactics, and alliances.

Key sources

1. Bazbauers, A. R. and Engel, S. (2021). *The Global Architecture of Multilateral Development Banks: A System of Debt or Development?* Abdington: Routledge.

This book offers the only overview of all 30 MDBs – how they evolved, what their developmental priorities are and their relationship to expanding debt relations. In addition to sections on each of the MDBs, the book offers chapters of three of the MDBs main areas of operation: infrastructure, human development, and climate finance.

2. Ben-Artzi, R. (2016). *Regional Development Banks in Comparison: Banking Strategies versus Development Goals.* Cambridge: Cambridge University Press.

Provides a detailed examination of the design and development strategy of four of the five major regional development banks: the African, Asian, European and Inter-American Development Banks.

3. Bond, P. (2016). BRICS banking and the debate over sub-imperialism. *Third World Quarterly,* 37: 611–62.

Outlines how the NDB as well as the BRICS Contingent Reserve Arrangements, a proposed alternative to the IMF, do not provide an alternative to the global debt-based financial order, rather support it.

4. Kapur, D., Lewis, J. P. and Webb, R. (1997). *The World Bank Its First Half Century, Volume 1: History.* Washington DC: Brookings Institution Press.

This second official history of the World Bank provides lots of vital detail on the policies, lending and organization history and politics of the first and still most influential MDB, the World Bank.

References

Banerjee, Abhijit, Angus Deaton, Nora Lustig, Ken Rogoff, and Edward Hsu. (2006). *An Evaluation of World Bank Research, 1998–2005.* September 24: World Bank.
Bazbauers, Adrian. (2016). The World Bank as a Development Teacher. *Global Governance* 22 (3):409–426.
Bazbauers, Adrian Robert, and Susan Engel. (2021). *The Global Architecture of Multilateral Development Banks: A System of Debt or Development?* Abdington: Routledge.

Ben-Artzi, Ruth. (2016). *Regional Development Banks in Comparison: Banking Strategies versus Development Goals.* Cambridge: Cambridge University Press.

Bond, Patrick. (2006). *Looting Africa.* London: Zed Books.

Bond, Patrick. (2016). BRICS banking and the debate over sub-imperialism. *Third World Quarterly* 37 (4):611–629.

Bond, Patrick. (2018). Ecological-economic narratives for resisting extractive industries in Africa. *Research in Political Economy* 33:73–110. doi: 10.1108/S0161-723020180000033004

Bond, Patrick. (2020). BRICS Banking and the Demise of Alternatives to the IMF and World Bank. In *International Development Assistance and the BRICS*, edited by Jose A. Puppim de Oliveira and Yijia Jing, 189–218. London: Palgrave Macmillan.

Bond, Patrick, and Dominic Brown. (2020). The World Bank's rating obsession will negate debt justice. Counterpunch 19 August, accessed 12 April 2021. www.counterpunch.org/2020/08/19/world-banks-rating-obsession-will-negate-debt-justice

Bracking, Sarah. (2009). *Money and Power: Great Predators in the Political Economy of Development.* London: Pluto Press.

Bretton Woods Observer. (2021). World Bank support to Covid-19 vaccination fails to address fundamental barriers to equitable access. 1–2, accessed 12 April 2021. www.brettonwoodsproject.org/wp-content/uploads/2021/03/bw_observer_spring_21_online.pdf

Broad, Robin. (2006). Research, knowledge, and the art of 'paradigm maintenance': the World Bank's Development Economics Vice-Presidency (DEC). *Review of International Political Economy* 13 (3):387–419.

Culpeper, Roy, and Nihal Kappagoda. (2016). The new face of developing country debt. *Third World Quarterly* 37 (6):951–974. doi: 10.1080/01436597.2016.1138844

Dossani, Sameer. (2014). BRICS Bank: New Bottle, How's the Wine? *Bretton Woods Bulletin* (February).

Engel, Susan. (2010). *The World Bank and the Post-Washington Consensus in Vietnam and Indonesia: Inheritance of Loss.* London: Routledge.

Engel, Susan, and Adrian Robert Bazbauers. (2020). Multilateral Development Banks: Washington Consensus, Beijing Consensus or banking consensus? In *Rethinking Multilateralism in Foreign Aid: Beyond the Neoliberal Hegemony*, edited by Viktor Jakupc, Max Kelly and Jonathan Makuwira, 113–131. Abingdon, Oxon, and New York: Routledge.

Engel, Susan, Nadeen Madkour, and Owain D. Williams. (2020). The World Bank's response to Covid-19. Occasional Paper No. 1 Jubilee Australia Research Centre.

Flyvbjerg, Bent. (2009). Survival of the unfittest: why the worst infrastructure gets built – and what we can do about it. *Oxford Review of Economic Policy* 25 (3):344–367.

George, Susan, and Fabrizio Sabelli. (1994). *Faith and Credit: The World Bank's Secular Empire.* London: Penguin Books.

Glennie, Jonathan. (2021). *The Future of Aid: Global Public Investment.* London and New York: Routledge.

Güven, Ali Burak. (2018). Whither the post-Washington Consensus? International financial institutions and development policy before and after the crisis. *Review of International Political Economy* 25 (3):392–417. doi: 10.1080/09692290.2018.1459781

Haas, Peter. (1992). Epistemic Communities and International Policy Coordination. *International Organization* 46 (1):1–35.

Humphrey, Chris. (2014). The politics of loan pricing in multilateral development banks. *Review of International Political Economy* 21 (3):611–639. doi: 10.1080/09692290.2013.858365

Humphrey, Chris. (2016). *The problem with development banks' environmental and social safeguards.* ODI, accessed 18 January 2020. www.odi.org/blogs/10379-problem-development-banks-environmental-social-safeguards-mdbs

Kose, M. Ayhan, Peter Nagle, Franziska Ohnsorge, and Naotaka Sugawara. (2020). *Global Waves of Debt: Causes and Consequences.* Washington DC: World Bank.

Maasdorp, Leslie. (2020). *The New Development Bank: What Washington should know about the newest Multilateral Development Bank.* Center for Global Development, 21 October, accessed 12 April 2021. www.cgdev.org/event/new-development-bank-what-washington-should-know-about-newest-multilateral-development-bank

Malpass, David. (2007). Don't panic about the credit market. *Wall Street Journal*, 7 August, accessed 12 October 2021. www.wsj.com/articles/SB118645120890190059

Malpass, David. (2020). Remarks by World Bank Group President David Malpass on G20 Finance Ministers Conference Call on COVID-19. World Bank, accessed 19 January 2021. www.worldbank.org/en/news/speech/2020/03/23/remarks-by-world-bank-group-president-david-malpass-on-g20-finance-ministers-conference-call-on-covid-19?cid=SHR_SitesShareTT_EN_EXT

Marini, Ruy Mauro. (1965). Brazilian Interdependence and Imperialist Integration. *Monthly Review* 17 (7):14–24.

Moody's Investors Service. (2020). Moody's affirms IBRD's Aaa rating, maintains stable outlook, accessed 12 April 2021. www.moodys.com/research/Moodys-affirms-IBRDs-Aaa-rating-maintains-stable-outlook--PR_416449

Oliver, Robert W. (1975). *International Economic Co-operation and the World Bank*. London and Basingstoke: Macmillan Press.

Oxfam. (2020). *From Catastrophe to Catalyst: Can the World Bank make COVID-19 a turning point for building universal and fair public healthcare systems?* Oxford: Oxfam GB.

Park, Susan. (2014). Institutional Isomorphism and the Asian Development Bank's Accountability Mechanism: Something Old, Something New, Something Borrowed, Something Blue? *The Pacific Review* 27 (2):217–239.

Robinson, William. (2014). *Global Capitalism and the Crisis of Humanity*. Cambridge: Cambridge University Press.

Schneider, Helen, Duane Blaauw, Lucy Gilson, Nzapfurundi Chabikuli, and Jane Goudge. (2006). Health Systems and Access to Antiretroviral Drugs for HIV in Southern Africa: Service Delivery and Human Resources Challenges. *Reproductive Health Matters* 14 (27):12–23.

Soederberg, Susan. (2006). *Global Governance in Question: Empire, Class and the New Common Sense in Managing North-South Relations*. London: Pluto Press.

Stern, Nicholas. (2013). Emerging Powers as Emerging Economies. accessed 6 April 2020. www.youtube.com/watch?v=4ZKQ6wQ-29w

Thakur, Ramesh. (2014). How Representative are BRICS? *Third World Quarterly* 35 (10):1791–1808.

Wade, Robert. (1996). Japan, the World Bank, and the Art of Paradigm Maintenance: *The East Asian Miracle* in Political Perspective. *New Left Review* 217:3–36.

Wade, Robert. (2006). Choking the South. *New Left Review* 38:115–127.

Williamson, John. (2004). A Short History of the Washington Consensus. Paper presented at a conference sponsored by Foundation CIDOB and the Initiative for Policy Dialogue, Barcelona.

World Bank. (2012). IBRD Articles of Agreement. Washington: World Bank.

World Bank. (2020a). Program Information Document: Ecuador Second Inclusive and Sustainable Growth DPL (P169822). Washington: World Bank, PIDA26827.

World Bank. (2020b). Program Information Document: Ecuador Third Inclusive and Sustainable Growth DPL (P174115). Washington: World Bank, PIDC30348.

9

NORTHERN AND SOUTHERN NON-GOVERNMENTAL ORGANIZATIONS

Nicola Banks and Badru Bukenya

Introduction

Non-governmental organizations (NGOs) have become increasingly important actors in global development. Research on NGOs has also increased alongside their expanding importance, with a recent systematic review of 35 years of scholarship on development NGOs highlighting its rich theoretical roots and diverse empirical insights (Brass et al. 2018). Understanding NGOs and their contributions means recognizing and exploring diversity within and across NGOs and NGO sectors. One key difference here, of course, is that between International NGOs (INGOs) – those rooted in Northern Headquarters[1] that collect funds from a variety of sources to spend in projects run in the Global South – and Southern NGOs, local and national organizations pursuing development goals in the Global South, often partnering with INGOs in their operations. There is, of course, great diversity among organizations within these broad categories, depending on an organization's size, type, and the broader NGO sector in which they find themselves. But for the purposes of this chapter, these broad categories serve a useful purpose in allowing us to explore the hugely different challenges facing NGOs in different parts of the world as a result of changing operational environments.

Recent evidence on NGO sectors in the Global North highlights that they are stable, if not growing rapidly (Banks and Brockington 2020; Davis 2019; Schulpen 2020), though this may be challenged by several high-profile scandals and the COVID-19 pandemic. Meanwhile, in the Global South, NGOs have experienced increasingly hostile environments, with shrinking civil society space restricting their operations and threatening their viability (Glasius et al. 2020). We also see that relationships between different development actors are showing signs of change. This includes increasingly vocal proponents for 'shifting the power' from North to South so that more effective, locally-led transformations can take place. It is these issues that we explore in this chapter – what we know about NGO sectors in the Global North and South, the different challenges they face, and whether and how development configurations may be changing across the aid chain and these geographic spaces to enable greater local ownership of global development agendas and greater transformative change on the ground.

These questions are important, because it has been long-recognized that the deeply-rooted managerial practices and philosophies that are institutionalized within the aid chain have constrained the ability of NGOs to promote genuinely transformative change on the ground.

DOI: 10.4324/9781003017653-11

Their position in the aid chain exposes NGOs (both International and Southern NGOs) to power asymmetries that encourage them to prioritize accountability to donors rather than those they serve: this often results in the reshaping of their strategies and activities to fit in with donor objectives instead of their own priorities, for example, or of prioritizing arduous reporting and accountability procedures to donors rather than the communities that they serve. Such relationships leave local NGOs with limited flexibility to prioritize local needs on the ground and hinders the ability of local communities to hold them to account. Herein lies the main criticism of NGOs: pressures in the aid chain mean that NGOs have become largely professionalized organizations that are proficient at service delivery, but have become detached from the civil society roots that let them represent their grassroots communities and pursue more transformative agendas (Banks et al. 2015).

One of the important questions we can ask then, is to what extent are these distortions being addressed in the aid chain? We first look at NGO sectors in the Global North and South to deepen our understanding of these distortions. We then investigate recent changes influencing the sector to see whether these changing development configurations or operational environments address these distortions. While multiple stakeholders are increasingly voicing the need for change, we see less evidence of tangible and positive widespread change in this direction.

Southern NGOs

We look first to the vast array of local and national NGOs that pursue development in the Global South (hereafter SNGOs). Obviously SNGOs are a diverse group differing across a range of dimensions, including their membership, geographical dispersion, motivation and values, leadership, methods of work, levels of funding and capacity. From the outset we make an important distinction between SNGOs and grassroots organizations also known as community-based organizations (CBOs). The latter are indigenous development organizations usually started on a self-help basis for the benefit of members, are managerially unstructured (but can grow to employ a few professional staff) and have activities that span across smaller sub-national areas. On their part, SNGOs are formal organizations formed for the purpose of benefiting the public (not just their members), and have defined officers as well as organizational permanence. While the majority of SNGOs originate from within their countries of operation, the number of foreign or International NGOs (INGOs) opening and operating local headquarters has increased over the years. Most of these originate from high-income European and North American countries, but a handful have also emerged from lower- and middle-income countries such as Bangladesh and South Africa.

There is wide consensus that SNGOs have exponentially grown over the last three decades across countries with varied social, economic and political contexts (Kellow and Murphy-Gregory 2018). This rapid expansion can be attributed to dynamic shifts occurring within and outside the countries they are located within, including the 1990s' struggles for democratization and aid donors' promotion of neoliberal policies and the 'good governance' agenda in the 1980s and 1990s (Bukenya and Hickey 2014). The growth is also a response of both the founders and donors to the alleged advantages of SNGOs.

SNGOs are widely acclaimed as best-positioned to undertake grassroots mobilization of citizens to challenge the social structures that keep people in poverty – and therefore the most reliable route for achieving inclusive and sustainable development (Banks et al. 2015; Mercer and Green 2013). Indeed, promoting a vibrant civil society globally is a key motivation for bilateral donors to fund NGOs (OECD 2020). Yet questions remain on whether donors are funding the right organizations for this. SNGOs proximity to the grassroots and deployment of participatory approaches in their work imply that they can identify the most deserving for support, are more innovative and have the flexibility to respond swiftly to local needs compared to the more distant and bureaucratic state agencies. SNGOs are also claimed to operate at relatively lower cost and are more sustainable due to their local rootedness and reliance on community voluntary input. While these qualities give SNGOs comparative advantages over the public sector (or even INGOs) with regards to delivering development – and while their link in the aid chain has the greatest potential for promoting transformative change – in reality and as discussed below, being at the 'bottom' of the aid chain implies that they possess the least power, resources and freedom to realize their full potential. In this chapter we explore three major issues facing SNGOs, namely: (i) shifts in SNGO financing and their implications on NGO-state relations; (ii) stringent regulatory frameworks that are being used to close civil society space; and (iii) relationships between SNGOs and INGOs.

Narrow donor funding channels and the volatile aid landscape

In their quest to promote the good governance agenda bilateral and multilateral donor agencies have long-championed aid modalities that encourage partnership-driven development approaches. Such arrangements expect NGOs to play a central role in rendering the state more effective through policy engagement, advocacy, ensuring accountability, and the provision of public services (Bukenya and Hickey 2014). Yet these expectations have implications on the nature of NGOs that can thrive in the Global South. Funding is steered towards formal organizations with demonstrated track records, reliable reporting systems and high degrees of professionalization. Such NGOs tend to be urban-based and staffed by middle-class professionals, a world away from the poor and marginalized communities that they are meant to represent. So while donors may recognize and value the role of civil society in inclusive development – evident from a switch away from using the term 'NGOs' to 'CSOs' in their policy language – they have not funded a more diverse range of civil society organizations beyond these (OECD 2020). Their spending patterns highlight their commitment to funding professionalized organizations that least resemble the civil society organizations who can represent local constituencies or offer genuine development alternatives (OECD 2020) and they rarely fund SNGOs directly. Instead, funding arrangements are based on 'complex transnational contracting chains' in which the relationship between donors and local NGOs is mediated by 'several sub-donors and sub-recipients' (Mercer and Green 2013). Thus, scope for funding organizations that are more suitably placed to understand local realities and pursue local priorities is remarkably limited.

Reductions in foreign aid have also created new difficulties for SNGOs. From the mid-2000s state-donor relationships were strained by concerns over progress on governance and corruption in several developing countries, causing reductions in foreign aid (Pallas and Sidel 2020). This was aggravated for NGOs by aid reductions in countries graduating from low to middle-income status (Kumi 2017). Such a volatile funding landscape has reignited debates around the need for SNGOs to venture into alternative revenue generating strategies, particularly those targeting domestic sources (Kumi and Hayman 2019). This strategy has allowed some large organizations in South Asia, like BRAC and Proshika in Bangladesh, to reduce their dependence on international donors. However, no similar examples of successful domestic resource mobilization have been documented among NGOs in sub-Saharan Africa, perhaps because high levels of poverty limit such possibilities (Kumi 2017). There is also a paucity of studies that systematically analyse SNGO's local fundraising strategies, leaving us with an insufficient understanding of 'best practices' in this respect, the potential benefits of such strategies and their ability to compensate for the shortcomings of donor funding, among others. A rare exception is

Smith's (2019) Gambian study of four NGOs' self-financing activities. While carried out with the intention of overcoming the obstacles and inconsistencies of donor funding channels, self-financing mechanisms offered little potential to transform SNGO fortunes or to challenge the distortions of the aid chain. Generating limited incomes, they were unable to compare with or compensate for the levels of finance available from donors.

Shrinking civil society space

SNGOs are disproportionately affected by the increasing adoption of measures by governments around the world that restrict the ability of NGOs to operate autonomously (CIVICUS 2019; Musila 2019; Glasius et al. 2020). Legal restrictions create barriers to NGOs' entry, funding, and advocacy activities. Between 2003 and 2012, 39 low and middle-income countries introduced laws limiting foreign funding for NGOs (Dupuy et al. 2016). New regulations and restrictions are not always brought in with the explicit intention of shrinking civil society space; governments claim that regulations are meant to streamline NGO operations and limit financial and political maleficence or external interference in domestic affairs. Though with the growth in authoritarian regimes, sometimes laws are clearly meant to control civil society. In addition to these legal restrictions, states in the Global South also restrict NGO activities through a wide array of political, administrative and extra-legal strategies, such as the stigmatization of civil society actors within public discourse and threats and the use of violence by state and non-state actors (Hossain et al. 2018).

These restrictions have negative implications on SNGO operations and wider development outcomes. In particular, such restrictions typically target NGOs working in advocacy activities that states consider to be politically sensitive; these are often funded by international agencies and/or have strong transnational links (CIVICUS 2019; Hossain et al. 2018; Musila 2019). In contrast, 'non-political' organizations and activities are encouraged to operate, predominantly those focusing on service delivery (Bukenya and Hickey 2014). While civic space is not necessarily shrinking for all SNGOs, it is changing in terms of who participates and on what terms. Again, here, we see opportunities for SNGOs seeking more transformative change particularly disadvantaged.

SNGO relationships with donors and INGOs

Bilateral donors continue to prefer funding INGOs rather than SNGOs directly. INGOs receive approximately ten times more funding from bilateral the donors than SNGOs (OECD 2020). Only 7 per cent of DAC donor funds for civil society goes directly to SNGOs. In comparison, donor-country based INGOs received 66 per cent of funds, and other Northern-based INGOs a further 27 per cent.[2] Further, most donor funds go to INGOs based in the donors' own countries, though an increasing proportion is channelled through INGOs based in other DAC-member countries (Figure 9.1).

Thus, despite most bilateral donors reporting that they support SNGOs, the reality is that they do so through heavily intermediated ways, primarily through INGOs headquartered in their own countries. There are clear benefits of this for donors. With limited capacity to administer financial support to large numbers of smaller NGOs in the Global South, there are much lower transactions costs associated with funding INGOs to carry out this contracting (OECD 2020). Donors also highlight the benefits that funding INGOs brings in terms of raising public awareness and engaging with their citizens to strengthen public belief in aid (OECD 2020). Tighter legal, regulatory, and administrative requirements in member countries also facilitate assumptions around

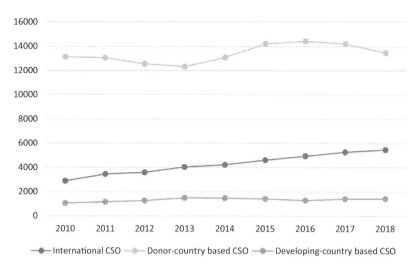

Figure 9.1 ODA allocations by type of CSO, in USD billions
Source: Adapted from OECD (2020)

'professionalism,' thereby generating familiarity and trust. Poole (2018) highlights that aid agencies perceive SNGOs as being more susceptible to corruption; thus contracting through INGOs is perceived to minimize financial risks. Issues of trust also affect the SNGOs that do receive donor funding. Among Cambodian INGOs and NGOs, for example, those staffed by foreigners have higher chances of accessing bilateral aid (Suarez and Gugerty 2016).[3]

If donors truly want to strengthen civil societies globally, channelling a greater proportion of funding directly to SNGOs is necessary. Some donors have used funding modalities that promote consortium arrangements, including NGO partners in the Global South. Yet INGOs tend to take the gate-keeping lead on such bids, leaving SNGOs at the bottom of the aid chain. Power and influence is, thus, in favour of INGOs and SNGOs get few opportunities to directly represent their work to donors, widen their funding networks or build technical capabilities and organizational sustainability. Antagonism between SNGOs and INGOs on these issues is becoming increasingly apparent, as the final section of this chapter discusses. First, however, we look in more detail at what we know about INGOs, these more powerful actors in the aid chain.

Northern NGOs

Recent research has highlighted the significance of INGOs to foreign aid efforts globally. In 2015, for example, British INGOs spent around £7 billion – the equivalent of 55 per cent of the UK's Official Development Assistance (ODA) that year (Banks and Brockington 2020). The Netherlands INGO sector is significantly smaller and has remained fairly consistent at around €2.3 billion between 2010 and 2017 (Schulpen and van Kempen 2020). Similarly, across 2011–2015 Canadian INGOs spent an average CA$3.4 billion annually, with 14 ODA-eligible countries receiving more aid from Canadian INGOs than they do from Canadian ODA (Davis 2019). They are major players in global development.

Many of us are familiar with International NGOs. These include the biggest household names, such as Oxfam, Save the Children or Action Aid, among others. Recent research has demonstrated the concentration of resources into a small number of dominant INGOs across Northern INGO sectors. Banks and Brockington (2020), for example, find that these 'mega'-NGOs dominate nearly 90 per cent of sectoral expenditure across British INGOs. Similarly in the Netherlands, 11 per cent of the largest INGOs disproportionately share around 82 per cent of the sector's income (Schulpen and van Kempen 2020). Yet these largest INGOs are by no means representative of INGOs more broadly, and we actually know surprisingly little on the more diverse range of smaller- and mid-sized INGOs that comprise the biggest number of Northern INGO sectors. A lack of research into this diversity thus makes it hard for pol-icymaking in the sector to be evidence-based beyond these heavy-hitters (Banks et al. 2020).

What do we know about funding for INGO sectors?

Asking who funds INGOs is an important question given that modalities and sources of funding are central to long-standing critiques of INGOs and NGOs. Of particular interest here is the volume and nature of funding INGOs receive from donor governments, who gen-erally offer the least flexible forms of funding, accompanied by the strictest reporting cri-teria. These funding characteristics effectively 'shackle' INGOs into being highly-professional service-providers and weaken their links to the grassroots constituencies they seek to represent (Banks et al. 2015). Yet because there have been few attempts to measure the extent to which INGOs depend upon government funding it has been hard to prove or disprove assumptions questioning INGO autonomy as a result of over-reliance on government funds.

   This inspired Davis (2019) to explore the income sources of Canadian INGOs. Government funds in fact constituted only 9 per cent of total INGO revenue across 2011–15 and was highly concentrated: 83 per cent of INGOs received no governmental funding (Davis 2019). In the UK, too, income from the general public (40 per cent of total funding across 2009–2016) vastly outweighed the 18 per cent contributed by the UK government, and these government funds are largely concentrated among the largest INGOs in the sector (Banks and Brockington 2020). This highlights again the importance of understanding diversity in the sector, rather than gener-alizing across the sector from the experiences of the largest influential players. Yet the depth of influence of donor agencies and their priorities remains strong on INGOs, even if the 'problem' of donor funding dependence is significantly less prevalent than commonly thought. Loman et al. (2011) highlight that in the Netherlands, even INGOs receiving limited ODA funds are still strongly inclined to follow governmental aid priorities. Likewise, the 'aid orphan-aid dar-ling' debate holds for INGOs as well as donors, with INGOs tightly clustering in donor priority countries (c.f. Koch et al. 2009).

   Donor governments channelled a total of US$21 billion – 15% of all bilateral ODA – through INGOs in 2018. This is clear illustration of the value they place on them as development actors. Donors highlight two clear objectives for doing so, namely i) to contribute to stronger, more diverse civil societies globally; and ii) to provide services by acting as implementing partners on behalf of donors (OECD 2020). The second – and instrumental – objective has taken clear pri-ority over the first, as evidenced by exploring funding channels. Programme or project-based support (i.e. funding for specific delivery targets and activities) constitutes 85 per cent of donor funding to NGOs.[4] Only 15 per cent is distributed more flexibly to INGOs as core support, despite being preferred by INGOs and being recognized as the 'most development-effective type of support' (OECD 2020: 19).

   Donors have clearly placed priority on the funding modality most conducive to dem-onstrating development 'results' and 'value for money.' *Outcomes* (i.e. service delivery targets and activities) continue to take precedence over *impact,* measured more broadly in terms of a strengthened, more diverse civil society and more inclusive development processes and outcomes. Tying funds so closely to pre-defined service delivery factors makes it difficult to steer development efforts in ways that respond to partner needs on the ground in the Global South and does little to promote a strengthened civil society in partner countries. In fact, it is more likely to weaken this. Focusing on service delivery alone is insufficient to reach the deeper goals of transformation and social justice upon which NGOs are praised as the best-positioned development alternative for achieving: this requires a concerted effort to invest in and fulfil NGOs' civil society functions alongside their more tangible service delivery accomplishments (Banks et al. 2015).

Despite these long-standing arguments, donor behaviours, preferences and funding channels remain frustratingly unwilling to change. The most recent study of donor funding for civil society concludes that, 'DAC Members do not yet appear to be offering effective development support for CSOs as part of the enabling environment for civil society' (OECD 2020, 17).

Changing operating environments, changing development configurations?

The previous sections illustrate the ways in which long-recognized power inequalities between different links in the aid chain have left SNGOs – those best-placed to pursue transformative change on the ground – with insufficient power or resources to take the driving seat of development locally. We now look at three areas displaying positive moves in this direction, including i) greater calls towards localization; ii) a radical Dutch civil society funding policy that prioritizes social transformation over managerial principals and systems; and iii) lessons learned in response to the COVID-19 pandemic.

Rebalancing systemic power inequalities? Localization and the #ShiftThePower Movement

The need to shift the distribution of power and funding across the aid chain is increasingly recognized by diverse development stakeholders. The CSO Partnership for Development Effectiveness, a platform uniting CSOs globally has published two sets of common principles that should govern CSO partnerships: 2010's Istanbul CSO Development Effectiveness Principles and 2011's Seam Reap CSO Consensus on aid effectiveness. These highlight the importance of CSOs promoting and pursuing equitable partnerships and solidarity in their relationships with other CSOs and identify the critical need for explicit steps being taken to counterbalance pre-existing inequalities in power across CSOs in the North and South in order to promote organizational autonomy (OFCDE 2011).

The language of 'localization' is now at the heart of growing demands for power and resources to be channelled directly to Southern NGOs, rather than through heavily intermediated ways. In the humanitarian field, localization commitments were formalized in 2016 when 61 of the largest donors, humanitarian agencies and NGOs committed to the *Grand Bargain*. This outlined a localization agenda seeking to enhance the effectiveness and accountability of humanitarian action by increasing the volume of global funding going directly to national and local responders to 25 per cent of total humanitarian funds. Progress has been slow, however, with only seven signatories reporting to meet or exceed the 25 per cent target (ODI 2019).

Beyond the humanitarian field, voices for a normative shift in this direction are also strong. The hash-tagged #ShiftThePower global campaign and movement initiated by the South Africa-based Global Fund for Community Foundations argues for a move away from top-down and bureaucratic aid chains, towards aid chains in which national and local organizations have local ownership over development agendas – and the power and resources to tackle these. What evidence of change do we see in the practices of donors and INGOs in this respect?

INGOs, too, have been vocal proponents of localization,[5] but it is hard to verify the extent to which this is being implemented in practice given a lack of formal commitment and subsequent reporting requirements. Nor do suggestions of *how* this process is taking place suggest much cause for optimism. Commentators ask whether INGOs are ready to give up the privileged power and position they hold in the aid chain and highlight the need for Southern organizations to *take* power rather than wait for it to be shifted to them.[6] The funding landscape of INGOs has shifted in ways that adds new obstacles to localization, including increased rules and regulations around reporting and criticisms of the sector in the wake of several high-profile scandals that reduced funding and public belief in INGOs. This increasingly risk-averse funding environment may encourage INGOs to exert more control over Southern 'partners,' rather than less.[7]

Despite progress in mind-sets and debates around localization,[8] there remains much work to be done to realize tangible and transformational change. An open letter to INGOs from SNGOs published in early 2020 highlights that despite their expressed desires to find a more Southern-centric aid system, INGOs are still approaching the topic from a Northern-focused perspective and through extractive channels.[9] These practices, they argue, reinforce the unequal power dynamics that #ShiftThePower is trying to rebalance, closing off, rather than strengthening and opening up, the space for domestic civil society. They finish their passionate letter with a clear plea, '… that you work with us, not against us. We need to be supported, not competed with, and certainly not replaced.'

Reorienting donor priorities: moving from managerial to social transformative approaches to civil society funding

In the second section we discussed the rigidity of donor funding systems and their clear orientation to managerial processes that prioritize top-down accountability, measurable outcomes and 'value for money' over longer-term, more transformative impacts. In response to these issues, the Dutch government's Ministry of Foreign Affairs launched its *Dialogue and Dissent* (D&D) policy framework (2016–2020) for funding civil society organizations. D&D represents a deep paradigmatic shift in policy away from the managerialist philosophies that have come to characterize the global aid system towards a social transformative approach to funding civil society organizations. In doing so, it directly seeks to tackle the deep-rooted constraints INGOs face to promoting more transformative development outcomes. Nearly €1 billion of Dutch ODA funds has been spent through 25 Strategic Partnerships comprising of lead Dutch INGOs and their global networks of Southern NGOs and CBO partners. Kamstra (2020, on which the remainder of this section is based) outlines three D&D innovations that have been designed explicitly to address common constraints in the system. These are i) stressing the political roles of NGOs; ii) cooperating in Strategic Partnerships; and iii) using flexible theories of change and learning.

First, D&D is unique in moving away from funding service delivery activities. Recognizing that issues of poverty, inequality and exclusion are political, not technical problems, D&D solely funds the political role of NGOs.[10] Secondly, the policy also recognizes the complexities and difficulties associated with promoting political change and recognizes that NGOs must be seen – and strengthened – as political actors in their own right (rather than being viewed and funded as project implementers of donor priorities and projects). To do this, D&D funds Strategic Partnerships – networks of collaborating NGOs in the North and South – to build the strength and support necessary for such an approach and to enhance the network's advocacy capacity. Moving towards political goals also requires a third innovation, a more flexible approach to monitoring and accountability. Working politically in fast-changing and often shrinking civil society spaces requires NGO partners to continuously adapt what they do and how they operate. Such flexibility is impossible in rigid managerial systems and D&D moves away from these towards a system of learning and adaptation based on theories of change.

This radical approach to civil society funding is not without challenges. While hinging on principles of social transformation, D&D exists and is implemented within a primarily managerial environment within the broader Ministry of Foreign Affairs. Financial and reporting requirements still exist, and departments outside those running D&D do not share (or cannot prioritize) new values to the same extent. Problems have also emerged in realizing transformative principles from the NGO side, with several lead-INGOs struggling to shift the same level of power and autonomy to partner NGOs and CBOs in the Global South that the Ministry has afforded to them. Examples include local partners not benefiting from the same security of contracts or flexible funding opportunities as the Ministry extended to lead-INGOs.

Two things here are pertinent. First, is that after operating along managerial principles for decades, expecting change overnight is overly ambitious for donors and NGO partners alike. Second, is that despite its innovations, D&D is an exception within the broader aid system that remains overtly managerial. It thus comes up against challenges that dilute its principles. Some INGOs, for example, reported that receiving funding from multiple donors meant that they could not pass down flexible funding to Southern partners: they streamlined reporting efforts across their donors by requiring partners to report along the most stringent of these.[11] Regardless of these tensions, D&D sets an important precedent for other donors in beginning to unpick the managerialist ideologies that continue to underpin – and undermine – civil society funding from other major donors. These Dutch efforts can only be enhanced by other bilateral donors following suit.

The COVID-19 pandemic and its lessons for localized action

As we wrote this chapter several months into COVID-19's global lockdown, the pandemic's estimated impact on the fight against global poverty was catastrophic. Estimates suggest that the combined health, economic and social crises accompanying the crisis could push an additional 420 to 580 million people into poverty (Sumner et al. 2020). Clearly, more resources will be required to meet additional short- and long-term needs, at the same time as these look like they are shrinking. INGO revenues dropped substantially with falls in donations, and in revenue from closed charity shops.[12] Future bilateral ODA funds (tied to GNP in donor countries) will likely fall too given the worldwide economic slow-down.

COVID-19 has also highlighted the front line of localized action, reinforcing the importance of shifting power and resources to Southern actors. National responses may be coordinated by governments and medical scientists, but it is the civil society organizations that are closest to communities who can best identify and support the most vulnerable members of society. Southern NGOs have played a central role in providing the vulnerable with health, hygiene, and food requirements. Some large, national SNGOs like BRAC have coordinated massive responses across Bangladesh.[13] But this applies equally to the most localized actors, such as community-based and grassroots organizations, who are well-placed to carry out awareness-raising activities and monitor local cases, particularly given the social distancing and lock-downs that the pandemic necessitated. Their closeness to communities makes them ideal to function as part of local health and surveillance clusters, registering data, tracing contacts and supporting people with confirmed or suspected infection (Torres 2020). In Kenya, for example, *Muungano wa Wanavijiji,* the local arm of Shack/Slum Dwellers International, initiated a COVID-19 tracker and situation monitoring system that follows testing, infection rates, deaths, treatment, availability of sanitation stations, face masks and other protective equipment and what is being done about the situation across urban informal settlements.[14]

Given the urgency with which the COVID-19 crisis is playing out in communities across the world, flexibility, dynamism, and local knowledge are required for timely and effective responses, especially in situations where public health systems are weak. This requires a localized response, and this need will not disappear once the immediate crisis is over. SNGOs will remain critical in supporting the recovery of communities and preparing for possible reoccurrences (Iroulo and Boateng 2020).

Conclusion

Power inequalities in the aid chain have long-been recognized to hinder progress towards more transformative and inclusive development outcomes by prioritizing certain types of NGOs (largely professionalized and Northern-based NGOs) and types of activity (primarily service delivery) when it comes to allocating power and resources in the aid chain. Southern NGOs, those actors most closely aligned to local needs and realities on the ground, have been under-resourced and unable to live up to their transformative potential. Globally, the aid system has achieved little in the way of fostering more vibrant, diverse, and strengthened civil societies, as key donor objectives proclaim they seek to do. Discussions around localization and the Southern-led #ShiftThePower campaign are indicative of a strong normative shift in this direction, but the extent to which this is leading to actual change in the practices of INGOs and donors is unclear. Leading the way in turning the tide towards change are new funding innovations for civil society from the Dutch government that offer a stronger partnership-based channel for funding political action through flexible approaches to funding and monitoring. Yet the tensions that this new policy confronted highlight the breadth and depth of change required in a managerial aid system that is so engrained in the mind-sets and practices of donors and INGOs alike. Innovations still exist within a broader aid system that shape and constrain the transformative potential of INGOs and SNGOs. Until more donors move towards flexible and non-project-based funding mechanisms, and start channelling a larger proportion of funds directly to NGOs and CSOs in the Global South, the potential for change remains limited. One clear lesson from the COVID-19 crisis is the importance of strong and well-resourced SNGOs that can operate quickly and effectively at the community-level. Reaching this, however, requires discussions and debates around localization to be accompanied by significant action from INGOs and donors alike.

Key sources

1. Brass, J., W. Longhofer, R. S. Robinson, and A. Schnable. (2018). NGOs and international development: A review of 35 years of scholarship *World Development* 112: 134–149.

Conducting a systematic review of 35 years of scholarship on development NGOs, this article is a must-read for teaching, research and policy-makers alike. Alongside analysing the core subjects, topics and debates of NGO research over this period, it also identifies a well thought-out research agenda for those interested in the sector.

2. Banks, N., D. Hulme, and M. Edwards. (2015). NGOs, states and donors revisited: Still too close for comfort? *World Development* 66: 707–718.

Provides an up-to-date synthesis and advances long-standing debates around NGOs' ability to live up to their comparative advantages given their positioning in the aid chain. It also reflects on how NGOs can 'return to their roots' to offer genuine development alternatives.

3. Kamstra, J. (2020). Civil society aid as balancing act: Navigating between managerial and social transformative principles. *Development in practice* 30: 763–773.

It is important to reflect on positive changes and this article gives unparalleled insight into the Dutch design and implementation of a new funding approach for civil society, highlighting how donors can reimagine funding systems to promote more inclusive, political, and transformative NGO action. It also highlights the tensions that emerge given that new socially-transformative ideals and priorities clash with the managerial constraints of the broader donor system.

Notes

1 An increase in South-South cooperation also means that we increasingly see examples of INGOs rooted in the Global South. Bangladesh's BRAC, which now operates across six sub-Saharan African countries, Afghanistan, Myanmar, Nepal and the Philippines is one important and notable example here. BRAC has been voted 'the best NGO in the world' for the past four years in a row.

2 These do not add up to 100 per cent, because a small proportion of total funds spent are categorized as 'undefined.' These have been removed from the figures for clarity.

3 An anonymous letter in the *Guardian* highlights the need for the aid sector globally to move away from these practices that disadvantage and discriminate against local hiring, among others. See: www. theguardian.com/global-development/2020/jun/15/the-aid-sector-must-do-more-to-tackle-its-white-supremacy-problem (accessed 15 July 2020).

4 The technical definition of 'aid *through* NGOs' is payments by the official sector for NGOs to implement projects and programmes which are developed by or known to and approved by the official sector, and which is ultimately responsible for them. The technical definition of 'aid *to* NGOs' is official contributions to programmes and activities which NGOs have developed themselves, and which they implement on their own authority and responsibility. This enables some level of discretion in allocation of funds once received (OECD 2020, 34). There is, however, recognition that there may be some overlap between aid through NGOs and NGOs' own priorities as independent development actors.

5 The Shift The Power movement has been led by the Global Fund for Community Foundations, a South-African organization working with like-minded partners around the world who believe in people-led development, community change from the ground up and shifting power locally to these organizations to facilitate that. They have organized events and roundtables in 2016 and 2018. You can learn more about the movement's history at their website: www.globalfundcommunityfoundations. org/what-we-stand-for/shiftthepower/

6 'Are INGOs ready to give up power?' by Deborah Doane, accessible at: www.opendemocracy.net/en/ transformation/are-ingos-ready-give-power/ (accessed 29 May2020).

7 'Are INGOs ready to give up power?' by Deborah Doane, accessible at: www.opendemocracy.net/en/ transformation/are-ingos-ready-give-power/ (accessed 29 May 2020).

8 www.globalfundcommunityfoundations.org/what-we-do/influencing-development-debates-and-practice/ (accessed 29 May 2020).

9 'An Open Letter to International NGOs who are looking to localise their operations' by #ShiftThePower: www.opendemocracy.net/en/transformation/an-open-letter-to-international-ngos-who-are-looking-to-localize-their-operations/?source=wa (accessed 10 June 2020).

10 ODA funds allocated to D&D represent around 25 per cent of all Dutch ODA funds that go to civil society organizations. The remaining civil society funds are still spent through more traditional channels that prioritize service delivery.

11 A successor policy, *The Power of Voices*, addresses lessons learned from D&D while staying true to the same principles and objectives. Seeking to strengthen some of the main transformative principles that were undermined under D&D, for example, the Power of Voices emphasizes the importance of Southern leadership and more balanced autonomy and complementarity in partnerships.
12 In the UK Oxfam alone, for example, has reported losing £5 million a month as a result of charity shop closures. See: www.theguardian.com/global-development/2020/may/20/oxfam-to-close-in-18-countries-and-cut-1500-staff-amid-coronavirus-pressures (accessed 12 June 2020).
13 As of 21 May 2020, BRAC had more than 100,000 workers on the ground covering all 64 districts of Bangladesh had distributed cash support to over 340,000 households, sanitation products to more than 1.5 million and food packages to over 10,000 households. See www.brac.net/covid19/res/sitrep/COVID-19%20Sitrep_21%20May%202020.pdf (accessed 12th June 2020).
14 See www.muungano.net/muunganos-covid-19-response for more details.

References

Banks, N., D. Hulme, and M. Edwards (2015). NGOs, states and donors revisited: Still too close for comfort? *World Development* 66: 707–718.
Banks, N., and D. Brockington (2020). Growth and change in Britain's Development NGO sector (2009–15). *Development in Practice* 30: 706–721.
Brass, J., W. Longhofer, R. S. Robinson, and A. Schnable (2018). NGOs and international development: A review of 35 years of scholarship *World Development* 112: 134–149.
Bukenya B., and Hickey S. (2014). NGOs, Civil Society, and Development. In: Obadare E. (eds.) The Handbook of Civil Society in Africa. *Nonprofit and Civil Society Studies* 20:311–335. New York: Springer. DOI:www.doi.org/10.1007/978-1-4614-8262-8_19
CIVICUS. (2019). People Power Under Attack: Findings from the CIVICUS Monitor. Available: www.civicus.contentfiles.net/media/assets/file/GlobalReport2019.pdf
Davis, J. M. (2019). Real 'non-governmental' aid and poverty: comparing privately and publicly financed NGOs in Canada. *Canadian Journal of Development Studies* 40(3): 369–386.
Dupuy, K., Ron, J., and Prakash, A. (2016). Hands Off My Regime! Governments' Restrictions on Foreign Aidto Non-Governmental Organizations in Poor and Middle-Income Countries. *World Development* 84: 299–311.
Glasius, M., Schalk, J., and De Lange, M. (2020). Illiberal Norm Diffusion: How Do Governments Learn to Restrict Nongovernmental Organizations? *International Studies Quarterly*, 64(2): 453–468.
Hossain, N., Khurana, N., Mohmand, S., Nazneen, S., Oosterom, M., Roberts, T., Santos, R. Shankland, A., and Schröder, P. (2018). What does closing civic space mean for development? A literature review and proposed conceptual framework. *IDS Working Paper* 515.
Iroulo, L. C., and Boateng, O. (2020). African States Must Localise Coronavirus Response. (GIGA Focus Afrika, 3). Hamburg: GIGA German Institute of Global and Area Studies = Leibniz-Institut für Globale und Regionale Studien, Institut für Afrika-Studien. www.ssoar.info/ssoar/bitstream/handle/document/67242/ssoar-2020-iroulo_et_al-African_States_Must_Localise_Coronavirus.pdf?sequence=1&isAllowed=y&lnkname=ssoar-2020-iroulo_et_al-African_States_Must_Localise_Coronavirus.pdf (accessed 12 June)
Kamstra, J. (2020). Civil society aid as balancing act: Navigating between managerial and social transformative principles. *Development in Practice* 30: 763–773.
Kellow, A., and Murphy-Gregory, H. (eds.) (2018). *Handbook of Research on NGOS*. Edward Elgar Publishing.
Koch, D. J., A. Dreher, P. Nunnekamp and R. Thiele (2009). Keeping a Low Profile: What Determines Aid Allocation of Aid by Non-Governmental Organizations? *World Development* 37: 902–918.
Kumi, E. (2017). Domestic resource mobilisation strategies of National Non-Governmental Development Organisations in Ghana. *Bath Papers in International Development and Wellbeing*. No: 52/2017.
Kumi, E., and Hayman, R. (2019). Analysing the relationship between domestic resource mobilisation and civic space. INTRAC.
Loman, B., I. Pop, and R. Ruben (2011). Follow the leader: How Dutch development NGOs allocate their resources – the contradictory influence of donor dependency. *Journal of International Development* 23(5): 641–655.
Mercer, C., and Green, M. (2013). Making civil society work: contracting, cosmopolitanism and community development in Tanzania. *Geoforum* 45: 106–115.

Musila, G.M. (2019). *Freedoms Under Threat: The Spread of Anti-NGO Measures in Africa*. Washington: Freedom House.

Organisation for Economic Cooperation and Development (OECD) (2020). *Development Assistance Committee Members and Civil Society*. The Development Dimension, OECD Publishing, Paris, https://doi.org/10.1787/51eb6df1-en.

Open Forum for CSO Development Effectiveness (OFCDE) (2011). *The Seam Reap CSO Consensus on the International Framework for CSO Development Effectiveness*. Seam Reap: Open Forum for CSO Development Effectiveness.

Overseas Development Institute (ODI) (2019). *Grand Bargain annual independent report 2019*. London: Overseas Development Institute.

Pallas, C.L., and Sidel, M. (2020). *Foreign Aid Reduction and Local Civil Society: Recent Research and Policy Guidance for Donors and International NGOs*. Nonprofit policy forum. 1–8.

Poole, L. (2018). *Turning Rhetoric into Resources: Transforming the Financing of Civil Society in the Global South*. NEAR Network.

Schulpen, S. and L. van Kempen (2020). Does the 'Dutch' INGO exist? Mapping a decade of financial and organisational change. *Development in Practice*. 30: 722–737.

Smith, H. E. (2019). *The Impact of Self-Financing on the Success and Sustainability of Non-Governmental Development Organisations in The Gambia*. Unpublished PhD thesis, Coventry University.

Suarez, D., and Gugerty, M.K. (2016). Funding Civil Society? Bilateral Government Support for Development NGOs, *Voluntas* 27: 2617–2640.

Sumner, A., C. Hoy, and E. Ortiz-Juarez (2020). *Estimates of the impact of COVID-19 on global poverty*. Helsinki, Finland: UNU-WIDER.

Torres, I. (2020). Localising an asset-based COVID-19 response in Ecuador. *The Lancet* 395(10233): 1339.

10

PHILANTHROPY

Linsey McGoey

Introduction

Private philanthropies play an instrumental role in development policies, setting agendas at the national and global level. Organizations like the Bill & Melinda Gates Foundation (BMGF or Gates Foundation) draw media headlines every day for their contributions to COVID-19 vaccine development, their financing of the World Health Organization (WHO), and their efforts to eradicate polio and malaria. What could be contentious about this? A number of things. As this chapter discusses, media soundbites often obscure complex realities around (1) the financial contributions that foundations make in proportion to government actors; (2) private foundations' capacity to undermine democratic decision-making, and (3) the tendency for billionaire's philanthropies to champion market-led solutions than can compound the very same health and economic inequalities that foundations purport to ameliorate (McGoey 2015).

In the US, there has been an explosive growth of new private foundations in recent years, doubling from approximately 40,000 in 1995 to over 90,000 by 2020. At the same time, the growth of individual foundations does not mean that overall giving levels have actually increased as a percentage of the US Gross Domestic Product (GDP). It also does not mean that government actors have ceased to be, by far, the largest funders of global health and development initiatives. Take the example of the Gates Foundation, in total, it has disbursed over £50 billion in grants since its establishment in 2000 (through the conjoining of two separate entities founded in 1994). Other organizations such as Bloomberg Philanthropies and the UK-based Wellcome Trust have also individually and collectively disbursed billions towards global development. In comparison, private philanthropic giving in the US has stayed flat at about two per cent of overall GDP since the 1960s (Soskis 2017).

The sums flowing from private foundations towards health and development initiatives are *much* smaller than what rich nations spend on global development through official development assistance (ODA), which, for OECD countries, is about $130 billion annually. Of that ODA funding, approximately $40 billion is earmarked for health initiatives (IMHE 2019). Those sums are significant, but importantly, they are dwarfed by what developing nations spend through their own budgets on different health and welfare programmes. As the US-based Institute for Health Metrics and Evaluation has reported:

DOI: 10.4324/9781003017653-12

Of worldwide health spending, 74.0% (72.5–75.5) was financed through governments, 18.6% (18.0–19.4) was spent out-of-pocket, 7.2% (6.7–7.8) was financed through private insurance, and 0.2% (0.2–0.2) was financed by donors. Still, despite accounting for less than 1% of global health spending, development assistance for health makes up 25.4% (23.9–26.8) of health spending in low-income countries, where 9.6% of the global population lives.

(IMHE 2019: 13)

Given that total contributions by private donors like the Gates Foundation only account for a tiny fraction of total health spending globally each year, why is there growing socio-logical attention to private foundations? It is partly because foundations have the discretion to put a significant amount of their total funding towards flagship programmes and organizations at their own prerogative, rather than at the will of an elected body of representatives. Of the Gates Foundation's total health spending each year (approximately $2 billion annually), nearly a quarter is directed to the WHO. This amount − $500 million − means that a private foun-dation has been among the top three donors to WHO, along with the US and the UK, for nearly 20 years. With this influence, the Gates Foundation can use relatively small amounts of money to stipulate that targeted grants must be spent on priority areas that they want emphasized, such as polio eradication. As a result of this conditionality, global health journalist Laurie Garrett has suggested that at the WHO, few major decisions take place without being 'casually, unofficially vetted by Gates Foundation staff' (Garrett 2012, see also McGoey 2015; Stuckler et al. 2011).

This means that organizations such as the WHO, which have a constitutional mandate to respond in a democratic manner to nation states, are increasingly beholden to private philan-thropic foundations (McGoey 2015). Another concern with private philanthropy is that it has not made a dent in narrowing, and may actually be compounding, economic inequality. This is partly because private gift-giving is an arbitrary and voluntary act, unlike mandatory taxes, and can be retracted at the whim of donors, leading to the problem of uncertain and non-sus-tainable funding flows. Private disbursements can even be used to quietly lobby for things like erosion of labour protections, thus entrenching wealth divides (Kohl-Arenas 2016, Mayer 2016, Reich 2018, Willoughby-Herard 2015).

Both in the past and today, wealthy donors like Bill Gates have earned billions through lobbying for corporate and tax policies that exacerbate the same social and economic inequal-ities that philanthropists purport to remedy, for example the practice of subcontracting which reduces labour protections for contract workers (McGoey 2015). In the early 20th century, business magnates such as Henry Ford and John D. Rockefeller used their political power to battle unions and worker power on the domestic front in the US, and the foundations they endowed, the Ford and Rockefeller Foundations, would later extend similar political power globally, quietly funding efforts to undermine socialist governments such as it the case of Chile in the 1970s (Parmar 2012, 2017; Sigmund 1983). What is different now is the explicit ways that newer philanthropic foundations are more outspoken about using philanthropy to enrich corporate actors through first, direct charitable grants to for-profit companies; and second, an ideological emphasis on entrepreneurship and private-sector involvement as a solution to devel-opment challenges.

This chapter offers an overview of the history and the contemporary role of private foundations in global development policies. It points out that while private foundations are non-government actors, they have political effects, making it possible to effect policy changes that might be less palatable to different constituencies if executed by state or corporate actors. They are aided by an appearance of being independent of the self-interests of a state or a donor. This chapter has two main sections, starting with a historical focus on philanthropy: the Rockefeller Foundation's role in helping to establish international health as an important, distinctive medical and foreign policy arena during the early to mid-20th century. This section also examines the mid-20th to late 20th century, when US foundations such as Ford, Rockefeller and Carnegie were seen as a form of 'soft power' in the ideological battle between market-oriented and socialist approaches to global development. In the second and final section, the recent emergence of philanthrocapitalism is discussed paying particular attention to the growing role of the Gates Foundation in advocating for more for-profit involvement in global development.

Philanthropy and global development: historical overview

Decades before philanthropic foundations such as the Gates Foundation became household names, earlier philanthropical organizations, such as the Rockefeller Foundation, contributed to significant advances in global development, in particular fighting infectious diseases and developing new statistical methods to understand growth and economic inequality around the world. The Rockefeller Foundation was established through the fortune of John D. Rockefeller, an American oil tycoon whose strong interest in international health sprang in part from his grandson's death from scarlet fever at a young age (McGoey 2015). From 1913 to 1951, the Rockefeller Foundation's health division operated in more than 80 countries, and its vast financial resources made up a substantial portion of budget of the League of Nations Health Organization (LNHO). Founded in 1922, the LNHO was a precursor to the WHO. The Rockefeller Foundation was responsible for funding over half of the LNHO's budget, and it continued to finance the WHO even after it gained a more secure funding base via UN member-state contributions (Youde 2013).

This philanthropic support was not simply financial, it was also ideational, in that the policies that the LNHO and later the WHO implemented tended to align with the priorities of Rockefeller Foundation staff. Scholars such as Anne-Emanuelle Birn and Jeremy Youde have noted that the bureaucratic structure of the LNHO replicated the Rockefeller Foundation's own operational structure, with shared expertise when it came to knowhow in disease control. A similar type of influence has emerged in the last decade with new philanthropic actors, and many observers think this is evidence that private philanthropies tend to exert ideological influence on the organizations that they financially support (Birn 2014; Birn and Fee 2013; Youde 2013).

This ideational influence has attracted significant criticism from observers in both the Majority and Minority world who were critical of the approach of Rockefeller and other foundations, especially the Ford Foundation and Carnegie Corporation. Informally, these three organizations were known together as the 'Big three' US philanthropic funders. Critics scorned what they saw as the blinkered approach of development strategies, based on Western approaches, without sufficiently understanding the culture of non-Western regions where philanthropic grants were targeted. Philanthropic donors were seen as funding experts who parachuted into new regions with a degree of hubris about the superiority of their homegrown methods (Willoughby-Herard 2015).

Take, for example, the Khanna study, a medical experiment in India in the late 1950s. The goal of the study was to reduce fertility rates in that nation at a time when many experts feared that explosive population growth would undermine India's economic development. Funded by the Rockefeller Foundation and developed in collaboration with the Harvard School of Public Health and India's Ministry of Health, the study involved teams of researchers compelling villagers to take contraceptive pills in a controlled scientific experiment. Informed consent procedures were minimal, and later the researchers learned that many villagers wrongly thought that the researchers had come to their home village to build a better road system. Worried their homes faced the prospect of demolition, they treated the researchers as kindly as possible, tolerating the presence of the researchers and receiving the contraceptive pills: they thought it would be impolite not to. When the experiment was over, birth rates were *higher* among participants given the contraceptive pills than among those who did not receive contraception. The villagers accepted the pills but had not consumed them. Further, the failed experiment led to mistrust of scientific experts among study participants, because when the underlying goal of the study became clear – to reduce fertility – the study participants were angry that their reproductive autonomy and capacity to have progeny to care for their long-term needs was undermined. They felt they were treated as guinea pigs in a study directly aimed at harming the long-term health and economic survival of their families. Decades later US historian Matthew Connolly described the study as an example of 'American social science at its most hubristic' (Connolly 2008, 171; see also McGoey 2015).

Another misguided or problematic funding decision was the Rockefeller's support for the Kaiser Wilhelm Institute of Anthropology, Human Heredity, and Eugenics, founded in Berlin in 1927. During this period, mainstream attitudes in the most nations in the West, including Canada, the US, Britain, and Western Europe, supported eugenics, defined as the use of selective breeding, and at its extreme, euthanasia programmes to cultivate 'optimal' population growth. The Rockefeller's support of the Kaiser Wilhelm Institute was in step with mainstream social science thought. Even after it was clear that the Kaiser Wilhelm Institute was using its research to demonize and inflict harm on Jews, the Rockefeller Foundation continued to fund its eugenics research, until as late as 1939 (Black 2012).

Philanthropic grants have both unwittingly and wittingly been used to support authoritarian rulers who supress dissent from discriminated groups and even to support genocidal activity against marginalized populations. Even when not as politicized as in these cases, philanthropy is not politically neutral. It helps to confer legitimacy on political actors, helping to veil the scale of state violence because external witnesses often presume that if an actor *was* acting in a morally egregious or illegal way, then well-meaning philanthropists would cut ties with them. In reality, philanthropic organizations have considerable discretion to channel support to different recipients even when local civil society groups call for the funding to cease. One current example is the Gates Foundation's continued praise for Indian Prime Minister Narendra Modi, long after civil society actors in India have condemned this support because of concern over Modi's divisive, punitive treatment of Muslim groups and other minorities in India. In September 2019, 37 Indian activists and scholars wrote a letter to *The Guardian* calling on Gates Foundation to halt its granting of a Goalkeepers Global Goals Award to Modi, arguing that the award 'serves to legitimise and embolden Modi and his supporters to intensify their divisive politics in a way that is in complete contradiction with the spirit and soul of sustainable, equitable and fair development.'

During the mid-20th century, criticism of the goals and methods of private philanthropic foundations also centred on the concern that they were playing a quiet, unspoken role in perpetuating the power of North American and European economic interests abroad at a time when formal empires were disintegrating. By the 1970s, many nations that had been colonized by European powers had gained political independence, but they were still economically dependent on European-American alliances for financing and structural support (Kvangraven 2020). In this climate of ongoing, non-territorial economic dependency, the Big 3 foundations – Ford, Carnegie, and Rockefeller – were seen as major forces in the development of imperial 'soft power,' a form of foreign policy that tries to achieve foreign policy objectives through persuasion and through the spreading of ideological influence rather than direct military conquest or formal colonization.

The US and its political rival, the USSR, were engaged in a fierce Cold War struggle for global supremacy, and both super-powers turned to private philanthropy and government aid to influence the internal decision-making of newly independent nations in sub-Saharan Africa and elsewhere. As well as seeking to influence newly independent former colonies of European imperial powers, nations such as Chile and other South American nations that had long enjoyed political independence were also the target of philanthropic investment by the Ford and Rockefeller Foundations. Afflicted by internal inequality, they faced strong grassroots pressure to redistribute wealth in more equitable ways. In Chile, an explicitly socialist president, Salvador Allende, was elected to office in 1970, but in 1973 a military coup backed by the US government deposed him. The Rockefeller and Ford Foundations had worked in collaboration with US Agency for International Development (USAID) to provide doctoral training for Chilean economists at universities such as Chicago, which was known as a hotbed of 'free market' economic theorizing developed by neoliberal thinkers such as Friedrich Hayek and Milton Friedman. After studying with Friedman and other Chicago School economists, they returned to Chile to work with the military junta that had deposed Allende, playing a close role in reversing various socialist policies introduced by Allende (Sigmund 1983, Parmar 2012, 2017). Historian Inderjeet Parmar (2017) argues that while the stated aim of USAID and the Ford and Rockefeller Foundations was to promote the social and economic 'development' of Chile, the practical results were very different, leading to military rule that supressed the liberty and freedom of Chileans.

By focusing their educational support towards institutions such as the University of Chicago that leaned ideologically towards anti-statist, pro-market approaches to economic development, philanthropic foundations quietly steered new cohorts of academic leaders away from socialist or statist economic blueprints towards market-oriented development solutions. These donors' emphasis on market-oriented development policies grew stronger over the 1990s, a time when, after the fall of Berlin Wall, pro-market economists like Milton Friedman claimed that their approach to political economic development was vindicated by the failure of Soviet centrally planned economies. Friedman once wrote that 'we have won the war of ideas' (quoted in McGoey 2019, 306). By this, he meant that various pro-market reforms, including restrictions of trade tariffs and private ownership over the means of production, were integral to higher employment and GDP growth across the world. By the 1990s, Friedman's belief in the value and efficiency of private-sector enterprise had grown so pervasive that many scholars at elite business schools in the US and elsewhere began to argue that the problem with non-profit organizations was their failure to replicate a business-like approach. This belief in the superiority of market-led solutions to global development has underpinned a major new trend in philanthropy: philanthrocapitalism.

The Gates Foundation and the philanthrocapitalist turn

The Gates Foundation has been even more outspoken than earlier foundations in asserting that direct subsidies to the private sector can help to approve development outcomes, taking the unusual step of disbursing non-repayable grants to for-profit companies like Mastercard and Vodacom, while counting these grants as 'charitable' in US tax filings. The Foundation's pro-market approach parallels Bill Gates's personal belief that for-profit companies are benevolent actors when it comes to health goals. Gates has made this view clear in speeches and published articles over the years. In 2008, in a speech at the World Economic Forum, he espoused the virtues of what he terms 'creative capitalism,' defined as 'an approach where governments, business, and nonprofits work together to stretch the reach of market forces so that more people can make a profit, or gain recognition, doing work that eases the world's inequities' (Gates 2008). Pro-corporate and pro-market philanthropy has been termed 'philanthrocapitalism,' coined in 2006 by *The Economist* magazine and later expanded in a book, by Matthew Bishop and Michael Green (2010), which describes philanthrocapitalism as a new approach to charity that seeks to emulate how business is carried out in the for-profit capitalist world. Because its adherents see capitalist enterprise as a 'naturally' altruistic mode of production because of its wealth creation, they were more explicit than earlier development experts in calling for the inclusion of corporate actors and for-profit companies in helping to achieve development goals (McGoey 2015, McGoey et al. 2018).

The Gates Foundation's novel practice of giving grants to corporations is driven by the belief that private profit-making and improving social welfare are naturally aligned goals (Gates 2008). But rather than being empirically proven, the belief that they go hand-in-hand is a controversial and a highly questionable one. It is a questionable because private profits can militate against redistributive policy aims by concentrating wealth in the hands of the few, rather than disbursing it to the many. The goals of private industry – to maximize private profits – often lead to diminished spending on wages and worker benefits, such as sick pay. Private rewards also incentivize companies to exaggerate internal production costs in order to justify charging exorbitant prices for goods and services, including pharmaceutical drugs and vaccines (Light and Warburton 2011; Lazonick and Mazzucato 2013). Because of long-standing recognition of the conflict between private profits and public welfare, US charity laws explicitly bar the use of private grants to further private enrichment. This prohibition is what makes the scale of the Gates Foundation's gifts to corporations an unusual precedent, and one that might breach US laws aimed at reducing conflicts of interests in giving practices (Schwab 2020). The Gates Foundation maintains that its gifts to corporations are legal because they serve charitable ends, but the positive effects of corporate gift-giving for marginalized and impoverished continents is unclear. By giving non-repayable grants to some of the world's wealthiest corporations, the Gates Foundation is exacerbating excessive returns to the private sectors at a time of stagnating levels of national growth levels, compounding the problem of economic inequality that scholars such as Thomas Piketty have highlighted in recent years (Piketty 2014).

Market-oriented solutions and their limits

Bill Gates' stated belief in the benevolent, pro-social nature of for-profit corporations has a long heritage in theories of political economy that stretch back 200 years to the ideas of foundational thinkers in political economy such as Adam Smith, author of *The Wealth of Nations* (1776). Smith posited that private enterprise and personal profit-seeking can have positive secondary effects on the general public by fuelling economic growth, employment, and tax revenue. But importantly, Smith included a series of caveats that have been side-lined by proponents of philanthrocapitalism today. Smith called, for example, for government regulations on predatory practices like usury, the lending of loans with extortionate interest terms. He insisted that a democratically representative legislature accountable to the public should and must regulate predatory business practices by treating industry groups with 'suspicious attention,' and implement checks on concentrated financial and political power (see McGoey 2019).

This emphasis on the need to restrict corporate predation and malfeasance through placing strict limits on private profit-making is missing in the philanthrocapitalist approach of the 21st century. Take, for example, microfinance initiatives that have been funded and championed by the Gates Foundation and other philanthropic and governmental funders over the past three decades. The initial impetus behind them was to extend credit to 'unbanked' individuals and families in order to stimulate economic investment, entrepreneurship, and growth. But today, many MFIs operate on a for-profit rather than non-profit basis, and both for-profit and for-profit MFIs charge excessively high interest rates. In other words, they perpetuate the same problem of usury that Adam Smith insisted needed to be better regulated. Today, punitive interest rates on micro-loans have created indebtedness among loan recipients rather than helped to reduce poverty, leading to crippling debt loads for many recipients (Engel and Pedersen 2019).

Despite increasing debt for many loan recipients and only very marginal gains when it comes to poverty alleviation, government aid donors and private foundations continue to heavily subsidize non-profit and for-profit microfinance providers (Cull et al. 2016). Why would different organizations in the Global North continue to offer grants to for-profit companies who offer microcredit loans on a for-profit basis? Definitive answers are hard to find, but it seems possible that financial benefits to for-profit companies have produced a powerful lobby that makes it hard for the World Bank, Gates Foundation and other donors to charge course, even though the harms of under-regulated microfinance lending are clear and widely known (Ghosh 2013). Microfinance can be profitable for affluent investors, which creates an incentive to perpetuate the practice, even if the gains for the explicit beneficiaries – economically marginalized groups – are unclear. A similar problem can be seen in the use of private philanthropy to subsidize for-profit companies that offer mobile money transfers in developing countries.

In 2010, the Gates Foundation offered a non-repayable grant of $4.8 million to Vodacom, a subsidiary of a British company Vodafone, to develop M-Pesa, a system that enables mobile phone users to pay bills through text messages. During this time, the parent company, Vodafone, was valued at £80 billion, despite this, they still received grant funding from DfID, the former foreign aid branch of the UK government, as well as private philanthropic grants from Gates, to develop a new mobile transfer scheme (McGoey 2015). Much like its grants to Mastercard, the Gates Foundation's gifts to Vodacom raise worrying questions. Given that Mastercard and Vodacom profit from the grants, shouldn't the company be legally beholden to repay it, especially given that the Gates Foundation confers tax benefits on its owners? Is it morally or ethically fair for a private philanthropy to subsidize the financial returns of for-profit companies?

For many years, organizations such as the Gates Foundation and DfID hailed M-Pesa as an unqualified success story, citing usage and business expansion as evidence of positive effects on the ground. But in recent years, problems similar to those that beset microfinance have become clear: the mobile apps often increase indebtedness rather than improving poverty or inequality in nations such as Kenya where usage has exploded. Development scholars Kevin Donovan, Emma Park, Serena Natile have studied mobile payments, and found that users speak of it like a type of albatross around their necks, and even referred to it as a type of slavery because it creates intractable obligations for recipients. The press in Kenya frequently report suicides associated with debt burdens (Donovan and Park 2019; Natile 2020). Some evidence suggests that mobile schemes can increase savings levels, while other studies show that under-regulated interest on loans is fuelling heavy debt burdens. Users who are unable to cover their monthly expenses through their meagre wages often rely on the apps for fast forms of credit, with some users see-sawing between different lending apps to cover debts to different lenders. Donovan and Park (2019) add that mobile money programmes 'valorizes the role of markets to improve people's lives – an ideology buoyed by the conservative turn in places such as the UK's Department for International Development and the outsized influence of the Gates Foundation' (see also Odundo Owuor 2019).

Conclusion

The fact that market-oriented development programmes like M-Pesa and microfinance have created at best mixed results for the global poor, and often compounded unbearable debt, points to a profound problem with the 'creative capitalism' ethos espoused by Bill Gates. In suggesting that it is feasible to easily marry the realms of for-profit business and global development, philanthrocapitalists skirt the problem of when profits and social purposes clash. This does not mean that the business sector and private philanthropy cannot play any positive role in development initiatives. But it does mean that there needs to be stronger corporate regulations in place, such as limits on the interest that lenders can charge on microcredit loans. It also highlights the need to rethink many of the core assumptions of the philanthrocapitalist model, where the inefficiency or a lack of 'results' in the non-profit sector are attributed to a failure to replicate a business approach to development. In reality, the opposite is more like to be true: thwarted efforts to narrow global and national inequality and eradicate poverty come from *too* much of a business mentality in development spheres, rather than too little of it.

A counter-mobilization against the philanthrocapitalist turn is growing. The fact that private profit-making can undermine and harm the public interest has once again come to the forefront of debates over development policy, especially as tragedies such as the COVID-19 pandemic expose failings. For example, many large corporations are not providing sick pay and other worker benefits, which has contributed to avoidable death and suffering. Recognizing the conflict between private gain and public benefits, some development scholars have called for what they term 'radical philanthropy' (Herro and Obeng-Odoom 2019). By this term, they mean that philanthropy should be directed at radically altering the status quo by redistributing wealth and power, rather than further concentrating it in large multinationals such as Mastercard. For this version of philanthropy to flourish, there needs to be shift away from the current corporate model encapsulated by the philanthrocapitalist shift, and more legal and political constraints on the ability of foundations such as the Gates to direct their giving in a way that enriches large corporate actors.

Key sources

1. Birn, A-E. (2014). Philanthrocapitalism, Past and Present: The Rockefeller Foundation, the Gates Foundation, and the Setting(s) of the International/Global Health Agenda. *Hypothesis* 12(1): e8, doi:10.5779/hypothesis.v12i1.229

A comparison between the Gates Foundation and Rockefeller Foundation, Birn offers an in-depth historical understanding of similarities and differences between the approaches of different funders, underscoring the novelty of the market-led approach of today's foundations and the impact for global health.

2. Connolly, M. (2008). *Fatal Misconception: The Struggle to Control World Population*. Cambridge, MA: Harvard University Press.

This book examines problems surrounding myopic attitudes to local cultural norms in regions where development practitioners and social scientists operate. As Connolly shows, various tacit and explicit neo-Malthusian population control efforts have been a key feature of much development policy over the 20th century. Connolly explores how different private foundations played an instrumental role in shaping neo-Malthusian policies.

3. McGoey, L. (2015). *No Such Thing as a Free Gift: The Gates Foundation and the Price of Philanthropy.* London: Verso.

McGoey explores the ideological origins of the philanthrocapitalist movement, showing how the practices of today's donors differ from earlier donors such as Carnegie, Rockefeller, and Ford at the turn of the 20th century. It is the first book-length examination of the Gates Foundation's influence on global health and development.

4. Natile, S. (2020). *The Exclusionary Politics of Digital Financial Inclusion: Mobile Money, Gendered Walls.* London and New York: Routledge.

This book offers a close examination of problems with microfinance, with specific attention to the role of new mobile payments systems. Natile's book is particularly useful for understanding the gendered implications of new systems of microlending and their impact on women globally.

5. Parmar, I. (2012). *Foundations of the American Century,* Colombia: Columbia University Press.

Parmar examines the history of the Big 3 foundations – Carnegie, Rockefeller, and Ford – and their quiet 'soft power' role in securing US economic dominance globally during the mid-20th-century.

References

Birn, A-E. (2014). Philanthrocapitalism, Past and Present: The Rockefeller Foundation, the Gates Foundation, and the Setting(s) of the International/Global Health Agenda. *Hypothesis* 12(1): e8, doi:10.5779/hypothesis.v12i1.229

Birn, A-E and Fee, E. (2013). The Rockefeller Foundation and the International Health Agenda. *The Lancet* 381: 1618–19.

Bishop, M. and Green, M. (2010). *Philanthrocapitalism: How Giving Can Save the World.* London: Bloomsbury.

Black, E. (2012). *War Against the Weak: Eugenics and America's Campaign to Create a Master Race.* New York: Dialog Press.

Connolly, M. (2008). *Fatal Misconception: The Struggle to Control World Population.* Cambridge, MA: Harvard University Press.

Cull, B., Demirgüc-Kunt, A. and Morduch, J. (2016). The microfinance business model: Enduring subsidy and modest profit. Policy Research Working Paper, No. 7786. World Bank, Washington, DC. URL: www.openknowledge.worldbank.org/handle/10986/24867

Donovan, K. and Park, E. (2019). Perpetual Debt in the Silicon Savannah. *Boston Review*, September 19, URL: http://bostonreview.net/class-inequality-global-justice/kevin-p-donovan-emma-park-perpetual-debt-silicon-savannah#:~:text=Perpetual%20Debt%20in%20the%20Silicon%20Savannah%20Kenya's%20poor,now%20they%20call%20it%20slavery.%20September%2020,%202019

Engel, S and Pedersen, D. (2019). Microfinance as Poverty-Shame Debt. *Emotions and Society* 1(2): 181–196.

Garrett, L. (2012). Money or Die: A Watershed Moment for Global Public Health. *Foreign Affairs*, March 6, URL: www.foreignaffairs.com/articles/2012–03–06/money-or-die

Gates, B. (2008). 'Creative Capitalism' Speech. *World Economic Forum*, URL: www.gatesfoundation.org/media-center/speeches/2008/01/bill-gates-2008-world-economic-forum

Ghosh, J. (2013). Microfinance and the challenge of financial inclusion for development. *Cambridge Journal of Economics* 37(6): 1203–1219.

Herro, A., and Obeng-Odoom, F. (2019). Foundations of Radical Philanthropy. *Voluntas: International Journal of Voluntary and Nonprofit Associations*, 30(4): 881–890.

Institute for Health Metrics and Evaluation (IMHE) (2019). *Financing Global Health 2018: Countries and Programs in Transition.* Seattle, WA: IHME, 2019.

Kvangraven, I. (2020). Beyond the Stereotype: Restating the Relevance of the Dependency Research Programme. *Development and Change* 52(1): 76–112.

Kohl-Arenas, E. (2016). *The Self-Help Myth*. Oakland: University of California Press.

Lazonick, W. and Mazzucato, M. (2013). The risk-reward nexus in the innovation-inequality Relationship. Industrial and Corporate Change 22(4): 1093–1128.

Light, D. and Warburton, R. (2011). Demythologizing the high costs of pharmaceutical research. *BioSocieties* 6: 34–50.

Mayer, J. (2016). *Dark Money: The hidden history of billionaires behind the radical right*. New York: Penguin Random House.

McGoey, L. (2015). *No Such Thing as a Free Gift: The Gates Foundation and the Price of Philanthropy*. London: Verso.

McGoey, L. (2019). *The Unknowers: How Strategic Ignorance Rules the World*. London: Zed Books.

McGoey, L., Thiel, D. and West, R. (2018). Philanthrocapitalism and Crimes of the Powerful. *Politix* 1: 29–54.

Natile, S. (2020). *The Exclusionary Politics of Digital Financial Inclusion: Mobile Money, Gendered Walls*. Oxon: Routledge.

Odundo Owuor, V. (2019). Mobile based lending is a double edged sword in Kenya = helping but also spiking personal debt. *Quartz Africa*, 5 October.

Parmar, I. (2012). *Foundations of the American Century*. Colombia: Columbia University Press.

Parmar, I. (2017). Corporate Foundations and Ideology. In: G. Baars and A. Spicer (eds.) *The Corporation*. Cambridge: Cambridge University Press, 434–447.

Piketty, T. (2014). *Capital in the 21st Century*. Cambridge: Harvard University Press.

Reich, R. (2018). *Just Giving: Why Philanthropy is Failing Democracy and How It Can Do Better*. Princeton: Princeton University Press.

Sigmund, P. (1983). The Rise and Fall of the Chicago Boys in Chile. *SAIS Review* 3(2): 41–58.

Schwab, T. (2020). Bill Gates Gives to the Rich (Including Himself). *The Nation*, March 17, URL: www.thenation.com/article/society/bill-gates-foundation-philanthropy/

Soskis, B. (2017). Giving Numbers: Reflections on Why, What, and How We are Counting. *NonProfit Quarterly,* November 1, URL: www.nonprofitquarterly.org/giving-numbers-reflections-counting

Stuckler, D, Basu, S. and McKee, M. (2011). Global Health Philanthropy and Institutional Relationships: How Should Conflicts of Interest Be Addressed? *PLoS Medicine* 8(4) e1001020:1–10.

Willoughby-Herard, T. (2015). *Waste of a White Skin: The Carnegie Corporation and the Racial Logic of White Vulnerability*. Berkeley: University of California Press.

Youde, J. (2013). The Rockefeller and Gates Foundations in Global Health Governance. *Global Society* 27(2): 139–58.

11

SOCIAL ENTERPRISE AND INCLUSIVE ECONOMIC DEVELOPMENT

Narayan Gopalkrishnan and Hurriyet Babacan

Introduction

As we embark on this great collective journey, we pledge that no one will be left behind. Recognizing that the dignity of the human person is fundamental, we wish to see the Goals and targets met for all nations and people and for all segments of society. And we will endeavor to reach the furthest behind first.

(UN 2015, 4)

The rallying call of the 2030 Agenda for Sustainable Development is that 'no one will be left behind.' This is a call to build a more inclusive world that is based on broadening the concepts of poverty and addressing inequality directly (Fukuda-Parr and Hegstad 2018). The Sustainable Development Goals (SDGs) accordingly provide a blueprint for action for a more sustainable and inclusive world. Several of the 17 goals are directly linked to economic outcomes such as poverty, work, hunger, consumption, education, inequality, and the remainder of the goals have cross-cutting and interconnecting links to economic development. And yet there is a clear recognition that sustained economic development is difficult to maintain without the implementation of inclusive economic development programs (Ostry et al. 2014). Across the world we see serious concern with uneven growth, arising from the development of global capitalist economies with tendencies that lead to growing inequality, particularly of income and wealth distribution over time (Piketty 2014).

Income inequality has increased in nearly all world regions in recent decades. For example, in 2016 the top 10% of earners took 37% of the national income in Europe, 41% in China, 46% in Russia, 47% in US-Canada, around 55% in sub-Saharan Africa, Brazil, and India and 61% in the Middle East. Global wealth disparity is also stark and exposes the nature of global economic inequality. In 2019, the richest 10% owned 82% of global wealth while the top 1% alone owned 45% of global wealth (Shorrocks et al. 2019, 13). Further, equal access to economic and social opportunities is still influenced by factors such as gender, race, ethnicity, migrant status, the socioeconomic status of parents, rurality, and disability (Babacan 2020; UN 2020). The trends towards economic exclusion continue even during times of major global economic growth as measured by Gross Domestic Product (GDP). The global growth rate in 2019–2020

DOI: 10.4324/9781003017653-13

(pre-pandemic rate) was approximately 3.2% (IMF 2020). During this period, the global wage growth declined to 1.8 % from 2.4 % in 2016 (ILO 2019), indicating that economic growth is not redistributed across productive units. Failure of growth to trickle down to the poor exacerbates inequality and exclusion (Van Gent 2017).

Promoting productive employment and decent work for all is Goal 8 of the SDGs, and yet many people in the world continue to be in precarious employment and insecure economic positions. Two billion people, more than 61% of the world's employed population, are employed in the informal economy where they are not covered by formal arrangements such as labour legislation (ILO 2018). Workers in the informal sector are vulnerable to casualized work, insecurity of livelihoods and income, increased poverty levels, lack of employment entitlements and heightened risk of occupational hazard and safety (OECD 2019). As argued by the International Labour Organization 'most people enter the informal economy not by choice but out of a need to survive' (ILO 2015, p. 2). Employment in the informal economy has adverse impacts on women, older workers, rural workers and the poor (Babacan 2020; OECD 2019). Data also indicates that countries with higher levels of total share of national employment in the informal economy have lower Human Development Index levels (ILO 2018, p. 44).

Many of the above issues are critical to the implementation of the SDGs. The metamorphosis of the Millennium Development Goals (MDGs) into the SDGs has been described as a process to 'carry forward the unfinished agenda of MDGs for continuity and sustain the momentum generated while addressing the additional challenges of inclusiveness, equity, and urbanization and further strengthening global partnership by including CSOs and private sector' (Kumar et al. 2016, p. 4). The importance of the additional SDGs that emphasize inclusion emerged from the demands of social movements across the world and other institutions such as the World Economic Forum that pointed out that the first major threat to social peace and economic stability was, and continues to be, inequality (Fukuda-Parr and Hegstad 2018). However, there are strong arguments to suggest that the SDG targets relating to economic outcomes will not be realized as per projections (UN 2020) and that alternative thinking is needed about the way we approach economic development. A development approach that encompasses an agenda beyond growth and income is needed to ensure that the benefits of growth are shared equitably across all parts of society (Van Gent 2017) and as envisioned in the SDGs. One such alternative paradigm is Inclusive Economic Development, an approach that addresses many of the issues of inequality and unsustainability that have emerged from more traditional economic development approaches. Inclusive Economic Development is discussed further in the next section and social enterprise is explored as one way in which inclusive economic development can be achieved.

Inclusive economic development

Much of the academic literature uses the terms 'inclusive growth' and 'inclusive development' interchangeably but they are quite different in terms of their focus and impacts. Inclusive growth or inclusive economic growth allows 'all members of a society to participate in, and contribute to, the growth process on an equal basis regardless of their individual circumstances' (Ali and Son 2007, 2). However, this continued emphasis on economic growth can still facilitate exclusion, and create further inequalities in society (Kanbur and Rauniyar 2010). Inclusive development, on the other hand, is about a process that 'generates broad-based participation, and specifically reduces poverty and social exclusions' (Chatterjee 2005, 3). The goal of inclusive development or inclusive economic development is to go past traditional and narrow understandings of development as a primarily economic process to more complex ones that incorporate notions of equity, rights and citizenship (Hickey et al. 2015). It is concerned with the distribution of social and material benefits across social groups and categories and also the structural factors that cause and sustain exclusion and marginalization of vulnerable groups in society (Van Gent 2017).

Some of the key aspects of inclusive economic development can be summarized as:

- Just economic growth, based across sectors, and inclusive of the larger part of the country's labour force.
- Equitable opportunities and equal participation in economic growth processes and benefits.
- Going beyond individual wealth accumulation strategies to focusing on societal or collective well-being.
- Focusing less on short-term gains and more on long-term sustainability.
- Changes in the economic structures of production enabling just redistribution.
- Progressive and redistributive economic policies including taxation, price regulation and macro-economic policy.
- Facilitative institutions for inclusive economic development, factoring in the role that formal and informal institutions play in economic development.
- Spatial equity, particularly to address negative economic impacts of urbanization or rurality.
- Improvement in social indicators that have a direct bearing on economic development such as education and health.
- Strengthening capacities and enablers to participate in economic processes including education, skills, digital connectivity, and infrastructure.
- Safety nets and social protections.
- Considering the aspirations of individuals, groups, and communities in development.

(Acemoglu and Robinson 2012; Donahue et al. 2017;
Gupta et al. 2015b; Kanbur and Rauniyar 2010)

The role of institutions is particularly important in inclusive economic development. Acemoglu and Robinson (2012) differentiate between inclusive institutions and extractive institutions in this context. They argue that inclusive institutions are those that allow the broad participation of the citizens of the country and uphold the rule of law, placing constraints and checks and balances in systems and supporting plurality. Extractive institutions, on the other hand, are different from inclusive institutions in every aspect and are designed by politically and economically powerful elites to extract resources from the rest of society. Extractive institutions can lead to insecure property rights, limit entry to markets and concentrate power in the hands of a few, with limited checks and balances, and are not likely to lead to broad-based and sustained economic development. The authors further point out that that growth is far more likely to happen with inclusive institutions rather than extractive ones (Acemoglu and Robinson 2012). The vulnerability context and erosion of trust in public institutions also impacts on the role of institutions. Increasing concentration of wealth and income affects trust in the role of politics and public institutions to address the needs of the majority. Lack of trust destabilizes political systems, hinders the functioning of democracy and threatens prosperity through its effect on the climate for investment and economic growth. It also threatens the underlying fabric that holds societies together (Babacan 2019). The expansion and growth of more inclusive organizations will help to build trust and reduce vulnerability.

Social enterprises are organizations that effectively respond to the issues of equity and exclusion discussed earlier, and play an important role in inclusive economic development. Unlike commercial enterprises which begin with the search for opportunities to value capture or make profits, the process of social enterprise begins with the identification of opportunities to create or enhance social value (Agrawal and Sahasranamam 2016). These opportunities can often emerge from social problems that are impacting negatively on the lives of people in the concerned communities. In Australia, for example, Castellas and Barraket (2017) found that social enterprises employed people with disabilities and female managers at twice the rate of mainstream small businesses. The authors also point out that 12% of jobs in social enterprises are held by previously long-term unemployed people, a significant issue in most countries across the world. The most exciting part of social enterprise projects is that while their primary stated focus might be in working with one kind of social issue, such as lack of health, they often help with a number of other issues such as youth crime, delinquency, or lack of education, to name a few. They can also emerge as a way to challenge institutional and social barriers. The nature and processes of social enterprises are further explored in the next section.

Social enterprise as an inclusive economic development approach

Social enterprises are a rapidly expanding phenomenon and play a key role in inclusive economic development globally. In Europe for example, social enterprise was given specific consideration as a vehicle for work creation as a response to the Global Financial Crisis, with a particular focus on supporting young people (Barraket et al. 2016). Social enterprises are already a significant part of the economies of many countries, with the Global Enterprise Monitor reporting in 2016 that social enterprise activity formed 3.7% of the broader entrepreneurial activity across 58 national economies, ranging from 4% in Iran to 14% in Senegal (Bosma et al. 2016). Generally, social enterprises operate at the intersection of the profit and the not-for-profit sectors and, as such, may involve stakeholders from either sector or both. They can be found in most sectors of the economy and serve a range of geographical markets, with a particular focus on the local (Barraket et al. 2017). Accordingly they can involve a range of activities such as community economic development, profit generation to support not-for-profit programs, cross-sectoral programs, corporate social programs as well as for-profit programs that prioritize social mission and impact (Gray et al. 2003). However, it must be acknowledged that while the term 'social enterprise' is relatively new, the phenomenon of combining social benefit and trading activity is not new, as cooperatives and friendly societies have been doing this over a long period of time (Eversole 2013).

Social enterprises are generally viewed as those organizations that:

1. Are led by an economic, social, cultural, or environmental mission consistent with a public or community benefit.
2. Trade to fulfil their mission.
3. Derive a substantial portion of their income from trade; and
4. Reinvest the majority of their profit/surplus in fulfilment of their mission.

(Barraket et al. 2016)

However, the boundaries of what constitutes a social enterprise are not as clear-cut as this definition implies and its details need to be further unpacked. The key component of every social enterprise is the nature of its primary purpose – which is that of creating or enhancing social and ecological value as against the generation of profit and wealth for individuals. This focus on social and/or ecological value enables social enterprises to 'fill the gap' where markets and extractive institutions have failed (Zahra et al. 2008). So while commercial enterprises are looking for economic opportunity and are driven by profit as purpose, social enterprise look for social opportunity or social purpose (Agrawal and Sahasranamam 2016). Some of the key identifiers of these social opportunities have been identified by Zahra et al. (2008) as the prevalence of the social issue, the relevance of the social issue to the background, values, skills, and resources of those involved in the social enterprise process, the urgency and accessibility of the social issue and the radicalness of the solutions. These authors suggest that the more urgent and inaccessible the social issue is and the more radical the solution is, the more likely it is that social enterprises will address it rather than traditional systems. This positioning of social enterprise is of particular significance in terms of the SDGs and addressing the challenges of inclusion and equity as well as the notion of strengthening global partnership (Kumar et al. 2016). Nevertheless, social enterprises are exposed to considerable market pressures and demands to demonstrate efficiency, all of which can impact on the primacy of their mission (Abbott et al. 2019).

The second component is that it has a focus on the need for organizational sustainability. Unlike traditional not-for-profit institutions, social enterprises work clearly towards economic returns that will ensure the long-term sustainability of the organization and the maintenance of their social purpose. Social enterprises trade goods and services for purpose, deriving a substantial portion of their income from this trading activity, and as such there is an emphasis on the utilization of business skills and processes towards building organizational sustainability but not at the cost of the mission of the organization (Weerawardena and Mort 2006). This balance between the mission of the organization and its sustainability is relevant to the notion of inclusive economic development in that it builds the capacity of the organization to be inclusive and reduces dependence on external funding sources, while keeping the organization in a position to continue to address its mission over the longer term. Eversole (2013, 577) also makes the point that social enterprises "meet needs on the ground and leverage local opportunities by working across, rather than within, established categories and sectors" and as such "leverage multiple resources and generate multiple forms of value". In addressing social issues, social enterprises do not try to predict an ideal solution and then try and bring together the resources to achieve it, but rather they often try to create a solution based on the resources at hand (Corner and Ho 2010). This is further supportive of the notion of inclusive and sustainable economic development by optimizing the use of available resources.

The third key component of any social enterprise is that a proportion of the profits that are generated through the enterprise would either be reinvested in it or applied towards building other forms of social value (Santos 2012). An example of this would be non-governmental organizations that invest money made from one social enterprise towards supporting other work with their client populations. Especially in the case of for-profit organizations, this dividing line has to be drawn, in terms of the utilization of profits, so as to ensure that the economic drivers do not overwhelm all other drivers of the process. The issue of reinvestment of profits remains a vexed subject, with social entrepreneurs like Mohammed Yunus, founder of Grameen Bank, arguing that there should be no value capture and that all profits should be reinvested, while others believe that as long as the primary aim is social value creation, only a proportion of the profits needs to be reinvested into achieving the overall goals of the enterprise. So while all social enterprises reinvest profits into their mission, there is differing practice globally in terms of the proportions of reinvestment (Bosma et al. 2016). Nevertheless, this reinvestment does bring in additional resources towards addressing social issues, resources that may have otherwise been redirected into extractive institutions.

In a study of over 2,000 very influential social entrepreneurs who form the Ashoka Foundation Fellows in over 70 countries, Chandra, Jiang, and Wang (2016) suggest that the strategies that social entrepreneurs adopt towards their primary goals can be examined as those that are 'materially focused' and those that are 'symbolically focused' as well those that are focused on both. As this classification suggests, there are a broad range of strategies that are part of social enterprise and that the choice of strategies often depends on the social issues being addressed and the context within which they are they being addressed. However, it is also fair to say that most social enterprise projects have elements of both material and symbolic focus. The same authors distill the different strategies down into six main meta-strategies as follows.

1. Individual empowerment that aims to enhance the quality of life at the individual and the community level by building skills, knowledge, and access to resources for the individuals. Examples of this would include small businesses like cafes and laundry services whose primary purpose is to develop skills among individuals who may be marginalized for some reason. Similarly micro credit and other financial support systems support individuals to develop their skills to run productive businesses.

2. Collective action that is about enabling communities to work together to achieve their shared goals, and working with different forms of collectivities towards dealing with social issues. An example of this form of strategy would be a local health collective that runs mental health support for people within a local geographical area.

3. Building physical capital that is about developing the physical (and virtual) spaces that will enable people and communities to achieve their social, ecological, and economic goals. This could be in the form of incubation buildings, community hubs and such physical space, or it could internet portals and hubs that enable people and communities to optimize time, acquire needed information, knowledge, and skills or to network effectively towards achieving shared goals.

4. Reforming the system, while widely used in High Income Countries, is often viewed with scepticism because it is about challenging the status quo. While the other meta-strategies tend to be more on the lines of consensual activity, this one is about structural change and can be 'contestual.' Social activism, advocacy and media campaigns can often be part of this strategy.

5. Evidence-based practice that is about building the evidence that the stated goals of social enterprises are being met and that the strategies adopted are actually working. As the field of social enterprise matures it becomes even more important to adopt this as a key strategy.

6. Prototyping is of course the process by which new social ventures are developed on a small scale and tested before being scaled up and replicated.

Of these, the first three are the most commonly used, while the others are less common but nevertheless central to the building of an effective social enterprise sector. Using these meta-strategies, social enterprises work across many key areas of inclusive and sustainable development, such as improving levels of employment especially among marginalized groups, improving access to basic services such as housing, health, and education, improving access to sustainable energy as well as reducing waste and other negative ecological impacts. Many of the organizations within the social enterprise sector work towards the achievement of the SDGs as part of their mission, especially addressing the issues of equity and inclusion that have been discussed earlier. As such, social enterprises present as a strong and viable alternative to the traditional for-profit/not-for-profit silos and an important driver of inclusive economic development.

However, social entrepreneurship is not a panacea for all the social and ecological issues that we face. One of the criticisms that is levelled at the sector is that social enterprise is essentially a transferring of traditional public responsibilities into the private sector. Nichols (2011) suggests that social enterprises privatize collective processes and replace them with individual notions of how social needs should be addressed. Further, social enterprises can easily remain focused on one social, economic, or ecological mission and can cause harm to other areas in the pursuit of this mission. As an example of this, one award-nominated social enterprise at the International Innovation and Social Entrepreneurship Awards (co-judged by one of the authors of this chapter) was attempting to alleviate poverty by intensively farming goats in a poor rural community. The negative effects of factory farming on the quality of life of the animals as well as the impacts on the environment had completely been ignored in the pursuit of the poverty alleviation agenda of this project. This kind of thinking is not uncommon in social enterprises unless careful thought and appropriate frameworks are put into place to ensure more holistic processes. Another very significant issue faced by social enterprises is that the focus on social/ecological missions can be overwhelmed by the financial drivers of the organization. As social enterprises are at the intersection of the for-profit and not-for-profit sectors and have to operate in an environment with private-sector investment, as well as government expectations of self-sustainability in relatively short time scales, decisions can easily be taken that increase profitability while not necessarily supporting the mission of the organization (Gupta et al. 2015a).[1]

Further, as Barraket et al. (2017) argue in the Australian context, while the practice of social enterprise is relatively mature, the language and processes of social enterprise and entrepreneurship remains contested and continues to evolve. This lack of clarity continues to be a complex issue in the sector, and has significant resource and process implications. An example of this is where Kernot and McNeill (2011) discuss the *Mars Hill Café* in Western Sydney, Australia, which identified as part non-profit and part business. The founders of this social enterprise found that they had difficulty in getting rental premises because of the non-profit focus, while they also had difficulty in accessing government grants because they were a trading entity and did not fit traditional funding guidelines. This 'lack of fit' is an ongoing problem that affects the ability of social enterprises to work effectively. Dacin et al. (2011) expand on this issue when they argue that the legitimacy of social enterprises is subject to their being able to satisfy the interests of stakeholders in both the for-profit and not-for-profit sectors, and in the process being subject to a number of operational tensions.

The social enterprise sector continues to evolve through all these tensions and contradictions and is becoming an important stakeholder in working towards inclusive development in many countries across the world. Through the innovative range of strategies that they adopt, they are increasingly able to address some of the gaps that have been left between the for-profit sector and the not-for-profit sectors and are beginning to make an impact in terms of achieving the SDGs. Two examples of social enterprises are discussed in the next section and some questions presented for consideration.

Case Study 1

The *Big Issue* is a street newspaper model of social enterprise that was first established in 1991 in London and has since been expanded to 35 other countries both in Low and Medium Income Countries (LMICs) and High Income Countries (HICs). The organization primarily works with people who are the most marginalized in society, especially those who are homeless, in dire poverty and perhaps impacted on by issues of addiction and/or mental health issues. This model enables people to earn a decent income while also developing themselves as micro-entrepreneurs. The model as established in most countries involves the vendors buying the newspaper at a particular price and selling it at approximately double that price, the difference being the margin that the vendor earns. *The Big Issue* has been extremely successful and has led to the establishment of other social enterprises such as the *Big Issue Shop* (an online marketing platform), the *Big Life Company* (involved in a number of areas including healthcare, childcare and employment training), *Women's Enterprise* (training and employment), and *Homes for Homes* (generating innovative funding for social and affordable housing) just to name a few. Additionally, the *Big Issue* has been able to establish foundations such as *Big Issue Invest* which finances social enterprises to further support marginalized groups in society. They also help develop the social enterprise ecosystem through investment in programs like the *Big Issue Classroom* which educates school groups about social issues like homelessness and The *Big Idea*, involving university students in planning and developing new social enterprises.

Case Study 2

Barefoot College International is a social enterprise that works in almost 100 countries globally with a particular focus on LMICs. While the present structure of the organization has been in place only since 2015, the *Barefoot College* itself began in 1972 in a small village in Rajasthan, India. The college worked with rural women, who were generally poorly educated and often illiterate, to enable them to become agents of change in their own communities. One of the key focus areas was the demystification and dissemination of solar technology, and over 3,000 women trained by the college have since become solar engineers (also known as solar mamas) establishing decentralized solar facilities in rural homes and rural institutions. Currently, the college works across a number of areas of inclusive development including provision of safe and reliable energy, improving access to education, clean and sustainable drinking water, health, and sustainable livelihoods. In the process they have been able to impact positively on millions of people and directly address 14 of the 17 SDGs. *Barefoot College International* continues to work with women across many countries by establishing Barefoot Regional Vocational Training Centres for Women which enable technology transfer, access to information, and the development of innovation at the local level. Besides government grant income and private philanthropy, the college is able to generate substantial income from within the range of goods and services that it provides both domestically in India as well as internationally.

Questions to consider

1. As both these case studies are of social enterprises that began at the local level and have since scaled up to work across a number of countries, what do you think could be the implications for the primary mission of the social enterprise?
2. What are the enablers to this kind of scaling up? What would/could be possible barriers?
3. What do you think about these organizations as drivers of inclusive economic development? How do these organizations address issues of equity, rights, and citizenship?
4. Unlike commercial institutions, economic productivity is not the primary focus of these organizations. What then needs to be considered in determining the productivity and impact of these organizations and why?

Key sources

1. Abbott, M., Barraket, K., Castellas, E. I. P., Hiruy, K., Suchowerska, R., and Ward-Christie, L. (2019). Evaluating the labour productivity of social enterprises in comparison to SMEs in Australia. *Social Enterprise Journal,* 15, 179–194.

An exploration of the complexity of evaluating the impacts of social enterprises.

2. Barraket, J., Mason, C., and Blain, B. (2016). Finding Australia's Social Enterprise Sector 2016: Final Report. Melbourne: Social Traders and Centre for Social Impact, Swinburne.

A useful report to understand the operating contexts of social enterprises in a HIC context.

3. Bosma, N., Schott, T., Terjesen, S., and Kew, P. (2016). Global Entrepreneurship Monitor 2015 to 2016: Special Report on Social Entrepreneurship. London: Global Entrepreneurship Research Association.

A report that presents the findings of the largest comparative study of social entrepreneurship in the world.

4. Kanbur, R. and Rauniyar, G. (2010). Conceptualizing inclusive development: with applications to rural infrastructure and development assistance. *Journal of the Asia Pacific Economy,* 15, 437–454.

A useful introduction to inclusive development.

5. Zahra, S. A., Rawhouser, H. N., Bhawe, N., Neubaum, D. O., and Hayton, J. C. (2008). Globalization of social entrepreneurship opportunities. *Strategic Entrepreneurship Journal,* 2, 117–131.

This article explores the globalization of social enterprises and discusses the nature of social opportunity.

Note

1 Social enterprises are not purely for-profit and can often access government funding that is not available to commercial enterprises. Many social enterprises actually emerge from the community sector and work towards self-sustainability over larger time frames.

References

Abbott, M., Barraket, J., Castellas, E. I. P., Hiruy, K., Suchowerska, R., and Ward-Christie, L. (2019). Evaluating the labour productivity of social enterprises in comparison to SMEs in Australia. *Social Enterprise Journal,* 15, 179–194.

Acemoglu, D. and Robinson, J. A. (2012). *Why Nations Fail: The Origins of Power, Prosperity and Poverty.* London: Profile Books.

Agrawal, A. and Sahasranamam, S. (2016). Corporate social entrepreneurship in India. *South Asian Journal of Global Business Research,* 5, 214–233.

Ali, I. and Son, H. (2007). *Defining and measuring inclusive growth: Application to the Philippines. ERD Working Paper 99*. Manila: Economics and Research Department, Asian Development Bank.

Babacan, H. (2019). Demonstrating Value for Human and Social Services Investment. *International Journal of Community and Social Development*, 1, 273–294.

Babacan, H. (2020). Women and Economic Dimensions of Climate Change. In: Chaiechi, T. (ed.) *Economic Effects of Natural Disasters*. London: Elsevier Publishers.

Barraket, J., Mason, C., and Blain, B. (2016). *Finding Australia's Social Enterprise Sector 2016: Final Report*. Melbourne: Social Traders and Centre for Social Impact, Swinburne.

Barraket, J., Douglas, H., Eversole, R., Mason, C., McNeill, J., and Morgan, B. (2017). Classifying social enterprise models in Australia. *Social Enterprise Journal*, 13, 345–361.

Bosma, N., Schott, T., Terjesen, S., and Kew, P. (2016). *Global Entrepreneurship Monitor 2015 to 2016: Special Report on Social Entrepreneurship*. London: Global Entrepreneurship Research Association.

Castellas, E. I. and Barraket, J. (2017). Social enterprises are building a more inclusive Australian economy. *The Conversation* [Online]. Available: www.smartcompany.com.au/growth/social-enterprises-building-inclusive-australian-economy/

Chandra, Y., Jiang, L. C., and Wang, C. J. (2016). Mining Social Entrepreneurship Strategies Using Topic Modeling. *PLoS ONE*, 11 (3), e0151342. https://doi.org/10.1371/journal.pone.0151342

Chatterjee, S. (2005). Poverty Reduction Strategies-Lessons from the Asian and Pacific Region on Inclusive Development. *Asian Development Review*, 22, 12–44.

Corner, P. D. and Ho, M. (2010). How Opportunities Develop in Social Entrepreneurship. *Entrepreneurship Theory and Practice*, 34, 635–659.

Dacin, M. T., Dacin, P. A., and Tracey, P. (2011). Social entrepreneurship: a critique and future directions. *Organization Science*, 22, 1203+.

Donahue, R., McDearman, B., and Barker, R. (2017). *Committing to Inclusive Growth: Lessons for Metro Areas from the Inclusive Economic Development Lab*. Washington DC: Metropolitan Policy Program, Brookings Institute.

Eversole, R. (2013). Social enterprises as local development actors: Insights from Tasmania. *Local Economy: The Journal of the Local Economy Policy Unit*, 28, 567–579.

Fukuda-Parr, S. and Hegstad, T. S. (2018). 'Leaving No One Behind' as a Site of Contestation and Reinterpretation. *Journal of Globalization and Development*, 9(2), 1–11.

Gray, M., Healy, K., and Crofts, P. (2003). Social enterprise: is it the business of social work? *Australian Social Work*, 56, 141–154.

Gupta, S., Beninger, S., and Ganesh, J. (2015a). A hybrid approach to innovation by social enterprises: lessons from Africa. *Social Enterprise Journal*, 11, 89–112.

Gupta, J., Pouw, N. R. M., and Ros-Tonen, M. (2015b). Towards an elaborate theory of inclusive development. *European Journal of Development Research*, 27(4), 541–559.

Hickey, S., Sen, K., and Bukenya, B. (2015). *The Politics of Inclusive Development*. Oxford: Oxford University Press.

ILO (2015). *Transition From the Informal to the Formal Economy: Report No.204*. Geneva: International Labour Organization.

ILO (2018). *Women and Men in the Informal Economy: A Statistical Picture*. 3 ed. Geneva: International Labour Organization.

ILO (2019). *Global Wage Report 2018/19*. Geneva: International Labour Organization.

IMF (2020). *A Crisis Like No Other, An Uncertain Recovery: World Economic Outlook Update, June 2020*. Washington DC: International Monetary Fund.

Kanbur, R. and Rauniyar, G. (2010). Conceptualizing inclusive development: with applications to rural infrastructure and development assistance. *Journal of the Asia Pacific Economy*, 15, 437–454.

Kernot, C. and McNeill, J. (2011). *Australian Stories of Social Enterprise*. Sydney: University of New South Wales.

Kumar, S., Kumar, N., and Vivekadhish, S. (2016). Millennium Development Goals (MDGs) to Sustainable Development Goals (SDGs): Addressing Unfinished Agenda and Strengthening Sustainable Development and Partnership. *Indian Journal of Community Medicine: Official Publication of Indian Association of Preventive & Social Medicine*, 41, 1–4.

Nichols, A. (2011). Social Enterprise and Social Entrepreneurship. *In:* Edwards, M. (ed.) *The Oxford Handbook of Civil Society*. Oxford: Oxford University Press.

OECD (2019). *Tackling Vulnerability in the Informal Economy*. Paris: Development Centre Studies, OECD Publishing.

Ostry, J. D., Berg, A., and Tsangarides, C. G. (2014). *Redistribution, Inequality and Growth, Note No:SDN/14/02*. Washington DC: International Monetary Fund.

Piketty, T. (2014). *Capital in the Twenty-First Century*. Cambridge, Belknap: Press of Harvard University Press.

Santos, F. M. (2012). A Positive Theory of Social Entrepreneurship. *Journal of Business Ethics*, 111, 335–351.

Shorrocks, A., Davies, J., and Lluberas, R. (2019). *Global Wealth Report 2019*. Zurich: Credit Suisse Research Institute.

UN (2015). *Transforming Our World: The 2030 Agenda for Sustainable Development*. New York: United Nations.

UN (2020). *World Economic Situation and Prospects*. New York: Department of Economic and Social Affairs, United Nations.

Van Gent, S. (2017). *Beyond Buzzwords: What is 'Inclusive' Development? Synthesis Report*. Leiden: Include Secretariat.

Weerawardena, J. and Mort, G. S. (2006). Investigating social entrepreneurship: A multidimensional model. *Journal of World Business*, 41(1), 21–35.

Zahra, S. A., Rawhouser, H. N., Bhawe, N., Neubaum, D. O., and Hayton, J. C. (2008). Globalization of social entrepreneurship opportunities. *Strategic Entrepreneurship Journal*, 2, 117–131.

PART 2

Sustainability and the environment

12

INTRODUCTION

Sustainability and development

Albert Salamanca and Pichamon Yeophantong

The relationship between environmental sustainability and development continues to be contested. There are those who think that the primary feature of development – industrial capitalism – 'can never be sustainable' (Jensen and McBay 2009 quoted in Foster et al. 2011, 8). But there are also others who believe that economic development can be rendered sustainable, leading to the well-known idea of sustainable development. This idea has been the guiding principle for development and environmental governance since the early 1990s, even though its roots can be traced back to the 1972 Stockholm Declaration.

Globally, the drive towards sustainable development was galvanized with Agenda 21, which advanced a list of goals critical to achieving development that effectively 'meets the needs of the present without compromising the ability of future generations to meet their own needs.' These were reshaped and incorporated into the United Nation's (UN) Millennium Development Goals (MDGs), formulated in 2000, which were then reconstituted and adopted as part of Agenda 2030 and the UN Sustainable Development Goals (SDGs) in 2015. Over time, the prominence of environmental issues has notably become more saturated within the global development agenda: compared to the MDGs, the SDGs – as the name already suggests – has environmental sustainability at its core and, with Goal 13, readily and explicitly acknowledges the need for urgent action to tackle climate change. Similarly, discussions remain ongoing about how existing tools and measures such as the Human Development Index can better incorporate environmental considerations and a 'univeralist' perspective on development, where human rights are assured and capabilities enhanced on the basis of 'justice, not charity' (Anand and Sen 1994, 6). More recently, in view of the growing threats and persecution faced by environmental human rights defenders (EHRDs), the global agenda has come to spotlight the need to secure peoples' right to a safe, clean, healthy and sustainable environment, which is afforded constitutional recognition and protection in over 100 states (Knox 2018).

That said, these permutations in global development can be taken collectively as either a positive sign of the global community 'getting serious' about environmental sustainability, or evidence of the inherent tensions between development on the one hand, and the natural environment on the other. What *is* clear, however, is that the last three decades still have not led to a state of greater sustainability. Why this is the case is largely a function of how we do 'development,' which is the focus of many – if not all – chapters in this Handbook.

DOI: 10.4324/9781003017653-15

With chapters penned by scholars and practitioners hailing from different disciplinary and professional backgrounds, this section unpacks the exigent issues, big debates and influential concepts in sustainability and development, and reflects on how these could translate in practice into potential pathways for sustainability. Even so, none of the chapters here can purport to offer clear-cut solutions to the current dilemmas involved in balancing the needs of development and human progress against those of environmental sustainability. Instead, they offer policy-relevant ideas and pedagogical ruminations on how to critically think, explore and understand – both in the real world and the classroom – the numerous sustainability challenges facing our global ecosystem.

The section opens with the notion of 'planetary boundaries,' where Ilan Kelman (Chapter 13) looks at how it has originated and evolved. Kelman observes how notions such as equity, justice, and fairness are excluded in the way in which planetary boundaries are drawn. Kelman suggests that the current conceptualization of planetary boundaries tends to restrict actions to technical interventions such as geoengineering. How the discourse on planetary boundaries emerged is a powerful example of how scientific paradigms come to gain traction.

Antonio G.M. La Viña and Jameela Joy M. Reyes (Chapter 14) focus on another term that has quickly gained resonance across the academic, policy and public realms: that is, the Anthropocene. They, however, question whether the term is adequate to the task, arguing that 'Capitalocene' is a more appropriate concept for capturing the role that capital accumulation plays in the 'ecological rift' between humans and nature (Foster, Clark and York 2011). These critical perspectives are further explored through Andrew McGregor and Ashraful Alam's foray into more-than-human approaches (Chapter 15). They tackle more-than-human subjects that look at the 'the spatial and temporal relations that constitute worlds.' The observations advanced in this chapter challenge human exceptionalism in the way we interrogate, act, and make sense of the world. In more-than-human approaches, the human is decentred towards the primacy of the multispecies, such that, as McGregor and Alam argue, 'more-than-human development is thick development.'

Susie Jolly (Chapter 16) traces the history of thinking on the linkages between gender and environment, arguing that gendered differences in our perceptions and experiences of the environment reflect the intersectional nature of identities. Thus, there are no such things as 'gender' and 'environment': all are social constructs embedded within power relations. Misogynist attitudes, Jolly suggests, can be seen reproduced in the manner that climate change deniers reflect the attitude of white, masculine industrialists who build their riches on fossil fuel extraction. This leads Jolly to propose a radical vision that eschews the market economy in favour of a caring economy.

Echoing Jolly's observations about the need for 'radical' changes to, and perspectives on, development, Henry Veltmeyer (Chapter 17) examines the concept of extractivism and its practical evolution and, in so doing, unpacks the key traits of the world capitalist system and its discontents. Like Jolly, he considers how a 'new world' based on more sustainable forms of inclusive and post-liberal development might be possible, putting forward the Indigenous notion of *buen vivir* as one potential pathway to social and environmental justice. A crucial factor in the 'politics of anti-extractivism,' according to Veltmeyer, is the resistance that emerges on the extractive frontier, which helps to prompt the search for alternatives.

Speaking to these extractivist dynamics, Feifei Cai and Pichamon Yeophantong (Chapter 18) focus their discussion on how resource conflicts impact development. Drawing on the case of Cambodia, they illustrate how ecological distribution conflicts, which can stem from environmental degradation, social injustice or, more specifically, the unfair distribution of environmental costs and benefits, constitute everyday realities for some communities. But while these conflicts can clearly constitute an impediment to sustainable development, Cai and Yeophantong also note how they have the potential to drive positive social and political change – if not transformation – through civic resistance and environmental peacebuilding.

Aside from causing social injustice and 'development in reverse,' the unbridled extraction of environmental resources can – and indeed, already has – resulted in species loss and mass extinction, which is the topic explored by Hilary Whitehouse (Chapter 19). She discusses the plummeting levels of biodiversity across the globe and the effect this is having on the functional wellbeing of humans and ecosystems. The chapter reviews the drivers of the ongoing extinction crisis before canvassing some of the conservation and policy approaches intended to restore long-term, ecological integrity. Whitehouse concludes by articulating the need for greater awareness of biodiversity loss as well as the importance of enhancing human relationships with the natural world.

The final set of chapters in this section turn from the big debates and development dilemmas to addressing more specific sustainability issues. Lorraine Elliott's chapter on transnational environmental crime (TEC) (Chapter 20) applies a development lens to understand the context and consequences of transnational environmental crime. In particular, she explores the many development challenges associated with *responding* to TEC and explains how these challenges often arise due to a combination of weak governance, ineffective conservation and environmental protection frameworks, and limited enforcement resources. According to Elliott, these challenges reflect the complexity of illicit transboundary transactions themselves, ranging as they do across extraction, production, transportation, and disposal practices, as well as associated illegal practices such as money-laundering, fraud, and corruption.

Mia Chung and Pyrou Chung (Chapter 21) analyse the challenges to realizing Indigenous environmental rights and how new technologies might be leveraged to assist Indigenous Peoples and ethnic minorities (IEM) with safeguarding their rights. Crucially, their chapter underscores the need for the cautious and context-sensitive application of technology-driven approaches. This is to ensure that these new technologies reflect the realities faced by IEM and local communities, and not the 'top-down' agendas of donor organizations.

'Securing equitable and sustainable food systems,' writes Sango Mahanty (Chapter 22), 'is a critical global challenge.' Her chapter provides an overview of the major challenges involved in food production and distribution, drawing on the experiences of small Cambodian and Vietnamese farmers. Although the drivers of food insecurity are complex and interconnected, Mahanty spotlights how existing agricultural interventions have also tended to sidestep important structural factors and neglect environmental concerns. Achieving sustainable food systems thus calls for a broader approach that addresses the intersections between agricultural markets and environmental change, and the diverse livelihoods of small-scale farmers.

Like food systems, energy security is critical to human development. Eko Priyo Purnomo, Aqil Teguh Fathani and Abitassha Az Zahra (Chapter 23) investigate the concept of renewable energy and its practical relevance to modern societies by examining over 100 peer-reviewed journal articles. Alongside a discussion of the negative externalities and decreasing supply of carbon-intensive energy sources, the authors set out the major barriers to the uptake and deployment of clean energy systems before proceeding to reflect on the implications for policy, practice, and pedagogy.

The section then turns our attention to yet another unfolding crisis. Danny Marks (Chapter 24) argues that the worsening plastic waste crisis amounts to a failure of transboundary environmental governance. The situation is especially severe in Asia, which is where '75 per cent of globally exported waste ends up.' Following this, Marks reveals how inequality and injustice are deeply embedded in this issue, specifically in relation to the unfair distribution of harm between those who contribute to the problem and those who suffer from it.

In view of these complex challenges, it is fitting for the section to conclude with a critical review of sustainable development discourses. Thomas McNamara (Chapter 25) discusses how sustainability is an excellent concept for considering the disjuncture between ideals of empowerment and local ownership, as well as between the practices and reporting structures that empower voices from the Global North. Using Malawi as a case study, the chapter levels critiques of sustainability to open up a space for students to engage with texts that are central to development apparatuses and discourses.

Taken together, these chapters speak to the importance of environmental considerations to sustainable human development, but also to how environmental challenges will often elude straightforward solutions and instead require contextually contingent remedies that pay heed to the local context. For students interested in exploring these issues further, our section's key message is a simple one: in the post-COVID world, the right to a safe, clean, healthy, and sustainable environment is one that all countries must readily accept – and one which we all must also strive to protect.

References

Anand, S. and Sen, A (1994). Sustainable Human Development: Concepts and Priorities. *Paper written in preparation for the 1994 Human Development Report.* United Nations Development Programme.

Foster, J.B., Clark, B. and York, R. (2011). *The Ecological Rift: Capitalism's War on the Earth.* New York: Monthly Review Press.

Jensen, D. and McBay, A. (2009). *What We Leave Behind.* New York: Seven Stories Press.

Knox, J.H. (2018). Report of the Special Rapporteur on the issue of human rights obligations relating to the enjoyment of a safe, clean, healthy and sustainable environment. *Seventy-third session of the UN General Assembly* (A/73/188).

13

PLANETARY BOUNDARIES

Ilan Kelman

What are planetary boundaries?

Rockström et al. (2009b; see also Rockström et al. 2009a) proposed planetary boundaries as a framework for indicating how humanity can live on Earth safely. The planetary boundaries framework is meant to support global sustainability and development endeavours by providing assumed thresholds beyond which large-scale, comparatively sudden environmental changes will produce major problems. The framework is defined in terms of each environmental process, its defining characteristic (i.e., control variable), the threshold, the uncertainties in play, and the state of knowledge. The first two in this list are in Table 13.1.

The ideas behind the framework have long been debated. For instance, fundaments of thresholds for carrying capacity are long-standing (e.g. Trumble and Fraser 1932) and the ecological footprint of humanity is presumed to have a threshold of the total biologically productive area available, above which consumption is not sustainable (Wackernagel and Rees 1996). Meanwhile the phrase 'planetary boundaries' had already appeared in fields ranging from pedagogy (Janik et al. 2005) to planetary physics (Remo 2007).

For Rockström et al. (2009b), thresholds are biophysical, in effect claimed to be part of nature and thus calculable, although it is unclear how immutable they view the final calculation to be. In contrast, such boundaries are more typically seen as subjective decisions by humanity for humanity, depending on the definition of 'safe' and how to determine the meaning of 'living safely.' It is unclear how variable the boundaries would be as societal values shift; nor who should have the right or authority to set and alter these boundaries.

Steffen et al. (2015) provide updates of the formulation, state, and expression of the planetary boundaries (Table 13.2). They also explore planetary boundaries at a sub-planetary scale on the premise that regional influences and changes can affect global Earth System processes and hence the global planetary boundaries. Steffen et al. (2015, p. 1259855-8) further layer the planetary boundaries, terming 'climate change' and 'change in biosphere integrity' as 'core planetary boundaries through which the other boundaries operate.' They suggest that crossing the other planetary boundaries 'does not by itself lead to a new state of the Earth system' (p. 1259855-8). This potentially contradicts Rockström et al. (2009b, 2) who view a planetary boundary as leading to a new state when crossed, since it produces 'unacceptable global environmental

DOI: 10.4324/9781003017653-16

Table 13.1 Proposed planetary boundaries (Rockström et al. 2009b)

Earth System process	Control variable
Climate change	Atmospheric CO_2 concentration, ppm
	Energy imbalance at Earth's surface, W m^{-2}
Ocean acidification	Carbonate ion concentration, average global surface ocean saturation state with respect to aragonite (Ω_{arag})
Stratospheric ozone depletion	Stratospheric O_3 concentration, DU
Atmospheric aerosol loading	Overall particulate concentration in the atmosphere, on a regional basis
Biogeochemical flows: interference with P and N cycles	P: inflow of phosphorus to ocean, increase compared with natural background weathering
	N: amount of N_2 removed from atmosphere for human use, Mt N yr^{-1}
Global freshwater use	Consumptive blue water use, km^3 yr^{-1}
Land-system change	Percentage of global land cover converted to cropland
Rate of biodiversity loss	Extinction rate, extinctions per million species per year (E/MSY)
Chemical pollution	For example, emissions, concentrations, or effects on ecosystem and Earth System functioning of persistent organic pollutants (POPs), plastics, endocrine disruptors, heavy metals, and nuclear wastes.

Table 13.2 Updated planetary boundaries (Steffen et al. 2015)

Earth System process from Rockström et al. (2009b)	Updated in Steffen et al. (2015)	Updated control variable in Steffen et al. (2015)
Atmospheric aerosol loading	(No change.)	Global: Aerosol Optical Depth (AOD), but much regional variation
		Regional: AOD as a seasonal average over a region.
Biogeochemical flows: interference with P and N cycles	Biogeochemical flows: (P and N cycles)	P Global: P flow from freshwater systems into the ocean
		P Regional: P flow from fertilizers to erodible soils
		N Global: Industrial and intentional biological fixation of N
Chemical pollution	Introduction of novel entities	None defined.
Climate change	(No change.)	Atmospheric CO_2 concentration, ppm
		Energy imbalance at top-of-atmosphere, W m^{-2}
Global freshwater use	Freshwater use	Global: Maximum amount of consumptive blue water use (km^3 yr^{-1})
		Basin: Blue water withdrawal as % of mean monthly river flow
Land-system change	(No change.)	Global: Area of forested land as % of original forest cover
		Biome: Area of forested land as % of potential forest
Rate of biodiversity loss	Change in biosphere integrity	None fully accepted.

Note: 'Ocean acidification' and 'Stratospheric ozone depletion' remain unchanged from previous framework in Table 1.

change … in relation to the risks humanity faces in the transition of the planet from the Holocene to the Anthropocene.'

Nonetheless, even crossing planetary boundaries seems to not necessarily yield the catastrophic change apparently defining the boundaries, because Rockström et al. (2009b, 21) suggest that humanity might be fine '[o]n condition that these are not transgressed for too long.' No specific timeframe is given for each boundary. The aim of staying within planetary boundaries is explained in terms of resilience (within their narrow definition of the term) whereupon stability is emphasized, even while seeming to imply that Earth System stability seen in recent millennia is the exception rather than the rule over longer time scales.

Critics of planetary boundaries

Critics of planetary boundaries tend to highlight three linked issues: concerns about the framework's technicalities, omissions, and poor engagement with previous and ongoing literature. The examples given in each sub-section are illustrative rather than comprehensive.

Technicalities

One overarching criticism of the planetary boundaries framework, as alluded to at the end of the last section, is the assumption that Earth System stability is an appropriate baseline to seek. Lewis (2012) describes how the focus on Holocene conditions distracts from process problems which did not exist until recently, such as plastic waste. This argument applies to chemical pollution which was included as a planetary boundary in Rockström et al. (2009b) and then changed to 'introduction of novel entities' in Steffen et al. (2015). Yet no control variable or boundary has been set, although 'stratospheric ozone depletion' has been defined by a parameter in Rockström et al. (2009b).

Fernández and Malwé (2019) lay out key technical critiques of planetary boundaries, notably outlining the drawbacks of thresholds for characterizing the environment and questioning the specific values or parameters selected. As an example of both problems, de Vries et al. (2013) deconstruct Rockström et al.'s (2009b) planetary boundary of the nitrogen cycle within biogeochemical flows, pointing out flaws and offering a revision. Their advice was taken for revising this boundary in Steffen et al. (2015). Similarly for the planetary boundary on 'rate of biodiversity loss,' Mace et al. (2014) and Samper (2009) have explained why extinction rate and species richness (Table 13.1) are inadequate indicators, with Steffen et al. (2015) then taking some of this advice in trying to revise the boundary (Table 13.2).

Another critique of the biodiversity loss planetary boundary and its revision as 'biosphere integrity' (Montoya et al. 2017a) raise numerous points of confusion regarding definitions of 'planetary boundary' and assumptions about species and ecosystems. Rockström et al. (2017) responded by suggesting errors in the initial critique and repeating material from the two core planetary boundaries publications. As Montoya et al.'s (2017b) reply indicates, two fundamental questions remain unanswered. First, what are the precise differences between 'tipping points, planetary boundaries, safe operating space, resilience, and irreversible changes' (Montoya et al. 2017b, 234) and what are the advantages of all these terms? This concern is more poignant given the various critiques of 'resilience' (e.g., Alexander 2013; Pugh 2014) which have been sidestepped in planetary boundaries proposals. Second, why does the biosphere integrity planetary boundary remain undefined, when Rockström et al. (2017) have had eight years from the initial publication?

Similar analyses and debates can be provided for each planetary boundary, including their revisions. Whether or not this task is worthwhile is an open question. Any definition and any parameter can be critiqued, with scientific consensus not necessarily being viable or desired. For example, for the 'climate change' planetary boundary, 'atmospheric CO_2 concentration, ppm' is not supported by everyone. Knutti et al. (2016) argue for global mean surface temperature, whereas some policy documents (e.g. UNFCCC 2015) promote a temperature-related goal despite calculation difficulties (Richardson et al. 2018). The choice of an environmental parameter, rather than a human one, such as total emissions or emissions reductions, further relates to criticisms of omissions in the literature.

Omissions

Downing et al. (2019) collate major concerns of omission within the planetary boundaries framework, all of which point out how much of a human perspective is missing from the framework. Many authors see planetary boundaries as being developed from physical science, for physical science. While accepting the subjective and political nature of determining the numerical values used as boundaries, the critics identify major gaps in how the planetary boundaries framework accounts for people, for human needs, and for principles of humanity, such as heterogeneity, contextualization, and intangibles, with each one summarized in the next paragraph.

Humanity's heterogeneity is missing from planetary boundaries which presumes generic people with generic and likely uniform needs (Downing et al. 2019). The critics lament what they see as an implied tension between social needs and the environmental boundaries, whereas contextualization of human needs might indicate how they could be attained without environmental harm. In addition to these tangible human elements, intangible needs and interests are said by some critics to be absent from planetary boundaries (Downing et al. 2019). These intangibles are concepts such as fairness, equity, and justice, embracing the long-established applied philosophy investigations of resource allocation, who decides and who pays, and generational disparities (e.g., Meadows et al. 1972; Parfit 1984). All these topics have an extensive literature which could have been used to inform planetary boundaries when the framework was being developed.

Ultimately, the critics highlighting omissions indicate that the planetary boundaries framework calculates that we are supposedly still operating within several of the planetary boundaries, yet we know that we are not operating safely in terms of foundational principles and, hence, basic humanity. When the conclusions from a systematic framework do not match what is observed empirically, the starting point for developing the framework might be questioned as inadequate.

Engagement with other science

Steffen et al. (2015, 1259855–1) state that 'The precautionary principle suggests that human societies would be unwise to drive the Earth system substantially away from a Holocene-like condition.' Aside from not accounting for critiques of the precautionary principle (e.g. Cross 1996; Holm and Harris 1999; Hughes 2006) nor justifying it as a starting point – also an issue in Rockström et al. (2009b) – this stance does not suggest what humanity should do when natural forces 'drive the Earth system substantially away from a Holocene-like condition.' The Earth's orbital cycles must eventually do so (Hodell 2019) and the planetary boundaries argument would seem to imply that we should geoengineer our way out of these millennial-scale natural changes.

Yet the precautionary principle has been invoked to indicate how transformations of the planet could be opposed in order to avoid further human-caused catastrophic change (Elliott 2010). Elliott (2010) also covers the diversity of understandings and implementations of the precautionary principle, again indicating aspects bypassed in the planetary boundaries framework.

Another body of literature which planetary boundaries might usefully have accounted for is 'creeping environmental problems' (CEPs), also termed 'creeping environmental changes' and 'creeping environmental phenomena' (Glantz 1994). CEPs refer to small, ongoing changes in environmental conditions which then add up to major problems, but these problems become apparent only after a threshold has been surpassed when it is typically too late. Examples of CEPs are human-caused climate change, ocean acidification, biodiversity loss, and pollution – in fact, all the Earth System processes in planetary boundaries are CEPs – and the notion has been applied in practice to specific sustainable development cases such as the Aral Sea (Glantz 1999).

CEPs are presented as concerns for society expressed through the environmental conditions. The literature on CEPs provides a powerful society-based impetus to consider more formalization and potentially systematic quantification (where appropriate), which is exactly the point of the planetary boundaries framework. Additionally, the multi-scalar nature and cross-scalar influences of CEPs provide a useful approach for connecting global and sub-global scales, which is one issue that Rockström et al. (2009b) had struggled with. Steffen et al. (2015) acknowledge this issue of scales and start to link different scales, yet still do not fully engage with prior work which could have advanced the resolution substantially. In placing the planetary boundaries framework within existing literature that supports its ethos, the framework could have avoided some of these pitfalls which are only now being corrected years after its first proposal.

Planetary boundaries for policy and practice

Irrespective of the critics of planetary boundaries, the idea has become entrenched within some scientific viewpoints, a situation which can influence policy and practice. Since science is meant to serve society, these outcomes are positive and should be encouraged. Even in common circumstances when science is not entirely robust or requires continuing investigations to overcome uncertainties and unknowns, the scientific process and results could and should improve policy and practice. Nonetheless, a lack of clarity is evident with respect to the framework's overall scientific grounding, and this raises questions for its use in policy and practice.

Downing et al. (2019) are impressed by the framework's high number of citations, but then explain that only 6 per cent of the citations, some of which are labelled as 'commentaries,' really connect significantly to planetary boundaries. This suggests that the vast majority of citations are unspecific rather than using or significantly linking with the framework, as confirmed by Downing et al. (2019, 3) explaining that most 'articles use the term 'planetary boundaries' as shorthand for issues of global unsustainability.' As well, 80 per cent of the citations they found were to the non-peer-reviewed Rockström et al. (2009a) rather than to the original Rockström et al. (2009b), suggesting that the publication venue generated more interest in the framework than the utility of the framework itself or its ability to withstand scrutiny.

But irrespective of scientific interest, the planetary boundaries framework has had some high-level penetration into policy and practice, with many examples compiled by Fernández and Malwé (2019). They focus on the United Nations (UN) listing some material from the UN Secretary General as well as for UN sustainable development conferences. They also describe opposition to the framework from national governments concerned that it has not been tested or critiqued enough, while potentially distracting from countries' views of core sustainability issues. Several European-level documents are mentioned as well, along with notes regarding reservations about planetary boundaries and differences between the original publications and the policy documents.

The creators of planetary boundaries actively seek this level of influence, with SEI (2017) explaining their views of the relevance to the 2030 sustainability agenda. They feature Sustainable Development Goal 12: 'Ensure sustainable consumption and production patterns' (UN 2015), as the importance of planetary boundaries is in quantification including scaling to sub-global levels such as national or regional delineations. Aside from the limits of quantification inherent in planetary boundaries, another difficulty emerges in Goal 12's indicators relying heavily on Gross Domestic Product (GDP).

GDP, in effect, is a metric for consumption, so increasing it is the antithesis of sustainable development and especially Goal 12, as somewhat indicated by Randers et al. (2019) who had tried to determine the impact of achieving the Sustainable Development Goals (SDGs) on planetary boundaries. Decreasing GDP to zero, though, is equally destructive, since then no purchases would be feasible, including clothes, food, water, electricity, or the computer and software with which I produced this chapter. It is thus remarkably fuzzy exactly what Goal 12 seeks regarding GDP. Yet planetary boundaries, as articulated by SEI (2017), adopts these vague indicators uncritically and without considering instead the many alternatives to GDP which have been long been explored in detail including deglobalization (Bello 2008), degrowth (Kallis 2018), doughnut economics (Raworth 2017), the circular economy (Lacy et al. 2019), and the steady state economy (Washington and Twomey 2016). The attempted marriage of planetary boundaries and Goal 12 is made more awkward by Target 8.1 (UN 2015) being 'at least 7 per cent gross domestic product growth per annum in the least developed countries' which cannot support sustainable development. Planetary boundaries had the opportunity to be a useful counterpoint to the policy difficulties posed by the SDGs, but this opportunity appears to have been missed by SEI (2017) and other observers.

Concerning efforts to implement the framework at the national level, Pisano and Berger (2013) also adopt an uncritical approach in trying to show the relevance of planetary boundaries for sustainable development in eight European countries. While Switzerland refers to the framework directly, Finland's statement that '[t]he carrying capacity of nature is not exceeded' is ascribed to planetary boundaries even though notions of carrying capacity long pre-date planetary boundaries (e.g. Trumble and Fraser 1932). Hazy statements from countries regarding the possibility for considering planetary boundaries at some point in the future are taken to have significant policy relevance. Pisano and Berger (2013) provide more substantive ongoing and hoped for implementation in Sweden, which makes sense given that the scientists leading the framework are based in Stockholm. They provide limited evidence or conclusions regarding the relevance of planetary boundaries for national policy.

More recent expressions of planetary boundaries in national or sub-national policies are scattered. One clear example is the Netherlands Environmental Assessment Agency (Lucas and Wilting 2018) which aims to apply the framework for developing sustainability policy targets in their country. The work was conducted under the auspices of the Planetary Boundaries Research Network which lacks recent activity and does not seem to have been impactful. Clift et al.'s (2017) critiques aptly explain the limited policy relevance and operationalization, notably vis-à-vis continuing issues with multiple spatial scales, equity, and international cooperation for action.

Interestingly, efforts at sub-planetary implementation are discouraged by Steffen et al. (2015, 1259855–8), who write that '[t]he PB framework is not designed to be "downscaled" or "disaggregated" to smaller levels, such as nations or local communities.' Consequently, the policy and practice relevance should apparently be determined only at the global level for which relevant, effective, legally binding regimes are limited. Progress in the SDGs, for instance, is mainly self-reported and voluntary (UN 2015, Clause 74).

Teaching and thinking about planetary boundaries

The problems described above, combined with the populism of planetary boundaries within some sectors, provide an excellent example for teaching how to think about and critically analyse scientific paradigms and attempts to create scientific paradigms. Why might scientists seek traction for, and policy and practice application of, a framework without first resolving its flaws? Or, by establishing its firm place within research, policy, and action, will the framework's difficulties be overcome? The conclusion might even be that, irrespective of the challenges, adopting and applying planetary boundaries leads to more good than harm.

The first step in the classroom is to explain planetary boundaries in terms of the phrase's meaning, origins, biases, problems, and possible applications. With this knowledge foundation, students could be asked to examine and explain the positive, negative, and neutral aspects. It could be through a critiquing prose analysis or a Strengths-Weaknesses-Opportunities-Threats/Constraints (SWOT/SWOC) table for organizing thoughts. At this point, broad agreement should be summarized regarding what the planetary boundaries framework means, what its limitations are, what they offer, and what they do not offer.

Students could then answer decision-related questions:

- Do planetary boundaries survive in-depth scrutiny?
- Irrespective of the problems, does the framework suffice to be accepted as robust science?
- Irrespective of the scientific worthiness, does the framework contribute to policy and action?

This stage could be framed as being about the class making an active choice to accept or reject the planetary boundaries framework. It could be explored on a case-by-case basis, accepting, rejecting, or seeking improvement for each individual boundary, characteristic, or threshold.

This decision-making exercise could be as a discussion, potentially seeking consensus or a majority opinion, or perhaps though a parliamentary debate with resolutions along the lines of 'This house adopts "Planetary Boundaries" for all sustainability legislation' or 'This house bans "Planetary Boundaries" from all government projects.' Fostering the extremes might be overzealous, setting up a false all-or-nothing choice, even if helping to bring out the core arguments and disagreements. An alternative would be describing what is agreed and what is not agreed. A more scientific or legalistic debating style called 'Cross-Examination' emulates a dissertation viva or exam. One person expounds their argument related to planetary boundaries, which could be for, against, or in between. Another person then cross-examines them, asking questions and letting them answer. The outcome could be an attempt to reach an agreement or an audience vote regarding who was most convincing.

Beyond agreeing to disagree or in-class referenda, the final stage of teaching planetary boundaries should be about interrogating the process of uptake of scientific thought and frameworks. The ideals behind basic principles embodied in the framework could be teased out with further examination of why those ideals and principles led to results achieving limited populism. Even where an idea or framework has scientific merit alongside general agreement about its relevance for policy and practice, there can be a lack of questioning, critique, and desire to improve. Where does this lacuna of challenging an apparent 'expert' or 'authority' come from? Could it be fear, ignorance, apathy, exhaustion, dictatorship, or other reasons which impede potential critics from raising their voices about scientific concerns and which stop those criticized from fully responding?

With some explanations, solutions are required. What do we need to better accept critique and to engage with it through sensible discussions? How could we convince ourselves and others to always undertake improvements? In a hyper-partisan world, where some with power and resources frequently lie, make up information, reject established science, and manufacture self-serving discourses explicitly lacking balance, how could we do better in bringing people together, not in spite of disagreement but because of disagreement.

Lessons for the classroom and for ourselves have long been accepted. Herman and Chomsky (1988; updated as Herman and Chomsky 2002) detail 'Manufacturing Consent' about how those with control of information select the agenda so that those receiving information are guided towards a pre-determined consensus. McLuhan (1964) explained that 'The medium is the message': how content is conveyed can be more important than the content itself. How do these analyses of communication apply to scientific constructs and their application? What could and should be done to break through the barriers erected by promoting and supporting a single framework? Are planetary boundaries an exception or a norm regarding how science operates?

As human beings, we are inherently biased. Scientists are no different and the process of science has fully acceptable biases which are often not admitted (Martin 1979). The planetary boundaries framework provides an excellent example for teaching students about the presence and nature of these biases, not only to make them explicit for critique but also to map out how to do better. Part of this work must include critiquing the critiques.

Key sources

1.   Bartlett, A.A. (2004). *The essential exponential!* Lincoln: University of Nebraska.

This book provides a quantitative approach to explaining the meaning of sustainability through challenging the growth paradigm. Going well beyond Malthusian constructs, this book examines the practical implications of ever-increasing human population numbers and ever-increasing consumption by people, compared with the actual resources available. Energy is a prominent example, especially fossil fuels. To provide the information, this book collects several papers spanning decades, giving an overview of how discussions evolved and the debates worthy of further engagement.

2.   Fuller, R.B. (1969). *Operating manual for spaceship earth.* Carbondale: Southern Illinois University Press.

This book analyses planetary resource limits and what could be done to avoid exceeding these limits. Topics cover poverty, technology, and pollution. Systems thinking is used to describe views of how to improve stewardship of the planet within the context of some historical reflections. The importance of resource limits is highlighted along with general advice and expectations regarding how to 'operate' our planetary 'spaceship.'

3. Glantz, MH 2003, *Climate affairs: a primer*. Covelo: Island Press.

This book provides baseline knowledge for understanding environmental and social impacts of climate change. It offers a more comprehensive and flexible approach than planetary boundaries for thinking from a systems perspective for action. The book brings together climate-relevant physical science, impacts, and effects including for economics, politics, law, policy, and ethics. Interactions among climate, society, and the environment are highlighted, demonstrating how complexities can be addressed in reality, especially in realms where misinformation is abundant.

4. Schumacher, E.F. (1973). *Small is beautiful: a study of economics as if people mattered.* London: Blond & Briggs.

This book explores scales of sustainability in order to highlight the advantages of smaller scales for living. As a counterpoint to mass production and mass consumerism, it explores people and resources, considering the problems caused by continually scaling up, expanding, and using more resources. Alternatives are offered, noting especially the finiteness of resources meaning that the focus ought to be on people's needs and interests.

5. WCED (World Commission on Environment and Development) (1987). *Our common future.* Oxford: Oxford University Press.

This report is core for following discussions about assessing and acting on humanity's global impacts. It provided a foundational and often-critiqued definition of 'sustainable development.' With 'common' as a motif throughout the report – covering common concerns, common challenges, and common endeavours – the report sought to fulfil the mandate of 'a global agenda for change.' It summarizes key aspects of the world's sustainability problems at the time, followed by specific action points for moving forward.

References

Alexander, D.E. (2013). Resilience and disaster risk reduction: an etymological journey. *Natural Hazards and Earth System Sciences*. vol. 13, pp. 2707–2716.

Bello, W. (2008). *Deglobalization: ideas for a new world economy*. London: Zed Books.

Clift, R., Sim, S., King, H., Chenoweth, J.L., Christie, I., Clavreul, J., Mueller, C., Posthuma, L., Boulay, A.M., Chaplin-Kramer, R., Chatterton, J., DeClerck, F., Druckman, A., France, C., Franco, A., Gerten, D., Goedkoop, M., Hauschild, M.Z., Huijbregts, M.A.J., Koellner, T., Lambin, E.F., Lee, J., Mair, S., Marshall, S., McLachlan, M.S., Milà i Canals, L., Mitchell, C., Price, E., Rockström, J., Suckling, J., and Murphy, R. (2017). The challenges of applying planetary boundaries as a basis for strategic decision-making in companies with global supply chains. *Sustainability* 9: article 279.

Cross, F.B. (1996). Paradoxical perils of the precautionary principle. *Washington and Lee Law Review*, vol. 53, pp. 851–928.

de Vries, W., Kros, J., Kroeze, C., and Seitzinger, S.P. (2013). Assessing planetary and regional nitrogen boundaries related to food security and adverse environmental impacts. *Current Opinion in Environmental Sustainability*, vol. 5, pp. 392–402.

Downing, A.S., Bhowmik, A., Collste, D., Cornell, S.E., Donges, J., Fetzer, I., Häyhä, T., Hinton, J., Lade, S., and Mooij, W.M. (2019). Matching scope, purpose and uses of planetary boundaries science. *Environmental Research Letters*, vol. 14: article 073005.

Elliott, K. (2010). Geoengineering and the precautionary principle. *International Journal of Applied Philosophy*, vol. 24, pp. 237–253.

Fernández, E.F. and Malwé, C. (2019). The emergence of the 'planetary boundaries' concept in international environmental law: a proposal for a framework convention. *Review of European, Comparative & International Environmental Law (RECIEL)*, vol. 28, pp. 48–56.

Glantz, M.H. (1994). Creeping environmental problems. *The World & I,* June, pp. 218–225.

Glantz, M.H. (1999). *Creeping environmental problems and sustainable development in the Aral Sea basin.* Cambridge: Cambridge University Press.

Herman, E.S. and Chomsky, N. (1988). *Manufacturing consent: the political economy of the mass media.* New York: Pantheon Books.

Herman, E.S. and Chomsky, N. (2002). *Manufacturing consent: the political economy of the mass media.* 2nd edn. New York: Pantheon Books.

Hodell, D.A. (2019). The smoking gun of the ice ages. *Science*, vol. 354, pp. 1235–1236.

Holm, S. and Harris, J. (1999). Precautionary principle stifles discovery. *Nature,* vol. 400, p. 398.

Hughes, J. (2006). How not to criticize the precautionary principle. *The Journal of Medicine and Philosophy*, vol. 31, no. 5, pp. 447–464.

Janik, D., Bills, M., Saito, H., and Widjaja, C. (2005). Neurobiological and transformational learning. *The IPSI BgD Transactions on Advanced Research*, vol. 1, pp. 19–22.

Kallis, G. (2018). *Degrowth*. Newcastle upon Tyne: Agenda.

Knutti, R., Rogelj, J., Sedláček, J., and Fischer, E.M. (2016). A scientific critique of the two-degree climate change target. *Nature Geoscience*, vol. 9, pp. 13–18.

Lacy, P., Long, J., and Spindler, W. (2019). *The circular economy handbook*. London: Palgrave Macmillan.

Lewis, S. (2012). We must set planetary boundaries wisely. *Nature*, vol. 485, p. 417.

Lucas, P. and Wilting, H. (2018). *Using planetary boundaries to support national implementation of environment-related sustainable development goals*. The Hague: PBL Netherlands Environmental Assessment Agency.

Mace, G.M., Reyers, B., Alkemade, R., Biggs, R., Chapin III, F.S., Cornell, S.E., Díaz, S., Jennings, S., Leadley, P., Mumby, P.J., Purvis, A., Scholes, R.J., Seddon, A.W.R., Solan, M., Steffen, W., and Woodward, G. (2014). Approaches to defining a planetary boundary for biodiversity. *Global Environmental Change*, vol. 28, pp. 289–297.

Martin, B. (1979). *The bias of science*. Canberra: Society for Social Responsibility in Science.

McLuhan, M. (1964). *Understanding media: the extensions of man*. New York: McGraw-Hill.

Meadows, D.H., Meadows, D.L., Randers, J., and Behrens III, W.W. (1972). *The limits to growth*. New York: Universe Books.

Montoya, J.M., Donohue, I., and Pimm, S.L. (2017a). Planetary boundaries for biodiversity: implausible science, pernicious policies. *Trends in Ecology & Evolution,* 33: 71–73.

Montoya, J.M., Donohue, I., and Pimm, S.L. (2017b). Why a planetary boundary, if it is not planetary, and the boundary is undefined? A reply to Rockström *et al. Trends in Ecology & Evolution,* vol. 33, p. 234.

Parfit, D. (1984). *Reasons and persons*. Oxford: Oxford University Press.

Pisano, U. and Berger, G. (2013). *Planetary boundaries for SD: from an international perspective to national applications*. Vienna: European Sustainable Development Network.

Pugh, J. (2014). Resilience, complexity and post-liberalism. *Area*, vol. 46, pp. 313–319.

Randers, J., Rockström, J., Stoknes, P.E., Goluke, U., Collste, D., Cornell, S.E., and Donges, J. (2019). Achieving the 17 Sustainable Development Goals within 9 planetary boundaries. *Global Sustainability*, vol. 2, article e24.

Raworth, K. (2017). *Doughnut economics*. White River Junction, Vermont: Chelsea Green Publishing.

Remo, J.L. (2007). Classifying solid planetary bodies. *AIP Conference Proceedings,* vol. 886, paper 284.

Richardson, M., Cowtan, K., and Millar, R.J. (2018). Global temperature definition affects achievement of long-term climate goals. *Environmental Research Letters,* vol. 13, article 054004.

Rockström, J., Steffen, W., Noone, K., Persson, Å., Chapin III, F.S., Lambin, E.F., Lenton, T.M., Scheffer, M., Folke, C., Schellnhuber, H.J., Nykvist, B., de Wit, C.A., Hughes, T., van der Leeuw, S., Rodhe, H., Sörlin, S., Snyder, P.K., Costanza, R., Svedin, U., Falkenmark, M., Karlberg, L., Corell, R.W., Fabry, V.J., Hansen, J., Walker, B., Liverman, D., Richardson, K., Crutzen, P., and Foley, J.A. (2009a). A safe operating space for humanity. *Nature*, vol. 461, pp. 472–475.

Rockström, J., Steffen, W., Noone, K., Persson, Å., Chapin III, F.S., Lambin, E.F., Lenton, T.M., Scheffer, M., Folke, C., Schellnhuber, H.J., Nykvist, B., de Wit, C.A., Hughes, T., van der Leeuw, S., Rodhe, H., Sörlin, S., Snyder, P.K., Costanza, R., Svedin, U., Falkenmark, M., Karlberg, L., Corell, R.W., Fabry, V.J., Hansen, J., Walker, B., Liverman, D., Richardson, K., Crutzen, P., and Foley, J.A. (2009b). Planetary boundaries: exploring the safe operating space for humanity. *Ecology and Society*, vol. 14, article 32.

Rockström, J., Richardson, K., Steffen, W., and Mace, G. (2017). Planetary boundaries: separating fact from fiction. A response to Montoya et al. *Trends in Ecology & Evolution,* vol. 33, pp. 232–233.

Samper, C. (2009). Rethinking biodiversity. *Nature Climate Change,* vol. 3, pp. 118–119.

SEI (2017). *How the Planetary Boundaries framework can support national implementation of the 2030 Agenda.* Stockholm: Stockholm Environment Institute.

Steffen, W., Richardson, K., Rockström, J., Cornell, S.E., Fetzer, I., Bennett, E.M., Biggs, R., Carpenter, S.R., de Vries, W., de Wit, C.A., Folke, C., Gerten, D., Heinke, J., Mace, G.M., Persson, L.M., Ramanathan, V., Reyers, B., and Sörlin, S. (2015). Planetary boundaries: guiding human development on a changing planet. *Science,* vol. 347, article 1259855.

Trumble, H.C. and Fraser, K.M. (1932). The effect of top-dressing with artificial fertilizers on the annual yield, botanical composition, and carrying capacity of a natural pasture over a period of seven years. *Journal of the Department of Agriculture of South Australia,* vol. 35, pp. 1341–1353.

UN (2015). *Resolution adopted by the General Assembly on 25 September 2015: 70/1. Transforming our world: the 2030 agenda for sustainable development.* New York: United Nations General Assembly.

UNFCCC (2015). *Conference of the parties, Twenty-first session, Paris, 30 November to 11 December 2015, FCCC/CP/2015/L.9/Rev.1.,* Bonn: UNFCCC (United Nations Framework Convention on Climate Change).

Wackernagel, M. and Rees, W.E. (1996). *Our ecological footprint.* Gabriola Island, BC: New Society Publishers.

Washington, H. and Twomey, P. (2016). *A future beyond growth: towards a steady state economy.* Abingdon: Routledge.

14

ANTHROPOCENE, CAPITALOCENE, AND CLIMATE CHANGE

Antonio G. M. La Viña and Jameela Joy M. Reyes

Introduction

Between late October and the first two weeks of November 2020, five typhoons made landfall in the Philippines, three of which heavily ravaged the country, leaving hundreds of thousands displaced. Typhoon Goni, known in the Philippines as Super Typhoon Rolly, made landfall in the southern region of Luzon as a Category 5 super typhoon, and became one of the strongest tropical cyclones on record, its strength comparable to Typhoon Haiyan in 2013 and Meranti in 2016, which also devastated the Philippines and other countries of Southeast Asia.

It is now widely accepted that these typhoons, with their increasing intensity and frequency, and their devastating consequences – inundation, displacement, and loss of lives and livelihoods – are a result of an ever-warming world. There is, moreover, no doubt that while climate change will be global in its effects, it will be felt differently in some areas and by some people more than others. Already, small island developing states (SIDS) and other countries in the Pacific are feeling the brunt of more intense and frequent typhoons. Many of these countries which have contributed little to anthropogenic climate change in terms of emissions, are likely to bear the brunt of its effects. In 2019, the Philippines emitted 1.39 tonnes of CO^2 emissions per capita as opposed to China's 8.12 tonnes of CO^2 emissions per capita and the USA's 15.52.[1]

In this entry in the Handbook, we explore climate change in the context of the Global North and South, using the lens of the Anthropocene, or the geological epoch that stems from the belief that humans have so significantly impacted the Earth's ecosystems, not least due to anthropogenic climate change, that naming a new period was necessary. We do this on three levels: first, by looking into the definition of Anthropocene, and the debates on whether or not we have actually entered the new geological Age of the Anthropocene; second, by exploring the responsibilities and duties of the Global North and Global South attendant by this new geological epoch and reflected in the loss and damage provisions of the Paris Climate Change Agreement; and third, by considering the policies that underpin the Paris Agreement and the most recent Conference of Parties (COP). The chapter will end by noting the gaps and disconnections in current practices and raise questions for consideration by students and practitioners.

DOI: 10.4324/9781003017653-17

Situated in time: the Anthropocene

The concept of the Anthropocene, while first mooted as early as the 1960s, did not come into full prominence until 2000, when the Nobel Prize-winning atmospheric chemist Paul Crutzen and limnologist Eugene Stoermer used the term in the newsletter of the International Geosphere-Biosphere Programme (IGBP). Two years later, Crutzen (2002) published his article the 'Geology of Mankind' in the journal *Nature*, writing:

> *Because of these anthropogenic emissions of carbon dioxide, global climate may depart significantly from natural behaviour for many millennia to come. It seems appropriate to assign the term 'Anthropocene' to the present, in many ways* **human-dominated, geological epoch,** *supplementing the Holocene – the warm period of the past 10–12 millennia. The Anthropocene could be said to have started in the latter part of the eighteenth century, when analyses of air trapped in polar ice showed the beginning of growing global concentrations of carbon dioxide and methane. This date also happens to coincide with James Watt's design of the steam engine in 1784.*
>
> (Crutzen 2002, 23) (Emphasis ours)

Crutzen proceeded to discuss the ways by which humanity has exploited the Earth's resources, including the rapid expansion of the human population over the last four centuries, the disappearance of tropical rainforests and the growing extinction of species, and the accelerating release of carbon dioxide. He clarified, too, that these effects have 'largely been caused by only 25% of the [world's] population.' He started his closing paragraph with this: 'Unless there is a global catastrophe – a meteorite impact, a world war or a pandemic – mankind will remain a major environmental force for many millennia' (Crutzen 2002, 23).

Of course, however, this does not mean that the Anthropocene was immediately accepted as a term to describe and denote the current global condition. Not only have there been debates on the agreed start date of the Anthropocene, but also on different interpretations of the concept and, even, on its utility.

Anthropocene arguments on the start date

While Crutzen situated the start date of the Anthropocene in the late 18th century, other scholars have proposed either a later or an earlier date. One of those who agreed with Crutzen was Jan Zalasiewicz, a British geologist. In 'Are we now living in the Anthropocene' (2008) published with members of the Stratigraphy Commission of the Geological Society of London, Zalasiewicz and his colleagues discussed the stratigraphic grounds on which new epochs can be designated. This is based on the selection of a Global Stratigraphic Section and Point (GSSP or the 'golden spike') and an agreement on the date for its inception. The paper's authors concurred with Crutzen, writing that a case can be made for the Anthropocene on the basis that the Earth has endured changes sufficient to leave a global stratigraphic signature distinct from the Holocene, and that this dates from the start of the Industrial Revolution.

In 'Defining the Anthropocene', Lewis and Maslin (2015) rejected the Industrial Revolution as marking the start of the Anthropocene, because it is 'not derived from a globally synchronous marker.' Instead, they recommended either the 1610 Orbis spike (beginning of colonialism, global trade, and coal) or 1964 (post-Great Acceleration). The reasons for the disagreement for the start date vary, as do their significance. Lewis and Maslin (2015) argue that:

*the event or date chosen as the inception of the Anthropocene will affect the stories people con-
struct about the ongoing development of human societies.*

*Past scientific discoveries have tended to shift perceptions away from a view of humanity as
occupying the centre of the Universe. In 1543 Copernicus's observation of the Earth revolving
around the Sun demonstrated that this is not the case. The implications of Darwin's 1859
discoveries then established that Homo sapiens is simply part of the tree of life with no special
origin.* **Adopting the Anthropocene may reverse this trend by asserting that humans
are not passive observers of Earth's functioning.** *To a large extent the future of the only
place where life is known to exist is being determined by the actions of humans. Yet, the power
that humans wield is unlike any other force of nature, because it is reflexive and therefore can
be used, withdrawn or modified.* **More widespread recognition that human actions are
driving far-reaching changes to the life-supporting infrastructure of Earth may well
have increasing philosophical, social, economic and political implications over the
coming decades.**

(Lewis and Maslin 2015, 178) (Emphasis ours)

Other researchers push back the start date of the Anthropocene even earlier, going as far back
as 3,000 BC, when agriculture and livestock cultivation grew extensively. It can be seen, then,
that knowing not only when the Anthropocene began, but also recognizing that it indeed *has*
begun, emphasizes the role that humans have played not only in its inception, but also in how
we should proceed with policies and laws moving forward.

The Anthropocene: not purely a geological debate

Aside from debates on when it began, many scholars have also looked into the definition of
Anthropocene, and the consequences of naming a new epoch after the workings of humanity.
This section will look into the debates beyond geology – including geopolitics and socio-eco-
nomic debates, especially in the context of climate change.

Jason Moore, an environmental historian and sociologist, wrote in 'Anthropocene or
Capitalocene?' (2016) that there is a need to also look into capitalism's role in driving climate
change and, more specifically, anthropogenic climate change. In his book, Moore makes a
case for the Capitalocene, writing that none of the other terms captures the 'basic historical
pattern modern of world history as the "Age of Capital" – and the era of capitalism as a world
ecology of power, capital, and nature' (Moore 2016, 6). In a later paper on 'The Capitalocene
and Platenary Justice' (2019), Moore echoes other radical thinkers and climate justice activists
in writing of the 'historical responsibility for climate change in a system committed to a sharply
unequal distribution of wealth and power' (Moore 2019, 50). Taking this standpoint, the reductionist phrase 'anthropogenic climate change' shifts the blame onto the victims of exploitation,
violence, and poverty:

*The Anthropocene (and before that, Spaceship Earth) tells us that planetary crisis is more or less
a natural consequence of human nature – as if today's climate crisis is a matter of humans being
humans, just as snakes will be snakes and zebras will be zebras. The truth is more nuanced,
identifiable, and actionable: we are living in the Capitalocene, the Age of Capital. We know –
historically and in the present crisis – who is responsible for the climate crisis. They have names
and addresses, starting with the eight richest men in the world with more wealth than the bottom
3.6 billion humans.*

(Moore 2019, 50)

With Moore's focus on the Capitalocene, as opposed to the more general Anthropocene, he clarified that there was a *Geological* Anthropocene (which is the concern of geologists and earth system scientists), and a *Popular* Anthropocene (which is much wider, and involves conversations across the humanities and social sciences encompassing historical development and the planetary crisis), the latter of which is confronted by Capitalocene. His approach pays attention to the fact that, historically, there has been little room for most of humanity – peoples of colour, Indigenous Peoples, and virtually all women – to shape global development. Therefore, he claims, '[w]hen the Popular Anthropocene refuses [sic] name *capitalogenic* climate change, it fails to see that the problem is not *Man* and Nature, but *certain* men committed to the profitable domination and destruction of most humans and the rest of nature' (Moore 2019, 53). Further, and corollary to this, he delves into the history of industrialization, the economy, globalization, white supremacy, and the culture of endless accumulation. For him, the Capitalocene started in 1492, when '[T]he Capitalocene thesis pursues analyses that link such consequences to the longer histories of class rule, racism and sexism, all of which form, in the modern sense, in 1492' (Moore 2019, 53).

Anthropocene, the Global North, and the Global South: a climate justice conversation

At the start of this chapter, we wrote that SIDS and developing countries, particularly in the Pacific, but also elsewhere in the world, are often the most vulnerable to climate change even though they emit the least greenhouse gases into the atmosphere. This section will explore further this great divide between the Global North and South, especially when it comes to duties and responsibilities, and responses to climate change.

In 'Worlds apart? The Global South and the Anthropocene', Jens Marquardt (2018) echoes Jason Moore and others, asking: is the Anthropocene an 'overarching paradigm for humanity that speaks to all humans in the same way' (Marquardt 2018, 201)? For Marquardt, there is a need to look at who is historically responsible for climate change and the global environmental crisis. Despite the 'universality' of the discourse on the Anthropocene, the concept still reflects a powerful Western discourse, and Marquardt, citing Bauriedl (2015), writes that it is 'even more striking that the ideal of modernization and industrialization driven by Western countries is reframed as the primary solution for global environmental threats, although the same industrialized countries are historically most responsible for these problems' (Marquardt 2018, 202). The first criterion Marquardt examines, representation, shows, unsurprisingly perhaps, the massive underrepresentation of writers from the Global South on the Anthropocene debate based on language – almost the entire scientific discourse is in English (97%), which inevitably erases the voices and ideas of non-English speakers, which can lead not just to disenfranchisement, but also the exclusion of their views in the discussion of problems and the brainstorming of potential results. The former being an issue of the psyche, the latter, an institutional concern.

On the criterion of contribution, Marquardt (2018, 208) writes that '[d]ealing with case studies in the Global South does not automatically mean the inclusion of (critical) voices from the Global South; rather, it can also mean jumping on the mainstream Anthropocene band-wagon'. As regards framing, Marquardt suggests that the Global South is usually framed as more passive in the discourse, dependent on Northern technology. In the book they edited on global environmental governance in the Anthropocene, Pattberg and Zelli (2016) talked about governance challenges in the Anthropocene, and mentioned the need for a more dynamic view of responsibility, considering the complexity of the debates, since they entail crucial questions including those on social behaviour. They ask 'Which actors gain responsibility? Which ones lose out?' (Pattberg and Zelli 2016, 5).

The UNDP (2019, 175) notes that inequalities permeate the Anthropocene, from cause to effect such that that 'inequality runs the gamut of climate change'. Countries with higher human development generally emit more carbon per person and have higher per capita ecological footprints. Moreover, richer countries and communities tend not to experience the full effects of their impacts on the environment, as a significant portion are felt by 'less-visible countries and communities elsewhere, including to [sic] those along global supply chains' (UNDP 2019, 176–177). Finally, the impacts are also shifted to future generations, exacerbating intergenerational economic inequality.

Practice and policy

Closing the North-South gap

Gonzalez (2015) looks into the priorities of both the Global South and North with regard to policies and laws put in place to address climate change, drawing out the differences. According to her, the:

> North has historically emphasized environmental problems of global concern (such as ozone depletion and species extinction), whereas the South has generally prioritized poverty alleviation and environmental problems with more direct impacts on vulnerable local populations (such as desertification, food security, the hazardous waste trade, and access to safe drinking water, sanitation, and energy). Southern countries have demanded that the North assume responsibility for its immense contribution to major environmental problems (such as climate change), but the North has only grudgingly accepted the principle of common, but differentiated, responsibility on the basis of its superior technical and financial resources while disavowing responsibility on the basis of its historic contributions to these crises. In almost every area of environmental concern, North-South negotiations have featured a deep and growing chasm between the call by some Northern states for collective action to protect the environment and the South's demand for social and economic justice.
>
> *(Gonzalez 2015, 409) (Citations omitted)*

Moreover, she argues that the North-South divide is not the only obstacle to environmental cooperation, as even members of the South are in conflict between and among one another, thus impeding finding compromises in climate change negotiations.

The UNDP (2019) paper, cited earlier, highlights the need for broadly shared approaches to resilience, focusing on inclusive and sustainable human development. This need for widely shared approaches is precisely the understanding that negotiating parties had when they entered into the Paris Agreement in 2015, particularly with regard to the provision on loss and damage.

The Paris Agreement and subsequent COPs

The Paris Agreement, which built upon The United Nations Framework Convention on Climate Change (UNFCCC), sought to fight climate change by keeping global temperature rise this century well below 2 degrees Celsius above pre-industrial levels and to pursue efforts to limit the temperature increase even further to 1.5 degrees Celsius. Among its provisions which was negotiated in the Conference of the Parties 21 (2015) is the setting of minimum obligations of countries, climate financing and the global stocktake, while recognizing the specific needs and special circumstances of developing countries in their fight to combat climate change. It also reflected the principles of equity and common but differentiated responsibilities considering different national circumstances.

Article 4 of the Agreement recognizes the link between development and poverty, accepting that the peaking of greenhouse gas emissions will take longer for developing country Parties. The rest of the Article further laid down the duties of developed and developing country Parties (par. 4 and 5), and indicated the work that least developed countries and SIDS can do in view of their special circumstances (par. 6). Article 9 also discusses the provision of financial resources to assist developing country Parties with respect to both mitigation and adaptation (par. 1).

Article 8 provides for the Warsaw International Mechanism for Loss and Damage (WIM) as the authority and guidance on loss and damage of the COP. Loss and damage refers to – and includes – effects associated with climate change impacts, and the Executive Committee of the WIM guides the implementation through five strategic workstreams: slow onset events (including sea level rise), non-economic losses, comprehensive risk management approaches, human mobility, and action and support. However, despite these, the Conference of Parties agreed in Paragraph 51 of their Report (2016) that Article 8 of the Agreement on the inclusion of the provision on loss and damage 'does not involve or provide a basis for any liability or compensation.'

The WIM was supposed to be reviewed during COP 22, but negotiators decided to adopt a five-year work plan and continue the review in COP 24 in Katowice, Poland, when the Paris Rulebook, as well as the Katowice Climate Package, was developed (Forest Foundation and Parabukas 2019). In COP 25 in 2019, held in Madrid, Spain, there was difficulty in reaching a consensus between the developed and developing countries, as again, because of their particularly vulnerable situation, developing countries had bigger – and more urgent – demands to 'significantly strengthen the ability of the WIM to facilitate work on-the-ground to address these impacts' (Pierre-Nathoniel et al. 2019). Unfortunately, many developed countries resisted this call.

Moving forward

Without bold and decisive collective action, the effects of climate change will exacerbate over time. It is therefore imperative that countries come together and concretely comply with their commitments, if not create more ambitious ones, especially developed countries.

In September 2020, Chinese President Xi Jinping surprised many climate change experts when he pledged that China would speed up reductions in emissions and reach carbon neutrality by 2060 (Myers 2020). As the world's top greenhouse gas (GHG) emitter, this was a welcome pronouncement; however, questions remain regarding the lack of detail in his plan. On his first day in office, President Joe Biden re-entered the Paris Agreement less than four months after the withdrawal by President Trump took effect, claiming that climate change posed an 'existential threat' to humanity (Milman 2021). For its part, the EU aims to be climate-neutral by 2050, and pledged an 'EU-wide net greenhouse gas emissions target of at least 55 percent by 2030, compared with 1990 levels' (Limam 2020). A carbon neutral goal by 2050 is also the target set by Japan (Dooley et al. 2020), Canada (Phys 2020), South Africa (Lo 2020), and the United Kingdom (The Climate Change Act 2008 (2050 Target Amendment Order) 2019). Achieving these goals, however, will require a drastic change of current policies, including, most importantly, turning away from coal, which a number of the aforementioned countries still use.

For their part, SIDS and other developing countries should continue to push for the continued technological and financial assistance of developed countries with regard to their policies on mitigation and adaptation, even as they work on strengthening their own internal policies. Peel and Lin (2019, 682) discuss the contribution of the Global South to 'transnational climate litigation and governance' and argue that the Global South's experience of climate litigation is 'essential if transnational climate jurisprudence is to contribute in a meaningful way to climate governance, and particularly, to ensuring just outcomes for the most climate-vulnerable.'

Pedagogy

Young people are at the forefront of climate change discussions. Their lives will be affected by climate change and as future leaders and policymakers they will be in a position to shape climate change policies. Therefore, it is imperative that they be taught about climate change, that their voices on the environment and just transition be amplified, and that the solutions they propose be listened to and meaningfully taken into account.

Climate change education must start at the earliest age possible. Moreover, it should continue throughout a person's life in a way that is appropriate for their age, especially since the discussion on climate change is wide, and encompasses not just the physical and environmental sciences, but the social science and humanities as well.

Age-appropriate content should not cause anxiety and potential paralysis for the student of climate change, but should instead raise awareness and propel the student towards action. Additionally, content should be solutions-oriented, interdisciplinary and multi-dimensional: while in-classroom discussions are necessary, there is also a need to venture outside to improve problem-solving skills, and to allow the student to see climate change as a function of many different parts, and therefore combating it will involve a holistic approach as well: nature trips, including reforestation; fora, discussions, and immersions with Indigenous Peoples and other environmental defenders; a historization of climate change through socio-economic and political lenses, these are all relevant and beneficial, and will inspire creative solutions to combat climate change.

Parents also have a responsibility, as climate change education – as earlier stated – should start at the earliest possible opportunity, and should not be limited to the confines of formal educational institutions. As the student grows older, more opportunities arise for the deepening (and broadening) of their knowledge on climate change. Where a toddler kindergartener can be taught to dispose of trash responsibly and not waste food, a grade schooler will be taught the natural sciences – the concepts of earth science and global warming and the consequence of climate change not just on human lives but on the world at large, and a teenage high schooler can widen their knowledge by being taught the politics that surround climate change. As high school is when a student becomes more in touch with their politics, and begin to form a deeper understanding of their personal advocacies, this can also be a time for them to explore environmental activism, get into entrepreneurship, and understand their role as changemakers.

In university and beyond, the student of climate change can look into economic analysis, and learn specific skills in relation to their field of study in a way that contributes to further understanding climate change. Throughout all these, the individual can also look into case studies, empirical studies, participate in local and international comparative exchanges, and join moot court and negotiation exercises, particularly those that relate to environmental law and policy.

UNESCO, for its part, has already included climate change education as part of its Education for Sustainable Development (ESD) programme, as there is a growing international recognition of ESD as necessary and integral to quality education and sustainable development. Goal 4 of the 2030 Agenda for Sustainable Development, adopted by all United Nations Member States in 2015, stipulates inclusive and equitable quality education and promote life-long learning opportunities for all. Specifically, the report of the Open Working Group of the General Assembly on Sustainable Development Goals includes Goal 4.7, which provides that by 2030, it should be ensured that 'all learners acquire knowledge and skills needed to promote sustainable development, including among others through education for sustainable development and sustainable lifestyles, human rights, gender equality, promotion of a culture of peace and non-violence, global citizenship, and appreciation of cultural diversity and of culture's contribution to sustainable development' ('Open Working Group Proposal for Sustainable Development Goals' 2014).

In the United States, former US Secretaries of Education Arne Duncan and John B. King Jr sent a letter to President Biden proposing that climate change education be included in the US public schools curriculum, claiming that it provides a way to prepare the students for a world drastically transformed by climate change, in addition to learning potential climate solutions to the crisis including renewable energy (Milman 2020). New Jersey was the first state to mandate climate change lessons at every grade level, including activities like learning climate science and writing climate change essays based on America the Not So Beautiful by Andy Rooney. In the Philippines, the Department of Education is aiming to strengthen the integration of Climate Change Education (CCE) in school curricula, with the goal of making learners become climate literate (Department of Education 2020), supplementing existing programs such as the Establishment of Youth for Environment in Schools Organization (YES-O).

Some of these initiatives are even youth-led, such as Teach The Future, which was launched in 2019 and is a UK-based organization led by students, who posit that current climate education is inadequate, and students are not well-equipped to either prepare for the effects of climate change or understand the solutions.

In the Philippines, some of the pioneering educational work has been done in the environmental science and political science departments of Ateneo de Manila University and in several law schools – the Ateneo Law School, De La Salle University College of Law, Lyceum Philippines University College of Law, Xavier University College of Law, Ateneo de Zamboanga College of Law, Far Eastern University Institute of Law, Polytechnic University of the Philippines College of Law, and University of Makati College of Law. Students in these schools have been exposed to climate change negotiation games that are based on what has happened in the conferences of the parties. Students have also participated in climate justice moot court competitions, imagining themselves as Judges of the International Court of Justice, agents (counsels) of applicant countries that are demanding compensation for loss and damage arising from climate change, and agents defending developed countries from such claims.

Postscript

Many experts see the 2020s as the critical decade for climate change, and for good reason, considering the urgency and reality of a warming world (Shukla et al. 2019; IPBES 2019). It is imperative that scientists, international law practitioners, and policymakers, including social scientists and economists come together to respond to this looming threat to the existence of humanity; equally – and perhaps more important – is the amplification of voices at the margins: those who live in SIDS and other archipelagic or island states, those who live below the poverty line, those who will potentially be displaced by natural hazards, Indigenous Peoples, women, and youth.

So have we officially transitioned to the Anthropocene? Geologists may yet disagree, but most environmental activists and scientists are more united in their response, which is a resounding 'yes.' The world has been significantly altered by human actions, reflected most clearly in climate change. However, while the concept of the Anthropocene is valuable, it is equally important to acknowledge that anthropogenic climate change will affect humanity differently. The reality is that communities which will feel the harshest effects are also those that have contributed the least to climate change. As we enter this new decade, it is of utmost importance that the conversation around climate justice be strengthened – climate change and its effects, and the policies which are supposed to combat it, for instance, should be heavily informed by a rights-based approach emphasizing climate justice.

Key sources

1. Ellis, M.A. and Trachtenberg, Z. (2014). Which Anthropocene is it to be? Beyond geology to a moral and public discourse. *Earth's Future*, vol. 2, no. 2:122–125.

In this essay, Ellis and Trachtenberg posit that the Anthropocene has, at its core, a moral content, which has important implications for its scientific study. First, understanding the Anthropocene will require not just input from the life sciences, but the social sciences as well, and theorizing its character will necessitate a collaborative effort across scholarly disciplines. Second, there is a need for strategic research when it comes to Anthropocene. Third, there is a need to challenge Earth scientists to acknowledge that 'their subject matter demands a kind of moral engagement' (Ellis and Trachtenberg 2014, 124).

2. Antadze, N. (2019). Who is the other in the age of the Anthropocene? Introducing the unknown other in climate justice discourse. *The Anthropocene Review, vol.* 6, no. 1–2:38–54.

In this essay that discusses both the Anthropocene and moral philosophy, Antadze introduces the notion of the Other in environmental and climate justice discourse, and concludes it with a discussion on the implication of this 'nature of togetherness in the age of the Anthropocene' (Antadze 2019, 40). He embarks on a discussion regarding the need to engage with the Unknown Other – through asymmetrical responsibility and through empathy.

3. Ruppel, O.C. (2013). Intersections of law and cooperative global climate governance – challenges in the Anthropocene. In *Climate change: international law and global governance: volume ii: policy, diplomacy and governance in a changing environment.* O.C. Ruppel, C. Roschmann, and K. Ruppel-Schlichting (ed.). Baden-Baden: Nomos Verlagsgesellschaft mbH.

Ruppel looked into the governance and legal challenges in the Anthropocene, mentioning the climate negotiations, the IPCC, the UN Security Council, international human rights law, the potential for a climate refugee law, and trade, among others. He noted that there is a need for these disciplines to become more integrated and coherent in order to respond to the challenges posed by climate change, positing that climate change can 'only be tackled through a combination of political, legal, and natural science tools' (Ruppel 2013, 44).

4. Simangan, D. (2020). Where is the Asia Pacific in mainstream international relations scholarship on the Anthropocene? *The Pacific Review*. www.doi.org/10.1080/09512748.2020.1732452.

Simangan took a regional-level analysis to look at implications of global environmental change for societies, with the intention to answer the question '[a]re Asia-Pacific experiences in the Anthropocene present in mainstream IR (international relations) scholarship?' She looked into bibliographic portfolio, bibliometric analysis or portfolio, systemic analysis, and definition of the research question and objectives. In her paper, she found three themes in locating the Asia-Pacific experience in the discourse surrounding Anthropocene. First, she found that Anthropocene has become a useful analytical lens for scholars to (re)examine old issues. Second, it shows the ontologies of nature found in the region. Third, she writes that the region is a site of 'Anthropocene paradox.' She concludes that the Anthropocene demands a radical rethinking of the current international system to eradicate the ecologically destructive modernist approaches to governance on a global level, international law, geopolitics, economic development, and sustainability.

5. Davis, H. and Todd, Z. (2017). On the importance of a date, or, decolonizing the Anthropocene. *ACME: An International Journal for Critical Geographies,* vol 16, no. 4:761–780.

Davis and Todd assert the importance of making the relations between the Anthropocene and colonialism explicit in order to better understand the current situation humanity is in, and move forward in a way that assures the sustainability of life. According to them, by linking Anthropocene with colonization, it calls for 'the consideration of Indigenous philosophies and processes of Indigenous self-governance as a necessary political corrective' (Davis and Todd 2017, 763). They make their case for colonialism as the start date of the Anthropocene for two reasons: first, to open up the implications of the Anthropocene beyond the realm of Western thought, and second, to argue that to use a date that coincides with colonialism in the Americas will show that the current ideology is defined by proto-capitalist logic. They argue that the Anthropocene 'as the extension and enactment of colonial logic systematically erases difference.' (Davis and Todd 2017, 769) While they focus on colonialism in the Americas, the paper is still relevant with regard to discussions on the implications of the Anthropocene and why it is important to contextualize, de-universalize, and decolonize the same. For the authors, 'if we are to adapt with any grace to what is coming, those with power … would do well to begin to listen to' the voices of Indigenous peoples, or those who have not forgotten our 'entwined relations and dependency on this body of the Earth' (Davis and Todd 2017, 776).

Note

1 www.knoema.com/atlas/Philippines/CO2-emissions-per-capita

References

Bauriedl, S. (2015). *Wörterbuch Klimadebatte*. Bielefeld: Transcript Verlag.
Conference of the Parties, Twenty-first session, Paris (2015). *FCCC/CP/2015/L.9/Rev.1,* 30 November to 11 December 2015. Bonn: UNFCCC.
Crutzen, P.J. (2002). Geology of mankind. *Nature*, vol. 415, 23.

Dooley, B., Inoue, M., and Hida, H. (2020). Japan's new leader sets ambitious goal of carbon neutrality by 2050. *The New York Times*, viewed 23 February 2021, www.nytimes.com/2020/10/26/business/japan-carbon-neutral.html

Forest Foundation and Parabukas (2019). UNFCCC Negotiations: A Resource Book. Makati City: Forest Foundation Philippines.

Gonzalez, C.G. (2015). Bridging the North-South divide: international environmental law in the Anthropocene. *Pace Environmental Law Review*, vol. 32, no. 2:407–433.

Intergovernmental Science-Policy Platform on Biodiversity and Ecosystem Services (2019). 'Report of the Plenary of the Intergovernmental Science-Policy Platform on Biodiversity and Ecosystem Services on the work of its seventh session (IPBES/7/10/Add.1)', viewed 23 February 2021, www.ipbes.net/sites/default/files/ipbes-7–10_en.pdf

Lewis, S. and Maslin, M. (2015). Defining the Anthropocene. *Nature*, vol. 519, no. 7542:171–180. www.doi.org/10.1038/nature14258

Limam, A. (2020). Carbon neutral goals: where does Europe stand?. *CGTN*, viewed 21 February 2021, www.newseu.cgtn.com/news/2020–09–24/Carbon-neutral-goals-Where-does-Europe-stand--U1QzCeaFxe/index.html

Lo, J. (2020). South Africa aims to reach net zero emissions in 2050 – while still burning coal. *Climate Home News*, viewed 23 February 2021, www.climatechangenews.com/2020/09/16/south-africa-aims-reach-net-zero-emissions-2050-still-burning-coal/

Marquardt, J. (2018). Worlds apart? The Global South and the Anthropocene. In T Hickmann, L Partzsch, P Pattberg & S Weiland (ed.), *The Anthropocene debate and political science,* 1st edn. London: Routledge.

Milman, O. (2020). Teaching climate crisis in classrooms critical for children, top educators say. *The Guardian*, viewed 29 March 2021, www.theguardian.com/education/2020/dec/03/teaching-climate-crisis-classrooms-critical-children

Milman, O. (2021). Biden signals radical shift from Trump era with executive orders on climate change. *The Guardian*, viewed 23 February 2021, www.theguardian.com/us-news/2021/jan/27/joe-biden-climate-change-executive-orders

Moore, J.W. (2016). Introduction: Anthropocene or Capitalocene? Nature, history, and the crisis of capitalism. In JW Moore (ed.). *Anthropocene or Capitalocene? Nature, history, and the crisis of capitalism.* Oakland: PM Press.

Moore, J.W. (2019). The Capitalocene and planetary justice. *Maize,* vol. 6:49–54.

Myers, S. (2020). China's pledge to be carbon neutral by 2060: what it means. *The New York Times*, viewed 21 February 2021, www.nytimes.com/2020/09/23/world/asia/china-climate-change.html

Open Working Group proposal for Sustainable Development Goals (2014). *Open Working Group of the General Assembly on Sustainable Development Goals*, Document A/68/970, viewed 29 March 2021, www.undocs.org/A/68/970

Peel, J. and Lin, J. (2019). Transnational climate litigation: the contribution of the Global South. *American Journal of International Law*, vol. 113, no. 4:679–726, www.doi.org/10.1017/ajil.2019.48

Pattberg, P. and Zelli, F. (2016). Global environmental governance in the Anthropocene: an introduction. In P. Pattberg and F. Zelli (ed.). *Environmental politics and governance in the Anthropocene: institutions and legitimacy in a complex world,* 1st ed., London: Routledge.

Pierre-Nathoniel, D., Siegele, L., Roper, L., and Menke, I. (2019). Loss and damage at COP25 – a hard fought step in the right direction, viewed 23 February 2021, www.climateanalytics.org/blog/2019/loss-and-damage-at-cop25-a-hard-fought-step-in-the-right-direction/

Phys.org (2020). Canada govt seeks carbon neutrality by 2050, viewed 23 February 2021, https://phys.org/news/2020-11-canada-govt-carbon-neutrality.html

Report of the Conference of the Parties on its twenty-first session (2016). *FCCC/CP/2015/10/Add.1*, 30 November to 13 December 2015, UNFCCC, Bonn.

Republic of the Philippines Department of Education (2020). DepEd eyes strengthening of climate change awareness in basic education, viewed 29 March 2021, www.deped.gov.ph/2020/12/16/deped-eyes-strengthening-of-climate-change-awareness-in-basic-education/

Shukla, P.R., Skea, J., Slade, R., van Diemen, R., Haughey, E., Malley, J., Pathak, M., and Portugal, P.J. (ed.) (2019). Technical summary. In P.R. Shukla, J. Skea, B.E. Calvo, V. Masson-Delmotte, H.O. Pörtner, D.C. Roberts, P. Zhai, R. Slade, S. Connors, R. van Diemen, M. Ferrat, E. Haughey, S. Luz, S. Neogi, M. Pathak, J. Petzold, P.J. Portugal, P. Vyas, E. Huntley, K. Kissick, M. Belkacemi, and J. Malley (ed.), *Climate Change and Land: an IPCC special report on climate change, desertification, land degradation, sustainable land management, food security, and greenhouse gas fluxes in terrestrial ecosystems.* Intergovernmental Panel on Climate Change (IPCC).

The Climate Change Act 2008 (2050 Target Amendment) (2019), Order 2019, No. 105. viewed 23 February 2021, www.legislation.gov.uk/uksi/2019/1056/contents/made

United Nations Development Programme (2019). In *Human Development Report 2019 Beyond Income, Beyond Averages, Beyond Today: Inequalities in Human Development in the 21st Century.* United Nations Development Programme. New York: United Nations.

Zalasiewicz, J., Williams, M., Smith, A., Barry, T.L., Coe, A.L., Brown, P.R., Brenchley, P., Cantrill, D., Gale, A., Gibbard, P., Gregory, F.J., Hounslow, M.W., Kerr, A.C., Pearson, P., Knox, R., Powell, J., Waters, C., Marshall, J., Oates, M., Rawson, R., and Stone, P. (2008). Are we now living in the Anthropocene? *GSA Today,* vol. 18, no. 2:4–8, www.doi.org/10.1130/GSAT01802A.1

15

MORE-THAN-HUMAN DEVELOPMENT

Andrew McGregor and Ashraful Alam

Introduction

More-than-human concepts are sweeping the social sciences and environmental humanities. Challenging the anthropocentrism embedded within much of the academy, more-than-human approaches adopt a relational lens, focusing attention upon the spatial and temporal relations that constitute worlds. Humans are seen as one actor among many, in constant negotiation with the agencies of non-human others, such as animals and plants, but also soils, winds, rivers, climates and much more. Human achievements are never purely human, instead they reflect collaborations with non-humans, where humans and non-humans come to be arranged in a particular way to enable development to occur. A farmer works with soils, worms, crops, and seasons to sustain their livelihood just as an academic may utilize energy, technology, coal and coffee to write their paper. Through more-than-human approaches such relations are brought into focus. Mirroring Indigenous ontologies, the world is seen as animated and alive, as opposed to a staid background for human achievement, or a set of resources for human exploitation.

In this chapter, we explore what more-than-human approaches can contribute to development research, teaching and practice. Such work is timely as development studies and practice have yet to engage with more-than-human insights in any significant way. We believe this is a regrettable oversight for at least three reasons. First, development policy is increasingly attempting to grapple with non-human agencies through initiatives like the Sustainable Development Goals; second, development practitioners have much to contribute to more-than-human debates and practice through their professional activities; and third, if we were to identify a core driver of the deteriorating planetary conditions that are changing the planet, it is surely the atomized focus on measurable human development outcomes above the broader more-than-human processes through which development occurs. In what follows, we first develop the concept of more-than-human development before analysing the challenges it poses to how we conceptualize and approach core development concerns such as community and empowerment. We then reflect on the ramifications of the concept for practice and policy, before finally exploring how to incorporate more-than-human approaches into pedagogy.

DOI: 10.4324/9781003017653-18

More-than-human development, or thick development

More-than-human approaches are most commonly associated with science and technology scholars such as Bruno Latour and Donna Haraway, but also social scientists including political theorist Jane Bennet, geographer Sarah Whatmore and anthropologist Anna Tsing. In different ways, each has challenged the notion of human exceptionalism and sought to foreground the role of non-humans in shaping worlds. Their collective efforts redistribute agency (the ability to affect change) away from the singular human actor, to instead recognize the more-than-human networks that must be in place for change to occur. Arguably, for example, rainfall patterns or the distribution of mineral resources, have had a lot more impact on human development patterns than any singular development policy. Under this approach humans are positioned as just one entity among many entangled together within complex more-than-human worlds. While linguistically clunky, the more-than-human descriptor serves as a lens in which to view human achievements as outcomes of much broader sets of relations involving living and non-living non-human parts.

Of course, many of these ideas go well beyond recent social theory and are reflected in long established relational concepts associated with many Indigenous cultures. Rather than seeing humans as separate from or superior to non-human actors, concepts like the Australian Aboriginal term Country, stress interconnection and embeddedness with place, where dynamic more-than-human relations provide the foundation for diverse multi-species societies (Suchet-Pearson et al. 2013). This contrasts sharply with the human focus of development studies and practice where development success is most often measured according to the wellbeing of humans, almost as if they exist independent of the relations that sustain them. Hence abstracted economic indices like Gross Domestic Product (GDP), or the health and education indices of the Human Development Index, sever connections with place, and in the process make those relations that contribute to GDP, health, and education, but also the multitude of other important place-based relations, invisible. When hidden from view such relations become under-valued and vulnerable to disruption.

In the context of these advances in social theory, we believe the concept of human development needs to be renovated or replaced. By emphasizing the human, the multitude of relations that sustain communities are deprioritized, often with destructive consequences, which are now evident at multiple scales. Human development logics, for example, are part of the rationale proponents draw upon to continue to promote fossil fuels, despite the overwhelming evidence that the long-term impacts of fossil fuel led development will far exceed any temporary benefits. Similarly, the destructive rationale to keep extending agriculture into forests is premised on the human development logic that it will boost jobs, incomes, GDP, etc, again with dire epidemiological, ecological and livelihood impacts. While the Sustainable Development Goals have sought to minimize the adverse outcomes that arise from a narrowly conceived focus on human development, they, like sustainable development discourses more generally, tend to ask how human development can be sustained without destroying the resource base (see below).

A more-than-human approach is fundamentally different. Following McGregor and Thomas (2018), we see development as a multi-species achievement, propelled by place-based assemblages comprised of actors such as humans, plants, animals, soils, landscapes, technologies and weather. In this sense, more-than-human development is *thick* development. In contrast to the narrow focus of human development (economic growth or human wellbeing), more-than-human development focuses on the different temporal and spatial rhythms of the multi-species communities through which spaces and places evolve. While most people would agree that human development is context dependent, context too often slips into the background. Within more-than-human development there is no separation between human and context, all are developing together. Humans, for example, are part of the context in which an avian community, or a plant, or a lake, is developing into new forms. In decentering the human, development practices and approaches must become thicker, more sensitive to, and responsible for, the broader array of biological and geophysical actors tied up in development processes.

Anna Tsing (2015), for example, has brilliantly exposed how matsutake mushrooms are entangled within human cultures, economies, migration patterns, politics, and empires. Her rich ethnographic work at once exposes the limitations and conceits of human development, and the opportunities a thicker more-than-human lens enables for exploring currently-hidden possibilities for doing development better. Through focusing on the mushroom new pathways become possible. In a similar way, Yeh and Lama (2013) highlight how caterpillar-fungus is intimately entangled with Tibetan livelihoods and communities, creating new opportunities and risks for rural development. Tibetan development is dependent not only on the agencies of wild growing caterpillar-fungus, which has so far defied farmed production, but the broader relations between the caterpillar-fungus, Chinese consumers, and the state. A more-than-human development approach would prioritize maintaining and improving these relations as a way of securing and improving the conditions for pickers, consumers, and caterpillars.

Taking a different tack, the hydrosocial cycle concept has proved popular in demonstrating how water and society are internally bound together in inseparable ways (Linton and Budds 2014). For example, a change in politics, policy or technology that results in the building of a dam, will affect the flow of water, which then enables and disables particular forms of development. Similarly, a storm, flood, or drought will affect politics, policy, and technology, also affecting how development occurs. In this sense, water is not a backdrop for development but an active agent in development, shaping and being shaped by the social organization of societies, politics, inequality, health, economies etc. McGregor et al. (2019) have similarly proposed a sociocarbon cycle to conceptualize the active agencies of carbon molecules in enabling and disabling different forms of development, from fossil fuel dependent economies to renewable energy and reforestation projects, as well as the emergence of climate refugees. Through both frameworks a thick lens is adopted to identify possibilities for working with the non-human agencies of water and carbon to bring about more just and convivial more-than-human societies.

Such approaches require rethinking how development is approached in at least two significant ways. First, development practitioners need to become skilled in what Anna Tsing refers to as the *arts of noticing*. Development studies currently train people to focus on issues relating to human development, primarily health, education, and economies, and how these aspects of society can be improved via internal and external inputs. 'Good change,' as Robert Chambers once famously described development, is when those dimensions improve as a result of development activities. These are laudable goals and we are not suggesting they should be replaced by a different set of environmentally-oriented goals. What we do suggest, however, is that good change need not be limited to the human, and it should be informed by the arts of noticing. This requires practitioners to become sensitive to and prioritize the everyday more-than-human relations that contribute to community wellbeing, where community is understood in a multi-species way (see below). This requires slowing down and 'learning to be affected' (Latour 2004) by the many different types of often mundane relations and non-human agencies that contribute to worlds and co-create places. Mushrooms, swamps, trees, and soils are all active agents, rather than inanimate objects, within development. Thick development is development that is attuned to these diverse more-than-human relations, and seeks to build from and co-become with them.

A second consequence of a more-than-human development framework is a heightened understanding of context that can spatialize and diversify how development is understood. In contrast to homogenizing all development as human development, a more-than-human approach recognizes that the development morphology of a place will reflect the biophysical contexts in which development occurs. For example, development in geographically dispersed mountainous regions of the world is likely to share particular development characteristics because of their biophysical conditions, just as coastal, riverine, and desert communities will be shaped by the similar types of more-than-human relations present within these bioregions. Without wanting to revert to a crass environmental determinism, there can be no doubt that the biophysical world influences the development morphologies that emerge. Such differences should be celebrated and development policy steered to nurture diverse development morphologies and expertise based on bioregion. Mountainous development, coastal development, and desert development, for example, explicitly recognize the influence of non-human actors in a way that a more universalized discourse of human development does not. To some extent, this already happens in practice but needs clearer articulation in theory, policy, and pedagogy.

Approaching more-than-human development

If human development, in its simplest sense, is about enhancing the quality of human life, more-than-human concepts unsettle traditional development thinking by highlighting how humans are positioned within the biophysical world in the company and modality of non-human others. For example, poverty or homelessness is inevitably influenced by the conditions of and access to non-humans, such as plants, animals, or water, in a particular setting. More-than-human perspectives provide a theoretical apparatus to help us become cognisant of such relationships in a 'thick way' (Stensöta 2015, 194). This includes a view of why non-humans are crucial to human wellbeing and survival as well as the reasons why a shared survival and wellbeing is necessary through nurturing, sustaining, and protecting relations of various kinds. In its focus on interdependence and context specificity, more-than-human development contradicts the immanent anthropocentrism and expert-centrism of the typical development approach where 'everything that is not expressed through human rules' is ruled out (ibid. 194).

How to lessen the human-centrism in everyday spaces has been a focus for some time in feminist geographies, environmental humanities, and critical social sciences. We borrow from these ideas to *thicken* some of the widely invoked concepts in development discourse, such as community, engagement, empowerment, and participation in order to experiment with pathways and practices of doing development differently. In the following subsections, we propose three prompts (Figure 15.1) to guide more-than-human development praxis.

Recognizing more-than-human communities

In the field of development, the term, community, assumes a collective actor (Titz, Cannon and Krüger 2018) composed of multiple *human* social groups who are usually from the same geographic locale, and bound together through assumed commonalities in goals and values, communication, comfort and familiarity (Bastian 2011). Terms, such as community-driven and community-led grant a moral licence to a development intervention, in the sense that the *human* beneficiaries are driving the project through inclusive practices (UNHCR 2008). Yet, scholars critique the often oversimplified mobilization of the term which may result in the

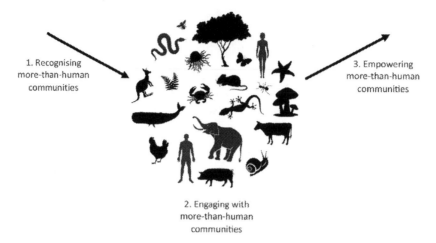

Figure 15.1   Approaching more-than-human development
Source: Author

failure to recognize local fissures and power imbalances (Adhikari and Goldey 2010; Titz et al. 2018) within and between human and more-than–human communities.

More-than–human approaches resist the temptation to recognize sameness as the basis or an aspirational ideal for community. Instead, it is focused on the difficult task of forming political partnerships with an openness towards diversity. This is an important ideological shift to recognize the *thickness* of communal affiliations that include diverse relations involving different types of humans and non-humans, such as chickens, crops and weeds. Recognizing multispecies or more-than–human communities challenges development trajectories that often erase not only human social differences such as gender, race, or ethnicity within communities, but also diverse non-human members of those communities. By imagining a more-than-human community, a more inclusive developmental ethos can evolve where, as Val Plumwood (1995) famously said, the human is understood in ecological terms and the non-human in ethical terms.

Engaging with more-than-human communities

Reconfiguring the traditional community inevitably requires development practitioners to experiment with ways of enabling non-humans to become recognized agents of development. This is a challenging task due to the traditional development repertoire not usually recognizing non-humans as having a role within development or a right to participate or flourish within multi-species communities. In this regard, Donna Haraway (2008, 88–89) encourages the term *response-ability* to emphasize both the responsibility and the ability of humans to notice and respond in ways that nurture and support more-than–human worlds. Cultivating a sense of response-ability enables deeper engagement and greater accountability to place-specific more-than–human interests.

For development practitioners to engage with more-than–human agencies and needs, they need to explore ways of going beyond expert frames (Dowling et al. 2016). One strategy is to establish localized *vantage points* (Haraway 1988, 583; 1991, 154), such as that of Tsing's mushroom (2015), or Yeh and Lama's (2013) caterpillar-fungus, from where localized multi-species performances and partnerships become noticeable. Unsettling the top-down, expert-lay power relationships, or even grassroots but anthropocentric perspectives, by explicitly focusing on the roles of non-human actors within development can enable ways of doing development differently. Rather than being exterior to development decision making, adopting, and responding to non-human vantage points provides a form of multi-species participation within development agendas.

Empowering more-than-human communities

Empowerment should be seen as a collective outcome involving more-than-human communities (Whatmore 2002, 3). However, the idea of empowering or giving agencies back to non-humans for their active participation is easier said than done. In this regard, feminist ethics of care (Tronto 1993) can help reimagine development as a multispecies reparation activity that includes everything that participates 'to maintain, continue, and repair our world so that we can live in it as well as possible' (Fisher and Tronto 1990, 40). Successful reparation (Tronto 2013; Power 2019) requires reciprocity from both the care-giver and care-receiver. Inspired by these works, Alam and Houston (2020) have conceptualized *care collectives* to show how various under-represented actors, women, children and birds collectively empower themselves, participate and generate alternate development spaces.

A different approach comes from Bruno Latour (2009, 64–65 and 70) who develops the idea of *spokespersons* who become sensitive to and learn to be affected by the multiple non-verbal ways non-humans communicate. Whether it is the brittle leaves of an undernourished tree, the acidification of tired soil, or the optimistic auto-rewilding of fast-growing species in degraded areas, non-humans and more-than-human spaces communicate in diverse ways. Spokespersons engage with these more-than-human forms of communication to articulate the interests of non-humans as best they can. Of course, speaking for non-humans is problematic in all sorts of ways, but it is surely better than current practices where non-humans are rarely provided with political spaces to speak at all.

Within the legal sphere there has been considerable progress in recognizing and empowering non-human perspectives. In 2017, the Whanganui River in the north Island in Aotearoa New Zealand has been granted the same legal right of a human being. Parliament passed legislation declaring that Te Awa Tupua, the river and all its physical and metaphysical elements, is an indivisible, living whole, and henceforth possesses all 'the rights, powers, duties, and liabilities' of a legal person (Charpleix 2018, 27). The sacred Ganges and Yamuna rivers in India, the Lake Erie in Toledo, Ohio, have gained similar legal status. This kind of legislation shifts away from the anthropocentric and resource-centric (Charpleix 2018) developmental ideology towards an eco-centric one in which the river or the lake system with all its human and non-human components is empowered to flourish in collective ways.

Practice and policy implications

At a policy level, more-than-human approaches have most in common with work on the Sustainable Development Goals (SDGs) and the ecosystem services (ES) concept championed through the Millennium Ecosystem Assessment. However, while the SDGs and ES bring more-than-human relations into view, both are overwhelmingly anthropocentric in their outlook, whereby the value of non-humans is assessed according to their contribution to human development. In the preamble to the SDGs, for example, there is the unusual reference to making '*human* habitats... safe, resilient and sustainable' (emphasis added) and offering up of the non-human world for consumption: 'A world in which consumption and production patterns and use of all natural resources – from air to land, from rivers, lakes and aquifers to oceans and seas – are sustainable' ('Transforming Our World' 2015, 3–4). Similarly, the term ecosystem services emphasizes that the non-human world should be understood and valued according to the services it provides for human development. Pushed by environmental economists, the overwhelming focus within ecosystem services research has been to place an economic value on those services (Chaudhary et al. 2015).

Despite these limitations, however, both initiatives do acknowledge that non-human agencies matter. For this reason, they may provide openings for more-than-human approaches. Policies informed by more-than-human ideas would seek to bring non-human agencies into the foreground of development processes and planning. This requires thick development, thinking about how places, and all the more-than-human relations that comprise them, can develop together. Of course, change means changing relations and some will inevitably be lost, however a focus on 'good change' would seek to identify the meaningful relations that contribute to the desirable qualities of a place and maintain or build from them. These are not just the relations that benefit humans, but the relations that benefit non-humans, such as non-human habitats, waterways, food sources, etc. Importantly, the assumption should not be that human and non-human interests are in conflict or that recognizing non-human interests may somehow threaten human interests (a recurring narrative with sustainable development that limits its radical potential). Instead, development policy should be framed around identifying shared interests, the types of more-than-human relations where human and non-human interests intersect. It is from this starting point, that *ecologically* sustainable development can be built and where ecosystem services discourse can be extended to recognize the *human services* we provide for non-humans.

Field-based development practitioners have much to offer in terms of the practices that could be utilized to recognize more-than-human relations. The types of activities outlined in Participatory Rural Appraisal and Assets Based Community Development, are good examples of grounded grassroots development practices that can be tailored towards recognition of more-than-human relations. Transect walks, for example, are ideal tools through which practitioners can become sensitive to place, as long as practitioners are engaged in the arts of noticing, actively seeking to observe and nurture more-than-human relations (Richardson-Ngwenya 2014). Developing forums where grassroots and expert knowledge, such as those of farmers, gardeners, biologists, and ecologists, can come together and learn from one another is an important step in the arts of noticing. Being in place so practitioners can walk with, talk about, be shown, and engage with places are essential components of learning about more-than-human worlds (Pitt 2015).

Through our own research and practice we have endeavoured to find ways to do development differently. In 2014–2015, we adopted two creative methodologies, namely a participatory photography exercise called *photo-response* (Alam et al. 2018 and 2020a) and a mobile mapping exercise that includes walking interviews (Alam et al. 2020b) to examine peri-urban development in the fringe ecologies of Khulna City in Bangladesh. The methodologies shone light on the more-than-human agencies that mattered to rural migrants, and the intimate reciprocal relationships that sustained families and places, differing radically from more conventional urban development strategies. For example, migrants secured their informal homes by caring for places through clearing weeds, collecting trash, and engaging in other forms of farm-based labour in order to avoid eviction from absentee landlords. Migrants developed reciprocal relationships of care and support with plants, animals, water, etc. and other humans for various services derived from local ecologies to form sustainable livelihoods (as described in Figure 15.2). These sorts of findings provoke a rethink of land beyond the economic value

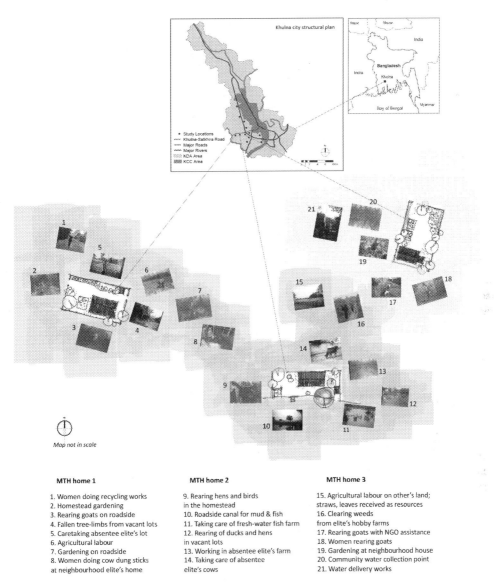

Figure 15.2 More-than-human relations around homes in peri-urban areas of Khulna city, Bangladesh
Source: Authors' fieldwork

as an integrated affective system, where the sustainability of the system is important for both human and non-human co-flourishing for a shared future. And yet we must acknowledge that the benefits or services received by migrant homemakers or their patrons and the ecology are not equal. Non-humans are considered as resources, but at least for migrant homemakers, the flourishing of non-humans was intimately connected to their own flourishing. Experiments like these provide important insights into how development practitioners can slow down to recognize, engage with and empower the various more-than-human agencies that comprise the *thickness* of developing areas.

Pedagogy

Teaching more–than–human approaches is similarly experimental. Following Whatmore and Landström (2011), who devised the *competency group* methodology as a means of exploring more–than–human responses to the flood problem in Pickering (UK), we have been experimenting with how to teach more–than–human approaches in the classroom. The key objective is to encourage students to approach particular development morphologies, in our case the bushfire landscape in Australia and the low–lying coastal plains in Dunedin, Aotearoa New Zealand, through a more–than–human development lens.

We used our own illustration of the competency platform (Figure 15.3) to show students how the development landscape is thickened when we include non–humans as agents within development strategies, rather than just humans, in the Pickering case. The illustration helped bring the diverse agencies, such as the scientists, local government agencies, and citizens, along-side maps, bikes, rivers, storms, flora, and fauna, into view. We explained how the competency group had particular knowledge of non–human agents and could bring these into the fore-ground when developing flood solutions.

Students were then asked to engage with either the bushfire context or the issues related to the flood incidents in Dunedin with a more–than–human approach. This involved mapping the different human and non–human actors and then developing a strategy that took into account the agencies of these actors in developing solutions. We observed that students quickly became more attuned to the *arts of noticing* in each case. Instead of immediately envisaging development solutions to the given problem, students were more interested in uncovering the diverse human and non–human relations that constituted the more–than–human communities in each case. This art of noticing resulted in a greater degree of responsiveness to the needs of both humans and non–humans. The exercise, according to the student feedback, has lasting impressions on them as they learned to slow down and envisage shared developmental outcomes that cater for a much broader array of agencies.

Getting students out of the classroom to engage directly with more–than–human relations is another important pedagogical tool. For example, at Macquarie University we have liaised with local D'harawal elders with particular botanical expertise to take students on guided experiential tours of the campus. During these walks, the non–human agencies of Country become more and more apparent as plants, insects and animals emerge from the background to become active subjects in the storying of place. The multiple properties and roles of native plants become active subjects in the telling of Indigenous histories and cosmologies. Similar to a transect walk, the embodied processes of listening, smelling, touching, and walking sensitize students to connections in which they are bound and through which the world is co–produced. Later, in class, concepts such as non–human Spirit Ancestors within D'harawal lore are discussed, and the ways in which such concepts structure Aboriginal societies and more–than–human relations. While walking tours with Indigenous knowledge holders are particularly valuable for their multigenerational and cosmologically challenging perspectives, walking with other local community groups or biological and ecological experts is another way through which more–than–human relations and concepts can be made visible and students can be trained in the arts of noticing.

Postscript

We end by acknowledging the limitations of a more–than–human approach and the need for more research in this space. Perhaps the greatest challenge for more–than–human development is how to move from a resource–centred view of non–humans to a different set of ethics that are

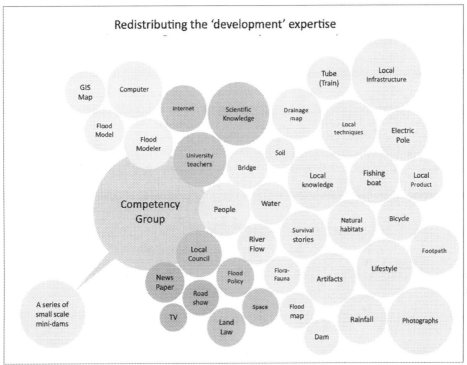

Authors' illustration of Whatmore and Landström's (2011) 'flood apprentices' exercise in Pickering, UK

Student task: How would you engage with the bushfire landscape in NSW rural areas?
Image: BLMIdaho, licensed with CC BY 2.0.

Student task: How would you engage with the flood landscape in South Dunedin?
Image: RNZ / Ian Telfer

Figure 15.3 Authors' teaching slides for in-class exercise; Top: authors' illustration of the 'competency group' platform; bottom left: class exercise at Macquarie University, Australia; bottom right: class exercise at the University of Otago, Aotearoa New Zealand.

still evolving but based more on care and co-becoming. As the scale of development projects increase the ability to recognize, engage and empower more-than-human relations becomes more difficult, however being aware of and caring for local environments through small scale grassroots initiatives, such as those with rivers, urban ecologies, and wildlife, is already being practised. Within such initiatives, more-than-human ideas can become conceptual assets for development practitioners interested in developing places, rather than only people, in the pursuit of good change.

The question of exactly how to pursue more-than-human development is unclear, but rather than lament the lack of a clear path forward we see this as an opportunity for creative experimentation. The challenge is to develop tools and ways of thinking that thickens development by foregrounding non-human agencies and recognizing humans as one actor among many. Here, we have suggested three key steps based on recognition, engagement, and empowerment, and see openings within policy, practice, and pedagogy where more-than-human ideas can have resonance. It requires slowing down and recognizing and appreciating the diverse development morphologies and opportunities that are co-created by place-based biophysical agencies. We imagine some of these ideas are already familiar to many grassroots development practitioners working on human-environment interfaces and are excited by not only what more-than-human approaches can bring to development, but by what development can bring to more-than human approaches.

Key sources

1. Alam, A, McGregor, A. and Houston, D. (2018). Photo-response: approaching participatory photography as a more-than-human research method. Area, vol. 50: 256–265.

Photo-response combines visual, verbal, and spatial responses to examine more-than-human dimensions of migrant homes in peri-urban Khulna. How can non-humans inform lives and livelihoods that shape particular development morphologies? What methods/methodologies can we use to recognize more-than-human agencies shaping development needs?

2. McGregor, A. and Thomas, A. (2018). Forest-led development? A more-than-human approach to forests in Southeast Asian development. In *Routledge Handbook of Southeast Asian Development*. London: Routledge.

In this chapter, McGregor and Thomas adopt a forest-centred perspective to demonstrate how forests have shaped development morphologies in Southeast Asia. What other development morphologies are there and what are the major more-than-human agencies shaping development in each morphology?

3. Tsing, A. (2015). *The mushroom at the end of the world: or the possibility of life in capitalist ruins*. Princeton: Princeton University Press.

In this masterful book, Tsing explores the agency of wild growing matsutake mushrooms in sustaining livelihoods and economies for a variety of social groups. What other non-human agencies sustain livelihoods in your country, and what sort of ethics should inform these more-than-human relations?

4. Yeh, E. and Lama, K. (2013). Following the caterpillar fungus: nature, commodity chains, and the place of Tibet in China's uneven geographies. *Social and Cultural Geography*, vol. 14, no. 3,: 318–340.

Yeh and Lama demonstrate how the relations between Tibetans, the Chinese state and wild growing caterpillar-fungus shape the livelihoods and forms of development that are possible. Can you think of similar more-than-human relations between states, ethnic groups, and non-humans?

5.   Suchet-Pearson, S., Wright, S., Lloyd, K., and Burarrwanga, L. (2013). Caring as country: towards an ontology of co-becoming in natural resource management. *Asia Pacific Viewpoint*, vol. 54, no. 2: 185–197.

The thought-provoking article draws on Australian Indigenous philosophies to show the ways humans can and should relate to the environment through processes of reciprocity and co-becoming. Co-becoming offers alternate natural resource management pathways through attending to vibrant more-than-human relations.

References

Adhikari, K.P. and Goldey, P. (2010). Social capital and its 'downside': the impact on sustainability of induced community-based organizations in Nepal. *World Development*, vol. 38, no. 2: 184–194.

Alam, A., McGregor, A. and Houston, D. (2018). Photo-response: approaching participatory photography as a more-than-human research method. *Area*, vol. 50: 256–265.

Alam, A. and Houston, D. (2020). Rethinking care as alternate infrastructure. *Cities,* vol. 100, 102662, doi: www.doi.org/10.1016/j.cities.2020.102662

Alam, A., McGregor, A., and Houston, D. (2020a). Neither sensibly homed nor homeless: re-imagining migrant homes through more-than-human relations. *Social & Cultural Geography*, vol. 21, no. 8: 1122–1145.

Alam, A., McGregor, A., and Houston, D. (2020b)., Women's mobility, neighbourhood socio-ecologies and homemaking in urban informal settlements. *Housing Studies*, vol. 35, no. 9: 1586–1606.

Bastian, M. (2011). The contradictory simultaneity of being with others: exploring concepts of time and community in the work of Gloria Anzaldúa. *Feminist Review*, vol. 97, no.1: 151–167.

Chaudhary, S., McGregor, A., Houston, D., and Chettri, N (2015). The evolution of ecosystem services: a time series and discourse-centred analysis. *Environmental Science and Policy,* vol. 54: 25–34.

Charpleix, L. (2018). The Whanganui River as Te Awa Tupua: place-based law in a legally pluralistic society. *The Geographical Journal*, vol. 184, no. 1: 19–30.

Dowling, R., Lloyd, K., and Suchet-Pearson, S. (2016). Qualitative methods II: 'more-than-human' methodologies and/in praxis. *Progress in Human Geography*, vol. 41, no. 6: 25–34.

Fisher, B. and Tronto, J. (1990). Toward a feminist theory of caring. In E.K. Abel and M.K. Nelson (ed.). *Circles of Care: Work and Identity in Women's Lives*. Albany: State University of New York.

Haraway, D. (1988). Situated knowledges: the science question in feminism and the privilege of partial perspective. *Feminist Studies*, vol.14, no. 3: 575–599.

Haraway, D. (1991). *Simians, cyborgs, and women: the reinvention of nature*. New York: Routledge.

Haraway, D. (2008). *When Species Meet,* vol 3. Minneapolis MN: University of Minnesota Press

Latour, B. (2004). How to talk about the body? The normative dimension of science studies. *Body & Society*, vol. 10, no. 2–3: 205–229.

Latour, B. (2009). *The politics of nature*. Cambridge, Massachusetts: .Harvard University Press.

Linton, J. and Budds, J. (2014). The hydrosocial cycle: defining and mobilizing a relational-dialectical approach to water. *Geoforum*, vol. 57: 170–180.

McGregor, A. and Thomas, A. (2018) Forest-led development? A more-than-human approach to forests in Southeast Asian development. In *Routledge Handbook of Southeast Asian Development*. London: Routledge.

McGregor, A., Challies, E., Thomas, A., Astuti, R., Howson, P., Afiff, S., Kindon, S. and Bond, S. (2019). Sociocarbon cycles: Assembling and governing forest carbon in Indonesia. *Geoforum,* vol. 99: 32–41.

Pitt, H. (2015). On showing and being shown plants: a guide to methods for more-than-human geography. *Area*, vol. 47, no. 1: 48–55.

Plumwood, V. (1995). Human vulnerability and the experience of being prey. *Quadrant*, vol. 39, no. 3: 29.

Power, E.R. (2019). Assembling the capacity to care: caring-with precarious housing. *Transactions: Institute of British Geographers,* vol. 44, no. 4: 763–777.

Richardson-Ngwenya, P. (2014). Performing a more-than-human material imagination during fieldwork: muddy boots, diarizing and putting vitalism on video. *Cultural Geographies*, vol. 21, no. 2: 293–299.

Stensöta, H.O. (2015). Public ethics of care – A general public ethics. *Ethics and Social Welfare*, vol. 9, no. 2: 183–200.

Suchet-Pearson, S., Wright, S., Lloyd, K., and Burarrwanga, L. (2013). Caring as country: towards an ontology of co-becoming in natural resource management. *Asia Pacific Viewpoint*, vol. 54, no. 2: 185–197.

Titz, A., Cannon, T., and Krüger, F. (2018). Uncovering 'community': challenging an elusive concept in development and disaster related work. *Societies,* vol. 8, no. 3, 71, www.doi.org/10.3390/soc8030071

Transforming Our World: The 2030 Agenda for Sustainable Development (2015). *A/RES/70/1,* UN General Assembly. New York: United Nations.

Tronto, J.C. (1993). *Moral boundaries: a political argument for an ethic of care.* London: Routledge.

Tronto, J.C. (2013). *Caring democracy: markets, equality, and justice.* New York: NYU Press

Tsing, A. (2015). *The mushroom at the end of the world: or the possibility of life in capitalist ruins.* Princeton: Princeton University Press.

UNHCR (2008). *A community-based approach in UNHCR operations.* Geneva: UNHCR.

Whatmore, S. (2002). *Hybrid geographies: Natures cultures spaces* London: Sage.

Whatmore, S.J., and Landström, C. (2011). Flood apprentices: an exercise in making things public. *Economy and Society,* vol. 40, no.4: 582–610.

Yeh, E. and Lama, K. (2013). Following the caterpillar fungus: nature, commodity chains, and the place of Tibet in China's uneven geographies. *Social and Cultural Geography,* vol. 14, no. 3: 318–340.

16

GENDER, SEXUALITY, AND ENVIRONMENT

Susie Jolly

Introduction

We are living in an environment of multiple crises including the climate emergency and climate denial. Covid has exacerbated gender inequalities and exclusions of people with non-conforming sexualities and gender identities. Development sectors and funding are under attack. Here in the UK where I write from, the international development budget is being slashed, and the UK Department for International Development is being absorbed into the Foreign and Commonwealth Office, meaning funding will be more closely tied to promoting UK interests. On the brighter side, Trump lost the US elections, COVID lockdowns reduced greenhouse gas emissions at least temporarily, and China recently announced they will become carbon neutral before 2060. I take these as glimmers of hope on which to build.

In these extreme times, what are the interconnections between gender, environment, and development? Overall, women have less control than men over natural resources such as land and water. LGBTQI+, people living with HIV, sex workers, and people with non-normative families can face greater vulnerability to impacts of environmental catastrophes, and exclusion from humanitarian aid (Root 2020; Devakula et al. 2018). Masculine elites make decisions on development trajectories, such as China's rapid economic development, which traded off air pollution and other environmental damage, and increasing inequalities, in exchange for growth. And international development sector institutions such as UNEP call for 'Green new deals' and women's participation in these projects (Wichterich 2012). How are such connections understood and elucidated? What solutions and ways forward have been proposed on these interconnected issues?

This chapter introduces conceptual tools and frameworks for addressing these connections and moving towards solutions. In this chapter, I first review the history of approaches to addressing the interconnections from '*Women Environment and Development*' and '*Gender Environment and Development*' to feminist political ecology. I discuss understandings of the terms 'gender' and related gendered understandings of 'environment,' which underlie the approaches. I then examine heteronormativity and homonormativity in the different approaches. I explore current strategies for change to avoid environmental catastrophe, and how gender relates to the solutions proposed. Finally, a section with suggestions on how to teach these issues is provided. I draw on a variety of sources, not just academic publications, but also news articles, panels and

DOI: 10.4324/9781003017653-19

interviews, podcasts, chatlogs and NGO webpages, to ensure that diverse and current debates are included.

Gender and the environment in global development discourse

'Gender' and '(the) environment' are relatively recent concepts. 'The environment' entered western discourses around the late 1960s. Gender only started to be commonly applied to human societies in the 1970s in English (MacGregor 2017), and at different times in different languages and places. In the 1990s in mainland China, the term *'shehui xingbie'* (gender) was coined and started to be used in the burgeoning feminist scene around the 1995 UN Fourth World Conference on Women held in Beijing.

Over recent decades, in the development sector, a shift from the framing of *'Women, Environment and Development'* to *'Gender, Environment and Development'* has mirrored the shift from *'Women in Development'* to *'Gender and Development'* approaches (Bradshaw and Linneker 2014). At the 1985 UN Third World Conference on Women in Nairobi, evidence was presented that women are worse affected by environmental crises and have a particular role to play in preventing it. This was discussed under the rubric of *'Women, Environment and Development.'* The shift to *'Gender, Environment and Development'* approaches in the 1990s served to emphasize that masculinity and femininity are constructed, women are not homogenous, men must be part of the analysis, and power relations between women, as well as between women and men, influence how they interact with the environment.

Different understandings of how far gender, sex and the environment are constructed underly major fissures in approaches to the interconnections. Some strands of ecofeminism take the environment to be a natural state which needs to be protected, and see women as having an essential connection to nature (Rocheleau et al. 1996). However, not all ecofeminism is essentialist, and it has been argued that even the earliest eco-feminists were aware of the potential for social construction of women's relationship with nature (MacGregor 2017).

In the 1990s, analyses were developed of gender (Rocheleau et al. 1996), sex (Butler 1990) and the environment (Escobar 1996) as constructed. Feminist political ecology emerging from development studies, and related disciplines of feminist geography and feminist political economy, emphasizes how power relations construct our gendered understandings and relationships with the environment, with these power relations mediating even the scientific understandings of environmental issues. A foundational text explains:

> *Our approach to feminist political ecology examines the very definition of 'environment' and the gendered discourse of environmental science, environmental rights and resources, and environmental movements, using feminist critiques of science...as well as the analyses and actions of feminist and environmental movements.*
>
> *(Rocheleau et al. 1996, 8)*

In parallel with the idea that how environments are understood is political, gender is understood to be a social construct which influences how environments are experienced:

> *We suggest that there are real, not imagined, gender differences in experiences of, responsibilities for, and interests in 'nature' and environments, but that these differences are not rooted in biology per se. Rather, they derive from the social interpretation of biology and social constructs of gender, which vary by culture, class, race, and place and are subject to individual and social change.*
>
> *(Rocheleau et al. 1996, 3)*

MacGregor describes how with a poststructuralist approach 'gender and environment are taken … in the field of gender and environment research … to be social constructions rather than empirical objects, there are no such things as "gender" or the "environment"' (MacGregor 2017, 2). She explains how things like climate change and trans-misogynist violence are real but the way that they are understood and the reasons they emerge depend on social contexts and power relations.

The evidence for and conceptualizations of the constructed nature of masculinities and femininities, and the emphasis on power, paved the way for applications of queer theory to the development arena from the late 1990s. These included the understanding of sex categories themselves as constructed and functioning as a means of discipline and social control (Mason 2018). The division of all people into categories of male and female has been criticized as a culturally specific western framing, at odds with the global multiplicity of sex identities repressed by colonialism (Campuzano 2006; Hines 2020) and with the biological diversities in sex characteristics (Roughgarden 2003).

In gender and development organizations, gender is generally taken to be socially constructed, but whether sex is also seen to be socially constructed or straightforwardly biological is subject to much heated debate. Many organizations recognize some mutability in sex categories. However, radical feminist 'sex-based rights' organizations and some Christian conservative groups, maintain that sex is fixed by biology, and cannot be contested (Hines 2020).

While academic discussions of the intersections ensued, in development policy in the noughties the attention to the environment decreased (Wichterich 2012; Leach 2020) and mainstream gender policies shifted to a new emphasis on a depoliticized instrumentalist version of women's economic empowerment (Kelleher 2017). With evidence emerging on the extent and speed of the climate crisis, the environment came more visibly back onto the agenda in the 2010s (Leach 2020). At the same time, the world has become more polarized between anti-gender movements working together with climate change denialists, and movements for more egalitarian relationships between people of different genders and between people and other species.

For today's 'anti-gender movement,' feeding into right wing populist regimes such as in Brazil, Poland, and Trumpian America, sex and gender are both nature, and gender is a dangerous plot to undermine the social order:

> *the religious right has turned its sights on a new secular bogeyman: 'gender ideology' … It has telescoped all the issues to which they have hewed, from abortion and gay marriage to sexuality education, into one single theory allegedly at the heart of the western secular experiment: that gender – as a mutable social construct – exists in the first place, as opposed to sex, which in their view is objective, natural and divinely ordained.*
>
> *(Gevisser 2020, 10)*

Climate denial and misogynist rhetoric interconnect. Older men sceptical of climate change may identify as the men who built the industrial world which is now called upon to change to avoid further environmental damage. They may take these calls as an attack on their masculine identities (Anshelm and Hultman 2014). Fossil fuels can contribute to identity creation. Daggett develops the concept of 'Petro masculinity' to explore the role of fossil fuels in sustaining white patriarchal rule, and contributing to the interlinked climate denial, misogyny and racism of new authoritarian movements in the West (Daggett 2018).

Both climate denialists and poststructuralist feminists have critiqued science. How do these critiques, coming from such different angles, relate to each other and to approaches such as

Greta Thunberg's call to 'unite behind the science' and 'listen to the scientists' (Milman and Smith 2020)? Some climate change sceptics or deniers explicitly critique constructivist theories such as poststructuralism and postmodernism: 'Many topics such as feminism, gender, socialism, postmodernism and climate change hysteria … they are all related' states the podcast by Naomi Seibt, a climate sceptic spokesperson linked to the Heartland Institute (a pro-Trump American thinktank) and the right wing AFD party in Germany (Cockburn 2020). However, it is also possible to use constructivist logic to critique 'the science,' which Thunberg advocates for, as constructed by social dynamics (Norgaard 2019).

Poststructuralist critiques of scientific knowledge have been charged with opening the way for relativism and a 'post-truth era' where leaders like Trump and Bolsonaro can deny the science on climate change and on COVID. How to address this pitfall? One of the leading thinkers in this domain, Donna Haraway, responds that feminism has always been 'anti-perspectival,' critiquing the post-truth idea that truth depends on which perspective you see it from. Her critiques of science as 'situated knowledges' were always 'anti-perspectival' (Haraway 2019, 41). They recognize the role of the observer in influencing the material results. She never said the results were 'made up,' but that they were 'made by the scientists' (Haraway 2019, 18). She sees the world as constituted by a series of relationships rather than a series of pre-existing units mediated by relationships. 'The layers are inherited from other layers, temporalities, scales of time and space, which don't nest neatly but have oddly configured geometries. Nothing starts from scratch' (Haraway 2019, 33). This view is entirely compatible with a materialist commitment to a better world, in contrast with 'Post-truth' which 'gives up on materialism.' And she will in a strategic essentialist move, defend the science in the current environment (Haraway 2019).

Majority and minority world power dynamics

Notwithstanding the vital contributions by majority world scholars such as most famously Vandana Shiva and Esther Wangari, scholarship on the gender-environment connections has been dominated by scholars from or in the minority world, due to colonialist and unequal means of academic production. But most explorations of climate change and gender are focussed on the majority world. Most incorporation of women or gender into environmental programmes and policy is done as part of sustainable development programming located in the majority world. While the minority world inflicts more environmental damage than the majority world, the majority world is being asked to change (MacGregor 2017). This parallels calls for poor men in the majority world to become less patriarchal and take on a greater share of reproductive labour, instead of the western dominated patriarchal systems that penalize poor people of all genders, and deplete systems of state and social care, being the ones that are asked to change (Bedford 2009). Climate finance is a key mechanism to address these inequalities. Gender analysis of climate finance is to date almost non-existent.

Population policy is a contentious example of where environment, development and gender intersect. 'Empowering poor women to limit their reproductivity in the service of global environmental policies that do little to limit the consumption patterns of the rich' (MacGregor 2017, 18). Wichterich (2012) decries the language of overpopulation in the majority world as depleting nature as racist when the regions and population groups with high birth rates have relatively low per capital resource and energy use as well as low emissions. However, others in gender and development circles, such on as the Population and Sustainability Network website, argue that 'family planning is a human-rights based climate adaptation,' and advocate better access to contraception to enable 'babies by choice not chance,' a slogan of MSI Reproductive

Choices (The Margaret Pyke Trust). The argument is made that people want choice over if and when to have babies, and that population policies need to be tethered to human rights approaches to ensure they offer genuine choice and not pressure from a top-down population control agenda.

One of the biggest changes in the international development sector in the new millennium is the rise in 'South-South Cooperation,' and international state and philanthropy and aid-like flows from majority world countries, such as China, Korea, India, and Brazil. Their changing role brings new gender and sexual orders to the development arena (Mawdsley 2020), as well as shifting environmental practices. Initial research and discussion explores such connections, for example on China's investment in the energy sector in Africa (Shen 2020), and gender and sexuality analyses of Chinese cultural narratives around becoming a world power as the US declines (Liu and Rofel 2018). However, to date exploration of gender *and* environment interconnections in South-South cooperation is a gap.

Heteronormativity and homonormativity in development approaches to gender and the environment

The term heteronormativity was originally coined by Berlant and Warner, and defined as 'the institutions (and) structures of understanding…that make heterosexuality seem not only coherent… but also privileged.' (Berlant and Warner 1998, 548). International development discourses and practices have been much critiqued as heteronormative. In regard to sexuality, men have been stereotyped as perpetrators, women as victims, and any other gender identities erased (Jolly 2007). Household models are often presented as centred around heterocouples with children, with a male head of household, regardless of reality (Jolly 2011).

Similar critiques can be applied to gender, environment, and development approaches. The persistent focus on women as victims/saviours, and men as damaging the environment, even though differences between women and men's care of the environment have been exaggerated (MacGregor 2017) echoes men as perpetrator/women as victim caricatures. Transgender people have made it into the picture in development discourses in relation to HIV, but not in relation to the environment, although evidence is beginning to be documented that they may face specific vulnerabilities and play particular roles in reconstruction, but are nonetheless are often excluded from humanitarian aid (Devakula et al. 2018).

Echoes of the heteronormative household model are found in calls to save the environment for the sake of 'future generations' (MacGregor 2017). In fact, many families do not have children, and have sometimes specifically decided not to have children for environmental reasons. On the front lines, in humanitarian aid, prioritizing families as defined as 'people with children' can in itself be understood as heteronormative as in some contexts LGBTQI+, particularly gay men, transgender and intersex people, may be less likely to have children, and face obstacles to doing so (Bailey 2020). Furthermore, the focus on future generations suggests that it's still about people rather than a less anthropocentric and speciest goal of saving animals and biodiversity more broadly (Bauman 2015).

Simply including LGBTQI+ in existing approaches may not address injustices in the development sector. A substantial body of scholarship has emerged critiquing the development sector for 'homonormativity' (Mason 2018). Where LGBTQI+ have been included, this is sometimes done using narrow understandings and problematic labels reflecting western framings of sexuality and presented as a way for the more backward majority world to catch up with the more civilized West (Gosine 2018). Note the deployment of feminism and LGBTQI+ rights by right wing nationalist parties in Europe to call for exclusion of immigrants who are posited to be more homophobic and sexist than indigenous populations. Mirroring colonialist deployments of gender and sexuality discourses, majority world governments and movements are able to deploy anti-gender narratives as a move against colonialism, and anti-colonialist narratives to combat gender equality.

Is homonormativity also present in gender, environment, and development discourses? While there may be the same potential for homonormativity in environment and development, as in the development sector more broadly, right now inclusion of gender and sexual minorities is often still seen as 'in the too hard basket' for the humanitarian sector (Humanitarian Advisory Group and VPride 2018), and funding for LGBTQI+-environment interconnections is almost non-existent. Hopefully the humanitarian sector can learn from the productive ways to tackle heteronormativity that have been created, for example the UHAI-East African Sexual Health and Rights Initiative, which facilitates participatory grantmaking for sexual and gender minorities and sex worker organizations in the region. Donors pool funds by transferring to UHAI to manage, and communities themselves decide through participatory mechanisms who gets the grants.

Ways forward for gender and the environment: a better market or beyond the market?

What solutions do feminist and gender advocates propose for the climate crisis? Economic systems are central to answering this question. From the perspective of feminist economics, the market-based view of natural resources such as land displays many similarities to the analysis of women's labour. 'Neoclassical economics bemoans the fact that both land and women are "under-utilized," lie fallow, and suffer from "underinvestment"' (Wichterich 2012, 27). Some studies apply similar logic to LGBT, that homophobia damages the economy by reducing the economic output and productivity of LGBT, and deters investment in their human capital (Badgett et al. 2019; Miller and Parker 2015).

Land is considered 'waste land,' even if in fact it's growing wildflowers which sustain vital biodiversity in insect populations or has become a squatters' community providing homes. 'In the same vein, women's labour for reproduction and subsistence farming are not perceived as productive or adding value but as external to the (market) economy, even though they secure social reproduction and sometimes also the regeneration of nature' (Wichterich 2012, 27). The monetary value of unpaid care work of women age 15 and over globally amounts to at least 10.8 trillion USD/year ('Time to Care: Unpaid and underpaid care work and the global inequality crisis' 2020). The monetary costs to society and economy of industrial air pollution in Europe alone were calculated to be up to EUR 189 billion already in 2012 (Adams 2014). These effectively constitute free subsidies for profit making enterprises which pay the costs of neither the pollution they cause, nor the regenerative labour which produces and sustains their workforces and environments. A report on the economics of biodiversity recently commissioned by the UK Department of Treasury concludes: 'The solution starts with understanding and accepting a simple truth: our economies are embedded within Nature, not external to it'(Dasgupta 2021, 5).

Economic models proposed to address the intertwined problems of climate crisis and misogyny range from limited reform of growth-oriented market economies to more radical change. Where do women's and LGBTQI+ organizations stand on these issues? Just as on other issues, the standpoints cover a range of positions, some examples of which are outlined below.

The Global Gender and Climate Alliance, a coalition of UN and civil society organizations seeks to mainstream gender to ensure women in the majority world benefit, for example by participation in commercialization of forest protection. They make gender equality the priority and remain agnostic on questions such as whether the market provides the best solution or not (Wichterich 2012, 22). In contrast, The Women's Global Call for Climate Justice, a campaign organized by regionally diverse women's rights organizations, calls for systems change and 'urgent reduction in production and consumption patterns by everyone, including and especially by those who have contributed the most to this problem from the developed world' (The Women's Global Call for Climate Justice, para. 21). There is some overlap between the two, for example the Women's Environment and Development Organization is on the organizational committees of both, suggesting a pragmatic approach by the organization to engage with different strategies.

Some LGBTQI+ organizers are engaging with environmental justice movements and calling for inclusion of LGBTQI+ and mainstreaming of gender and sexual rights into these arenas. They also take different positions on the market and how to achieve justice and inclusion. Queer members of Young Friends of the Earth Scotland have posted a blog arguing:

> *Capitalism and colonialism fuel climate change but are also a part of LGBTQ+ oppression. The same colonising forces which brought environmental degradation and extractivism also repressed and attempted to wipe out the diversity of sexuality and gender in indigenous communities across the world.* (Randall 2020)

This position contrasts with that of Queersxclimate.org, an international NGO based in Mexico describing itself as 'a queer led initiative promoting climate activism among the LGBTQ+ community.' They argue that LGBTQ+ are among the most vulnerable communities which will be further marginalized by climate crisis and 'All the development accomplishments, including the achievements on LGBTQ+ rights, could be erased within barely a decade by exacerbating resource scarcity and social unrest.' They seek to catalyse LGBTQ+ activists to contribute and lend power to environmental activism (Queers X Climate). They do not take an explicit position on the market in their website, but the consumer power of the pink economy in Mexico and beyond features as a theme, and they invite members to reduce and offset their carbon footprints, rather than calling for structural change.

Regulating the market?

Will for profit enterprises pay for the environmental damage they cause and the regenerative labour which sustains their work force and environments? But should we be 'selling nature to save it'? (Wichterich 2012, 19). 'Within the globalized model of efficiency, nature has been transformed into a subsystem of the market – supposedly with the aim of protecting it. To this end, nature has been quantified, priced, privatized, and traded …' to give it value in the market economy. In parallel, should we be calling for monetary remuneration for regenerative labour and informal care work (Federici 2018)? Should there be compensation for the emotional work sexual minorities do sustaining themselves and their communities in hostile environments? Should a 'care income' be paid for people caring for households and/or the environment? (*Blueprint for Europe's Just Transition* 2019)

Can these strategies solve the intertwined problem of gender and environment? Marxist feminist Federici reflects that 'wages for housework' is an interim goal to visibilize this labour, and end its subsidy of the capitalist economy, but not the end goal of a more fundamental transformation of capitalism and gender inequality (Federici 2018). Wichterich argues that the marketization of nature is 'diametrically opposed to sufficiency and climate justice' (Wichterich 2012: 19).

A more rational market could however reduce some of the harms. Some forms of solar power and onshore wind projects are now the cheapest form of new power generation for a majority of the world's population. Soon building new solar and wind plants will be cheaper than continuing to operate existing coal plants. Electric cars and buses will soon be as cheap as their polluting alternatives. Perverse subsidies on fossil fuels continue and are boosted by COVID recovery measures, but if the longer-term costs of climate change were factored in, these subsidies would be revealed to be irrational and causing economic disaster rather than recovery ('Making peace with nature: a scientific blueprint to tackle the climate, biodiversity and pollution emergencies', 2021):

> *this is not simply a market failure … Governments almost everywhere exacerbate the problem by paying people more to exploit Nature than to protect it … A conservative estimate of the total cost globally of subsidies that damage Nature is around US$4 to 6 trillion per year.* (Dasgupta 2021, 5)

Can the markets be 'redirected' and 'tamed,' with green new deals bringing structural transformation linking social justice and livelihoods, as well as investment in green technologies designed and deployed in the social contexts of their applications, and linked to citizen action, instead of top-down regulations which sideline human rights? (Leach 2020; Mehta et al. 2019). Can markets be regulated when, as Shiva declares 'the state has been hijacked' so is no longer holding the corporates to account? How to shift power and interests to enable regulation and a reverse of subsidies of fossil fuels and other climate damaging sectors?

And is continued growth compatible with sustainability? One broad set of approaches seeks to reform and regulate the capitalist market, without challenging the fundamental logic of growth. Initiatives such as UNEP's Green New Deal attempt to sustain the neoliberal capitalist market and growth principles while regulating to ensure environmental principles. OECD proposes 'green growth' (Wichterich 2012, 38). Others argue that the focus on endless growth is part of the problem. Growth is the main driver of the rise in greenhouse gas emissions, which have risen 1.5 per cent annually on average over the past decade (Leach 2020). '… capitalism works on the false assumption of limitless growth. There is no such thing as limitless growth – except perhaps for the cancer cell' (Vandana Shiva, cited in Ismi 2012: 10) .

Beyond the market …

Do we need more radical change to bring us back from climate crisis? An Extinction Rebellion founder argues that what we need to rebel about is not the climate crisis, which is only a symptom of the current order. The cause of climate crisis is the neocolonialist, white supremacist, heteronormative, patriarchal and class hierarchical current order, which treats the world, land, minerals and certain populations within it, as mere resources which are endlessly exploitable and from which value can be extracted for consumption by a Euro-American minority (Basden 2019).

Some visions of more radical solutions are emerging.

> *Feminist economists are critical of ecological modernization and have advanced the notion of a triple crisis – of capitalism, ecology, and care – that leads to holistic policy solutions. These solutions include investment in a caring economy based on meeting human needs through*

collectivization and redistribution rather than promoting endless growth by greening the cap-
italist market, which only serves to sustain the ecologically destructive neoliberal status quo.
(MacGregor 2017, 17)

Wichterich cites several examples of alternative measures of progress that go beyond growth, GDP and the market, already developed and in some cases implemented: the Stiglitz-Sen-Fitoussi commission in France, Bhutan's *Gross National Happiness*, and the *Genuine Progress Indicator* (GPI) developed by the NGO *Redefining Progress*. Yanis Varoufakis envisions a demo-cratic economic system, retaining a market for goods and services but not for land and labour, which reorients incentives towards a green transition and gender and other equalities rather than short term exploitation (Varoufakis 2020).

What is needed, together with a new economic order, is changing the relationship between people and the rest of the natural world (Haraway 2019). The current period has been labelled 'anthropocene' to draw attention to this being the era during which humans fundamentally change the global environment. Di Chiro critiques this label in its acceptance of neoliberal capitalism as a given, and in its obscuring of differences between which humans are causing this impact and benefiting from the damage. She proposes the 'White M(A)ntrhopocene' would be a more appropriate label. She calls for a future that departs from the gendered, racist, and ethno-centric environmental politics of the Anthropocene (Di Chiro 2017). Similar ideas are discussed in chapter 14 on the Anthropocene, Capitalocene and Climate Change.

Bauman likewise challenges the usefulness of the concept Anthropocene, in part because it neglects the differences in responsibility for environmental damage between different sets of humans and suggests it would be more accurate to call it 'capitolocene.' She proposes that the term is also problematic because it suggests the continued illusion of human exceptionalism and mastery when what is needed is for us to realize our place in the larger scheme of life and the world and that we need to stop trying to master it and treat it as of service to us. Queer theory can help us think our way away from the current hierarchizations, into this new future, by envisioning 'nature' as queer, climate as weird, ourselves as not so different from animals, and abandoning 'Habits such as purity laws regarding production and reproduction' which 'help to maintain the boundary between humans and the rest of the natural world, and boundaries between male and female' (Bauman 2015, 748).

In her vision, non-human life and life systems such as the climate are understood to have some kind of agency. She explains Barad's ideas that not only humans are engaged in performativity but the whole world: 'The planetary community is not a soundboard that can be controlled by a DJ, but rather it is like the whole rave in which ravers, DJ, and music are in a constant performance together, acting and re-acting to one another' (Bauman 2015, 750). This echoes Haraway's argument that 'its relations all the way down' (Haraway 2019, 42)

Moving forward together?

How to bring about the changes needed? Both Wichterich and Shiva see coalitions of different progressive causes as having potential to shift the system. 'Sociological and economic trans-formation requires … Large coalitions and networking between all social groups that want to abandon current forms of development' (Wichterich 2012, 46). Climate justice cannot be attained through single issue politics (Basden 2019).

There are different understandings of concepts such as sex, gender, environment, and of who is included and excluded by these terms. Goals also differ: what kind of development model to aim for; growth or not; to what degree the market should be regulated or superseded by other systems; and how far it's about saving people and how much weight should be given to non-human species. But, thankfully, different groupings share some common ground in understanding the urgency of addressing climate change. On a personal note, I spent a fraught evening with a friend of several decades who has now become an organizer in the sex-based rights movement, in contrast to my intersectional feminism. We tried to identify common ground on which we could agree and move forward on as feminists. There was precious little that we could agree on relating to sex or gender. However, we both agreed these are times of crisis, and that while gender and the environment interconnect, climate change is a more urgent issue than gender. Maybe there is hope for the collective action needed.

Suggestions for teaching: analyse the present, envision the future

Students/participants can draw on their experience and knowledge of climate and/or gender activism, policymaking, research, reading etc. Materials to browse are also suggested below. These exercises could be run as virtual or face-to-face discussions, or be done individually or in small groups, and/or become topics for a term paper or blog.

Gender and heteronormativity analysis:

- Khan, S and Jolly, S (2010), *Sex, gender and development: challenging heteronormativity,* Naz Foundation, London. This picture book presents a short, accessible yet nuanced overview of how international development work is heteronormative, and what needs to change. Do you think these insights and critique apply to international development work on climate change?
- 42D library (n.d.), 42d.org on inclusion of people with diverse sexual orientations, gender identities and expressions, and sex characteristics (SOGIESC) in humanitarian and development work
- Online dialogue 'From Words to Action: What More Should Be Done to Ensure Inclusive Resilience for All in Asia and the Pacific' (Asia Pacific Adaptation Network 2020), or read summary 'Resilience For All: Key Messages From The Virtual Dialogue on Inclusive Resilience' (Anschell and Tran 2020)
- ELDIS Key issue guide on 'Heteronormativity' (Wood 2017)

Then pick a question to address and analyse from the perspective of gender and heteronormativity, including in your answers attention to gender, sexuality and intersecting inequalities, e.g., Who causes climate change? Who is most affected by climate change? Or a topic to address, e.g., water/forest management in a particular context, asking: who has decision-making power? Who benefits? Who pays the costs? How do their interests influence what happens?

Envisioning a different future

Even more challenging than analysing the current situation, is to develop visions, plans and models for a better future. Imagining this future is not just daydreaming, but an essential solutions systems thinking ('Radical futures roundtable: Young climate feminists in conversation' 2020). What does this future look like? And what how does gender relate to it?

- Wichterich, C 2012, '*The future we want: a feminist perspective,*' Heinrich Boll Foundation, Berlin.

This report advocates for a fundamental shift in the economic order, critiquing how the market economy treats care and the environment as free resources, and seeks constant growth and short term profit. If short of time, focus on chapters 5 and 6.

- 'Radical futures roundtable: Young climate feminists in conversation' 2020, Women's Environment and Development Organization, viewed 23 April 2021, www.wedo. org/radical-futures-roundtable-young-climate-feminists-in-conversation/

This podcast with young activists from majority and minority world, lays out general principles for a new economic order and a new relationship between humans and nature, beyond cisheteropatriarchal colonialist monocropping, monocultures and extractivism. Start listening at 39 minutes to go straight to the question of what the future they envision looks like.

- Bauman, W (2015), 'Climate weirding and queering nature: getting beyond the Anthropocene,' *Religions*, vol. 6: 742–754.

This philosophical article urges us to get beyond the illusion of human mastery and exceptionalism which is part of what caused climate weirding in the first place. Instead, humans should be one part of a planetary community which respects the agency of non-human beings and bodies.

- 'Making peace with nature: a scientific blueprint to tackle the climate, biodiversity and pollution emergencies' (2021), United Nations Environment Programme, www. unep.org/resources/making-peace-nature

This report proposes transforming economic, financial, and productive systems, with specific actions proposed for a wide range of actors including governments, intergovernmental organizations, financial and private sectors, NGOs, communities, and individuals, and scientific and educational organizations. Read the index, the executive summary, and search for terms such as 'gender,' 'women,' 'sex,' LGBT etc.

Consider the strengths and weaknesses of the different texts. Explore the value of the texts and how they have been or can be used productively. Critique the texts looking at what they fail to do, and who they exclude. What do these four pieces have in common and how are they different? How do they complement or contradict each other? Students can also discuss and develop their own ideas for future models, for how a particular sector, farm, forest, energy producer, community, economy, or world could function.

Students can consider the models they themselves develop, or those outlined in the texts above, and drawing on the previous section, ask if the solutions proposed are heteronormative or homonormative. Put to them the questions that Wichterich (2012, 38) proposes:

Each and every new model of prosperity and sustainability, in theory and in practice … has to answer a few crucial questions regarding gender: How does it deal with making human rights, global social rights, and decision-making rights a reality for women, minorities, indigenous peoples, migrants, etc.? Which concept of justice does the new model rely on? How does it respond to the asymmetries of power between the North and the South, between social classes, between men and women? How does it deal with the division between care economy and market economy as well as the gender-hierarchical division of labor? Which relationship to nature is it based upon?

References

42 Degrees Library n.d., Homepage, viewed 12 February 2021, www.42d.org

Adams, M (2014), 'How much does industrial air pollution cost Europe?', *EEA newsletter*, vol. 2014, no. 4, viewed 20 April 2021, www.eea.europa.eu/articles/how-much-does-industrial-air

Anschell, N and Tran, M (2020), 'Resilience for all: key messages from the Virtual Dialogue on Inclusive Resilience,' Stockholm Environment Institute: perspectives, viewed 12 February 2021, www.sei.org/perspectives/resilience-for-all-key-messages-from-the-virtual-dialogue-on-inclusive-resilience/

Anshelm, J and Hultman, M (2014), 'A green fatwā? Climate change as a threat to the masculinity of industrial modernity,' *NORMA*, vol.9, pp. 84–96.

Asia Pacific Adaptation Network (2020), *Stream 1: inclusive resilience = from words to action: what more should be done to ensure inclusive resilience for all in Asia and the Pacific*, online video, viewed 12 February 2021, https://youtu.be/iJHZew5P7m0

Badgett, MVL, Waaldijk, K and Rodgers, YV (2019), 'The relationship between LGBT inclusion and economic development: macro-level evidence', *World Development*, vol. 120, pp. 1–14.

Bailey, M (2020), 'Leaving no-one behind: including LGBTQ in responses to climate disasters,' paper presented at the *International Federation of Red Cross and Red Crescent Societies Climate: Red Summit*, 9 September 2020.

Basden, S (2019), 'Extinction rebellion isn't about the climate,' viewed 12 February 2021, https://medium.com/extinction-rebellion/extinction-rebellion-isnt-about-the-climate-42a0a73d9d49

Bauman, W (2015), 'Climate weirding and queering nature: getting beyond the Anthropocene' *Religions*, vol. 6, pp. 742–754.

Bedford, K (2009), *Developing partnerships: gender, sexuality, and the reformed world bank*, University of Minnesota Press, Minneapolis.

Berlant, L and Warner, M (1998), 'Sex in public', *Critical Inquiry*, vol. 24, pp. 547–566.

Blueprint for Europe's just transition (2019), 2nd edn, The Green New Deal for Europe, viewed 23 April 2021, https://report.gndforeurope.com/

Bradshaw, S and Linneker, B (2014), *IIED Working Paper. 'Gender and environmental change in the developing world,'* IIED, London.

Butler, J (1990), *Gender trouble: feminism and the subversion of identity*, Routledge, London.

Campuzano, G (2006), 'Reclaiming travesti histories', *IDS Bulletin*, vol. 37, pp. 34–39.

Cockburn, H (2020), 'Anti-Greta: far-right groups trying to turn teenager into climate change-denying version of Greta Thunberg,' *The Independent*, viewed 24 Feb 2020.

Daggett, C (2018), 'Petro-masculinity: fossil fuels and authoritarian desire', *Millennium-Journal of International Studies*, vol. 47, pp. 25–44.

Dasgupta, P (2021), *The economics of biodiversity: the Dasgupta review = headline messages,* HM Treasury, London.

Devakula, D, Dotter, E, Dwyer, E and Holtsberg, M (2018), *Pride in the Humanitarian System*, Edge Effect.

Di Chiro, G (2017), 'Welcome to the white (M)anthropocene? A feminist-environmentalist critique,' in S MacGregor (ed.), *Routledge Handbook of Gender and Environment*, Routledge, London.

Escobar, A (1996), 'Construction nature: Elements for a post-structuralist political ecology', *Futures*, vol. 28, Issue 4, pp. 325–343.

Federici, S (2018), 'Silvia Federici reflects on wages for housework,' in J Hoffmann and D Yudacufski (ed.), *Feminisms in motion: voices for justice, liberation, and transformation*, AK Press, Chico, CA.

Gevisser, M (2020), 'The front line of the new gender wars,' *Financial Times*, viewed 27 June 2020.

Gosine, A (2018), 'Rescue and real love: same-sex desire in international development,' in CL Mason (ed.), *Routledge handbook of queer development studies*, Routledge, Abingdon.

Haraway, D (2019), 'A giant bumptious litter: Donna Haraway on truth, technology, and resisting extinction,' in M Weigel (ed.), *Logic Magazine*, Issue 9, viewed 23 April 2021, https://logicmag.io/nature/a-giant-bumptious-litter/

Hines, S (2020), 'Sex wars and (trans) gender panics: identity and body politics in contemporary UK feminism,' *The Sociological Review*, vol. 68, pp. 699–717.

Humanitarian Advisory Group and VPride (2018), *Humanitarian Horizons Practice Paper Series. 'Taking sexual and gender minorities out of the too-hard basket,'* Humanitarian Advisory Group, Melbourne.

Ismi, A (2012), 'Capitalism is the Crisis (Part III): time to sow seeds of a better world, says Vandana Shiva,' *The Monitor,* 1 November 2012, Canadian Centre for Policy Alternatives, Ottawa.

Jolly, S (2007), 'Why the development industry should get over its obsession with bad sex and start to think about pleasure,' *IDS Working Paper,* 283.

Jolly, S (2011), 'Why is development work so straight? Heteronormativity in the international development industry,' *Development in Practice*, vol. 21, pp. 18–28.

Kelleher, F (2017), 'Disrupting orthodoxies in economic development = an African feminist perspective,' *Feminist Africa*, vol. I22, pp. 128–138.

Khan, S and Jolly, S (2010), *Sex, gender and development: challenging heteronormativity*, Naz Foundation, London.

Leach, M (2020), 'The 2020s is the decade that demands more from environmental politics,' *IDS Opinions*, viewed 12 February 2021, www.ids.ac.uk/opinions/the-2020s-the-decade-that-demands-more-from-environmental-politics/

Liu, P and Rofel, L (2018) 'Wolf Warrior II: the rise of China and gender/sexual politics,' viewed 12 February 2021, https://u.osu.edu/mclc/online-series/liu-rofel/

MacGregor, S (2017), 'Introduction,' in S MacGregor (ed.), *Routledge handbook of gender and environment*, Routledge, Abingdon.

'Making peace with nature: a scientific blueprint to tackle the climate, biodiversity and pollution emergencies' 2021, United Nations Environment Programme, www.unep.org/resources/making-peace-nature

Mason, CL (2018), 'Introduction,' in CL Mason (ed.), *Routledge handbook of queer development studies*, Routledge, Abingdon.

Mawdsley, E. (2020), 'Queering Development? The Unsettling Geographies of South-South Cooperation.' *Antipode,* vol. 52, pp. 227–245.

Mehta, L, Srivastava, S, Adam, H, Alankar, N, Bose, S, Ghosh, U and Kumar, VV (2019), 'Climate change and uncertainty from 'above' and 'below': perspectives from India,' *Regional Environmental Change*, vol. 19, pp. 1533–1547.

Miller, J and Parker, L (2015), *Open for Business: The economic and business case for global LGB&T inclusion*, Open for Business: a coalition of companies supporting global LGB&T inclusion, London.

Milman, O and Smith, D (2020) '"Listen to the scientists" Greta Thunberg urges Congress to take action,' *The Guardian*, viewed 19 September 2019.

Norgaard, KM (2019), 'Making sense of the spectrum of climate denial,' *Critical Policy Studies*, vol. 13, pp. 437–441.

Queers X Climate n.d., *About*, 12 February 2021, www.queersxclimate.org/about

'Radical futures roundtable: Young climate feminists in conversation' (2020), Women's Environment and Development Organization, viewed 23 April 2021, https://wedo.org/radical-futures-roundtable-young-climate-feminists-in-conversation/

Randall, C (2020) 'Why climate change is an LGBTQ+ issue,' viewed 12 February 2021, https://foe.scot/why-climate-change-is-an-lgbtq-issue/

Rocheleau, D, Thomas-Slayter, B and Wangari, E (1996), 'Introduction to feminist political ecology: global issues and local experience,' in D Rocheleau, B Thomas-Slayter and E Wangari (ed.), *Feminist Political Ecology: Global Issues and Local Experience*, Routledge, Abingdon.

Root, R (2020), 'Responses to climate disasters must be LGBTQ-inclusive, experts say,' viewed 12 February 2021, www.devex.com/news/responses-to-climate-disasters-must-be-lgbtq-inclusive-experts-say-98062

Roughgarden, J (2003), *Evolution's rainbow: diversity, gender, and sexuality in nature and people*, University of California Press, Oakland.

Shen, W (2020), 'China's role in Africa's energy transition: a critical review of its intensity, institutions, and impacts,' *Energy Research & Social Science*, vol. 68, 101578.

The Margaret Pyke Trust n.d., *Homepage*, viewed 1 November 2020, https://populationandsustainability.org/

The Women's Global Call for Climate Justice n.d., *The call English*, viewed 12 February 2021, https://womenclimatejustice.nationbuilder.com/the_call_english

'Time to Care: Unpaid and underpaid care work and the global inequality crisis' (2020), *Oxfam Briefing Papers*, Oxfam International, Oxford.

Varoufakis, Y (2020), *Another now: dispatches from an alternative present*, Penguin, London.

Wichterich, C (2012), *The future we want: a feminist perspective*, Heinrich Boll Foundation, Berlin.

Wood, S (2017), ELDIS Key issue guide 'heteronormativity,' viewed 12 February 2021, www.eldis.org/keyissues/heteronormativity#top

17

EXTRACTIVISM

Henry Veltmeyer

Introduction

Nature in the Global South, and the habitat of communities that are close to and dependent on it, in recent years have been degraded and overwhelmed at a shocking rate. In many places this is due to ventures such as large-scale open-pit mining, oil extraction in tropical areas, and the spread of monoculture and agro-food extraction. These and other such forms of natural resource appropriation are usually known as 'extractivisms' (Gudynas 2013). The incidence of extractivism so understood has multiplied, as reflected in the huge volume of studies, reports, and analyses, both scholarly and from within civil society, that it has generated. Many reports have documented the destructive impacts of extractivisms on local communities and the environment. Others have defended them by proclaiming their economic benefits. All of this has led this mode of appropriation of natural resources to become the centre of political debate and social mobilization. This has been so pronounced in many countries in the Global South, particularly in Latin America, that it has become a critical issue for understanding development strategies and politics – and acting on this understanding.

Extractivism as a pathway towards development
The concept of extractivism

The term extractivism, as Gudynas (2010) uses the term, can be traced back at least as far as the 1970s as a means of describing developments in the mining and oil export sectors. Although Chiasson (2016), in a systematic review of the literature, notes that one of its very first appearances in academic and political discourse was in a paper written by Bunker in 1984. In any case, Gudynas (2018) notes that the term *extractivism* was promoted by large transnational corporations in what we would today describe as the 'extractive sector' of capitalist development (mining, fossil fuels), as well as by multilateral banks and governments. However, groups and organizations and civil society, as well as environmentalists and political activists opposed to extractivism or different 'modes of extraction,' for their negative social and environmental impacts, also use the term. This scenario became more complex as of the mid-2000s, particularly in South America where a series of left-leaning post-neoliberal regimes formed in what has been described as a 'progressive cycle' (or a 'pink tide' of regime change) in Latin American

DOI: 10.4324/9781003017653-20

politics – also opted for the extraction of minerals and metals, hydrocarbons (oil and gas) and agro-food products, and the export of these 'natural resources' in primary commodity form (Veltmeyer 2020).

In the context of these developments – i.e. the advance of resource-seeking *extractive* capital (investments in the extraction of natural resources in order to export them into the expanding markets for these commodities); a primary commodities boom in capitalist markets; and the emergence of a progressive cycle of left-leaning 'progressive' regimes in the search for add-itional revenues to finance the poverty reduction programs of inclusive development – the literature on extractivism and neoextractivism grew very rapidly. At first and for some years, these studies focused mostly on the 'extractive industries' of mining or fossil fuels (mainly oil and gas), but in recent years the study of extractivism has been extended to the agricultural sector (agriculture understood broadly to include not only the extraction of agro-food products but also the harvesting of forest and seafood products) – hence the recent concern with the development and resistance dynamics of agro-extraction or agro-extractivism (McKay 2020; Veltmeyer 2019). Thus, mining (of fossil fuels or hydrocarbon forms of energy) and agro-food products constitute what economists regard as the extractive sector of the economy, a sector whose development and resistance dynamics have been the object of study and major concern of scholars and activists in the field of international development studies as well as political economy and social ecology. The primary concern of these scholars and activists is with the development and resistance dynamics of extractivism and capitalist development in the Global South on the periphery of what has been described as the 'world capitalist system' (for example, see Robinson 2007).

Extractivism(s) as capitalism and imperialism

In the recent and ongoing debates, extractivism has been associated with capitalism as a phase in the evolution of or as a particular form of capital(ism), as opposed to, e.g., industrial capital(ism); and also imperialism, with reference here to a long history of European colo-nialism and 500 years of resistance and struggle of aboriginal peoples, who in Canada and elsewhere are described as 'first nations' and indigenous communities. In these debates there has been and remains considerable confusion as to what these terms, imperialism, capitalism, extractivism mean and how they are related. Part of the confusion results from the inter-changeable use of these terms in describing the social condition of economic exploitation and colonial rule, as well as the globalizing dynamics of accumulation and the capitalist develop-ment process – expansion into non-capitalist areas of the world, the looting and appropriation of wealth, subjugation of the owners of this wealth. Another source of confusion relates to the insistence of Gudynas, among others, that both capitalism and extractivism should be thought of in the plural – as varieties of capitalism and diverse modalities of accumulation and extractivism.

What has added to the confusion and complicated the debate is that the terms capit-alism and imperialism are often either conflated, the one reduced to the other, or not clearly distinguished. In the Marxist tradition, for example, imperialism is often understood in terms established by Lenin – as a particular phase in the evolution of capitalism, the most advanced stage of capitalism characterized by the export of capital (or, in the contemporary conjuncture of the development process and context, mobilization of the free flow of capital in the form of foreign direct investment (FDI) and the formation of an international division of labour between countries on the periphery of the system (as exporters of raw materials and natural resources in primary commodity form) and the centre (as exporters of capital and value added industrial products). In this context, different types of capital have been mobilized in diverse contexts, and capitalism has assumed different forms, albeit always exhibiting the same incessant drive to accumulate. As for imperialism in its diverse forms it has always exhibited an extractivist logic as well as a logic of domination within the centre-periphery framework of the world capitalist system.

In the neoliberal era, which can be traced back to the establishment in the 1980s of a new world order that reflects a commitment to free market capitalism, the intersection or intertwining of capitalism and imperialism was given a new twist by proponents of a theory of 'the new imperialism,' who argued that with the advent of globalization and the subsequent weakening of the State, the capitalist class no longer had to rely on the powers of the State to advance their economic interest (i.e. the expansion and accumulation of capital). Freed from regulatory constraints as well as reliance on state power, the guardians of the capitalist system and other agents and members of the international capitalist class could rely instead on the economic power embodied in their multinational corporations and financial institutions (Foster 2003; Panitch and Leys 2004; Robinson 2007). Whereas imperialism hitherto had been understood as political power exercised by the State in the service of Capital, it was now described in terms of the agency of unregulated and untrammelled global economic power embodied in institutions. As for extractivism – or extractivisms, as per the insistence of Gudynas – in this context it is understood not as a mode of production (capitalism) or an agency of state power (imperialism), but as a particular form of productive or economic activity or even as an 'industry' that can be associated with either or both capitalism and imperialism, and more precisely, as an economic model or strategy – a particular way of understanding development.

Extractivism, development and the world capitalist system

Extractivism, both as a mode of accumulation and way of understanding development, has had a long and sordid history, particularly in the context of the history of European colonialism in the Americas, where 'development' from the beginning meant both the subjugation and exploitation of the indigenous population and, what in contemporary development discourse, would be described as the violation of the fundamental human and the territorial rights of this population to the land and the commons.

In this historical context of European colonialism and imperialist exploitation extractivism referred primarily to the dynamics associated with mining, the mining of gold and silver for the purpose not so much of accumulation (in the form of merchant capital) as personal enrichment, and even more so, to sustain the treasuries of the European monarchy, depleted by the forces of intra-class warfare in the struggle to maintain the privileges of the landed gentry and the aristocracy. The collateral effects and consequences of what turned out to be a 500-year struggle against colonial rule and the advance of extractive imperialism and capitalism included looting and subjugation of the indigenous population and exploitation of both their labour power and vital productive resources, separating the communities of peasant farmers and the direct producers on the land from their means of production and access to the commons livelihoods. And neither these development dynamics, nor the forces of resistance that they generated, came to an end with the transition, in the 19th century, towards capitalism understood as a system of commodity production based on the capital-labour relation, which is to say, the exchange of labour against capital for a living wage. Indeed, by numerous accounts and narratives these development and resistance dynamics continued apace and acquired a new meaning in subsequent periods of capitalist development, that included what historians such as Eric Hobsbawm termed the 'age of imperialism' (1870–1914), and also the subsequent interwar years of British imperialism or Pax Britannica that saw the unfolding of a systemic crisis and construction of the welfare state, followed by three decades of state-led 'development' dubbed by some historians as 'Pax Americana,' and then the 'neoliberal era' – three decades of experimentation with an economic model based on a commitment to the supposed virtues of free market capitalism (Harvey 2003).

In addition to mining and the development of fossil fuels capitalism (the extraction first of coal and then oil and gas as sources of energy to fuel the modern economy), extractivism in this historical context of capitalist development entailed what today would be described as agro-extractivism – the extraction for the purpose of exports to the world market of agro-food and consumer products in the form and context of plantation agriculture, in conditions of both slavery and agricultural labour (in the production of sugar, rum, palm oil, tropical fruit, cotton, and so forth).

In the late 19th century and the early decades of the 20th century, both capitalism and extractivism – extractive capitalism, we might argue – expanded into non-capitalist areas and macro-regions on the periphery of what evolved into a world system, a system based on the capital-labour relation (the exchange of labour power for a living wage) and the accumulation of industrial capital (modern industry) in the centre of the system, and the accumulation of extractive capital (investments in the extraction of natural resources for the purpose of export-ation in primary commodity form) on the extractive frontier on the periphery of the world system.

By 1945, at the end of World War II, the so-called 'age of imperialism' came to an end in the context of a series of anti-imperialist struggles and colonial wars that challenged the hegemony of British-led colonial rule and imperial power. Within the confines and institutional dynamics of a new world order established in the wake of the collapse of the British Empire and the world order established to maintain it over the years, the states that had emerged victorious from the war convened to establish the principle and institutions of a new world order based on capitalism and a system designed to reactivate the economic development process based on the accumulation of capital, modern industry and a multilateral form of international or global gov-ernance rooted in the Bretton Woods system (the World Bank, the IMF, and the GATT, what would eventually evolve into and be replaced by the WTO) and the United Nations system of multilateral governance.

This system established the foundations of a new dynamic of capitalist development – that is, six decades of 'development,' a project or idea based on international cooperation designed to improve the social condition of the population in more 'economically backward' areas of the world – that can be traced out in the form of three development-resistance cycles.

In the first cycle of this new era, 'development' was advanced by the agency of the state – the welfare-development state that had been established in the ashes of the dying British empire/European colonialism – and a series of international organizations that served as an adjunct to the state and a surrogate of this development project. Development in this context was basic-ally understood as urbanization and labourforce participation, i.e., employment in the modern sector of industry and services, productive investments in the expansion of modern industry, and improvement in both the wages and working conditions. With this perspective 'devel-opment' took the following forms: (i) absorbing the supply of surplus agriculture labour in the workforce; (ii) increasing the rate of savings (from economic activity) and investment in industrial development to boost the dynamic of capital accumulation – to activate the devel-opment impulse and expand opportunities for employment and income in the modern sector; (iii) boosting investment in productive enterprise and industrial employment on the periphery of the system, where the economy was mired in rural poverty resulting from low productivity agriculture; (iv) incorporating and capacitating rural migrants and surplus agricultural labour for entry into the labour market in the modern sector with the assistance of social welfare policies; and (v) promoting domestic industry with an industrial policy designed to protect local firms from undue competition on the world market, and regulating the flow of capital in the form of foreign investment and corporate enterprise.

Development so understood entailed a series of reforms – social reforms regarding the expansion of social policy and a structural reform of macroeconomic policy, which had the effect of changing the structure of international economic relations in which countries on the periphery were embedded, and that had them serve as exporters of raw materials and commodities required by industry and consumers at the centre of the system. In other words, development at the centre of the system was predicated on extractivism – the extraction of natural resources in the enclave economies constructed on the periphery of the system.

Peripheral capitalism and extractivism in the neoliberal era

The second cycle of development and resistance unfolded in the 1980s and 1990s within the institutional and policy framework of a new world order (neoliberal globalization) designed to liberate the 'forces of economic freedom' (investment capital, the market, private enterprise) from the regulatory constraints of the welfare-development state. A major outcome of the installation of this new world order – apart from the destruction of forces of production in agriculture and industry built up under the aegis of the development state – was the massive increase in the flow of capital in the form of foreign direct investment (FDI), particularly as regards resource-seeking extractive capital (i.e. investments in the extraction of natural resources, such as fossil fuels, industrial metals and minerals and agro-food products in high demand in capitalist markets) (Veltmeyer 2013). Latin America, in particular, was a major beneficiary or recipient of increased capital flows and the accumulation of extractive capital, particularly in South America, where a series of left-leaning 'progressive' governments sought to finance their programs of poverty reduction with the additional fiscal revenues derived from commodity exports. One result of this increased flow of extractive capital was a deepening of a regional pattern of export primarization – the reliance on commodity exports to fuel the governments' programs of economic development and poverty reduction. Another result was a transmutation of the classical or traditional form of extractivism in the region into what Gudynas and others (see Svampa 2019, in particular) have described as *neoextractivism*: the use of fiscal revenues derived from commodity exports to finance programs of poverty reduction.

Resistance on the extractive frontier

A political truism confirmed by research into the dynamics of extractivism is that each phase and advance of capitalism in the development process generates and is associated with the emergence of new forces of resistance. In the first post-war cycle of capitalist development, the resistance predominantly took the form of a class struggle for improved wages and working conditions in the cities, and in rural areas as a struggle for land, a struggle led by dispossessed peasant farmers and largely landless rural workers, a struggle that took form as an armed insurrectional movement for revolutionary change ('armies of national liberation,' as these movements styled themselves). In the second development cycle that unfolded in the 1980s and 1990s amid conditions of the new neoliberal world order and free market capitalism, the resistance took the form of a sociopolitical peasant movement that confronted the neoliberal policy agenda of governments at the time within the framework of a post-Washington consensus that sought to bring the State back into the development process in the search for a more inclusive form of development. In the third 'development' cycle that coincided with a primary commodities boom and a progressive cycle of regimes, mostly formed in South America, the epicentre of a wave of extractive capital that hit Latin America in the 1990s and the first decade of the new millennium, the resistance took the form of a post-neoliberal policy regime oriented towards the neodevelopmentalism, or 'inclusive development' (Zibechi 2012) and neoextractivism (channelling commodity export revenues into programs of poverty reduction). But on the extractive frontier, the resistance has taken different very diverse forms, including violent confrontations between mining companies and the local communities negatively impacted by their operations; protests against the destructive impacts of extractive activities on both livelihoods and the environment; an eco-territorial movement of these communities to reclaim their territorial rights; proposals of more sustainable and inclusive forms of resource development; and construction of a social and solidarity economy based on cooperativism and local community-based development. Some authors (for example, Barkin and Sánchez 2017) even view the communities on the extractive frontier as the 'new collective revolutionary subject,' an agency of transformative and revolutionary change.

Towards a new world: the politics of anti-extractivism

One of the conclusions derived from the study of extractivism is that each advance of capital in the development process, each phase in the evolution of the capitalist system, generates or leads to new forces of resistance. By the same token, we can conclude that just as development leads to resistance, resistance leads to a search for alternatives. Thus, as noted above, the latest advance of capital on the extractive frontier, and the form taken by extractivism in the current context of the development process, has not only led to powerful forces of resistance against capitalism in its current extractivist form, but to a search for alternative (post-neoliberal) and more sustainable forms of inclusive development, including alternatives to development itself. A notable example of this is the proposal by indigenous people in the highlands of Ecuador and Bolivia to replace capitalist development and extractivism with a society in which people can 'live well' in social solidarity and harmony with nature (Acosta and Martínez 2009).

This indigenous notion of 'buen vivir' (living well) has generated an extended and heated debate about alternative development pathways and how to move forward towards 'another world' of social and environmental justice. Among the unresolved issues in this debate is whether the way forward includes extractivism. There are essentially three answers to this development question. One is the view argued by mainstream development economists and organizations such as the World Bank, which is that the wealth of natural resources with which so many developing countries in the Global South are endowed provides an unparalleled economic opportunity for these countries to advance and make substantive economic progress. From this perspective extractivism is a vital component of a successful development strategy. A second position is one that is argued, inter alia, by Alvaro Garcia Linera, until recently the vice-president of Bolivia but also a well-known sociologist regarded by some as one of Latin America's leading intellectuals. The position is that just as there can be no sustained development without an effective industrial policy or a process of industrial development, there can be no way forward for developing countries on the periphery of the world system without extractivism, i.e., without the extraction and exportation of the country's natural resource wealth and with governments taking action to ensure that everyone benefits from this development process. This means that the government must appropriate its due share of resource rents in the form of royalties and taxes, and requires that any foreign investments and operational engagement in the extraction and export of these resources include a commitment of the corporations in the extractive sector to process and industrialize these resources before they are exported.

A third position, one firmly held by many indigenous organization and communities as well as environmental activists, critics, and opponents of capitalist development in all forms, but particularly in the form of extractivism, is that a country's patrimony of natural resources is more of curse than a blessing (Acosta 2013). It is not that these critics and activists are necessarily opposed to the extraction of natural resources as such. The issue is with extractivism as a modality of capital accumulation, i.e., resource development based on a reliance on foreign direct investment and the control exercised by Capital over the development process. That is, the problem is capitalism. The view of many critics and activists is that capitalism by its very nature is exploitative both of Labour and Nature and beset by contradictions that cannot be overcome within the system (Foster 2002; Harvey 2014). A sustainable development strategy would require a socialist regime wherein the decisions regarding natural resource development and the social distribution of any economic surplus would be made by the owners of these natural resources who, in many cases, are the indigenous communities.

A fourth position on the politics of extractivism and natural resource development is one that is highly contested at the level of politics and policymaking in many developed and developing countries with both a capitalist state, i.e., a state that is designed or serves above all to advance and protect the interests of the capitalist class, and indigenous communities with a territorial claim on the resources slated for extraction and development. A typical example here would be Canada, a country that is highly dependent on its wealth of natural resources to maintain its advanced level of national development, but where decision-making regarding the disposition of these resources is highly contested and subject to powerful forces of resistance mobilized by indigenous organizations and communities (see Bowles and Veltmeyer 2013).

The politics of extractivism in these countries – particularly in the mining sector – is fraught with conflict,[1] but the dominant modality is a model of private sector-led development where the sub-soil natural resources are technically 'owned' by the State, and the privately-held capitalist corporations in the extractive sector are required to advance any extractive development project within a regulatory framework that requires the companies to secure not only a political license or permit to explore and operate but also a social license from the communities implicated or affected. While a government sanctioned permit presupposes a demonstration of Corporate Social Responsibility (CSR),[2] this social licence is conditioned on the companies submitting an environmental impact assessment and securing the informed consent of the local communities. It is here where the rubber hits the road, as it were, when it comes to the politics of extractivism. We take this point up in an outline of a proposed action that could be taken at the level of education.

Extractivisms, alternatives

To explore further and more fully understand the dynamics of extractivism as a modality of accumulation and development, we need to seek the answers to questions raised by the above discussion but which are yet to be satisfactorily addressed. To move in this direction, we could use the methodology of a webinar that can serve as the basis for an intensive workshop, and which is incorporated into course curricula.

The workshop that I propose is composed of between 4 and 6 webinar sessions organized in a virtual space (e.g. via Zoom) to discuss the development and resistance dynamics of what has been described as 'extractivism,' a modality of capitalist development based on foreign investments in the large-scale acquisition of land and the extraction of natural resources and the exportation of these resources in primary commodity form (unprocessed, with little to no value added by labour). An example of the structure of such a webinar is as follows:

Session 1:   What is extractivism?

Session 2:   The economics of extractivism: development dynamics.

Session 3:   The politics of extractivism: resistance dynamics.

Sessions 1–3 will encompass an introduction to the theme, with an emphasis on the development and resistance dynamics of extractivism, specifically (i) its negative social and environmental impacts, and (ii) the alternatives advanced by grassroots organizations, social movements, and activist scholars.

Session 4:   From resistance to alternatives: post-development and 'buen vivir',

In this session, the discussion will be focused on the dynamics involved in moving from resistance to the search for alternatives to extractivism. The working idea is that alternatives are born from resistance. At issue here, with particular reference to the Global South (countries, people and communities located on the periphery of the world capitalist system), are (i) the negative social and environmental impacts of extractivism, a modality of development based on the extraction of natural resources for export in primary commodity form; and (ii) the emergence of social movements that not only protest the destructive impacts of extractivism on the environment and livelihoods, but have experimented with concrete alternative pathways towards development.

Key sources

1.   Chu Chuji, M, Rengifo, G and Gudynas, E (2019), 'Buen vivir,' in A Kothari, A Salleh, A Escobar, F Demaria and A Acosta (ed.), *Pluriverse. A post-development dictionary*, Upfront, New Delhi.

The authors elaborate on the notion of 'buen vivir,' the notion of an alternate post-development future rooted in the cosmovision of indigenous peoples in the Andean highlands (Bolivia, Ecuador). In the imaginary of this notion, post-development implies a new form of society in which people can 'live well' in social solidarity and harmony with nature. Other chapters of this volume explore the implications of this notion of 'buen vivir' and other dimensions of 'post-development' in the search for an alternative to the development model of extractive capitalism currently in vogue in Latin America and much of the world.

2.   Deonandan, K and Dougherty, M (2016), *Mining in Latin America: critical approaches to the new extractivism*. London: Routledge.

The last two decades have witnessed a dramatic expansion and intensification of mineral resource exploitation and development across the Global South, especially in Latin America. This shift has brought mining more visibly into global public debates, spurring a great deal of controversy and conflict. This volume assembles new scholarship that provides critical perspectives on these issues.

3. Gudynas, E (2021). *Extractivisms: politics, economy and ecology,* Fernwood Publications, Halifax.

The author is a social ecologist and a major theorist and analyst of the development and post-development dynamics of extractivism and neoextractivism in Latin America. This introductory book adopts an interdisciplinary and critical perspective on extractivism, incorporating contributions from economics, politics and social ecology. Gudynas explores the negative local impacts of extractivism such as ecological and health degradation or violence, along with spillover effects that redefine democracy and justice in the context of capitalist development.

4. Svampa, M (2019), *Development in Latin America: toward a new future.*

Svampa explores the contemporary development and resistance dynamics of the workings of capitalism in the context of Latin America, where these dynamics have had their most notable outcomes. She focuses on the phenomenon of 'neoextractivism,' the combination of the global advance of resource-seeking extractive capital (foreign investments in the extraction of natural resources) and the commodities consensus (export of raw materials), among both neoliberal and progressive governments. She explores the complex dynamics of socio-environmental conflict associated with neoextractivism, as well as what she refers to as the 'eco-territorial turn.' In her analysis of these dynamics, she includes both the territorial, ecological and gender dimensions of extractive capitalism.

5. Veltmeyer, H and Petras, J (2014), *The new extractivism: a model for Latin America?* Zed books, London.

Veltmeyer and Petras have edited a series of analytical probes into the development and resistance dynamics of neoextractivism, which is the national development model implemented by the left-leaning 'progressive' regimes in South America and which was formed in the context of a primary commodities boom and 'pink tide' of regime change in the first two decades of the 21st century.

Notes

1   In the mining sector, there are records of different types of protests and resistance movements since the early years of the twentieth century. But in recent years, there has been a significant escalation in the number and intensity of conflicts in the mining sector (De Echave 2008; Lucio 2016; OCMAL 2018). OCMAL, and observatory of mining conflicts for 2016 documents 219 socio-environmental conflicts in the mining sector, affecting 234 communities in 20 countries.
2   On the notion and practice related to CSR see, inter alia, Sagebien and Lindsay (2011).

References

Acosta, A (2013), 'Extractivism and neo extractivism: two sides of the same curse', in M Lang and D Mokrani (ed.), *Beyond development: alternative visions from Latin America*, Transnational Institute, Amsterdam, pp. 61–86.

Acosta, A and Martínez, E (2009), *El buen vivir. Una vía para el Desarrollo,* Abya Yala, Quito.

Barkin, D and Sánchez, A (2017), 'The collective revolutionary subject: new forms of social transformation,' paper presented at the *Revolutions: A Conference,* September, Winnipeg.

Bowles, P and Veltmeyer, H (2014), *The answer is still no: voices of resistance to the Enbridge Northern Gateway Pipeline,* Fernwood Publishing, Black rock NS.

Bunker, SG (1984), 'Modes of extraction, unequal exchange, and the progressive underdevelopment of an extreme periphery: the Brazilian Amazon, 1600–1980', *American Journal of Sociology,* vol. 89, no. 5, pp. 1017–1064.

Chiasson, T (2016), 'Neo-extractivism in Venezuela and Ecuador: a weapon of class conflict', *The Extractive Industries and Society*, vol. 3, no. 4, pp. 888–901.

De Echave, J (2008), *Diez años de minería en el Perú*, CooperAcción, Lima.

Foster, JB (2002), 'Capitalism and ecology: the nature of the contradiction', *Monthly Review*, vol. 54, no. 4, pp. 6–16.

Foster, JB (2003), 'The new age of imperialism', *Monthly Review,* vol. 55, no. 3, pp. 1–14.

Gudynas, E (2010), *Americas Program Report. The new extractivism of the 21st Century: ten urgent theses about extractivism in relation to current South American progressivism*, January 21, pp. 1–14.

Gudynas, E (2013), 'Transitions to post-extractivism: directions, options, areas of action', in *Beyond development: Alternative visions from Latin America*, Transnational Institute, Amsterdam, pp. 165–188.

Gudynas, E (2018), 'Development and nature. Modes of appropriation and Latin American extractivisms', in J Cupples, M Palomino-Schalscha and M Prieto, (ed.), *The Routledge Handbook of Latin American Development*, Routledge, New York, pp. 389–399.

Harvey, D (2003), *The new imperialism*, Oxford University Press, Oxford.

Harvey, D (2014), *Seventeen contradictions and the end of capitalism*, Profile Books, London.

Lucio, C (2016), Conflictos socioambientales, derechos humanos y movimiento indígena en el Istmo de Tehuantepec, Universidad Autónoma de Zacatecas, México.

McKay, B (2020), *The politics of control: new dynamics of agro-extractivism in Bolivia*, Fernwood publications, Black Point NS.

OCMAL (2018), Conflictos mineros en América Latina: extracción, saqueo y agresión. Observatorio de conflictos mineros de América Latina, www.ocmal.org

Panitch, L and Leys, C (2004), *The new imperial challenge*, Monthly Review Press, New York.

Robinson, W (2007) 'Beyond the theory of imperialism: global capitalism and the transnational state,' *Societies Without Borders* vol. 2, pp. 5–26.

Sagebien, J and Lindsay, N (2011), *Governance ecosystems: CSR in the Latin American mining sector*, Palgrave Macmillan, Basingstoke and New York.

Svampa, M (2019), *Las fronteras del neoextractivismo en América Latina: conflictos socioambientales, giro ecoterritorial y nuevas dependencias*, Bielefeld University Press, Bielefeld.

Veltmeyer, H (2013), 'The political economy of natural resource extraction: a new model or extractive imperialism?,' *Canadian Journal of Development Studies*, vol. 34, no. 1, pp. 79–95.

Veltmeyer, H (2019), 'Resistance, class struggle and social movements in Latin America: contemporary dynamics,' *Journal of Peasant Studies*, vol. 46, no. 6, pp. 1264–1285.

Veltmeyer, H (2020), 'Latin America in the vortex of social change: development and social movement dynamics', *World Development*, vol. 140, June.

Veltmeyer, H and Petras, J (2014), *Neoextractivism: a new model for Latin America?* Zed Books, London.

Zibechi, R (2012), *Territories in resistance: a cartography of Latin American social movements*, AK Press, Oakland.

18

RESOURCE CONFLICT

Feifei Cai and Pichamon Yeophantong

Introduction

How does natural resource exploitation impact development? Neoclassical economic arguments contend that resource extraction contributes positively to economic growth by creating employment opportunities for local people, generating revenue for the government to use for public services, and fostering local businesses and infrastructure development (Frickel and Freudenburg 1996). This line of argument has, however, been empirically and normatively challenged by those who, through such concepts as 'extractivism' (see Chapter 17 of this Handbook) and 'ecologically unequal exchange' (Rice 2007), explore the conflicts and attendant challenges to institutional quality, governance, and human development induced by unbridled and unjust resource extraction.

Although natural resources can bring prosperity to some countries, they have also been linked to poverty, corruption and conflict in many developing contexts (Brabant and Gramling 1997; Reed 2002; Karl 2004; Sawyer and Gomez 2008; Rob 2013; Miller 2015; De Haas and Poelhekke 2019), contributing to what is popularly known as the 'resource curse' (i.e. the phenomenon where a country's abundance in natural resources results in adverse economic, governance and sustainable development impacts). While scholars have identified various factors that can contribute to the emergence of such a curse, including the type of resources involved (Bulte et al. 2005), sunk costs (Barham and Coomes 2005), institutional weakness (Robinson et al. 2006), firm-state relations (Bridge 2008), and rent cycling (Auty 2010), debate persists over how the curse takes effect and how it might be avoided in different socio-political settings.

Following this, much of the scholarship on this topic has been dedicated to unravelling the potential pathways for countries to escape from the negative impacts caused by resource extraction. One approach involves drawing upon international regulations and strengthening the rule of law (Collier 2008; McGee 2009; Papillon and Rodon 2017); other approaches have tended to centre on improving natural resource management models or promoting 'environmental peacebuilding' processes (Tengbe 2001; Söderholm 2006; Travers et al. 2011; Revollo-Fernández et al. 2016). Another less conventional pathway considers the role of civic resistance in generating change: as local opposition to resource extraction – especially those related to the fossil fuel industry – has grown on a global scale, their intended and unintended effects on domestic institutions and corporate behaviour have attracted increasing scholarly attention.

DOI: 10.4324/9781003017653-21

Further to analyses that view grassroots struggles as a potential solution to address deep-seated problems of unequal and unjust resource distribution (Bebbington et al. 2008b), this chapter will reflect on how civic resistance can act as a trigger to transform resource conflicts. While the likelihood that local opposition and activism will secure peoples' environmental and human rights is determined by many complex variables including, for example, a community's capacity to demand policy accountability from governments and companies, it is imperative that we do not underestimate the potential for resistance to effect meaningful – even if incremental – social and political change.

What are resource conflicts?

Resource conflicts refer to situations where the perceived incompatibility among two or more parties results in competing interests, goals, and aspirations over the control and use of natural resources[1] (Walker and Daniels 1997). Here, natural resources are treated as the likely trigger, target, or channel for conflicts (Baechler 1998). These conflicts can be categorized by intensity as non-violent disputes, non-violent and violent crises, or wars ('Conflict Barometer 2019' 2020); by geographical scope (i.e. global, international, regional, national, or local); and by actors (e.g. as a state or subnational conflict, business-community conflict) (Vargas-Hernández and Noruzi 2009). Most studies tend to be concerned with two overarching forms of conflict: those pertaining to non-renewable resources that are crucial to 'national or military resource security', and those 'livelihood conflicts pertaining to mostly renewable resources' (Le Billon 2008, 349). The former is usually associated with armed violence, whereas the latter is often described in terms of 'ecological distribution conflicts.' This section will analyse the drivers and consequences of these two types of resource conflict.

Armed conflicts related to non-renewable resources can encompass resource-induced war, group identity conflict, as well as civil strife and insurgency. But while some have argued that resource scarcity is a key driver of this type of conflict, others have demonstrated how resource abundance can also be a cause in certain contexts. The effects of resource scarcity on armed conflicts are notably explored by the Toronto Group, who posit that environmental scarcity[2] is likely to lead to civil conflict because it induces population movement, economic decline, and the weakening of states (Homer-Dixon 1994). This argument has, however, been challenged by other researchers who claim the limited impacts of resource scarcity on the emergence of armed violence (Homer-Dixon 1995; Percival and Homer-Dixon 1996; De Soysa 2002; Magnus Theisen 2008). They instead suggest that environmental scarcity is neither a necessary nor sufficient condition for civil conflict (Dessler 1999), and that its casual effects will rely more on the mediation of social, political, and economic factors (Hauge and Ellingsen 1998; De Soysa 2000; Suliman 2005; Magnus Theisen 2008).

With respect to how resource wealth can stoke conflict, Collier and Hoeffler (1998) had identified in their seminal study how a country's reliance on primary commodity exports could be positively linked to civil strife in a curvilinear relationship (Collier and Hoeffler 1998; Collier 1999; Collier et al. 2003; Collier and Hoeffler 2004). Here, the risk of conflict – specifically, civil war – increases when a higher percentage of national income is derived from primary commodity exports, as resources can provide finance and motive to insurgents or weaken state capacity through the 'Dutch disease.' Even so, debate persists over this association, with other scholars having attempted to replicate the study but achieving partially compatible or totally contradictory outcomes (Ross 2004b).

Fearon and Laitin (2003; Fearon 2005), for example, find little evidence to support Collier and Hoeffler's argument despite using the same measure of primary commodity exports and a similar dataset. Likewise, other studies argue that the total volume of resources is not necessarily associated with the onset of civil war (De Soysa 2002; Ross 2004b), noting how 'natural resources' is defined can also impact the strength of the correlation between resources and social conflict (De Soysa and Neumayer 2007). The contention here is that the characteristics of a resource may be a more compelling determinant of social conflict due to the political processes that produce and control them (Samset 2009; Mildner et al. 2011). For instance, an abundance in oil, gas, and diamonds is often identified as a source of heightened conflict risk (De Soysa 2000; Ross 2004b; Fearon 2005; Kok et al. 2009; Carlos José Crêspo 2015; Humphreys 2005; Le Billon 2008). Some scholars have also argued that lootable[3] resources (e.g. minerals) are more likely to motivate non-separatist conflict, whereas unlootable resources (e.g. oil and gas) tend to spark separatist conflict (Ross 2004a). Still, others note that geographical factors (e.g. the level of power concentration and distance from the centre/state) may also contribute to certain resources catalysing different types of violence (Le Billon 2001).

By comparison, the nexus between natural resources and ecological distribution conflicts[4] (EDCs) is more straightforward. EDCs are 'social conflicts born from the unfair access to natural resources and the unjust burdens of pollution' (Martinez-Alier 2018, 187). The emergence of EDCs is seen across a variety of sectors, including land, water, forest, mining, biomass, and even waste disposal (Gerber 2011; Özkaynak et al. 2012; Pérez-Rincón et al. 2019). These conflicts arise from the uneven and unfair distribution not only of environmental harm and benefits, but also of the social and environmental rights that impinge on peoples' livelihoods, sense of cultural security, and ability to participate in decision-making processes (Schlosberg 2003; 2007; see also Özkaynak et al. 2012). In this way, EDCs can – and frequently do – cut across other social fault lines (e.g. class, ethnicity, and religion), which can contribute to exacerbating the intensity and duration of such conflicts (Martinez-Alier 2018).

Resource-induced armed conflicts and EDCs generate varying consequences for human security, development, and sustainability. Violent conflicts like civil wars have observable, detrimental impacts on human welfare and development through loss of life, poverty, forced migration, the destruction of infrastructure and the natural environment, as well as less 'visible' implications for state capacity, accountability, and transparency (Collier 1999; Gates et al. 2010; Matthew et al. 2009). Identifying the impacts of EDCs, by contrast, is less straightforward as they may evolve or escalate in different ways and result in a spectrum of development and governance outcomes. In the absence of effective complaint mechanisms, vulnerable groups can mobilize and fight to protect their livelihoods, culture, and/or the natural environment by adopting different methods: they might use peaceful tactics (e.g. awareness-raising campaigns, petitions, media activism, lawsuits) to defend their rights (Pérez-Rincón et al. 2019); or they might turn to more confrontational approaches (e.g. street demonstrations and marches), even when these come at the risk of repression or violent confrontation with the state's law enforcement officials (e.g. the police) or a company's security guards.

As mentioned above, the broader development outcomes of civic resistance can and do vary across different contexts. It is, of course, possible to judge the effectiveness of a resistance movement on the basis of whether the targeted extractive project continues despite local concerns and opposition; whether governments or companies are pushed to improve their social and environmental performance as a result of the resistance; or if the project in question is cancelled altogether (Bebbington et al. 2008a; Bebbington et al. 2008b). But beyond such project-level outcomes, it is equally important to acknowledge how EDCs can catalyse deeper change at the social institutional level by bringing about more inclusive and accountable governing arrangements – and thus 'sustainable transformation' (Scheidel et al. 2018; Bebbington and Bebbington 2010).

Social transformation itself refers to 'radical, systemic shifts in deeply held values and beliefs, patterns of social behaviour, and multi-level governance and management regimes' (Westley et al. 2011, 762). When applied to conflicts, social transformation as an approach does not necessarily view conflicts like EDCs as something to avoid; instead, it emphasizes the importance of addressing their root causes as well as the need to problematize the sociocultural, economic, and ecological norms and values inherent in traditional models of (unsustainable) resource use, so as to engender durable solutions within the local context (Temper et al. 2018). Accordingly, while the cumulative effects of transformation will be fundamentally radical and systemic, in practice the shifts generated will often be incremental and might not always manifest in the immediate term. That said, more studies are still needed to better document and understand the effects of EDCs on institutional arrangements and how transformation might be achieved, while ensuring that resistors are not merely co-opted by 'status quo actors' (e.g. the government or mining companies) and that changes effected in public debates on resource governance translate into actual impact on government and corporate policies (Bebbington et al. 2008b). The next section draws on the Cambodian experience to explore how EDCs can potentially contribute to more profound institutional and social change.

Ecological distribution conflicts and their policy implications for Cambodia

Since Cambodia transformed into a market economy, the commodification of natural resources – especially of forests, fisheries, water, and land – has fuelled many EDCs across the country, as the social and environmental rights of vulnerable and marginalized groups are increasingly threatened (De Lopez 2002). It warrants note here, though, that the primary source of these conflicts has shifted from forest exploitation in the 1990s to land dispossession in the 2000s. Whereas the 1990s saw forests at the centre of escalating tensions among political factions as well as large-scale private businesses and subsistence farmers (Le Billon 2000; Bottomley 2002; Le Billon 2002; Stanley 2006; Chanthy and Schweithelm 2015), the early 2000s saw the Cambodian government issuing a new *Land Law* that led the unequal distribution of land to become the most significant cause of resource conflicts in the country over the past two decades (Norén-Nilsson and Bourdier 2019).

This law replaces forest concessions with controversial economic land concessions (ELCs) which allow the Cambodian government to lease 'state private lands'[5] to domestic and foreign companies for agricultural, mineral, infrastructure, hydropower, and tourism development purposes. However, these ELCs tend to overlap with the traditional land and natural resource rights of local and Indigenous communities, often resulting in the latter's loss of access to these lands and the resources needed to sustain their livelihoods, cultural traditions, and identity. Consequently, thousands of land- and resource-related disputes have surfaced over time, with more than half a million people in both rural and urban areas adversely affected ('Land Situation in Cambodia in 2013' 2014).

To push back on the 'land-grabbing' behaviour of local authorities and companies, affected communities are seen taking up 'everyday resistance' to challenge their dispossession and displacement (Schneider 2011). Usually with assistance from local and transnational NGOs (Swift 2015), they adopt tactics that range from filing complaints, starting petitions and setting up road blockades, to public demonstrations and direct confrontation with security guards and the police to resist encroachment by the Cambodian state and concessionaires (Touch and

Neef 2015). Although such attempts are normally met with one of the following categories of response from the state agencies or companies involved – that is, repression (e.g. violent or forceful evictions), tolerance, concession with discipline, or full concession (see Cai 2010) – the outcomes of resistance can be more mixed and variegated. Indeed, there is emerging evidence of how EDCs in Cambodia may have longer term impacts on state and corporate behaviours, as government and corporate strategies vacillate between the use of explicit violence or threats of violence and more 'constructive' responses, such as the reform of regulatory rules, engagement in compensation negotiations, and implementation of more responsible investment practices (Beban et al. 2017).

Even so, such outcomes are obviously not assured and may be impacted by a plethora of factors within the Cambodian context, including the political opportunity structures that local communities and activists engage with or the degree of solidarity and unity exhibited by an affected village (Diepart et al. 2019). Accordingly, just as one hears accounts of companies being compelled, through various channels, to adopt corporate social responsibility practices or comply with global standards and guidelines (Yeophantong 2020) – for instance, the Mong Reththy Investment Cambodia Oil Palm Co. Ltd (MRICOP) became the first Cambodian company to receive the Roundtable on Sustainable Palm Oil certification in an attempt to enhance its competitiveness in the European market (where social and environmental per-formance matters to consumers) – other examples abound of concessionaires and businesses continuing to behave in a 'business as usual' manner. A case in point is the Tianjin-based Union Development Group's continued development of the controversial Dara Sakor Resort despite considerable local opposition.

At the state policy level, Cambodian Prime Minister Hun Sen issued in 2012 *Directive 01BB* as a response to the rise in land-related EDCs. Comprised of measures for 'strengthening and increasing the effectiveness of ELC management,' the directive was followed by the launch of a land titling campaign that sought to mitigate land-related conflicts ('Land tenure and land titling' 2015). The process saw over 700,000 plots in 357 communes surveyed and 3.8 million titles reportedly issued; and by the end of November 2014, more than one hundred ELCs were downsized or cancelled outright due to breach of contract ('Notification' 2014). Moreover, in 2018, the Cambodian government reorganized the country's land dispute resolution body – the National Authority for Land Dispute Resolution[6] (NALDR) – by appointing the Minister of Land Management, Urban Planning and Construction as its new president to accelerate the resolution of land conflicts (Sotheary 2018).

Critics have, however, suggested that the titles issued to villagers were likely for small land parcels (Grimsditch and Schoenberger 2015). Observations were also made about how the land titling program was notably active in the lead-up to the 2013 general election – presum-ably in a bid by the government to win public approval – but with its activities slowing down post-election. Aside from the politicization of the initiative, there were reports of problem-atic instances where the collective ownership rights of Indigenous communities were getting suspended or jeopardized, and where certificates of title were not being delivered (Council for Land Policy 2012; Rabe 2013; 'Cambodia: Land in Conflict' 2013). Furthermore, concerns have been raised over how the effects of such reform programs are largely limited due to the endemic corruption, clientelism, and government-business collusion that characterize Cambodia's neo-patrimonial politics (Un and So 2011; Dwyer 2015; Schoenberger 2015).

But while these challenges have proved enduring, it is critical that we do not overlook the positive implications of certain policy developments with respect to mitigating conflict and expanding the public sphere. For instance, in Kratie province's Snuol district, where most land concessions are located, villagers were able to oppose the questionable implementation of *Directive 01BB* by local authorities, whereupon 312 families eventually received their titles and those affected were able to exert 'their right to have rights' (Schoenberger 2017). Other researchers have likewise observed how, despite obvious flaws, the land titling campaign contributed to a shift from 'fear of authorities to a demand for greater accountability and responsiveness' (Beban et al. 2017). Indeed, the formalization of this process has proved useful to communities when local authorities were supportive of equitable distribution, and when the communities themselves were supported by strong networks that assisted with monitoring the survey and registration processes (Beban et al. 2017). To discuss the broader dynamics of positive change further, the next section turns to the notion of environmental peacebuilding and considers how it might be leveraged to address resource conflicts in a more systematic manner.

Environmental peacebuilding as a solution?

Environmental peacebuilding has been increasingly touted as a means to mitigate resource-related conflicts by focusing on how environmental conditions can foster peace as opposed to violent competition. Defined as a subcategory of peacebuilding (Harari and Roseman 2008), it underscores the necessity of comprehensive and inclusive processes and approaches to address the root causes of conflicts, specifically those with an environmental or resource dimension, and to transform relations between hostile parties (see Lederach 1997). Environmental peacebuilding is thus viewed as an instrument of conflict transformation (Carius 2006; Conca and Dabelko 2002) and operates on the premise that environmental challenges extend beyond political boundaries, requiring long-term cooperation and people-to-people interactions to resolve (Dabelko 2006). From this perspective, natural resources and the environment, more broadly, can support peacebuilding by encouraging economic recovery and sustainable livelihoods, as well as by enabling dialogue, cooperation, and confidence-building (Matthew et al. 2009; Bruch et al. 2009).

It is possible to identify three main trajectories or pathways of environmental peacebuilding (Dresse et al. 2019; Carius 2006). The first is the technical pathway that addresses the environmental causes of conflicts by reducing resource scarcity and degradation. This pathway draws attention to innovative technological solutions; but due to difficulties in coordination and resource access among stakeholders, the effectiveness of this approach vis-à-vis broader peacebuilding efforts tends to be rather constrained. The second pathway is restorative dialogue that 'provid[es] neutral and shared spaces to acknowledge past injustice and recognize the other as a legitimate interlocutor' (Dresse et al. 2019, 12). Advocates of this approach contend that environmental dialogue has the potential to change competing actors' behaviours and perceptions, which then enables cooperation in the longer term. The third pathway concerns sustainable environmental peacebuilding, which stresses the need to address deep-seated power asymmetries in order to achieve equitable ecological distribution. This approach relies heavily on there being collective action and a high degree of institutionalization which can, however, limit opportunities for conflict transformation. That said, these three pathways are not mutually exclusive; rather, they speak to two interrelated frameworks that undergird peacebuilding efforts – environmental cooperation and environmental governance.

Environmental cooperation is a central tenet of environmental peacebuilding, along with disarmament, demobilization, reintegration, and revenue-sharing from resource extraction (Dresse et al. 2016). It is viewed as 'a catalyst for reducing tension, broadening cooperation, fostering demilitarization, and promoting peace' (Conca and Dabelko 2002, 9), which is achievable through such means as cultivating interdependence between conflicting parties, fostering new norms, deepening (transnational) civil society, and promoting inclusive institutions that allow for greater transparency and accountability (Conca 2001). Much of the focus in environmental cooperation is on transforming the perceptions and behaviours of conflicting parties through dialogue and confidence-building measures. By contrast, environmental governance places stronger emphasis on 'the establishment, reaffirmation or change of institutions to resolve conflicts over environmental resources' (Paavola 2007, 94). Here, institutions are understood to encompass not only formal rules (e.g. laws, policies) but also norms, customs, and values (Nelson 2010). These institutions work to shape the strategies of key stakeholders through a range of state-, market-, and community-based governance processes which, taken together, present the possibility for hybrid approaches that see the co-management of the natural environment by the state and communities, as well as by public-private and private-social partnerships (Lemos and Agrawal 2006), as the basis for peace.

Even so, questions remain as to the conditions under which environmental peacebuilding is feasible and most effective. Especially in developing, fragile and conflict-affected situations, burgeoning empirical evidence illustrates how the feasibility, suitability and effectiveness of peacebuilding mechanisms are largely contingent on contextual variables, such as the absence of high-intensity conflict and presence of external support (Ide 2019). Moreover, it is unclear as to what extent environmental peacebuilding approaches are suited to addressing EDCs that are rooted in deep distrust between affected communities and the government and/or companies, and which essentially stem from corrupt systems of governance that exacerbate asymmetrical power relationships. What these unresolved questions reflect is how theoretical perspectives on environmental peacebuilding are still emergent, but with the consensus in academic and policy circles being that it cannot – and should not – be treated as a 'one-size-fits-all' tool or concept. For efforts at environmental peacebuilding to work, they must respect the local context and be attuned to the nature of the conflict at hand, as well as the underlying political dynamics that inform peoples' lived experiences.

Conclusion

To explain resource conflicts and how they impact development, this chapter unpacked the nexus between the environment, natural resources, and two types of conflicts – armed conflict and ecological distribution conflicts. Scholarship on this topic highlights the wide-ranging implications that such conflicts hold for development in practice, underscoring the necessity of inclusive approaches to environmental governance and resource management. The chapter also offered examples from Cambodia to illustrate how, for some communities there, EDCs constitute everyday realities that impede sustainable development.

But although resource conflicts are oftentimes perceived as negative social phenomena with profound and long-lasting ramifications, it is important to appreciate how certain conflicts, specifically EDCs, do have the potential to become catalysts for positive social and political change – if not transformation – through civic resistance that challenges the status quo and brings into relief the sources of environmental and social injustice. As seen from the Cambodian experience, even though community-based resistance has not always yielded immediate or major changes in the practices and attitudes of state agencies or companies, there are cases where gradual shifts in the behaviours of these dominant actors towards greater responsiveness to local grievances have been perceptible and are, thus, promising. Contentious politics, in this regard, can contribute to instigating conflict-transformative approaches which, in turn, can help to mediate the protracted disputes over land and other natural resources that colour the lived realities of affected communities.

Pedagogical reflections

In the classroom, it is crucial that students appreciate the complexity of resource conflict dynamics and understand their cascading negative ramifications for development. But students must also be prompted to think critically about how such conflicts can enable social change and sustainable transformation. A comparative approach is useful in this respect, whereupon students can be tasked with discussing in breakout groups how certain resource conflicts play out in different country and local contexts. Each group should be assigned a particular conflict to explore and provided with the following set of questions to help ground their discussions:

- What does resource conflict look like in your chosen case? What are the key drivers of the conflict? Who are the key actors and how are their interests defined?
- What are the observable and/or 'less visible' impacts of this conflict?
- How have the key parties (e.g. affected communities, government, business) in the conflict responded? What kind of strategies did they leverage in their responses? And how did these strategies mitigate or aggravate the local situation?
- What kind of social, policy, and/or institutional shifts (if any) do you see emerging from the conflict? Has conflict been 'transformed' in your case?

After the breakout group discussions, students can then reconvene to compare their insights with other groups. Further to this activity, we would recommend that students be given a follow-up exercise where they can choose to role-play one of these stakeholders – government advisor, company manager, development consultant, NGO representative, or affected citizen – who has been requested to provide a debrief on an actual (past or ongoing) conflict. In the debrief, they will have to come up with a response strategy that seeks to resolve, mitigate, and/or contain the problems that arise, as well as identify opportunities for (positive) change by mapping out potential leverage and pressure points within their operational environment and the local context. The debrief will need to be developed realistically – that is, in accordance with the interests and objectives of their chosen stakeholder 'identity' – and presented to the rest of the class for discussion.

Key sources

1. Le Billon, P (2008), 'Diamond wars? Conflict diamonds and geographies of resource wars,' *Annals of the Association of American Geographers*, vol. 98, no. 2, pp. 345–372.

How does geography impact resource-related conflicts? Drawing on the case of conflict diamonds, Le Billon discusses in this article the shortcomings of existing perspectives on 'resource wars' and instead advances an insightful conceptual framework that addresses the 'resource-related spaces of vulnerability, risk, and opportunities for conflict.'

2. Temper, L, Walter, M, Rodriguez, I, Kothari, A and Turhan, E (2018), 'A perspective on radical transformations to sustainability: resistances, movements and alternatives,' *Sustainability Science*, vol. 13, no. 3, pp. 747–764.

Temper et al. adopt the perspective of conflict as a productive and transformative force that can address the issue of unfair ecological distribution to achieve greater sustainability. Through a power analysis, they reflect on how sustainable societies can emerge from social movements engaged in ecological conflicts by delving into how such movements challenge hegemonic power relations and serve to create 'new subjectivities, power relations, values and institutions.'

3. Harari, N and Roseman, J (2008), *Environmental peacebuilding theory and practice*, Friends of the Earth Middle East, Amman, Jordan.

In their article, Harari and Roseman review different theories and concepts relevant to environmental peacebuilding and link theory to practice with real-world examples – namely, the Good Water Neighbors Project – to evaluate the potential and real impacts of environmental cooperation to support peacebuilding in the Middle East.

4. Diepart, JC, Ngin, C and Oeur, I (2019), 'Struggles for life: smallholder farmers' resistance and state land relations in contemporary Cambodia,' *Journal of Current Southeast Asian Affairs*, vol. 38, no. 1, pp. 10–32.

This article examines three cases of struggle against economic land concessions within the context of unequal power distribution in Cambodia. It considers how the 'peasantry' resists and negotiates with state authorities and market actors, as well as how conflict management can result in contingent rules that allow smallholder farmers to better protect their land resources.

5. Martínez-Alier, J (2002), *The environmentalism of the poor: a study of ecological conflicts and valuation*, Edward Elgar Publishing Limited, Cheltenham, UK and Northampton, MA, USA.

In this book, Martínez-Alier demonstrates the diversity and geographical breadth of contemporary and historical movements that have transformed environmental injustices into ecological distribution conflicts. Aside from pointing out how the poor are often on the side of those seeking to conserve natural resources and preserve a clean environment in these conflicts, the book raises some salient questions: who is entitled to use natural resources and determine their appropriation? How are environmental values determined in ecological conflicts? And who has the right to impose such values?

Notes

1 The UN's definition of natural resources is used here, wherein 'natural resources are natural assets (raw materials) occurring in nature that can be used for economic production or consumption' ('Glossary of environment statistics' 1997).
2 Environmental scarcity derives from the degradation and depletion of environmental resources, population growth, and unequal resource distribution (Homer-Dixon 1994).
3 The lootability of a resource refers to whether a resource can be 'easily' acquired and 'appropriated by individuals or small groups of unskilled workers' (Ross 2003, 47).
4 Ecological distribution conflicts also refer to the social, spatial, and temporal asymmetries or inequalities in how different stakeholders use environmental resources and services (Martinez-Alier and O'Connor 1999).
5 According to 2001 *Land Law*, 'state land' is categorized into state public land and state private land, with the latter able to be leased to private businesses for various purposes.
6 Five formal land dispute resolution mechanisms exist in Cambodia: the Commune Councils, Administrative Commissions, Cadastral Commission, NALDR, and the judiciary (Khuneary and Nan 2016). The duties of the NALDR include investigating and resolving land complaints across the country and overseeing the work of lower-level agencies.

References

Auty, R (2010), 'Links between resource extraction, governance and development: African experience (ARI)', *Elcano Newsletter*, vol. 72, no. 9.

Baechler, G (1998), *'Environmental Change and Security Project Report. Why Environmental Transformation Causes Violence: A synthesis,'* Woodrow Wilson Center, Washington, DC.

Barham, BL and Coomes, OT (2005), 'Sunk costs, resource extractive industries, and development outcomes', in PS Ciccantell, DA Smith and G Seidman (ed.), *Nature, raw materials, and political economy (Research in Rural Sociology and Development, Vol. 10)*, Emerald Group Publishing Limited, Bingley.

Beban, A, So, S and Un, K (2017), 'From force to legitimation: Rethinking land grabs in Cambodia', *Development and Change*, vol. 48, no. 3, pp. 590–612.

Bebbington, A, Bebbington, D, Bury, J, Lingan, J, Muñoz, J and Scurrah, M (2008a), 'Mining and social movements: Struggles over livelihood and rural territorial development in the andes', *World Development*, vol. 36, no. 12, p. 2888.

Bebbington, A, Hinojosa, L, Bebbington, DH, Burneo, ML and Warnaars, X (2008b), 'Contention and ambiguity: Mining and the possibilities of development', *Development and Change*, vol. 39, no. 6, pp. 887–914.

Bebbington, DH and Bebbington, AJ (2010), 'Extraction, territory, and inequalities: gas in the Bolivian Chaco', *Canadian Journal of Development Studies (Revue canadienne d'études du développemen)*, vol. 30, no. 1–2, pp. 259–280.

Bottomley, R (2002), 'Contested forests: an analysis of the highlander response to logging, Ratanakiri province, Northeast Cambodia', *Critical Asian Studies*, vol. 34, no. 4, pp. 587–606.

Brabant, S and Gramling, R (1997), 'Resource extraction and fluctuations in poverty: a case study', *Society & Natural Resources*, vol. 10, no. 1, pp. 97–106.

Bridge, G (2008), 'Global production networks and the extractive sector: governing resource-based development', *Journal of Economic Geography*, vol. 8, no. 3, pp. 389–419.

Bruch, C, Jensen, D, Nakayama, M, Unruh, J and Gruby, R (2009), 'Post-conflict peace building and natural resources', *Yearbook of International Environmental Law*, vol. 19, no. 1, pp. 58–96.

Bulte, E, Damania, R and Deacon, R (2005), 'Resource intensity, institutions, and development', *World Development*, vol. 33, no. 7, pp. 1029–1044.

Cai, Y (2010), *Collective resistance in China: why popular protests succeed or fail,* Stanford University Press, Stanford.

Cambodian Ministry of Land Management (2014), *Notification*, Phnom Penh.

'Cambodia: Land in Conflict – An Overview of the Land Situation' (2013), Cambodian Center for Human Rights, Phnom Penh.

Carius, A (2006), 'Environmental peacebuilding: conditions for success', *Environmental Change and Security Program Report*, no. 12, pp. 59–75.

Carlos José Crêspo, S (2015), 'Resource wars: the new landscape of global conflict', *Monções*, vol. 4, no. 8, pp. 233–236.

Chanthy, S and Schweithelm, J (2015), 'Forest resources in Cambodia's transition to peace: lessons for peacebuilding,' in H Young and L Goldman (ed.), *Livelihoods, natural resources, and post-conflict peacebuilding*, Routledge, USA and Canada.

Collier, P (1999), 'Doing well out of war,' paper presented at the *Conference on Economic Agendas in Civil Wars*, London.

Collier, P (2008), 'Laws and codes for the resource curse', *Yale Human Rights and Development Law Journal*, vol. 11, p. 9.

Collier, P, Elliott, VL, Hegre, H, Hoeffler, A, Reynal-Querol, M and Sambanis, N (2003), *Breaking the conflict trap: civil war and development policy*, Oxford University Press Washington, DC.

Collier, P and Hoeffler, A (1998), 'On economic causes of civil war', *Oxford Economic Papers*, vol. 50, no. 4, pp. 563–573.

Collier, P and Hoeffler, A (2004), 'Greed and grievance in civil war', *Oxford Economic Papers*, vol. 56, no. 4, pp. 563–595.

Conca, K (2001), 'Environmental cooperation and international peace,' in P Diehl (ed.), *Environmental Conflict: An Anthology*, Westview Press, Boulder.

Conca, K and Dabelko, GD (2002), *Environmental Peacemaking*, Johns Hopkins University Press, Baltimore.

'Conflict Barometer 2019' (2020), Heidelberg Institution for International Conflict Research, Heidelberg.

Council for Land Policy (2012), *Instruction No. 20 on the Implementation of the Royal Government's Directive No. 1 Bor.Bor*, 26 July.

Dabelko, GD (2006), 'From threat to opportunity: exploiting environmental pathways to peace,' paper presented at the *Environment, Peace and the Dialogue Among Civilizations and Cultures*, Tehran, Islamic Republic of Iran.

De Haas, R and Poelhekke, S (2019), 'Mining matters: natural resource extraction and firm-level constraints,' *Journal of International Economics*, vol. 117, pp. 109–124.

De Lopez, TT (2002), 'Natural Resource Exploitation in Cambodia: An Examination of Use, Appropriation, and Exclusion,' *Journal of Environment & Development*, vol. 11, no. 4, pp. 355–379.

De Soysa, I (2000), 'The resource curse: are civil wars driven by rapacity or paucity?' in M Berdal and DM Malone (ed.), *Greed & griecance: economic agendas in civil wars*, Lynne Rienner Publishers, Colorado.

De Soysa, I (2002), 'Paradise is a bazaar? Greed, creed, and governance in civil war, 1989–99,' *Journal of Peace Research,* vol. 39, no. 4, pp. 395–416.

De Soysa, I and Neumayer, E (2007), 'Resource wealth and the risk of civil war onset: results from a new dataset of natural resource rents, 1970–1999,' *Conflict Management and Peace Science*, vol. 24, no. 3, pp. 201–218.

Dessler, D (1999), 'Environmental Change and Security Project Report. Review of the environment, scarcity, and violence,' Woodrow Wilson Center, Washington, DC.

Diepart, JC, Ngin, C and Oeur, I (2019), 'Struggles for life: smallholder farmers' resistance and state land relations in contemporary Cambodia,' *Journal of Current Southeast Asian Affairs*, vol. 38, no. 1, pp. 10–32.

Dresse, A, Fischhendler, I, Nielsen, JØ and Zikos, D (2019), 'Environmental peacebuilding: towards a theoretical framework,' *Cooperation and Conflict*, vol. 54, no. 1, pp. 99–119.

Dresse, A, Nielsen, JØ and Zikos, D (2016), *THESys Discussion Paper No. 2016–2. Moving beyond natural resources as a source of conflict: exploring the human-environment nexus of environmental peacebuilding*, Humboldt-Universität zu Berlin, Berlin.

Dwyer, MB (2015), 'The formalization fix? Land titling, land concessions and the politics of spatial transparency in Cambodia,' *The Journal of Peasant Studies*, vol. 42, no. 5, pp. 903–928.

Fearon, JD (2005), 'Primary commodity exports and civil war,' *Journal of Conflict Resolution*, vol. 49, no. 4, pp. 483–507.

Fearon, JD and Laitin, DD (2003), 'Ethnicity, insurgency, and civil war', *The American Political Science Review*, vol. 97, no. 1, pp. 75–90.

Frickel, S and Freudenburg, WR (1996), 'Mining the past: historical context and the changing implications of natural resource extraction,' *Social Problems*, vol. 43, no. 4, pp. 444–466.

Gates, S, Hegre, H, Nygård, H and Strand, H (2010), 'WDR Background Paper. Consequences of civil conflict,' World Bank.

Gerber, J (2011), 'Conflicts over industrial tree plantations in the South: who, how and why?' *Global Environmental Change*, vol. 21, no. 1, pp. 165–176.

'Glossary of environment statistics: studies in methods' (1997), United Nations, New York.

Grimsditch, M and Schoenberger, L (2015), *New actions and existing policies: the implementation and impacts of Order 01*, The NGO Forum on Cambodia, Phnom Penh.

Harari, N and Roseman, J (2008), *Environmental peacebuilding theory and practice*, Friends of the Earth Middle East, Amman.

Hauge, W and Ellingsen, T (1998), Beyond environmental scarcity: causal pathways to conflict, *Journal of Peace Research*, vol. 35, no. 3, pp. 299–317.

Homer-Dixon, TF (1994), 'Environmental scarcities and violent conflict: evidence from cases,' *International Security*, vol. 19, no. 1, pp. 5–40.

Homer-Dixon, T (1995), 'The ingenuity gap: can poor countries adapt to resource scarcity?' *Population & Development Review*, vol. 21, no. 3, pp. 587–612.

Humphreys, M (2005), 'Natural resources, conflict, and conflict resolution: uncovering the mechanisms,' *Journal of Conflict Resolution*, vol. 49, no. 4, pp. 508–537.

Ide, T (2019), 'The impact of environmental cooperation on peacemaking: definitions, mechanisms, and empirical evidence,' *International Studies Review*, vol. 21, no. 3, pp. 327–346.

Karl, TL (2004), Oil-led development: social, political, and economic consequences. CDDRL Working Papers, 80, p. 36.

Khuneary, H and Nan, T (2016), *Land dispute resolution mechanisms in Cambodia*, Parliamentary Institute of Cambodia, Phnom Penh.

Kok, A, Lotze, W and Van Jaarsveld, S (2009), 'Natural resources, the environment and conflict,' ACCORD, Durban.

'Land Situation in Cambodia in 2013' (2014), ADHOC, Phnom Penh.

'Land tenure and land titling' (2015), Open Development Cambodia.

Le Billon, P (2000), 'The political ecology of transition in Cambodia 1989–1999: war, peace and forest exploitation,' *Development and Change*, vol. 31, no. 4, pp. 785–805.

Le Billon, P (2001), 'The political ecology of war: natural resources and armed conflicts,' *Political Geography*, vol. 20, no. 5, pp. 561–584.

Le Billon, P (2002), 'Logging in muddy waters = the politics of forest exploitation in Cambodia,' *Critical Asian Studies*, vol. 34, no. 4, pp. 563–586.

Le Billon, P (2008), 'Diamond wars? Conflict diamonds and geographies of resource wars,' *Annals of the Association of American Geographers*, vol. 98, no. 2, pp. 345–372.

Lederach, JP (1997), *Building peace: sustainable reconciliation in divided societies*, United States Institute of Peace Press, Washington.

Lemos, MC and Agrawal, A (2006), 'Environmental governance,' *Annual Review of Environment and Resources*, vol. 31, no. 1, pp. 297–325.

Magnus Theisen, O (2008), 'Blood and soil? Resource scarcity and internal armed conflict revisited,' *Journal of Peace Research*, vol. 45, no. 1, pp. 801–818.

Martínez-Alier, J (2002), *The environmentalism of the poor: a study of ecological conflicts and valuation*, Edward Elgar, Northhampton.

Martínez-Alier, J (2018), 'Ecological distribution conflicts and the vocabulary of environmental justice,' in V Dayal, A Duraiappah and N Nawn (ed.), *Ecology, Economy and Society*, Springer, Singapore.

Martínez-Alier, J and O'Connor, M (1999), 'Distributional issues: an overview,' in J van den Bergh (ed.), *Handbook of Environmental and Resource Economics*, Edward Elgar, Cheltenham.

Matthew, R, Brown, O and Jensen, D (2009), 'From conflict to peacebuilding: the role of natural resources and the environment,' in S Halle (ed.), UNEP, Nairobi.

McGee, B (2009), 'The community referendum: participatory democracy and the right to free, prior and informed consent to development,' *Berkeley Journal of International Law*, vol. 27, no. 2, p. 635.

Mildner, SA, Lauster, G and Wodni, W (2011), 'Scarcity and abundance revisited: a literature review on natural resources and conflict,' *International Journal of Conflict and Violence*, vol. 5, no. 1, pp. 156–172.

Miller, R (2015), 'Natural resource extraction and political trust,' *Resources Policy*, vol. 45, pp. 165–172.

Nelson, F (2010), 'Introduction: the politics of natural resource governance in Africa,' in F Nelson (ed.), *Community rights, conservation and contested land: the politics of natural resource governance in Africa*, Earthscan, UK.

Norén-Nilsson, A and Bourdier, F (2019), 'Introduction: social movements in Cambodia,' *Journal of Current Southeast Asian Affairs*, vol. 38, no. 1, pp. 3–9.

Özkaynak, B, Rodríguez-Labajos, B, Chicaiza, G, Conde, M, Kohrs, B, Raeva, D, Yánez, I and Walter, M (2012), 'Mining conflicts around the world: common grounds from environmental justice perspective,' in B Rodríguez-labajos (ed.), *EJOLT Report*.

Paavola, J (2007), 'Institutions and environmental governance: a reconceptualization,' *Ecological Economics*, vol. 63, no. 1, pp. 93–103.

Papillon, M and Rodon, T (2017), 'IRPP Insight: Indigenous Consent and Natural Resource Extraction,' no. 16.

Percival, V and Homer-Dixon, T (1996), 'Environmental scarcity and violent conflict: the case of Rwanda,' *The Journal of Environment & Development*, vol. 5, no. 3, pp. 270–291.

Pérez-Rincón, M, Vargas-Morales, J and Martínez-Alier, J (2019), 'Mapping and analyzing ecological distribution conflicts in Andean Countries,' *Ecological Economics*, vol. 157, pp. 80–91.

Rabe, A (2013), *'Directive 01BB in Ratanakiri Province, Cambodia: Issues and impacts of private land titling in indigenous communities,'* Asia Indigenous Peoples Pact.

Reed, D (2002), 'Resource extraction industries in developing countries,' *Journal of Business Ethics*, vol. 39, no. 3, pp. 199–226.

Revollo-Fernández, D, Aguilar-Ibarra, A, Micheli, F and Sáenz-Arroyo, A (2016), 'Exploring the role of gender in common-pool resource extraction: evidence from laboratory and field experiments in fisheries,' *Applied Economics Letters*, vol. 23, no. 13, pp. 912–920.

Rice, J (2007), 'Ecological Unequal Exchange: International Trade and Uneven Utilization of Environmental Space in the World System', *Social Forces*, vol. 85, no. 3, pp. 1369–1392.

Rob, W (2013), 'Resource extraction leaves something behind: environmental justice and mining', *International Journal for Crime, Justice and Social Democracy*, vol. 2, no. 1, pp. 50–64.

Robinson, J, Torvik, R and Verdier, T (2006), 'Political foundations of the resource curse', *Journal of Development Economics*, vol. 79, no. 2, pp. 447–468.

Ross, ML (2003), 'Oil, drugs, and diamonds: how do natural resources vary in their impact on civil war?', in K Ballentine and J Sherman (ed.), *Beyond greed and grievance: The political economy of armed conflict*, Lynne Rienner, Boulder.

Ross, ML (2004a), 'How do natural resources influence civil war? Evidence from Thirteen cases', *International Organization*, vol. 58, no. 1, pp. 35–67.

Ross, ML (2004b), 'What do we know about natural resources and civil war?', *Journal of Peace Research*, vol. 41, no. 3, pp. 337–356.

Samset, I (2009), *'Natural resource wealth, conflict, and peacebuilding,'* Ralph Bunche Institute for International Studies, New York.

Sawyer, S and Gomez, ET (2008), 'Transnational governmentality and resource extraction: indigenous peoples, multinational corporations, multilateral institutions and the state,' in UNRISD (ed.), *Identities, Conflict and Cohesion Programme Paper Number 13*.

Scheidel, A, Temper, L, Demaria, F and Martínez-Alier, J (2018), 'Ecological distribution conflicts as forces for sustainability: an overview and conceptual framework,' *Sustainability Science*, vol. 13, pp. 585–598.

Schlosberg, D (2003), 'The justice of environmental justice: reconciling equity, recognition, and participation in a political movement,' in A Light and A De-Shalit (ed.), *Moral and Political Reasoning in Environmental Practice*, The MIT Press, Cambridge.

Schlosberg, D (2007), 'Defining environmental justice: theories, movements, and nature,' paper presented at the *Annual Meeting of the Association of the American Geographers*, 17–21 April, San Francisco.

Schneider, AE (2011), 'What shall we do without our land? Land grabs and resistance in rural Cambodia,' Paper presented at the *International Conference on Global Land Grabbing*, Land Deals Politics Initiative, University of Sussex.

Schoenberger, L (2015), 'Winning back land in Cambodia: community work to navigate state land titling campaigns and large land deals', paper presented at the *Land grabbing, conflict and agrarian-environmental transformations: perspectives from East and Southeast Asia*, 5–6 June, Chiang Mai University.

Schoenberger, L (2017), 'Struggling against excuses: winning back land in Cambodia,' *Journal of Peasant Studies*, 44, no. 4, pp. 870–890.

Söderholm, P (2006), 'Environmental taxation in the natural resource extraction sector: is it a good idea?' *European Environment*, vol. 16, no. 4, pp. 232–245.

Sotheary, P (2018), 'Government reorganises land dispute resolution body', *Khmer Times*, viewed 10 June 2020, www.khmertimeskh.com/561777/government-reorganises-land-dispute-resolution-body/

Stanley, B (2006), 'Maps, not guns, resolve resource conflicts in Cambodia,' International Development Research Centre, Ottawa.

Suliman, M (2005), 'Ecology, politics and violent conflict,' *Sudanese Journal for Human Rights' Culture*, no. 1.

Swift, P (2015), 'Transnationalization of resistance to economic land concessions in Cambodia,' paper presented at the *Land grabbing, conflict and agrarian-environmental transformations: perspectives from East and Southeast Asia*, 5–6 June, Chiang Mai University.

Temper, L, Walter, M, Rodriguez, I, Kothari, A and Turhan, E (2018), 'A perspective on radical transformations to sustainability: resistances, movements and alternatives,' *Sustainability Science*, vol. 13, no. 3, pp. 747–764.

Tengbe, JB (2001), 'Simulation modelling in resource management: a sustainable development approach to resource extraction in Sierra Leone,' *Journal of Environmental Planning and Management*, vol. 44, no. 6, pp. 783–802.

Touch, S and Neef, A (2015), 'Resistance to land grabbing and displacement in rural Cambodia,' paper presented at the *Land grabbing, conflict and agrarian-environmental transformations: perspectives from East and Southeast Asia*, 5–6 June, Chiang Mai University.

Travers, H, Clements, T, Keane, A and Milner-Gulland, EJ (2011), 'Incentives for cooperation: the effects of institutional controls on common pool resource extraction in Cambodia,' *Ecological Economics*, vol. 71, no. 1, pp. 151–161.

Un, K and So, S (2011), 'Land rights in Cambodia: how neopatrimonial politics restricts land policy reform,' *Pacific Affairs,* vol. 84, no. 2, pp. 289–308.

Vargas-Hernández, J and Noruzi, M (2009), 'Scale of conflicts between firms, communities, new social movements and the role of government,' *International Journal of Management and Innovation*, vol. 1, no. 2, pp. 15–42.

Walker, GB and Daniels, SE (1997), 'Foundation of natural resource conflict,' in B Solberg and S Miina (ed.), *Conflict management and public participation in land management*, European Forest Institute, Finland.

Westley, F, Olsson, P, Folke, C, Homer-Dixon, T, Vredenburg, H, Loorbach, D, Thompson, J, Nilsson, M, Lambin, E, Sendzimir, J, Banerjee, B, Galaz, V and Leeuw, S (2011), 'Tipping toward sustainability: emerging pathways of transformation,' *Journal of the Human Environment*, vol. 40, no. 7, pp. 762–780.

Yeophantong, P (2020), 'China and the Accountability Politics of Hydropower Development: How Effective are Transnational Advocacy Networks in the Mekong Region?', *Contemporary Southeast Asia*, vol. 42, no. 1, pp. 85–117.

19

THE EXTINCTION CRISIS

Hilary Whitehouse

Introduction

This decade may prove to be one of the most important in the entirety of human history. As discussed in this Handbook, humans are currently facing a number of crises. The crisis discussed in this chapter is that of plummeting biodiversity across the globe and the effect this is having on the functional wellbeing of humans and ecosystems. An existential crisis is described as an overwhelmingly complex situation that materially affects life and the continuity of existence. Accelerating loss of biodiversity – referred to as the extinction crisis – is often positioned as an existential crisis; not only because millions of species are in immediate danger of extinction, but also because their loss will have catastrophic effects on human societies and cultures. What is at stake here is the integrity of meaningful human life on Earth, and what we do today and tomorrow will make our futures.

With that in mind, the first part of this chapter, titled 'Recklessness', summarizes some of the leading evidence for ecological devastation and its drivers. Numerous scientific reports by research organizations, the United Nations, national governments, global think tanks and respected non-government organizations (NGOs) illuminate the extent of the extinction crisis. The reports discussed in this chapter are reputable sources of knowledge on the severity of the extinction crises and its many, structural causes. Of considerable concern is the trillions of dollars expended on financing destruction of the natural world each year and concomitantly, the trillions of dollars spent annually on fossil fuel subsidies. Surprisingly little global financial resources are currently spent on protecting and conserving nature and natural capital. Speaking to this, Antonio Guterres, the Secretary General of the United Nations, has described global financial settings as 'suicidal,' and in urging change, declared 2021 the first year of the UN Decade on Ecosystem Restoration (Guterres 2020). The implications of this declaration for shifts in financing remains to be seen.

The second part of the chapter, titled 'Reconstruction', canvases a few of the many international conservation ideas and policy settings devised to directly address the extinction crisis and promote actions to restore long-term, ecological integrity. The desirable remedy in the face of any crisis is to take positive, transformative action, at whatever scale is possible. This is where education has a significant role in leveraging change. While existential crises elicit feelings of great sadness, prolonged grief, anger and despair, extensive social science research illuminates

DOI: 10.4324/9781003017653-22

that the healthiest human response is to take remedial action and materially address any crisis situation.

There are multiple strategies to turn around biodiversity loss, too many to discuss in this one chapter. Consequently, it is up to every development practitioner to keep themselves informed of biodiversity matters in their own sites of practice, including local initiatives that are reversing biodiversity loss in material ways. Knowledge is important, as is continued learning. We do not know whether global transformation can be achieved in a timely manner – as halting the extinction crisis is not yet close to the top of any nation's political or economic priorities (Bradshaw et al. 2021). However, millions of individuals, groups and communities are acting to conserve and restore biodiversity at multiple levels and scales, and in doing so, are demonstrating how relationships between humans and the other-than-human world can be better adjusted and reconstructed.

The third part of this chapter is titled 'Relationships'. One of the major causes of the extinction crisis is a dominant, dysfunctional relationship between global systems of human organization and endeavour (finance, politics, manufacturing, education, transport etc. – see Dasgupta 2021) and the earthly beings living in the 'natural' world on which we completely rely. Insights from the field of environmental education are drawn upon to consider how pedagogies of care can be implemented within development practice. In relation to biodiversity, Bradshaw et al. (2021, in abstract) summarized that, while the science is strong, awareness is weak, and that, 'without fully appreciating and broadcasting the scale of the problems and the enormity of the solutions required, society will fail to achieve even modest sustainability goals.'

The educational idea is that bringing awareness of the importance of biodiversity and enhancing kin relationships with the natural world will generate localized momentum for enabling positive actions for ecological restoration and enhanced human wellbeing. Without a significant shift in our ecological relations and enhanced consideration for the value of the lives of others, human societies will continue to exceed ecological limits meaning predicted short to medium term futures may prove most undesirable.

Recklessness

Being a naturalist in the 21st century is like being an art enthusiast in a world where an art museum burns to the ground every year.
(Entomologist) Alex Wild, December 23, 2013, Twitter post

Human actions are causing the fabric of life to unravel, posing serious threats to the quality of life of people
Díaz et al. (2020)

In the history of human thought, extinction is a comparatively recent idea (Kolbert, 2014). Yet in the 21st century, extinction is becoming the dominant threat shadowing human life. Having achieved the Anthropocene (see Chapter 14 on the Anthropocene and Climate Change), where human activities totally overwhelm the functions of the Earth system, a general, human recklessness is placing multitudes of life forms in peril. Recent quantitative analysis indicates the present mass of humans on the planet is an order of magnitude higher than that of all wild animals combined (Bar-On et al. 2018). The consequences of the accelerating biodiversity crisis are well documented, but underacknowledged outside of academia (see Diaz et al. 2020; Kolbert 2014; MacKinnon 2013). In 2018, Brooke Jarvis introduced readers of the New York Times to the term 'Insect Apocalypse' with a warning the consequences of rapidly falling insect numbers will be dire (Jarvis 2018). The crash in insect numbers is a worrying phenomenon observed worldwide, and studies show over 40 per cent of known species are at risk of extinction (Newbold et al. 2016; Sanchez-Bayo and Wyckhuys 2019). The tragedy of this immense loss of natural life prompted ecologist E.O. Wilson (2016) to name our age the Eremocine, the age of loneliness.

The independent, Intergovernmental Science-Policy Platform on Biodiversity and Ecosystem Services (IPBES) was established in 2012 to assess the state of planetary biodiversity using methods founded on the observation of biological phenomena. Its purview includes innovative methodology for the collection of evidence on a global scale, biological conservation, sustainable use of biodiversity, long-term human wellbeing and sustainable development. The IPBES is not a United Nations body, but secretariat services are provided by the United Nations Environment Programme (UNEP). The IPBES Global Assessment Report on Biodiversity and Ecosystem Services Summary for Policymakers (IPBES 2019) was approved by a UN plenary and released in May 2019. This is the first intergovernmental report of its kind and builds on the landmark Millennium Ecosystem Assessment report published in 2005.

In evaluating the evidence, the Global Assessment Report warns that, on a planetary scale, biodiversity is declining at rates unprecedented in human history. Most worrying is that the rate of species extinctions is accelerating. One million species are in danger of extinction, most within decades, unless action is taken. The current global rate of extinction is tens to hundreds of times higher than the estimated rate over the last 10 million years. Such evidence, collected from a wide range of disciplines, indicates it is now time to put biodiversity conservation at the centre of human development initiatives. The biosphere is being altered 'to an unparalleled degree across all spatial scales' (IPBES 2019, 10). Ninety-six per cent of the mass of mammals on the planet are livestock raised by humans. Every other mammal now alive falls into the 4 per cent. Similarly, 70 per cent of all birds alive today are chickens (Bar-on et al. 2018). Even with these extraordinary imbalances, the IPBES concludes it is not too late to act rapidly and systematically to transform to the ways in which we organize our activities on Earth, with the caveat warning that failure to act will cause massive problems for humanity within near decades.

Reversal of the extinction crisis requires system-wide reorganization of human activities across economic, technological, military, political, scientific, educational, social, and cultural domains, from the personal and family level to global scale arrangements. Such dramatic change will not be easy. Yet, according to the Dasgupta Review on The Economics of Biodiversity (2021), governments already possess the necessary tools at their disposal to enable effective transformation. Studies reveal that time and again humans have proven themselves to be highly adaptable and innovative in response to difficult circumstances (Degroot et al. 2021), and we are quite capable of transformation. Returning to the IPBES report, this too argues that change can be achieved through deploying existing policy instruments to remediate biodiversity loss, even as populations of individual species and animal groups are down to 5 per cent or fewer of their estimated original numbers. Few things could be more irrational than to destroy the entire fabric of planetary life for shareholder profit and private enrichment – particularly when analyses repeatedly demonstrate it's much cheaper to preserve and conserve remaining ecosystems than to reconstruct them (Dasgupta 2021).

No realistic attempts at transformation will be successful until the mass means for organizing money are reorganized, and what is known as 'natural capital' is wholly and materially accounted for in global financing (Atkins and Atkins 2019; Dasgupta 2021). According to the United Nations, US$500 billion of public sector, (taxpayer) government money is spent annually on subsidizing 'environmental harm' (UNCBD 2020, 1). This figure excludes the trillions of dollars of private sector expenditure that is annually spent on environmental destruction. The Bankrolling Extinction Report, released in 2020 by Portfolio Earth (a NGO with a directorate in London), calculates that the 50 largest investment banks provided more than US$2.7 trillion to directly finance ecosystem destruction in 2019 alone. Sectors funded included agriculture, mariculture, forestry, mining, fossil fuel extraction, urban and rural infrastructure, tourism, transport, and logistics: all of which are identified as drivers of biodiversity loss by the IPBES.

The Bankrolling Extinction Report is critical of banking practices that support a global financial system, 'that free rides on biodiversity, and the regulators and rules which govern banks currently protect them from any consequences' (Portfolio Earth 2020, 6). Compare the trillions of dollars spent every year to destroy the fabric of planetary life (which undergirds biodiversity loss), to the just US$24 billion dollars is spent annually on managing and preserving natural reserves. In order to desirably protect 30 per cent of lands and oceans by 2030, it is estimated that just US$140 billion dollars a year will be needed – this represents only 0.16 per cent of the world's GDP (CFN 2020; Waldron et al. 2020). Similarly, according to the Paulson Institute (a privately funded, independent think tank with offices in Chicago, Washington and Beijing), expenditure required to reverse biodiversity collapse by 2030 will cost between US$700 billion and a trillion dollars a year (Deutz et al. 2020). This figure is less than one third of what is known to be expended on environmental destruction, and less than a quarter of global fossil fuel subsidies, estimated to be about US$4.7 billion annually (UNEP 2021).

The good news is that sustainable finance, that is, the concept of using investments and finance instruments to create a more sustainable earth system, has grown 'from niche to mainstream' (Credit Suisse AG and McKinsey Center for Business and Environment, 2016). In her writings on systems thinking, Donella Meadows (1999, 3) argued the most effective intervention leading to positive transformational change is changing 'the mindset or paradigm out of which the system – its goals, structure, rules, delays, parameters – arises.' Investing for sustainability and reconstruction, rather than destruction, is a paradigm that's time has arrived. Blueprints for change are progressively being researched and applied. For example, Diaz et al. (2019, 1) argue that multiple opportunities exist to 'change future trajectories through transformative action.' The extinction crisis has been manufactured by and through human actions and human-made systems and hierarchies of values. The reversal of this crisis can also be manufactured. There is a great, intrinsic resilience in natural systems, and the known and emerging leverage points for positive change are actionable from the smallest scale to the largest scale. For the estimated one million species of animals and plants on the edge of extinction every leverage point counts.

Reconstruction

It'd be nice to live in a world where everything isn't a damn crisis.
Meteorologist Nick Humphrey October 18, 2020, Facebook post

Biodiversity reconstruction actively includes protection, preservation, conservation, and restoration (including rewilding) with no limit on scale. Restoration can take place at the smallest of scales to the largest. At large scale, international and national governance, and policy mechanisms in place to protect nature include the Convention on Biological Diversity (CBD), the United Nations Convention to Combat Desertification (UNCCD), and the United Nations

Framework Convention on Climate Change (UNFCCC). From 1993 to 2021 the original Convention on Biological Diversity was the legal, international instrument for conservation and sustainable use of biological diversity including the fair and equitable sharing of benefits. The Convention covered ecosystems, species, genetic resources, biotechnology (through the Cartagena Protocol on Biosafety), science, politics, education, agriculture, business, and culture. The United Nations Decade of Biodiversity produced the Strategic Plan for Biodiversity 2011–2020 as an attempt to reverse biodiversity freefall. Twenty biodiversity targets were developed as hopeful mechanisms for international planning. These are known as the Aichi Biodiversity Targets (and can be viewed at www.cbd.int/aichi-targets/). These targets and conventions have yielded positive outcomes (see Secretariat of the Convention on Biological Diversity 2020), though it must also be acknowledged that they have also been progressively undermined over the last decades, and have not prevented global scale ecocide (Bradshaw et al. 2021).

As part of efforts to head off predicted disaster, in 2021 the Campaign for Nature – a partnership of the Wyss Campaign for Nature, the National Geographic Society, and more than 100 conservation organizations around the world – called on governments around the world to actively commit to the (financed) protection of at least 30 per cent of the planet's ecosystems by 2030 (see CFN 2020). The concept of reserving 30 per cent of land masses and oceans for nature alone is one that is fast gaining global traction, with benefits for climate mitigation as well as reconstructing biodiversity. Previously, land-based protected areas have covered just 10 to 15 per cent of the globe, while only 7 per cent of oceans are currently protected – most neglectful given oceans provide 80 per cent of our planet's oxygen. In the words of Tom Lovejoy (quoted by CFN 2020, n.p.), this level of ambition is needed because 'this is the last chance to secure a functional living planet for people and other forms of life.'

Indeed, it may be that as much as 50 per cent of the planet needs permanent protection to stave off mass extinction (Locke 2014). The recently launched Global Safety Net, a spatial analysis tool for land use planning, identifies that reserving 50 per cent of terrestrial ecosystems will have the greatest effect on reversing biodiversity loss, reducing carbon emissions, promoting natural carbon capture, and providing increased protection from zoonotic diseases (Dinerstein et al. 2020). Whatever levels are politically achievable, expert agreement is that much more of the planetary surface must be reserved, and that this protection must be properly financed (Dasgupta 2021). The parallel crises of biodiversity loss and climate change can be addressed through sufficient conservation of enough nature in the right places. As the ecologist E.O. Wilson has said of the 'half earth' idea, the real challenge is that 'people haven't been thinking big enough' (in Hiss 2014).

A related concept to large scale protection, and one requiring significant on-ground action, is land degradation neutrality, where degraded lands are restored to ecological function with concomitant benefits to human communities. This idea is promoted through the Bonn Challenge, launched in 2011, and sponsored by the International Union for the Conservation of Nature (IUCN) and the German government. The Bonn Challenge seeks to recreate landscapes to sustainably protect wildlife, regenerate forests, create ecological corridors, manage water quality, grow agroforestry and food and fibre plantations, and support ecologically productive agriculture at a scale where ecological, social, and economic priorities can be balanced.[1]

Turning to the United Nations Sustainable Development Goals (SDGs), there are seven goals directly relating to biodiversity: Goals 2, 3, 6, 11, 13, 14 and 15. In terms of social mechanisms, there are significant anti-extinction leverage points located within SDG 4 – Quality Education, and SDG 5 – Gender Equality. There is a wide body of evidence from social research disciplines that achieving greater gender equality is a highly effective social strategy for addressing the extinction crisis (UN Women 2018a; UN Women 2018b). Gender is a system for structuring unequal power relations where people designated as female have been and are still denied the same freedoms and opportunities as people designated as male. Addressing pervasive gender inequalities is necessary to solving the biodiversity crisis, particularly with respect to key matters of equality of income, food security, health, education, economic opportunity, governance and decision-making (UN Women 2018b). Women are continually struggling for recognition that the impacts of pollution, biodiversity loss and environmental degradation have on their livelihoods and their children's lives. Women and their children are also far more vulnerable to environmental perils (UNDP 2013). Social justice is inextricably linked with environmental justice.

From 2014, parties to the Convention on Biological Diversity, the International Panel for Climate Change, the Green Climate Fund, and the Global Environment Facility have presented action plans to address gender equity in biodiversity conservation and anti-extinction projects. Lau (2020) warns that simplistic approaches to addressing inequalities can have unintended consequences, and approaching gender as a power-laden, sociocultural construct also requires investigating intersectionality to understand gendered subjectivities and the materialities of bio-diversity reconstruction. Anti-extinction goals are unlikely to be achieved without directly and fully addressing women's concerns in development contexts, and there is much to be done here. According to the Global Gender Gap Report 2017 (WEF 2017), at present rates of social change, global gender equality is still over two hundred years away. If we are to successfully tackle the climate crisis and biodiversity loss we can't wait that long.

With respect to quality education, the turn to 'transformation' requires significantly greater educational investment to address social inequalities and further the agenda for conservation education. As the multi-country, Transforming Education for Sustainable Futures (TESF) pro-ject recently demonstrated, greater educational investment is a profound leverage point for global change toward fairer and more equal economies that recognize the vital integrity of nat-ural systems (TESF 2020). Transformations in education mean that our common understandings of what education is and how it is practiced are also changing. One of the most interesting, and necessary, ideas to emerge is that environmentally focused education and positive ecological actions are mutually intertwined. In the past, forms of learning and education that supported action, activism and personal agency were either undervalued, faintly acknowledged or rigor-ously despised. However, the third decade of the 21st century has seen a shift in how we see formal, informal, and community sustainability education, at all levels, from kindergarten to postdoctoral studies. Included within this shift is that biodiversity concerns can no longer sit at the margins of educational practice, and neither can climate concerns. Contemporary environ-mental education, and its corollaries, education for sustainability and education for sustainable development are discipline sites where learning, agency and meaningful action for future sus-tainability powerfully combine.

At the international level, the United Nations Educational, Scientific and Cultural Organization (UNESCO) launched a roadmap in 2020 for education for sustainable devel-opment (ESD) that prescribes the responsibilities of all signatory nations across five priority action areas. Priority action area 1 states that education for sustainable development (ESD) *must* be integrated into all national and state educational policies. Education policy makers and practitioners must assume responsibility towards bringing about global transformation, and policy and practice are instrumental to scaling up ESD in all education institutions, commu-nities, and informal learning settings. Nations are asked to make ecological and socio-envir-onmental learning take their full place alongside traditional curriculum studies. The rationale is, that without greater awareness and understanding of the biodiversity crisis – and the sub-sequent need for reform in structural arrangements and priorities – life opportunities for young people and future generations will be greatly circumscribed. Indeed, in the absence of deep change, young people's futures have been scientifically described as 'ghastly' (Bradshaw et al. 2021). Raising awareness through education is the first of the Aichi Targets under the Convention on Biological Diversity, and the Dasgupta Review also places high value on ecological education.

Priority action area 2 of the UNESCO ESD roadmap encourages learners to become agents of change, re-visioning learners as active and agentic (rather than passive) and as people able to take action in their own interests and that of others, including the more-than-human others. UNESCO's vision is that every educational institution and organization will align itself with ESD principles and practices. To quote the roadmap; 'this whole-institution approach to ESD calls for learning environments where learners learn what they live and live what they learn' (p. 28). This shift in thinking no longer sees a conceptual gap between formal learning and taking positive environmental actions, thereby opening up possibilities for productive partnerships (such as between educational institutions, researchers, educators, students and communities) where learning for conservation work is actively enabled (Ardoin et al. 2020; Toomey et al. 2017).

Priority action area 3 focuses on building the capacity of educators to better understand Agenda 2030 and the United Nations Sustainable Development Goals. Educators are asked to be much better prepared in terms of enacting the knowledge, skills, values, and actions possible within robust ESD programmes – which include development programs. There is a large body of published evidence on what effective ESD looks like across different nations and priority 3 draws on that evidence. Priority action area 4 also connects strongly with ESD research that consistently shows empowering and mobilizing young people is vital to transformation practice. The roadmap recognises young people 'continue to envision the most creative and ingenious solutions to sustainability challenges' (p. 32). This clear focus on intergenerational justice acknowledges the capabilities of young people and their decision-making capacities.

Priority action area 5 is also significant for development practitioners in identifying the importance of community-scale actions for enabling learning partnerships for change. The roadmap promotes partnerships for learning and active cooperation between learning institutions, the community, and business enterprises. Promoting partnerships ensures the latest knowledge and practices can be utilized to advance local and regional sustainability agendas. Priority action area 5 recognizes the leverage power of community as the scale at which environmental education and social transformation gains meaningful traction in anti-extinction work.

All of the above mechanisms for protection, reconstruction and education have value. Without action, accelerating cascades of trophic collapse will bring human societies to the brink of wholescale disaster (Laybourn-Langton et al. 2019). Biodiversity protection contributes to climate mitigation, quality of life, human wellbeing, food and water security, and psychological and physical benefits (Naeem et al. 2016). Thankfully, many individuals, conservation groups, networks, and increasingly businesses and investors are acting in favour of life affirming paradigm change. In an interconnected world, every environmentally positive action counts and all actions are accumulative.

Relationships

In this time of extinctions, we are going to be asked again and again to take a stand for life.
(Multi-species researcher) (Deborah Bird Rose, 2017, G61)

It is important to recognize that with matters of biodiversity, the emotions we deal with when confronting the matter of looming extinction can be overwhelming. Christy Clark (2020), writing about the disastrous 2019–2020 Australian summer, said:

> *there are so many details about these unprecedented bushfires that I have no idea how to process. Like so many in Australia, I have spent hours staring in disbelief at the sheer number of fires raging … and grieving the millions of acres of forest laid to waste in their wake … How do we even measure the cost of such a loss? … Each of those animals had a life of value and experienced fear and pain as it was taken from them.*

Similarly, the blogger Umir Haque (2020) turned his attention to the Australian fires and asked, 'how do you grieve … for tiny defenceless creatures perishing in walls of flame the size of skyscrapers?' The overwhelming emotion was grief, not only for lost homes and burned farms and domestic animals, also for burned forests and the billions of dead and suffering wild animals. It is important to grieve. To grieve, gives to animals and every other living being 'the necessary gift of inherent and inalienable worth' (Haque 2020). To bear witness to terrible eco-cide, is an act of humanity and also an act of terrible shame because, if koalas and wombats and lizards 'always had inherent worth, as in the rights and resources to live … Australia's country probably wouldn't have burned' (Haque 2020; see also Anderson et al. 2020; Steffensen 2020, WWF 2020).

The central question is always one of value. As the late Deborah Bird-Rose (2017, G56) saw it, one of:

> *the most interesting things about humans is our remarkable plasticity as individuals and as a species. While cruelty is indeed one of the great insignia of a distinctly human way, there are other sides of our capacity that help bring ourselves into fellowship with others.*

The fellowship between humans and others is a characteristic of our humaneness. For so many scientists, scholars, writers, artists, activists and wildlife carers, the ways to transformation lie within renewed fellowship and kinship with what is known as the natural world. The heterogeneous groups of people who work tirelessly for wild animal, wild plant (see Magdalena 2018) and fungi conservation are too often volunteers who spend their own money and time to save lives nominally under the protection of the state. So many people care, doing what they can, where and how they can, finding themselves emotionally wobbling between distress and inspiration. Global finance pours billions into exacerbating death; the heated atmosphere turns normal fires into conflagrations and precipitation into dangerous floods; and then, caring individuals and groups rally to the rescue, coordinating with each other to scrape together the resources necessary to keep the remaining creatures and habitats alive.

There is no disputing that the living world is in trouble, and with it ourselves. Consequently, grief and hope co-exist. For Rebecca Solnit (2016, iv), hope 'locates itself in the premise that we don't know what will happen and in that spaciousness of uncertainty is room to act.' It is time to act to save our kin and in doing so offer ourselves forms of redemption. There is no other rational choice; it's clear that biodiversity protection needs massive global and national investment if anti-extinction measures at the scale required are going to be effective (CFN 2020; Paulson Institute 2020; WEF 2020). The current settings for protection are far from systematic, coherent, or robust. However, the good news is that this can change, and that settings for destruction are and can be changed by people who care. All over the globe, there are people whose care matters in the profoundest sense.

Caring counts. Indeed, it is the most important of personal human acts. Caring is a deep form of attention – what Nel Noddings (1984) called 'engrossment'. A number of pedagogical practices in environmental education begin with teaching and learning how to be more greatly attentive to the natural world, not only to observe who is missing, but also who is still there. Actions as simple as asking people to notice bird song begin the shift in attention to their noticing birds' plight (Gaskell 2020). Inattentiveness is a common failing. To live in the infinitely beautiful world and yet not know of whom the world is composed is a chastening critique of contemporary human life. Attentiveness is an invitation to kinship, and to kindship. It is for this reason that recent research has focused on affective domain learning, taking into account the strong emotions that arise when learning about the extinction crisis.

Caring is at the foundation of any pedagogy of conservation development work, no matter where this work takes place (Heise 2016). So much is already being done, from building nest boxes, devising a rescue program to halt a species from sliding into extinction, regenerating and expanding local habitat, planting food forests for landless farmers, remaking urban biodiversity tracts, financing water and food security projects, seeking meetings with policy makers to strengthen environmental protection laws, and taking direct action to protect a remnant swampland from the bulldozers; all these are positive material actions to reconstruct biodiversity and each depends on people actioning care. Care expresses value, assumes an agentic position and has material effects. Care exists and can be defined, appreciated, taught, learned, and modelled.

For the esteemed educator and philosopher Nel Noddings (2015), care is relational and transactional, receptive and reciprocal. In order for there to be care, there has to be someone who cares and someone or some ones (including an interlinked collection of beings, such as an ecosystem or special place) who is (and are) cared for. Care lies within the relationship of who or what is cared for and cared about. Material expressions of care are constituted from within that caring relationship. Every carer is attentive, open to experiencing deep emotion in the act of caring, and able to demonstrate responsiveness to the needs of those cared for. There are over one million species currently in need of urgent care. No person or group can take on the totality of the world's biodiversity crisis. However, every person and group can make a material difference to the lives (human or non-human) of those for whom you chose to care. Caring is a moral choice, and wherever you work as a development practitioner and educator, you will find a species in need of your care. Choose your animal, plant, or fungi, and pay them the respect of your attention. What needs to be done to serve your conscience and protect their lives will soon become clear enough.

Key sources

1.   Atkins, J. and Atkins, B. (Eds.) (2019) *Around the World in Eighty Species: Exploring the Business of Extinction*. New York: Routledge.

Most causes of extinction are directly or indirectly linked to corporate and financial activities, political and business corruption, and weakened regulatory settings. This collection explores the many ways in which accounting, business, politics, and finance can be used to prevent species extinctions, and how transformed finance arrangements can fund restoration and life enhancing practices. The basic argument is that extinction has been paid for, and from the ruins and remnants it's time to pay for biological and ecological restoration.

2. Kolbert, E. (2014) *The Sixth Extinction. An Unnatural History,* New York: Henry Holt and Company.

There is a reason this book won the Pulitzer Prize for general nonfiction in 2015 and was listed as the best nonfiction book of all time by *The Guardian* in 2016. Following on from her 2006 book, *Field Notes from a Catastrophe: A Frontline Report on Climate Change,* Kolbert again travels around the globe to document extinctions in oceans and on land. Her conclusion is that, should we not do enough to stem the loss, between 20 and 50 per cent 'of all living species on earth' will be extinct by the end of the 21st century. If so, this may mean the end for us too, and we may well deserve our own extinction.

3. Stanford, C. (2012) *Planet Without Apes,* Boston: Harvard University Press.

A deeply considered book about interspecies genocide. Stanford documents and analyses the tragedy of forest destruction and the systematic killing of the four types of great apes, gorillas, chimpanzees, bonobos, and orangutans – humans are considered the fifth great ape. It's a brilliant book about mostly sad events. Stanford concludes that, should we choose the right policies and financial settings, viable wild populations of remaining great apes can be preserved in their reduced numbers. He is not overly optimistic this will happen as the slaughter continues apace.

4. Steffensen, V. (2020). *Fire Country: How Indigenous Fire Management Could Help Save Australia,* Melbourne: Hardie Grant

The Australian continent is blessed with rich biodiversity that is fast becoming lost. The 2019–2020 bushfires killed billions of animals. Billions more are at risk every hot summer. This is a sensitive and highly informative book about Indigenous Australian traditional cultural and ecological knowledge. What has become known as 'cultural burning' is a means for healing county and restoring landscapes, and Steffensen carefully explains how and why Indigenous caring works.

5. Tsing, A., Swanson, H. Gan, E. and Bubandt, N. (Eds.) (2017). *Arts of Living on a Damaged Planet,* Minneapolis: University of Minnesota Press.

This is a collection of challenging and inspiring scholarly articles on interspecies sociality working at the intersections between the social and natural sciences, the arts, and humanities. The work is divided into two parts, Ghosts and Monsters, and the contributors are leaders in their fields. This is a book you can return to many times for inspiration and remembrance.

Acknowledgements

I would like to sincerely thank the editors, and especially Dr Kearrin Sims, for their most helpful feedback and continued support in preparing this chapter. It isn't easy to write about extinction.

Note

1 See www.bonnchallenge.org/about-flr

References

Anderson, P., James, P. Komesaroff, P.A. (Eds.) (2020) *Continent Aflame: Responses to an Australian Catastrophe*. Sydney: Palaver.

Ardoin, N.W., Bowers, A.W. and Gaillard, E. (2020) 'Environmental education outcomes for conservation: A systematic review.' *Biological Conservation*, 241, 108224. www.sciencedirect.com/science/article/pii/S0006320719307116

Atkins, J. and Atkins, B. (Eds.) (2019) *Around the World in Eighty Species: Exploring the Business of Extinction*. New York: Routledge.

Bar-On, Y.M., Phillips, R. and Milo, R. (2018) *The biomass distribution on Earth*. PNAS, 115 (25), 6065–6511: https://doi.org/10.1073/pnas.1711842115

Bird-Rose, D. (2017) 'Shimmer: When all you love is being trashed.' Chapter 3 in Tsing, A., Swanson, H. Gan, H. and Bubandt, N. (eds.) *The Arts of Living on a Damaged Planet*, Minneapolis: University of Minnesota Press (pages G51–G61).

Bradshaw, C.J.A. et al. (2021) 'Underestimating the challenges of avoiding a ghastly future.' Frontiers in Conservation Science, January 13, 2021: https://doi.org/10.3389/fcosc.2020.615419

Campaign for Nature (CFN) (2020). 'Highlights and policy implications of new economic report: Protecting 30 per cent of the planet for nature: costs, benefits and economic implication': www.campaignfornature.org/

Clark, C. (2020). 'Biodiversity loss is a flaming tragedy.' Eureka Street. 13 January, 2020: www.eurekastreet.com.au/article/biodiversity-loss-is-a-flaming-tragedy

Dasgupta, P. (2021) *The Economics of Biodiversity: The Dasgupta Review*. London: HM Treasury.

Degroot, D. et al. (2021) 'Towards a rigorous understanding of societal responses to climate change'. *Nature*, 591, 539–550, www.nature.com/articles/s41586-021-03190-2

Deutz, A., Heal, G. M., Niu, R., Swanson, E., Townshend, T., Zhu, L., Delmar, A., Meghji, A., Sethi, S. A., and Tobinde la Puente, J. (2020) *Financing Nature: Closing the Global Biodiversity Financing Gap*. Chicago and New York: The Paulson Institute, The Nature Conservancy, and the Cornell Atkinson Center for Sustainability.

Díaz, S. et al. (2019). 'Pervasive human-driven decline of life on Earth points to the need for transformative change.' *Science*, 366, 6471–6476. https://doi.org/10.1126/science.aax3100

Dinerstein, E., Joshi, A.R., Vynne, C., Le, T.L., Pharand-Deschenes, F., Franca, M., Fernando, S., Birch, T., Burkart, K., Asner, G.P. and Olson, D. (2020) 'A Global Safety Net to reverse biodiversity loss and stabilize Earth's climate.' *Science Advances*, 6(36). https://advances.sciencemag.org/content/6/36/eabb2824

Gaskell, D. G. (2020) 'The Voices of Birds and the Language of Belonging', *Emergence Magazine*, Issue 5, Inverness, CA: Kalliopeia Foundation. https://emergencemagazine.org/issue/language/#stories

Guterres, A. (2020) 'Secretary-General's address at Columbia University: The State of the Planet.' December 2, 2020. www.un.org/sg/en/content/sg/speeches/2020-12-02/address-columbia-university-the-state-of-the-planet

Heise, U. K. (2016) *Imagining Extinction. The Cultural Meanings of Endangered Species*. Chicago and London: The University of Chicago Press.

Haque, U. (2020) 'How Do You Grieve for a Billion Lost Lives?' Medium: Eudaimonia and Co Blog post. January 15, 2020. https://eand.co/how-do-you-grieve-for-a-billion-lost-lives-ac0c488024b5

Hiss, T. (2014) 'Can the World Really Set Aside Half of the Planet for Wildlife?' *Smithsonian Magazine*. www.smithsonianmag.com/science-nature/can-world-really-set-aside-half-planet-wildlife-180952379/?no-ist

IPBES (2019) *Global Assessment Report on Biodiversity and Ecosystem Services of the Intergovernmental Science-Policy Platform on Biodiversity and Ecosystem Services*. E. S. Brondizio, J. Settele, S. Díaz, and H. T. Ngo (eds.) Bonn: Intergovernmental Science-Policy Platform on Biodiversity and Ecosystem Services.

Jarvis, B. (2018). 'The insect apocalypse is here. What does this mean for the rest of life on Earth?' New York Times Magazine. November 27, 2018. www.nytimes.com/2018/11/27/magazine/insect-apocalypse.html

Laybourn-Langton, L., Rankin, L. and Baxter, D. (2019) *This Is A Crisis: Facing Up to the Age of Environmental Breakdown*. London: Institute for Policy Research.

Lau, J.D. (2020) 'Three lessons or gender equality in biodiversity conservation.' *Conservation Biology*, 34(6), 1589–1591.https://conbio.onlinelibrary.wiley.com/doi/full/10.1111/cobi.13487

Locke, H. (2014) 'Nature Needs Half: A Necessary and Hopeful New Agenda for Protected Areas in North America and around the World.' *The George Wright Forum*, 31(3), 359–371. www.jstor.org/stable/43598390

MacKinnon, J.B. (2013). *The Once and Future World. Nature as It Was and Could Be.* Boston: Houghton, Mifflin, Harcourt.

Magdalena, C. (2018) *The Plant Messiah. Adventures in Search of the World's Rarest Species.* New York: Doubleday.

Meadows, D. (1999). 'Leverage points: Places to intervene in a system.' Academy for System Change: http://donellameadows.org/archives/leverage-points-places-to-intervene-in-a-system/

Naeem, S., Chazdon, R., Dufy, E., Prager, C. and Worm, B. (2016) 'Biodiversity and Human Well-Being: An Essential link for Sustainable Development.' *Proceedings of the Royal Society B*, (Special feature on the value of biodiversity) 283: 20162091. https://royalsocietypublishing.org/doi/pdf/10.1098/rspb.2016.2091

Newbold, T. et al. (2016). 'Has land use pushed terrestrial biodiversity beyond the planetary boundary? A global assessment.' *Science*, 353(6296), 288–291, DOI: 10.1126/science.aaf2201

Noddings, N. (1984) *Caring. A Feminine Approach to Ethics and Moral Education.* Berkeley: University of California Press.

Noddings, N. (2015) 'Care ethics and "caring" organizations.' Chapter 5 in Engster, D. and Hamington, M. (eds.) *Care Ethics and Political Theory.* Oxford: Oxford University Press (p. 72–86).

Paulson Institute (2020) *Financing Nature: Closing the Global Biodiversity Financing Gap.* New York: Paulson Institute, The Nature Conservancy, and Cornell Atkinson Centre for Sustainability. www.paulsoninstitute.org/key-initiatives/financing-nature-report/

Portfolio Earth (2020). *Bankrolling Extinction. The Banking Sector's Role in the Biodiversity Crisis.* London: Portfolio Earth. https://portfolio.earth/

Sanchez-Bayo, F. and Wyckhuys, K.A.G. (2019) 'Worldwide decline of the entomofauna: A review of its drivers'. *Biological Conservation*, 323, 8–27. www.sciencedirect.com/science/article/pii/S0006320718313636#!

Secretariat of the Convention on Biological Diversity (2020) *Global Biodiversity Outlook 5 – Summary for Policy Makers.* Montréal. www.cbd.int/gbo/gbo5/publication/gbo-5-spm-en.pdf

Solnit, R. (2016) *Hope in the Dark: Untold Histories, Wild Possibilities.* Edinburgh: Canongate.

Steffensen, V. (2020) *Fire Country: How Indigenous Fire Management Could Help Save Australia.* Melbourne: Hardie Grant.

Toomey, A.H., Knight, A. T. and Barlow, J. (2017) 'Navigating the space between research and implementation in conservation'. *Conservation Letters* 10(5), 619–625. https://conbio.onlinelibrary.wiley.com/doi/epdf/10.1111/conl.12315

TESF (2020) 'Transforming Education for Sustainable Futures': Foundations Paper London: Transforming Education for Sustainable Futures. https://tesf.network/resource/tesf-foundations-paper/

United Nations Convention on Biodiversity (UNCBD) (2020). 'Biodiversity Numbers Fact Sheet': www.cbd.int/events/unbiodiversitysummit/factsheet-numbers.pdf

United Nations Development Programme (UNDP) (2013) 'Gender and Disaster Risk Reduction.' Gender and Climate Change: Asia and the Pacific Policy Brief 3. New York: UNDP. www.undp.org/

United Nations Environment Programme (UNEP) (2021) 'Making Peace with Nature'. A scientific blueprint to tackle the climate, biodiversity and pollution emergencies. Oslo: Norwegian Ministry of Climate and Environment. www.unep.org/resources/making-peace-nature

UN Women (2018a) 'Turning Promises into Action'. Gender Equality in the 2030 Agenda for Sustainable Development. www.unwomen.org/sdg-report

UN Women (2018b) 'Towards a gender-responsive implementation of the Convention on Biological Diversity'. November 2018. New York: UN Women. www.unwomen.org/en/digital-library/publications/2018/11/towards-a-gender-responsive-implementation-of-the-convention-on-biological-diversity

UNESCO (2020) 'Education for Sustainable Development. A Roadmap'. https://unesdoc.unesco.org/ark:/48223/pf0000374802

Waldron, A. et al. (2020) *Protecting 30 per cent of the planet for nature: costs, benefits and economic implications.* Working paper analysing the economic implications of the proposed 30 per cent target for areal protection in the draft post-2020 Global Biodiversity Framework. London: Campaign for Nature. www.campaignfornature.org/protecting-30-of-the-planet-for-nature-economic-analysis

Wilson, E.O. (2016) *Half-Earth: Out Planet's Fight for Life.* New York: W.W. Norton.

WEF (2017) *The Global Gender Gap Report 2017.* New York: World Economic Forum. www.weforum.org/reports/the-global-gender-gap-report-2017

WEF (2020) *The Global Risks Report 2020,* New York: World Economic Forum. www.oliverwyman.com/content/dam/oliver-wyman/v2/publications/2020/January/Global_Risks_Report_2020.pdf

WWF (2020) *Another 37 Million Animals Could Be Lost Next Decade If Government Fails to Properly Enforce National Environment Laws.* Sydney: WWF Australia.

20

TRANSNATIONAL ENVIRONMENTAL CRIME AND DEVELOPMENT

Lorraine Elliott

Introduction

The core of activities that constitute 'transnational environmental crime' (TEC) includes the illegal wildlife trade, timber trafficking and the trade in stolen timber, and the black market in ozone-depleting substances. As global awareness of these kinds of illegal transboundary transactions has increased, the TEC agenda has expanded to include the illegal movement across borders of hazardous and toxic wastes (including electronic wastes), fisheries crime and, in some cases, the transboundary dimensions of illegal mining, and fraud executed in the context of emissions trading schemes (sometimes referred to as carbon crime). TEC is reputed to be one of the fastest growing areas of criminal activity. Figures on the exact size and value of these kinds of illicit markets have been estimated at anything between US$70 billion and US$213 billion annually (Nellemann et al. 2014, 13), with other sources putting the value as high as US$1 trillion (Jose et al. 2019, 8).

Illicit criminal endeavours such as the illegal wildlife trade, timber trafficking and fisheries crime rely on a political economy of 'lootable commodities' – those that are 'high in value but have low economic barriers to extraction' (Farah 2010, 2) – and 'uncritical markets [that] ensure that there are buyers for goods at the right price, regardless of how they are obtained, processed or transported' (Nellemann et al. 2010, 34). Those markets are driven by both price and cost differentials – in the former situation where expected returns are higher than for analogue legal trade and the latter when the costs of compliance with regulations can be avoided through illegal practices. Illegal trade also arises when 'demand exceeds the supply of legal products' in the case of timber or ozone-depleting chemicals, for example (OECD 2011, 7).

Developing countries have been the focus of much of the global debate about TEC sectors such as the illegal wildlife trade and timber trafficking – both as source and destination countries and markets. There is a strong presumption in the policy and commentary literature that developing countries are 'more likely' to function as key TEC nodes. In part, particularly for sectors such as the illegal wildlife trade and timber trafficking, this is a function of high levels of biodiversity and the kinds of species that are in demand for consumptive purposes and as status goods. It also arises through the narrative lens of 'weak states' and those characterized by what Wennmann (2004, 105) refers to as 'socio-economic destitution,' which are assumed more likely to offer the opportunities that attract criminal groups of the kind involved in TEC.

DOI: 10.4324/9781003017653-23

However, developed countries are not absolved of complicity in the supply chains and markets of demand and supply that underpin these forms of illegal and environmentally unsustainable endeavours. The second and third largest markets for illegal wildlife (flora and fauna) are the United States (US) and the European Union, including for species and specimens that are sourced from other developed countries. About 20 per cent of the timber imported into the European Union is thought to come from illegal sources. Developed countries are also more likely to be the source of illegally traded and dumped wastes, including e-waste, and the black market in ozone-depleting substances was driven initially by demand in the US, Europe, and Japan.

These are not occasional movements of goods. In the environmental sphere of the illicit, smuggling of timber, wildlife, pollutants, and waste is a daily occurrence. At least some of the activity that constitutes TEC at a global and regional level is opportunistic, or undertaken by individuals or by operators who work at a small scale. It is also an area that has become increasingly systematic and well-financed, involving organized crime groups, sophisticated smuggling chains and complex trade routes. Large quantities of environmental contraband are moved across borders, sometimes by individual smugglers and rather ordinary forms of concealment (in cars, luggage, express post-bags and, as with drugs, hidden on the person) but often in bulk consignments dispatched by ship, barge, truck, and plane.

The development challenges associated with *responding* to TEC are many. They can arise in the context of weak governance, ineffective conservation and environmental protection frameworks, and limited enforcement resources. But they are also a function of the complexity of illicit transboundary transactions themselves, ranging as they do across extraction, production, transportation, and disposal practices, as well as associated illegal practices or cross-over crimes such as money-laundering, fraud, and corruption. As a result, TEC has been analysed as a problem of criminality and criminal justice, governance and regulation, international political economy, international law, and even security studies (for an overview, see Elliott and Schaedla 2016). This chapter extends this analysis by applying a development studies lens to understand the context and consequences of TEC. From a development perspective, TEC 'undermines sustainable economic growth, equitable development, and environmental conservation' (Kishor and Oksanen 2006, 12). It 'deprive[s] governments of much-needed revenues, ... undermine[s] legal businesses' (Nellemann et al. 2016, 7) and incurs development and ecological costs to future generations.

Implications for sustainable development

As the Global Initiative Against Transnational Organized Crime (2014, 1) puts it, organized criminality across multiple sectors is a 'spoiler for development.' Transnational environmental crime is no exception. It has often severe consequences for developing countries and their development prospects as well as for the sustainable livelihoods of many of the world's poorer and vulnerable peoples. The pathways and negative feedbacks that connect TEC and development impacts are complex. For the purposes of this chapter, these fall into three broad, overlapping areas of inquiry that reflect key pillars of sustainable development: environmental impacts, economic consequences and human (in)security.

Environmental impacts

Various forms of TEC are complicit in species endangerment, an unsustainable loss of living resources, and the destabilization of ecosystem services (see also Chapter 19 of this Handbook).

The environmental and conservation impacts of TEC can be extensive and have been well documented. The damage to charismatic and apex species – elephant, rhinoceros, big cats, and great apes in particular – is perhaps the most prominent of these broader sustainability impacts. For example, elephant numbers in both Asia and Africa continue to be under pressure from poaching with a 30 per cent reduction in population numbers across Africa. Poaching has reduced the global population of tigers in the wild to number no more than 4000 individuals (Musing 2020, 6). Despite a slow and uneven reduction in the absolute number of animals killed, rhino numbers have declined over the last two decades to the extent that African black rhinos are now listed as critically endangered and white rhino as 'near threatened' (Emslie et al. 2016). The illegal wildlife trade, driven by pharmacopeia demands in traditional medicines, by protein demands in the bushmeat trade, by growing middle class demands for collectible high-status specimens, items, and ornamentals, by the global pet trade, and even by biomedical research demands, has also resulted in severe impacts on wild populations of less-well known species of fauna and flora. Examples (a very small subset) include Indian Star Tortoises (Stoner and Shepherd 2020), pangolin (Nijman, Zhang and Shepherd 2016), totoaba (Crosta et al. 2018), the Spix's Macaw (Dale 2018), and cactus and cyad species (Margulies et al. 2019).

The ecological impacts of the illegal and predatory forms of forest extraction that are crucial to illicit transnational timber supply chains can include biodiversity loss, land degradation, disruption to carbon cycles and an increase in the potential for soil erosion and water runoff (Pacheo 2016). The suite of activities that constitute fisheries crime (de Coning and Witbooi 2015) contributes to the unsustainable harvesting of fish stocks, disruption of marine food chains and threats to marine biodiversity. The illegal production and consumption of ozone-depleting substances undermines efforts to contain damage to the ozone layer with consequences for public health (such as an increase in skin cancers, cataracts, and greater vulnerability to infectious diseases) and productivity in plants and phytoplankton. The covert and illegal dumping of hazardous and toxic wastes results in contamination of water tables, river systems and local ecosystems, and compromises animal, plant, and human health, often in the world's poorest countries.

Economic consequences

TEC imposes severe economic consequences on developing countries through the depletion of natural resources, the distortion of markets, the diversion of funds and assets through money-laundering and rent-seeking, through the loss of government revenue that might otherwise have been available through taxes and licence fees, and through the economic impact of corruption and associated practices. Criminal enterprise and illicit economies create what the Global Initiative Against Transnational Organized Crime (2014, 5), in its work on development responses to organized crime, refers to as a 'crime trap' in which the future is stolen as funds are diverted from the legitimate economy which is then 'crowd[ed] out' by the illicit economy. The World Bank has estimated that because of environmental crimes, '[g]overnments in source countries forego an estimated $7–12 billion each year in potential fiscal revenues,' a shortfall that 'hinders economic growth … and increases development risks and vulnerabilities beyond national borders' (Jose et al. 2019, 8). These are revenues that could have been deployed in key development sectors such as public health, food security, social welfare, education, and infrastructure. The overall annual economic costs to developing countries through the loss of resources and legal commerce associated with TEC is much higher though the estimates are quite wide – between US$91 billion and US$259 billion according to Nellemann et al. (2016, 15), though these figures include illegal minerals extraction as well.

The pursuit of economic development can also generate opportunity structures for extraction and production that feed into TEC. Growing affluence within developing countries has been identified as a factor exacerbating pressure on natural resources and creating global markets for illicitly harvested or produced consumptive goods (TRAFFIC 2008). The expansion of human settlements, mining, forestry, agriculture, and industry, puts existing habitat areas and species under increased pressure. The development of forest plantations and associated road networks facilitates easier access for illegal logging and for the illegal hunting and extraction of already endangered species of fauna and flora. Special economic zones and free trade zones, intended to attract investment and increase the presence of developing countries in global value chains, also offer opportunities for illicit trade and trans-shipment with minimal customs interventions and little fear of interception (OECD 2018, 116).

The bribery, corruption and fraud that has become a deeply embedded feature of TEC has both economic and non-economic consequences that undermine development efforts. Corruption can take place at all stages in illicit chains of custody and at any point at which illegal or criminal activity is involved. This can include extraction and harvest through hunting, illegal logging as well as other forms of capture such as theft, the use of fraudulent logging permits or CITES documentation. It includes production, not only of prohibited or controlled chemicals such as ozone-depleting substances, but also the transformation of wildlife and timber-related products into, for example, carved items and tanned skins, traditional medicines, fashion items, and furniture. It involves transportation crimes associated with the movement of specimens, chemicals and wastes from source to market: this can include physical transportation, management of warehousing and logistics, and disposal through illicit sales and dumping. Zain (2020) reports traders shipping wildlife and timber easily through air and sea ports with the payment of bribes to officials. The use of fraudulent documentation is a common feature of the black market in ozone-depleting substances and the illegal wildlife trade. Illegal logging operations function on the back of bribes to forest officials, police and military, and extensive opportunities for tax fraud (Nellemann/INTERPOL 2012).

Corruption is also implicit in the profit-related practices such as money-laundering that enable TEC. Bribery and coercion have hampered efforts to bring perpetrators to justice. In this regard, the kinds of corruption that enable TEC also undermine attempts to strengthen the kinds of good governance that are often argued to be at the core of development. It corrodes the institutions of the state and compromises core values such as the rule of law (Elliott 2007). In the most extreme cases of high-level corruption, fraud, and personal patronage as core to TEC activities and illicit profits, the state no longer functions in the Weberian sense as a provider and guarantor of public goods, including those associated with human and sustainable development, but more as a 'protection racket' or kleptocracy that sustains private appropriation, resource asset stripping and rent-seeking.

Human security

TEC threatens not just the security of habitat, ecosystems and species, and economies but also the security of those people and communities who are most vulnerable to the consequences of illegally sponsored environmental degradation, to the violence that can accompany the demand for illegal resources and, often, to the prosecution of those whose participation is driven not by greed and profit but by subsistence needs and survival (Duffy et al. 2016). The depletion of species and associated biological resources through illegal hunting and extraction can have an impact on poorer communities for whom subsistence and 'safety-net' access to fauna and flora can be important for wellbeing in physical, economic, and socio-cultural terms. The ecologically and economically unsustainable rise in trafficking of wildlife products and the illegal clearance of forest lands also threaten the livelihood of communities that derive economic and development gains from tourism (OECD 2018, 64). Human insecurity can drive participation in TEC. As Rademeyer (2016, 29) points out, with respect to the illegal wildlife trade, '[p]overty, inequality and limited opportunities in economically underdeveloped rural communities ... create livelihoods and opportunities that are sustained through wildlife trafficking and poaching.' Yet poor communities are rarely significantly complicit in the illegal trade. As a World Bank/TRAFFIC report argues, 'the links between wealth, poverty and engagement in the wildlife trade are complex: people involved in the trade are not necessarily poor, and the poor who are involved usually do not drive the trade [and] ... they do not capture the majority of the trade's monetary value' (TRAFFIC 2008, xii).

The human insecurities associated with TEC, especially the illegal wildlife trade, are often embedded in much longer colonial histories that removed hunting rights from local populations, thus criminalizing what would otherwise have been subsistence practices (see Duffy et al. 2016). They are also a function of protection and conservation efforts that ignored or overlooked historical and contemporary occupation of areas deemed wilderness or excluded local communities from negotiations and practices of declaring areas to be wildlife reserves, protected areas, and national parks.

Some human security threats are very direct. As Nellemann et al. (2014, 49) point out, the illegal extraction of natural resources creates conditions that can lead to 'severe human rights abuses,' 'threaten human populations located close to valuable wildlife resources,' and in worse case scenarios result in situations in which 'local communities are subject to threats, intimidation, forced labour ...human trafficking ... sexual exploitation, and murder.' In already vulnerable societies and polities, criminal groups and the victimization and violence that can accompany activities associated with the poaching and theft of live animals and animal parts can pose a serious threat to individual security and to the security of local communities. Violence in support of such activity often functions at a local level to intimidate villagers into compliance or, occasionally, to eliminate those who resist environmental exploitation. Each year, people from local communities and national agencies involved in protecting wildlife, preventing poaching, fighting against illegal logging and timber trafficking, or defending against the hazards of illegal waste and chemical trade are the victims of violence and murder. More than 1000 wildlife and anti-poaching rangers lost their lives in the decade to 2019, with many more suffering injury and harm (Belecky, Singh and Moreto 2019).

TEC also creates a range of insecurities for those who constitute the labour force for such illegal activities, such as those who work as chain-saw operators, illegal fishing crew, or as labourers in the production or management of illegal wastes and chemicals. These insecurities arise through loss of subsistence resources and income generating opportunities, or through the vulnerability of having to rely on low levels of pay in illicit extractive and production sectors characterized by little if any form of worker safety or protection. Studies on fisheries crime, now recognized as a form of TEC by agencies such as INTERPOL, have drawn attention to problems of bonded and child labour, sexual exploitation, and physical abuse (see UNODC, 2011). Reports on illegal logging and forest exploitation also document forced labour, sexual slavery and violence (Urrunaga et al. 2012).

People and communities are also made vulnerable and insecure in the face of social and political conflicts (see also chapter 18) that are increasingly associated with some forms of TEC, particularly those that involve supply-side pressure on renewable and non–renewable natural and living resources. Gore et al. (2019, 784) suggest that TEC is 'directly stimulated by continued

Lorraine Elliott

or renewed conflict in many of the world's most deadly contexts.' These TEC-conflict pathways can arise through disputes over territory spilling into protected areas, national parks, or biodiversity hotspots. However, they are more likely to involve illegal resource extraction and harvesting on the part of government and non-government actors to fund civil conflict. Studies reveal that a range of recent and contemporary conflicts, such as those in Angola, Democratic Republic of Congo, Nepal, Liberia, and Cambodia, have been characterized by high levels of illegal and unsustainable poaching and timber harvesting (see Elliott 2016 for a summary).

Policy implications

Calls for effective multilateral action to deal with the multiple development-related challenges of environmental crime have gained policy traction with greater awareness of the ecological, economic, social, political and, indeed, security consequences of the activities involved in the illegal wildlife trade, timber trafficking, black market in ozone-depleting substances, illegal waste dumping and fisheries crime. Some of these calls, such as those embedded in Sustainable Development Goal (SDG) 15.7 and 15.C that identify the need for 'urgent action to end poaching and trafficking of protected species of flora and fauna,' are directed at specific sectors. Others are more wide-ranging, a product of institutional settings in development, law enforcement, conservation, and heritage.

Yet TEC governance and policy development is also increasingly polycentric and fragmented, characterized by a multiplicity of rules and rule-making venues that operate within non-hierarchical policy spaces. There is no single international agreement that addresses the challenges of TEC or that provides an overarching framework to guide governments and other actors in the development of policy responses across prevention, protection, and law enforcement. Key multilateral environmental agreements (MEAs) include the 1973 Convention on International Trade in Endangered Species of Wild Fauna and Flora (known as CITES), the 1987 Montreal Protocol on Substances that Deplete the Ozone Layer (the Montreal Protocol), and the 1989 Basel Convention on the Control of Transboundary Movements of Hazardous Wastes and their Disposal (the Basel Convention). These are not transnational crime or law enforcement agreements. Their purpose is to enhance conservation and environmental protection efforts through establishing guidelines on the production, use of and trade in specific substances or species. The growing recognition that TEC is a challenge not just for those with a focus on environmental protection but also for those fighting transnational crime has drawn international agencies with a mandate for crime prevention, border protection, and law enforcement into the policy and operational milieu, including the World Customs Organization (WCO), INTERPOL and the UN Office on Drugs and Crime. International and local nongovernmental organizations have also become key policy, advocacy, capacity-building and (at times) enforcement actors in exposing and seeking to overcome TEC, working in both formal and informal partnerships with governments and intergovernmental organizations. Key among these (although the list would be a long one) are the Environmental Investigation Agency, TRAFFIC and WWF, the International Fund for Animal Welfare, and the Wildlife Justice Commission.

One of the themes driving global policy responses to TEC has been the need to strengthen 'international cooperation, capacity-building, criminal justice responses and law enforcement efforts.'[1] Governments are urged to define natural resource crimes as serious organized crime and to implement multiagency strategies that, in efforts to 'follow the money,' include financial intelligence units, anti-corruption agencies, and customs and tax units (Jose et al. 2019, 30). Intergovernmental and national policy and practice on TEC-related law enforcement and

238

criminal justice has ranged widely across global and regional operations, training and capacity development for enforcement best practice, technical assistance to strengthen legal frameworks, improved technology for knowledge exchange and intelligence gathering, national capacity and cooperation among law enforcement agencies, and awareness raising for national customs agencies around illegal environmental trade.

A second field of policy effort has focused on ensuring that trade – in wildlife, timber, and fish, for example – is not only legal but also functions within sustainable limits. Market-based incentive structures that rely on certification, investment, and procurement policies, such as those that underpin the European Union Forest Law Enforcement, Governance and Trade Programme or the prior informed consent procedures instigated by the Montreal Protocol and the Basel Convention, have sought to encourage national policies on sustainability and legal verification to address both demand and supply-side motivations. The policy approaches are designed to promote transparent and responsible supply chains for products that can be traded legally on the basis that licensing and permitting protocols can also help trading countries and companies identify possible illegality. These market-driven policies are supplemented with approaches that rely on financial support. Key MEAs such as the Montreal Protocol have also instituted funding mechanisms to assist developing countries meet their trade and protection obligations. The Global Environment Facility funds the Global Wildlife Program (managed by the World Bank) with the goal of promoting sustainable development by curbing the illegal wildlife trade and promoting local wildlife-based economies.

This emphasis on local economies and questions of scale is reflected in policies that account for the ways in which local communities are affected by TEC-related activities, particularly with respect to extractive illegalities in the wildlife, timber and fisheries sectors, and that also recognize that '[c]ommunity-based considerations are an integral part of sustainable, long-term solutions' (Jose et al. 2019, 28). SDG 15.C specifically calls for efforts to 'increas[e] the capacity of local communities to pursue sustainable livelihood opportunities.' Proposals for effective management strategies have traversed a range of possibilities around ecosystem management, the development of local nature-based solutions, securing more sustainable sources of extractive or harvested products, engaging local communities in law enforcement activities, and improving the role, status and capacity of local enforcement actors.

Creating opportunities for local input into TEC policy also demands that communities are engaged as stakeholders through transparent, effective, and inclusive decision-making and implementation mechanisms. Policy efforts also need to focus more specifically on overcoming the economic disincentives that might impel local participation in illegal resource and environmental extraction, production, and distribution (though they rarely benefit in any meaningful way from the profits that are generated in these sectors) and the constraints that TEC sectors impose on local development opportunities. This demands that TEC policy engages with broader social considerations, is grounded in pro-poor approaches that can strengthen alternative sustainable livelihood opportunities, and empower vulnerable populations in both rural and urban areas (Duffy et al. 2016). Poaching, illegal logging and fisheries crime can be better managed in a context in which land tenure and property rights for local communities and indigenous peoples, including those relating to protected areas, are enhanced (see for example CITES 2019), or in which artisanal fishing practices are recognized and protected. The illegal trade in ozone-depleting substance or in toxic wastes requires that better economic choices are available to those who would otherwise have little choice but to derive income from transporting cylinders of illegal ozone-depleting substances or other chemicals, selling them in local markets or working in hazardous production and disposal conditions.

Pedagogical reflections

In both policy reports and academic literature, TEC has been examined in the context of criminology, conservation and environmental protection, law enforcement, regulatory theory, global governance, and development. This means that teaching on TEC is likely to involve interdisciplinary sources and approaches regardless of the specific discipline or field of study within which that teaching occurs. There are various ways in which TEC can be introduced to students in a development studies context. This could focus on specific case studies such as the illegal wildlife trade, fisheries crime, timber trafficking, or the black market in ozone-depleting substances. It could focus on how regulatory responses account for the development impacts of these sectors, breaking this down into the various activities such as poaching, harvesting, smuggling, transporting, and marketing that constitute an illegal chain of custody. It could explore the impact on local communities through an evaluation of how these kinds of activities are implicated in violence and human rights abuses, the extent to which poverty might be a factor in illegal harvesting, extraction and production, and the relevance of alternative livelihood and community conservation strategies. Finally, teaching on TEC could take a more meta-level approach to explain how these various sectors affect the economic security of communities and states through corruption, fraud, and money-laundering, through the loss of natural resources and the damage to habitat, and through the impact on national income through the non-payment of fees, taxes, and excise.

When working with students who are new to these issue-areas and themes, a case-study approach is likely to offer a useful teaching modality that will encourage them to identify and understand the key development themes involved in a specific case or sector of transnational environmental crime. This can be achieved through a teaching presentation, class-based discussion, assessment tasks such as essays and policy reports, or a combination of these. This kind of investigation could also be pursued through a country-case-study, particularly if the focus is on the economic development consequences at state or government level of TEC sectors.

Two issues that might excercise those who are leading teaching-and-learning on TEC in a development studies context is how to approach the selection of cases, and how to identify relevant sources. The selection of cases can be independent of the analytical framework outlined above but some TEC sectors will be more relevant than others depending on the geographic location in which these pedagogical exercises are being conducted. The illegal wildlife trade (IWT) is the most extensively covered form of TEC in the policy and academic literature. For wildlife trade cases, I would encourage moving beyond studies of charismatic mega-fauna (elephants, rhinos, great apes, big cats) to non-charismatic species such as reptiles, birds, eels, tortoises, geckoes, and smaller species (the slow loris and pangolin as just two examples). The choice of IWT cases should also recognize the extent to which plant species are harvested and traded illegally, though they rarely attract the kind of international media attention that mega-fauna does. Case studies of timber trafficking – that is the movement of 'stolen' timber across borders in which illegal logging is just the first phase – can likewise move beyond high-value species such as rosewood or mahogany to consider other forms of illegal trade that might include timber for construction purposes, garden furniture, and the like. Cases of the black market in ozone-depleting substances are partly time-constrained in that certain ODS chemicals are no longer in demand and therefore no longer produced (even illegally), although historic case studies will provide some useful insight into the development context for this trade.

In terms of sources, for teaching purposes and particularly for class-based discussion and analysis, primary case-study sources, and materials, including those based on investigations, are a good place to start. This primary material is produced by both non-governmental and inter-governmental organizations. A preliminary 'go-to' list would include the NGOs TRAFFIC, the Environmental Investigation Agency, the International Fund for Animal Welfare, WWF, and the Wildlife Conservation Society. Key international organizations whose reports can be included in case-study investigations include INTERPOL's Environmental Security Unit, IUCN, the UN Office on Drugs and Crime, the UN Environment Programme, as well as relevant conventions and treaties such as CITES, the Basel Convention and the Montreal Protocol. Research reports and academic literatures can then be deployed in a teaching situation as useful supplementary sources.

Conclusion

As the discussion above indicates, TEC covers a range of sectors of both living and non-living resources, illicit chemicals such as ozone-depleting substances that are deliberately produced, and the unwanted toxic and hazardous outputs of production and consumption. The illicit trade in these sectors is driven by a complexity of supply and demand, being marked by both opportunistic and organized extraction and smuggling, and often involves sophisticated chains of custody and supply. From a development perspective, TEC affects all three pillars of sustainable development: environmental, social, and economic. This makes them difficult challenges to solve but it also opens scope for moving beyond regulatory and policy responses that focus on TEC primarily in trade, law enforcement or even conservation terms. As suggested here, a development-focused approach to fighting transnational environmental crime involves policy-activity at a national level but also requires further attention to the ways that local communities are affected by the activities that make TEC possible and can be involved in strategies to overcome them.

Key sources

1. Elliott, L and Schaedla, WH (ed.) (2016), *Handbook of transnational environmental crime*, Edward Elgar, Cheltenham.

This edited volume compiles a wealth of interdisciplinary expertise across international relations and global governance, security studies, international law, criminology, development studies and conservation biology. Across a range of TEC case studies, the authors examine the drivers of illegal environmental activity, the environmental and development consequences, and institutional and regulatory responses. The volume is enhanced by shorter observations written by enforcement and policy practitioners.

2. Felbab-Brown, V (2018), *The extinction market: wildlife trafficking and how to counter it*, Hurst Publishers, London.

This book examines the causes and modalities of poaching and wildlife trafficking, and analyses policy tools to respond to these challenges. In the context of regulatory theory and debates about tensions between human welfare, animal welfare and conservation, the author explores trade bans, alternative livelihood programs, anti-money-laundering and demand reduction.

3. Wyatt, T (ed.) (2016), *Hazardous waste and pollution: detecting and preventing green crimes*, Springer, London.

This volume takes a green criminological approach to 'brown' crimes in TEC to provide insights into how such crimes can be conceptualized, controlled and regulated. It includes case studies of the black market in ozone-depleting substances and the illegal trade in e-waste (electronic waste) as well as other forms of pollution crime including pollution from sea-going vessels and chemical accidents.

4. Duffy, R, Freya, AV, John, S, Buscher, B and Brockington, D (2016), 'Toward a new understanding of the links between poverty and illegal wildlife hunting,' *Conservation Biology*, vol. 30, No. 1, pp. 14–22.

In this article, the authors examine illegal wildlife hunting, particularly of species that enter the illegal wildlife trade, as an issue of poverty and development rather than simply one of conservation. They explore the nature of poverty to understand what motivates people to hunt illegally. They do so against the backdrop of an analysis of the complex history of wildlife laws and the criminalization of hunting practices.

5. Stoett, P and Omrow, DA (2021), *Spheres of transnational ecoviolence: environmental crime, human security, and justice*, Palgrave, London.

In this book the authors examine the extent and perpetrators of various forms of TEC, including wildlife and plant crime, illegal fisheries, toxic waste trade, and climate-related crimes. They analyse the impact on development opportunity costs including through forms of violence and coerced labour. In their discussion of policy responses, they argue for the importance of taking justice and human security into account.

Note

1 Doha Declaration on Integrating Crime Prevention and Criminal Justice into the Wider United Nations Agenda to Address Social and Economic Challenges and to Promote the Rule of Law at the National and International Levels, and Public Participation (UN Doc. A/CONF.222/L.6, 2015), at paragraph 9(e).

References

Belecky, M, Singh, R and Moreto, W (2019), *Life on the frontline 2019: a global survey of the working conditions of rangers*, WWF, Geneva.

CITES (2019), 'CITES Strategic Vision 2021–2030, Resolution Conf. 18.3,' presented at the *18th Conference of Parties*, 17–28 August, Geneva, Switzerland.

Crosta, A, Sutherland, K, Talerico, C, Layolle, I and Fantacci, B (2018), *Operation fake gold*, Elephant Action League, Los Angeles, CA.

Dale, A (2018), *Spix's Macaw heads list of first bird extinctions confirmed this decade*, Birdlife International, viewed 10 November 2020, www.birdlife.org/worldwide/news/spixs-macaw-heads-list-first-bird-extinctions-set-be-confirmed-decade

de Coning, E and Witbooi, E (2015), 'Towards a new 'fisheries crime' paradigm: South Africa as an illustrative example', *Marine Policy*, vol. 60, pp. 208–215.

Duffy, R, Freya, AV, John, S, Buscher, B and Brockington, D (2016), 'Toward a new understanding of the links between poverty and illegal wildlife hunting', *Conservation Biology*, vol. 30, no. 1, pp. 14–22.

Elliott, L (2007), 'Transnational environmental crime in the Asia Pacific: an un(der)-securitised security problem?', *The Pacific Review*, vol. 20, no. 4, pp. 499–522.

Elliott, L (2016), 'The securitisation of transnational environmental crime and the militarisation of conservation,' in L Elliott and WH Schaedla (ed.), *Handbook of transnational environmental crime*, Edward Elgar, Cheltenham.

Elliott, L and Schaedla, WH (2016), 'Transnational environmental crime: excavating the complexities – an introduction,' in L Elliott and WH Schaedla (ed.), *Handbook of transnational environmental crime*, Edward Elgar, Cheltenham.

Emslie, RH, Milliken, T, Talukdar, B, Ellis, S, Adcock, K and Knight, MH (2016), *African and Asian rhinoceroses – status, conservation and trade*, IUCN Species Survival Commission African, Asian Rhino Specialist Groups and TRAFFIC. CoP17 Doc. 68 Annex 5.

Farah, D (2010), *Transnational crime, social networks and forests: using natural resources to finance conflicts and post-conflict violence*, Program on Forests, Washington DC.

Global Initiative Against Transnational Organised Crime (2014) *Improving development responses to organized crime*, Global Initiative against Transnational Organized Crime, Geneva.

Gore, ML et al. (2019) 'Transnational environmental crime threatens sustainable development', *Nature Sustainability*, vol 2, pp. 784–6.

Jose, J, Montero, M, Wright, E and Khan, MN (2019), *Illegal logging, fishing and wildlife trade: the costs and how to combat it*, The World Bank Group, Washington DC.

Kishor, N and Oksanen, T (2006), 'Combating illegal logging and corruption in the forestry sector: strengthening forest law enforcement and governance', *Environment Matters*, Annual Review July 2005–June 2006, pp. 12–15.

Margulies, JD, Bullough, LA, Hinsley, A, Ingram, DJ, Cowell, C, Goettsch, B, Klitgård, BB, Lavorgna, A, Sinovas, P and Phelps, J (2019), 'Illegal wildlife trade and the persistence of "plant blindness"', *Plants, People, Planet*, vol. 1, no. 3, pp. 173–82.

Musing, L (2020), *Falling through the system: The role of the European Union captive tiger population in the trade in tigers*, TRAFFIC and WWF, Cambridge.

Nellemann, C, Redmond, I and Refisch, J (2010), *The last stand of the gorilla: environmental crime and conflict in the Congo basin*, UNEP and GRID-Arendal, Norway.

Nellemann, C and INTERPOL Environmental Crime Programme (ed.) (2012), *Green carbon, black trade: illegal logging, tax fraud and laundering in the world's tropical forests. A rapid response assessment*, United Nations Environment Programme and GRID-Arendal, Nairobi and Arendal.

Nellemann, C, Henriksen, R, Baxter, P, Ash, N and Mrema, E (ed.) (2014), *The environmental crime crisis: threats to sustainable development from illegal exploitation and trade in wildlife and forest resources – A rapid response assessment*, United Nations Environment Programme and GRID-Arendal, Nairobi and Arendal.

Nellemann, C, Henriksen, R, Kreilhuber, A, Stewart, D, Kotsovou, M, Raxter, P, Mrema, E, and Barrat, S (ed.) (2016), *The rise of environmental crime – A growing threat to natural resources peace, development and security. A UNEP/INTERPOL rapid response assessment*, United Nations Environment Programme/ RHIPTO Rapid Response–Norwegian Center for Global Analyses, Geneva.

Nijman, V, Zhang, MX and Shepherd, CR (2016), 'Pangolin trade in the Mong La wildlife market and the role of Myanmar in the smuggling of pangolins into China,' *Global Ecology and Conservation*, vol. 5, pp.118–26, https://doi.org/10.1016/j.gecco.2015.12.003

OECD (2011) 'Illegal trade in environmentally sensitive goods: draft synthesis report' 2011, *COM/ TAD/ENV/JWPTE (2011) 45*, OECD Trade and Agriculture Directorate/Environment Directorate, OECD, Paris.

OECD (2018), *Governance Frameworks to counter illicit trade*, OECD, OECD Publishing, Paris.

Pacheo, P (2016), 'Multiple and intertwined impacts of illegal forest activities,' in D Kleinschmit, S Mansourian, C Wildburger and A Purret (ed.), *Illegal logging and related timber trade – dimensions, drivers, impacts and responses. A global scientific rapid response assessment report*, International Union of Forest Research Organizations, Vienna.

Rademeyer, J (2016), *Beyond borders: crime, conservation and criminal networks in the illicit rhino horn trade*, Global Initiative Against Transnational Organized Crime, Geneva.

Stoner, SA and Shepherd C (2020), 'Using intelligence to tackle the criminal elements of the illegal trade in Indian Star Tortoises Geochelone elegans in Asia', *Global Ecology and Conservation*, vol. 23, pp. 1–9, https://doi.org/10.1016/j.gecco.2020.e01097

TRAFFIC (2008) *What's driving the wildlife trade? A review of expert opinion on economic and social drivers of the wildlife trade and trade controls in Cambodia, Indonesia, Lao PDR and Vietnam*, East Asia and Pacific Region Sustainable Development Discussion Papers, The World Bank, Washington DC.

UNODC (2011) *Transnational organized crime in the fishing industry*, UN Office on Drugs and Crime, Vienna.

Urrunaga, JM, Johnson, A, Orbegozo, ID and Mulligan, F (2012), *The laundering machine*, Environmental Investigation Agency, London.

Wennmann, A (2004), 'The political economy of transnational crime and its implications for armed violence in Georgia', *CP 6: The Illusions of Transition: which perspectives for Central Asia and the Caucasus*, Graduate Institute of International Studies/CIMERA, Geneva.

Zain, S (2020), *Corrupting trade: An overview of corruption issues in illicit wildlife trade*, WWF/Targeting Natural Resource Corruption Programme, Gland, Switzerland.

21

INDIGENOUS RIGHTS, NEW TECHNOLOGY AND THE ENVIRONMENT

Pyrou Chung and Mia Chung

Introduction

Environmental governance and natural resource management have often been removed from Indigenous stewardship. This stems from the long and violent history of colonialism. Despite a period of 'decolonization' during which countries in Southeast Asia (including four of the five Mekong sub-region countries, i.e. Cambodia, Lao PDR,[1] Myanmar, and Vietnam, with Thailand never having formally been colonized) declared national independence, the power structures inherent in past colonization processes are replicated and perpetuated through the application of traditional models of economic development (Rutazibwa 2018; Rutazibwa 2019; Ziai 2015). For Indigenous Peoples and ethnic minorities (IEM), these conditions have been further exacerbated by the paternalistic attitudes of national governments toward them. The result: continued disenfranchisement of Indigenous Peoples and ethnic minorities.

Drawing on the Mekong sub-region experience, this chapter explores how over 100 Indigenous and ethnically distinct communities in the region have struggled to retain their autonomy, while taking note of the significant barriers they face in asserting their rights to their identity – as well as over their land and the natural resources located therein. We discuss, in particular, how new technology could be inclusively leveraged to centre the needs of IEM communities while protecting their rights and connection to the natural environment.

Before proceeding further, a caveat warrants note here. In Southeast Asia, IEM are frequently referred to by many names by outsiders, including 'Indigenous Peoples' (Baird 2014) and 'tribes,' even though tribal boundaries may not have been how IEM originally grouped themselves (Scott 2009). The term 'ethnic minority' is also widely used and can be understood as encompassing more than Indigenous Peoples, though political appropriation in national contexts has complicated the term. Further, the description of 'minority' is in some cases inaccurate: many IEM in the region have culturally contiguous enclaves straddled across country borders, such that they may be a minority in one state but constitute a majority as a community (this is the case for the Hmong people, whose numbers equal the population of Laos). Colonial external observers have also originally referred to IEM groups in the region's mountainous regions as 'savages' (*Moi* in Vietnamese, *Kha* in Laotian, or *Phnong* in Khmer), as well as the still-used 'Montagnards' (mountain dwellers) and 'hill tribes' – labels which often imply inferiority.

DOI: 10.4324/9781003017653-24

Of course, IEM groups have names by which they refer to themselves that constitute the most accurate way of referring to these communities. This chapter endeavours to use these local-level names when describing the experiences of a specific IEM group. However, for ease of reference, when speaking of the shared experiences of marginalization faced by IEM in the region, we use the term 'Indigenous Peoples and ethnic minorities,' while acknowledging that the use of such overarching terms as 'Indigenous' or 'ethnic minorities' suggests a level of cohesion and unity within and between groups that does not exist in reality. It is beyond the scope of this chapter to detail how these naming and definitional complexities have evolved into an issue of significant contention for IEM in Southeast Asia; instead, suffice it to note here that these issues have invariably impacted IEM's ability to demand state and international recognition of their rights.

Defining IEM rights and recognition

IEM in the Mekong sub-region have a varied history. While some groups, like the Mon and the Kayah or Karenni in Myanmar and the Lahu in Laos migrated into the region, others are considered the original inhabitants of the land, such as the Khmu in Northern Laos (Battachan et al. 2019). Despite this diversity, the IEM in the region do share common experiences. They have historically been dominated by others – through colonization as well as state-building – and at present, remain marginalized economically, politically, and culturally ('Managing forests, sustaining lives, improving livelihoods of indigenous peoples and ethnic groups in the Mekong region, Asia' 2013).

The United Nations (UN) began discussions on how to include Indigenous rights in the international agenda in the early 1980s, but it was not until 2007 – when the United Nations Declaration on the Right of Indigenous Peoples (UNDRIP) was finally agreed – that they were able to agree on how this was to be achieved. This voluntary, non-binding document represents global recognition of the importance of protecting Indigenous Peoples' rights. Self-determination is a key component of UNDRIP, forming part of the preamble as well as articles in the declaration ('UNDRIP,' article 34). Land and natural resources also underpin the framework, which sees the preamble including specific articles against the forcible removal of Indigenous Peoples from their lands and territories ('UNDRIP,' article 10), as well as confirming their right to the lands, territories, and resources which they have traditionally owned, occupied or otherwise used or acquired ('UNDRIP,' article 26). These provisions, in effect, set out IEM's right to own, use, develop, and control these lands, territories and resources, and acknowledge the inextricable connection between Indigenous cultures, land and natural resources ('UNDRIP,' article 25).

While UNDRIP is meant to represent the most comprehensive global document on Indigenous rights, it still defines indigeneity by reference to first colonial contact by Westerners (rather than, for example, by Chinese or Mongolian peoples). This Western-centred perspective is rooted in settler colonialism and is inherently problematic: while most countries in the region were colonized by European powers, the colonial experience of IEM in the Mekong sub-region stretches back beyond the relatively recent European incursions into the region. Moreover, although all governments in the Mekong sub-region are signatories to UNDRIP, they have seized upon this 'first Western contact' definition of indigeneity to maintain the status quo with respect to Indigenous rights. In particular, the region's governments have tended to either claim that there are no Indigenous Peoples within their borders, or that all their peoples are Indigenous (Baird 2020).

Table 21.1 IEM counts by country

Country	# of IEM groups recognized by national government	Population (2020)	% of total population	Self-identified groups (2014)	Self-identified population estimate (2014)
Cambodia	24 (2020)	400,000 (2020)	2–3%	19–21 (2014)	101,000–190,000 (2014)
Laos	50 (2020)	2.4–4.8 million (2018)	35–70%	unknown	unknown
Vietnam	53 (54 including Kinh) (2020)	13.4 million (2020)	14.6%	90+ (2014)	10 million (2014)
Thailand	9 (2020)	900,000 (2020)	1%	25+ (2014)	5 million (2020)
Myanmar	8 (2020)	unknown	unknown	16.5 million (2014)	32% (2014)

Source: IWGIA 2020; AIPP 2014; Sengdara, 2018.

Indeed, rather than using 'Indigenous Peoples,' the governments of Vietnam, Cambodia, Laos, and Thailand use the term 'ethnic minorities.' In Myanmar, 'national races' is used. Each of the countries also apply these terms in slightly different ways. The Laos government considers all citizens to be a member of an ethnic minority ('Ethnic minorities and indigenous peoples' 2019), whereas in Myanmar 'national races' are defined quite narrowly, impacting citizenship ('The Republic of the Union of Myanmar' 1982). In practical terms, the complication that arises from such naming is that IEM in these countries have not been given access to the rights delineated in the UNDRIP, with the most notable being the lack of implementation of the right to self-determination. Following this, even basic processes such as enumeration by national census are rendered complicated, exemplified by the discrepancies that arise as between self-identified population counts and nationally recognized numbers (see Table 21.1).

IEM and environmental governance

Regardless of the idiosyncrasies in cultural practices across the different IEM groups in the Mekong sub-region, reverence towards the environment remains a unifying theme for all of them. IEM groups have lived in partnership with their lands and territories for years. This relationship uniquely defines their indigeneity and sovereignty as peoples united by cultural beliefs, which promote holistic understandings of nature that allow for mutual respect, responsibility, and reciprocity. As a result, livelihood practices reflecting IEM ways of knowing are often recognized as being well-attuned to the intrinsic connectivity of natural environments. This translates into practices such as swidden (or rotational) agriculture that respect natural cycles and better conserve the environment.

Yet, such traditional ecological knowledge (TEK) has been largely dismissed by both national governments and international bodies supporting conservation, which typically rely on a Western-centric scientific model of conservation that promotes disciplinary silos and knowledge 'exclusivity'. This in turn perpetuate discourses of economic development and frameworks of environmental conservation predicated on the absence of human influences on ecosystems – that is, where humans and nature are viewed as separate entities (known as 'fortress' or 'colonial conservation' (Adams and Mulligan 2003)). These mainstream conservationist narratives have a tendency to stigmatize TEK and cultural practices as 'primitive' or 'harmful.' Indigenous agricultural practices[2] – in particular, migratory and subsistence farming – are demonized as responsible for low productivity and environmental degradation, and are often conflated with negative portrayals of IEM themselves. Together, such 'decontextualizing' of Indigenous knowledge has been used to justify the replacement of IEM by others as stewards of the land and natural environment (Fox et al. 2009; Dressler et al. 2010; 'Shifting cultivation: livelihood and food security' 2015). In the Mekong region, IEM communities have been removed and forcibly prevented from re-entering their traditional territories by park and forest rangers charged with protecting a protected area (Domínguez and Luoma 2020). According to Robin Roth, such strict delineation of protected areas and exclusion zones in Thailand's forests has produced the most conflict between IEM and the state (Roth 2008). Conflict in the Mekong region is ongoing in this regard (Pye 2016; Phromma et al. 2019; Allendorf et al. 2006; Wageningen University and Research n.d.; Dwyer 2017).

Aside from environmental conservation, agricultural industrialization is another major justification for dispossessing IEM of their lands. In a bid to reduce poverty and hunger by intensifying agricultural production with technological innovations (Harwood 2018), governments in the Mekong region have focused on redistributing land to increase agricultural productivity. In Vietnam, the 1967 'Land to the Tiller' Law, which gave all land that was not directly cultivated to farmers, stripped away the land rights of primarily IEM participating in swidden agriculture, due to the perceived unproductivity (Le Coq et al. 2001). The government at the time prioritized rice production, using new foreign technologies to develop irrigation, making available modified seed varieties imported from the United States, (Hazell 2009), as well importing the necessary pesticides and chemicals to grow them (Le Coq et al. 2001). These developments not only had negative environmental impacts, but also further alienated IEM from their land – a situation aggravated by the introduction of land rights systems that did not recognize the IEM focus on community-managed land and non-commoditized agriculture.

Since the 1970s, the co-management of natural resources gradually emerged as a means for environmental protection – one considered more equitable, sustainable, just, and democratic (Borrini et al. 2007). It includes IEM communities as a key stakeholder, alongside central and local government authorities, in environmental governance (Ballet et al. 2009). Even so, while this model at first glance appears to move away from more top-down approaches, it actually remains a variation of the 'colonial conservation' approach (Dominguez and Luoma 2020; Fa et al. 2020; Lang 2020). In practice, IEM knowledge continues to be overlooked (Stevenson 2004). In many cases, restrictions on land use are still being imposed on IEM communities, even as the data suggests that doing so is ineffective and may even be detrimental to conservation (Ballet et al. 2009; Armitage 2002; Roth 2008). IEM livelihoods, cultures and perspectives remain largely absent from this model (Stevenson 2004), even though IEM are disproportionately impacted.

Traditional ecological knowledge versus technology for environmental protection?

Traditional ecological knowledge defines the relationship between IEM and their environment. The term was originally coined by Fikret Berkes in 1987 as 'a cumulative body of knowledge and beliefs, handed down through generations by cultural transmission, about the relationship of living beings (including humans) with one another and with their environment' (Berkes 1993, 3). TEK does not seek to control nature and constitutes an integrated system of knowledge, practices and beliefs (Berkes 1993).

As the above discussion reveals, IEM perspectives – particularly as channelled through TEK – are notably absent from models of environmental protection that remain wedded to Western scientific knowledge and top-down management limiting IEM participation. This omission of TEK from environmental governance serves as one of the primary means through which IEM continue to be divested of their rights. One key example has been in regard to the introduction of information and communication technologies (ICT). Here, TEK has frequently been depicted as 'prelogical' or 'irrational' and thus incompatible with new technology (Berkes 1993). For this reason, TEK is ignored when ICT are developed uncritically for environmental governance. Thus, these tools can widen the inequality gap between IEM and mainstream communities, ultimately further disenfranchising IEM.

Because ICT is fundamentally built upon a Western-centric and siloed perspective on scientific knowledge, the push to use these tools in environmental governance and natural resource management can have adverse implications for IEM and their rights. Without meaningful input from IEM and inclusion of TEK, ICT development will invariably be non-inclusive. Indeed, English remains the main language of ICT, which naturally excludes a broad demographic of users. ICT also requires basic literacy (i.e. reading and writing) as well as digital literacy and numeracy, which also limits usage by IEM who may not have access to the relevant education or the necessary technological infrastructure (e.g., electricity). As such, technological tools that have been developed with a Western cultural context in mind are unlikely to be relevant or helpful to other cultures. Similarly, considerations of data protection and security, which are normally intrinsic to software development, have yet to take into proper account TEK requirements and political sensitivities, with data management protocols not incorporating points of reference that matter to IEM communities.

Indeed, as a further consequence of the way ICT is developed, these tools now block IEM access to fundamental land and other rights. E-government and online registry tools, often developed with the support of international funders, use Western-centric scientific models to define land and natural resources. IEM are then forced to prove their claims to the land and natural resources within a framework that is both foreign and irrelevant to them, and thus, many IEM encounter difficulties in proving their claim to traditional lands. This impacts the ability of IEM individuals to access the health system, pay taxes, attend school, and further (Chen et al. 2013), while simultaneously divesting them of the ability to access rights pursuant to national legal frameworks. In this way, IEM face an uphill battle in accessing basic rights, from land to education.

ICT-driven approaches continue to be pushed by multilateral institutions like the Asian Development Bank, backed by development agencies and conservation organizations, especially those from developed nations. Where TEK is incorporated, it is arguably done in a superficial and tokenistic way, used to vindicate certain approaches or mentioned only specifically in relation to disaster adaptation and mitigation. For example, consider the applications of geospatial mapping (using GIS technology to digitize the natural environment). TEK typically does not differentiate between different categories of the environment, but instead, considers knowledge of the 'land' as including all living things (inclusive of humans) as well as the physical geographies of the immediate environment. In this sense, the understanding of 'ecosystems' is implicit, subsumed into a broader knowledge of different plants and animals that co-exist, types of weather, and other interrelated aspects of their environment. However, Western scientific knowledge, implicit in GIS technology, takes a more reductionist view, one that defines environmental structure and their boundaries distinctly. The use of GIS in integrated participatory modalities with TEK has thus become a tool of epistemological assimilation that erodes Indigenous cultures (Rundstrom 1995). Participatory GIS approaches, heavily funded by international development agencies, have been used extensively in blended knowledge models to aid IEM to lobby governments and industry towards greater land claims. However, in most circumstances in the Mekong sub-region, such use of technology has de-legitimatized TEK, instead reaffirming Western scientific power structures that are essentially urban, high-tech, capital-intensive, and 'expert' (Dunn et al.1997). The result: further divestment of Indigenous rights through forced evictions and relocation to make way for conservation reserves, parks, or state-private land development schemes proposed as conservation solutions, all derived from GIS 'knowledge' systems (Martinez 2018).

Little to no time is spent on ensuring that new technologies reflect the realities faced by IEM and local communities, such that they do not always understand how to use these technologies or the purpose they are meant to serve. Training for such ICT tools may be provided, but this often lasts only for the length of the project and may only be specific to the tool itself, rather than addressing more basic skill requirements like digital literacy. Moreover, projects sponsored by development agencies typically happen within a timeframe set by the funding schedule and implemented pursuant to the goals and objectives of the agencies rather than those of the communities impacted. It can be unclear as to what long-term benefits are to accrue for the communities themselves. When GIS systems are established externally apart from communities with little to no opportunities for long-term funding support, the primary factor for sustainability is money (Chapin et al. 2005).

Implications for practice

Increasingly, the ICT that is being introduced to IEM communities in development projects is digital. As these tools are dependent on the internet – commonly considered to be a democratizer of technology – they are presumed to be accessible by all. Yet, as noted by Meinrath, Losey and Pickard in their 2011 article, participation in internet technology is based on the internet as a 'common' resource, which in turn enables non-discriminatory access and use of the network (Meinrath et al. 2011). However, the reality is that digital access is not equal, and IEM are ill-prepared and poorly supported to participate in it. Nor are they necessarily given a choice to participate or not. Critics have compared the current digital ecosystem with the medieval system of feudalism, in which the ruling (kings) and managing (lords) classes extracted value from the producer class (serfs), who had little choice and influence over the system and did not benefit. Digital feudalism sees kings and lords replaced by investors and entrepreneurs, with users as serfs. The value being extracted is data (Owyang 2019). The uncritical use of digital technologies by development agencies further perpetuates this extractive framework.

Thus, tools to support IEM in accessing and protecting their rights regarding natural resources must also be sensitive to their digital rights. Already considered an extension of human rights in the internet era (Hutt 2015), these rights include:

- Internet access for all;
- Freedom of expression and association;
- Access to knowledge;
- Shared learning and creation through the use of free and open-source software and technology development;
- Privacy, surveillance, and encryption;
- Governance of the internet; and
- Awareness, protection, and realization of rights ('APC Internet Rights Charter' 2006).

In the Mekong sub-region, inequalities in digital access remains a key barrier to accessing internet access for all, while low digital literacy underlies inequalities in accessing the other six digital rights. To this end, there are a number of rights-based approaches and tools that have proven useful. Indigenous Data Sovereignty (IDS) can be used as a framework to consider how to centre IEM perspectives in data, alongside the CARE principles for Indigenous data governance (both are defined in the next section).

IDS refers to 'Indigenous Peoples' possession of the locus of authority over the management of data about their communities, their territories, and their ways of life' (Kukutai and Taylor 2016). In other words, it refers to when IEM have control over how such data is collected, manipulated, managed, and used by themselves as well as by governments, corporations, and development agencies. IDS complicates current understandings of data, which are based on Western-developed notions of individual ownership and use through copyright and licensing and directs attention to the power relationships and postcolonial dynamics in existing data agendas.

Following this, the open data movement has been considered an inclusive way of approaching data ownership and use. The Open Knowledge Foundation defines 'open data' as data and content that can be freely used, modified, and shared by anyone for any purpose. In furtherance of this, the FAIR principles were developed, with each of the letters in the acronym standing for findable, accessible, interoperable and reusable ('Fair Principles' 2016). However, in the context of IEM data, the open data movement is problematic. It reflects international processes that are exclusive of Indigenous voices, circumventing the 'free, prior, and informed consent' (FPIC) principles outlined in UNDRIP, such that the pursuit of open data may be in direct tension with the rights of Indigenous Peoples to govern their data.

In recognition of these issues, the CARE principles for Indigenous data governance were developed. CARE stands for 'collective benefit, authority to control, responsibility, and ethics.' These principles are people- and purpose-oriented and are intended to complement the use of the FAIR principles. Notably, these principles are not dependent on digital skill, and highlight the importance of security and governance, along with the awareness and protection of rights. Digital skills and data literacy training should, therefore, be considered a fundamental way to support IEM in accessing rights both regarding natural resources as well as in the digital realm. Such training supports communities in their freedom of expression and association, access to knowledge, as well as shared learning and creation.

The CARE principles reflect the global Indigenous context. Originally drafted in English, they have been translated into some national languages. However, for Indigenous Peoples of Southeast Asia, a direct translation of the concepts has limited relevance given differing contexts, and the specialized nature of the concepts mean that few local words suffice, such that much of the meaning is lost. There are currently initiatives underway to adapt the principles to the local context using localized concepts and language. Yet even with these limitations, these principles can be a central tool for IEM to use to implement IDS or access their rights, since they allow the user or community of users to be at the centre of their data governance. Coupled with the use of decentralized tools tailored to the digital capacities and access levels of the communities themselves, it creates opportunities for greater Indigenous control over data about themselves. These principles and tools can also be a stepping-stone to improving internet access and, again, are a means to support the realization of all other digital rights.

Recommendations for pedagogy

This section sets out example lesson plans that could be used in a course to support students to further explore the issues and challenges discussed above. We would suggest that these plans be structured around three thematic areas: local adaptation of the IDS concept and CARE principles; data literacy training; and the use of decentralized tools.

Plan 1: discussing IDS recognition and adaptation of CARE principles to the local context

Students need to first understand the key concepts of 'Indigenous knowledge' and 'data,' after which discussions can shift towards the types of knowledge and information that Indigenous Peoples hold, as well as how such knowledge and information ought to be governed. During these discussions, students can be asked the following question prompts: Who are the stewards of this data/knowledge? How is data/knowledge shared among the community? How is this data/knowledge stored? And how is data/knowledge communicated externally?

Students must also comprehend IDS within a rights framework where Indigenous Peoples exercise sovereignty and are able to self-determine that sovereignty through rights to land and territories. To this end, in follow-up discussions, students can be asked to reassess the levels of authority that determine indigeneity, as well as discuss how the CARE principles might be contextualized within specific local contexts, especially those in which Indigenous communities are exceptionally diverse, to ensure that the modalities of communication are accessible to the most marginalized within these communities.

By the end of the lesson, students should understand how:

- IDS deliberately repositions control of data back to Indigenous Peoples and that 'Indigenous data governance is decision making. It is the power to decide how and when Indigenous data are gathered, analysed, accessed and used' (Walter 2018, 3).
- Governance of data is the ownership and control over Indigenous data. At the same time, data for governance speaks to the availability of quality, relevance, and access to data. Both are necessary to ensure the reciprocal relationship between Indigenous sovereignty and IDS.
- Strengthening Indigenous institutions increases the capacity to govern data and, in turn, facilitates stronger evidence-based decision-making to ensure Indigenous nation-rebuilding, which is integral to better positioning Indigenous nations to negotiate as sovereign governments with national governments (Carroll et al. 2019).

Plan 2: data literacy training

In this lesson, students should explore how training in data literacy (i.e., the ability to read, understand, create, and communicate with data) constitutes an important means to support IEM in accessing their rights and realizing IDS frameworks in the digital space. In the context of natural resource management, data produced by IEM may include the geolocation of sites of cultural significance, traditional names and uses of medicinal plants, agricultural practices and patterns, amongst many others. With respect to accessing their rights, students can explore in a seminar format how data can be used to support Indigenous land rights claims and leveraged as evidence to combat illegal encroachment on their land. In class discussions, students can also be asked to deliberate the ways in which data literacy training can support more effective participation in data collection, aggregation, and dissemination, and assist activists supporting IEM or forest communities to advocate for greater control over land and natural resource governance.

Students could also be asked to prepare a concept note on an issue related to Indigenous rights and environmental governance that they would like to advocate for. They would then be asked to present their note and reflect on the fundamental principles of digital and data literacy.

By the end of the lesson, students should understand how:

- The ability to understand data as well as how it is used to further certain interests and agendas is of profound importance to Indigenous communities who have, for generations, not been represented in data and/or have had their data extracted from them.
- Just as IDS is a powerful tool for increasing Indigenous Peoples' agency over their rights to their data and data governance, data literacy is fundamental to facilitating their ability to exercise these rights.

Plan 3: using decentralized tools

Decentralized tools can be a useful tool for IEM to implement IDS or access their rights. But because IEM's level of digital and technological skills is generally low, this raises the question of how and what kind of technological tools can be designed and developed to help IEM document their natural resource rights.

For this lesson, students could be tasked to explore the Mapeo application and whether it is fit-for-purpose in different local contexts. Mapeo is an open-source, offline-first technology for documenting, monitoring, and mapping various types of data, including natural resource ownership. It is accessible to many IEM communities because it does not require internet or any special hardware to collect, view, or share the data. Originally developed with input from IEM in the Amazon, this tool features IEM needs at its core. However, when introduced to IEM in the Mekong sub-region, users identified that the tool misses some functionality, including multi-language capacity, sufficient security protocols, and support for lower literate communities. Accordingly, students can be asked to assess the tool's utility in different IEM communities and reflect more broadly on the following set of questions: What are the current risks to Indigenous communities within existing national data ecosystems? Is it possible to employ new technologies to preserve traditional knowledge and still be scalable? Where is data stored and housed and who has access to this? How can we increase Indigenous representation within the ICT sector to ensure that technological tools are created within the IDS framework?

By the end of the lesson, students should understand how:

- Technological tools like the Mapeo app should be designed and developed to help communities document their rights to and ownership of natural resources.
- Building a fit-for-purpose application that allows IEM to govern their own data within a truly Indigenous data ecosystem remains challenging. The Mapeo app represents a small step to introduce a technical solution that offers greater control over Indigenous data.

Conclusion

ICT can offer new and innovative solutions to managing natural resources. Yet, the existing, predominant frameworks of conservation and ICT development remains vested in siloed, Western-centric knowledge systems and practices. ICT tools uncritically developed can perpetuate inequalities and be misused to further the epistemological assimilation of Indigenous knowledge and value systems.

Indigenous Data Sovereignty offers an alternative approach that challenges traditional schools of thought about how knowledge, comprising both data and information, is framed within Indigenous worldviews. It questions the status quo on data governance and privacy, and offers a modality for IEM to assert control over data about themselves and their lives. The application of IDS can thus aid IEM within the communities of Southeast Asia and beyond by not only furnishing them with ways to access their right to self-determination, but also imbuing them with the ability to exert agency over environmental governance and their lands.

Key sources

1. Nelson, KM and Shilling, D (ed.) (2018), *Traditional ecological knowledge: learning from Indigenous practices for environmental sustainability*, Cambridge University Press, Oxford.

This book offers a collection of compelling cases of environmental sustainability through the lens of traditional ecological knowledge from diverse backgrounds, tribal groups and geographical locations. Grounded in the understanding that biological and cultural diversity are intertwined, the book defines, interrogates and problematizes the many definitions of traditional ecological knowledge and sustainability.

2. Kukutai, T and Taylor, J (ed.) (2016), *Research Monograph No. 38. Indigenous data sovereignty: towards an agenda,* The Australian National University, Canberra.

As the global 'data revolution' accelerates, how can the data rights and interests of Indigenous Peoples be secured? Premised on the UNDRIP, this book argues that Indigenous Peoples have inherent and inalienable rights relating to the collection, ownership, and application of data about them, and about their lifeways and territories. It asks: what does data sovereignty mean for Indigenous Peoples, and how is it being used in their pursuit of self-determination? The varied group of mostly Indigenous contributors theorize and conceptualize this fast-emerging field and present case studies that illustrate the challenges and opportunities involved. These range from Indigenous communities grappling with issues of identity, governance and development to national governments and NGOs seeking to formulate a response to Indigenous demands for data ownership. While the book is focused on Canada, Australia, Aotearoa/New Zealand, and the United States, much of the content and discussion will be of interest and practical value to a broader global audience.

3. Rainie, S, Kukutai, T, Walter, M, Figueroa-Rodriguez, O, Walker, J and Axelsson, P (2019), 'Issues in open data – Indigenous data sovereignty,' in T Davies, S Walker, M Rubinstein and F Perini (Ed.), *The State of Open Data: Histories and Horizons*, African Minds and International Development Research Centre, Cape Town and Ottawa.

Rainie et. al. raise questions regarding the validity of open data applications within Indigenous contexts by observing fundamental questions about assumptions of ownership, representation, and control in open data communities. Because IDS refers to the right of Indigenous Peoples to control data from and about their communities and lands, the authors reflect on how the IDS framework thus challenges concepts of open data by drawing attention to the power and postcolonial dynamics within many data agendas.

4. Dyson, LE, Hendriks, M and Grant, S (ed.) (2007), *Information technology and Indigenous people*, Idea Group Inc., Calgary.

This collection of articles represents a broad cross-section of the research community and practitioners and explores many interesting ideas and opinions that give voice to Indigenous Peoples on the role of information technology (IT) in their lives. It is structured with specific case studies where applications of technology have been used both successfully and unsuccessfully. The book will enable the reader to make informed decisions for planning and action in Indigenous IT-related areas.

5. Gómez-Baggethun, E, Corbera, E and Reyes-García, V (2013), 'Traditional ecological knowledge and global environmental change: research findings and policy implications,' *Ecology and Society*, vol. 18, no. 4, p. 72, http://dx.doi.org/10.5751/ES-06288-180472

This article introduces the theme of TEK and its ability to strengthen the resilience of communities in the face of multiple stressors brought about by global environmental change. It examines the hybridization of TEK with new technical knowledge systems and offers insights into how to promote the maintenance and restoration of living TEK systems as sources of social-ecological resilience.

Notes

1 Hereafter Laos.
2 Like the FAO in 1957, calling shifting agriculture the most serious land-use problem in the tropical world.

References

Adams, WM and Mulligan, M (2003), *Decolonizing nature: strategies for conservation in a postcolonial era*, Earthscan Publications, London.

Allendorf, T, Swe, KK, Oo, T, Htut, Y, Aung, M, Aung, M, Allendorf, K, Hayek LA, Leimgruber, P and Wemmer, C (2006), 'Community attitudes toward three protected areas in Upper Myanmar (Burma)', *Environmental Conservation*, vol. 33, no. 4, pp. 344–352.

'APC Internet Rights Charter' (2006), Association for Progressive Communications, viewed 11 November 2020.

Armitage, D (2002), 'Socio-institutional dynamics and the political ecology of mangrove forest conservation in Central Sulawesi, Indonesia', *Global Environmental Change*, vol. 12, no. 3, pp. 203–217.

Baird, IG (2014), 'Translocal assemblages and the circulation of the concept of "Indigenous Peoples" in Laos', *Political Geography*, vol. 46, pp. 54–64.

Baird, IG (2020), 'Thinking about indigeneity with respect to time and space: reflections from Southeast Asia,' *Space, Population, Society,* https://journals.openedition.org/eps/9628#bibliography

Ballet, J, Koffi, KJM and Komena, KB (2009), 'Co-management of natural resources in developing countries: The importance of context', *Économie internationale*, vol. 4, no. 120, pp. 53–76.

Battachan, K, Su, L, Mali, AL, Ahuan, K, Liu, P, Dacquigan, E, Coleman, M, Bataclao, M and Galagal C (2019), 'Situation on lands, territories and resources of indigenous peoples in Asia: Bangladesh, China, Japan, Laos, Myanmar, Nepal, Sri Lanka, Taiwan, Timor Leste and Vietnam', in G Shimray, JAL Guillao and G Gangmei (ed.), Chiang Mai, Thailand: AIPP.

Berkes, F (1993), 'Traditional ecological knowledge in perspective,' in JT Inglis (ed.), *Traditional ecological knowledge concepts and cases*, IDRC Books/Les Éditions du CRDI, Ottawa, CA.

Borrini, G, Farvar, MT, Nguinguiri, JC and Ndangang, VA (2007), *Co-management of natural resources: Organizing, negotiating, and learning-by-doing*, GTZ, Germany Ministry for Economic Co-operation and Development and IUCN.

Carroll, S.R., Rodriguez-Lonebear, D. and Martinez, A., (2019). Indigenous Data Governance: Strategies from United States Native Nations. *Data Science Journal*, vol. 18, no. 1, p. 31. DOI: http://doi.org/10.5334/dsj-2019-031

Chapin, M, Lamb, Z and Threlkeld, B (2005), 'Mapping indigenous lands', *The Annual Review of Anthropology*, vol. 34, pp. 619–38.

Chen, S, Francois, F, Jütting, J and Klansen, S (2013), *PARIS 21 Discussion Paper No. 1. Towards a Post-2015: framework that counts: developing national statistical capacity*, PARIS 21.

Domínguez, L and Luoma, C (2020), 'Decolonizing conservation policy: how colonial land and conservation ideologies persist and perpetuate indigenous injustices at the expense of the environment', *Land*, vol. 9, no. 3, p. 65.

Dressler, W and Roth, R (2010), The good, the bad, and the contradictory: neoliberal conservation governance in rural Southeast Asia, *World Development*, vol. 39, no. 5, pp. 851–862.

Dunn, CE, Atkins, PJ and Townsend, JG (1997), 'GIS for development: a contradiction in terms?', *Area (London 1969)*, vol. 29, no. 2, pp. 151–159.

Dwyer, M (2017), *UN-REDD Program. Land and forest tenure in Laos: Baseline overview 2016 with options for community participation in forest management*, FAO, UNDP and UNEP.

'Fair Principles' (2016), *Home*, Go Fair, www.go-fair.org/fair-principles/

Fox, J, Fujita, Y, Ngidang, D, Peluso, N, Potter, L, Sakuntaladewi, N, Sturgeon, J and Thomas, D (2009), 'Policies, political-economy, and swidden in Southeast Asia.' *Human Ecology: An Interdisciplinary Journal*, vol. 37, no. 3, pp. 305–322.

Fa, JE, Watson, JEM, Leiper, I, Potapov, P, Evans, TD, Burgess, ND, Molnár, Z, Fernández-Llamazares, Á, Duncan, T, Wang, S, Austin, BJ, Jonas, H, Robinson, CJ, Malmer, P, Zander, KK, Jackson, MV, Ellis, E, Brondizio, ES and Garnett, ST (2020), 'Importance of indigenous peoples' lands for the conservation of intact forest landscapes,' *Frontiers in ecology and the environment*, vol. 18, no. 3, pp. 135–140.

Hazell, P (2009), *IFPRI Discussion Paper 00911. The Asian Green Revolution*, IFPRI, www.ifpri.org/publication/asian-green-revolution

Harwood, J (2018), 'Green Revolution,' in H Callan and S Coleman (ed.), *International Encyclopedia of Anthropology*, John Wiley, New York, viewed 11 January 2021, www.academia.edu/37659199/Green_Revolution

Hutt, R (2015). 'What are your digital rights?' *World Economic Forum*, 13 November, viewed 11 November 2020.

Lang, C (2020), 'The EU's NaturAfrica must avoid colonialism in conservation: protected areas should be managed by Indigenous Peoples themselves,' REDD, 12 June, https://redd-monitor.org/2020/06/12/the-eus-naturafrica-must-avoid-colonialism-in-conservation-protected-areas-should-be-managed-by-Indigenous-peoples-themselves/

Le Coq, JF, Dufumier, M and Trébuil, G (2001), 'History of rice production in the Mekong Delta,' paper presented at the *Third Euroseas Conference*, 6–8 September 2001, London.

Kukutai, T and Taylor, J (ed.) (2016), *Research Monograph No. 38. Indigenous data sovereignty: towards an agenda*, The Australian National University, Canberra.

'Managing forests, sustaining lives, improving livelihoods of indigenous peoples and ethnic groups in the Mekong region, Asia: Lessons learned from the Learning Route' (2013), IFAD, Procasur and AIPP.

Martinez, D (2018), 'Redefining Sustainability through Kincentric Ecology: reclaiming indigenous lands, knowledge, and ethics,' in KM Nelson and D Shilling (ed.), *Traditional ecological knowledge*, Cambridge University Press, pp. 139–174.

Meinrath, SD, Losey, JW and Pickard, VW (2011), 'Digital feudalism: enclosures and erasures from digital rights management to the digital divide,' *CommLaw conspectus*, vol. 19, no. 2, p. 423.

Nelson, KM and Shilling, D (ed.) (2018), *Traditional ecological knowledge: learning from Indigenous practices for environmental sustainability*, Cambridge University Press, Oxford.

Owyang, J (2019), 'The rise of digital feudalism: chances are, you're a serf,' *Jeremiah Owyang's Blog*, 14 September, https://web-strategist.com/blog/2019/09/14/the-rise-of-digital-feudalism-chances-are-youre-a-serf/

Phromma, I, Pagdee, A, Popradit, A, Ishida, A and Uttaranakorn, S (2019), 'Protected area co-management and land use conflicts adjacent to Phu Kao-Phu Phan Kham National Park, Thailand,' *Journal of Sustainable Forestry*, vol. 38, no. 9, pp. 486–507.

Pye, D (2016), 'Cambodia declares protected area in hotly contested Prey Lang Forest,' *Mongabay*, 2 May, https://news.mongabay.com/2016/05/cambodia-protected-areas/

Roth, RJ (2008), 'Fixing the forest: the spatiality of conservation conflict in Thailand,' *Annals of the Association of American Geographers*, vol. 98, no. 2, pp. 373–391.

Rundstrom, RA (1995), 'GIS, Indigenous Peoples, and Epistemological Diversity,' *Cartography and Geographic Information Systems*, vol. 22, no. 1, pp. 45–57.

Rutazibwa, OU (2018), 'On babies and bathwater: decolonizing international development studies,' in S de Jong, R Icaza and OU Rutazibwa (ed.), *Decolonization and feminisms in global teaching and learning*, Routledge, London.

Rutazibwa, OU (2019), 'What's there to mourn? Decolonial reflections on (the end of) liberal humanitarianism,' *Journal of Humanitarian Affairs*, vol. 1, no. 1, pp. 65–67.

Scott, JC (2009), *The art of not being governed: an anarchist history of upland Southeast Asia*, Yale University Press, London.

Sengdara, S (2018), 'NA approves Brou as official Lao ethnic group', *Vientiane Times*, http://annx. asianews.network/content/na-approves-brou-official-lao-ethnic-group-87342

'Shifting cultivation: livelihood and food security: new and old challenges for Indigenous Peoples in Asia' (2015), AIPP and IWGIA, www.academia.edu/15275628/Shifting_Cultivation_Livelihood_and_ Food_Security_New_and_Old_Challenges_for_Indigenous_Peoples_in_Asia

Stevenson, MG (2004), 'Decolonizing co-management in Northern Canada,' *Cultural Survival Quarterly*, vol. 28, no. 1, pp. 68–71.

'The Republic of the Union of Myanmar' (1982), *Burma Citizenship Law*, viewed 24 November 2020.

'UNDRIP' (United Nations Declaration on the Rights of Indigenous Peoples) (2007), UN, www.un.org/ development/desa/indigenouspeoples/wp-content/uploads/sites/19/2018/11/UNDRIP_E_web.pdf

Wageningen University & Research n.d., *Co-management in protected areas of Vietnam,* www.wur.nl/en/ show/Co-management-in-protected-areas-of-Vietnam.htm

Walter, M (2018), 'The voice of Indigenous data: Beyond the markers of disadvantage' in *First Things First: Griffith Review (60)*, The Text Publishing Company, Australia.

Ziai, A (2015), *Development discourse and global history: from colonialism to the sustainable development goals*, Routledge, London.

22

SUSTAINABLE FOOD SYSTEMS

Sango Mahanty

Introduction

Securing equitable and sustainable food systems is a critical global challenge. Although food security is a key sustainable development goal (consider Sustainable Development Goal 2 'Zero Hunger'), the United Nations' latest food security assessment shows that 'we are still off track' in ending hunger and food insecurity (FAO et al. 2020, viii). The evidence is concerning. About one quarter of the world's population experience hunger, with many of these food insecure populations concentrated in Asia and Africa (ibid. xix). During 2020, this situation was compounded by COVID-19's disruptions to farm production and supply chains. This chapter reviews some major challenges for food production and distribution, drawing on the experience of small-scale Cambodian and Vietnamese farmers and their engagements in agricultural markets. Although the drivers of food insecurity are complex and interconnected, many agricultural interventions tend to sidestep important structural factors and neglect environmental concerns. Achieving sustainable food systems therefore calls for a broader approach that addresses land and resource security. Greater attention is also needed to the intersections between agricultural markets and environmental change, and the diverse livelihoods of small-scale farmers (smallholders).

Perspectives on sustainable food systems

The FAO defines a sustainable food system as one that 'delivers food security and nutrition for all in such a way that the economic, social and environmental bases to generate food security and nutrition for future generations are not compromised' (FAO 2018, 2). Food systems sustainability therefore goes well beyond the quantity of food being produced, to also incorporate strong financial returns, broadly based social benefits and neutral or positive environmental impacts (ibid.). Yet diverse actors, such as international agencies, businesses and farmers may have very different interests and levels of power, whether they are working towards capital accumulation or to secure local control over food resources (food sovereignty). We therefore need to consider whose interests are served by current food systems, and to recognize potential frictions between different goals and interests among those involved in food systems.

DOI: 10.4324/9781003017653-25

Although most researchers view food systems as multi-dimensional, a broad distinction can be made between pro-market or neoliberal approaches, and critical perspectives. Key points of difference include problem definition with current food systems, perceived developmental benefits of the market, and methodological orientations. These points of difference in turn have important implications for the kinds of policy or governance responses that are advocated within these perspectives.

Neoliberal approaches advocate for markets as a pathway to poverty alleviation. Their analytical focus falls upon factors that weaken the potential operation and benefits of markets, such as outmoded agricultural technologies, market access and profitability for different value chain actors. Proposed interventions then aim to address these weaknesses while strengthening market engagement among smallholders. An example is FAO's campaign to 'end hunger' through sustainable food systems. Advocating for a systems approach, FAO guidance explains that interventions in food systems need to consider the institutions, resource access, labour, inputs, knowledge, finance, production to consumption systems, and environmental relationships (FAO 2018, 2). Value chain interventions are another example within this neoliberal perspective, aiming to improve returns to various value chain actors through market-oriented support. This approach has gained wide purchase among donor agencies and is discussed in my case study below. Ironically, while attempting to support smallholders within markets, some interventions in this vein have instead tied farmers into potentially adverse forms of market engagement, such as contract farming (Mahanty 2022).

In contrast, critical perspectives identify historical-structural causes for the current inequalities within food systems, as these forces have shaped 'what is being produced and for whom' (Flachs 2019, 2; Akram-Lodhi 2013). My own study of cassava farmers in the Cambodia-Vietnam border region, for instance, relates contemporary market engagements back to colonial state formation, and subsequent land and labour transformations to understand the current configuration of power and benefits within contemporary, globally-connected market networks (Mahanty 2022). Unlike the faith in markets that underpins neoliberal perspectives, critical scholars emphasize that sustainable food systems cannot be achieved without social justice. Pathways towards justice may at times involve challenging the market by working at its interstices to achieve local goals and collective action. Locally-driven food sovereignty movements are one such example (Galt 2013; Borras, Franco and Suarez 2015). Another example is the push for distributive reforms in fundamental assets such as land (Akram-Lodhi 2013).

To explore these issues further, I focus here on smallholder producers who have continuing importance for global food systems, even as their rates of migration and uptake of off-farm work has grown (Rigg et al. 2016). The cases below show that sustainable food systems are inseparable from the struggles and livelihoods of smallholders, and the need to enable diverse livelihood opportunities in this highly differentiated group.

Challenges: the case of smallholders in Cambodia and Vietnam

Case studies from my ongoing research in Cambodia and Vietnam illustrate three of these major food system sustainability challenges. First, I discuss unsustainable debt which has emerged from state and donor interventions to promote land commodification together with the flourishing of an unregulated credit sector. Second, I consider how agricultural industrialization works against smallholder interests in Vietnam's reforms to enable land aggregation by corporate actors. Finally, I show that market-oriented interventions that do not address structural inequities can further entrench these. Ultimately, I argue for interventions that are sensitive to the diverse livelihoods and inequalities among smallholders, who are key players in food systems sustainability in Southeast Asia and beyond (Akram-Lodhi 2013).

Debt

In line with the ideas of Hernando de Soto and other neoliberal thinkers (1989), interventions to formalize rural land titles have been a focal area for donors in recent decades. The logic of this intervention model is that formal titles provide land security and can also be used as collateral for loans, enabling smallholders to invest in farming improvements (ibid.). This is expected to drive more efficient and productive modes of agriculture and to reduce rural poverty (ibid.). Although the strengthening of land security ostensibly aims to improve smallholders' opportunities, this is often not realized in practice. The Cambodian case is illustrative.

Although Cambodia's *2001 Land Law* attempted systematic land registration, titling was never completed (Biddulph 2010). After 2010, many rural people still lacked secure land access especially in frontier provinces where market opportunities as well as low population densities in frontier provinces provided a magnet for migration and informal land-claiming. In 2012, the government deployed teams of students working with state officials to such regions to measure, map and title smallholdings of up to 5 hectares (Grimsditch and Schoenberger 2015). The initiative was known as Order 01 (specifically, *Order 01 on Measures Strengthening and Increasing Effectiveness of the Economic Land Concessions (ELC) Management*, a declaration under the existing Land Law). Uncleared lands and lands under conflict were supposed to be excluded from titling, but well-connected landholders could often gain land titles for such lands (ibid.). The policy had important anticipatory effects, producing a flurry of land clearing, small-scale land-grabbing, and cultivation of the fast-growing crop cassava. These farmers aimed to demonstrate that land was in active use before land titling teams arrived (Mahanty and Milne 2016). Many newly cleared lands were titled during the Order 01 period, making this intervention a key contributor to land clearance as well as land grabbing in agricultural frontiers.

Aside from the landscape change and accumulation catalysed by Order 01, the mobilization of capital with formal titles has also been problematic. The lending opportunities created by Order 01, together with microfinance institutions' move towards commercialized operations, saw the proliferation of new microfinance institutions (MFIs) in Cambodia (Green 2019). In 2019, the Cambodian NGO LICADHO estimated that the country's MFIs held about 1 million land titles through secured loans, placing a sizeable population at risk of land dispossession.

We know from a long tradition of critical agrarian studies and critical geography that the outcomes of rural market formation are socially and geographically differentiated. Those with assets and connections are usually the first to capitalize and enhance their positions while poorer households tend to lag or fall behind (Nevins and Peluso 2008; Bernstein 2010). In Cambodia, for instance, tenant farmers and landless labourers have seen few tangible benefits from land formalization and productivity-oriented interventions, while well-connected elites have been able to garner significant tracts of land. Undeniably, a proportion of farmers have been able to scale up their agricultural production, and taken on perennial crops such as fruit trees, pepper or rubber (Mahanty 2022). Yet my research shows that those beneficiaries were usually well endowed with land and had the resources to acquire equipment and pay for labour. Others, especially farmers in the middle to low-income range, showed clear signs of livelihood and social stress. Debt was contributing to impoverishment, and reinforced pre-existing inequalities in land holdings, assets, and connections. Land dispossession has now become a critical concern for this group of smallholders (Mahanty 2022; Green 2019).

Agricultural industrialization

From a neoliberal perspective, technological improvements to improve agricultural efficiency and productivity are a key pathway to better market performance. Mechanization is often advocated as a key component of this approach, often requiring some consolidation of land holdings to enable economies of scale. The challenges this poses for smallholders is illustrated by the case of Vietnam's land consolidation policy debate (drawing on To et al. 2019).

During the French colonial period, large tracts of Vietnam's farmland were owned by French plantation owners and large Vietnamese landlords. After independence in 1953, Northern Vietnam adopted radical land reforms to collectivize farming, while South Vietnam retained highly unequal land holdings. When the country was reunified in 1975, a socialist land tenure system applied across the board, seeing the distribution of 'land to the tiller.'

After a period of collectivization, food deficits and local unrest emerged. This prompted the government to change its approach in 1981. Under the government's new *Directive 100*, farmers could enter contracts with their cooperatives to retain a proportion of their harvest after meeting their collective quota (Rambo 1973). However, farmers still had little choice over land use and crops, which left them dissatisfied (Kerkvliet 2005). Furthermore, the cooperatives maintained tight control over inputs and were often slow to pay farmers for their produce. Farmers also lacked individual tenure security and were reluctant to make land improvements (ibid.).

These developments led to the 1993 *Land Law*, the country's third major land reform. The new law distinguished between land ownership, which rested with the government, and 20-year use rights, which farmers could hold, transfer, lease, mortgage and inherit. These use rights were formalized with land use certificates known as 'red books.' Equality was a key principle in this next wave of land-rights distributions, in relation to land size and fertility (Thuan 2018). The size of the holding given to each household was determined based on the number of household residents. However, this approach contributed to land fragmentation over time, as holdings were progressively divided between family members.

In 2003, the government revised the *Land Law* to provide more secure tenure and to facilitate land markets. Red book holders could now sell their land rights to others. This commercialization of land use rights drove a new land market and ownership quickly started to concentrate in fewer hands. Between 2001 and 2010, almost one million hectares of agricultural land were converted to industrial or commercial economic development projects for products such as rubber, coffee, and timber. This also had the effect of diverting land from food production. The social impacts were also significant: an estimated 11 per cent of the Vietnamese population has been displaced by land conversion (Garrido et al. 2011). Unsurprisingly, conversion of land from agriculture to other uses has been a primary cause for most major land disputes in recent years (Gillespie et al. 2014).

The most recent amendments to the *Land Law* in 2013 further broadened the scope and duration of land rights. Specifically, new additions included the option for red book holders to deploy their land use rights as a share in joint ventures. For example, households could use their land for a joint venture with a rubber or coffee company to establish a plantation. A cap of 30 hectares was applied for annual crops and 300 hectares for perennial crops, and the duration of rights was changed from 20 to 50 years for both categories of crops. The government actively pursued a policy of land consolidation, encouraging farmers to exchange land parcels with each other to promote greater efficiency – a policy that has shown mixed results (Wells-Dang et al. 2016).

The most recent proposed changes to the 2013 *Land Law* were still undergoing revisions at the time of writing. If, as is widely expected, the revisions pass, private sector investors will find it very easy to concentrate land for industrial agriculture. Government advocates argue that smallholders lack the skills and resources to efficiently utilize their lands, which are also becoming increasingly fragmented. They contrast this with the financial resources of the private sector, their greater market access and capacity to adopt high-tech production systems if they could achieve economies of scale in farm production.

The government's departure from the pro-poor 'land to the tiller' principle has catalysed heated debate in Vietnam. Opponents to the policy emphasize its detrimental impacts on the livelihoods of small and poor farmers. A particular point of concern has been the risk that large-scale private holdings will return Vietnam to the French era of landlords or promote corruption in government. So far this counter-narrative has been unable to gather sufficient momentum to shift the land concentration agenda, creating a very real risk that the land concentration policy may indeed facilitate the formation of a new landed elite in Vietnam, accumulated from smallholders' land.

This case highlights that the temporality of land reforms needs to be considered. Land redistribution to smallholders is one of the key intervention spaces flagged by critical agrarian scholars along with collective action to elevate the rights of local producers (Akram-Lodhi 2013). In the Vietnamese case, though, we see that land reforms continue to evolve over time, and pro-smallholder policies can be overturned by logics that prioritize technical and economic efficiency.

Market interventions

Value chain interventions have gained wide purchase as a pathway to strengthen agrarian livelihoods. These programs follow the logic that market incorporation provides the greatest hope to improve farmers' livelihoods. These approaches emerged after the technocratic green revolution reforms of the 1960s and 1970s, which gave a more emphasis to market mechanisms (Moseley 2016). Importantly, although the language of value chains resembles the notion of commodity chains that grew from Marxist world systems theory, the goals of commodity chain analysis and value chain interventions differ significantly (Neilson, Pritchard and Yeung 2014). The latter embrace market formation and assume that market incorporation delivers better welfare for rural populations. They also reflect the trickle-down assumption that a flourishing private sector will enable rural development (ibid.). In contrast, commodity chain analysis exposes the unequal structures of global capitalism in relation to labour, production and consumption (Bernstein 2010).

In Cambodia, agricultural value chain development has been integral to the government's Trade Sector-Wide Approach (Trade SWAP), an umbrella initiative led by the Ministry of Commerce (Royal Government of Cambodia 2014). With donor support, the government mapped a suite of interventions that prioritized cassava because of the emerging domestic boom in this crop to supply Vietnamese and Thai processing industries. The first intervention discussed here, supported by a multilateral donor, the United Nations Development Programme (UNDP), initially studied value chain functioning and supported new farmers' associations ('Cambodia CEDEP' 2013).

During 2016, there was a major downturn in cassava prices that impacted Cambodian smallholders severely. A mid-term review ('Mid-term review of the Cambodian CEDEP' 2017), which came after the 2016 cassava price slump, warned of the deterioration of the 'cassava sector business environment.' It noted that the market had become volatile, and the profitability of cassava had slipped, leaving 'farmers either locked in a downward spiral of debt and poverty or going into crop diversification' (ibid. 19). The problems were in part attributed to Cambodian famers' being held captive by well-established cassava processing industries in Thailand and Vietnam. The mid-term review called for a realignment of project activities to address this issue (ibid. 7). Yet the review did not fundamentally question the capacity of value chain interventions to manage global market volatility, beyond creating a 'better business environment' and helping the government to develop a dedicated cassava policy (ibid. 49).

The UNDP subsequently advocated for contract farming as the best means to assist farmers by strengthening value addition within Cambodia. Annual cropping contracts were viewed as a means to provide farmers with a steady seasonal price, while guaranteeing cassava inputs for fledgling processing industries (UNDP 2018). Here, UNDP personnel were repeating the logic that has underpinned the growth of contract farming across the developing world, where contract farming was often embraced as a 'win-win' approach that would support farmer incomes (Eaton and Shepherd 2001) *and* provide certainty for start-up firms (Humphrey and Navas-Alemán 2010). In Cambodia, the UNDP put their support behind the Chinese company Green Leader, which already had global operations in energy and land-based commodities (Green Leader 2017). The two parties signed a memorandum of understanding in April 2018, which the UNDP Cambodia country director lauded as a much needed 'private catalyst to a new phase in which more value is generated for the cassava industry, redistributing value-added downwards to different actors and towards the end of the value chain' (Chin 2018). His statement reflected the trickle-down assumption that value creation at the company level would flow through to actors lower down on the value chain.

It was too early at the time of writing to predict the results of the UNDP's cassava experiment, but Thailand's mixed experience with three decades of contract farming is instructive (Glover 1992). In Thailand, many of the problems that cropping contracts were intended to solve have been left unresolved. Farmers and companies often reneged on agreed production targets and prices (Singh 2005). Over time, many farmers abandoned farming contracts to regain flexibility and independence, including the opportunity to negotiate better prices. At the same time, some farmers stayed with farming contracts, because they thought these provided better access to markets, updated knowledge and farming inputs (Schipmann and Qaim 2011; Sriboonchitta and Wiboonpoongse 2008). The mixed outcomes of contract farming were causing some firms to venture into direct production in peri-urban areas in order to guarantee supply to their factories (Schipmann). If Cambodian farmers were to renege on their contracts, processing companies might similarly acquire land for cassava production as a logical next step; an example of this already existed in Tbong Khmum province. For many smallholders, being locked into cassava production could translate to higher risks of debt and land dispossession.

In contrast with the private sector and contract farming focus of the UNDP project, a second value chain intervention by the Australian Centre for International Agricultural Research (ACIAR) focused on farmer-level interventions. This project aimed to strengthen linkages between value chain actors at the farm level, in government, in research institutions and from industry. Farm-level interventions targeted cassava cultivation practices and disease management through extension agencies. It also contributed to the development of a national cassava policy in Cambodia – all in support of smallholder-based cassava markets. According to project documents, smallholder markets provided the opportunity to shift agro-industries away from large-scale land concessions, with their well-documented and serious impacts on communities and environments (ACIAR 2014). The intervention's focus on cassava was explained in terms of cassava's growing economic and livelihood importance (ibid.). As such, like the UNDP intervention, the ACIAR project aimed to understand and influence market conditions, but ACIAR's approach had a greater emphasis on smallholder livelihoods than was evident in the later phases of the UNDP project. In this sense, the ACIAR value chain intervention appeared more sensitive to social conditions.

Since 2016, however, the project's efforts were diverted to addressing market volatility and, especially, disease. The Cassava Mosaic Disease, a virus transmitted by the whitefly (*Bemisia Tabaci*), entered Cambodia in 2015 (Wang et al. 2015). Other diseases that have affected cassava cultivation include Cassava Witches Broom disease, and mealybug (Smith, Newby and Cramb 2018). By 2018, stemming the spread of these diseases was viewed as an urgent priority to secure cassava production. The ACIAR project team acknowledged, however, that this was challenged by the weak 'incentive structures' around disease management and the complex social networks that share and sell seed stock (ibid. 16). Collaborations between this project and agricultural agencies now centre on quelling the spread of disease, which is proving a major challenge, and highlights the risks of smallholder reliance on one crop.

Ultimately, both the ACIAR and UNDP value chain interventions discussed here are premised on the view that smallholder market incorporation produces favourable development outcomes. UNDP's pathway was through private sector investment and industry development, while ACIAR's centred on technical support to producers. Yet both interventions, with their single-commodity emphasis, had the effect of entrenching and fetishizing one crop – cassava – with its associated risks of market volatility and disease. They ultimately applied simplified ideas on how specific crops fit in with smallholders' livelihoods. The (likely) risks of the value chain interventions, especially strategies such as contract farming, were not fully considered and exposed farmers with fewer assets and safety nets to greater precarity and debt.

Conclusion

Smallholder producers are key players in global food networks, yet their market engagements have diverse implications. While some benefit from the income and opportunities these networks provide, many producers also face the risks of debt, land dispossession and the volatility caused by factors such as demand and price shifts, disease, and a changing climate. The cases in this chapter illustrate the broader issue whereby interventions to support market-oriented production in agricultural frontiers often promote land clearance and environmental change. Frictions between smallholders and industrial agriculture are also a source of social conflict in frontier areas.

At the same time, neoliberal scholars propose that smallholder market engagement is an effective pathway to rural development. Greater market efficiency is also flagged as a key requirement for sustainable food systems, leading to interventions of the kind discussed in this chapter: land formalization initiatives, technological innovations, and value chain projects. Yet the evidence so far shows mixed outcomes from such interventions. They usually assist better-off farmers, while poorer and middle-income farmers may not benefit or may even experience new risks. In contrast, critical scholars attribute the problems with food systems to historical and structural terms, resulting in calls for more radical interventions such as major land redistributions and collective action by smallholders, such as movements for food sovereignty (Akram-Lodhi 2013). The Vietnamese case study illustrated that even pro-poor land distributions can shift direction over time.

Overall, this chapter highlights that we need to pay attention to the diverse geographies and circumstances of smallholder producers. In policy terms, the example of agricultural value chain interventions highlights the need to look beyond a 'one-size-fits-all' approach.

Notes for pedagogy

This chapter starts by introducing debates around the concept of sustainable food systems, which is foundational for this topic. One useful way to engage students in exploring this concept and its contested nature is by setting up a class debate around controversial statement that speaks to key contested issues. Teams could present arguments for and against the statement, or they could work in groups to document arguments for and against the statement. Here are two examples of potential debating topics, but others could be developed from the content in the section titled 'Perspectives on Sustainable Food Systems':

- Sustainable food systems are a matter of making farming systems more efficient through the use of technology.
- Inequality is the greatest barrier to sustainable food systems.

Another way to explore the topic of sustainable food systems is through case studies. This approach invites students to explore what kinds of interventions are currently underway to promote better farming systems, and to draw lessons from these. This would work well for a written assignment, for instance, tasking students to explore how a particular project or intervention has responded to the challenge of achieving sustainable food systems. Students can explore the strengths and weaknesses of the approach used in the case study, much as this chapter has done with value chain interventions.

Suggested questions for group reflection and discussion

1. What does sustainability mean in the context of food systems?
2. Why are smallholders relevant to discussions of sustainable food systems?
3. Given the limitations of the various intervention strategies discussed in this chapter, how can governments, donors and civil society groups do better?
4. Can smallholders collaborate in productive ways with large-scale industrial producers? Explore this in the context of palm oil production.

Key sources

1. Akram-Lodhi, H (2013), *Hungry for change: farmers, food justice and the agrarian question.* Fernwood Publishing, Nova Scotia.

In this book, Akram-Lodhi provides an accessible and useful overview of why global food systems are currently unsustainable. He particularly explores the disadvantages faced by small farmers in market exchanges, and the historical political and economic roots of this problem. A detailed understanding of these drivers of unsustainable agriculture is a crucial foundation if we are to move towards more equitable and sustainable food systems.

2. FAO, IFAD, UNICEF, WFP and WHO (2020), 'The state of food security and nutrition in the world 2020: transforming food systems for affordable healthy diets,' FAO, Rome.

This is a key resource on international efforts towards Sustainable Development Goal 2, which is integral to sustainable food systems. This report outlines the current state of play and the key challenges we face globally in addressing food security. It also addresses the specific impacts of COVID-19 on global food security. While Akram-Lodhi's work emphasizes structural change, this FAO report emphasizes the importance of making nutritious food more accessible and affordable, and proposes economic instruments and markets to improve food production and distribution systems.

3. Mahanty, S (2022), *Unsettled frontiers: market formation in the Cambodia-Vietnam borderlands.* Cornell University Press, Ithaca.

This book includes the Cambodian case studies included in this chapter. It provides a grounded exploration of how contemporary farming systems are shaped by a range of factors such as migration, land security and social networks that mediate farming practices and markets. As discussed in the case studies presented here, Mahanty argues in this book that smallholder farming interventions often try to influence small parts of this complex whole, which limits their effectiveness and value.

4. Neilson, J, Bill, P and Yeung, HW (2014), 'Global value chains and global production networks in the changing international political economy: an introduction,' *Review of International Political Economy*, vol. 21, no. 1, pp. 1–8, doi: 10.1080/09692290.2013.873369

This paper by Neilson et al. provides a useful review of the literature on global value chains, which are an integral component of global food systems. Neilson et al. summarize key approaches to analysing global value chains, current debates on how they function and the roles of different value chain actors. The paper is a well-researched introduction to this topic.

References

ACIAR (Australian Centre for International Agricultural Research) (2014), *Project Proposal: Developing Cassava Production and Marketing Systems to Enhance Smallholder Livelihoods in Cambodia and Laos.* Bruce, Australia: ACIAR.

Akram-Lodhi, H (2013), *Hungry for change: farmers, food justice and the agrarian question,* Fernwood Publishing, Halifax.

Bernstein, H (2010), *Class dynamics of agrarian change,* Fernwood Publishing, Halifax.

Biddulph, R (2010), Geographies of evasion: the development industry and property rights interventions in early 21st century Cambodia, doctoral thesis, University of Gothenburg, Gothenburg.

Borras, SM, Franco, JC and Suarez, SM (2015), 'Land and food sovereignty', *Third World Quarterly,* vol. 36, no. 3, pp. 600–617.

Chin, S (2018). 'Cambodian Cassava Catches Mosaic Virus', *The Asean Post,* viewed 20 August 2018, https://theaseanpost.com/article/cambodian-cassava-catches-mosaic-virus.

De Soto, H (2000), *The mystery of capital: why capitalism triumphs in the west and fails everywhere else,* Basic books, New York.

FAO (2018), 'Sustainable food systems: concept and framework', *Policy brief,* viewed 1 August 2020, www.fao.org/sustainable-food-value-chain

FAO, IFAD, UNICEF, WFP and WHO (2020), 'The state of food security and nutrition in the world 2020: transforming food systems for affordable healthy diets', FAO, Rome.

Eaton, C and Shepherd, WA (2001), *FAO agricultural services bulletin 145. Contract farming: partnerships for growth,* FAO, Rome.

Flachs, A (2019), 'Political ecology and the industrial food system,' *Physiology and Behaviour,* vol. 220, 112872.

Galt, RE (2013), Placing food systems in first world ecology: a review and research agenda, *Geography Compass,* vol. 7, no. 9, pp. 637–658.

Garrido, ADM, Anderson, JH, Davidsen, S, Vo, DH, Dinh, DN and Tran, HTL (2011), *Recognizing and reducing corruption risks in land management in Vietnam*, National Political Publishing House Hanoi, Vietnam.

Gillespie, J, Hualing, F and Pham, DN (2014), *Land-taking disputes in East Asia*, United Nations Development Program, Hanoi.

Glover, D (1992), 'Introduction,' in D Glover and LT Ghee (ed.), *Contract farming in South-East Asia: three country studies*, University of Malaya, Kuala Lumpur, pp. 1–9.

Green, WN (2019), From Rice Fields to Financial Assets: Valuing Land for Microfinance in Cambodia. *Transactions of the Institute of British Geographers* vol. 44, no. 4, pp. 749–76262.

Green Leader (2017), *Corporate overview*, viewed 10 May 2019, www.greenleader.hk/EN/about1_1

Grimsditch, M and Schoenberger, L (2015), *Land and livelihoods program. New actions and existing policies: the implementation and impacts of Order 01*, The NGO Forum on Cambodia, Phnom Penh.

Humphrey, J and Navas-Alemán, L (2010), *IDS Research Report 63. Value chains, donor interventions and poverty reduction: a review of donor practice*, Institute of Development Studies, Brighton.

Kerkvliet, B (2005), *The power of everyday politics: how Vietnamese peasants transformed national policy*, Cornell University Press, New York.

Mahanty, S (2022), *Unsettled Frontiers: market formation in the Cambodia-Vietnam borderlands*, Cornell University Press, Ithaca.

Mahanty, S and Milne, S (2016), 'Anatomy of a boom: cassava as a 'gateway' crop in Cambodia's North-eastern borderland,' *Asia Pacific Viewpoint*, vol. 57, no. 2, pp. 180–193, doi: https://doi.org/10.1111/apv.12122

UNDP (2013) 'Cambodia Export Diversification and Expansion Program (CEDEP) II: cassava, marine fisheries products, the Royal Academy of Culinary Arts, and Program Evaluation Function', UNDP, New York.

UNDP (2017), 'Mid-term review of the Cambodian Export Diversification and Expansion Programme (CEDEP II cassava component)', UNDP, New York.

UNDP Cambodia (2018), 'Cassava in Cambodia: Sustainable industrial development', viewed 13 March 2021, https://undpcambodia.exposure.co/cassava-in-cambodia

Moseley, WG (2016), 'The new green revolution for Africa: a political ecology critique,' *Brown Journal of World Affairs*, vol. 23, no. 2, pp. 177–190.

Neilson, J, Pritchard, B and Yeung, HW (2014), 'Global value chains and global production networks in the changing international political economy: an introduction,' *Review of International Political Economy*, vol. 21, no. 1, pp. 1–8, doi: 10.1080/09692290.2013.873369

Nevins, J and Peluso, NL (2008), 'Introduction: commoditization in Southeast Asia,' in J Nevins and NL Peluso (ed.), *Taking Southeast Asia to market: commodities, nature, and people in the neoliberal age*, Cornell University Press, New York, pp. 1–26.

Rambo, TA (1973), *Comparison of peasant social systems of Northern and Southern Vietnam*, Center for Vietnamese Studies, Carbondale.

Rigg, J, Salamanca, A and Thompson, E.C (2016) 'The Puzzle of East and Southeast Asia's Persistent Smallholder.' *Journal of Rural Studies*, vol. 43, pp. 118–133, doi: https://doi.org/10.1016/j.jrurstud.2015.11.003

Royal Government of Cambodia (2014), *Cambodia's Diagnostic Trade Integration Strategy 2014–2018*, Viewed 12 June 2019, www.kh.undp.org/content/cambodia/en/home/library/poverty/cambodia-trade-integration-strategy-2014–2018/

Schipmann, C and Qaim, M (2011), 'Supply chain differentiation, contract agriculture, and farmers' marketing preferences: the case of sweet pepper in Thailand,' *Food Policy*, vol. 36, no. 5, pp. 667–677, doi: https://doi.org/10.1016/j.foodpol.2011.07.004

Singh, S (2005), 'Contract farming system in Thailand,' *Economic and Political Political Weekly*, vol. 40, no. 53, pp. 5578–5586.

Smith, D, Newby, J and Cramb, R (2018), *Cassava Program Discussion Papers. Developing value-chain linkages to improve smallholder cassava production in Southeast Asia*, University of Queensland, Brisbane.

Sriboonchitta, S and Wiboonpoongse, A (2008), *ADBI Discussion Paper 112. Overview of contract farming in Thailand: lessons learned*, Asian Development Bank Institute, Tokyo.

Thuan, NQ (2018), *Tích tõ, tập trung _Şt _ai cho phát tri'n nông nghi»p › Vi»t Nam trong _i·u ki»n mĩi* (land consolidation and concentration for agricultural development in Vietnam in the new environment), viewed 15 December 2018, www.vanhoahoc.vn/nghien-cuu/van-hoa-viet-nam/van-hoa-ung-xu-voimoi-truong-tu-nhien/3486-nguyen-quang-thuan-tich-tu-tap-trung-dat-dai-cho-phat-trien-nong-nghiepo-viet-nam-trong-dieu-kien-moi.html

To, X, Mahanty, S and Wells-Dang, A (2019), 'From "land to the tiller" to the "new landlords"? the debate over Vietnam's latest land reforms,' *Land*, vol. 8, no. 120, pp. 1–19.

Wang, HL, Cui, XY, Wang, XW, Liu, SS, Zhang, ZH and Zhou. XP (2015), 'First report of Sri Lankan cassava mosaic virus infecting cassava in Cambodia,' *Plant Disease*, vol. 100, no. 5, pp. 1029–1029, doi: 10.1094/PDIS-10–15–1228-PDN

Wells-Dang, A, Pham, QT, Burke, A (2016), 'Conversion of land use in Vietnam through a political economy lens,' *Journal of Social Sciences and Humanities*, vol. 2, no. 2, pp. 131–146.

23

RENEWABLE ENERGY

Eko Priyo Purnomo, Aqil Teguh Fathani, and Abitassha Az Zahra

Introduction

Energy is crucial to the existence and growth of human societies (Azam et al. 2015). As the world faces an impending energy crisis due to a variety of factors including overpopulation, poor infrastructure, climate change, and energy waste (Poudyal et al. 2019; Parks 2020; Charles Rajesh Kumar and Majid 2020; Poudyal et al. 2019), the capacity to develop and supply renewable fuels has also become closely linked to sustainable economic growth and has emerged as a cornerstone of national security strategies (Bo et al. 2015). This chapter examines the challenges and opportunities presented by renewable energy, while examining energy consumption trends and efforts at transitioning from conventional to renewable energy sources in the Asia-Pacific region. It also conducts a review of existing literature (published between 2015 and 2020) on this topic: using the following keywords —'energy consumption,' 'energy supply,' 'energy policy,' 'energy planning,' 'conventional energy,' 'renewable energy,' and 'sustainable energy' – it examines the results in the form of more than 100 articles in journals indexed by Scopus. In so doing, aside from discussing the state of existing knowledge on the necessity of renewable energy (Güney 2020), it also seeks to illustrate the key themes addressed in the literature on clean energy and energy supply.

Why renewable energy matters

In these uncertain times, where disruptive events like the COVID-19 pandemic coalesce with the ongoing ramifications of anthropogenic climate change, energy security has emerged as a primary concern for governments everywhere (Kumar and Majid 2020). There is an urgent need for countries to bolster their energy capacity with the production and use of renewable energies, and for national energy policies to become more aligned with sustainable energy guidelines (Ilbahar et al. 2019) as well as with the concepts of scientific development and cyclical economy. Renewables are currently the cleanest and most promising energy source for future generations, with their stable prices and promise of zero carbon emissions (Lau et al. 2010; Dent 2015; Griffiths 2017). It is, of course, customary to note that an increased supply of renewable energy could serve to replace carbon-intensive energy sources and significantly reduce global warming emissions (Boukli Hacene et al. 2020; Pradhan et al. 2020). Renewable

DOI: 10.4324/9781003017653-26

energy facilities also require less maintenance than traditional generators. Their fuel, being derived from available natural resources, can reduce the cost of operation and a country's dependence on fuels and energy from offshore sources (Kumar and Majid 2020).

Indeed, these past few decades have witnessed considerable transformation and innovation in the areas of energy access and low carbon transition (Kuamoah 2020). In the Asia-Pacific, a growing number of governments have begun to focus more on developing 'greener,' alternative energy sources such as solar, biomass and wind energy (Huang and Liu 2017; Ilbahar et al. 2019), as the negative externalities of fossil fuels (e.g. toxic air pollution) become increasingly visible and severe. For countries like Japan and China, renewables are now recognized as a strategic source of energy, delivering cleaner, more secure, and accessible electricity to millions of people (Azam et al. 2015; Garvey 2015; Hufen and Koppenjan 2015). In this way, the development of renewable energies would serve to not only enhance resource use and reduce reliance on fossil fuels, but also optimize energy demand systems and allow for more equitable access to electricity, especially in industrializing societies (Erdiwansyah et al. 2019). At the same time, compared with fossil fuel technologies, which are typically mechanized and capital intensive, the renewable energy industry is more labour-intensive (Liu and Zeng 2017). This means that, on average, more jobs are created for each unit of electricity generated from renewable sources than from fossil fuels.

But because renewable technologies require a high level of investment, private sector capital, technology and innovation are all required to supplement restricted public sector funding for these necessary services through public–private partnerships (Solaun and Cerdá 2019). Thus, renewable energy projects often face difficulties in generating significant economic benefits due to market constraints that cause entrepreneurs to be less incentivized to pursue such projects (Díaz-Cuevas et al. 2019; Parks 2020). To attract more foreign and social capital, existing financing channels that limit rural energy development in developing countries need to be actively addressed and expanded by local governments. One way to do this is through the establishment of investment diversification schemes, where state investment is complemented by multi-channel investment (Kuleli Pak et al. 2015; Mosobi et al. 2015; Doukas et al. 2017). Certainly, increased investment by governments is known for being central to developing the renewable energy sector, especially in rural areas which are abundant in renewable energy sources (Amuzu-Sefordzi et al. 2018). Here, governments would need to focus on converting 'policy incentives' into 'technological innovation' by, for example, providing financial subsidies as well as technical support and guarantees (Kardooni et al. 2015).

While renewable energy encompasses a broad range of sources and technologies, some are more favoured for their 'green-ness' than others. For example, some consider solar and wind power as preferable to hydropower which requires the construction of large and potentially destructive dams, which can release greenhouse gases (methane), foster diseases related to stagnant water, displace populations, and contribute to drought (Madan et al. 2020; Fraundorfer, M and Rabitz, F 2020). Moreover, wind and solar energy require virtually no water to operate and thus do not pollute water resources or strain supply by competing with agriculture, drinking water systems, or other essential water needs. Even so, hydropower is still praised for its ability to increase electricity production dramatically, which was the justification for the Three Gorges Dam in China, and is deemed to be more 'sustainable' than fossil fuels as it cannot be depleted like the latter (Kuamoah 2020).

Trends in renewable energy research

Why, you might ask, is the increase in research publications on the topic of renewable energy noteworthy? The state of knowledge on renewable energy can impact as well as reflect the ways in which policy practitioners think about sustainable policymaking in this space. As previously mentioned, we identified and evaluated over 100 research articles that focused on the topic of renewable energy through a keyword search of the Scopus journal database. Overall, there has been a clear increase in the number of studies that seek to apply renewable energy-related theories or frameworks, with many focusing on how to theorize the concept of sustainability (Lélé 2017) and others looking at how to develop alternative sources of energy that are safe for the environment (Afonso et al. 2017; Balakrishnan et al. 2020; Ghose et al. 2019). This growth mirrors broader global and regional trends in Asia that see significant yearly increases in energy consumption, which has in turn raised concerns about how non-renewable energy sources may be exhausted within the next 100 years, if global consumption were to be left unabated (Pickl 2019).

Based on our review of the literature, it is also notable how research on renewable energy has become more diverse in terms of questions asked and approaches used, compared to a decade ago. Since 2015, it has also become more inextricably connected to broader discussions about sustainability, with a reinvigorated emphasis on how future energy use might be transformed from conventional to renewable energy. Over time, there is also a sharpened focus on how sustainability approaches to energy transition must be rooted in and applied to multi-actor networks in order to facilitate policy uptake within governments (Tongsopit et al. 2016; Sunjoo et al. 2017; Liu 2019).

Furthermore, it deserves note how the production and publication of renewable energy research is still unevenly distributed in geographical terms, with certain countries producing significantly more research than others. Here, most of the published research comes from authors based in the United States, with China-based researchers coming in second place. It is also interesting to observe how, as countries with high conventional energy consumption, research from China, Malaysia and India has tended to focus on the practical application and implementation of renewable energy theories (Azam et al. 2015; Cedrick and Long 2017; Ghose et al. 2019). This not only reflects the growing concern among Asian countries of

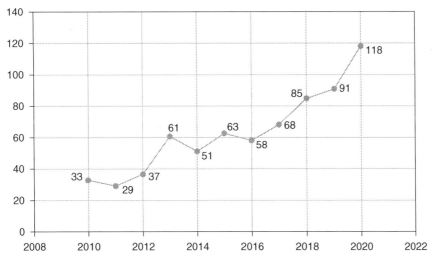

Figure 23.1   Total number of documents per year related to renewable energy and sustainability

Source: Scopus.com

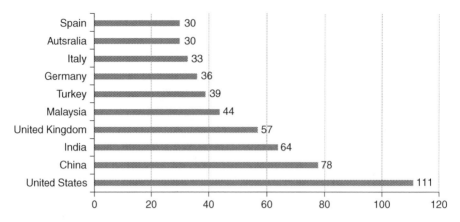

Figure 23.2 Country affiliation of authors

decreasing energy supplies (Purnomo et al. 2020; Sharvini et al. 2018), but it also underscores the need for greater collaboration between researchers in the Global North and Global South to encourage knowledge transfers and innovation on this vital topic.

Renewable energy in policy and practice

The unfolding energy crisis is largely caused by rapid industrialization and urbanization processes in countries such as the Australia, Brazil, China, Russia, and the United States (Eitan et al. 2019; Azam et al. 2015; Kuamoah 2020; Liu et al. 2018; Poudyal et al. 2019; Manasseh et al. 2017). Specifically, the Asia-Pacific region has recorded the highest growth in energy consumption for the period between 2010 and 2019, with the average rise being equivalent to 5–10 exajoules per year. Energy consumption in this region is driven by a range of end users – but namely, those in the industrial, transport and residential sectors (see Figure 23.3) where consumption has increased dramatically over the course of nearly three decades in line with the growing complexity of human needs (Erdiwansyah et al. 2019; Purnomo et al. 2018).

These high levels of energy consumption stem from the fact that most countries in the region are still developing and have, thus far, pursued policy agendas that prioritize economic growth and industrialization. East Asian countries (i.e., China, India, Japan, and South Korea), in particular, have adopted development pathways that entail large-scale energy consumption (Ahmed et al. 2017; Tongsopit et al. 2016; Kuleli Pak et al. 2015; Vijayalakshmi et al. 2017; Dent 2015; Rachmawatie et al. 2020). But for most countries in the region, the accessibility of energy-saving and greener technologies remains largely limited.

These realities, in effect, raise the question of how can developing countries and governments promote and facilitate the expansion of renewable energy to ensure future sustainability? One example can be taken from the European Union's (EU) policies on renewable energy, which has helped to initiate a shift from fossil fuels to biofuels. Under the direction of the European Commission Directorate General for Energy, the EU started to implement biofuel-related targets in 2003 with the issuance of the Fuel Quality Directives (2003 and 2009), and the subsequent introduction of the Renewable Energy Directive (RED) in 2009. RED mandated EU member-states to ensure that by 2020, 20 per cent of all energy usage were being derived from renewable sources (Monti and Martinez Romera 2020). This legally binding obligation was expanded in 2018 with the revision of RED (RED II) and includes a new renewable energy

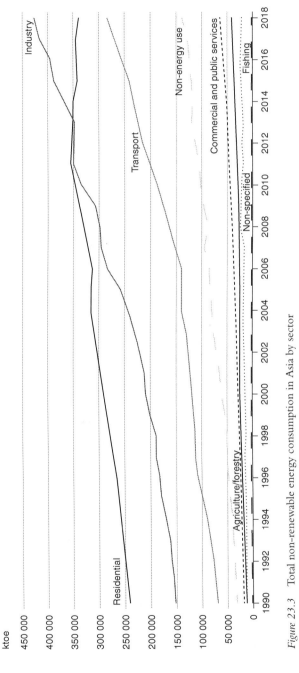

ktoe

Figure 23.3 Total non-renewable energy consumption in Asia by sector

Source: IEA 2020

target of at least 32 per cent by 2030, along with the requirement that member-states transpose the new provisions of RED II into their national laws.

Other emerging examples are found across the Asia-Pacific region, which is seeing more governments adopt alternative energy strategies. In view of the government's proposal to reduce carbon dioxide emissions by 37 per cent or around 315 million tons by 2030 (Han and Baek, 2017), South Korea is now in the middle of stepping up its energy policy. The South Korean government is focused more on financing the development of renewable energy technologies and supporting the Korea Electric Power Corporation (KEPCO), the country's largest electric utility, to increase the market penetration and integration of renewable energy into Korea's electric power system. In addition, the government has announced a guaranteed interconnection policy for renewable energy sources below 1MW. This effort is a response to the 2015 Paris Agreement and the 23rd Conference of the Parties 2017, which aims to increase renewable energy from 7 per cent to 20 per cent of the total power generated in South Korea by 2030 (Kim et al. 2018).

Similarly, in Southeast Asia, the Indonesian government enacted a national energy policy in 2006, which established the various laws, regulations, and targets to ensure that energy production and use are effective and adheres to sustainability principles by 2025 (Purba et al. 2015). The increased use of renewable energy comes at a time when reliance on oil, gas, and coal is already decreasing year by year. Singapore has likewise been making progress in the development of its renewable energy capacity, with the Singapore government having aimed for at least a 16 per cent reduction in greenhouse gas (GHG) emissions (Gasbarro, Iraldo, and Daddi 2017). The government has also facilitated and funded research and innovation in the renewable energy sector by, for instance, holding dialogue with various energy market authorities and inking international research cooperation agreements (Ali, Khan, and Khan 2018; de Andres, MacGillivray, Roberts, Guanche, and Jeffrey 2017).

Conclusion

As global energy consumption continues to increase, the depletion of energy resources – combined with the acute environmental problems that stem from carbon-intensive economic activity – renders it imperative for countries to prioritize the use and production of renewable energy, especially wind and solar power, to secure a more sustainable future (Gürtler et al. 2019; Masoud 2020). Certainly, the plethora of risks and vulnerabilities posed by climate change has brought this requirement into sharp relief (Colomco et al. 2015; Han 2020).

In public and policy debates about global climate change and how to best address it, a major concern centres on the need to ensure intergenerational equity – or put differently, the recognition that climate change stands to have the most impact on the welfare of future generations. But whereas the drivers of the problem are largely well understood by the general public, greater awareness is still needed on how it might be tackled, together with a more nuanced appreciation of the current limitations to renewable energy technology and policies. It is especially important for students to understand these constraints – alongside new opportunities – and arrive at constructive ideas on how to either work with or around them.

Teaching and researching renewable energy

Following this, we suggest that a class on renewable energy involve two types of activities: individual and group work. These suggestions are based on our own experiences teaching this topic in Indonesia. For the individual task, it should focus on getting students to reflect critically about the promise and potential pitfalls of renewable energy. Here, students could be asked to write short reflection pieces on what they think about renewable energy: how can it be implemented in their community or country? Can renewable energy actually be implemented in their community/country right now? Or what are the barriers to implementation (e.g. is it too costly)?

Regarding the group task, students should be asked to divide themselves into smaller groups and then asked to discuss a range of topics, including the advantages and disadvantages of renewable energy as well as what kind of policies are needed to support its use. We would also normally include questions that get students to discuss what renewable policies look like in the Global North compared to the Global South.

Students should also be encouraged to think about renewable energy policy and planning from different stakeholders' perspectives and at multiple scales. To this end, they could be asked to engage in class discussions from the standpoint of a government official trying to implement energy-related reforms in a particular country; or they could be tasked with thinking through the geopolitics of negotiating energy policy at the global level. Further to this, incorporating a 'light' fieldwork component into the course can be useful in getting students to explore alternative perspectives and contexts. For example, students could undertake a half- or one-day visit to a government agency or company, where they determine whether clean energy is being used at the organization and why this is or is not the case. In doing so, they could speak to managerial staff to determine what incentives, if any, exist to support the transition to renewable energy and reflect, in a more practical setting, what kind of strategies might be needed to (further) incentivize 'greener' energy consumption.

The benefit of this type of activity also has the potential to be two-fold, in terms of raising the awareness of students about renewable energy use in their community/country as well as of those working in the government agency or company visited. Indeed, to reform energy use and transition to more sustainable energy sources, we all must play an active role not only in engaging with the policymaking process, but also in advancing the important research and pedagogical innovations currently underway in the field.

Key sources

1. Afonso, TL, Marques, AC and Fuinhas, JA (2017), 'Strategies to make renewable energy sources compatible with economic growth,' *Energy Strategy Reviews*, vol. 18, pp. 121–126.

This article discusses how renewable energy has not contributed to economic growth, whereas non-renewable energy has. It makes the case for why and how these findings are to be incorporated into the formulation of more effective energy strategies.

2. Azam, M, Khan, AQ, Zaman, K and Ahmad, M (2015), 'Factors determining energy consumption: Evidence from Indonesia, Malaysia and Thailand,' *Renewable and Sustainable Energy Reviews*, vol. 42, pp. 1123–1131.

This study explains how a country's foreign direct investment inflows, economic growth, trade openness, and Human Development Index have a positive and statistically significant impact on energy consumption. Here, the authors posit that energy policy should thus be oriented to support the expansion of renewable energy, which can help with sustaining economic growth and development.

3. Balakrishnan, P, Shabbir, MS, Siddiqi, AF and Wang, X (2020), 'Current status and future prospects of renewable energy: A case study,' *Energy Sources, Part A: Recovery, Utilization and Environmental Effects*, vol. 42, no. 21, pp. 2698–2703.

This article examines the installed capacity of renewable energy in developing countries, with a focus on wind, solar, and hydropower energy. According to its findings, the most significant barrier to the further development of renewable energy in such countries is the private sector's reluctance to invest due to the large expenditure involved and late returns on capital.

4. Charles Rajesh Kumar, J and Majid, MA (2020), 'Renewable energy for sustainable development in India: Current status, future prospects, challenges, employment, and investment opportunities,' *Energy, Sustainability and Society*, vol. 10, no. 1, pp. 1–36.

In this article, the authors discuss how sustainable development can be made possible by ensuring peoples' access to affordable, dependable, and sustainable energy. They argue that, in the Indian case, strong government support for renewable energy, combined with an improved economic situation, has propelled India into becoming one of the world's most promising and rapidly expanding renewable energy markets.

5. Erdiwansyah, Mahidin, M., Mamat, R., Sani, M. S. M., Khoerunniza, F., and Kadarohman, A. (2019), 'Target and demand for renewable energy across 10 ASEAN countries by 2040,' *Electricity Journal*, vol. 32, no. 10, p. 106670.

This article illustrates how innovative grid technology, together with a renewable energy portfolio, can provide considerable benefits to countries in Southeast Asia. Accordingly, 'good' national policies are key to the development of the region's renewable energy capacity. The authors posit, however, that the region is yet to be optimally managed to maximize its renewable energy potential. How can the countries in this region learn to implement sustainable energy better?

References

Afonso, TL, Marques, AC and Fuinhas, JA (2017), 'Strategies to make renewable energy sources compatible with economic growth', *Energy Strategy Reviews*, vol. 18, pp. 121–126.

Ahmed, K, Bhattacharya, M., Shaikh, Z, Ramzan, M and Ozturk, I (2017), 'Emission intensive growth and trade in the era of the Association of Southeast Asian Nations (ASEAN) integration: An empirical investigation from ASEAN-8', *Journal of Cleaner Production*, vol. 154, pp. 530–540.

Ali, Q., Khan, M. T. I., and Khan, M. N. I. (2018), 'Dynamics between financial development, tourism, sanitation, renewable energy, trade and total reserves in 19 Asia cooperation dialogue members', *Journal of Cleaner Production*, vol. 179, pp. 114–131. https://doi.org/10.1016/j.jclepro.2018.01.066

Amuzu-Sefordzi, B, Martinus, K, Tschakert, P and Wills, R (2018), 'Disruptive innovations and decentralized renewable energy systems in Africa: A socio-technical review', *Energy Research and Social Science*, vol. 46(July), pp. 140–154.

Azam, M, Khan, AQ, Zaman, K and Ahmad, M (2015), 'Factors determining energy consumption: Evidence from Indonesia, Malaysia and Thailand', *Renewable and Sustainable Energy Reviews*, vol. 42, pp. 1123–1131.

Balakrishnan, P, Shabbir, MS, Siddiqi, AF and Wang, X (2020), 'Current status and future prospects of renewable energy: A case study', *Energy Sources, Part A: Recovery, Utilization and Environmental Effects*, vol. 42, no. 21, pp. 2698–2703.

Bo, Y, Bao-Hua, W and Fei-Ling, S (2015), 'The problems facing renewable energy use in rural China', *Energy & Environment*, vol. 26, no. 3, pp. 437–443.

Boukli Hacene, MA, Lucache, D, Rozale, H and Chahed, A (2020), 'Renewable energy in Algeria: desire and possibilities', *Journal of Asian and African Studies*, vol. 55, no. 7, pp. 947–964.

Cedrick, BZE and Long, PW (2017), 'Investment motivation in renewable energy: a PPP approach', *Energy Procedia*, vol. 115, pp. 229–238.

Charles Rajesh Kumar, J and Majid, MA (2020), 'Renewable energy for sustainable development in India: Current status, future prospects, challenges, employment, and investment opportunities', *Energy, Sustainability and Society*, vol. 10, no. 1, pp. 1–36.

Colomco, E, Bologna, S, and Masera, D (2015), 'Renewable energy and sustainable development', *Renewable Energy and Sustainable Development*, vol. 31, no. 1, pp. 7–34.

de Andres, A., MacGillivray, A., Roberts, O., Guanche, R., and Jeffrey, H (2017). 'Beyond LCOE: A study of ocean energy technology development and deployment attractiveness,' *Sustainable Energy Technologies and Assessments*, vol. 19, pp. 1–16. https://doi.org/10.1016/j.seta.2016.11.001

Dent, CM (2015), 'China's renewable energy development: policy, industry and business perspectives', *Asia Pacific Business Review*, vol. 21, no. 1, 26–43.

Díaz-Cuevas, P, Domínguez-Bravo, J and Prieto-Campos, A (2019), 'Correction to: integrating MCDM and GIS for renewable energy spatial models: assessing the individual and combined potential for wind, solar and biomass energy in Southern Spain', *Clean Technologies and Environmental Policy*, vol. 21, no. 9, p. 1871.

Doukas, H, Karakosta, C and Eichhammer, W (2017), 'Renewable energy policy dialogue towards 2030 – Editorial of the special issue', *Energy and Environment*, vol. 28, no. 1–2, pp. 5–10.

Eitan, A, Herman, L, Fischhendler, I and Rosen, G (2019), 'Community–private sector partnerships in renewable energy', *Renewable and Sustainable Energy Reviews*, vol. 105, pp. 95–104.

Erdiwansyah, M, Mamat, R, Sani, MSM, Khoerunniza, F and Kadarohman, A (2019), 'Target and demand for renewable energy across 10 ASEAN countries by 2040', *Electricity Journal*, vol. 32, no. 10, p. 106670.

Fraundorfer, M and Rabitz, F (2020), 'The Brazilian renewable energy policy framework: instrument design and coherence', *Climate Policy*, vol. 20, no. 5, pp. 652–660.

Garvey, SD (2015), 'Integrating energy storage with renewable energy generation', *Wind Engineering*, vol. 39, no. 2, pp. 129–140.

Gasbarro, F., Iraldo, F., and Daddi, T. (2017). 'The drivers of multinational enterprises' climate change strategies: A quantitative study on climate-related risks and opportunities', *Journal of Cleaner Production*, vol. 160, pp. 8–26. https://doi.org/10.1016/j.jclepro.2017.03.018

Ghose, D, Pradhan, S and Shabbiruddin (2019), 'Development of model for assessment of renewable energy sources: a case study on Gujarat, India', *International Journal of Ambient Energy*, vol. 0, no. 0, pp. 1–25.

Griffiths, S (2017), 'Renewable energy policy trends and recommendations for GCC countries', *Energy Transitions*, vol. 1, no. 1, pp. 1–15.

Güney, T (2020), 'Renewable energy consumption and sustainable development in high-income countries', *International Journal of Sustainable Development and World Ecology*, vol. 00, no. 00, pp. 1–10.

Gürtler, K, Postpischil, R and Quitzow, R (2019), 'The dismantling of renewable energy policies: The cases of Spain and the Czech Republic', *Energy Policy*, vol. 133, p. 110881.

Han, D., and Baek, S. (2017). 'Status of renewable capacity for electricity generation and future prospects in Korea: Global trends and domestic strategies,' *Renewable and Sustainable Energy Reviews*, vol. 76 (November 2015), pp. 1524–1533. https://doi.org/10.1016/j.rser.2016.11.193

Han, H (2020), 'Energy cooperation with North Korea: conditions making renewable energy appropriate', *Journal of Environment and Development*, vol. 29, no. 4, pp. 449–468.

Huang, P and Liu, Y (2017), 'Renewable energy development in China: spatial clustering and socio-spatial embeddedness', *Current Sustainable/Renewable Energy Reports*, vol. 4, no. 2, pp. 38–43.

Hufen, JAM and Koppenjan, JFM (2015), 'Local renewable energy cooperatives: revolution in disguise?', *Energy, Sustainability and Society*, vol. 5, no. 1, pp. 1–14.

International Energy Agency (2020), *Data and Statistics*, www.iea.org/data-and-statistics/?country= WORLD&fuel=Energy consumption&indicator=TFCShareBySector

Ilbahar, E, Cebi, S and Kahraman, C (2019), 'A state-of-the-art review on multi-attribute renewable energy decision making', *Energy Strategy Reviews*, vol. 25, pp. 18–33.

Kardooni, R, Yusoff, S B and Kari, FB (2015), 'Barriers to renewable energy development: five fuel policy in Malaysia', *Energy and Environment*, vol. 26, no. 8, pp. 1353–1361.

Kim, S., Lee, H., Kim, H., Jang, D. H., Kim, H. J., Hur, J., … Hur, K. (2018). 'Improvement in policy and proactive interconnection procedure for renewable energy expansion in South Korea', *Renewable and Sustainable Energy Reviews*, vol. 98 (January), pp. 150–162. https://doi.org/10.1016/j.rser.2018.09.013

Kuamoah, C (2020), 'Renewable energy deployment in Ghana: the hype, hope and reality', *Insight on Africa*, vol. 12, no. 1, pp. 45–64.

Kuleli Pak, B, Albayrak, YE and Erensal, YC (2015), 'Renewable energy perspective for Turkey using sustainability indicators', *International Journal of Computational Intelligence Systems*, vol. 8, no. 1, pp. 187–197.

Lau, KY, Yousof, MFM, Arshad, SNM, Anwari, M and Yatim, AHM (2010), 'Performance analysis of hybrid photovoltaic/diesel energy system under Malaysian conditions', *Energy*, vol. 35, no. 8, pp. 3245–3255.

Lélé, SM (2017), 'Sustainable development: a critical review', *World Development*, vol. 19, no. 6, pp. 607–621.

Liu, J (2019), 'China's renewable energy law and policy: a critical review', *Renewable and Sustainable Energy Reviews*, vol. 99, pp. 212–219.

Liu, X and Zeng, M (2017), 'Renewable energy investment risk evaluation model based on system dynamics', *Renewable and Sustainable Energy Reviews*, vol. 73, pp. 782–788.

Liu, X, Zhang, S and Bae, J (2018), 'Renewable energy, trade, and economic growth in the Asia-Pacific region', *Energy Sources, Part B: Economics, Planning and Policy*, vol. 13, no. 2, pp. 96–102.

Madan, D, Mallesham, P, Sagadevan, S and Veeramani, C (2020), 'Renewable energy scenario in Telangana', *International Journal of Ambient Energy*, vol. 41, no. 10, pp. 1110–1117.

Manasseh, R, McInnes, KL and Hemer, MA (2017), 'Pioneering developments of marine renewable energy in Australia', *The International Journal of Ocean and Climate Systems*, vol. 8, no. 1, pp. 50–67.

Masoud, AA (2020), 'Renewable energy and water sustainability: lessons learnt from TUISR19', *Environmental Science and Pollution Research*, vol. 27, no. 26, pp. 32153–32156.

Mosobi, RW, Chichi, T and Gao, S (2015), 'Power quality analysis of hybrid renewable energy system', *Cogent Engineering*, vol. 2, no. 1, pp. 1–15.

Monti, A and Martinez Romera, B (2020), 'Fifty shades of binding: Appraising the enforcement toolkit for the EU's 2030 renewable energy targets', *Review of European, Comparative and International Environmental Law*, vol. 29, pp. 221–231.

Parks, D (2020), 'Promises and techno-politics: renewable energy and Malmö's vision of a climate-smart city', *Science as Culture*, vol. 29, no. 3, pp. 388–409.

Pickl, MJ (2019), 'The renewable energy strategies of oil majors – From oil to energy?', *Energy Strategy Reviews*, vol. 26, p. 100370.

Poudyal, R, Loskot, P, Nepal, R, Parajuli, R and Khadka, SK (2019), 'Mitigating the current energy crisis in Nepal with renewable energy sources', *Renewable and Sustainable Energy Reviews*, vol. 116, p. 109388.

Pradhan, S, Ghose, D and Shabbiruddi (2020), 'Present and future impact of COVID-19 in the renewable energy sector: a case study on India', *Energy Sources, Part A: Recovery, Utilization and Environmental Effects*, pp. 1–11. DOI: https://doi.org/10.1080/15567036.2020.1801902

Purba, N. P., Kelvin, J., Sandro, R., Gibran, S., Permata, R. A. I., Maulida, F., and Martasuganda, M. K. (2015). 'Suitable Locations of Ocean Renewable Energy (ORE) in Indonesia Region-GIS Approached', *Energy Procedia*, vol. 65, pp. 230–238. https://doi.org/10.1016/j.egypro.2015.01.035

Purnomo, EP, Anand, PB and Choi, JW (2018), 'The complexity and consequences of the policy implementation dealing with sustainable ideas', *Journal of Sustainable Forestry*, vol. 37, no. 3, pp. 270–285.

Purnomo, EP, Ramdani, R, Salsabila, L and Choi, JW (2020), 'Challenges of community-based forest management with local institutional differences between South Korea and Indonesia', *Development in Practice*, vol. 30, no. 8, pp. 1082–1093.

Rachmawatie, D, Rustiadi, E, Fauzi, A and Juanda, B (2020), 'Driving factors of community empowerment and development through renewable energy for electricity in Indonesia', *International Journal of Energy Economics and Policy*, vol. 11, no. 1, pp. 326–332.

Sharvini, SR, Noor, ZZ, Chong, CS, Stringer, LC and Yusuf, RO (2018), 'Energy consumption trends and their linkages with renewable energy policies in East and Southeast Asian countries: challenges and opportunities', *Sustainable Environment Research*, vol. 28, pp. 257–266.

Solaun, K and Cerdá, E (2019), 'Climate change impacts on renewable energy generation: a review of quantitative projections', *Renewable and Sustainable Energy Reviews*, vol. 116, p. 109415.

Sunjoo, P, William, B, Taekyoung, L and Junseop, S (2017), 'Challenges in large-scale government-led sustainable development: the case of the Four Major Rivers Restoration Project', *International Development Planning Review*, vol. 39, no. 4, pp. 399–421.

Syaifuddin, M., Purnomo, E.P., Salsabila, L., Fathani, A.T., and Mitra Adrian, M (2021), 'Development of Aerotropolis in Kulon Progo with Green Infrastructure Concept', *IOP Conference Series: Earth and Environmental Science,* vol. 837, no. 1, 012014. https://doi.org/10.1088/1755-1315/837/1/012014

Tongsopit, S, Kittner, N, Chang, Y, Aksornkij, A and Wangjiraniran, W (2016), 'Energy security in ASEAN: a quantitative approach for sustainable energy policy', *Energy Policy*, vol. 90, pp. 60–72.

Vijayalakshmi, S, Girish, GP and Singhania, K (2017), 'Role of renewable energy in Indian power sector', *Energy Procedia*, vol. 138, pp. 1073–1078.

24

TRANSBOUNDARY GOVERNANCE FAILURES AND SOUTHEAST ASIA'S PLASTIC POLLUTION

Danny Marks

Introduction

Waste, particularly single-use plastic, is a worsening externality of the current neoliberal political-economic system. Consumption is essential for capitalism to work effectively. The good life is often defined by material abundance. From its inception, modern-day capitalism has promised to spread this abundance wide and far. Its incessant cycle of accumulation and production leads to the proliferation of goods (Dauvergne 2010). Final consumption expenditures have skyrocketed from US$2 trillion in 1970 to US$64 trillion in 2019. However, goods cannot be used forever, particularly given that producers plan for them to become obsolete quickly (Iles 2004). Hence, they must be discarded and will then become waste. The packaging that comes with the production and consumption of goods also adds to the amount of waste produced. Only 9 per cent of all plastic packaging is recycled; about 12 per cent is incinerated. The remaining 79 per cent accumulates in landfills, dumps and the natural environment (McVeigh 2020).

Much of this accumulated waste, particularly plastic, finds its way into our oceans. As much as 13 million tons of plastic enters our oceans annually, according to a 2018 study by UN Environment. Over 60–80 per cent of all marine litter, plastic pollution takes either the form of user-end products (plastic bags, bottles, and packaging) or industry raw materials (resin, granules and pellets) (Marks, Miller and Vassanadumrongdee 2020). While 10–20 per cent of plastic waste is generated by ocean fishing and aquaculture, such as discarded fishing gear, the majority of plastic litter, around 80 per cent is land-based (Clapp 2012). Plastic debris makes its way to the ocean through rivers, drainage systems, storm runoff, industrial processes, beach visitors, ineffective waste management in land and illegal dumping.

Marine plastic pollution is a major transboundary problem because it causes social and environmental damage on a global scale, with an estimated cost of US$2.5 trillion per year (Beaumont et al. 2019). Many marine animals, such as turtles, fish, whales, and seabirds, have been found entangled in plastic debris or to have ingested plastic found in their stomachs. Some 267 species of marine animals have been adversely impacted by plastic debris through entanglement or ingestion, although this number will invariably increase as smaller species are studied.

DOI: 10.4324/9781003017653-27

Humans also have consumed plastic by eating these animals, which contributes to health risks such as cancer and infertility (Sharma and Chatterjee 2017). This debris is creating huge ocean patches and plastics are also washing onto shores. The vast majority of the waste, around 80 per cent, is land-based (Clapp 2012) and will have made its way to the ocean through rivers and other waterways. Researchers have calculated that if current trends persist, by 2050, plastic will outweigh fish in the oceans (Ellen Macarthur Foundation 2017).

Further, increased marine plastic pollution has been an externality of our response to a global pandemic. COVID-19 has triggered an estimated use of 129 billion face masks and 65 billion every month. This has led to more than 1.5 billion face masks entering the waterways and oceans, thereby increasing the amount of marine plastic pollution ('6000 metric tons of COVID-19 masks have entered oceans' 2020). Further, COVID-19 has contributed to a global crash in oil markets. Because oil is a key raw material to make plastic, using virgin plastic for packing is at an all-time low cost. These trends have contributed to an increase of 30 per cent waste production in 2020 compared to 2019 (Ford 2020).

In this chapter, I argue that the worsening plastic waste crisis is a failure of transboundary environmental governance. But also, I will show that inequality and injustice are deeply embedded in this sector, in terms of those who contribute to the problem and those who suffer from it. I use Southeast Asia as a case study given that it is home to four of the world's worst six plastic polluting countries – Thailand, Indonesia, the Philippines, and Vietnam (Jambeck et al. 2015), and has become the new dumping ground of the world's plastic waste. I will also discuss the impact of COVID-19 in the region and show how injustice in relation to plastic waste has arisen in the region. I will then conclude by discussing how I have approached this topic pedagogically to offer insights and approaches on how the topic can be taught.

Unequal and unjust transboundary governance of plastic waste

More specifically, since plastic waste can transcend national borders through its mobility, it is a challenge of governing transboundary environmental commons. As (Miller 2020) defines it, transboundary environmental governance is the *collective* of state, societal and private sector decision-making, norms, and practices that shape the formal and informal (re)distribution of environmental costs and benefits across territories and timeframes. As political spaces for governing common pool resources across administrative borders within and between countries, transboundary commons require geographically dispersed communities of environmental practice who either come together to protect a particular environmental good or to respond to a cross-border environmental threat or crisis (Miller, Middleton, Rigg and Taylor 2020). Because plastic waste can be a 'fugitive' or 'non-stationary' resource, multi-sited commoning relations are needed to coordinate the governance of them and other similar resources (Miller 2020). Also, because plastic like other transboundary resources cannot easily be regulated within a country's border, the challenge of sovereignty makes it more difficult for transnational regimes to govern it.

However, the transboundary governance of marine plastic pollution is failing for a number of reasons. First, at the international level, there is no plastics treaty with binding targets and timelines (Borrelle et al. 2017). For example, the United Nations Convention on the Law of the Sea (UNCLOS) sought to create a legal framework to protect and preserve the marine environment, but it has not sufficiently addressed the key sources, types, and entry points of marine pollution. For example, it does not distinguish between sea-based and land-based pollution (Vince and Hardesty 2018). More positively, by defining regimes of responsibilities and characteristics of bodies of water and the responsibilities of coastal and littoral states, UNCLOS has provided a framework to potentially resolve the issue (Duvic-Paoli 2020). Second, ethical consumer norms are neither strong nor comprehensive enough to address this threat (ibid.). Third, both the fossil fuel and plastic industries has been effective in pushing back against policies which would curb plastic consumption, such as plastic bag and import bans (Clapp and Swanston 2009; Tabuchi, Corkery and Mureithi 2020). Instead, it has invested in marketing strategies aimed at convincing consumers to take responsibility for their own waste (Fuhr and Patton 2019). Overall, governance of both the plastic and fossil fuel sectors are fragmented both horizontally – between sectors and product lines – as well as vertically, with loopholes and limited implementation at the national and subnational levels (Dauvergne 2018).

At the supranational or regional level of governance among blocs of nation-states, there has been some success in coordinating transboundary governance efforts, but not in Asia. The European Parliament voted in October 2018 to ban single-use plastics by 2021 (Yeginsu 2018). Caribbean countries have also implemented a Regional Action Plan on Marine Litter Management (Vince and Hardesty, 2018). Across Asia, however, collective action remains limited. In Southeast Asia, ASEAN countries agreed in January 2019 to the Bangkok Declaration on Combating Marine Debris in the ASEAN Region, which will serve as a guide to tackling plastic pollution. Yet ASEAN itself acknowledges that the challenges of addressing marine plastic pollution are tremendous and difficult to achieve, especially in the context of its own geopolitical culture of non-interference in the domestic affairs of individual countries and non-confrontational approach to addressing transboundary environmental issues (Miller, Middleton, Rigg and Taylor 2020).

Given these weaknesses at the international level and regional level, there is a need to 'ratchet up national policies' (Dauvergne 2018, 29). At an Inter-Parliamentary Union hearing to plan for the Ocean Conference in February 2017, a number of countries stated that they wanted to address this problem. However, they admitted that they lacked enforceable legislation and supporting infrastructures to build compliance with sustainability measures across scales and sectors of governance (Borrelle et al. 2017).

Concurrently, our system of consumption and waste are unequal and unjust in terms of those who profit and suffer from this system. As Dauvergne (2010) argues in *The Shadows of Consumption*, the most deleterious effects of consumption (and waste) are borne by groups and places which can least afford to bear these costs. Our globalized system deflects the socioenvironmental costs of consumption to distant places. Even within countries and cities, these costs are directed to places to poorer areas where low-income communities or indigenous communities live, thereby reinforcing pre-existing inequality patterns. Examples in terms of waste include landfills and incinerators close to poor neighbourhoods or marine plastic pollution which adversely affects subsistence fishers in Indonesia (see Nash 1992). Overall, as Dauvergne (2010) asserts, these costs often gravitate far away from centres of power. Examples he gives include the Amazon rainforests, marine life of the Arctic, the Pacific Ocean depths, and the atmosphere (ibid.).

Further, similar to the electronic waste sector (see Iles 2004), environmental injustice arises in the process of plastic usage and waste. Governments and corporations in developed countries benefit the most since they do not have to deal with the waste. Further, within developing countries, designers, oil producers, manufacturers, and recyclers also benefit. They do not closely monitor what happens to their products and the latter profit from recycling them. Producers rarely bear extended responsibility for their products. Instead, it is the workers, such as waste collectors and pickers, factory workers in the recycling sector, and even nearby communities who must bear the brunt of toxic contamination and other adverse environmental effects of waste processing and disposal. As I will show, the Southeast Asia experience is a telling one.

Southeast Asia as the new dumping ground

Seventy-five per cent of globally exported waste ends up in Asia. For example, the United Kingdom exports about 70 per cent of its plastic due to not having the processing capacity to handle it (McVeigh 2020). Since July 2017, when China began to ban imports of plastic waste, Southeast Asia has become a dumping ground for wealthier countries' waste. After China's ban, the amount of plastic waste imported to countries like the Philippines, Malaysia, and Indonesia more than doubled (Marks 2019). In Thailand, the increase was many-fold: 2018, Thailand received 481,000 tons of plastic waste imports, compared with 70,000 tons in 2016 (Macan-Markar 2019). Overall, between 2016 and 2018, regional imports of plastic waste rose by 171 per cent ('Southeast Asia's struggle against the plastic waste trade' 2019).

As the amount of foreign waste accumulated and resentment grew among local populations, Southeast Asian governments began to refuse to act as the world's dumpsite. For example, Malaysia and the Philippines have already returned waste that had improper labelling to Spain and South Korea, respectively (Daniele and Regan 2019). Malaysia, Thailand, and Vietnam also restricted plastic waste imports, with a complete ban planned in the coming years.

These return shipments and restrictions, however, only somewhat signal that waste management is changing and improving in the region. There is increasing awareness of the environmental and social problems of waste, particularly plastic. As highlighted in a report by Global Alliance for Incinerator Alternatives (2019), across Southeast Asia, waste is causing contaminated water, failed crops, and respiratory illnesses. Fish are ingesting plastic. Dead whales are turning up in Thailand and Indonesia with many kilograms of plastics in their stomachs. These various factors contribute to the recent refusals or restrictions in accepting additional waste from high-income countries.

Less positively, some Asian countries still do not have strict restrictions on waste. Indonesia still permits the importation of plastic waste to support industrial operations. Statistics reveal that the amount of plastic waste and scrap the country imported in 2018 jumped by 141 per cent (Siddharta 2019). A lot of waste is also illegally entering Southeast Asian countries. An audit found that almost one-third of waste imported into East Java, for instance, was labelled as paper scraps despite actually being illegal scrap plastic (Graham 2019). This means that these piecemeal bans could have counterproductive effects. For example, Indonesia's waste processing is worse than in many other places and it is projected the country will soon become the largest importer of waste. Further, some materials that have become contaminated cannot be used again, including those that are mislabelled, mixed with non-recyclables, or cleaned improperly, and consequently, often wind up in landfills or the oceans (Jain 2020).

A larger problem is that the changes that are needed to drastically improve these countries' plastic management have yet to occur. Single-use plastic consumption is still worryingly high in these countries. And comprehensive bans or taxes, such as on single-use bags, are few to non-existent. Voluntary measures have often been promoted, but they still have limited effectiveness (Vassanadumrongdee and Marks 2020). These countries' waste management is also woefully inadequate. Recycling rates throughout the world, but especially in Southeast Asia, remain low. In many places, there is no separation of household waste. Littering remains pervasive. At the household and community scales, inadequate infrastructure contributes significantly to the plastic pollution problem. Rubbish bins are often too small, uncovered and infrequently collected (Yukalang et al. 2017)

Many Southeast Asian dumpsites are unprepared to deal with the burgeoning volumes of plastic waste. Of Thailand's 27.8 million tons of plastic waste in 2018, at least 27 per cent was improperly disposed, including via open dumping ('Booklet on Thailand state of pollution 2018' 2019). Much of this plastic ends up in waterways, then flowing into oceans. More than half of Indonesia's landfills are open dumpsites (Kahfi, Putra and Dipa 2019). In these sites, waste is piled improperly – increasing the risk of floods, fires, and trash avalanches. This has led to deaths in the Philippines, Indonesia, and India (Marks 2019). Some waste is also illegally incinerated, releasing toxic gases that harm human health. Policymakers in Southeast Asia have yet to prioritize waste management. They still need to significantly invest in improving waste infrastructure and facilities.

COVID-19 has caused plastic waste to surge in Southeast Asia. This is particularly due to the widespread disposal of single-use face masks, takeaway food containers, and packaging from online shopping. For example, in April 2002, Bangkok's daily average of 2,115 tons of single-use plastics waste per day grew to more than 3,400 tons a day (Sukumaran 2020). Lockdowns halted more than 80 per cent of the recycling value chain in Southeast Asian countries, such as the Philippines and Vietnam (Kaplan and Stuchtey 2020). Moreover, due to COVID-19, there has been a 50 per cent reduction in demand for recycled plastic in Southeast Asia. This is because the economic recession has dampened demand for oil which in turn has made the price of new plastic cheaper (Brock 2020).

Thailand is a prominent example of a country where the increased waste imports have had significant impacts upon segments of the population, particularly low-income groups. Overall, Thailand produced two million tons of plastic waste in 2018, but only one-quarter (500,000 tons) were recycled, mostly plastic bottles (Sukumaran 2020). The country, like others in East and Southeast Asia, has also struggled to expand its domestic capacity to keep pace with surging waste imports that China had previously absorbed (Marks, Miller and Vassanadumrongdee and Marks 2020). Many of these recycling firms which processed the waste were discharging untreated wastewater to save money. As an example, in 2018, police raided Dexin Industries in Samut Prakarn after the government found that the company was illegally dumping wastewater from its waste recycling process (Nanuam 2018). This plastic waste processing has contributed to worsening wastewater in the past few years. This increased wastewater has adversely affected the livelihoods of aquaculture farmers in the southern peri-urban area of Bangkok, such as Bang Khun Thian and parts of Samut Prakarn. They pointed to wastewater intrusion which has caused disease outbreaks as a major source of vulnerability to their livelihoods. The frequency of polluted water entering their ponds has increased in recent years (Marks and Breen 2021). But it is not only in Bangkok and Samut Prakarn where smallholder farmers are suffering from wastewater intrusion, but in other areas of the country as well. As an example, in June 2020, smallholders in Ban Khai district in Rayong province complained to the media that wastewater had leaked into a local rubber plantation and ruined 10 rai of crops (Kongrut 2020). Further, according to the head of a local NGO (personal communication, 6 November 2020), many of these recycling firms which processed the waste were emitting 'all sorts of pollution' and are a key contributor to increased industrial air pollution in Bangkok since 2018.

Injustice is deeply embedded in the flows of plastic in Southeast Asia. As the example from Thailand shows, those profiting the most from plastic usage and waste are consumers as well as the factories' owners and shareholders but, since they neither live nearby or work in the factories, they do not have to breathe the factories' toxic pollutants. Instead, factory works or communities, often lower-income ones residing in areas nearby, suffer health problems. For example, in Samut Sakhon Province, which has the fourth-highest concentration of factories in Thailand, PM2.5 levels severely exceeded Thailand's standard but also the majority of blood samples taken by the local government contained lead at dangerous levels. School students also suffered respiratory problems ('Clean air blue paper: insights on the impact of air pollution and its root causes' 2020). Additionally, smallholder farmers suffer from the negative externalities produced by waste processing, namely wastewater intrusion. Other groups which suffer are waste collectors and those who live near incinerators.

The way forward

If Southeast Asian countries no longer accept waste from high-income countries, where will the waste go? Only 9 per cent of plastic waste worldwide is recycled. Western countries have few easy solutions to deal with plastic waste, as it is often too costly for them to recycle it themselves. Unlike China, they cannot readily convert waste into new products. Given this lower demand plus the numerous deleterious effects of waste recycling, it would be sensible for Southeast Asian countries to follow China's lead and adopt an all-waste import ban too.

Ultimately, manufacturers need to make products that can be better recycled. But some materials, such as plastic wrap film and composite materials, cannot be reprocessed easily. Western countries must also find ways to reduce their consumption of single-use plastic.

Grassroots environmental collectives can also help to mitigate the transboundary spread of plastic litter. In 2016, when Chulalongkorn University launched its Zero Waste Program, students learned during their orientation week how to reduce plastic consumption, supported by the availability of water refilling stations on campus. Students also quickly adjusted to paying two baht (USD$0.06) for bioplastic cups made from sugarcane and plastic bags purchased from campus shops, including from the Japanese-owned 7-Eleven convenience store. Less than a year after the program's launch, the number of plastic bags consumed on campus had dropped by 90 per cent, from 132,000 per month down to less than 13,000 bags per month (Marks, Miller and Vassanadumrongdee 2020).

Political will is generated by these sorts of success stories, which serve as examples of best practice for emulation and replication across borders. From a transboundary governance perspective, political will is crucial to the mobilization of communal activities that change our collective relationship with plastic. When political will is low among governments and big businesses, then external pressure from NGOs, financial institutions, and the mass media can help to turn public opinion toward participatory pathways to environmental reform (ibid.).

At the regional level, bodies such as ASEAN have an instrumental role to play in supporting civil society, plastic producers, retail businesses and governments across the region. With its non-interventionist political culture, ASEAN's emphasis on protecting regional common goods through sustainable development strategies is not only palatable but appealing to member countries as it emphasizes collective economic, health and social rewards while avoiding apportioning blame to individual governments. In March 2019, ASEAN's environment ministers took a positive first step to laying the groundwork for such transboundary cooperation by approving in principle the aforementioned Bangkok Declaration (Gong 2019). While considerable work remains to be done in translating this framework into actionable policies, region-wide consensus about the shared threat posed by marine plastic pollution represents a necessary starting point for thinking through collective forms of policy redress (Marks, Miller and Vassanadumrongdee 2020).

Pedagogical recommendations

I would suggest that when crafting a lesson plan and teaching about waste, with a focus on plastic waste, one should start by asking students where they think their waste today went and what will happen to it eventually, such as whether it will stay in landfills, be recycled, or end up in the ocean. The next topic to be discussed could then be the various waste disposal options, including open burning, landfills, and communal open dumping, and the risks involved in each option. I would also suggest talking about how the amount of global waste has increased significantly in recent years and ask students why they think this has occurred. Following this, students could be asked about which countries produce the most waste – a topic that could be linked to increased consumption.

Before moving into a discussion about the different types of plastic, I would recommend undertaking a group exercise, which would see different groups exploring the unequal costs, benefits, and environmental injustices in four different industries, with each group to discuss a different industry. The groups could then be asked to identify winners and losers, including non-humans (see Chapter 15 of this Handbook), and to explain why and how environmental justice has arisen in their given industry. Industries to be discussed could include electronic waste, single-use plastic (e.g., bags, straws, and bottles), the fast fashion industry, and soft and bottled drinks.

After this discussion, and before moving to the topic of plastic waste, I would recommend discussing electronic waste and asking students what would make their electronic devices more sustainable. A good reference is the Greenpeace Guide (2017) which grades the major electronics companies and reveals how many of these companies do a poor job in terms of making their products sustainable. This would present an opportunity to introduce to students the concept of 'planned obsolescence' (see Iles 2004), which could then segue into a discussion of how injustice and inequality is embedded within the electronic waste sector and asking students to identify the winners and losers in this sector. I also suggest linking this discussion of injustice to that of neoliberalism (see Chapter 17 of this Handbook).

Next, I suggest moving to the topic of plastic waste, microplastics, and marine plastic pollution. Students can be introduced to the problems of marine plastic pollution – about how we now eat microplastic, which has been found in our stool. A one-minute clip could be shown here of a diver off the coast of Bali who has filmed a deluge of plastic with only the occasional fish being seen (see Lamb 2018). Then I would hone in on two major contributors to this problem: the fast fashion industry and single-use plastic usage. For the discussion of the fast fashion industry, I recommend discussing the issue of microplastics, throwaway clothes, and the numerous ways this industry is unsustainable. For the discussion of microplastics, a clip 'The Story of Microplastics' (www.storyofstuff.org/movies/story-of-microfibers/) is a useful visual aid. The eye-opening documentary, 'The True Cost' (https://truecostmovie.com/), which discusses the hidden social and environmental costs of the fast fashion industry could also be assigned.

The lesson can conclude with a discussion of single-use plastic, with a focus on Southeast Asia as a dumping ground of developed countries' waste and how these countries also contribute so much marine plastic pollution themselves. I would include a discussion of poor waste management practices of land-based sources to help students understand how these countries contribute to the problem. Topics to explore further could include the lack of household separation of waste, open dumping, and the poor state of landfills. The discussion could then end with a reflective session on what can be done to address this problem and ask students for suggestions.

Such a lesson plan can help to inform students about the key social and environmental issues related to waste, particularly the issue of marine plastic pollution and microfibers, but would also make the discussion more personalized. Further, it can help encourage students to think more about how inequality and injustice have arisen in this sector, their own waste and consumption practices, and what they can do to affect change. Moreover, students would learn about the governance issues related to plastic pollution, which can help encourage the identification of potential solutions.

Key sources

1. Borrelle, SB, Rochman, CM, Liboiron, M, Bond, AL, Lusher, A, Bradshaw, H and Provencher, JF (2017), 'Opinion: why we need an international agreement on marine plastic pollution,' *Proceedings of the National Academy of Sciences*, vol.114, no. 38, pp. 9994–9997.

This article argues that international collaboration is needed to reduce the demand for single-use plastic products and enable a shift to a sustainable plastics economy. In particular, the international community must commit to quantitative targets to reduce plastic emissions into our oceans.

2. Dauvergne, P (2018), 'Why is the global governance of plastic failing the oceans?' *Global Environmental Change*, vol. 51, pp. 22–31.

This article argues that the governance of plastic is failing due to industry efforts to resist government regulation and deflect accountability combined with the fragmented and inefficient governance at both the international and regional levels.

3. Jambeck, JR, Geyer, R, Wilcox, C, Siegler, TR, Perryman, M, Andrady, A, Narayan, R and Law, KL (2015), 'Plastic waste inputs from land into the ocean,' *Science*, vol. 347, no. 6223, pp. 768–771.

By linking worldwide data on solid waste, population density, and economic status, this article is the first one to estimate the mass of land-based plastic waste entering the ocean, calculating that 4.8 to 12.7 million metric tons entered the ocean in 2010. It also found that population size and the quality of waste management systems were the greatest determinants of which countries contribute the most plastic marine debris.

4. Marks, D, Miller, MA and Vassanadumrongdee, S (2020), 'The (geo)political economy of Thailand's marine plastic pollution crisis,' *Asia Pacific Viewpoint*, vol. 61, no. 2, pp. 266–282.

This article on Thailand presents a good case study of why the governance of marine plastic pollution is failing in a Southeast Asian country. It finds that insufficient incentives, scalar disconnects, and inadequate ownership over waste reduction are major barriers to reduce marine plastic pollution.

5. Vince, J and Hardesty, BD (2018), 'Governance solutions to the tragedy of the commons that marine plastics have become,' *Frontiers in Marine Science*, vol. 5, www.frontiersin.org/articles/10.3389/fmars.2018.00214/full

This article reviews the challenge of global plastic pollution in the context of governance and policy, providing examples of successes, opportunities, and levers for change. It also discusses the role of regulation, public perception, and social license to operate (SLO) in managing waste that enters the ocean.

References

Beaumont, NJ, Aanesen, M, Austen, MC, Börger, T, Clark, JR, Cole, M, Hooper, T, Lindeque, PK, Pascoe, C and Wyles, KJ (2019), 'Global ecological, social and economic impacts of marine plastic', *Marine Pollution Bulletin,* vol. 142, pp. 189–195.

'Booklet on Thailand state of pollution 2018' (2019), *Pollution Control Department,* viewed 31 January 2021, www.pcd.go.th/file/Booklet per cent20on per cent20Thailand per cent20State per cent20of per cent20Pollution per cent202018.pdf

Brock, J (2020), 'Special report: plastic pandemic – COVID-19 trashed the recycling dream,' *Reuters,* viewed 31 January 2021, www.reuters.com/article/health-coronavirus-plastic-recycling-spe-idUSKBN26Q1LO

Clapp, J (2012), 'The rising tide against plastic waste: unpacking industry attempts to influence the debate', in S Foote and E Mazzolini (ed.), *Histories of the dustheap: waste, material cultures, social justice,* pp. 199–225, MIT Press, Cambridge, MA.

Clapp, J and Swanston, L (2009), 'Doing away with plastic shopping bags: international patterns of norm emergence and policy implementation', *Environmental Politics,* vol. 18, no. 3, pp. 315–332.

'Clean air blue paper: insights on the impact of air pollution and its root causes' (2020), *Thailand Clean Air Network,* viewed 31 January 2021, https://thailandcan.org/Clean_Air_Blue_Paper_EN.pdf

Daniele, U and Regan, H (2019), 'Plastic waste dumped in Malaysia will be returned to UK, US and others,' *CNN,* viewed 31 January 2021, www.cnn.com/2019/05/28/asia/malaysia-plastic-waste-return-intl/index.html

Dauvergne, P (2010), *The shadows of consumption: consequences for the global environment,* MIT Press, Cambridge, MA.

'Discarded: communities on the frontlines of the global plastic crisis' (2019), *Global Alliance for Incinerator Alternatives,* Berkeley, https://wastetradestories.org/wp-content/uploads/2019/04/Discarded-Report-April-22.pdf

Duvic-Paoli, L-A (2020), 'Fighting plastics with environmental principles? The relevance of the prevention principle in the global governance of plastics', *AJIL Unbound,* vol. 114, pp. 195–199.

Ellen Macarthur Foundation (2017), 'The New Plastics Economy: Catalysing action,' viewed 20 October 2018, www.ellenmacarthurfoundation.org/publications/new-plastics-economy-catalysing-action

Ford, D (2020), 'COVID-19 has worsened the ocean plastic pollution problem,' *Scientific American,* viewed 31 January 2021, www.scientificamerican.com/article/covid-19-has-worsened-the-ocean-plastic-pollution-problem/

Fuhr, L and Patton, J (2019), 'Plastic production is the problem, and not plastic waste,' *Irish Examiner,* viewed 31 January 2021, www.irishexaminer.com/breakingnews/views/analysis/plastic-production-is-the-problem-and-not-plastic-waste-910047.html

Gong, L (2019), 'More plastic bags than fish: East Asia's new environmental threat,' *East Asia Forum,* viewed 22 April 2019, www.eastasiaforum.org/2019/04/20/more-plastic-bags-than-fish-east-asias-new-environmental-threat/

Graham, B (2019), 'Australia rubbish: Indonesian village being inundated with illegal plastic waste from Australia,' *News.Com.Au,* viewed 31 January 2021, www.news.com.au/technology/environment/indonesian-village-being-inundated-with-illegal-plastic-waste-from-australia/news-story/9a03e5a78d67f25994f4176731a30822

Greenpeace Guide, (2017), 'Guide to Greener Electronics', retrieved from www.greenpeace.org/usa/reports/greener-electronics-2017/

Iles, A (2004), 'Mapping environmental justice in technology flows: computer waste impacts in Asia', *Global Environmental Politics,* vol. 4, no. 4, pp. 76–107.

Jain, A (2020), 'Trash trade wars: Southeast Asia's problem with the world's waste,' *Council on Foreign Relations,* viewed 31 January 2021, www.cfr.org/in-brief/trash-trade-wars-southeast-asias-problem-worlds-waste

Kahfi, K, Putra, N and Dipa, A (2019), 'Inadequate landfills worsen Indonesia's waste problems', *The Jakarta Post,* viewed 31 January 2021, www.thejakartapost.com/news/2019/03/03/inadequate-landfills-worsen-indonesias-waste-problems.html

Kaplan, R and Stuchtey, M (2020), 'Collateral damage: COVID-19's impact on ocean plastic pollution Greenbiz', *GreenBiz*, viewed 31 January 2021, www.greenbiz.com/article/collateral-damage-covid-19s-impact-ocean-plastic-pollution

Kongrut, A (2020), 'Wastewater sparks local ire,' *Bangkok Post*, viewed 31 January 2021, www.bangkokpost.com/thailand/general/1930900/wastewater-sparks-local-ire

Lamb, K (2018) '"Plastic, plastic, plastic": British diver films sea of rubbish off Bali,' *The Guardian*, viewed 1 February 2021, www.theguardian.com/world/2018/mar/06/plastic-british-diver-films-sea-rubbish-bali-indonesia

Macan-Markar, M (2019), 'New law in Thailand risks drawing an avalanche of plastic waste,' *Nikkei Asian Review*, viewed 31 January 2021, asia.nikkei.com/Spotlight/Environment/New-law-in-Thailand-risks-drawing-an-avalanche-of-plastic-waste

Marks, D (2019), 'Southeast Asia's plastic waste problem,' *East Asia Forum*, viewed 31 January 2021, www.eastasiaforum.org/2019/06/26/southeast-asias-plastic-waste-problem/

Marks, D and Breen, M (2021), 'The Political Economy of Corruption and Unequal Gains and Losses in Water and Sanitation Services: Experiences from Bangkok,' *Water Alternatives*, vol. 14, no. 3, pp. 795–818.

McVeigh, K (2020), 'New rules to tackle "wild west" of plastic waste dumped on poorer countries,' *The Guardian*, viewed 31 January 2021, www.theguardian.com/environment/2020/dec/29/new-rules-to-tackle-wild-west-of-plastic-waste-dumped-on-poorer-countries

'6000 metric tons of COVID-19 masks have entered oceans' (2020), *Marine Insight*, viewed 31 January 2021, www.marineinsight.com/shipping-news/6000-metric-tons-of-covid-19-masks-have-entered-oceans/

Miller, MA (2020), 'Bordering the environmental commons', *Progress in Human Geography*, vol. 44, no. 3, pp. 473–491.

Miller, MA, Middleton, C, Rigg, J and Taylor, D (2020), 'Hybrid governance of transboundary commons: insights from Southeast Asia', *Annals of the American Association of Geographers*, vol. 110, no. 1, pp. 297–313.

Nanuam, W (2018), 'Prawit orders end to imports of hazardous waste,' *Bangkok Post*, viewed 31 January 2021, www.bangkokpost.com/thailand/general/1489106/prawit-orders-end-to-imports-of-hazardous-waste

Nash, AD (1992), 'Impacts of marine debris on subsistence fishermen: an exploratory study', *Marine Pollution Bulletin*, vol. 24, no. 3, pp.150–156.

Sharma, S and Chatterjee, S, 2017, 'Microplastic pollution, a threat to marine ecosystem and human health: a short review', *Environmental Science and Pollution Research*, vol. 24, no. 27, pp. 21530–21547.

Siddharta, A (2019), 'Indonesia vows to send back illegal plastic waste,' *Voice of America*, viewed 31 January 2021, www.voanews.com/east-asia/indonesia-vows-send-back-illegal-plastic-waste

'Southeast Asia's struggle against the plastic waste trade' (2019), *Greenpeace Southeast Asia*, viewed 31 January 2021, www.greenpeace.org/southeastasia/publication/2559/southeast-asias-struggle-against-the-plastic-waste-trade/

Sukumaran, T (2020), 'Plastic pollution plagues Southeast Asia amid Covid-19 lockdowns,' *South China Morning Post*, viewed 31 January 2021, www.scmp.com/week-asia/health-environment/article/3096554/plastic-pollution-plagues-southeast-asia-amid-covid-19

Tabuchi, H, Corkery, M and Mureithi, C (2020), 'Big oil is in trouble. Its plan: flood Africa with plastic', *The New York Times*, viewed 31 January 2021, www.nytimes.com/2020/08/30/climate/oil-kenya-africa-plastics-trade.html

Vassanadumrongdee, S and Marks, D (2020), 'Thailand takes action on plastic,' *East Asia Forum*, viewed 30 January 2021, www.eastasiaforum.org/2020/01/27/thailand-takes-action-on-plastic/

Yeginsu, C, 2018, November 28, 'European Parliament Approves Ban on Single-Use Plastics', The New York Times, Retrieved from www.nytimes.com/2018/10/25/world/europe/european-parliament-plastic-ban.html.

Yukalang, N, Clarke, B, Ross, K, Yukalang, N, Clarke, B and Ross, K (2017), 'Barriers to effective municipal solid waste management in a rapidly urbanizing area in Thailand', *International Journal of Environmental Research and Public Health*, vol. 14, no. 9, pp. 1013.

25

SUSTAINABLE DEVELOPMENT DISCOURSE

Thomas McNamara

Introduction

A constant struggle of teaching global development, particularly at Master level, is encouraging students to consider development workers' positions in the structures and power dynamics that stabilize global injustices. Even students who are passionate about the role of Global North institutions and politics in underdevelopment are often confronted by or resistant to works that foreground development practice's unjust power structures or that demonstrate how project reports discount local ontology (classically Ferguson 1994 or Mosse 2004). In responding to such works students often unconsciously invoke common rejoinders to post-and-critical development theorists. Some draw-up what Pieterse (2010) describes as a 'counterpoint paradigm,' claiming as obvious the superiority of projects that are small, ecologically sound and that are guided by local values, and stating that development, or at least the development they will practice, is moving in this direction. Other students follow Yarrow (2010) in arguing for analysing development less through political structures and more through the personal moral projects of development workers. There is nothing inherently unconvincing about these arguments, but they are often presented by students, and in development textbooks, as providing a complete refutation of enduring criticisms of development discourse and procedures (Ziai 2017). I aim to, and have struggled with my attempts to, encourage students to appreciate how difficult it is to fully incorporate local critiques into development practice and to consider the relationship between development workers' moral projects and the material and political systems that development can entrench.

Sustainability is an excellent concept for considering the disjunctures between development's ideals of empowerment and local ownership; and the practices and reporting structures that empower voices from the Global North and development professionals. Sustainability and sustainable development have deeply contested meanings, yet their ambiguity and positive valence make it difficult to object to values and mechanisms that claim the label 'sustainable.' Ideally, debate centred around sustainability should therefore enable overlap and exchange between the histories, cultural norms and moral projects of donors, practitioners, and recipients, allowing each to challenge their interlocutors without questioning their intentions. However, I found that in rural Malawi, the norms of sustainable development practice, and the values and power

DOI: 10.4324/9781003017653-28

dynamics it renders invisible, quietened local critiques of projects, while enabling development workers to ensure coherence between policy and practice.

In teaching introductory Masters-level courses on development theory and practice, I utilize a case study from Malawi to critique practices justified through sustainability and to introduce post-development theory. This builds upon a semester-long activity into development buzzwords (detailed in the final section of this chapter) used to illustrate how development's key terms are created through compromise and consensus (Cornwall and Eade 2010; Mosse 2004). Late in the semester I use critiques of sustainability to open space for students to engage with texts that are particularly critical of development apparatuses and discourse. Ziai (2017) argues that many development studies courses simultaneously incorporate specific post-development claims, while rejecting post-development politics. Through encouraging students to consider how they will be incentivized to downplay local critiques, I aim to highlight to them the inherently political nature of development as a project and their own role in maintaining or challenging this.

Critical development studies, sustainability, and development practice

Sustainability is one of global development's most ubiquitous and contested terms. Its advocates argue that this is a strength, with its flexibility creating space for agreement and alliances between actors with diverse interests (Cornwall and Eade 2010). Brutland's classic definition of sustainable development as 'development that meets the needs of the present without compromising the ability of future generations to meet their own needs' can be deployed to make claims based upon the immediate needs of those in the Global South, while also respecting concerns about long-term self-reliance and the environmental impact of economic growth (Mazibuko 2013). The Sustainable Development Goals build upon Brutland's legacy, and the mainstreaming of sustainability in development, creating a framework that affirms the equal significance of economic growth, societal progress and environmental maintenance (Holden et al. 2017). They are similarly lauded for encouraging compromise, partnership, and the inclusion of diverse viewpoints.

Critics argue that instead of sustainability enabling alliances through uncertainty, the positive resonance of the term impedes debate, with criticisms of a project or practice labelled 'unsustainable.' Sustainability's ambiguity, combined with development's power dynamics, privileges Global North and pro-market voices. Critiquing Brutland, Mazibuko (2013) questions why the desires of future generations should have the same moral significance as the immediate needs of the world's poor. Alternatively, Weber (2017) claims that the SDGs consolidate a 'market episteme' crowding out space for non–capitalist approaches to development and for concepts of sustainability that question the possibility of ever-increasing growth.

More significantly for my work, the meaning of concepts like sustainability, development and even one's needs are necessarily embedded in local value systems, yet sustainable development is too often treated as a universal concept. Insights derived from the Global South are typically used to support or distinguish between Global North understandings, rather than as coming from separate ontologies with differing meaning for needs, aspirations and moral responsibilities (Herrera 2019). For example, the Andean concept of Beun Vivir – a discourse of living well that activity rejects commodification- is often framed as an expansion of strong sustainability;[1] while Maori development narratives are presented as a 'quadruple bottom line' – the triple bottom line[2] enhanced by cultural continuity (Scrimgeour and Iremonger 2011). Similarly, advocates of the SDGS frequently describe the need to find partners, who can adapt a global aspiration to a local context. Sustainable development, in the form espoused in the SDGs and in many development projects, reflects a northern discourse, which is framed as 'global', and which produces a cluster of norms, practices and accountability structures. When local definitions of sustainability clash with this global culture of sustainability, they are at most understood as criticisms of specific projects, rather than seen through genuine differences in understandings of the good life and of moral interconnection (McNamara 2017). In the specific case of Malawi, sustainable development is translated into *chitukuko kulutilisya* – literally 'change for the better that is continuing.' In the context of a moral and economic system built through intra-communal interdependence, this continuing improvement is often understood to be antithetical to attempts to create independent consumers and producers through 'sustainable development' projects.

Sustainability is not the only buzzword that justifies specific development projects and entwines them with the practices and power dynamics of international development. There are many others. Identifying the relationship between these terms is crucial to understanding how development norms emerge and are stabilized. Cornwall and Eade (2010) note that sustainability and other terms such as empowerment, participation and capacity building, appear in 'chains of equivalence,' which link development project documents and norms at a local, national and international level. This means that interpretations of buzzwords that are consistent with each other take precedent over discordant meanings (Cornwall and Eade 2010). For example, sustainability's frequent entwinement with participation and empowerment encourages understandings which posit sustainability within a geographically defined community, creating 'local solutions' that do not hoist ongoing demands on a development agency (Swilder and Watkins 2017). Development funders will therefore describe a project as sustainable if it has an environmental focus; if it will generate income; or if it will utilize the labour or resources of its recipients (which serve as proxies for the community's enthusiasm for the project and therefore as evidence that it will continue after the development agency departs) (Mazibuko 2013). They will also reject projects requiring ongoing transfers of resources or long-term engagement with a community as unsustainable, in doing so reducing space for arguing if this is morally righteous or good development practice.

Critiques of sustainability (and this chapter) link and occasionally entwine post-development and critical approaches to development (Ziai 2017). This is slightly disingenuous: some post-development thinkers explicitly reject the constructive criticism of alternative and critical approaches to development (Escobar 2010; Ziai 2017); while certain critical development theorists argue that post-development ignores developments' material achievements and regular reforms (Mosse 2013; Pieterse 2010). However, many of the same key academics are drawn upon by both post-development and critical development works, some but not all of whom would identify with post-development (c.f. Ferguson 1994; Li 2014). Similarly, most international development courses draw from critiques that are made by both post-and-critical development thinkers, and acknowledge disagreement over the implications of these claims (Harcourt 2017; Ziai 2017).

Both post-development and critical development theorists claim to champion local critiques of development in their responses to and utilizations of sustainability. The latter often sees participatory research techniques and advances in monitoring and evaluation as somewhat enabling local critiques and ontology (Cornwall and Eade 2010; Pieterse 2010). The former presents this as ventriloquizing the poor, with unjust power dynamics forcing local communities to agree to and accept responsibility for external projects (Escobar 2010). Ziai (2017) argues that this has a created a new middle-ground in development studies pedagogy, whereby those teaching development courses often state, 'I'm not post-development but,' then give students important critiques of the development apparatus that they desire to work within. However, these lecturers then quickly explain that the critiques are extreme and offer no way forward. By contrast, I instead attempt to foreground to students that, in the development projects I have witnessed, these critiques were legitimate and that the structures of development work means that they will likely be legitimate elsewhere. Criticizing sustainability through these frameworks provides a practical example for students to grapple with in their studies and in their later practice.

Building upon Mosse (2004), I explore how development actors maintain policy coherence – protecting development models, strategies and projects from the critiques of recipients, through describing these as sustainable development and as necessary for sustainability. These criticisms ring particularly true in nations like Malawi, where there is a vast power asymmetry between international development overseers and local development practitioners and between local professionals and recipients (Swilder and Watkins 2017). Beneficiaries often participate in projects because they represent their best, but still meagre, chance of personal reward, and development structures depoliticize the wealth of a national development elite and an unjust global system (Cammack 2017).

In Malawi, sustainable development is a key concern of many small NGOs, who compete for non-recurrent funds from institutional donors. These organizations run trainings and workshops, organizes volunteer infrastructure projects or facilitates income generating activities (Swilder and Watkins 2009). There is substantial literature arguing that sustainability measures which place costs on a community do not work; with poorer recipients abandoning projects they value highly when a price is attached to development. Yet both in Malawi and elsewhere, employees throughout the development industry work to fashion narratives of success, attesting to the sustainability of the project they worked for, and affirming the importance of sustainability more generally (Swilder and Watkins 2017). If students desire to work in such environments, it is likely they will find themselves professionally incentivized to create coherence between their best-practice models and the consistently worsening material conditions of most recipients (Tawfik and Watkins 2007). They will be encouraged to discount some community critiques as a 'lack of understanding' and to see projects that almost directly match donor priorities as co-created by the community (Swilder and Watkins 2017).

Contested meanings of sustainability in rural Northern Malawi

To illustrate how development norms and practices downplay local critiques, I recount a vignette from my fieldwork in Northern Malawi.[3] This story details how, in an environment of extreme poverty, sustainability's moral valance encouraged NGO workers to discount villagers' critiques of their projects, even when these critiques manifested through a local sustainability discourse. This story is set in Vsawa, a collection of approximately 80 villages along a 30 km stretch of fishing and farming villages along the Northern Malawian lakeside. Vsawa was poor, even by Malawian standards, with the average Vsawan spending under 200MK(0.58USD) per day.[4] Only ten villages had any electricity, and only one flat-bed truck and a small number of motor-boats connected Vsawans to the rest of the nation. All Vsawans farmed, some fished, and a tiny minority were formally employed as teachers, health workers and, increasingly, as development practitioners.

Mbwezi was an NGO that ran health, education, agriculture and micro-finance programs, provisioning whatever they could garner from grants or donor funds. It had a staff of about 30 people, mainly Malawians. Tertiary educated workers stayed in Mzuzu (Northern Malawi's capital), searching for grants, and engaging in monitoring and evaluation, while high-school educated staff lived in Vsawa and facilitated projects. *Nkuvira* was a small development organization founded by a British development worker, who oversaw a staff of about 20 Vsawans and a stream of international volunteers. This organization ran a library, a community centre, a pesticide-free garden with chickens, and an 'electrification room' where mobile phones could be charged using solar panels.

More important to Vsawans than their development projects were the symbols that both NGOs emitted, linking development to wealth, and presenting it as gifted to the community by the outside world. *Mbwezi's* internal walls were covered in illustrations of two-parent, two-children, white families. One of the external walls had a large, graffiti-text sign that read 'To the people of Vsawa … development.' The NGO owned two 4WDs and its staff carried laptops and phones that were unlike those owned by anyone in Vsawa. These paled in comparison to the wealth of the international volunteers and Masters students who worked at *Nkuvira,* and who would occasionally leave their phones, clothes and iPods as gifts for villages with whom they felt an affinity. Further, while not officially linked to *Nkuvira,* the only other substantive business in this village was a luxury travel lodge that tourists accessed by boat from Nkhata Bay. *Nkuvira's* manager and international volunteers stayed at this lodge.

While modest, *Mbwezi* and *Nkuvira's* set-up were relatively common in Malawi, a 'donor darling' that combines extreme rural poverty, English language penetration and peace (Morfit 2011). The nation's poverty, rurality and lack of geostrategic significance means that many of the power dynamics that are either contested or covertly present in other aid industries are overt in Malawi (Swilder and Watkins 2017). Donors provide 40 per cent of Malawi's recurrent funds and attempt to direct these to internationally-aligned NGOs, due to their distrust of the nation's government (Page 2019). These donors are known to be particularly feckless, moving their support between organizations based upon trends in the development industry and their perceptions of each NGO's monitoring and evaluation (Page 2019). The power relation between intra-national development staff and project recipients is equally discordant. Less than four-per cent of Malawians complete high-school and less than one-per cent go to university (Frye 2012). Almost all elites work in the development sector, which is a key provider of stable employment and personal wealth (Swilder and Watkins 2017). Long-held beliefs conflate wealth, education and intelligence and these have been exacerbated by a development industry where almost every project depicts development as a transfer of knowledge and resources from an urban elite to the rural poor (Page 2019). Local development staff, even gardeners and nightwatchmen, will almost always be among a village's richest residents (Page 2019). Sitting-fees from a day of training are more than a month's salary and microloans or seed provisions can significantly shorten a family's hungry season (Swilder and Watkins 2017).

For most development agencies operating in Malawi, sustainability means combating rural Malawians' 'dependence,' on the aid industry. This typically occurred through 'education' and 'empowerment,' which appear in chains of equivalence (Swilder and Watkins 2009). Every NGO I interacted with had sustainability in their mission statement and *Mbwezi's* management pledged not to pay villagers per-diems, because they were not sustainable. Some, including *Nkuvira,* expanded sustainability to include financial self-sufficiency, which necessitated that they receive funds from inside Vsawa. Neither of these understandings resonated with rural Malawians' concept of *chitutuko kulutilisya.* 'Change for the better which continues' was built upon intra-national interdependence between the rich and the poor, and Malawi would need assistance from the international rich for the foreseeable future.

Nkuvira's desire for financial self-sufficiency was a long-standing source of tension between the organization and Vsawans. *Nkuvira's* Tumbuka staff received 14,000MK (40USD) per month, more than the nurse who was the only medical practitioner within four hours walk.

Over time, their salaries (which were paid by donors in 2012) would supposedly come from selling eggs and vegetables to Vsawans, charging mobile phones on *Nkuvira's* solar panels, and from villagers' donations. *Nkuvira's* founder told village headmen when she arrived that *Nkuvira* would not be giving away things for free. Thinking that any development was better than nothing, the headmen agreed. However, villagers came to resent paying *Nkuvira* for assistance, believing that development should be provided by donors.

Nkuvira's donors would occasionally organize clothing drives from British schools. They would send *Nkuvira* bales of secondhand clothing and old mobile phones. These would be sold to villagers for about 500MK ($1.42) each and the money used to pay staff wages and maintain the community centre. Villagers' resentment of this process eventually led to an ugly clash between two of the Malawian staff and some fishermen in a local pub. These fishermen angrily proclaimed that *Nkuvira's* manager was 'living off us' by charging for clothes and services. After this, *Nkuvira* held a series of meetings to re-sensitize the community to the advantages of sustainability. This was to occur through an interactive storytelling method, to encourage participation from the community.

Nkuvira's meetings took place in villages near its youth centre and comprised a local staff member, an international volunteer, and any villagers who chose to attend. The staff member told a fable, co-written by the international staff, about a fisherman who gave away all his fish for free. This fisherman was popular in his village, but he could not afford to have his net repaired when it was damaged. The community had reduced their farming in response to his gifting, so they starved when he was no longer able to provide fish.

This story had little resonance to a Northern Malawian audience. Fishermen gave away fish all the time, through complex peer-and-patron-client networks (McNamara 2019). These same networks repaired a fisherman's nets, which was not normally a form of paid labour. Those who had received fish would not reduce their farming. At a pragmatic level they would need to give the fisherman farm products to thank him for his generosity. At an existential level receiving a gift entwined one's personhood with the gifter, providing a moral obligation to work harder and improve oneself, one's family and one's village (McNamara 2019). Further, in rural Malawi, one's wealth shapes one's responsibilities to peers, family, and community. Even at 14,000MK (40USD) per month, *Nkuvira's* Malawian staff were some of the village's highest earners; its youth centre contained multiple computers in a community where no one else had regular electricity; and each of its stream of volunteers had spent many years' worth of any villager's cash income on flights. If *Nkuvira* was a fisherman, then a 'fisherman' at a similar level of wealth would probably be expected to gift fish to the entire community. Villagers knew that such a fisherman's wealth would not be seriously impacted by gifting fish and that *Nkuvira's* donors were not financially dependent upon the 500MK (1.42USD) they paid for a bundle of clothes.

People spoke through the story, to the frustration of the speaker and volunteer. After he finished his tale, the staff member attempted to question the uncooperative villagers. He began by asking, 'What should the net owner have done?'

An older man answered, 'We cannot judge the net owner, we do not know what was in his heart.'

The staff member repeated himself and the other man, ignoring the question, stated:

> *Last week I went to your clothing market and bought two pants and some underwear. You took a photo of me when I received the clothes, but not one when I was paying. This is so that you could show the donors back in England that you had given me the things, which they gave you for free.*

There was a rumble of agreement among the onlookers, and the speaker continued, 'However, we need more of these things [clothing markets] in the community, and they should happen more often.'

The international volunteer spoke, claiming that the donors knew they were selling the clothes, as *Nkuvira's* donors approved of sustainability. This was translated into chiTumbuka. Villagers protested that the donors must provide ongoing assistance if the project was to *kulutilisya* (continue/be sustainable). The *Nkuvira* employee claimed that this was outside *Nkuvira's* mission statement. He then asked, 'What is sustainable development?'

The village's headman replied:

> We know that other NGOs have been shut down because they failed to deliver the development they promised...you should just take photos and show them to donors, and then they will help us, why do you need to find money here?

The staff member re-joined, 'We will take comments at the end' in English, and then in chiTumbuka, 'What is sustainable development?' Villagers turned their back on him, and the group disbursed.

This experience inspired me to find out what sustainable development meant to rural Malawians. The villagers provided definitions emphasizing ongoing assistance from the rich to the poor, the moral entwinement of donors and recipients and the primacy of one's daily needs. Their definition was vastly different to my own, and to that taught in many development programs. It was more consistent with Malawian moral norms, which were based upon personhoods embedded in mutual dependence and upon the rich having moral responsibility for the wellbeing of the poor. It also more accurately reflected the material conditions of life in Malawi than what I had learnt about sustainability as a development student, foregrounding that development projects had costs (material and economic) and that Malawians' ability or inability to meet these costs was not an indicator of their desire for the project.

Relatively succinctly, I was frequently told:

> For development to be sustainable, it just needs money. It is better that this money is found outside, where there are many donors, than here where everyone is poor.

Translating sustainable development to *chitukuko kulutilisya* (change for the better that continued), Vsawans described those which provided current and ongoing assistance, or which would help children, who were the community's future.

> Sustainable means things change if development is continuing, but they may revert if it is not. People can always find help if development is continuing but not if it isn't... . Sustainable development is schools, roads, and hospitals. Development is better if it is sustainable/continuing because cultivating the road will make [the village] better, and it provides jobs. If we had just built it [the road], it would be overrun by now.

Sustainability was therefore not signified though independence, but through extra-community linkages that generated continuous assistance. Linked to this realistic assessment of Malawi's economy was a belief in the moral righteousness of interdependence at a local and global level. A Malawian bar owner, who was frustrated that *Mbwezi* was considering charging interest on its microloans (to make them sustainable) explained:

I did a lot for Mbwezi; I helped them build the school and the hall and now it is my turn to receive. [Mbwezi had initially used volunteers, including Ceazie to build its office]. When Mbwezi arrived, it was like a child, but now it is like a parent; it is here to take care of us… Malawi is poor because of slavery, so the British people who got rich because of slavery have to give to us now. That is why they are giving seventy million dollars [a reference to British development aid] and why they are giving to Mbwezi.

The combination of these criticisms created an internally consistent definition of sustainability. Continuing change for the better entailed both costs and benefits, a group of exceptionally poor people could desire the benefits without being able to pay the costs. Nor should Malawians be expected to pay these costs – development was typically a gift provided by rich donors to poor Malawians, in the manner that the rich commonly gift to the poor in Malawi. The wealth that manifested through development organizations should provide obligations to the poor especially when one considered the colonial histories and ongoing injustices of Malawi's relationships with donor nations.

These critiques were not recorded in *Mbwezi* and *Nkuvira's* Monitoring and Evaluation documents, which attested that the NGOs were engaging in best-practice. I asked a staff member from *Nkuvira* and one from Mbwezi about the sensitization meetings that *Nkuvira* had held and about their broader experience encouraging sustainability in Vsawa. *Nkuvira's* employee claimed, 'I can tell by people's actions and their unwillingness to be involved with the project that they are not understanding sustainability.' The Mbwezi employee disagreed. He argued that around half of the villagers understood and valued sustainability: 'If I tell them there is training, maybe half will say "give me money" and half will say "we want training and we want to learn."' For this employee, understanding sustainability meant that villagers attended training without wanting to be paid.

These statements reflect a narrative about donor dependency inhibiting sustainable development that was common among Malawi's elite (and that had encouraged the fable about the fisherman). However, this narrative did not faithfully describe Malawi's economy and social norms. Malawi's poverty is foremost the ongoing result of deliberate policies to create a resource-scarce labour reserve for Rhodesia, combined with the legacy of an apartheid-aligned dictator (c.f. Page 2019). The nation cannot and should not be expected to become self-sufficient. Similarly, Malawians did not learn 'dependency' from donors, but instead have long created personhood through mutual obligations between the rich and the poor (c.f. Swilder and Watkins 2017). When NGO staff described villagers as 'understanding sustainability' they were referring to these villagers learning an external moral and economic system and accepting their limited agency within it.

Over time villagers increasingly re-engaged with the NGOs. They desired the commodities Nkuvira bought which, because they had been donated, were vastly cheaper than those obtained from other providers and the loans *Mbwezi* offered at below bank interest rates. They also desired the development the NGOs signified. Its computers, salaried work, and international connections, all of which it overtly linked to development. Vsawans participated in projects with the desire to use a computer or phone, the hope that they would find work as an NGOs gardener or that a staff member or volunteer would gift them a t-shirt and the dream that the organization would either employ them for a project or would sponsor them for tertiary study. It was these dreams and aspirations that represented sustainable development to Vsawans, and they would be incorporated in their local sustainability narrative. In contrast, the project reports that travelled up from the village created by both local staff and volunteers undertaking internships affirmed that the community had initially been resistant to sustainability, due to Malawians' culture of dependence. However, they had been properly sensitized through participatory techniques like storytelling. A key learning for future projects was that additional sensitization was necessary when sustainable projects were to be delivered.

This silencing of local critique is not idiosyncratic of these organizations but is crucial to the relationship between local development projects and broader development knowledge. A cycle took place between donor demands for a sustainable project, based on their readings of best-practice documents and their previous engagement with development. These were carried out by field staff under conditions of profound inequality between themselves, donors, and the local community. Villagers' resistance was interpreted as ignorance, to be cured through practices drawn from best-practice documents, which detailed methods for sensitization. When these resource-scarce villagers then re-engaged with the project, this affirmed the righteousness of sustainability. When the local and international staff and donor agencies wrote case studies, this provided more evidence to a literature that places sustainability in chains of equivalence with empowerment and participation, further impeding local sustainability narratives when they challenge this chain.

Lessons and teaching techniques

Due to the way introductory development studies subjects are structured, overtly critical and post-development approaches are often taught late in these courses (c.f. Harcourt 2018). I use this story to build upon a semester-long activity that I conduct to teach students about how key terms are constructed in international development. To illustrate to students how development is rendered coherent and technical through buzzwords, the first time a student uses any of a number of key terms during a semester,[5] each student individually writes the meaning of that term, they then create a small group definition, and we then negotiate a class definition. Students are encouraged to consider how their own definitions are incorporated or silenced in this process. As different development projects are introduced into the classes, we discuss how their definitions of development's key terms are different to our own, the commonalities and differences between them and how (and if) class members would alter their definitions to work with these agencies.

Students are only introduced to rural Malawian development narratives late in the course. They therefore investigate this critique after they have spent several weeks exploring how usable definitions of terms like empowerment and sustainability are shaped by consensus between development workers and coherence within and between projects. This activity allows students to see how this consensus-based process creates a coherence that does not incorporate local critiques and to explore how local's participation in projects is shaped by unjust power dynamics and unachievable dreams of development (c.f. De Vries 2007 and Mosse 2004).

When faced with criticisms of a supposedly sustainable development project's efficacy or politics, some students argue for focusing on development work as a moral ideology and see this as reconciling the critiques made by local sustainability narratives. Their claims build upon ethnographic explorations of development workers, which have been used to both justify and rebut critiques of development power structures (Mosse 2013). Yarrow (2010) argues that we should study development 'beyond politics' through the aspirations of development workers. Harrison (2013) is concerned that ethnographies of development work will undermine the legitimacy of aid as an ethical project. She claims that such studies divert attention from development projects.' Other students attempt to explore how their own narratives of sustainability are embedded in moral ontologies, material circumstances and power dynamics, and consider how this creates tensions and misunderstandings between development's providers and recipients. While sustainability's meaning is deeply contested, few students are prepared to question its importance, and most are supportive of the SDGs. Combining this vignette with their own experiences and values then enables them to appreciate how the moral valence of sustainability and narratives where development occurred through combating villagers' 'lack of understanding'; as well as how this enabled a coherence between the daily activities of a development project and Malawi's broader development norms, silencing intra-village critiques in the process.

More broadly, this focus on local sustainability narratives is not used to discount the importance of the person moral projects described by Yarrow (2010) and Harrison (2013), but to foreground that development workers' moral projects have material and political implications, that every act necessarily either challenges or cements development norms. This approach to teaching post-development and local critiques of development responds to Ziai's (2017) criticisms, where critical approaches to development are simultaneously integrated and downplayed in development studies courses. Local meanings of sustainability are different to those used by the development industry, and students almost inevitably both value local knowledge and sustainability. This provides a place for them to consider both the positive and negative material effects of development and how facilitating projects with a positive material affect contributes to a regime of knowledge that rarely fully incorporates local critique.

Key sources

1. McNamara, T (2017) They Are Not Understanding Sustainability: Contested Sustainability Narratives at a Northern Malawian Development Interface *Human Organisation* 76(2):121–130.

McNamara argues that the meaning of sustainability is determined through personal experiences, cultural norms, and political economies. What does sustainability mean for you? How is this similar or different to Brutland and to the SDGs? Why do you think your understanding of sustainability is how it is?

2. Swilder, A and Watkins, S (2009) Teach a Man to Fish: The Sustainability Doctrine and its Social Consequences *World Development* 37(7): 1182–1196.

Swilder and Watkins argue that narratives of sustainability in Malawi limit organizations to running trainings and workshops, organizing volunteer infrastructure projects or facilitating income generating activities. Why do these say this? What aspect of sustainability is focused upon and how is this sustainability narrative generated?

3. Cornwall, A and Eade, D (2010) *Deconstructing Development: Buzzwords and Fuzzwords* Warwickshire, United Kingdom. Practical Action Publishing.

Cornwall and Eade describe sustainability as both a 'buzzword' and a 'boundary term.' What is the different between the two terms? Under what circumstances have you seen sustainability serve as a 'buzzword' or a 'boundary term'?

4. Mosse, D (2004) Is Good Policy Unimplementable? Reflections on the Ethnography of Aid Policy and Practice *Development and Change* 35(4)639–671.

Mosse claims that development actors maintain policy coherence – protecting development models, strategies from realities on the ground. How does this chapter argue that sustainability does this? Do you believe this is a fair criticism? Which other terms have you seen being used in a similar manner?

> 5. De Vries, P (2007) 'Don't Compromise Your Desire for Development!' A Lacanian/ Deluzian Rethinking of the Anti-Politics Machine *Third World Quarterly* 28(1):25–43.

De Vries depicts development as a desiring machine, where small amounts of externally provisioned development, and the presence of signifiers of external opulence shape intra-community aspirations for development. How does this interact with what you see as sustainable development? How does a desire for development entwined with material advancement shape the possibilities of sustainable development?

Notes

1 'Strong sustainability' sees the economy encompassed in society, which is in term encompassed within a finite ecology.
2 The equal importance of the environment, society, and the economy.
3 I spent about a year doing ethnography in Vsawa, Northern Malawi between 2012 and 2015. This vignette is explored in more detail in McNamara 2017, see pages 125–128 of McNamara 2017 & pages 295–297 of this chapter.
4 Malawi's currency lost over half its value after being de-pegged from the dollar in 2012. I have placed the conversation rate at 350MK to the USD in this article, but it oscillated between 100 and 500MK to the USD during the time I was working in Malawi. In 2012, Malawi's GDP per person was 380 USD, a touch more than a dollar a day per person.
5 Sustainability, empowerment, education, local, capability, and (inevitably in the first lesson) development.

References

Cammack, D (2017) Malawi's Political Settlement: Crafting Poverty and Peace, 1994–2014. *Journal of International Development* 29(5):661–677.
Cornwall, A and Eade, D (2010) *Deconstructing Development: Buzzwords and Fuzzwords* Warwickshire. Practical Action Publishing.
De Vries, P (2007) 'Don't Compromise Your Desire for Development!' A Lacanian/Deluzian Rethinking of the Anti-Politics Machine. *Third World Quarterly* 28(1):25–43.
Escobar, A (2010) Latin America at a Crossroads: alternative modernizations, post-liberalism, or post-development? *Cultural Studies* 24(1):1–64.
Ferguson, J (1994) *The Antipolitics machine: Development, Depoliticisation and Bureaucratic Power in Lesotho.* University of Minnesota Press.
Frye, M (2012) Bright Futures in Malawi's New Dawn: Educational Aspirations as Assertions of Identity. *American Journal of Sociology* 117(6):1565–1624.
Harcourt, W (2017) The making and unmaking of development: using Post-Development as a tool in teaching development studies. *Third World Quarterly* 38(12):2703–2718.
Harcourt, W (2018) 'People and Personal Projects': a rejoinder on the challenge of teaching development studies. *Third World Quarterly* 39(11):2203–2205. DOI: https://doi.org/10.1080/01436597.2018.1460595
Harrison, E (2013) Beyond the looking glass? 'Aidland' reconsidered. *Critique of Anthropology* 33(3):263–279.
Herrera, V (2019) Reconciling global aspirations and local realities: Challenges facing the Sustainable Development Goals for water and sanitation. *World Development* 1(18):106–117.
Holden, E, Linnerud, K and Banister, D (2017) The imperatives of sustainable development. *Sustainable Development* 25:213–226.
Li, T (2014) *Land's End: Capitalist Relations on an Indigenous Frontier* Duke University Press.

Mazibuko, S (2013) Understanding underdevelopment through the sustainable livelihoods approach. *Community Development* 44(2):173–187.

McNamara, T (2017) 'They Are Not Understanding Sustainability': Contested Sustainability Narratives at a Northern Malawian Development Interface. *Human Organisation* 76(2):121–130.

McNamara, T (2019) 'Me and the NGO Staff, We Live Like Azungu': Malawian Moral Economies of Development. *Human Organisation* 78(1):43–53.

Morfit, S (2011) AIDS is Money: How Donor Preferences Reconfigure Local Realities. *World Development* 39(1):64–76.

Mosse, D (2004) Is Good Policy Unimplementable? Reflections on the Ethnography of Aid Policy and Practice. *Development and Change* 35(4):639–671.

Mosse, D (2013) The anthropology of international development. *Annual Review of Anthropology* 42:227–246.

Page, S (2019) *Development, Sexual Cultural Practices and HIV/AIDS in Africa*. Springer.

Pieterse, J (2010) *Development Theory: Deconstructions/Reconstructions*. 2nd Edition. London, Sage.

Scrimgeour, F and Iremonger, C (2011) *Maori Sustainable Economic Development in New Zealand: Indigenous Practices for the Quadruple Bottom Line*. University of Waikato, Hamilton, New Zealand.

Swilder, A and Watkins, S (2009) 'Teach a Man to Fish': The Sustainability Doctrine and its Social Consequences. *World Development* 37(7): 1182–1196.

Swilder, A and Watkins, S (2017) *A Fraught Embrace: The Romance and Reality of AIDs Altruism in Africa*. Princeton University Press.

Tawfik, L and Watkins, S (2007) Sex in Geneva, sex in Lilongwe, and sex in Balaka. *Social Science and Medicine* 65(5):1090–1101.

Weber, H (2017) Politics of 'Leaving No One Behind': Contesting the 2030 Sustainable Development Goals Agenda. *Globalizations* 14(3):399–414.

Yarrow, T (2010) *Development beyond Politics: Aid, Activism and NGOs in Ghana*. Palgrave Macmillan.

Ziai, A (2017) 'I am not a Post-Developmentalist, but…' The influence of Post-Development on development studies. *Third World Quarterly* 38(12):2719–2734.

PART 3

Inequality and inequitable development

26

INTRODUCTION

Inequality and inequitable development

Kearrin Sims and Jonathan Rigg

Inequality is a critical multi-scalar and multi-sector challenge for global development. Inequalities exist at the local, national, and global scale, and across and within different genders, classes, ethnicities, religions, age groups, levels of education, and other demographic cohorts. In terms of economic inequality, it is now estimated that the world's richest one per cent own 44 per cent of global wealth (inequality.org). There are more billionaires than ever before, yet hundreds of millions continue to live in extreme poverty (see Rigg and Sims, Chapter 27). As the wages and wealth of elites have multiplied, corporate profit shifting and other forms of tax avoidance have also grown, with an estimated 100 to 500 billion USD of tax revenue lost each year (see Etter-Phoya et al., Chapter 28). Inequalities in wealth and income also correlate with carbon footprints, with the world's wealthiest being the largest contributors to the global climate crisis, and are thus critically important to tackling climate change.

Inequality exists between and within countries, and includes both uneven standards of living and unequal access to opportunity. According to Oxfam International (2020), men own 50 per cent more of the world's wealth than women, while the value of global unpaid care work undertaken by women is US$10.8 trillion. Women also outnumber men in paid care work, such as healthcare and social work, placing them at heightened risk of exposure to communicable diseases such as COVID-19. Indeed, the COVID-19 pandemic highlighted existing and forged new inequalities with a sharply uneven distribution of care and harm, seeing mass-scale unemployment alongside uneven burdens of infection across classes and ethnicities. Inequalities shaped by gender, race, class, and other factors intersect, leading to multi-layered forms of disadvantage and injustice.

Recognizing that inequality takes many guises and has manifold consequences, the chapters in this section offer different entry points for examining the relationships between poverty, inequality, and development. Traversing a range of topics, theoretical positions, and case studies, they demonstrate the need for continued efforts to understand and address inequality in all its forms.

The opening chapter by **Jonathan Rigg** and **Kearrin Sims** provides an overview of the polarized debates regarding if, where, how, and by how much poverty has decreased. Rather than taking a position on these debates, Rigg and Sims work on unpacking and explaining

DOI: 10.4324/9781003017653-30

the differences in understandings of poverty and its prevalence. Attention is given to poverty lines and trends, poverty's multidimensional features, poverty drivers, the spatial distributions of poverty, and the heterogeneous ways in which poverty is experienced by different cohorts of people. Regardless of which side of the debate one favours, Rigg and Sims also note that one undisputable point regarding the current prevalence of poverty in the world today is that more could be done – and that the persistence of poverty is a consequence of policy and power imbalances.

Questions regarding persistent poverty and inequality are also tackled by **Etter-Phoya et al.**, who explore the crucially important issue of how inequality is structurally reproduced within the global financial system. Critical of what they describe as a biased and largely dysfunctional global financial and tax architecture and a global financial system that is deeply biased against the interests of the majority world, Etter-Phoya et. al discuss the politics and uneven power relations behind global tax regulation and illicit financial flows. Arguing that the paradigmatic shift within development studies from international to global development must encompass addressing and redressing the structural inequities in the global financial and tax systems that have been established, and are maintained by, minority world countries, the authors then propose a number of policy solutions. These include the need for greater (enforced) transparency within the global financial system, automatic exchange of information between tax authorities, the identification, registration, and disclosure of the beneficial owners of companies, partnerships, trusts and private foundations, and public country by country financial reporting. Global finance and tax regulation are topics that require far greater attention within development studies, and the authors of this chapter – affiliates of the Tax Justice Network (TJN) – provide a contribution towards this end.

Interwoven with the structural inequalities of the global financial system are the vast injustices embedded within global extractive economies. For **Etienne Roy Grégoire and Pascale Hatcher**, Extractive Industries (EIs) – industries where the intensity of exploitation exceeds the rate of renewal of natural resources, ecosystemic resilience capacities, or the viability of alternative development models and life projects – have been a key vector of inequality and vulnerability across resource-rich countries in the Global South and in colonial contexts in the North. In particular, and as their chapter elaborates, EIs have been responsible for the expansion and entrenchment of political economies of accumulation by dispossession that have exacerbated pre-existing tensions around sovereignty and self-determination; a catalyser of political violence against human rights defenders (including land and environmental defenders) and already marginalized populations; and a site of struggle in the political economy of climate change.

With EIs playing a dominant economic, social, and political role in more than 81 developing countries, where an estimated 3.5 billion people live, finding alternatives and better means of regulation is a critically important challenge for social and environmental justice. One common means by which practitioners have sought to 'clean up' EIs, is to enhance corporate social responsibility (CSR). Under the right circumstances, CSR can provide important improvements in industry practice. Yet, as Grégoire and Hatcher summarize, research also shows serious limitations in relying too heavily on corporate good behaviour. Accordingly, they argue that what is required are forms of co-production of knowledge with affected communities, transnational solidarity networks and, perhaps most importantly of all, the rethinking of our relationship with nature and the reconciling of our economies with the biophysical limits of planet Earth. This requires pedagogical approaches that both inform students of current sectoral/policy norms, and which seek to radically transcend such norms.

Displacement and spatial inequalities

Turning to thematic inequality challenges, **Edo Andriesse and Kristian Saguin** focus on within-country spatial inequalities of development. With a particular focus on South, East and Southeast Asia, they demonstrate that spatial inequalities remain a persistent challenge within many countries that are otherwise lauded as development success stories. Attention is given to the 'profound' spatial disparities that continue to exist between core cities and peripheral regions, spatial targeting strategies to address such disparities, and some of the key theoretical lenses that have been used to make sense of these challenges. Regarding spatial targeting, the authors summarize debates surrounding cash-transfers to poor households, area-based approaches (that emphasize context specificity), remote ethnic minorities, periphery and semi-periphery lowland farming communities, and coastal areas. Theoretical topics discussed in advance of these spatial inequalities include circular and cumulative causation, growth pole strategies, and the role of institutions and localized responses in tackling spatial inequality.

In addition to covering all of the above themes and issues, Andriesse and Saguin also provide the reader with a valuable case study of decentralization in the Philippines, where they argue decentralization has empowered communities to deliver better and more contextually-specific and locally-responsive services, but which has shown little correlation with overall trends in reducing spatial inequality. This case study offers a number of useful insights for development practice, complemented by the chapter's closing section on pedagogy and its call for the use of different mapping tools and techniques in the analysis of spatial inequalities, as well as harnessing of the 'affective power' that underlies learning and teaching on inequality.

Evident in many of the chapters within this section is the importance of being attentive to how power dynamics shape development. In **Philip Hirsch**'s chapter on land grabbing and exclusion, for example, he examines land access and land politics before outlining a framework for explaining processes of dispossession and accumulation associated with land grabbing. This framework, the 'powers of exclusion approach,' provides a way to assess the relative justice or injustice in land grabbing and other land-related policies and processes, not through a critique of the fact of exclusion *per se*, but rather through an analysis of four main powers that enable exclusion to occur: the power of regulation, the power of the market, the power of force and the power of legitimation. In explaining each of these powers, Hirsch works his way through different attempts to define land grabbing and associated concepts and provides a nuanced reading of land grab debates which shows the centrality of land to much contemporary inequality. Writing on pedagogy, he emphasizes the value of case-study analysis and a nuanced discussion of land grabbing terminology.

Complementing Hirsch's chapter, is **Diana Suhardiman's** analysis of forced displacement and resettlement caused by development projects and activities. Globally, it is estimated that almost one million people are displaced by development projects every year, bringing widely documented hardships including loss of homes, livelihoods, and culturally embedded place-attachments. As a result, much attention has been given to how development-induced displacement and resettlement (DIDR) can be managed more effectively, in order to reduce harmful effects. Through the use of a case study analysis of DIDR in Laos – which occurred as part of the Government of Laos's (GoL) widespread and sometimes forced resettlement of minority groups living in remote upland areas – Suhardiman seeks to advance debates by demonstrating how displaced and resettled peoples can be active agents for change, adapting and adjusting their livelihood strategies, and navigating power struggles to enact their collective interests and resistance. What this chapter tells us, is that while it is important to interrogate the harm wrought by DIDR, we should not focus only on harm. Rather, Suhardiman reminds us of the need to also look at how communities respond to and cope with change, while trying to improve their livelihoods. Regarding pedagogy and practice, this requires direct conversations with displaced communities and the addition of grassroots activities alongside efforts to improve DIDR legislation.

Displacement and resettlement are both longstanding and increasing challenges for global development. Indeed, as climate change and its effects worsen, we are now entering a new phase in resettlement management, where entire countries and communities totalling in the hundreds of millions of people (Bangladesh is a prominent example) will be displaced as a result of rising sea levels and other climatic effects. Resettlement for climate change is/will be both forced and voluntary, and as **Andreas Neef and Lucy Benge** explain in their chapter, when it comes to mobility and migration, how a person's movement is categorized shapes both how they are defined and the solutions available to them.

Recognizing that there is currently little international political appetite to create protection policies for climate-related migration across borders, Neef and Benge unpack the different labels that are placed on people moving as a consequence of climate change in order to provide an understanding of the key debates and narratives regarding how climate-related migration is conceptualized. They ask: who has created these narratives? Who is depicted in them? And whose voices are missing? Each of these questions is discussed both at a conceptual level, and through a case-study analysis of the ways in which policy discourse in Fiji often places more emphasis on physical risk factors and economic opportunities propelling movement, rather than the cultural or socio-psychological factors that contribute to decisions to migrate or to remain in place. Regarding the question of voice and missing voices, the authors argue that both practice and pedagogy around climate migration needs to be attentive to whose voices are being prioritized, whose stories or perspectives are obscured, what the political implications of this might be, and how climate-induced human movement can be simultaneously both an opportunity *and* a loss.

Gender, education, and culture

Education is widely considered as one of the most important catalysts for upward social mobility and the alleviation of inequality. Globally, there are strong correlations between impoverishment and low levels of education, as well as between wealth and high levels of education. Yet, as **Youyenn Teo** argues in her chapter on Singapore's 'world class' education system, the empirical evidence on the relationship between education and inequality is not always so straightforward.

Singapore is a country that is often placed on a pedestal for its successes across a range of development issues – not least its world class education. It performs superbly on global indicators of school attendance and student performance, including the prized Programme for International Student Assessment (PISA) rankings, where it places second globally. Less considered, however, is that differences in scores between socioeconomically advantaged and disadvantaged Singapore students are higher than the OECD average. Teo shows that such differences in educational performance are structurally embedded within the Singaporean state's focus on 'talent' – and the consequent examination-based 'sorting' of students from 9 years old onwards. Drawing on extensive interviewing of parents as well as her own personal experiences, Teo reveals the highly competitive and stressful nature of schooling, associated financial burdens on families due to private tutoring expenses, and the uneven access to educational opportunity and success relative to family wealth. Painting a picture of an education system that has become somewhat beholden to examination-based learning, she argues that, even within 'top-performing' systems, education does not automatically contribute to reducing inequality. Indeed, Singapore's highly competitive education system creates and reinforces inequality. If this is to be resolved, Teo argues that addressing inequality needs to be a serious aim of education, rather than an expected (naturally occurring) outcome.

Turning to gender inequality, **Archana Preeti Voola and Bina Fernanadez** deliver a synoptic review of shifting trends in gender and development, calling for an intersectional feminist approach within contemporary research, policy, practice and pedagogy initiatives. Commencing with Boserup's seminal work (1970) on women's role(s) in economic development, they note how studies of gender have proliferated and broadened to include the analysis of men and masculinities, gender non-conforming groups, as well as towards gender 'mainstreaming' within policy and practice. After tracing key shifts in ideas, Voola and Fernandez then turn their attention to different domains of gender inequalities, discussing gendered divisions of labour, violence against women and inequalities in healthcare. To carry gender-responsive policies in development forward, Voola and Fernandez call for the systematic collection and analysis of gender disaggregated data. Key gender analysis tools and frameworks are summarized, and emphasis is given to the need to keep a feminist perspective in focus for gender and development practice and pedagogy.

Gender-based violence (GBV), including violence against women, is a priority target of Sustainable Development Goal 5 (Gender Equality) and is a core theme taken up by **Sara Amin et al.**, who provide a rich overview of key GBV debates, issues, and policy shifts. In addition to discussing GBV, Amin et al. examine three other key domains for gender equity – education, employment, and political representation – discussing key trends and priorities. This conceptual and issue-based grounding is then followed by two detailed case studies of pedagogical practices around gender and development in two Southern universities – The Asian University for Women (AUW) and The University of the South Pacific (USP). Drawing from these experiences, the authors argue that teaching programs which aim to promote gender equality must ensure that women from marginalized groups and gender and sexual minorities are proactively recruited and supported, and that men and analysis of masculinities are involved in discussions and initiatives; curricular and extracurricular content and practices are developed in dialogue with grassroots organizations, civil society actors, governments and private sector actors advocating for gender equality; and, that context-responsive pedagogical practices are provided, whereby students and communities are linked and put into conversation with each other.

This section closes with a final chapter by **Anna Hayes and Kearrin Sims**, that explores how expanding South-South cooperation is bringing changes to global development which include new forms of violence against ethnic minority communities. Noting that no country has been more influential in reworking global development norms, ontologies, and practices than China, the authors explore the violent ties between the Chinese Communist Party's (CCP) political oppression and human rights abuses in Xinjiang Uyghur Autonomous Region (XUAR) and the CCP's ambitious Belt and Road Initiative (BRI). Due to its geographic location in China's north-western border region, XUAR is widely perceived as of critical importance to the BRI, and thus offers a useful case study for examining how the CCP has wedded development to state-expansionism and efforts to erase cultural diversity. Tracking the history of development interventions in XUAR and the extreme state violence that has grown over the past decade – whereby over 1 million Uyghurs and other Muslim minority nationalities have been imprisoned in detention camps – the chapter argues that the violence unfolding is directly linked to the development goals of the BRI, and the CCP's attempts to bring modernization and security to the ethnic minority Uyghurs who inhabit this 'gateway' to Central Asia and beyond. Accordingly, Hayes and Sims call for new empirical research and new (and increased) engagement with 'Southern' actors to better understand how shifts in the global development arena are producing new forms of violence, as well as reiterating the longstanding need to take culture more seriously when thinking about development and its effects. As they conclude, each of these priorities require attentiveness to development's shifting materialities, and its shifting epistemic features.

Conclusion

Inequalities in wealth, power and knowledge are impossible either to escape or to erase. They operate at all scales, from the intimate to the transnational, and from the economic to the social and the political. Despite inequality's ubiquity, the contributions to this section show that state actions and government policies can play a role in reducing its more egregious forms while ameliorating its worst effects.

27

POVERTY

No meeting of minds

Jonathan Rigg and Kearrin Sims

What has been happening to poverty?

In 1969 Dudley Seers, a founding scholar of development economics, argued that rather than focusing only on economic growth, development economists should be asking:

> *What has been happening to poverty? What has been happening to unemployment? What has been happening to inequality? … If one or two of these central problems have been growing worse, especially if all three have, it would be strange to call the result 'development,' even if per capita income doubled.*
>
> *(Seers 1970, 354)*

Half a century later, on 19 January 2019, Bill Gates tweeted a graph of global extreme poverty (Figure 27.1) from 1820 to 2015, writing: 'This is one of my favourite infographics. A lot of people underestimate just how much life has improved over the last two centuries.'[1]

This graph draws on data from a piece by Max Roser and Esteban Ortiz-Ospina on 'Global Extreme Poverty.' In that article, Roser and Ortiz-Ospina argue that:

> *The available long-run evidence shows that in the past, only a small elite enjoyed living conditions that would not be described as 'extreme poverty' today. But with the onset of industrialization and rising productivity, the share of people living in extreme poverty started to decrease. Accordingly, the share of people in extreme poverty has decreased continuously over the course of the last two centuries. This is surely one of the most remarkable achievements of humankind.*
>
> *(Roser and Esteban Ortiz-Ospina 2019 https://ourworldindata.org/ extreme-poverty#the-mis-perceptions-about-poverty-trends)*

In response to Gates' tweet, the anthropologist Jason Hickel wrote a piece in *The Guardian* with the title 'Bill Gates says poverty is decreasing. He couldn't be more wrong.' Hickel expanded:

> *What Roser's numbers actually reveal is that the world went from a situation where most of humanity had no need of money at all to one where today most of humanity struggles to*

DOI: 10.4324/9781003017653-31

Beating back poverty

Two hundred years ago, virtually everyone lived in extreme poverty.
Today fewer than 10% of people do.

gates
notes

▓ % in extreme poverty ▓ % not in extreme poverty

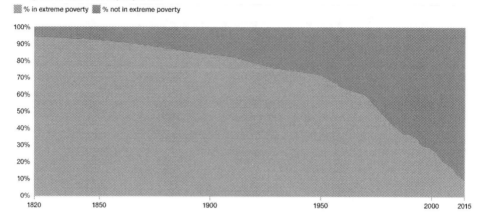

Figure 27.1   Extreme poverty 1820–2015

Source: https://twitter.com/BillGates/status/1011660648231518208/photo/1

> *survive on extremely small amounts of money. The graph casts this as a decline in poverty, but in reality what was going on was a process of dispossession that bulldozed people into the capitalist labour system, during the enclosure movements in Europe and the colonisation of the global south.*
>
> *(Hickel 2019a)[2]*

This debate brings together many of the issues that have confounded the analysis of poverty from the very start: matters of definition, measurement, methodology, process, history – and of ideology. As such, it provides a particularly fertile stage on which to think about some of the key fault lines in development as a process of transformation and development studies as a field of scholarship.

We are not taking any position here on these debates – siding either with the poverty eradicators such as Gates or the poverty persisters like Hickel – but rather drawing attention to the reasons why, and the grounds on which, these differences of opinion emerge. This debate, while it can sometimes seem arcane, even esoteric, is profoundly important. It is not, in other words, 'just academic.'

Understanding and alleviating poverty lies at the very heart of what most of those working in the development field seek to achieve. As Seers made clear, poverty is the lynchpin of development; without a decline in poverty, any claim to development is hopelessly compromised. The first of the World Bank's two headline goals is to end extreme poverty by 2030;[3] while Oxfam International's 'purpose' is to 'create lasting solutions to the injustice of poverty.' This leads relentlessly to the questions that this entry poses: Who are the poor? How is poverty defined and measured? Where do the poor live? What has happened to poverty over the development era? And, most importantly of all, why has poverty either declined or persisted?

Questioning poverty

Who are the poor?

The Roser graph (Figure 27.1) shows poverty declining from nine-in-ten of the world's inhabitants in 1820 to one-in-ten in 2015, based on the $1.90 International Poverty Line. In other words, in 1820 90 per cent of the globe were getting by on less than this income figure; in 2015, it was just 10 per cent.

But does this $1.90 poverty line capture the experience of poverty and adequately identify the poor? Or should the line be drawn rather higher? And if so, where?

It is broadly accepted by development economists that $1.90 is a *very* low threshold – Hickel goes so far as to write that 'it is an insult to humanity' while Alston, the UN's Special Rapporteur on extreme poverty argues that the International Poverty Line 'is explicitly designed to reflect a *staggeringly low* standard of living, *well below any reasonable conception of a life with dignity*' (Alston 2020, 5 [emphases added]). The exercise of drawing a poverty line rests on empirical data, but these data cannot, in themselves, arrive at the line. In the final analysis, it is a judgement as to where such a line should be drawn – in other words, deciding what are the 'necessaries' for a 'creditable' existence (Smith 2007 [1776], 676). This makes it, while not arbitrary, at least subjective. The scope for drawing different lines comes, not least, from the work of the World Bank itself.

In 2017, the World Bank announced[4] that it would begin to report poverty rates for countries using two new international poverty lines, along with its $1.90 line: a lower middle-income International Poverty Line of $3.20/day, and an upper middle-income International Poverty Line of $5.50/day. Roser's graph would look rather different using these two lines, with the proportion of the world living in poverty in 2015 rising from one-in-ten on the basis of the $1.90 line, to one-in-four for the $3.20 line, to almost one-in-two for the $5.50 poverty line (Figure 27.2).[5] It is significant that the World Bank's $1.90 extreme poverty definition is 'taken as a "governmental standard" [and] reflects a *political judgment* about the extent of ambition on

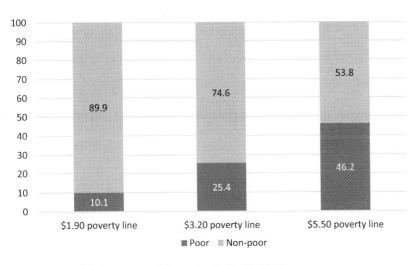

Figure 27.2 Global poverty at different thresholds (%, 2015)

Source: World Bank databank https://data.worldbank.org/indicator/

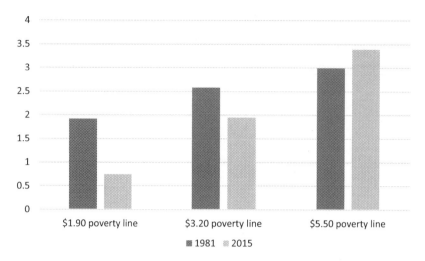

Figure 27.3 Numbers living in poverty at $1.90, $3.20 and $5.50 poverty lines (1981 and 2015, billions)
Source: World Bank databank

the part of the Member States of the United Nations' [emphasis added]. 'An alternative poverty line,' the report states, 'may be more intrinsically defensible, but it does not have the same claim on political leaders' (World Bank 2017, 3). With these words, the World Bank seems to accept that while the $1.90 line is not arbitrary, its use is justified – and defended – on political rather than on academic, methodological, or even developmental grounds.

While where we draw the line matters, it is also true that whatever line we select, global poverty has declined as a proportion of the world's population. That said, if we focus on numbers of poor and not on the proportion of poor as a measure of development achievements, then this claim for success is less certain. At $1.90 a day, poverty in both absolute (the total number of poor people) and relative (the poor as a proportion of the total population) terms has declined, and dramatically so. This is the story and the lesson of Figure 27.1. But using the $3.20 line and the number of poor has shown only a modest decrease, while the $5.50 poverty line reveals a growing number of poor between 1981 and 2015, from just under 3 billion in 1981 to 3.4 billion in 2015 (Figure 27.3). Percentage differences in poverty rates are important for tracking change, but the sheer number of poor matters too.

Do we have the data to identify the poor?

A related question is whether – and this is the case wherever we draw the line – the data are available to estimate the incidence and depth of poverty and its evolution over time. Generally, the poorer the country the worse (in terms of quantity *and* quality) the data. And the further we go back in time, the greater these problems become. So, the places where poverty is most pronounced – and therefore the places in which we should be most interested – are also those places where we are least well informed about the extent and nature of that poverty.

Accordingly, it is doubtful that we can accurately assess trends in global poverty from 1820 to 2015. Income poverty data were only systematically collected from 1981, and even this was highly uneven with large parts of the world essentially constituting data free zones. The extrapolation back to 1820 raises additional statistical challenges. These early data are based on Angus Maddison's (see www.rug.nl/ggdc/historicaldevelopment/maddison/releases/maddison-project-database-2018?lang=en) long-run historical data set of per capita GDP. This has been used to estimate per capita income and from this to derive rates of poverty. But caution is necessary. In Bolt et al.'s (2018) extension of the Maddison dataset, for example, they accept that:

> *Despite our inclusion and estimation of numerous historical benchmarks, our understanding of comparative income levels becomes based on sparser data as we move back further in time. This is particularly pressing in regions such as Africa and large parts of Asia, but there [are] also important gaps in 19th century Latin America.*
>
> (Bolt et al. 2018, 29)

For some scholars, combining data collected systematically and at an increasingly granular level since 1981 to estimate living standards with disparate historical data assembled retrospectively and with uneven and often very thin geographical coverage produces a graph that 'might make for nice social media, but [is] not rooted in science' (Hickel 2019a). For others, however, it is good enough to show the progress of poverty even given the gaps to which Bolt et al. allude. The tension here concerns whether we are content to accept such statistical gaps and inconsistencies in order to shed some light on the big picture.

Finally, this debate over the raw data overlooks a broader and more important issue: is income – even if we had accurate, broad-based data – the key metric that we should use to measure and evaluate poverty? What does income capture and what does it miss?

What do we measure?

In almost all studies and reports, 'poverty' is shorthand for 'income poverty.' While the received wisdom among both scholars and practitioners may be that poverty is multi-dimensional, 'income-poverty measures still predominate, or at the very least are implicitly accorded the position of "first among equals"' (Sumner 2007, 5). This is partly for practical reasons: such data are cheap and easy to collect, widely available, and internationally comparable. There are also two further – and important – underlying assumption held by many development economists and international development agencies: first, that income data are objective and therefore less susceptible to the ambiguities associated with other data; and second that, in the final analysis, development and poverty eradication *are* about economic and income growth. This means that notwithstanding the rhetoric, income remains both the lodestar and the best surrogate for poverty.

In the 1990s, associated with the wider participatory turn in development, there was a move to adopt the then unorthodox approach of asking the poor (and non-poor) about their views and experiences of poverty, leading to the production of a slew of Participatory Poverty Assessments or PPAs. This culminated in the 2000/2001 *World Development Report* 'attacking poverty' (World Bank 2001), which drew on the 'realities' of over 60,000 poor women and men across 60 countries, the 'voices of the poor' project (World Bank 2001, 3). Using these grounded, personal, and subjective vignettes of poverty, the report made a case for poverty being about much more than inadequate income or even low levels of human development. It is 'also vulnerability and a lack of voice, power and representation' (2001, 12). Notably, even in this report which was so committed to broadening the view of poverty and giving the poor a human face, it was recognized that 'much' of the difference in poverty performance across the world could be attributed to differences in economic growth (2001, 29). Once again, then, the primacy of economic forces and measures (GDP growth and income) are in the ascendant.

What was the past like?

The point that the experience of poverty is not just – or perhaps even mainly – about a lack of income is particularly pertinent when an attempt is made to trace poverty back over the *longue durée*. Not only are the data lacking or uncertain (see above), but even attempting to put an income value on lives, living and livelihoods, when these are subsistence or semi-subsistence, is questionable. Hickel writes:

> *Prior to colonisation, most people lived in subsistence economies where they enjoyed access to abundant commons – land, water, forests, livestock and robust systems of sharing and reciprocity. They had little if any money, but then they didn't need it in order to live well – so it makes little sense to claim that they were poor.*
>
> (Hickel 2019a)

In writing this, Hickel is paying homage to the work of the anthropologist Marshall Sahlins in the opening chapter to his influential book *Stone Age Economics,* entitled 'The original affluent society':

> *The world's most primitive people have few possessions, but they are not poor. Poverty is not a certain small amount of goods, nor is it just a relation between means and ends; above all it is a relation between people. Poverty is a social status. As such it is the invention of civilization. It has grown with civilization, at once as an invidious distinction between classes and more importantly as a tributary relation—that can render agrarian peasants more susceptible to natural catastrophes than any winter camp of Alaskan Eskimo.*
>
> (Sahlins 2004 [1972], 36)

An awareness of being 'poor' arises when the development project seeks to make people non-poor. The identification of the problem creates the context – and the impetus – for its amelioration. As Norberg-Hodge similarly writes in her analysis of poverty and development discourse in Ladakh, in north-western India 'I am convinced that the people of Ladakh were significantly happier before development than they are today' (1992, 27). That said, it is also the case that life expectancy and other indices of health status in these 'primitive' but 'affluent' societies were low. Sahlins' notional Alaskan 'Eskimo' (Inuit) may have had no conception of being poor, they had to wait until they were touched by development for that to emerge, but average life expectancy was probably between 35 and 45 years and infant and maternal mortality rates far higher than would be deemed acceptable today.[6]

What happened to poverty under colonialism and what is happening to the poor under capitalism?

If conditions are worse – or not significantly better – today than they were in the past, we need to ask 'why?' Two historical forces are seen to have been central to the 'underdevelopment' of the 'Third World': colonialism and capitalism.

Beginning with the former, the most central question is whether the inequalities and inequities of today are a legacy of historical processes, that shaped and created the modern, industrialized world. As Hickel states:

> *We cannot ignore the fact that the period 1820 to circa 1950 was one of violent dispossession across much of the global South. Colonizers...[coerced] people into the labour market: imposing taxes, enclosing commons and constraining access to food, or just outright forcing people off their land. The process of forcibly integrating colonized peoples into the capitalist labour system caused widespread dislocation.*
>
> (Hickel 2019a)

The point here, and this is particularly the case for backward-looking estimations of national income, is that while colonialism may have increased national wealth it also, and for many more than a handful of colonial subjects, at the same time pushed individuals and sometimes countries into an increasingly precarious existence. Amartya Sen's work (1981) on the Bengal Famine of 1943 and Stephen Platt's (2018) history of China and the Opium Wars of 1839–1842 provide compelling insights into the effects of colonialism in particular places, and at particular historical junctures. It is hard to see colonialism as 'developmental' in these contexts and it is equally very easy to see such historical processes driving people into increasingly tenuous existences, thus creating poverty.

Debates over the effects of colonialism finds their echo in disputes over the effects of late capitalism. Scholars write of 'immiserising growth' and 'adverse incorporation' to highlight the ways in which the poor may not just fail to receive their 'fair' share of a growing economic pie, but may actually be harmed by such growth. For Mosse (2010, 1161), 'the point is that the poverty of certain categories of people is not just unimproved by growth or integration into (global) markets, but deepened by it.' This, furthermore, is a product of a particular *type* of growth which might be labelled 'free-market' or 'neo-liberal.' Joseph Stiglitz (2013, 64), former Chief Economist at the World Bank, has argued that those at the bottom of the income pyramid have been losers from globalization and that policies need to rectify this tendency. While adverse incorporation may be a quasi-Marxian notion, the idea that we see around the globe a 'race to bottom' under the forces of globalization is entertained even by those in the mainstream, like Stiglitz.

Even if we accept that poverty has declined, there is no doubt that things could be better and the fact that they are not is due to policy decisions taken, nationally and internationally. It is this that explains debates over and concern for 'pro-poor growth' (Son and Kakwani 2008), 'shared growth' (Page 2006), 'inclusive growth' (Ali and Son 2007) and 'inclusive development' (Kanbur and Rauniyar 2010). To illustrate how pro-poor growth should function, Martin Ravallion provides a helpful example based on two notional country cases. Both experience a two per cent rate of growth in per capita income, see no change in income distribution, and have a poverty rate of 40 per cent. The only difference is their rate of inequality. One is a low inequality country (Gini coefficient of inequality of 0.30) and the other a high inequality country (Gini coefficient of inequality of 0.60). The former would experience a decline in poverty of 6.4 per cent per year, halving in little more than ten years. The latter would see poverty falling at 1.2 per cent per year and would take 57 years to halve the initial poverty rate (Ravallion 2004, 16).

Where are the poor?

Identifying who are the poor permits us to place them geographically, whether intra-nationally or inter-nationally.

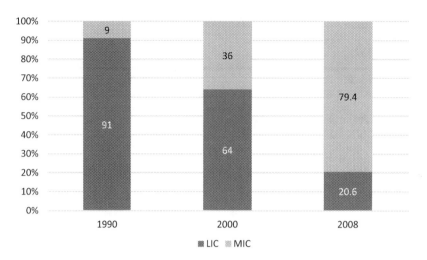

Figure 27.4 The distribution of the poor between low-income and middle-income countries ($2.00 per day)

Source: data from Sumner 2012: 2

Until around 1990, the extreme poor – those living on less $1.90 per day – were to be found in poor countries. Indeed, in 1990, 91 per cent of the extreme poor were living in low-income countries (Figure 27.4). This meant that a concern for poor people led, inexorably, to an interest in poor countries, their progress and development. The not unreasonable assumption was that promoting growth in such countries (but see above), and especially pro-poor growth, would also address extreme poverty at a global level. Replicated across the world, this would in turn achieve the eradication of extreme poverty, the first of both the UN's Millennium Development Goals (2000–2015) and Sustainable Development Goals (2015–2030). (But again, using the $1.90 per day measure.) By 2008, however, geographies of poverty had shifted dramatically, such that just 21 per cent of the extreme poor were in low-income countries. To address global poverty, in other words, we have to focus on middle-income as much as low-income countries.

Of course, the cause of this shift is not that poor people moved from low- to middle-income countries between 1990 and 2008, taking their poverty with them. Rather, the countries them-selves have progressed from low- to middle-income, while their extreme poor have remained behind. A few countries, notably China and India, make up for a good deal of this shift in the poor between categories. What this means for addressing poverty is that while the extreme pov-erty rate (the incidence of poverty) may remain highest in low-income countries, the scale of poverty – in terms of numbers of poor – has become a middle-income country problem. The danger is that because countries are no longer poor it is assumed that their populations have also escaped poverty. Some have; but many millions have not.

Significantly, the growth of the countries of East and Southeast Asia – not least, China, Indonesia, and Vietnam – has meant that the extreme poor have become increasingly concentrated in the countries of South Asia and, especially, Sub-Saharan Africa. In 1990, these two regions accounted for 44 per cent of those living on less than $1.90 a day; in 2018, it was 89 per cent (Figure 27.5). What has not changed, however, is the concentration of the poor in rural areas. The large majority of the extreme poor are to be found in the countryside – four

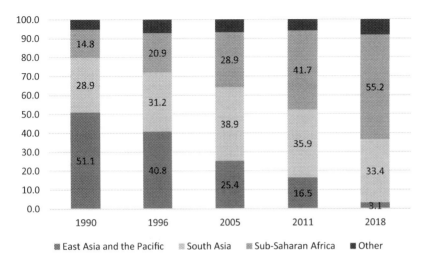

Figure 27.5 The regional distribution of the poor ($1.90)
Source: data from World Bank (2020)

out of every five people in 2018 – and most of these are smallholders, tenant farmers or the landless. The world is becoming ever more urban, and extreme poverty ever more rural.

A singular body of 'the poor'?

Sen's (1981) study of famines makes the point that standard approaches to poverty are reductionist in the sense of making us think that there is a body of people who can be labelled '*the* poor.' The sources and experiences of one person's or one household's poverty are very different. Poverty is plural even though poverty lines may reduce populations to one of two conditions: poor and non-poor. Sen writes:

> *A small peasant and a landless labourer may both be poor, but their fortunes are not tied together. In understanding the proneness to starvation of either we have to view them not as members of the huge army of 'the poor', but as members of particular classes, belonging to particular occupational groups, having different ownership endowments, and being governed by rather different entitlement relations. Classifying the population into the rich and the poor may serve some purpose in some context, but it is far too undiscriminating to be helpful in analysing starvation, famines, or even poverty.*
>
> *(Sen 1981, 156)*

There is, as Sen (1979) suggests, both an identification problem (who are the poor?) and an aggregation problem (how can the very different characteristics of these poor be aggregated into a single overall measure?). The importance of knowing *who* we are talking about has policy ramifications because interventions to address poverty must start with a full understanding of why people – individuals, households, and communities – are poor in the first place. Like Sen, Martin Ravallion, who proposed the original one-dollar-a-day International Poverty Line in 1990, also highlights 'the importance of more, micro, country-specific, research on the factors determining why some poor people are able to take up the opportunities afforded by an expanding economy – and so add to its expansion – while others are not' (Ravallion 2001, 1813). Dealing in averages and aggregates creates the image of poverty as a singular condition with an equally singular solution, preventing this sort of forensic and granular view of poverty. When poverty is also reduced to a matter of income this becomes yet more acute.

A matter of justice?

The UN's Special Rapporteur on extreme poverty and human rights is given the mandate to highlight the plight of those living in extreme poverty and to 'highlight the human rights consequences of the systematic neglect to which [the extreme poor] are subjected.'[7] In 2018 and 2019 the Special Rapporteur (at the time Professor Philip Alston) visited the UK (November 2018) and Laos (March 2019) to undertake country studies.[8] Table 27.1 picks out extracts from the two reports which highlight that the countries – despite the UK and Laos being so different in wealth and much else – share many similarities when it comes to Alston's assessment of poverty. In both places, economic growth has not fed through into substantial poverty alleviation; measurement, data and definitional issues persist; officials seem to have little idea – or little concern – for what it means to be poor; and poverty 'justice' is unfulfilled. Time-and-again, Alston returns to questions of the rights of the poor. As Easterly similarly writes in *The Tyranny of Experts: The Forgotten Rights of the Poor*:

> *The conventional approach to economic development, to making poor countries rich, is based on a technocratic illusion: the belief that poverty is a purely technical problem amenable to such technical solutions as fertilizers, antibiotics, or nutritional supplements. … The technocratic illusion is that poverty results from a shortage of expertise, whereas poverty is really about a shortage of rights.*
>
> *(Easterly 2013, 6–7)*

Notwithstanding the 'unprecedented opulence' (Sen 1999, xi) of the world, there are hundreds of millions of people in rich countries as well as poor who live frugal, meagre lives, often without dignity, their rights undermined, and for whom economic growth has delivered few lasting benefits. Understanding who these people are, and in all their complexity, remains a central task for development practitioners, policy makers and scholars.

A pedagogy of poverty

This entry in the *Routledge Handbook of Global Development* is structured around a dispute as to whether global poverty is close to being eradicated or has persisted even while global wealth has accumulated. It forces a consideration of the assumptions that lie behind the figures we use to substantiate our arguments and positions. This opens up a range of questions pertaining not just to poverty and its amelioration, but development more generally. Some of these are quite technical questions; others rather more conceptual. Together, they emphasize that poverty – like development – requires a consideration of the following:

- Can the poverty data be trusted? Are they reliable and robust? Where are any shortcomings concentrated?
- Is the poverty line drawn correctly?
- More widely, does the definition of poverty employed in such poverty lines adequately capture destitution?

Table 27.1 Poverty in Laos and the UK

	UK	Laos
GDP per capita (2019)	US$42,300	US$2,535
Poverty rate at $1.90	0.2%	21.2%
Poverty rate at $5.50	0.5%	84.4%
Definition and measurement of poverty	'To address poverty systematically and effectively it is essential to know its extent and character. Yet the United Kingdom does not have an official measure of poverty.' (page 15)	'Very limited data and a lack of transparency around what data does exist make it difficult to accurately assess the current state of poverty in Lao PDR, and mean that programs and responses are being designed around information that may not reflect the actual situation.' (page 4)
Pro-poor growth	'The UK is the world's fifth largest economy, it contains many areas of immense wealth, its capital is a leading centre of global finance, its entrepreneurs are innovative and agile... It thus seems patently unjust and contrary to British values that so many people are living in poverty.' (page 1)	'...Lao PDR's economic growth strategies have too often destroyed livelihoods, created or exacerbated vulnerability, and lead to impoverishment for many groups. ... By emphasizing aggregate economic growth over poverty reduction...the government has achieved impressive GDP numbers but at times has failed to make meaningful changes in the lives of a very large number of people in poverty.' (page 1)
Politics and poverty	'...poverty is a political choice. Austerity could easily have spared the poor, if the political will had existed to do so' (page 22)	'One minister told me [the Special Rapporteur], "For me, poverty is natural, a force of nature, not a man made calamity."' (page 4)
Disconnect between government and people	'...there is a striking and almost complete disconnect between what I heard from the government and what I consistently heard from many people directly, across the country' (page 16	'...government officials were eager to share ambitious targets and the creation of new committees, but were only rarely able to provide evidence that their policies had benefited people in the real world... .' (page 4)
Human rights and poverty	'Abandoning people to the private market in relation to a service [transport] that affects every dimension of their basic well-being is incompatible with human rights requirements.' (page 24)	'Some approaches to poverty alleviation have...prejudiced the human rights of poor and marginalized people. ... Deep structural barriers prevent the full realization of human rights by people in poverty.' (page 1)

Sources: quotations extracted from UNHCR 2018 (www.ohchr.org/Documents/Issues/Poverty/EOM_GB_16Nov2018.pdf) and UNHCR 2019 (www.ohchr.org/Documents/Issues/EPoverty/EOSVisitToLao28Mar2019_EN.pdf); data on poverty and GDP per capita from World Bank 2020

- Goal 1 of the United Nations Sustainable Development Goals is 'No Poverty.' How do the SDGs define and measure poverty? What does the SDG understanding of poverty overlook?
- What is lost when we aggregate poverty to the global level?
- Is it reasonable to make cross-country comparisons?
- Does a focus on income poverty capture the multi-dimensional nature of poverty – is it an acceptable proxy?
- What variables should be included in a multi-dimensional poverty index? How should these be weighted? Examine the relevant literature to interrogate this question.
- What processes produce (and reduce) poverty in the 21st century? How are these different from processes that operated in the past?
- In October 2020, the World Bank estimated that the COVID-19 pandemic could push as many as 150 million people into extreme poverty. How has COVID-19 disproportionately affected the poor, in both the rich(er) and poor(er) worlds? What has connected poverty in these worlds during the pandemic? And what structural changes are needed to pull these people back out of poverty?
- Hickel and Kallis have argued that 'there are no scientific grounds upon which we should not question growth, if our goal is to avoid dangerous climate change and ecological breakdown' (2019, 15). How might poverty alleviation be best achieved alongside reductions in economic growth?
- How does the 'texture' of poverty change across different places? What does poverty look like in your home town/state? Who, predominantly, makes up the poor in your local area?
- A challenge, particularly for students (and scholars) who themselves may never have been poor, is to understand what poverty is like. Can we ever truly understand what it means to be poor if poverty has not been our own lived experience? What methods might we use to try and better understand poverty as experience? Narayan et al. (2000, xvii) quote the words of a poor Ghanaian man: 'Poverty is like heat; you cannot see it; you can only feel it; so to know poverty you have to go through it.'

Key sources

1. Hickel, J. (2019a). 'Bill Gates says poverty is decreasing. He couldn't be more wrong.' *The Guardian*. London.
2. Roser, M. and Ortiz-Ospina, E. (2019). Global extreme poverty. Our World in Data. Downloaded from: https://ourworldindata.org/extreme-poverty#the-mis-perceptions-about-poverty-trends.

These two sources are a good place to start because they underscore why there is such a gulf of views regarding what might seem to be a simple question, namely: is poverty declining? Both are accessible and couched in non-technical language.

3. Hickel, J. (2019b). 'Progress and its Discontents.' *The New Internationalist*, 7th August. https://newint.org/features/2019/07/01/long-read-progress-and-its-discontents.

This is a longer intervention by Hickel, but covering similar points and arguments. Again, it is accessible.

4. World Bank (2001). *World Development Report 2000/2001: Attacking Poverty*. New York: Oxford University Press. https://openknowledge.worldbank.org/handle/10986/11856.

This is a key publication – available for free download – that shows the World Bank 'discovering' a different way of thinking about and measuring poverty. Whether the ethos here has been fully embraced by the Bank is another matter.

5. Sumner, A. (2007). 'Meaning versus Measurement: Why Do "Economic" Indicators of Poverty Still Predominate?' *Development in Practice* 17(1): 4–13.

This paper can be quite nicely paired with the 2000/2001 World Development Report of the World Bank. Here Sumner explores why economic indicators of poverty remain the 'first among equals' notwithstanding the broad acceptance that we need to take a multi-dimensional approach to poverty and its measurement.

6. Mosse, D. (2010). 'A relational approach to durable poverty, inequality and power.' *The Journal of Development Studies* 46(7): 1156–1178.

This paper, drawing mainly on evidence from India, argues that poverty persists, first, because it emerges from economic and political relations rooted in historical processes and, second, due to deep-seated social categorizations.

7. Ravallion, M. (2001). 'Growth, Inequality and Poverty: Looking Beyond Averages.' *World Development* 29(11): 1803–1815.

This paper is important partly because of who has written it – Martin Ravallion, one of the architects of the original $1-a-day International Poverty Line – but also because it recognizes that poverty should not be aggregated and reduced to averages but recognized as highly diverse in its origins, nature and effects.

8. Alston, P. (2020). The parlous state of poverty eradication: report of the Special Rapporteur on extreme poverty and human rights. New York, Human Rights Council, United Nations. www.ohchr.org/EN/Issues/Poverty/Pages/parlous.aspx.

This paper echoes some of Hickel's concerns. Its importance comes from the fact that it is written by the UN's Special Rapporteur on extreme poverty and human rights.

9. Sen, A. (1981). *Poverty and famines: an essay on entitlement and deprivation*. Oxford: Oxford University Press.
10. Sen, A. (1999). *Development as freedom*. New York: Oxford University Press.

Of scholars writing about poverty, Novel Prize laureate Amartya Sen is the most prominent and influential. These two books, separated by almost two decades, are representative of his work on poverty and development, providing beautifully written and academically persuasive cases for asking the question: what is constitutive of poverty and development?

Notes

1 https://twitter.com/billgates/status/1086662632587907072?lang=en.
2 See Hickel 2019b for a longer discussion of the issues and the case he makes.
3 The other is to 'promote shared prosperity.'
4 https://blogs.worldbank.org/developmenttalk/richer-array-international-poverty-lines#:~:text=Start
ing%20this%20month%2C%20the%20World,%2C%20set%20at%20%245.50%2Fday.
5 The blog also stated: 'Let us be completely clear: The World Bank's headline threshold to define
extreme global poverty is unchanged, at $1.90/day. The Bank's goal of ending poverty by 2030, and the
United Nations Sustainable Development Goal 1.1, are both set with respect to this line.'
6 Statistics Canada 'Life expectancy at birth for Inuit of the former Northwest Territories rose from
29 years in 1941 to 1950 (38 years less than for Canada overall), to 37 years in 1951 to 1960 (33 years
less), to 51 years in 1963 to 1966 (21 years less), and to 66 years in 1978 to 1982 (19 years less).'
7 www.ohchr.org/en/issues/poverty/pages/srextremepovertyindex.aspx.
8 The country reports can be accessed here: www.ohchr.org/EN/Issues/Poverty/Pages/CountryVisits.
aspx.

References

Ali, I. and Son, H. H. (2007), 'Measuring inclusive growth', *Asian Development Review*, vol. 24, no.1,
pp. 11–31.
Alston, P. (2020), 'The parlous state of poverty eradication': report of the Special Rapporteur on extreme
poverty and human rights, Human Rights Council, United Nations New York.
Bolt, J., R. Inklaar, H. de Jong, and J. Luiten van Zanden (2018). *Rebasing 'Maddison': New income comparisons
and the shape of long-run economic development*. Groningen, Groningen Growth and Development Centre,
University of Groningen.
Easterly, W. (2013), *The tyranny of experts: economists, dictators, and the forgotten rights of the poor*, Basic Books,
New York.
Hickel, J. (2019a), 'Bill Gates says poverty is decreasing. He couldn't be more wrong,' *The Guardian*
London, www.theguardian.com/commentisfree/2019/jan/29/bill-gates-davos-global-poverty-
infographic-neoliberal.
Hickel, J (2019b), 'Progress and its Discontents,' *The New Internationalist*, 7th August, https://newint.org/
features/2019/07/01/long-read-progress-and-its-discontents.
Hickel, J. and Kallis, G. (2019), 'Is Green Growth Possible?', *New Political Economy*, vol. 25, no. 4,
pp. 469–486.
Kanbur, R. and Rauniyar, G. (2010), 'Conceptualizing inclusive development: with applications to rural
infrastructure and development assistance', *Journal of the Asia Pacific Economy*, vol. 15, no. 4, pp. 437–454.
Mosse, D. (2010), 'A relational approach to durable poverty, inequality and power', *The Journal of
Development Studies,* vol. 46, no. 7, pp. 1156–1178.
Narayan, D., Chambers, R., Shah, M.K., Petesch, P. (2000), *Voices of the Poor: Crying Out for Change*,
Oxford University Press for the World Bank, New York.
Norberg-Hodge, H (1992), 'Learning From Ladakh: A Passionate Appeal for 'Counter-Development'',
Earth Island Journal, vol. 7, no. 2, pp. 27–28.
Page, J. (2006), 'Strategies for Pro-poor Growth: Pro-poor, Pro-growth or Both?', *Journal of African
Economies,* vol. 15, no. 4, pp. 510–542.
Platt, S. (2018), *Imperial Twilight: The Opium War and the End of China's Last Golden Age*, Atlantic Books,
London.
Ravallion, M. (2001), 'Growth, Inequality and Poverty: Looking Beyond Averages', *World Development,*
vol. 29, no. 11, pp. 1803–1815.
Ravallion, M. (2004), *Pro-Poor Growth: A Primer*, World Bank, Washington DC.
Roser, M. and Ortiz-Ospina, E. (2019), *Global extreme poverty. Our World in Data*, viewed, https://ourworl
dindata.org/extreme-poverty#the-mis-perceptions-about-poverty-trends.
Sahlins, M. (2004), *Stone Age economics*, Routledge, London.
Seers, D. (1970), 'The meaning of development', *Ekistics*, vol. 30, no. 180, pp. 353–355.
Sen, A. (1979), 'Issues in the Measurement of Poverty', *The Scandinavian Journal of Economics*, vol. 81,
no. 2, pp. 285–307.
Sen, A. (1981), *Poverty and famines: an essay on entitlement and deprivation*, Oxford University Press, Oxford.

Sen, A. (1999), *Development as freedom*, Oxford University Press, New York.

Smith, A. (2007), *An inquiry into the nature and causes of the wealth of nations (books I, II, III, IV and V)*, Metalibri, Amsterdam.

Son, H. H. and Kakwani, N. (2008), 'Global Estimates of Pro-Poor Growth', *World Development,* vol. 36, no. 6, pp. 1048–1066.

Stiglitz, J. E. (2013), *The price of inequality*, W.W. Norton & Company, New York.

Sumner, A. (2007), 'Meaning versus Measurement: Why Do 'Economic' Indicators of Poverty Still Predominate?', *Development in Practice,* vol. 17, no. 1, pp. 4–13.

Sumner, A. (2012), 'Where Do The Poor Live?', *World Development,* vol. 40, no. 5, pp. 865–877.

World Bank (2001), *World Development Report 2000/2001: Attacking Poverty*, University Press, New York: Oxford.

World Bank (2017), *Monitoring Global Poverty: Report of the Commission on Global Poverty*, World Bank, Washington DC.

World Bank (2020), *Poverty and Shared Prosperity 2020: Reversals of Fortune,*, World Bank, Washington DC.

28

GLOBAL FINANCIAL SYSTEMS AND TAX AVOIDANCE

Rachel Etter-Phoya, Moran Harari, Markus Meinzer, and Miroslav Palanský

Illicit financial flows and revenue losses in majority countries

Half a century after the end of most formal colonial oppression, the global financial and tax systems remain deeply biased against the interests of majority world ('developing') countries. This bias is salient in the worldwide impact of policies shaped by the key organization in global tax governance, the Organization for Economic Co-operation and Development (OECD), as well as data gaps around tax and financial matters the OECD presides over (Ates 2020; Cobham 2020a). This club of minority world ('developed') countries designs policies that cover most of global economic activity and thus heavily impact the distribution of tax revenues and the shape and magnitude of tax avoidance.

One result of this biased and largely dysfunctional global financial and tax architecture are large volumes of illicit financial flows and associated tax revenue losses disproportionately affecting majority world country economies and public coffers. Another result is the undermining of the state building function of taxation, which is one of the 4 'Rs' functions of taxation: revenue, redistribution, repricing and representation (Cobham 2020b, 14). Representation is vital for sound and accountable institutions and good governance, as citizens hold governments to account over spending of tax money as one dimension of the social contract. Under the current global system, majority world country elites can too easily escape paying taxes, allowing their governments to shift their allegiance away from their citizens to foreign donors and aid agencies or to the natural resource sector. The challenge is how to reconfigure this global system to make it more inclusive for greater social and economic equity. This chapter reviews the politics behind the current global setup for regulating tax and illicit financial flows, and discusses policy solutions and how progressive transformation could be supported.

Two related debates are important to the subjects discussed in this chapter, yet were beyond its scope. The first relates to the scope of the definition of illicit financial flows and particularly if tax abuses by multinational companies are part of such flows (for an overview of the debate, see Cobham and Janský 2020, 81–83). The second related debate concerns the various methodologies to estimate the scale of illicit financial flows and revenue losses (for a review of this literature and its varying methodologies, see Cobham and Janský 2020). Globally, it is estimated that between 100 and 500 billion USD is lost in tax revenue annually as a result of corporate

DOI: 10.4324/9781003017653-32

profit shifting; while between 6 trillion and 30 trillion USD of global private financial wealth is held offshore.

The Political economy of global tax governance

The current international tax system has its roots in the period of the demise of the British Empire and the end of political colonialism. The key principles of international taxation were negotiated in the 1920s under the auspices of the League of Nations, and the current international tax system continues to largely reflect the interests of former colonial powers. For example, many British dependencies located in the Caribbean are famous corporate tax havens and secrecy jurisdictions. Their extraordinary role in the global financial system has been and continues to be possible through the explicit or implicit support from the City of London. During the 1960s and 1970s, parts of the British government encouraged or condoned the enactment of laws conducive to secretive banking, shell company formation and trust administration in these remaining outposts of the British empire. This happened in order to reduce reliance of these crown dependencies and overseas territories on British taxpayer monies and to feed funds to London (Ogle 2020; Shaxson 2011).

Since the 1960s, the OECD has taken over the role of managing and shaping international tax rules. Its founding members in 1961 were 20 rich countries, including all Western European former colonial powers. Today, its membership still comprises almost exclusively minority world countries. The design of its international tax policies is where these countries create an unlevel playing field, with asymmetric benefits for minority world countries clearly discernible. These policies include, among others, those around blacklisting tax havens, bilateral tax treaties, tax information exchange, accounting, and financial reporting standards for multinational companies (namely, country by country reporting), and the calculation and distribution of taxable profits within a multinational company.

This section focuses on three of these policies and discusses the way minority world countries continue to shift rights to tax and revenues their way, at the expense of majority world countries.

Blacklisting tax havens

Despite the lack of a clear and agreed definition, the term 'tax haven' has been widely used in research and international policy-making to designate countries that were threatened with countermeasures, required to change domestic policies and asked to sign up to international agreements to counter tax evasion and avoidance. Historically, tax haven lists have been highly political and lacked any transparent criteria for inclusion (Meinzer 2016).

The OECD published its first blacklist with 35 tax havens in 2000 as part of its Harmful Tax Competition project. After the new US administration withdrew its support in 2001, the project largely stalled due to its vulnerability to accusations of hypocrisy, illegitimacy, and political bias. This included, for example, its exclusion of the United Kingdom's overseas territories like Bermuda and Cayman Islands, and its omission of OECD members (chiefly the US) and their role in hiding dirty money (Sharman and Rawlings 2006). Subsequently, the tax havens were invited to the negotiation table for shaping the future 'international standard' for countering tax evasion and avoidance through bilateral information exchange agreements.

Following the financial crisis, this initiative was revived at the G20 summit in London 2009, when the OECD published a white-grey-black list of jurisdictions refusing these bilateral information exchange treaties (OECD 2009). Within a week of publication, no country was left on the blacklist because commitment to the treaties was sufficient to be removed. While the OECD introduced a peer review process to entice and monitor compliance with their standard, its blacklist remained empty until much later. The key problem of the standard and peer review process was its limitation to 'upon request' information exchange. In order to successfully request information under this system, jurisdictions have to provide in advance evidence about individual taxpayers evading taxes abroad ('a smoking gun'). This requirement proved – and continues to prove – to be an immense obstacle especially for majority world governments whose administrations operate with less capacity, not least because evidence obtained through whistle-blowers and dataleaks would not be accepted as legitimate evidence for requesting information. Hence, no effective deterrent against tax evasion was created and secrecy jurisdictions continued to provide only a trickle of information. They were largely able to adapt their business model by engaging in 'mock compliance' (Woodward 2016).

In 2016, the OECD defined three criteria to assess non-cooperative jurisdictions, two of which need to be met by a jurisdiction in order to escape the OECD's blacklist. Due to the flaws of these criteria, it was not surprising that a year later, in June 2017, the OECD published a tax haven list which contained only one tiny jurisdiction, Trinidad and Tobago. In December of the same year, the European Union released its own blacklist of non-cooperative jurisdictions, largely relying on OECD's standards. However, so far not even one OECD or EU country was included in the blacklist.

Evidence from the last two decades shows that blacklists have been ineffective in attempts to combat tax avoidance and evasion. For example, Johannesen and Zucman's econometric analysis shows that signing bilateral information exchange treaties, as guided by the OECD, has not changed much of the total amount of wealth managed offshore and at best only led the relocation of deposits to the least compliant havens (2014, 89; 65). The ineffectiveness of the black-grey-and-white listing approach rests on the difficulty of insulating lists from political manipulation. This is related to a lack of a consistent and objective definition of the term 'tax haven' (Meinzer 2016; Cobham et al. 2015). Further, it is not legitimate to apply a dichotomous approach to tax systems because in most cases, 'tax havenry' is not a binary division but rather a matter of degree of particular financial or legal measures abused by individuals and companies for their own benefit.

In line with the notion that all jurisdictions can be placed on a spectrum based on the intensity of their corporate tax aggressiveness and financial secrecy, the Tax Justice Network publishes the Corporate Tax Haven Index and the Financial Secrecy Index, which measure the contribution of each jurisdiction to the global problems of corporate tax abuse and financial secrecy, respectively (see a more detailed explanation on the indices in next section). These indices have revealed that, contrary to the picture painted by most blacklists, some of the world's wealthiest countries like OECD and EU member states are the main perpetrators, recipients, or conduits of illicit financial flows. In other words, dominant narratives that present majority world countries as corrupt ought to be considered in the context of similarly corrupt and unlawful practices by minority world countries.

As presented in Table 28.1, the results of the latest editions of the two indices indicate that minority world countries ranked among the top ten countries. These are the countries with tax and financial systems and regulations that received the highest scores for providing financial secrecy and enabling corporate tax avoidance.

Furthermore, several minority world countries, like the United Kingdom, are also known to 'outsource' corporate tax avoidance and some of the worst forms of financial secrecy to their dependencies. As a result, British overseas territories and crown dependencies (e.g. Cayman

Table 28.1 Top ten countries in the Corporate Tax Haven Index 2019 and Financial Secrecy Index 2020

Jurisdiction	Rank in CTHI 2019	CTHI 2019 value	Rank in FSI 2020	FSI 2020 value
British Virgin Islands	1	2769	9	619
Bermuda	2	2653	40	289
Cayman Islands	3	2534	1	1575
Netherlands	4	2391	8	682
Switzerland	5	1875	3	1402
Luxembourg	6	1795	6	849
Jersey	7	1541	16	467
Singapore	8	1489	5	1022
Bahamas	9	1378	22	407
Hong Kong	10	1372	4	1035
United Arab Emirates	12	1245	10	605
United States	25	408	2	1487
Japan	Not assessed	/	7	696

Source: Corporate Tax Haven Index (Tax Justice Network 2019); *Financial Secrecy Index* (Tax Justice Network 2020).

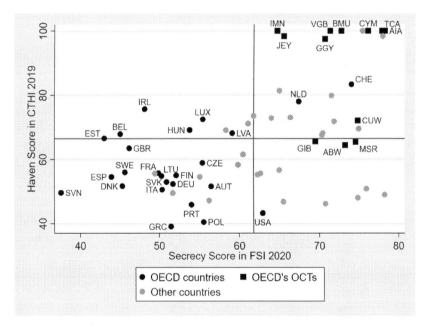

Figure 28.1 Haven Scores of the Corporate Tax Haven Index 2019 and Secrecy Scores of the Financial Secrecy Index 2020, OECD vs. OECD's Overseas Countries and Territories vs. non–OECD countries. Lines show the means of the two samples.

Source: Authors based on Janský and Palanský (2019)

Islands and British Virgin Islands) dominate the top of the two indices (see Figure 28.1) and obtain higher haven and secrecy scores than the United Kingdom. If the scores of the controlled dependencies were combined with the United Kingdom's score, the United Kingdom would rank first on both indices and is thus considered to be the world's greatest enabler of corporate

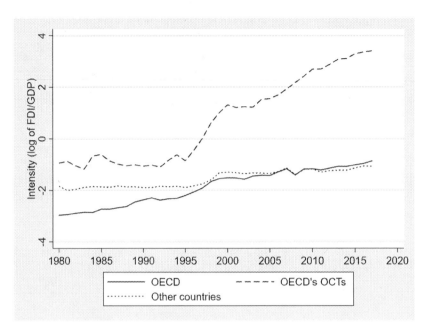

Figure 28.2 Development of the intensity indicator (Foreign Direct Investment as a share of GDP), OECD vs. OECD's Overseas Countries and Territories vs. Non–OECD countries.

Source: Authors

tax avoidance. A similar pattern of more secretive tax rules in dependencies can also be observed in the remnants of the colonial empire of the Netherlands (Curacao, Aruba) and for the United States (US Virgin Islands, Puerto Rico). The results of the Financial Secrecy Index 2020 illustrate the hypocrisy of minority world countries in curbing financial secrecy: compared with majority world countries (excluding OECD dependencies) that had an average secrecy score of 67 out of 100, minority world countries had an average secrecy score of 54 out of 100 while their dependencies had an average secrecy score of 73.

As a result of such outsourcing of financial secrecy, the disproportional relationship between the international activity of OECD's dependency jurisdictions – as measured by Foreign Direct Investment (FDI) – and the size of their economy (as measure by their GDP) has consistently increased from 1995, as indicated in Figure 28.2. In the same time period, large scale profit misalignment (i.e. tax avoidance) of US multinationals began to take off. These excessive levels of FDI growth in the OECD's dependencies suggest that much or most of this FDI is phantom FDI which exists on paper but does not consist of new factories, mines or infrastructure (Damgaard et al. 2019). Phantom FDI may indicate, for example, roundtripping of capital where domestic investors hide their identities behind shell companies in dependencies and then invest in the local economy pretending to be foreign, all to take advantage of tax incentives for foreign entities (Meinzer et al. 2019; Millán-Narotzky et al., Forthcoming).

Double tax treaties

These indices epitomize the power of the OECD's approach to blacklisting, which excludes the minority world countries from taking responsibility for tax abuse and profit shifting. Yet, as

Meinzer argues (2019, 23), blacklisting is but one of several examples of the way the OECD's tax policy design and decision-making primarily serve the interests of minority world countries, often at the expense of majority world countries. Another important example with strong historical roots in colonial exploitation is the signing of bilateral tax treaties which are designed to divide taxing rights between two jurisdictions over income arising from cross-border investment (Picciotto 1992). Over 3800 treaties in force in 2019 regulate the taxation of cross-border investment. Many of these are based on the OECD's model tax convention, which is like a template used for most treaty negotiations. The highly complex and detailed rules for deciding how to calculate the appropriate income are laid down in OECD's transfer pricing guidelines which are used far beyond the OECD.

The era of tax treaties commenced in the 1920s in the lead up to decolonization. Mitigation of double taxation is achieved by limiting the taxing rights of the source country (where multinational subsidiaries usually operate factories or mines and sell their produce), thereby distributing the tax revenues to the residency countries of multinational companies. The distributional conflict inherent in the allocation of taxing rights in double tax treaties goes back to the League of Nations when the first model for a double tax treaty was negotiated (Picciotto 1992, 49–60).

At the inception of this international system of taxation, concerns were raised about tax avoidance and evasion, but the issue was left unaddressed for fear of 'placing any obstacles in the way of the international circulation of capital, which is one of the conditions of public prosperity and world economic reconstruction' (League of Nations 1927). With this framing of tax obstacles preventing prosperity, minority world countries succeeded in curtailing the taxing rights in former colonies through tax treaties for the benefit of their multinational companies. It is widely accepted that the most extensively followed OECD model convention attributes most taxing rights from the source countries – which are usually majority world countries – to the capital exporting country, i.e. the jurisdiction of residence of the investor, which is often a minority world country. In addition, research has shown that majority world countries lose billions in tax revenues due to profit shifting strategies of multinational corporations engaging in 'treaty shopping' where they artificially structure their corporate group in order to invest through treaties with the least possible tax (Hearson 2018).

One of the specific avenues where taxing rights of majority world countries are limited in treaties are withholding taxes. These taxes are important backstops against illicit financial flows and tax avoidance especially for majority world countries as they are relatively easy to administer. Upper limits are imposed on the withholding tax rates a country can levy on outbound payments, such as dividends, interest, royalties, or management/technical service fees. A source country is most disadvantaged when the withholding tax rate is set to 0 per cent, because it prevents the source country from levying any withholding tax, resulting in a loss of tax revenues (Millán-Narotzky et al. forthcoming). The results of the 2019 edition of the Corporate Tax Haven Index indicate that several minority world countries are responsible for aggressively driving down the withholding tax rates in majority world countries through their treaties, eroding their already limited defences against illicit financial flows. According to the Corporate Tax Haven Index, compared with other countries, minority world countries were on average 41 per cent more aggressive towards majority world countries. Among minority world countries, France, and the United Kingdom, both former colonial empires, were the most aggressive. The treaties these countries negotiated with majority world countries included withholding tax rates that were on average 8 and 7 percentage points, respectively, lower than the average withholding tax rates offered by those countries.

Country by country reporting

The final example of OECD policies favoring minority world countries is the case of country by country reporting. Tax avoidance exposés about multinational companies may be commonplace today, but the first major story was run by *The Guardian* in 2007 following a six-month investigation into the banana trade. According to Global Justice Now, 133 of the top 200 economic entities in the world by revenue generation are not countries but companies. Yet the information multinationals are required to disclose to governments and the public is far narrower than the requirements on governments and companies that only operate in a single jurisdiction. This opacity means that for many multinationals which are mostly resident in minority world countries, their sales, subsidiaries, profits, staff, assets, and taxes paid per country are unknown, making it hard especially for majority world revenue authorities to identify and check suspicious activity.

The diplomatic struggle between minority and majority world countries to tackle tax and other abuses of multinational companies began with the ending of colonialism. Since the 1970s, this debate has crystallized around improving corporate tax transparency in the form of public country by country reporting in order to enhance accountability of multinational companies, to level the playing field between domestic and multinational corporate groups, and to address base erosion and profit shifting (Cobham et al. 2018). In 1977, a Group of Experts on International Standards of Accounting and Reporting within the UN Commission for Transnational Corporations (UNCTC) recommended a multinational company to publish financial statements for each company that is part of that multinational group, including information on intra-group trade. The recommendation reflected continuing efforts of the G77 countries to improve their terms of trade and ensure sovereignty over their natural resources by initiating transparency regulations for transnational corporations in the interest of the source countries. However, following pressure by lobbyists, in 1978 the recommendation was blocked by the OECD, and the OECD forced the UNCTC to replace majority voting with unanimous decision-making. As a result, and for the next 15 years, until the dissolution of the UNCTC, minority world countries rejected such disclosure proposals from majority world countries and no consensus on binding standards for transnational corporations was reached (Cobham et al. 2018).

In 2003, the Tax Justice Network published a first draft accounting standard for a country by country reporting requirement which set out the basis to ensure that multinational corporations provide the public with effective disclosure about their activities and risks at the country level. Despite the efforts of the tax justice movement to advocate for this initiative in the subsequent years, the OECD continued to resist such proposals. It was ten years later, following the 2008 financial crisis, when the interests between activists in majority and minority countries converged around the fiscal risks of multinationals and led to country by country reporting making its way onto the G8 and G20 agendas. Finally, in 2013, this set the direction of the OECD's Base Erosion and Profit Shifting (BEPS) Action Plan to require multinational companies to report their economic activity and taxes paid according to a common template. After massive pressure from private sector lobbyists, the adopted OECD's standard for country by country reporting was a watered-down version because, among others, it prescribed only secretive, non-public country by country reporting for very large multinational companies (Meinzer et al. 2019, 121–122).

In this weak version of country by country reporting rules, disclosure is only to the tax authorities in the jurisdiction where the company is headquartered. Tax authorities in other jurisdictions where the company has subsidiaries may request information through a complex system, requiring bilateral negotiations. This further exacerbates inequalities in global taxing rights and has resulted in a relative decline in the capacity of majority world countries to tax multinational companies. While as of November 2019, middle-income countries had 933 bilateral relationships for exchanging country by country reporting information, none of the so-called 'least-developed countries' receive such data (United Nations Department for Economic and Social Affairs, 2020, 45).

As of January 2020, four years into the implementation of OECD's country by country reporting standard, out of the 119 countries participating in the exchange system for country by country reports specifically, 62 countries have no access to those reports (they can only send but not receive the reports from other countries); most of these are majority world countries. In comparison, of the 57 countries that already have access, 38 countries are minority world countries (Civil Society Organisations 2020). As of June 2020, there were no indications that the OECD might make the county by country reporting data publicly available or otherwise increase the accessibility or robustness of its country by country reporting standard.

The continued dominant and decisive role of the OECD in governing global tax regulations and financial and economic systems is summarized by the wide remit of its policies across trade, banking, and investment, as illustrated in Figure 28.3. The OECD tax treaties and transfer pricing guidelines jointly regulate the intra-group trade portion of global trade, where there are high risks for profit shifting, while the tax treaties also directly regulate the foreign direct and financial investment components of the global financial system. At the same time, the OECD excludes majority world voice and interests, as acknowledged implicitly in the United Nations' Financing for Sustainable Development Report 2020:

While significant progress has been made in international tax cooperation, the interests and voice of developing economies require greater priority and attention. The global community could better ensure effective inclusion in tax norm-setting processes; adaptation of tax norms and practices to

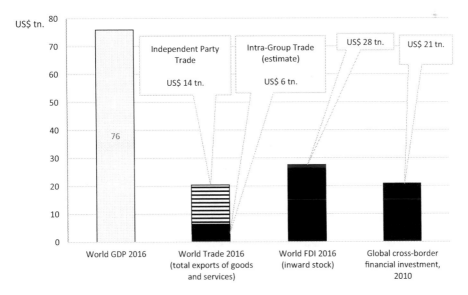

Figure 28.3 Volumes of international cross-border economic activity subject to OECD tax policies (in black)

Source: Authors based on (Meinzer 2019, p. 2)

the realities and needs of developing countries; and greater investment in capacity-building from development partners.

(United Nations Department for Economic and Social Affairs 2020, 37)

The ABC of tax transparency: transforming global tax politics

Radical tax transparency is central to transforming the structural injustices of the global financial system, which currently prioritizes the economies of former imperial powers and other countries in the minority world (Ogle 2020; Oswald and Christensen 2017; Shaxson 2011). The Tax Justice Network, established in 2002, has identified and advocates for a number of measures to address illicit financial flows; three reinforcing, key policy changes have become known as the ABC of tax transparency (Cobham 2020a, 112–116). The ABC of tax transparency – automatic exchange of information, beneficial ownership disclosure and country by country reporting – when first introduced into policy discussions by the Tax Justice Network was dismissed as utopian. This section considers what these interventions look like at their best and the state of current practice.

When assets and accounts are moved beyond the reach and oversight of tax authorities, typically by the wealthy in society, inequality grows as the progressive nature of personal income tax is undermined, the function of tax for redistribution is eroded and general tax compliance across society is impacted. This is facilitated by secrecy jurisdictions where non-residents can move their assets and thus are incentivized not to declare assets with domestic tax authorities. All countries are affected and while difficult to assess given their hidden nature, estimates suggest that about US$190 billion global tax revenue is lost annually due to undeclared offshore assets (Zucman 2014). African nations particularly suffer. Ndikumana and Boyce (2018) estimate that 30 African nations suffered US$1.8 trillion in accumulated stock of capital flight between 1970 and 2015, while the stock of debt owed by these countries as of 2015 was just over a quarter of the stock of capital flight assets (US$496.9 billion) held abroad, making the continent a net creditor to the world.

To counter this offshore tax evasion, the automatic exchange of information for tax purposes provides tax authorities with information about the financial accounts and assets held in other countries (or 'offshore') by residents in their country and can serve as a deterrent to misreporting. Some progress has been made to date with the OECD's Common Reporting Standard (CRS) for the automatic exchange of financial account information between tax authorities and as of December 2019, 108 jurisdictions had signed the Multilateral Competent Authority Agreement to implement the standard (OECD 2019). However, majority world countries face structural and regulatory barriers to implementation and many are not in reciprocal relationships with some of the most secretive jurisdictions that contribute more to the problem of offshore wealth. As a result, none of the 'least developed countries,' as defined by the UN, receive information through automatic exchanges (United Nations Department for Economic and Social Affairs 2020, 44). In addition, the United States is not a signatory and opted to implement a domestic exchange of information regime through the Foreign Account Tax Compliance Act of 2010, which has very limited reciprocity but includes sanctions for non-compliant financial institutions which face a 30 per cent withholding tax on all US-source payments for non-compliance. In contrast, the OECD's CRS lacks sufficient sanctions for non-compliance and wilful misreporting does not carry significant prison sentences or fines for individuals in most countries.

A second key policy is the identification, registration, and disclosure of the beneficial owners of legal vehicles, including companies, partnerships, trusts and private foundations. Beneficial ownership disclosure aids in investigating cross-border financial transactions in efforts to stem money laundering, tax evasion, corruption, and terrorist financing. Correct ownership information is also a prerequisite for a robust automatic exchange of information system to cross-check information or provide information when financial institutions have not been required to collect ownership information.

There is a global movement towards improved beneficial ownership transparency. Countries are recommended to ensure domestic financial institutions have access to beneficial ownership information and for exchange in the Financial Action Task Force's Recommendations on Anti-Money Laundering and Combating the Financing of Terrorism and the OECD's Global Forum on Transparency and Exchange of Information. In the European Union, measures introduced in the fourth and fifth Anti-Money Laundering Directives in 2015 and 2018, respectively, go much further than international recommendations and require member states to set up public central registries for companies and some trusts.

Several voluntary initiatives are also advocating for public beneficial ownership disclosure – most notably through the Open Government Partnership's coalition of governments leading on the implementation of public central beneficial ownership registries and the Extractive Industries Transparency Initiative requirement for over 50 participating countries to introduce public beneficial ownership registries for the extractive sector in 2020.

An assessment of beneficial ownership transparency in 133 jurisdictions, revealed that as of February 2020, 81 jurisdictions have introduced beneficial ownership legislation (Harari et al. 2020). Just two years earlier, 37 of jurisdictions assessed at the time did not have legislation in place. However, beneficial ownership legislation is not all created equal. In short, effective beneficial ownership transparency should (1) include all legal vehicles in a country, (2) be triggered for a legal vehicle to legally exist or operate in an economy or where a resident is a participant in a legal vehicle, (3) require the registration of all individuals rather than set a threshold of ownership, (4) include the identity, name, address, identification number, date of birth and tax identification number of the beneficial owners, (5) require information to be verified and validated such as with other records, through notarization, and making information public for citizen groups or journalists to investigate, (6) set criminal sanctions for non-compliance, and (7) require all information to be public, for free, in open data format (Knobel et al. 2017). At present, no jurisdiction requires the registration of both beneficial and legal ownership information for all legal vehicles and makes this available in open data format to the public (Harari et al. 2020). Ecuador currently leads the way in beneficial ownership transparency as along with effective registration, legal and beneficial ownership information for companies and partnerships and the ownership information of domestic law trusts are available online and for free (Harari et al. 2020, 25).

The third policy is public country by country reporting. As mentioned above, early predecessors of current proposals for country by country reporting date back to the 1970s and the United Nations. The current, most robust standard for country by country financial reporting requires multinational companies to publish data on the global allocation of income, profit, taxes paid and other payments to government and economic activity in all the jurisdictions where there is a subsidiary and to reconcile this with financial statements (Tax Justice Network 2020, 92–107). As was explained above, by declaring the accounting data to be tax data, and introducing a complex system for accessing it, the OECD version of country by country reporting exacerbates information and thus power asymmetries.

As long as these data will not be made public, majority world countries can legislate to partially overcoming this asymmetry by requiring any multinationals operating within their borders to file local country by country reports. Concurrently, some limited sectoral measures have been taken for public reporting. These were introduced specifically with varying scope for the extractive sector in the US in 2010 with rules introduced in 2016 (subsequently repealed in 2018), the European Union in 2013, Canada and Norway in 2014 and the Ukraine in 2018 and for financial institutions in the European Union in 2014 and 2015 (Tax Justice Network 2020, 92–107). Evidence from these relatively recent public reporting requirements suggests that public country by country reporting can have a deterrent effect on tax avoidances practices. In the banking sector, public country by country reporting significantly increased tax payments made by reporting banks compared to non-affected banks and this effect was stronger for those banks operating in tax havens (Overesch and Wolff 2018). There are also voluntary approaches, including the tax reporting standard introduced in 2019 by the Global Reporting Initiative which leads in global sustainability standard reporting, and the aforementioned Extractive Industries Transparency Initiative, yet these are inevitably limited by their voluntary nature so companies can opt out and face no sanctions for non-compliance.

The more a country opts out of or implements watered-down versions of the ABC of tax transparency and similar transparency policies, the higher its financial secrecy will be and therefore the higher the risk for it to facilitate illicit financial flows. The degree to which countries are vulnerable to highly secretive jurisdictions can be identified in a country's external economic relationships. One way to assess the risk of a country's external economic relationships, as explained earlier, is to examine its vulnerability and exposure to illicit financial flows. The Illicit Financial Flows vulnerability tracker, a new online tool published by the Tax Justice Network in June 2020, measures each country's vulnerability to various forms of illicit financial flows (trade, FDI, portfolio investment and banking positions) over different periods of time. For example, Figure 28.4 provides the top ten sources of Nigeria's vulnerability to illicit financial flows in inward FDI. This reveals the importance of British controlled dependencies, such as Bermuda, the British Virgin Islands and the Cayman Islands. These insights could be used by Nigerian authorities to focus on auditing risky investments and transactions for illicit financial flows and to review its tax treaties for instance with the Netherlands.

Conclusion

The structural imbalances and biases in the global tax and financial systems are not challenges of the past, but continue to facilitate asymmetric illicit financial flows harming predominantly the citizens, societies, institutions, and economies of majority world countries. In 2020, these ongoing struggles can be observed in the negotiations on the reform of the taxation of the digital economy 2019–2020 at the OECD. These negotiations resemble past patterns whereby powerful OECD members ultimately decide on the policy outcomes and fail to effectively take into account voices and interests of majority world countries. After having set the rules in the BEPS project, the OECD created the so-called Inclusive Framework to help rolling out its new policies, yet promising all non-OECD members an equal say. However, when majority world countries voiced dissent on the proposed next round of policy reforms in 2019–2020, their concerns were largely ignored (Cobham 2020b). Thus, a pertinent question for future research in international tax governance and development is how to design and implement international instruments and organizations that better represent majority world country's interests. For example, the United Nations High Level Panel on International Financial Accountability, Transparency and Integrity for Achieving the 2030 Agenda (2020) could establish a new global Convention at the United Nations to counter illicit financial flows (Meinzer et al. 2019).

To transform these deeply entrenched structures that protect privileges enjoyed predominantly by minority world countries will require sustained engagement in political struggles at

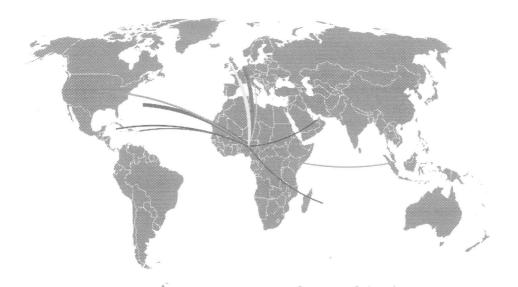

	Partner country	Vulnerability share	Secrecy score
1	Netherlands	26%	67
2	Bermuda	19%	73
3	United States	10%	63
4	France	9%	50
5	British Virgin Islands	8%	71
6	Cayman Islands	7%	76
7	United Kingdom	5%	46
8	Mauritius	4%	72
9	Singapore	2%	65
10	China	1%	60

Figure 28.4 Top ten origins of Nigeria's vulnerability to Illicit Financial Flows in inward Foreign Direct Investment (FDI), 2017

Source: *https://iff.taxjustice.net/#/profile/NGA*

all levels over many years and decades to come. Future practitioners for supporting progressive global transformation will need to combine robust knowledge in technical domains such as international accounting rules, tax law and financial regulation with softer skills such as a firm commitment to social justice, capabilities for team work and courage if necessary to blow the whistle.

The current state of affairs in the global financial and tax architecture illustrates the case for a paradigm shift (not only) in the field of development studies which has been foreshadowed by the shift from the MDGs to the SDGs. Development scholars have recently suggested shifting from 'international development' to 'global development,' looking no longer primarily or only at 'low income' or 'developing' countries, but at challenges faced across both the 'north' and 'south' (Oldekop et al. 2020). To be complete, the new global development paradigm has to encompass addressing and redressing the structural inequities in the global financial and tax systems that have been established, and are maintained by, minority world countries. To transform lives and politics, this paradigm shift must include multidisciplinary analysis, including by academic disciplines that remain somewhat inconsiderate to historical and structural power imbalances, such as public economics, law, and some strands of political science.

Key sources

1. Cobham, A. (2020a). *The Uncounted* (1 edition). Polity (Cambridge, UK; Medford, MA, USA, 2020).

This book is accessibly written and sharply diagnoses 'Development's data problem' consisting of data shortages and blindspots, resulting in gravely distorted policy decisions as 'unpeople' at the bottom of the wealth distribution and 'unmoney' at the top go unnoticed.

2. Cobham, A., and Janský, P. (2020). *Estimating illicit financial flows: A critical guide to the data, methodologies and findings.* Oxford University Press. https://global.oup. com/academic/product/estimating-illicit-financial-flows-9780198854418?q= 9780198854418&cc=gb&lang=en#

This book represents the state of the art in methodologies for estimating illicit financial flows and is available in open access for free download. It embeds technical econometrics in key political debates and connects it to the Sustainable Development Goals.

3. Meinzer, M. (2019). Countering cross-border tax evasion and avoidance: An assessment of OECD policy design from 2008 to 2018 [Utrecht University]. http://coffers.eu/wp-content/uploads/2019/10/1910-Meinzer-PhD-Dissertation-OECD-Tax-Policies.pdf

This doctoral dissertation analyses the bias against majority world countries in the policy design of the OECD and is available in open access for free download. The four case studies review tax haven blacklisting, automatic exchange of information, country by country reporting and the requirements around beneficial ownership of legal entities.

4. Shaxson, N. (2011). *Treasure Islands. Tax Havens and the Men Who Stole the World.* Bodley Head.

This book provides a deeply researched historical account of linkages between colonial exploitation, high level corruption and the rise of tax havens around the world. It meticulously traces the actors and decisions in minority countries such as the United States and the United Kingdom to welcome dirty money.

5. Unger, Brigitte, Lucia Rossel, and Joras *Combating Fiscal Fraud and Empowering Regulators* Ferwerda, eds., (COFFERS) (Oxford, 2021)

This edited volume provides an overview of current cutting-edge research on tax avoidance and evasion and will be available in open access form January 2021. It synthesizes results from a three year pan-European research project whose other outputs are available here: http://coffers. eu/publications/

References

Ates, L. (2020), Tax Information Production, Sharing Use and Publication, Viewed 11 November 2020, www.factipanel.org. https://assets-global.website-files.com/5e0bd9edab846816e263d633/5f15fc4abad3db9e9f105889_FACTI%20BP2%20Tax%20information.pdf

Civil Society Organisations (2020), Joint response to OECD public consultation document on the review of Country-by-Country Reporting, Viewed 11 November 2020, https://financialtransparency.org/wp-content/uploads/2020/03/Submission-letter-OECD-consultation-on-CBCR.pdf

Cobham, A. (2020a), US blows up global project to tax multinational corporations. What now? Viewed 11 November 2020, www.taxjustice.net/2020/06/19/us-blows-up-global-project-to-tax-multinational-corporations-what-now/

Cobham, A. (2020b), *The Uncounted*, Polity Press, Cambridge.

Cobham, A., Janský, P., and Meinzer, M. (2015), The Financial Secrecy Index: Shedding New Light on the Geography of Secrecy, *Economic Geography*, vol. 91, no. 3, pp. 281–303.

Cobham, A., Janský, P., and Meinzer, M. (2018), A half-century of resistance to corporate disclosure, *Transnational Corporations = Investment and Development*, vol. 25, no. 3, pp. 160.

Damgaard, J., Elkjaer, T., and Johannesen, N. (2019), *What Is Real and What Is Not in the Global FDI Network?* International Monetary Fund.

Harari, M., Knobel, A., Meinzer, M., and Palanský, M. (2020), *Ownership registration of different types of legal structures from an international comparative perspective: State of play of beneficial ownership – Update 2020*, Tax Justice Network.

Hearson, M. (2018), *The European Union's tax treaties with developing Countries: Leading by Example?* [Report for the European United Left/Nordic Green Left (GUE/NGL) in the European Parliament].

Janský, P., and Palanský, M. (2019), The most important tax havens and the role of intermediaries, *COFFERS Working Paper*.

Johannesen, N., and Zucman, G. (2014), The End of Bank Secrecy? An Evaluation of the G20 Tax Haven Crackdown, *American Economic Journal: Economic Policy*, vol. 6, no. 1, pp. 65–91.

Knobel, A., Meinzer, M., and Harari, M. (2017), What Should Be Included in Corporate Registries? A Data Checklist-Part 1: Beneficial Ownership Information, *A Data Checklist-Part 1*

League of Nations (1927), Double Taxation and Tax Evasion: Report (C. 216. M. 85), Viewed 11 November 2020. http://adc.library.usyd.edu.au/view?docId=split/law/xml-main-texts/brulegi-source-bibl-3.xml;collection=;database=;query=;brand=default

Meinzer, M. (2016), Towards a Common Yardstick to Identify Tax Havens and to Facilitate Reform. In T. Rixen and P. Dietsch (Eds.), *Global Tax Governance – What is Wrong with it, and How to Fix it*, pp. 255–288.

Meinzer, M., Ndajiwo, M., Etter-Phoya, R., and Diakité, M. (2019), Comparing tax incentives across jurisdictions: A pilot study, *Tax Justice Network*, pp. 43. https://dx.doi.org/10.2139/ssrn.3483437

Millán-Narotzky, L., Garcia-Bernardo, J., Diakité, M., and Meinzer, M. (Forthcoming), Double Tax Treaty Aggressiveness: Who is bringing down taxing rights in Africa?, *Tax Justice Network*.

Ndikumana, L., and Boyce, J. K. (2018), Capital Flight From Africa: Updated Methodology and New Estimates [PERI Research Report], *Political Economy Research Institute (PERI)*.

OECD (2009), A progress report on the jurisdictions surveyed by the OECD Global Forum in implementing the internationally agreed tax standard.

OECD (2019), Signatories of the Multilateral Competent Authority Agreement on Automatic Exchange of Financial Account Information and Intended First Information Exchange Date.

Ogle, V. (2020), 'Funk Money': The End of Empires, The Expansion of Tax Havens, and Decolonization as an Economic and Financial Event, *Past & Present*.

Oldekop, J. A., Horner, R., Hulme, D., Adhikari, R., Agarwal, B., Alford, M., Bakewell, O., Banks, N., Barrientos, S., Bastia, T., Bebbington, A. J., Das, U., Dimova, R., Duncombe, R., Enns, C., Fielding, D., Foster, C., Foster, T., Frederiksen, T., … Zhang, Y.-F. (2020), COVID-19 and the case for global development, *World Development*, vol. 134.

Oswald, M., and Christensen, J. (2017), The Spider's Web: Britain's Second Empire (Documentary), Viewed 11 November 2020, www.youtube.com/watch?v=np_ylvc8Zj8

Overesch, M., and Wolff, H. (2018), Does Country-by-Country Reporting Alleviate Corporate Tax Avoidance? Evidence from the European Banking Sector (SSRN Scholarly Paper ID 3075784), *Social Science Research Network*.

Picciotto, S. (1992), *International Business Taxation*, Weidenfeld and Nicolson.

Sharman, J. C., and Rawlings, G. (2006), National Tax Blacklists. A Comparative Analysis, *Journal of International Taxation*, vol. 17, no. 9, pp. 38–54.

Tax Justice Network (2019), Corporate Tax Haven Index (CTHI) (2019) Methodology, Viewed 11 November 2020, www.corporatetaxhavenindex.org/PDF/CTHI-Methodology.pdf.

Tax Justice Network (2020), Financial Secrecy Index 2020 Methodology, Viewed 11 November 2020, https://fsi.taxjustice.net/PDF/FSI-Methodology.pdf

United Nations Department for Economic and Social Affairs (2020), *Financing for Sustainable Development Report 2020: Inter-agency Task Force on Financing for Development*, United Nations, Viewed 11 November 2020.

United Nations High Level Panel on International Financial Accountability, Transparency and Integrity for Achieving the 2030 Agenda (2020), *Overview of Existing International Institutional and Legal Frameworks Related to Financial Accountability, Transparency and Integrity. FACTI Panel Background Paper*, Viewed 11 November 2020.

Woodward, R. (2016), A Strange Revolution: Mock Compliance and the Failure of the OECD's International Tax Transparency Regime. In P. Dietsch and T. Rixen (Eds.), *Global Tax Governance. What is Wrong With It and How to Fix It*, pp. 103–121.

Zucman, G. (2014), Taxing across Borders: Tracking Personal Wealth and Corporate Profits, *Journal of Economic Perspectives*, vol. 28, no. 4, pp. 121–148.

29

GLOBAL EXTRACTIVISM AND INEQUALITY

Etienne Roy Grégoire and Pascale Hatcher

Introduction

The promises of extractivism have fed development discourses across the Global South, but as opposed to more sustainable models of natural resources stewardship, or alternative models of nature-society interaction, Extractive Industries (EIs) imply, by definition, an intensity of exploitation that exceeds the rate of renewal of natural resources, ecosystemic resilience capacities, or the viability of alternative development models and life projects (Gudynas 2015). This chapter analyses how and why EIs have been a vector of inequality and vulnerability across resource-rich countries in the Global South and in colonial contexts in the North.

This chapter has three sections. First, it provides a brief overview of the main approaches around these challenges. It then provides a discussion on contemporary policy and practices around extractive industries and how these may be ill-equipped to tackle the sector's socio-environmental and human rights record. Lastly, it offers some suggestions on teaching and learning practices to better understand, and act upon, the challenge that extractivism poses to students, researchers, and practitioners.

Global extractivism and the politics of exclusion: an overview

This section provides an overview of a range of multi-disciplinary approaches that explain why extractivism is a vector of inequality and vulnerability. Three complementary entry points are addressed. EIs as: (1) A political economy of accumulation by dispossession, exacerbating pre-existing tensions between sovereignty and self-determination; (2) a catalyser of political violence against human rights defenders (including land and environmental defenders) and already marginalized populations; and (3) a site of struggle in the political economy of climate change.

Accumulation by dispossession

In describing the current global political economy as involving processes of 'accumulation by dispossession,' David Harvey (2004) has linked growing wealth inequalities to the production of devastated territories, increased levels of exclusion and vulnerability, and harsher responses to social unrest. Global EIs are perhaps the best illustration of this process.

DOI: 10.4324/9781003017653-33

EIs play a dominant economic, social and political role in more than 81 developing countries where an estimated 3.5 billion people live (World Bank 2020a). The recent commodity boom, driven by increased demand for emerging markets, intensified capital accumulation, the rapid depletion of known reserves and the development of new technology, have propelled EIs to venture into the world's most remote areas. EIs now represent a significant part of the GDP for many countries rich in natural resources across the Global South (see Figure 29.1). According to data from the United States Geological Surveys, in the Asia-Pacific region, which accounts for more than half of the world's total production of alumina, bauxite, iron ore, refined copper and steel, Indonesia, Papua Guinea and the Philippines are top exporting developing countries that also account for more than 80 per cent of the total mineral exploration budget for the region (cited in Hatcher and Roy Grégoire forthcoming). Africa accounts for an estimated 30 per cent of the world's total mineral reserves (UNECA 2017, 1), with key exporting countries such as the Democratic Republic of the Congo, one of the richest countries on the continent and yet, one of the world's lowest performing countries on the Human Development Index. Latin America, which notably produces 45 per cent of the world's silver, 39 per cent of its copper, 19 per cent of its gold and 44 per cent of its lithium, has long been the main destination for exploration capital in mining, oil and gas in the Global South, with a total of about 2.5 billion USD in 2015 alone (Hatcher and Roy Grégoire forthcoming). Figure 29.1 summarizes the share of natural resources as a percentage of the GDP by region (1970–2018).

Yet as early as the 1990s, it had already become obvious that despite increased investments in the sector, developing countries were struggling to harness EIs revenues and to meet expectations around EI-led development. Proponents of the resource curse thesis point out that some of the world's worst performers in terms of development outcomes are also some of the countries most endowed in natural riches (Sachs and Warner 1995). This 'paradox of plenty' is linked to a range of factors, including the volatility of revenues, governments over-reliance

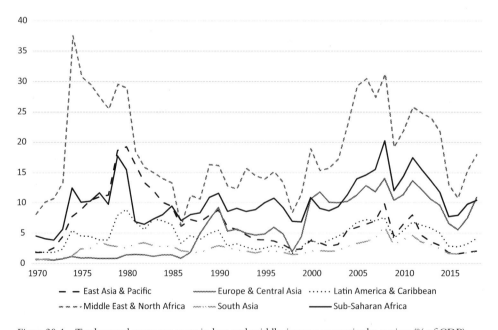

Figure 29.1   Total natural resources rents in low and middle-income countries by region (% of GDP)

on EIs revenues in times of high commodity boom, and deleterious macro-economic impacts such as the 'Dutch-disease'. Crucially, the abundance of wealth generated by EIs tends to be concentrated into few hands, leading to corruption and rent seeking, and in turn, increased inequalities and even violent conflicts (Rosser 2006).

While better governance has since been advocated as a solution to the resource curse, heterodox political economists stress that the sector's development outcomes will continue to disappoint unless policy makers account for the structural relations of power and influence that shape and govern EIs within globalizing capitalism (Hatcher and Roy Grégoire forthcoming). Spearheaded by international financial institutions, waves of neoliberal reforms to national mining codes initiated in the 1980s and 1990s delivered the mining sector into private, often foreign, hands, hence shifting the state away from its traditional developmental/regulator roles (Campbell 2009; EIR 2003). In conjunction with wider structural reforms, these new extractive regimes shifted parts of the state's power and authority to the private sector, redefining the relations between elites and communities, as well as reducing institutional capacity for development and environmental protection (Campbell and Hatcher 2019).

These reforms have further pitted governments against each other in a race to attract foreign direct investment, a process that has noticeably undermined the fiscal capacity of states to realize social and economic human rights (Campbell 2009). This reconfiguration of the state facilitated net flows of capital (legal and illegal) that are clearly detrimental to countries in the Global South. For example, while more than 70 per cent of African exports stem from the extractive sector, an estimated US$25 billion of extractive revenues leaves the continent annually in the form of illicit individual and corporate flows (UNECA 2017, 1–2). Conservative estimates suggest that between 1990 and 2008, the continent lost approximately US$170 billion in illicit flows (UNDP 2015, 12). Meanwhile, the latest 'commodities super-cycle' has produced tremendous returns for increasingly consolidated transnational networks of private actors.

Overall, extractivism has had a profound impact on the world's political institutions inasmuch as it exacerbates some of the tensions that are central to post-colonial statehood – the contradiction, for example, between the principle of 'sovereignty over natural resources' and the principle of 'self-determination and people's right to natural resources'. This is especially the case where Indigenous self-determination is concerned, but also, more largely, wherever states' actions depart from a fulfillment of their populations' common good (see Gilbert 2018). Some states – mainly under left-leaning governments in Latin-American countries – have resorted to a form of 'neo-extractivism.' Sometimes nationalizing mining and oil and gas assets, these states have used mobilized extractive rent in anti-poverty programs to widen social participation and shore up support for extractive policies; but as an emerging literature in the field of political ecology shows, neo-extractivism only produces a new geography of inequality and vulnerability for groups – mainly Indigenous – whose alternative development models are excluded (Burchardt and Dietz 2014, p. 479). Neo-extractivism thus reproduces the tension between sovereignty and self-determination, albeit through discursive shift that pits 'the common good' against the claims of the communities most affected by resource extraction (Svampa 2015).

Marginalization and violence against populations and human rights defenders

Extractivism has severe socio-environmental impacts, including – as per a review commissioned by the World Bank in 2000: deforestation, loss of biodiversity, water pollution, industrial poisoning, landscape alteration, impacts on water availability, displacement of population (often forced) and the disruption of their ways of life; the reconfiguration of local social structures; sexual violence; loss of livelihoods, collective identities and spiritualities; economical dependency and deepening of inequalities; and frequent association with authoritarianism, gross human rights violations and armed conflict (EIR 2003).

Accordingly, the extension of the 'extractive frontier' has led to a steady increase in extractive conflicts (Scheidel et al. 2020). In a context where domestic political and legal environments cater to (mainly foreign) extractive investors political dissent is increasingly stigmatized – it is illustrative to note that 2019 saw a record number of environmental activists killed globally, with activists protesting against projects in the mining sector specifically being the most at risk (Global Witness 2020). Criminalization of dissent is also a common issue around extractive projects. Challenges to the state's exclusive sovereignty – whether through competing claims to self-determination, alternative modes of territorial occupation or development models incompatible with extractivism – tend to be targeted as threats to national security or confronted with terrorism and other related charges (Global Witness 2020).

Part of this narrowing of local political spaces is directly linked to international pressure on resource-rich but cash strapped states in the Global South to implement disciplinary mechanisms for debt financing and foreign investment attractiveness. For Szablowski, such global pressures have led the state to adopt a strategy of 'selective absence' whereby it strategically transfers its authority to mediate aspirations and grievances unto the extractive company (2007, 27), hence dislocating communities from local/national political spaces.

These encounters between global institutions and local political spaces produce new types of claims as well as new subjectivities which reflect the social governance model of the institution involved (see, e.g., Hatcher and Lander forthcoming).

While jurists document and discuss legal remedies for murders, collective rapes, abusive detentions, misuse of the criminal system and arbitrary suspensions of civil liberties accumulating around extractive projects (see Roy Grégoire 2019), extractive companies have also developed increasingly sophisticated strategies and multi-scalar instruments to obtain support or manage conflict around their projects (Hatcher 2020). Critical socio-legal scholars and criminologists argue that far from simply compensating for 'weak' jurisdictions, these strategies are often irremediably entangled with historical dynamics of exclusion: racism, patriarchy, counter-insurgency doctrines and colonialism (Pichler 2016; Roy Grégoire 2019).

The political economy of climate change and the 'new economy'

EIs have long been acknowledged as a main contributor to global warming (EIR 2003). The political clout of carbon-intensive industries nevertheless induces G20 governments to provide substantial public finance to coal mining and oil and gas production despite these governments international climate commitments. Both as a carbon-intensive industry and as an impediment to decisive climate action, extractivism deepens the inequalities and vulnerabilities already associated with climate change.

In keeping with its long-standing claims as a vector of progress and technological advance, extractivism has also found opportunities for accumulation in 'the green transition' – transport electrification and the development of more efficient batteries, for example, have sparked intense interest for minerals such as lithium, graphite and cobalt (World Bank 2020b). Emerging environmental justice literature suggests, however, that green extractivism fails to meet the scope and scale of politico-economical transformations necessary for human survival, while reproducing the same mechanisms of inequality and exclusion in extractive territories (Brand et al. 2020). The production of devastated landscapes is also intricately linked to the rise of the 'new economy': minerals that are necessary to the functioning of electronic devices, columbite-tantalite (coltan) for example, have long been procured in the midst of one of the worst humanitarian catastrophes in recent decades – the war-torn Democratic Republic of Congo. Other studies show that 84 per cent of platinum and 70 per cent of cobalt resources, considered key minerals for 'low-carbon energy technologies' are found in 'high-risk contexts' (Lèbre et al. 2020).

Overcoming these challenges demands a full consideration of the three interrelated dynamics outlined above and, according to radical democracy theorists, a recuperation of a form of democratic control over the reciprocal relationship between human societies and nature. For Pichler (2016), a radical democratic approach would:

> highlight the political and specifically democratic dimensions of self-determination and equality in the access to and control over nature and natural resources [,] the recognition of diverse modes of living, the inclusion in and the control over decision-making power and problem-solving [,] and the democratic allocation of responsibility to challenge the existing order, enlarge the public sphere, and only then enable the redistribution of nature and natural resources – among humans, social groups, and states.
>
> *(2016, 47)*

Practice and policy

The acute socio-environmental challenges linked to EIs have led to a wide range of multi-scalar policy initiatives and practices. However, it is relevant to recall that these are deployed within a larger framework that saw the retreat of the state from the sector and that has, in parallel, invested transnational corporations with unprecedented authority to mitigate the sector's mounting socio-environmental impacts.

While newer generations of mining regimes have enhanced socio-environmental impact assessments provisions, these regimes are nonetheless having equivocal consequences as private operators are being called upon to manage local politics, territorial organization, environmental protection, and access to justice (Campbell 2009; Szablowski 2007). Under this model of governance, the aspirations and grievances of local populations are to be resolved directly by operators through their Corporate Social Responsibility (CSR) instruments. In the process, however, these aspirations and grievances are often reconstructed in a manner that is consistent with the operator's interest. Revindications that are not compatible with the operators' interests tend to be stigmatized and subjected to political violence; the political logic that emerges from this governance model, the attributes of sovereignty are shared between states and transnational companies while self-determination and access to justice is rendered increasingly elusive (Roy Grégoire and Monzón 2017).

The surge of CSR stems from the mid-1990s when international civil society movements began to be successful in attracting global attention onto the stark socio-environmental impacts of large-scale extractive projects across the Global South (Dashwood 2011). In their attempts to secure a 'social licence to operate,' transnational corporations established a wide range of corporate voluntary guidelines and initiatives to address the sector's social, environmental, and human rights record. These global regimes rely mainly on informal law and as such, independent monitoring and public reporting remains voluntary and discretionary, leading to dubious results (see Dashwood 2011; Hatcher and Lander forthcoming). Hamann and Acutt (2003) further underline that these corporate practices often serve to pre-empt and/or preclude compliance with state-sponsored regulations and standards.

Stricter standards for the industry do exist, mainly those tied to international investment banks. The International Financial Corporation's (the World Bank's private arm) Performance Standards on Environmental and Social Sustainability, are arguably the most stringent in the industry, including Performance Standard 7, which in certain circumstances, can trigger the principle of Free, Prior, and Informed Consent (FPIC) for Indigenous Peoples. However, numerous case studies have again highlighted the limitations of these schemes, notably their failure to address the core problems tied to the sector's practices, including the confidentiality of contracts which is key to keeping taxation and royalty rates to a minimum, as well as the massive illicit flows that continue to characterize the sector (Campbell 2013; UNDP 2015). In parallel, global grievance mechanisms tied to the presence of international investors have also increasingly been recognized as important vehicle to report non-compliance to regulatory norms. But here again these transnational instruments often fail to deliver proper recourses, notably because they are characterized by serious asymmetries in accessing power and resources – money, information, technical expertise, and time – disadvantaging communities seeking redress (Altholz and Sullivan 2017; Simons and Macklin 2014).

Rethinking CSR: the African mining vision

In an attempt to challenge technocratic takes on CSR, novel initiatives have recently been pursued, most notably the ones proposed by the African Union and UNECA in 2009. The 'African Mining Vision' (AMV) emphasizes that CSR instrument 'should not be considered a substitute for government responsibility towards its citizens in providing basic infrastructure and other public goods' and further argued that it should rather 'complement government efforts through local government institutions and local authorities'. AMV is groundbreaking as in a decisive departure from mainstream approaches to CSR, it makes the case for an overarching approach to EIs policy that:

> The different types of frameworks should be considered part of a national policy debate on the mining industry's obligations regarding social development objectives. Without such debate, there is danger that the CSR requirements in a jurisdiction will be left to the industry to determine. This ad hoc approach can lead to uncertainty of how much should be spent on CSR and what types of CSR projects should be developed as well as the mechanisms for their development. Indicators around assessing the impact of good CSR projects must be built into the framework and applied by a range of stakeholders, such as civil society. The framework must focus on stakeholder consultation and allow for review of obligations and commitments. This review must be based on reporting requirements that should be part of the CSR framework.
>
> (AMV, 2011: 88-89)

In 2011, the United Nations Human Rights Council unanimously endorsed the Guiding Principles on Business and Human Rights (UN 2011), also known as the 'Ruggie Principles.' The latter, which were subsequently widely endorsed by governments and corporations, ostensibly do not create new obligations for states or corporations; they outline, rather, a hybrid governance framework that includes existing binding and non-binding norms for states and non-binding 'responsibilities' for corporate actors. In such light, they do little to overcome the limitations of the above-mentioned CSR instruments and regulation schemes; they also leave open the debate as to which policy instruments should be put in place by governments to regulate the harmful activities of corporations operating in other jurisdictions (Simons and Macklin

2014). The Ruggie Principles have thus been criticized for legitimizing ineffective private-led remedies and for delaying decisive governmental action (Coumans 2017). Under pressure from civil society organizations, nevertheless, some jurisdictions are moving forward with mandatory human rights due diligence laws to hold corporations accountable for their activities overseas, especially in the European Union (Hatcher and Roy Grégoire forthcoming). Other international schemes have been driven by civil society, notably the now well-established Extractive Industries Transparency Initiative, a voluntary initiative to encourage the transparency and disclosure of revenues in EIs (Klein 2017).

While the involvement of global actors – corporate, governmental, or non-profit – in EIs has seen the recent rise of multiple initiatives to tackle the sector's socio-environmental legacy, these schemes have often led to policies and practices that continue to give precedence to market-oriented norms. This points to the existence of a patent asymmetry in norm domestication whereby 'weaker' norms, the ones geared towards the protection of local communities, their livelihood, rights and environment, remain unevenly implemented (Hatcher 2020).

Pedagogy

A critical appraisal of the 'policy solutions' put forward with regards to EIs demands a higher degree of reflectiveness in knowledge production and knowledge mobilization. In 2007, for example, the United Nations adopted the Declaration on the Rights of Indigenous Peoples (UNDRIP), which for the first time enshrined the right of Indigenous Peoples to self-determination over their territories and natural resources in international law (UN 2007, 3, 4, 26.2, and 36.1). This direct challenge to state's unilateral sovereignty claims has not put an end to the debates around the regulation of EIs. These, rather, have switched towards competing conceptions and mechanisms for the implementation of the right to FPIC (UN 2007, 10, 11, 19, 28 and 29). In these ongoing struggles, most of the above-mentioned instruments and governance schemes are being mobilized, and new ones are being put forward. In this section we discuss how research and teaching can help assess these governance proposals and intervene in those debates.

As Indigenous legal scholars have argued, the practices of exclusion and marginalization discussed here point to deeper contradictions: on the one hand, they express the difficulty of the modern conception of sovereignty to accommodate alternate worldviews, development models or 'lifeworlds' (Ash et al. 2018); and on the other hand, as anthropologists have shown, they reflect the difficulty of euro-centric ontologies to acknowledge the relation of mutual dependency that exists between humans and the rest of nature (Sahlins 2008).

Both contradictions can be expressed as the ontological exclusion and violence of extractivism. Indeed, the difficult reconciliation of sovereignty with plural claims to self-determination and alternative development models is closely linked to the commodification of nature exacerbated in extractivism. As feminist and post-colonial literature has shown, the euro-centric ontological divide between nature and culture is accompanied by other binaries in constructing overarching ideological structures that justify violent hierarchies by ascribing citizenship, humanness, morality, desirability, reason, etc. (Ashcroft et al. 2008, 1–18).

As long as dominant societies refuse to apprehend nature as something else than a source of rent, alternative ontologies will be met with exclusion and violence. Critical research in the field of EIs thus implies a form of co-production of knowledge with affected communities. Indigenous and de-colonizing pedagogical tools, such as Smith's classic '*Decolonizing Methodologies: Research and Indigenous Peoples*' (2012), written for both Indigenous and non-Indigenous audiences, are useful starting points.

Transnational solidarity networks and their multi-level impact on governance are another important area of fieldwork, one that is linked to extractivism's impact on the reconfiguration of political subjectivities and political belonging in extractive territories.

Lastly, as climate change obliges societies around the globe to rethink their relationship with nature and to reconcile their economies with the biophysical limits of planet Earth, practical investigations into transitioning out of the current political economy are much needed. To that end, pedagogy should help illuminate the mutual entanglement of our current political structures and extractivism, an entanglement that make those changes difficult.

Key sources

1. Asch M, J Borrows and J Tully (eds.) (2018) *Resurgence and Reconciliation: Indigenous-Settler Relations and Earth Teachings*, Toronto, Buffalo, London: University of Toronto Press.

A delicate exploration of the ontological underpinnings of extractive conflicts and ways to overcome them. The discussion is centred around the colonial context, but it is relevant to the global extractive contexts.

2. Campbell B (ed.) (2013) *Modes of Governance and Revenue Flows in African Mining.* Basingstoke, Hampshire: Palgrave Macmillan.

A political economy analysis of mining and revenues in Africa based on cases studies from Ghana, Mali, and the Democratic Republic of the Congo. It offers perspectives from academics, policymakers, and practitioners from Africa and beyond.

3. Katz-Rosene R and M Paterson (2018) *Thinking Ecologically About the Global Political Economy,* London: Routledge.

A guide to the necessary re-conceptualization of the international political economy. The book offers avenues out of global capitalism's reliance on extractivism by re-integrating socio-ecological processes into how we think about trade, finance, production, interstate competition, globalization and inequalities.

4. Sawyer S and Gomez ET (eds.) (2012) *The Politics of Resource Extraction: Indigenous Peoples, Multinational Capital, and the State.* Basingstoke: Palgrave Macmillan.

This edited piece on Indigenous Peoples and extractivism examines mega resource extraction projects in Australia, Bolivia, Canada, Chad, Cameroon, India, Nigeria, Peru and the Philippines. It offers an overview of the intricacies of a multi-level governance system involving Indigenous communities, multinational corporations, and domestic and multilateral government institutions. It also illuminates opportunities for efficient activism and policy intervention.

5. Simons P and Macklin A (2014) *The Governance Gap: Extractive Industries, Human Rights, and the Home State Advantage.* London: Routledge.

A comprehensive and critical discussion of governance instruments in the Business and Human rights field. The authors convincingly argue that for voluntary measures to effectively prevent human rights abuses, they must be backed by coercive normative instruments.

References

Altholz, R and Sullivan, C (2017), 'Accountability & International Financial 99 Institutions: Community perspectives on the World Bank's Office of the 100 Compliance Advisor Ombudsman', *Accountability & International Financial Institutions.*

Asch M, J Borrows and J Tully (eds) (2018), *Resurgence and Reconciliation: Indigenous-Settler Relations and Earth Teachings*, University of Toronto Press, Toronto.

Ashcroft B, Griffiths G and Tiffin H (eds) (2008), *Post-Colonial Studies. The Key Concepts* 2nd ed., Routledge, New York.

Brand U, Görg C and Wissen M (2020), 'Overcoming neoliberal globalization: social-ecological transformation from a Polanyian perspective and beyond', *Globalizations,* vol. 17, no. 1, pp. 161–76.

Burchardt H-J and Dietz K (2014), '(Neo-)extractivism – a new challenge for development theory from Latin America', *Third World Quarterly*, vol. 35, no. 3, pp. 468–86.

Campbell B (ed.) (2009), *Mining in Africa*, Pluto Press, London, Ottawa.

Campbell B (ed.) (2013), *Modes of Governance and Revenue Flows in African Mining*, Palgrave Macmillan, Basingstoke, Hampshire.

Campbell B and Hatcher P (2019), 'Neoliberal reform, contestation and relations of power in mining: Observations from Guinea and Mongolia', *The extractive industries and society*, vol. 6, no. 3, pp. 642–53.

Coumans C (2017), 'Do no harm? Mining industry responses to the responsibility to respect human rights', *Canadian Journal of Development Studies / Revue canadienne d'études du développement*, vol. 38, no. 2, pp. 272–90.

Dashwood H S (2011), 'Sustainable development norms and CSR in the global mining sector', Palgrave Macmillan, New York, pp. 31–46.

EIR (Extractive Industries Review) (2003), *Striking a Better Balance. Final Report of the Extractive Industries Review. Volume 1: The World Bank Group and Extractive Industries*, Jakarta; Washington D.C, Viewed 28 July 2020, http://documents1.worldbank.org/curated/en/222871468331889018/pdf/842860v1 0WP0St00Box382152B00PUBLIC0.pdf.

Gilbert J (2018), 'Sovereignty, Self-Determination, and Natural Resources: Reclaiming Peoples' Rights', *Natural Resources and Human Rights: An Appraisal*, Oxford University Press, Oxford, pp. 13–33.

Global Witness (2020), *Defending Tomorrow. The Climate Crisis and Threats Against Land and Environmental Defenders*, London, Brussels, Washington DC: Global Witness, Viewed 28 July 2020, www.globalwitness.org/documents/19939/Defending_Tomorrow_EN_low_res_-_July_2020.pdf.

Gudynas E (2015), *Extractivismos: Ecología, Economía y Política de Un Modo de Entender El Desarrollo y La Naturaleza*, Cochabamba: CEDIB/CLAES.

Hamann R and N Acutt (2003), 'How should civil society (and the government) respond to 'corporate social responsibility'? A critique of business motivations and the potential for partnerships', *Development Southern Africa*, vol. 20, no. 2, pp. 255–270.

Harvey D (2004), 'The 'New' Imperialism: Accumulation by Dispossession', *Socialist Register*, vol. 40, pp. 63–87.

Hatcher P (2020), 'Global Norm Domestication and Selective Compliance: the case of Mongolia's Oyu Tolgoi Mine', *Environmental Policy and Governance*, vol. 30, pp. 252–262.

Hatcher P and J Lander (forthcoming) 'Negotiating Political Spaces on the 'Final Frontier': Extractive Development and Transformations in the Practice and Governance of Citizenship in Mongolia', *Journal of Contemporary Asia.*

Hatcher P and Roy Grégoire E (forthcoming) 'Governance of Extractive Industries', *Elgar Handbook on Governance and Development*, Edward Elgar Publishing: Cheltenham.

Klein, A (2017), 'Pioneering extractive sector transparency: A PWYP perspective on 15 Years of EITI', *The Extractive Industries and Society*, vol. 4, no. 4, pp. 771–4.

Lèbre, É, Stringer, M, Svobodova, K, Owen, JR, Kemp, D, Côte, C, Arratia-Solar, A and Valenta, RK (2020), 'The social and environmental complexities of extracting energy transition metals', *Nature Communications*, vol. 11, no. 1, pp. 1–8.

Pichler M (2016), 'What's democracy got to do with it? A political ecology perspective on socio- ecological justice', in Pichler M, Staritz C, Küblböck K, Plank C, Raza W, Ruiz Peyré F (eds) (2017) *Fairness and Justice in Natural Resource Politics*, Routledge, London.

Rosser A (2006), *The political Economy of the Resource Curse: A Literature Survey*, Institute of Development Studies, Brighton.

Roy Grégoire E (2019), 'Dialogue as racism? The promotion of 'Canadian dialogue' in Guatemala's extractive sector', *The Extractive Industries and Society*, vol. 6, no. 3, pp. 688–701.

Roy Grégoire E and Monzón LM (2017), 'Institutionalising CSR in Colombia's extractive sector: disciplining society, destabilising enforcement?', *Canadian Journal of Development Studies / Revue canadienne d'études du développement,* vol. 38, no. 2, pp. 253–71.

Sachs JD and AM Warner (1995), *Natural resource abundance and economic growth*, National Bureau of Economic Research Working Paper, Cambridge.

Sahlins M (2008), *The Western Illusion of Human Nature: With Reflections on the Long History of Hierarchy, Equality, and the Sublimation of Anarchy in the West, and Comparative Notes on Other Conceptions of the Human Condition*, Prickly Paradigm Press, Chicago.

Scheidel, A., Del Bene, D., Liu, J., Navas, G., Mingorría, S., Demaria, F., Avila, S., Roy, B., Ertör, I., Temper, L. and Martínez-Alier, J (2020), 'Environmental conflicts and defenders: A global overview', *Global Environmental Change*, vol. 63, pp. 1–12.

Simons P and Macklin A (2014), *The Governance Gap: Extractive Industries, Human Rights, and the Home State Advantage*, Taylor & Francis Group, London.

Smith LT (2012), *Decolonizing Methodologies: Research and Indigenous Peoples*, Zed Books, London.

Svampa M (2015), 'Commodities Consensus: Neoextractivism and Enclosure of the Commons in Latin America', *South Atlantic Quarterly*, vol. 114, no. 1, pp. 65–82.

Szablowski D (2007), *Transnational Law and Local Struggles: Mining Communities and the World Bank*, Hart Publishing, Oxford; Portland.

UN (United Nations) (2007), *United Nations Declaration on the Rights of Indigenous Peoples*, A/RES/61/295, New York: United Nations, General Assembly, www.un.org/Docs/journal/asp/ws.asp?m=A/RES/61/295

UN (2011), *Guiding Principles on Business and Human Rights: Implementing the United Nations 'Protect, Respect and Remedy' Framework*, HR/PUB/11/9. New York: United Nations, Viewed 11 August 2017, www.ohchr.org/documents/publications/guidingprinciplesbusinesshr_en.pdf (accessed 11 August 2017).

UNDP (United Nations Development Program) (2015), *Illicit Financial Flows from the Least Developed Countries 1990–2008*, Discussion Paper. United Nations, United Nations Development Program.

UNECA (United nations Economic Commission for Africa) (2017), *Impact of Illicit Financial Flows on Domestic Resource Mobilization*, Addis Ababa: UNECA Conference of Ministers of Finance, Planning and Economic Development. Available at: www.uneca.org/sites/default/files/PublicationFiles/impact_of_illicit_financial_flows_-_april_2017_-_web.pdf

World Bank (2020a), Extractive Industries, Viewed 6 November 2020, www.worldbank.org/en/topic/extractiveindustries

World Bank (2020b), Climate-Smart Mining. Minerals for Climate Action: The Mineral Intensity of the Clean Energy Transition, Viewed 24 September 2020, http://pubdocs.worldbank.org/en/9617115 88875536384/Minerals-for-Climate-Action-The-Mineral-Intensity-of-the-Clean-Energy-Transition.pdf.

30

SPATIAL INEQUALITY AND DEVELOPMENT

Edo Andriesse and Kristian Saguin

Introduction

Despite the rise of the middle class in many countries across the world, inequality remains a pressing global problem. It is unclear when the majority of Global South countries will approach average living standards enjoyed in the Global North, and within countries, economic inequality has seen enduring poverty, social unrest and political extremism (Stiglitz 2013; Piketty 2014; Wuthnow 2018). Due to increasing disparities between average and median incomes, the middle class in many countries face increasing financial pressures.[1] As Oxfam International (2021) note:

> The world's ten richest men have seen their combined wealth increase by half a trillion dollars since the pandemic began – more than enough to pay for a COVID-19 vaccine for everyone and to ensure no one is pushed into poverty by the pandemic.

Overall, it has become clear that coupling economic growth with sustained reductions in poverty and inequality is a daunting task. Inequality remains an important development challenge for the 21st century, and has multiple dimensions including inequalities in access to land, health, education, and opportunities to participate in decision making, among others. In this chapter, we focus on spatial inequality and socioeconomic disparities within countries, with special reference to South, East and Southeast Asia. This part of the world has been lauded for its economic performance and remarkable reductions in absolute poverty. Notable 'success stories' include Japan, the four Asian Tigers of South Korea, Taiwan, Singapore and Hong Kong, several Southeast Asian countries, Sri Lanka, the Maldives, and (since the 1990s) the powerhouses of China and India.

Income inequality *can* decline. As Wan and Wang (2018, 32) note, 'for Asia as a whole, the Gini index increased from 38.43 per cent in 1965 to 42.80 per cent in 2006 before declining in recent years.' Similarly, in his research on inequality trends in East Asia and Latin America in the 2000s, Stewart (2016, 70) also notes that 'rising inequality is not inevitable even in the context of increasing globalization and competitive pressures' and that 'while there are strong tendencies making for rising inequality in a globalizing capitalist world, governments can counter this by well-designed policies.' Nevertheless, structural transformation from agriculture

DOI: 10.4324/9781003017653-34

to manufacturing and services accompanied by massive processes of rural-urban migration as well as prioritizing urban areas in public policy (so-called urban bias) has culminated in profound spatial disparities between core cities and peripheral regions. China and India have several core areas of high economic productivity (e.g. Shanghai, Yangtze River Delta megalopolis, Guangzhou, Pearl River Delta megalopolis, Mumbai, Bengaluru) but inequalities between urban and rural settings remain high (Krishna 2017; Yang 2020). Other countries face a disproportionate share of economic activity in just one core city with insufficient attention being paid to secondary cities and continued hardships in the countryside (Rigg 2012). In the following section, we provide an overview of theories and approaches to within-country spatial inequality. This is followed by a discussion of policy and practice in Southeast Asia before turning attention to potential pedagogical methods in teaching spatial inequality.

An overview of within-country spatial inequality

Core-periphery relations

A suitable starting point to discuss spatial inequality is the concept of circular and cumulative causation, advanced by Gunnar Myrdal (1957). This concept can be defined as a process whereby 'once particular regions have by virtue of some initial advantage moved ahead of others, new increments of activity and growth will tend to be concentrated in already expanding regions because of their derived advantages, rather than in other areas of the country' (cited in Binns 2014, 102). This explains why many contemporary core areas in the Global South are situated in areas that were already the most developed – in terms of the availability of skilled labour, technology, infrastructure and connectivity, cosmopolitan mindset, and finance. Good examples are Rio de Janeiro, Nairobi, and Jakarta. In order to contribute to nation-building and promote economic development various countries have also built new capital and national administrative cities, such as Brasilia in Brazil, Abuja in Nigeria, and Putrajaya in Malaysia. Before the start of the COVID-19 pandemic, Indonesia had plans to build a brand-new capital city in Kalimantan, but it is not certain if this plan will go ahead.

Academics and policymakers continue to disagree about the implications – indeed, even the presence – of circular and cumulative causation. Does economic activity in core areas automatically trickle down to semi-peripheral and eventually even peripheral areas without substantial government intervention? Some like Albert Hirschman (1958) have contended it does, as demand for products and services moves to semi-peripheral areas once a certain saturation point has been reached in core areas, similar to the spread effects as formulated by Myrdal. Others, however, stress that government intervention remains necessary to support the weakest areas in a country. Higgens et al. (2010), Stiglitz (2013), Berdegué et al. (2015) and others claim that the spread effects are insufficient to attain spatially balanced development and to significantly reduce spatial inequality.

For Myrdal (1957, 13), national economies do not, by themselves, tend to move to equilibria and convergence of regional endogenous capacities, as is widely predicted within neoliberal economic models. In fact, he stressed the possibility of backwash effects. These pertain to all the assets that peripheral regions lose to the core regions within the process of circular and cumulative causation (Binns 2014, p. 102; Andriesse and Kang 2017). Investors prefer to open businesses in attractive cities and not in rural, peripheral, difficult-to-reach places. Consequently, the most skilled people migrate from peripheral to core areas where they can benefit from higher wages, more opportunities, and better amenities; more popularly termed a brain drain. These trends can certainly be observed in Asia. Spatial inequality has been an important component of overall inequality in both larger and smaller countries including China, India, Vietnam and Bhutan. In these four countries spatial inequality accounted for more than 30 per cent of overall inequality (Kanbur et al. 2014, 53).[2] Based on an analysis of incomes, human development, and infrastructure in India, Jose (2019, 8) concluded that regional inequality is 'alarming' and should be scrutinized at multiple scales, noting how 'not only interstate disparities but also intrastate disparities are on the rise.'

To counter backwash effects, create new spread effects, and support non-core areas, economists and geographers have explored different spatial strategies and policy interventions. For example, in the 1950s French economists studied spatial dimensions of economic development. The most famous was Francois Perroux (1955) who wrote about the need to establish new growth poles. He defined growth poles as centres (poles or foci) from which centrifugal forces emanate and to which centripetal forces are attracted. Each centre being a centre of attraction and repulsion has its proper field which is set in the field of all other centres (Andriesse and Kang 2017). Similarly, Parr (1999) argued that growth-pole strategies were employed to achieve four different but interrelated objectives: reviving depressed areas, encouraging regional de-concentration, modifying the national urban system and attaining inter-regional balance. The newly built capital and administrative capitals mentioned above, for example, can be considered as nodes designed to modify the national urban system. Successes have, however, been mixed. It is hard to defy the logic of circular and cumulative causation. Traditional commercial centres are highly inventive in reproducing their competitive edge through innovation.

Also important for examining spatial inequality is the work of economic geographers focusing on institutions, institutional change, and regional economic evolution (Clark et al. 2018). One of the main objectives of economic geography is to explain why some regions outperform others. Many economic geographers have focused on the role of formal and informal institutions (written and unwritten rules of the game), as well as path dependency, regional cultures of knowledge production and innovation seeking, and the changing nature of the relationships between firms, government, universities, and civil society actors in regional economies. For regions to perform better, it is argued that institutions should be as enabling as possible and actors should have enough freedom and flexibility to bring about process and product innovation so as to continuously adapt to the changing global competitive environment. These lines of inquiry make much sense for urban metropolises in the Global South that aim to reduce gaps in innovative capacity relative to the Global North. For example, due to innovative industries like information technology, cities such as Hyderabad and Bengaluru in India are developing as innovation hubs similar to well-established areas in the Global North, like Silicon Valley. However, these developments simultaneously widen socioeconomic inequalities within countries as hinterlands fail to benefit from growth in core areas. In other words, in the absence of spread effects, processes of uneven development are likely to intensify. This fits the description of India as a 'fast-growing, yet largely agrarian, developing economy' (Krishna 2017).

Thus, while this body of knowledge has significant explanatory power in regions presently seeking to benefit from the ongoing fourth industrial revolution, it is less adequate in peripheral areas predominantly focused on agriculture in the Global South; areas that usually see very few spread effects (see for some recent changes Song et al. 2020). Here, the work of development geographers appears to be somewhat more relevant as the primary foci of analysis have been poverty reduction, addressing environmental pressures (climate change impacts, air and water pollution), gender imbalances, empowering the marginalized such as indigenous communities, and indeed improving the inter-regional balance as proposed by Myrdal and Perroux.

An example of work in this vein investigates how marginalized communities improve their well-being by identifying local endogenous capabilities and reducing exposure to shocks such as commodity price busts or natural disasters (see Miller 2021). Nevertheless, whereas peripheral areas might be able to improve living standards, there is no guarantee that within-country socioeconomic inequalities will decline. When core areas grow faster than the rest of the country, peripheral areas do not catch up to the national average; indeed, they fall even further behind. Thus, inter-regional balance remains extremely hard to achieve, and pro-poor growth does not necessarily reduce spatial inequality. Whereas secondary cities and tourist hotspots in Thailand have grown significantly since the 1990s, for instance, the Bangkok Metropolitan Region accounts for approximately 40 per cent of Thailand's gross domestic product, and firms and investors rarely consider any alternatives outside this region. Before moving to policy and practice, the remainder of this section discusses spatial targeting for poverty reduction as researched by development geography and cognate disciplines.

Spatial targeting for poverty reduction

It is a widely held view that 'the rise in spatial inequality [in Asia and the Pacific] is not a reason to reverse openness and technological progress, or stop the reform process, but rather to reorient infrastructure investment to lagging regions, and to remove barriers to migration to the fast-growing regions' (Kanbur et al. 2014, 53). Yet, such views are also deeply contested. Some key questions include: Should scarce funds be allocated to lagging regions without meaningful prospects of development? Does it make more sense to stimulate employment in lagging regions or facilitate labour migration to core areas? Does the proliferation of slums in the Global South constitute evidence that rural-urban migration has merely relocated, rather than addressed, poverty?

One common means to address poverty has been to provide cash transfers to poor households. Such transfers can be spatially targeted. Evidence from Brazil has shown that spatially blind conditional cash transfers to the poorest households during the presidency of Lula da Silva have contributed to rapid poverty reduction, with the percentage of people living below 5.50 USD decreasing from 41.7 per cent in 2003 to 17.1 per cent in 2014. But conditional cash transfers do not guarantee long-term success. They do little to ensure long-term productive investments in agriculture, manufacturing, and services, and often do not sufficiently improve livelihood stability. Between 2015 and 2018 the Brazilian poverty rate increased, reflecting trends seen in other countries. During disasters and pandemics many households fall back into poverty and resort to dramatic measures such as returning to their ancestral, rural villages or cutting down on nutritious (but expensive) food. Therefore, some sort of spatial targeting is still warranted to foster longer-term spatial stability. This is particularly relevant for localities that are locked into a spatial poverty trap: where there are few opportunities to reduce poverty due to a bad location (slum or periphery) and high levels of chronic poverty (many households are not able to escape poverty for a long time).

Higgens et al. (2010, 22) propose that universal or sectoral policies should be matched by area-based approaches that emphasize context specificity, which entails 'a layering of policy instruments, with interventions and policies seeking to address the needs of specific life-cycle groups…overlaid with those focusing on the needs of other categories or groups (e.g. pastoralists, small farmers, microentrepreneurs, slum dwellers, widows, orphans, people living with HIV-AIDS, conflict-affected people…)' (see Berdegué et al. 2015 for the Latin American context). With respect to slum dwellers, one could think of practical issues like housing conditions, flood prevention, and waste management, but also blending these with notions and agendas of reducing spatial injustice (Soja 2013). In rural areas, widening geographical acting spaces of households by external actors is necessary in case villages are not able to provide sufficient livelihood stability, increase social capital and trust, and improve opportunities for marginalized groups (Pain and Hansen 2019, 50–58).[3]

As poverty is predominantly a rural phenomenon, it makes sense to focus on three specific geographical landscapes. First, many mountainous and remote areas in the Global South, often inhabited by indigenous communities, experience high levels of poverty. Due to ethnic and linguistic differences, such communities often face, if not overt discrimination, at least various forms of social exclusion on the part of both the majority populations and by national government and its agencies. These differences make it hard for such populations to benefit from employment opportunities outside their home villages. In these cases, facilitating labour migration to core areas is not likely to work without first improving levels of human capital (learning a new language, skills formation, etc.). For example, in the hills of West Java many residents find it hard to land decent well-paying jobs because they are not fluent in Bahasa Indonesia. In remote villages, Sundanese remains the common language at elementary school. Only when children attend middle school are they more fully exposed to Bahasa Indonesia.

Besides the persistence of inequalities between mountainous and core areas, another challenge is how to manage capitalist development in the uplands. Li (2014) reveals how the infiltration of capitalist arrangements into the uplands in Sulawesi, Indonesia, has led to new forms of intra-village inequality as well as the erosion of traditional ways of living. The regional economy of Sulawesi might have grown but many upland people have lost their livelihoods and their lives are now more precarious than before. Another layer of uncertainty is continuing environmental pressures (such as landslides, erosion, deforestation, droughts) and natural hazards (volcanic eruptions).

A second type of distinctive landscape is valleys and plains in the semi-periphery or periphery. Population densities are higher than in the mountains, there are more opportunities to make a living – in farming, manufacturing, and services – yet there is a great deal of variety in living standards. Smallholders in irrigated farming areas tend to do well while people practicing dryland agriculture suffer from droughts and floods. Some farmers own their land, others work as landless labourers whether on the land of smallholders or large corporate plantations. Many households have relatives working and living in large urban centres, sometimes abroad. In countries like Malaysia and Thailand, ageing has started to transform rural, semi-peripheral areas in significant ways. Senior citizens increasingly tend to the farm while people in their thirties and forties work in Penang, Kuala Lumpur, Bangkok, other cities, and abroad. Furthermore, livelihoods trajectories often are the result of changes in the immediate circle around households rather than the result of socioeconomic planning at the national level (Rigg 2012). Weddings, divorces, sickness, or having a bright child landing a white-collar job at a bank have a more profound impact on individual households than do national policies that may or may not be implemented effectively.

The third type of landscape with distinctive features is coastal areas. Those close to urban areas may have relatively high population densities; those more distant may be less densely settled but still attract migrants from inland areas. While a coastal location may bestow some locational benefits, they can also face multiple stresses: land degradation, coastal erosion, salinization, overfishing, illegal fishing, unemployment, typhoons, and persistent. Reducing inequality – relative to the average living standards of farmers in valleys for instance – has proven to be a daunting task. A significant complication is that fishing-based households occupy a marginal position within national political economies (Andriesse et al. 2020). This group is smaller in number than farmers and is much less influence in shaping policies concerning nation-building, food security, and national prosperity. The rice farmer has been an important part of the national identity of many countries in East and Southeast Asia, receiving attention during election campaigns, and protection at times of crises. But fisherfolk have received relatively little support from politicians and policymakers. Even in archipelagic Southeast Asia, the political economic position of coastal communities has remained marginal.

Policy and practice

Since the 2007–2008 Asian financial crisis, international organizations have started to pay more attention to the negative consequences of capitalism, including spatial inequality. For example, the Asian Development Bank has started to support interventions to assist lagging regions such as improving regional connectivity, developing new growth poles, strengthening fiscal transfers and removing barriers to migration to more prosperous areas (Kanbur et al. 2014). Nevertheless, the discussion above shows that reducing inequality is difficult, with differing opinions on the future trends of spatial inequalities ranging from optimistic (Wan and Wang 2018, 32) to pessimistic (Walker 2008; Apostolopoulou 2020). This section uses the Philippines as a case study to discuss decentralization – an important variable in the spatial targeting of wealth and resources.

The effectiveness of spatial targeting depends to a significant extent on who has the power to decide and who has the power to spend money. In each country, centre-local relations are different. In some, it is the national government; in others, local government agencies have relatively more influence. Also, de-facto political influence and financial strength often matters more than the official system as codified in the constitution. For example, in some federations the federal government is more influential than one would expect based on the constitution, whereas in others local and provincial agencies exercise considerable influence. Centre-local relations can change over time, thereby also changing the options available to reduce spatial inequality and improve the inter-regional balance.

Looney (2020, 157), in an insightful comparison of rural mobilization and development in China, Taiwan, and South Korea concluded that not only are formal institutions important, but also the degree of autonomy of local stakeholders to get things done. Successes can be achieved if there is 'central-local government coherence, controlled decentralization,' and 'an implementing coalition that includes the intended beneficiaries of the policy.' As China aims to be a fully modernized country by 2035, the urban-rural divide needs to be addressed as well (Yang 2020; Wei 2017). Similar to Higgens et al. (2010), Yang advocates for a policy mix of universal coverage of certain social security policies and more targeted efforts to support peripheral areas. A strong focus on facilitating rural-urban migration will be ineffective in addressing those aspects of development that are spatially situated in rural areas.

Post-Suharto Indonesia is another interesting case. The country embarked on a big bang decentralization in 1999. But it turned out to be too loose with too many opportunities for local elites to enrich themselves, so the national government partially recentralized certain laws and regulation (most notably to prevent environmental degradation) and partially further decentralized (setting up a village fund). This recent recalibration appears to have had positive outcomes: 'Although most autonomous regions still have high dependency on central government, they have improved the regional development indicators gradually. Particularly, regional disparity has started to decline since 2010, and local government proliferation seems to be more controllable due to the stricter procedures' (Talitha et al. 2020, 705). Nevertheless, for a country of approximately 6,000 inhabited islands spanning a distance equivalent of London to Doha, it is not easy to find the optimal level of decentralization.

The Philippine experience with decentralization is similarly instructive as a country characterized by a colonially inherited centralized state power persisting amid the historical primacy of its capital region – Metro Manila – and the marginality of much of its archipelagic periphery. The country embarked on one of the first decentralization projects in the Global South after the shift from a decades-long authoritarian regime. Through the passage of the Local Government Code in 1991, the national government transferred certain services and functions to local government units at the provincial, municipal/city and *barangay* (village) levels. Recent proposals have aimed to deepen local autonomy and further encourage development in peripheral regions through a shift towards a federal system.

This decentralization was accompanied by an ambitious industrial de-concentration project that sought to spur growth away from Metro Manila to its adjacent regions. The result has been increased economic activity outside of the capital, but collectively these adjacent regions continue to receive the bulk of investments, producing nearly two-thirds of national GDP. It has also led to dramatic social and environmental transformations, such as increased migration and population growth, urban sprawl, conversion of productive agricultural lands, displacement of farmers, fisherfolk and indigenous groups, and environmental degradation (Saguin 2021). An expanded mega-urban corridor centred around Metro Manila has emerged, containing master-planned cities and private-led developments that serve as hubs connected by vital infrastructure networks. Relying on agglomeration economies, these developments reinforce an urban-centred regional growth approach often promoted through the development of special economic zones (SEZs) and similar strategies (Ortega et al. 2015).

The country's spatial strategy, emphasized in the most recent medium-term national development plan, continues to recognize cities as primary drivers of economic growth that need to be linked to lagging peripheral regions through improved connectivity. The urban-centred growth approach promotes regional development in a manner similar to a hub-and-spoke model, where enhancing urban-rural linkages becomes key to addressing spatial inequality. In practice, however, backward linkages to the surrounding hinterlands continue to be limited, and their development potential depend on the type of economic sector that dominates, such as for example whether manufacturing or services (business process outsourcing, tourism) are the primary drivers of growth in regional centres (Balisacan et al. 2008; Clausen 2010). With a holistic approach absent, such a strategy also fails to account for other fundamental issues, such as the ongoing conflict in the southern island of Mindanao where intra-regional inequality is most pronounced.

Studies that assess the impacts of decentralization policies in reducing spatial inequality and bringing development in the Philippines paint a mixed picture (Akita and Pagulayan 2014). While empowering communities to deliver better and more contextually-specific and locally-responsive services, there has been little correlation with overall trends in reducing spatial inequality. Local government units remain marginal players in delivering economic development to their respective localities and often lack resources to adequately provide for basic public services: two conditions that are compounded by incoherent inter-level governance coordination (Balisacan et al. 2008). Issues of poor agricultural productivity, infrastructural deficits in a fragmented archipelago, limited investments in human capital and the concentrated power of local elites in peripheral areas constrain endogenous development in poorer regions (Clausen 2010; Andriesse 2017). In the case of the Philippines, decentralization alone as a policy to reduce spatial inequality and distribute benefits to peripheral areas has been insufficient and incomplete.

Pedagogy/teaching spatial inequality: mapping and affect

The spatial component of inequality lends itself to a variety of pedagogical questions and techniques to enhance student learning about their complex characteristics. For example, it highlights questions of representations of space and patterns in space that make maps particularly useful visual tools in teaching. Maps and other forms of geo-visualization can be used to demonstrate inequality across various scales and serve as a starting point to initiate conversations about the roots, effects, and solutions to inequality.

Techniques such as geographic information system (GIS) become an accessible and adaptable tool for illustrating and engaging with the spaces of inequality. Mapping through these techniques aids cross-country and inter-regional comparison of the spatiality and temporality of development and inequality. These provide not only visual tools for demonstrating inequality in a classroom setting but also create hands-on opportunities for students to directly engage with data from the ground. Through laboratory exercises, several dimensions of inequality may be identified, mapped and overlaid, including individual economic and development indicators, or aggregated indices. GIS also allows users to perform methods of spatial analysis with varying degree of complexity to illustrate patterns and connections not easily visible through simple thematic mapping.

The availability of volunteered geographic information (VGI) in the geoweb may also be harnessed as a visual methodology for pedagogical purposes in creative ways. For example, geotagged tweets or images posted through social media may be analysed and mapped to show alternative patterns of inequality apart from those collected through census and other more conventional methods. These spatialized data may then be visualized through techniques such as cartograms and other creative forms of 'situated mapping' that go beyond representations of the absolute spaces of inequality (Jung and Anderson 2017).

Various types of maps may be produced in class as a student learning activity, but these may also be co-produced with communities if such partnerships exist or are possible to forge. This can enrich service learning and development-related field activities beyond just visiting communities to see and encounter inequality first-hand. For example, university students and faculty in the Philippines have partnered with indigenous, farmer and urban poor groups in a series of 'counter-mapping' activities that aim to co-produce maps for communities campaigning against forms of development-related dispossession (Ortega et al. 2017). Apart from the practical use of maps that through geospatial, narrative, and performative techniques produce alternative representations of space, these activities aim to promote mutual learning about processes that drive inequality and their spatialized manifestations.

Network mapping may also be a useful learning exercise in illustrating the interconnections across space of various actors, places and relations, and their implications for local development. For example, activities that involve tracing value chains and global production networks enable students to stitch together the stories behind products and commodities that they encounter on a daily basis. Identifying governance and power relations within these chains and networks also provide a window into the points for intervention (through for example, upgrading) to improve the well-being, poverty reduction and development potentials of participation in particular value chains and production networks.

Inequality also holds a strong affective power that may be harnessed for pedagogy. Highlighting the emotional, embodied, and sensory experiences of inequality through various visual and literary media, field encounters and self-reflection articulates forms of empathy and an ethics of care that transcends space. More than just the feeling of suffering, tapping into the emotional geographies of development involves acknowledging how complex emotions and messy experiences of inequality may connect at a deeper level with people (Wright 2012). This focus on affective dimensions of what it means to experience development presents a different way to engage with spatial inequality beyond the visual as represented by mapping.

The persistence of inequality continues to be a major challenge in development planning across the world. This chapter presented an overview of the diverse approaches to understanding and addressing spatial inequality in their analytical, policy and pedagogical diversity. Based on lessons from the experiences of dynamic regions in Asia and elsewhere, there is no one-size-fits-all model or strategy to spurring development and reducing spatial inequality. However, it can be argued that policies and programmes to target specific areas or regions for development will not succeed without a holistic approach that incorporates the multiple dimensions of inequality and the contextual particularities of places that contribute to the uneven geographies of well-being.

Key sources

1. Miller, M. (2021), *Economic Development at the Community Level. Creating Local Wealth and Resilience in Developing Countries*. Abingdon and New York: Routledge.

This book provides a good overview of the magnitude of options local rural and urban communities have in order to improve living standards in the Global South.

2. Kanbur, R.; Rhee, C. and Zhuang, J. (eds.) (2014), *Inequality in Asia and the Pacific: trends, drivers, and policy implications*. Mandaluyong City and Abingdon: ADB and Routledge.

This book offers an insightful analysis of mostly economic inequalities in Asia. It can be used as a starting point from which to deepen investigations on spatial and other inequalities.

3. Krishna, A. (2017), *The Broken Ladder. The Paradox and Potential of India's One-Billion*. Cambridge: Cambridge University Press.

This book explains how it is possible that India is an emerging, fast-growing economy, yet simultaneously a country where millions of people essentially remain rural and impoverished.

4. Looney, K (2020), *Mobilizing for Development. The Modernization of Rural East Asia*. Ithaca and London: Cornell University Press.

This book compares rural modernization in China, Korea, and Taiwan, combining a focus on enabling institutions and state campaigns. As such it contributes significantly to explaining the relatively lower inequalities as experienced in East Asia.

5. Soja, E. W. (2013), *Seeking spatial justice*. Minneapolis: University of Minnesota Press.

This book applies a spatial approach to questions of urban inequality and strategies toward justice. Its focus on labour-community coalitions in Los Angeles would also be a fruitful perspective for analyses of addressing urban inequality in the Global South.

Notes

1 For example, in the Philippines the 2015 average family income was 267,000 Philippine Pesos, while 58.4 per cent of families earned less than 228,480 Pesos in 2017. And the percentage of families belonging to the (near) poor has remained quite stable implying that the relative size of the middle class has not expanded despites periods of economic growth (Albert et al. 2018).
2 In addition to spatial inequality Kanbur et al. (2014, 42) identify 'shifts in income distribution between skilled and unskilled labour [by examining returns to human capital and the skill premium] and between labour and capital [by analysing labour and capital income shares]' as channels through which income inequality can be measured. In turn these channels are driven by technological progress, globalization, and market-oriented reform.
3 Geographical acting spaces refer to the daily socio-spatial reach of households (Pain and Hansen 2019). Do they do have everything in their village? Do they have many contacts with the outside world in cities or even abroad? Generally, people in the most peripheral provinces and indigenous communities in mountainous areas have limited geographical acting spaces, but due to information and communication technologies, things are changing.

References

Albert, J., Santos, A. and Vizmanos, J. (2018), *Defining and profiling the middle class. Policy Notes No. 2018–18*. Manila: Philippine Institute for Development Studies. https://pidswebs.pids.gov.ph/CDN/PUBLICATIONS/pidspn1818.pdf

Akita, T., and Pagulayan, M. S. (2014), Structural Changes and Interregional Income Inequality in the Philippines, 1975–2009. *Review of Urban & Regional Development Studies: Journal of the Applied Regional Science Conference* 26 (2), 135–154.

Andriesse, E. (2017), Regional disparities in the Philippines: Structural drivers and policy considerations. *Erdkunde*, 97–110.

Andriesse, E. and J. Kang. (2017), An alternative to 'exodus capitalism'? Offshore services in Iloilo City, Philippines. *International Journal of Urban Sustainable Development* 9 (3), 333–345.

Andriesse, E., Kittitornkool, J. Saguin, K. and Kongkaew, C. (2020), Can fishing communities escape marginalization? Comparing overfishing, environmental pressures, and adaptation in Thailand and the Philippines. *Asia Pacific Viewpoint,* early view: https://onlinelibrary.wiley.com/doi/abs/10.1111/apv.12270

Apostolopoulou, E. (2020), Tracing the links between infrastructure-led development, urban transformation, and inequality in China's Belt and Road Initiative. *Antipode* early view: https://onlinelibrary.wiley.com/doi/10.1111/anti.12699

Balisacan, A. M., Hill, H., and Piza, S. F. (2008), Regional development dynamics and decentralization in the Philippines: ten lessons from a 'fast starter'. *ASEAN Economic Bulletin*, 293–315.

Berdegué, J., Escobal, J. and Bebbington, A. (2015), Explaining spatial diversity in Latin American Rural Development: Structures, institutions and coalitions. *World Development* 73, 129–137.

Binns T. (2014), Dualistic and unilinear concepts of development. In: Desai V., Potter R., editors. *The Companion to Development Studies*. 3rd ed. Abingdon: Routledge, p. 100–105.

Clark, G., Feldman, M., Gertler, M., and Wojcik, D. (2018), *The New Oxford Handbook of Economic Geography*. Oxford: Oxford University Press.

Clausen, A. (2010), Economic globalization and regional disparities in the Philippines. *Singapore Journal of Tropical Geography*, 31(3), 299–316.

Higgens, K., Bird, K., Harris, D. (2010), Policy responses to the spatial dimensions of poverty. ODI working paper 328/CPRC working paper 168. www.odi.org/sites/odi.org.uk/files/odi-assets/publications-opinion-files/5518.pdf

Hirschman, A. (1958), *Economic Theories of Development*. New Haven: Yale University Press.

Jose, A. (2019), India's regional disparity and its policy responses. *Journal of Public Affairs* 19, e1933.

Jung, J. K., and Anderson, C. (2017), Extending the conversation on socially engaged geographic visualization: Representing spatial inequality in Buffalo, New York. *Urban Geography*, 38(6), 903–926.

Li, T. (2014), *Land's End. Capitalist Relations on an Indigenous Frontier. Durham and London*: Duke University Press.

Myrdal, G. (1957), *Economic theory and underdeveloped regions*. New York, NY: Harper and Row.

Ortega, A. A., Acielo, J. M. A. E., and Hermida, M. C. H. (2015), Mega-regions in the Philippines: Accounting for special economic zones and global-local dynamics. *Cities*, 48, 130–139.

Ortega, A. A. C., Martinez, M. S. M., Dayrit, C., and Saguin, K. K. C. (2017), Counter-mapping for resistance and solidarity in the Philippines: Between art, pedagogy and community. In Kollektiv Orangotango+ (Ed.), *This is not an atlas: Global collection of counter cartographies* (pp. 144–151). Transcript Verlag.

Oxfam International (2021), Mega-rich recoup COVID-losses in record-time yet billions will live in poverty for at least a decade. www.oxfam.org/en/press-releases/mega-rich-recoup-covid-losses-record-time-yet-billions-will-live-poverty-least

Pain, A. and Hansen, K. (2019) *Rural Development*, Abingdon, Oxon: Routledge.

Parr J. (1999), Growth-pole strategies in regional economic planning: a retrospective view (Part 1. origins and advocacy). *Urban Studies* 36, 1195–1215.

Perroux F. (1955), Note sur la notion de pôle de croissance. *Economie Appliquée.* 8, 307–320 (in French).

Piketty, T. (2014), Capital in the Twenty-First Century (translated by Arthur Goldhammer). Cambridge and London: The Belknap Press of Harvard University Press.

Rigg, J. (2012), *Unplanned development: tracking change in South-East Asia.* London and New York: Zed Books.

Saguin, K.K. (2021), *Urban Ecologies on the Edge: Making Manila's Resource Frontier.* Berkeley: University of California Press.

Song, E., Gress, DR., Andriesse, E. (2020), Global production networks and (distributional) regional development: The cinnamon industry in Karandeniya and Matale, Sri Lanka. *Journal of South Asian Development* 15(2), 209–237.

Stewart, F. (2016), Changing perspectives on inequality and development. *Studies in Comparative International Development* 51, 60–80.

Stiglitz, J. (2013), *The Prize of Inequality*. London: Penguin Books.

Talitha, T., Firman, T. and Hudalah, D. (2020), Welcoming two decades of decentralization in Indonesia: a regional development perspective. *Territory, Politics, Governance* 8(5), 690–708.

Walker, K. (2008), Neoliberalism on the ground in rural India: Predatory growth, agrarian crisis, internal colonization, and the intensification of class struggle. *The Journal of Peasant Studies* 35 (4): 557–620.

Wan, G. and Wang, C. (2018), *Poverty and inequality in Asia 1965–2014. WIDER working paper 2018/121*. Helsinki: UNU-WIDER. www.wider.unu.edu/sites/default/files/Publications/Working-paper/PDF/wp2018–121.pdf

Wei, Y. (2017), Geography of Inequality in Asia. *Geographical Review* 107 (2), 263–275.

Wright, S. (2012), Emotional geographies of development. *Third World Quarterly*, 33(6), 1113–1127.

Wuthnow, R. (2018), *Decline and rage in rural America*. Princeton: Princeton University Press.

Yang, Y. (2020), China's bold new five-year plan. 13 December. www.eastasiaforum.org/2020/12/13/chinas-bold-new-five-year-plan/

31

LAND GRABBING AND EXCLUSION

Philip Hirsch

Introduction

Land is a central issue in global development. As most of the rural poor in the majority world rely mainly on farming for food and income, access to land is central to life and livelihood. Unequal access to land has for long been a key part of wider social and economic inequalities and hence an object of redress, and land politics have often been integral to debates over the dominant direction of development within capitalist economic systems. Land reform, community rights to land and related resources, recognition of customary tenure, and other attempts to push back against prevailing development models based on marketized land relations, have served as the basis for counter-movements against prevailing means of dispossession, concentration, and elite capture.

While land is a longstanding and continuing concern for those advocating for more equitable development paths, debates around land have changed over time. In part this has to do with changing circumstances from the post-colonial Cold War era to more recent times dominated by neoliberal paths of economic development. In part it has to do with the urbanization and industrialization that have diminished the relative significance of farming relative to the rest of the economy. In part it also has to do with changing conceptual lenses through which we understand the significance of land and the means by which people gain access to or are excluded from it.

During the early stages of the post-Second World War development era, two main concerns dominated debates over land in Asia, Africa, and Latin America. The first had to do with correcting the alienation of large numbers of rural poor from their lands as a legacy of colonialism and associated patterns of commodity-oriented economic development. The second had to do with making land more productive, particularly under the so-called Green Revolution that sought to employ packages of hybrid seeds, fertilizers, chemical applications and access to irrigation as well as agricultural machinery in order to lift living standards in rural areas, as well as to help serve the food needs of growing cities and increase the export base of developing countries. Both of these concerns lay behind agrarian reform agendas, and they were roped in to the politics of land reform and agricultural support programmes that were integral parts of the armoury of both sides in the Cold War as it played out in the majority world in Africa, Asia and Latin America.

DOI: 10.4324/9781003017653-35

Following the end of the Cold War, and with the rapid industrialization of many middle-income countries, land became a less critical feature of developmental and political agendas during the 1990s in particular. Despite the fact that land never lost its importance for the majority world's rural poor, it declined as a core object of development studies and practice. However, the first two decades of the 21st century have seen a remarkable revival of academic, programmatic, and political attention to land as a central theme of global development. While there are many facets to this renewed interest, much of the discourse and debate have centred on the so-called global land grab, a process that agrarian political economists refer to as 'the large-scale acquisition of land or land-related rights and resources by corporate (business, non-profit [including pension funds] or public) entities' (White et al. 2012, 619).

This chapter sets out the basic contours of land grabbing, showing how the phenomenon has been contested both as a concept and through scholarly and activist challenges to changes in control over land. It then outlines a framework that seeks to nuance and explain processes of dispossession and accumulation associated with land grabbing, employing the related concepts of access and exclusion. The chapter concludes with a suggested pedagogy of land debates relevant to contemporary circumstances that employs the lens of exclusion, suggesting that such an approach links land-specific issues to many other concepts, themes, and debates in global development.

Land grabbing and its discontents

The land grabbing phenomenon has many antecedents in colonial and post-colonial agrarian processes, and even further back in English enclosures, Scottish highland clearances, and dispossession of Indigenous people by settler societies in the Americas, southern Africa and Australasia. However, new mechanisms, justifications and contexts differentiate it from earlier manifestations (Peluso and Lund. 2011; GRAIN 2008). A distinct discourse centred on the so-called global land grab emerged during the first decade of the 21st century. Four main triggers helped generate the land grabbing phenomenon and debates around it at this time. In 2007–8, global food prices spiked, raising the spectre of food shortages and associated hunger for the first time since the 1970s. The World Bank's annual *World Development Report* for 2008 called for a revitalization of agriculture based on large scale investment, out of concern that farming was under-capitalized and hence not achieving its potential (World Bank 2007). Meanwhile, the 2008 Global Financial Crisis (GFC) prompted global capital to search for secure investment options, which in many cases were found in farmland investments that yielded returns both from agricultural products and from appreciation of land values. Finally, the rise of biofuels as an economically profitable outlet for crops in the context of a search for alternatives to fossil fuels raised the prospect of almost unlimited demand for agricultural products beyond their role in global food systems.

The discourse of the global land grab includes a range of terms that seek to nuance, sharpen, or otherwise qualify what some see as a the more emotive and accusatory term 'grabbing.' One of the more common terms is large scale land acquisitions (LSLAs), implying that the scale of enterprise is important, and often connoting the seizure, purchase, or negotiation of multiple smallholdings for consolidation into larger plots under lease or ownership by wealthy and powerful players. Another common term is land deals, which leaves open the question of whether the pre-existing land users have been party to the deals in question. A frequent assumption is that most such acquisitions and deals are transnational in nature, involving large agro-investors negotiating with host governments for access to land, but many studies have highlighted the important role of domestic capital in LSLAs, either as the sole or dominant investors, or in partnership with foreign players. As a consequence, a more encompassing definition of land grabbing is one that accounts for a wider range of actors, processes and contingent conditions:

> *Land grabbing is the control — whether through ownership, lease, concession, contracts, quotas, or general power — of larger than locally-typical amounts of land by any persons or entities — public or private, foreign or domestic — via any means — 'legal' or 'illegal' — for purposes of speculation, extraction, resource control or commodification at the expense of peasant farmers, agroecology, land stewardship, food sovereignty and human rights.*
>
> *(European Coordination Via Campesina.[1])*

Whichever definition is applied, several key empirical questions arise around the global land grab. How much land has been grabbed? Who or what is behind the phenomenon? What kinds of land are particularly prone to grabbing? What have been the consequences for those who have seen their land appropriated? The answers to these questions remain open and hotly debated.

Quantifying the global land grab is a bit like measuring the length of the proverbial piece of string. Much depends on how grabbing is defined. Clearly not all land transactions involve grabbing. Despite the inclusive definition above, some measures only count transnational deals, while others include domestic acquisitions. Whether or not there are lower size limits to what counts as a land grab affects what is counted. And so on. Early estimates in the land grabbing literature suggest that anything from 43 million (World Bank figures) to 227 million (Oxfam figures) hectares had been subject to large scale acquisitions by 2011 (White et al. 2012, 620 at footnote 2). A GRAIN report in 2016 indicated that land grabbing had continued at a slower pace and that increasing numbers of deals were failing (GRAIN 2016), whereas the multi-partner Land Matrix Initiative suggests that by 2020 land deals had been concluded for some 52 million hectares, while another 29 million hectares of intended or failed deals were recorded (https://landmatrix.org/global/ accessed 12 August 2020).

Corporate actors are normally associated with the global land grab, particularly given its association with the neoliberal turn in development. However, government policy and enabling legislation is also fundamental. In many countries, the granting of land concessions is key to patronage politics, for example the crony deals prevalent in Myanmar and Cambodia (Woods 2014; Un and So 2011). Critics also accuse international institutions including the World Bank of promoting unjust land appropriation through spurious arguments that land privatization unlocks its economic potential (Mousseau et al. 2020). In reality, the global land grab is a consequence of a 'web' of private and public actors and interests rather than any one acting in isolation (Borras et al. 2020).

Public policy that favours private land investments draws on some key — and contested — arguments. The 2008 World Development Report, *Agriculture for Development* (World Bank 2007) suggested that large areas of agricultural land are ripe for development through investment, either because they are unused ('wasteland') or because land is under-utilized. Sometimes the notion of a 'yield gap' (the difference between actual and potential output from given areas of land) has been employed to make the case for large scale land acquisitions (Deininger et al. 2010, 54). However, an empirical global study that looks more carefully at the kinds of land acquired under various land deals finds that in fact about one-third of large scale acquisitions target already more productive and accessible land that is being alienated from small scale users (Messerli et al. 2014, 457).

Numerous studies and reports document the dislocation and hardship among those who have had their land taken through land grabs (e.g. GRAIN 2008; Mousseau et al. 2020; Dell'Angelo, D'Odorico, and Rulli 2017; Global Witness 2020). Ethnic minorities, including Indigenous groups, are particularly hard hit, both because of their marginalized status and because their farming practices – including shifting cultivation, pastoralism or semi-nomadic hunting and gathering – depend on customary land practices not recognized by authorities who are party to land deals with corporate investors. When governments allocate so-called wastelands to investors, authorities often neglect or criminalize pre-existing use of such land by subsistence-oriented smallholders, employing discourses that portray previous users as backward and wasteful.

Debates over land grabbing thus involve both terminology and the phenomenon itself. Other terms associated with land grabbing reflect different resources that are grabbed (water grabbing); different pretexts under which alienation takes place, for example where environmental protection is the legitimizing argument for appropriation (green grabbing); and different means by which control over land is achieved, where the grab does not necessarily give the more powerful actor ownership or long-term lease rights over the land in question, but rather involves surrendering control through labour and financial processes of the ways in which land is used and surplus is extracted from it (control grabbing).

Contestation of land grabbing has entailed an intertwined set of academic and activist critiques. Academic critique is channelled in particular through some key journals, prominent among which is the *Journal of Peasant Studies*. The critique takes many forms. Justice is the central theme in much of the theoretical and empirically-based discussion. Other critiques seek to show that the land grab phenomenon is incompatible with the Sustainable Development Goals (Dell'Angelo, D'Odorico, and Rulli 2017) or that it is environmentally destructive (Lazarus 2014). Other critiques challenge the arguments employed to legitimize large scale land acquisitions, for example by showing that small farmers feed the world's population more efficiently, not to mention equitably, than does large scale agribusiness (GRAIN 2014). But such studies have in turn generated discomfort with overly generalized or simplified accounts of land grabbing, producing calls for a more contingent analysis of land transactions and developments in land relations. Pedersen and Buur (2016) identify what they refer to as a more nuanced second wave of literature on the global land grab, suggesting three key ways to take a more analytical approach and, by implication, one driven less by moral indignation. They suggest that new forms of commodification, changing structures of authority, and reconfiguration of rights offer perspectives on recent land investments that allow research to explore this contingency in different country cases. Even neighbouring countries that are known for land grabbing experience what have been termed 'variegated transitions,' based on different political economies of land relations (Kenney-Lazar and Siusue 2020).

Beyond academic critique, there has been a strident and varied set of activist responses to land grabbing as an unjust practice on the part of more powerful groups at the expense of the less powerful. Responses range from local challenges to global movements such as *La Via Campesina*. They include outright resistance, negotiation for better terms of incorporation on the part of the dispossessed, and acquiescence (Hall et al. 2015). The Land Research Action Network has published an excellent compilation of case studies presenting activist perspectives on dispossession and ways in which farmers' movements have been defending their land (LRAN 2018).

The movement against land grabbing is not entirely a reactive one. Alternatives are framed under the banners of agro-ecology and food sovereignty, both of which represent pushbacks on the global food regime based in large scale monocropping, agribusiness and dominant global commodity chains (McMichael 2015). Land grabbing has also evoked policy responses by governments and international organizations. The Food and Agricultural Organisation of the United Nations assembled guidelines for responsible practice in large scale land and other resource investments in direct response to concerns over large scale land acquisitions (FAO 2012). Some governments have issued moratoria on new concessions, and many have placed areal size limits on those granted.

Access and exclusion

Mark Twain famously advised, 'Buy land, they're not making it anymore!' By definition, land is a finite good. Unlike most commodities that can be produced in response to demand, land is what Karl Polanyi termed a 'fictitious commodity' (Polanyi 1944). This is not to imply that land has not been commodified, as the emergence of land markets attests. Rather, it suggests, first, that while its use can be intensified and its cultivability affected by climatic shifts, land ultimately cannot be produced. Second, treating something so fundamental to life and livelihood as a commodity misses its innate value beyond tradeable worth. Access to and control over land is socially determined, by markets and by other means. Similarly, exclusion and exclusive rights over land are determined in multiple ways. Moreover, while land is one of a number of factors of production that determine the ability of the world's rural poor and others to produce agricultural goods for consumption and sale, land also has multiple meanings beyond its significance as a productive resource. These include sources of identity, places of refuge in times of economic hardship, and inter-generational assets.

There is a very large literature on the meaning of land, its role in development and the policy shifts that are required for the world's rural poor to achieve more equitable access to land and its benefits. These range from Hernando De Soto's neoliberal approach to transforming land from 'dead capital' to 'live capital' through the formalization of property rights (De Soto 2000), to radical calls for land justice as outlined above. Derek Hall's volume titled simply *Land* (Hall 2012) provides a particularly readable and thoughtful discussion of its multiple meanings. For present purposes, however, this chapter outlines two related approaches to understanding social relations around land: Ribot and Peluso's (2003) theory of access and Hall, Hirsch and Li's (2011) powers of exclusion.

Access to land is achieved through multiple means, both formal and informal. Property is sometimes equated with sanctioned rights to land, whether enshrined in law or in custom. As a noun, *a* property is often understood colloquially to refer to a piece of land (and sometimes also to the built structures contained therein). Ribot and Peluso suggest that property, as the *right* to benefit from things (including land), is too limiting. Instead, they employ the notion of access as the *ability* to derive benefit from things, including the power, knowledge and various dimensions of social relations that go beyond rights per se.

The concept of access is helpful in understanding land grabbing in a number of respects. Some land grabs occur within the purview of the law, while others are extra-legal. Determining what gives a large scale operator the ability to access land previously held and used by poorer groups requires a contextual understanding that ranges from the legal regime, policy environment, structures of authority that constrain resistance, ideological environment that prioritizes certain uses and users over others, to cultural and other means by which access is achieved.

In some ways a mirror image to Ribot and Peluso's access approach to land relations, Hall et. al interrogate the means by which people are excluded from land. Exclusion may mean being prevented from gaining access, being deprived of previous access, or preventing others from accessing land. However, rather than mirroring access as a desirable or positive outcome, the powers of exclusion approach does not take a normative stance that sees exclusion as something to be regretted or avoided. Rather, it sees exclusion as a fundamental and universal feature of land relations, and it uses the concept to explain how people are excluded, and are able to exclude others, from land. Exclusion therefore has a double edge: for small and large farmers alike, a degree of exclusive control over land is a requirement for farming.

It is therefore the means by which exclusion occurs, rather than whether or not it exists, that determines who can access land at whose expense. The powers of exclusion approach provides a way to assess the relative justice or injustice in land grabbing and other land-related policies and processes, not through a critique of the fact of exclusion *per se*, but rather through an analysis of four main powers that enable exclusion to occur: the power of regulation, the power of the market, the power of force and the power of legitimation.

Regulation refers to the legal and institutional framework that determines how boundaries are drawn and rights within them assigned. It is mainly, but not exclusively, associated with the power and actions of the state. Land titling is a common means by which regulation gives exclusive – but not necessarily unconditional – rights over land to individuals, in the name of enhanced security and ability to profit from land that can be used as security in accessing loans necessary for putting the land in question to work. Regulatory powers are also applied to the delimitation and granting of concessions, providing the means for foreign investors in particular to exclude others from their land-based operations.

Markets refer to the exclusionary power of pricing that shuts out those unable to afford commodified land. As land becomes more readily alienable, so access is lost to those who have been forced by circumstance or who have chosen to sell land, often only to find that its speculative value over and above productive returns means that they and their children are forever shut out of re-acquiring land. Increasingly, moreover, land is being financialized, meaning that its purchase, sale and ultimate ownership is modulated through various financial instruments including hedge funds and superannuation investments (Ouma 2014). New approaches match land registries with block chain technologies to further facilitate land transactions, for example in World Bank projects in Zambia and elsewhere (Mousseau et al. 2020, 10–12).

Force is the third power of exclusion. At one level, the power of the state behind regulatory measures is a kind of force, but in many instances extra-legal measures also lie behind exclusion of less powerful farmers from land. The very notion of land grabbing implies a degree of forced dispossession, whether or not the formal powers of the state lie behind it. In many cases, the threat or possibility of violence, rather than its enactment in every case, serves to exclude. Furthermore, the use of force may not necessarily be applied or implied directly on those who have their land grabbed. Rather, it may be used to cower or silence journalists, legal activists and other civil society actors who seek to intervene on behalf of the dispossessed. A global report on killings in 2019 names the Philippines as the country with most agribusiness-related violence against land and environmental defenders, but such extra-judicial violence extends to the Congo, Colombia, Brazil, Cambodia and many other countries, while land continues to be grabbed not only for farming but also for mines, reservoirs and other extractive activities (Global Witness 2020).

Legitimation is the fourth power, serving as an ideological backstop to exclusion that allows different actors to promote it. The so-called reverse land reform that appears to mark 21st century land grabbing (Byerlee 2014) is based in part on a position that large scale investments – and investors – are necessary to modernize farming and allow land use to realize its potential. Similarly, the neoliberal position associated with Hernando De Soto's notion of untitled land as 'dead capital' promotes the land titling policies and programmes that, in turn, facilitate alienation of land through marketized relations.

All four powers of exclusion are applicable in a range of contexts. Indeed, for the authors of the 2011 book, it was primarily the hermeneutic usefulness of the approach applied to a range of actual processes and illustrative cases that gave it value and that set up its potential for a more generalizable way to investigate various dimensions of land relations, including land grabbing. Six sets of processes help to illustrate this, and all can help explain, nuance, and question the land grabbing phenomenon in one way or another.

The first process is formalization, or the application of state recognition to underpin exclusive rights over land. Titling programmes, land reform schemes, and zoning by demarcation are examples of such formalization in contexts where small farmers had previously enacted their own delimitation of rights to use land, forests and other resources. In some cases, such formalization is legitimized as a protection against land grabbing by legalizing smallholder rights over parcels of land. However, the opposite effect is also prevalent in many cases, as the bounding of defined areas of land through titling or other means by implication allows state agencies to allocate farmland and forests outside such boundaries to large scale investors – which has happened in Laos and Cambodia, for example (Hirsch 2011; Schönweger et al. 2012). Furthermore, titling strengthens the exclusionary power of the market as land becomes a more readily tradeable asset.

Conservation is the second process through which exclusions play out. This is associated not only with the more obvious power of legitimized exclusion in the name of the global common good of environmental protection, but also with corporate schemes that seek to 'offset' their environmental effects with exclusionary protected area financing. Fairhead, Leach, and Scoones (2012) coined the term 'green grabbing' to refer to the appropriation of land for various asserted environmental ends. These range from 'offset' projects, to bio-fuel agricultural leases, to privatized conservation initiatives such as those prevalent in southern Africa.

Expansion of areas planted for boom crops is the third process examined through the *Powers* lens. In many ways, this is a core process in the popular imagination of the land grabbing phenomenon, as it sees entire landscapes transformed through wholesale planting of a particular monoculture crop, whether it be perennials such as rubber or bananas, or annuals such as cassava or maize. Global commodity markets are the primary contextual driver of such crop booms, but specific tenure relations and exclusive access given to agribusiness through national policy approaches is what allows them to happen on the ground. Sugar estates in Brazil, rubber plantations in Vietnam, Laos, Cambodia and Thailand, oil-palm estates in Indonesia or the Congo all illustrate the crop boom phenomenon that excludes shifting cultivators and other smallholders from land that previously provided the basis for their livelihoods.

While land grabbing has been associated with the concentration of agricultural enterprises, many other demands on land also displace smallholders in more permanent processes of land conversion away from farming altogether. The fourth process in which powers of exclusion are manifest is post-agrarian land alienation, including energy projects such as hydro-electric dams whose reservoirs flood tens of thousands of hectares of prime farmland. Other post-agrarian land grabs include peri-urban land conversion for housing or industrial estates and tourism or recreational uses such as golf courses. All these have in common with agricultural land grabs that they rest on claims that new land usage practices will generate greater economic value than the previous farming practices.

Not all exclusions are large scale, and not all involve non-local actors – whether foreign or domestic in origin. The fifth context of exclusion is termed 'intimate exclusions,' in which the commodification of agriculture and land is associated with land lost to kin and other neighbours. This raises interesting and often difficult questions about the delimitation of land grabbing. In communities marked by significant differences in wealth, domination by particular clans or other groups, concentration of land in the hands of more powerful actors may be just as significant in effect as land grabs effected by external forces. Furthermore, gender-based exclusions are often closely tied up with intra-familiar relations and cultural norms that close off land control by women.

The sixth context of exclusion involves counter-movements including those pushing back against land grabbing. While such movements seek to resist or ameliorate the effects of exclusionary land policies, the law of unintended consequences can, ironically, generate other types of exclusion. For example, the framing of many counter-movements in ethno-territorial claims has led to exclusionary consequences for minority groups, such as the eviction of Madurese migrants by Dayaks in Kalimantan and of Javanese in Aceh. These kinds of ethnically-based evictions are not normally classified as land grabs, but the effect on the dispossessed can be just as traumatic (Hall, Hirsch, and Li 2011, 176).

A pedagogy of land grabbing through the lens of exclusion

Land grabbing is a rich context for teaching and learning not only about land relations, but also about many other themes and debates in global development. Justice (development for whom), efficiency (of factor inputs), (in)equality, sustainable development, the role of discourse in legitimizing and contesting mainstream development, financialization, food regimes, agro-ecology and globalization can all be taught by working backwards from the specific issue of land grabbing to the more generic concepts that lie behind the phenomenon and the debates that it generates. Similarly, concepts like property, entitlements and sustainability can be accessed in a more concrete manner by working to unpack land grabbing, both in conceptual terms and through empirical examples. Alternatively, key theoretical positions on enclosures, primitive accumulation, accumulation by dispossession and other fundamental themes in agrarian studies are made more accessible and immediate to students through discussion around the global land grab.

It is important to take a critical approach, not only to alert students to injustices innate in the land grabbing phenomenon, but also by encouraging critical thinking through serious consideration of challenges to the moniker of land grabbing as a way of framing recent developments in agriculture and land control. Such critique is also a means to sensitize students to the differences between analytical and normative approaches to development studies, and indeed to engaged social science more generally.

More concretely, a suggested pedagogy of the local land grab that takes access and exclusion as core concepts provides an analytical frame for students themselves to do this unpacking. Two fundamental challenges in the pedagogy of global development can also be addressed through this approach. The first of these is to apply concepts to real world situations through case studies, allowing students to interpret particular examples through an analytical lens while at the same time testing and questioning conceptual schema. Exercises and class discussion can be enhanced through a case study approach, whereby students are assigned the task of researching a geographically and historically specific land dispute commonly identified with land grabbing, identifying the actors and processes involved, and applying the concepts of access and exclusion to analysis of the case in question. Among the references below, LRAN (2018), Global Witness (2020) and a report by the Oakland Institute (Mousseau et al. 2020) provide recent and accessible such cases, while others can be sought online and within students' own countries' current and historical experience of land grabbing.

The second, related, challenge is to match locally specific examples to more universalistic trends, and hence to isolate contextually relevant conditions from conceptually more universal generalizations. Land grabbing can also be nuanced through a discussion of terminology. This can include different kinds of 'grabs' referred to in the literature, including land grabs, water grabs, green grabs and control grabs. Discussion and literature review on the part of students can also dissect the different terms that have been applied to nuance the normativity in the grabbing discourse, including large scale land acquisitions, land deals and the global land rush. Discussion of the more specific denotations and connotations of the different terms can sensitize students to the assumptions behind them.

Further reading

The following key references are recommended for further reading. Ribot and Peluso (2003) outlines the authors' theory of access. Hall, Hirsch and Li (2011) gives detailed treatment to the *Powers of Exclusion* framework as applied to Southeast Asia, and it can be used to consider applicability to other world regions. White et al. (2012) provides an excellent purview of the political economic critique of land deals by key activist scholars associated with the *Journal of Peasant Studies* analysis of the land grabbing phenomenon. Hall's very readable (2012) monograph challenges readers to expand their conceptualization of land, land relations and – by implication – land grabbing. FAO (2012) shows the governance response to land grabbing that brings together mainstream and critical voices.

Note

1 ECVC is the European arm of the global movement of peasants and other small farmers who resist the control of global agriculture by agribusiness and who promote food sovereignty over financialized and corporatized farming and land control. See www.eurovia.org/wp-content/uploads/2016/11/defining-land-grabs.pdf (downloaded 27 July 2020).

References

Borras Jr, SM, Mills, EN, Seufert, P, Backes, S, Fyfe, D, Herre, R, and Michéle, L (2020), 'Transnational Land Investment Web: Land Grabs, TNCs, and the Challenge of Global Governance' *Globalizations,* vol. 17, no. 4, pp. 608–28.

Byerlee, D (2014), 'The Fall and Rise Again of Plantations in Tropical Asia: History Repeated?' *Land,* vol. 3, no. 3, pp. 574–97.

Deininger, K, Jonathan L, Andrew N, Harris S and Mercedes S (2010), *Rising Global Interest in Farmland: Can It Yield Sustainable and Equitable Results?,* World Bank Publications, Washington.

Dell'Angelo, J, D'Odorico, P and Cristina Rulli, M (2017), 'Threats to Sustainable Development Posed by Land and Water Grabbing' *Current Opinion in Environmental Sustainability,* vol. 26–27, pp. 120–28.

De Soto, H (2000), *The Mystery of Capital: Why Capitalism Triumphs in the West and Fails Everywhere Else,* Civitas Books, New York.

Fairhead, J, Leach, M and Scoones, I (2012), 'Green Grabbing: A New Appropriation of Nature?', *The Journal of Peasant Studies,* vol. 39, pp. 237–61.

FAO (2012), *Voluntary Guidelines on the Responsible Governance of Tenure of Land, Fisheries and Forests in the Context of National Food Security,* Rome: Food and Agricultural Organisation of the United Nations (FAO).

Global Witness (2020), 'Defending Tomorrow: The Climate Crisis and Threats against Land and Environmental Defenders', viewed at www.globalwitness.org/en/campaigns/environmental-activists/defending-tomorrow/

GRAIN (2008), 'SEIZED! The 2008 Land Grab for Food and Financial Security', viewed at www.grain.org/go/landgrab

GRAIN (2014), 'Hungry for Land: Small Farmers Feed the World with Less than a Quarter of All Farmland', viewed at, www.grain.org/en/article/4929-hungry-for-land-small-farmers-feed-the-world-with-less-than-a-quarter-of-all-farmland#sdfootnote36sym

GRAIN (2016), 'The Global Farmland Grab in 2016: How Big, How Bad?' *Against the Grain*, viewed at www.grain.org/article/entries/5492-the-global-farmland-grab-in-2016-how-big-how-bad#_edn7

Hall, D (2012), *Land*, London: Polity.

Hall, D, Hirsch, P and Li, T (2011), *Powers of Exclusion: Land Dilemmas in Southeast Asia*, Singapore University Press, Singapore.

Hall, R, Edelman, M, Saturnino, B, Scoones, I, White, B and Wolford, W (2015), 'Resistance, Acquiescence or Incorporation? An Introduction to Land Grabbing and Political Reactions', *Journal of Peasant Studies*, vol. 42, no. 3–4, pp. 467–88.

Hirsch, P (2011), 'Titling against Grabbing? Critiques and Conundrums around Land Formalisation in Southeast Asia Global Land Grabbing', *LDPI*, Viewed at, https://landportal.org/library/resources/titling-against-grabbing-critiques-and-conundrums-around-land-formalisation.

Kenney-Lazar, M and Siusue, M (2020), 'Variegated Transitions: Emerging Forms of Land and Resource Capitalism in Laos and Myanmar', *Economy and Space*, pp. 1–19.

Lazarus, E (2014), 'Land Grabbing as a Driver of Environmental Change', *Area*, vol. 46, no. 1.

LRAN (2018), 'New Challenges and Strategies in the Defense of Land and Territory', viewed at, https://focusweb.org/system/files/landresearchactionnetwork_web.pdf

McMichael, P (2015), 'The Land Question in the Food Sovereignty Project', *Globalizations*, vol. 12, pp. 434–51.

Messerli, P, Giger, M, Dwyer, M.B, Breu, T and Eckert, S (2014), 'The geography of large-scale land acquisitions: Analysing socio-ecological patterns of target contexts in the global South', *Applied Geography*, vol. 53, pp. 449–459.

Mousseau, F, Currier, A Fraser, E and Green, J (2020), *Driving Dispossession: The Global Push to 'Unlock the Economic Potential of Land*, Oakland.

Ouma, S (2014), 'Situating Global Finance in the Land Rush Debate: A Critical Review', *Geoforum*, vol. 57, pp. 162–66.

Pedersen, RH and Buur, L (2016), 'Beyond Land Grabbing. Old Morals and New Perspectives on Contemporary Investments', *Geoforum*, vol. 72, pp. 77–81.

Peluso, NL and Lund, C (2011), 'New Frontiers of Land Control: Introduction', *Journal of Peasant Studies*, vol. 38, no. 4, pp. 667–81.

Polanyi, K. (1944), *The Great Transformation*. New York: Farrar and Rinehart.

Ribot, J C and Peluso, N L (2003), 'A Theory of Access', *Rural Sociology*, vol. 68, no. 2, pp. 153–81.

Schönweger, O, Heinimann, A, Epprecht, M, Lu, J and Thalongsengchanh, P (2012), *Concessions and Leases in the Lao PDR: Taking Stock of Land Investments*, University of Bern, Vientiane.

Un, K and So, S (2011), 'Land Rights in Cambodia: How Neopatrimonial Politics Restricts Land Policy Reform', *Pacific Affairs*, vol. 84, no. 2, pp. 289–308.

White, B, Borras, S M, Hall, R, Scoones, I and Wolford, W (2012), 'The New Enclosures: Critical Perspectives on Corporate Land Deals', *Journal of Peasant Studies*, vol. 39, no. 3–4, pp. 619–47.

Woods, K (2014), 'A Political Anatomy of Land Grabs', *Myanmar Times*, March.

World Bank (2007), *Agriculture for Development. World Development Report 2008*, World Bank, Washington DC.

32

FORCED DISPLACEMENT AND RESETTLEMENT

Diana Suhardiman

Introduction

Forced displacement and resettlement, including forms of forced migration, have taken centre stage in development debates globally (Price 2015; Hathaway 2007; Kibreab 1999). Driven by conflict that urges domestic and transnational migrations (Hepner and Tecle 2013), government policies on internal resettlement (Evrard and Goudineau 2004), large-scale infrastructure development and climate emergencies (Piquet et al. 2011), as well as their interactions, forced displacement and resettlement have taken numerous forms and configurations. This chapter focuses on development induced displacement and resettlement (DIDR), or displacement and resettlement that is caused by development projects and activities. Globally, it is estimated that almost one million people are displaced by development projects every year (Picciotto 2013).

In most cases, though not always, DIDR is a consequence of large-scale infrastructure development and resource extraction. While early work and scholarship on DIDR focused mainly on large-scale hydropower development, more recent literature has examined numerous other drivers including large-scale land concession for plantation agriculture and forest conservation, mining, and special economic zone development (Diepart and Sem 2018). Conceptually, these works cover a wide range of disciplinary background and approaches, including geography, development studies, anthropology, sociology, economics, and political science (Cernea 1997; Neef and Singer 2015). Consequently, displacement and resettlement are viewed, analysed and discussed with reference to various theoretical framings ranging from policy and institutional analysis, legal pluralism, livelihood options and strategies, transboundary water governance, water and land rights, to social and environmental impact assessment.

Despite these new directions, however, DIDR research remains largely disconnected from the study of other types of displacement, such as those induced by conflict and natural disasters (e.g. refugee studies, climate-induced displacement and resettlement) (Wilmsen and Webber 2015; Hathaway 2007). Further, while current literature looks at both benefits and detrimental impacts from DIDR, it rarely does so comparatively. For example, while scholars have introduced the idea of benefit sharing (Sadoff and Grey 2002) and corporate social responsibility (Utting 2005) in hydropower dam debates, these works continue to stand disconnected from other scholarly works that show how DIDR has resulted in massive and further impoverishment of the poor, how this contributes to the shaping of social (transnational) movements

DOI: 10.4324/9781003017653-36

as a response to DIDR, while also calling for stronger incorporation of social and political dimension in DIDR studies.

This chapter looks at current approaches to understanding displacement and resettlement and common (policy) responses to DIDR, including through Social and Environmental Impact Assessments (SIA/EIA). Placing DIDR within the broader context of natural resource governance and resource politics, it outlines the value of power analysis and gives particular attention to processes of reterritorialization in the overall (re)shaping of DIDR processes and outcomes. Viewing development as a discourse (Crewe and Harrison 1998), it presents DIDR as a product of power struggles and to a certain extent the manifestation of existing power asymmetry between powerful (e.g. international donors, national government, private developers) and less powerful actors (e.g. different groups of affected people). It links displacement and resettlement with the (re)shaping of state spaces, how the latter emerge through various forms of territorialization, and how they resemble both state's territorialization attempt and local community's ability to resist and/or adjust their livelihoods options and strategies through reterritorialization or territorialization from the ground up (Kramp et al. 2020), including within state spaces (Kenney-Lazar et al. 2018).

Moving beyond technocratic, growth-led development and displacement practices

In the increasingly integrated world economy, infrastructure developments (roads, bridges, dams, railways, etc.) are viewed and oftentimes presented as key elements to achieve economic and social progress (Picciotto 2013). Consequently, displacement and resettlement are viewed and treated as byproducts of modernization, common consequences of development and inevitable sacrifices to the achievement of this 'progress.' Here, DIDR is justified as part of the state's economic development measures to promote growth (e.g. GDPs) through large-scale infrastructure development and related land acquisitions that require displacement and resettlement, regardless of whether benefits from economic growth are distributed equally among the wider society in general, and with regard to affected communities in particular.

When displacement and resettlement are treated merely as byproducts, sacrifices, or even opportunities for development, DIDR studies neglect and ignore any alternative visions of development beyond dominant narratives of economic growth – usually presented by the state, private sector actors, or international financial institutions. Cernea's Impoverishment Risks and Reconstruction framework focuses on identifying risks of DIDR, which includes loss of land and jobs, homelessness, marginalization, increased morbidity and mortality, food insecurity, loss of access to common property resources, and the weakening of social and community ties as key risks of development induced displacement and resettlement (Cernea 1997). Similarly, Scudder and Colson (1982) offer four different stages (recruitment of affected residents, transition, community formation and economic development, and incorporation) to conceptualize the DIDR processes. Later, Scudder (2009) also provided additional key characteristics to conceptualize the processes. These include: the accelerated rate of social change, the predominantly involuntary nature of resettlement; resettlement as by product of a different development initiative, and the complexity associated with DIDR processes and outcomes. While useful, these frameworks do not question the need for development in relation to DIDR, which is perceived as key to promoting economic growth and reducing poverty. Technocratic and managerial approaches to large-scale infrastructure development reduce DIDR merely into apolitical, procedural, financial issues, as is demonstrated by the centrality of cost benefit analysis (CBA) in DIDR frameworks. As stated by Wilmsen and Webber (2015, 78), 'CBA was introduced to determine what is in the public's interest. However, the determination of the public and of who can speak on its behalf are questions of politics and power.'

With the need to involve community views in ongoing discussions on displacement and resettlement being driven by procedural steps embedded in existing policy guidelines, meaningful participation, or participation beyond the mere provision of information, remains limited. Scholars have emphasized the complexity of displacement and resettlement and how they often result in the further impoverishment and marginalization of the poor and most vulnerable (Bui et al. 2013), and thus the need to understand affected people's views in (re)shaping DIDR processes and outcomes. Research has also shown local communities' abilities to organize and resist proposed developments. Nonetheless, meaningful participation and empowerment continue to be hijacked by vested interests of DIDR industries and assemblages (de Landa 2006), as the latter take centre stage in DIDR policy guideline formulation, procedural arrangement and project implementation.

This highlights the needs for scholars to take into consideration the entire spectrums of displacement and resettlement in order to unpack how dominant power structures and relations have helped create, (re)construct, preserve, and reproduce processes of inclusion and exclusion. For example, bringing to light the politics of inclusion and exclusion in DIDR processes, scholars have shown how IFIs in collaboration with state actors and private developers construct their own definitions as to whom are entitled (or not) for displacement and resettlement assistance (Gupte and Mehta 2007). Vandergeest (2006) argues that DIDR is inherently a reorganization of people, place, and identity, intertwined through complex socio-economic and political arrangements. Or as stated by Wilmsen and Webber (2015, 81):

> *In praxis, resettlement rarely occurs according to the preferences of displaced people for self-determined 'community' resettlement. Instead, the 'communities' as depicted in a resettlement plan is a grouping of the 'units of entitlement' within administrative boundaries.*

Understanding livelihood reconstruction within the context of DIDR requires further unpacking of community strategies, as well as how such strategies are defined, adjusted, and reproduced based on household perceptions of space, their territorial rights, how these have been affected by development activities, and how this change (re)shapes their access to resources, beyond household income and assets.

Policy and institutional landscape of DIDR

Following the adoption of the first guidelines for DIDR by the World Bank in 1980, the majority of governments (including within developing countries) have formulated resettlement policies and guidelines at both national and project levels, while also being attentive to international standards. In 2000, the World Commission on Dams published a framework for decision-making for hydropower dam projects that sought to ensure local community participation in decision-making processes (WCD 2000). At the global policy level, the UN Guiding Principles on Internal Displacement (UNHCR 2004) has also underlined the rights of project-affected people to free prior and informed consent (FPIC) with regard to planned development projects, as well as participation in decision-making processes pertaining to their displacement and resettlement. Similarly, the UN's Guiding Principles on Business and Human Rights (UN Global Compact 2011) endorses the integration of human rights to businesses through the incorporation of the above principles and general guidelines to be followed by corporate businesses. Moreover, the need to protect the rights of indigenous people from DIDR has made it to the global policy discussions (UNHCR 2004).

In line with these policy guidelines and principles, Social and Environmental Impact Assessments (SIA/EIA) have also been developed as tools for identifying longer term project impacts, as well as to ensure transparent consultation and public hearings. Originating in the passing of the US National Environmental Policy Act (NEPA) in 1970, EIA spread rapidly across the globe and is now practiced in more than 100 countries, including by IFIs as their main indicator for socio-environmental safeguards. In 1989, the World Bank introduced a comprehensive environmental assessment policy requiring an EIA to be undertaken for major projects by the borrowing countries with World Bank supervision. Similarly, the ADB's environmental policy also recognizes the need to incorporate environmental considerations into national and sub-national development planning.

At the regional level, the Mekong River Commission has introduced Strategic Environmental Assessment as a means to foster open discussion and decision-making processes pertaining to hydropower development on transboundary rivers. At the national level, developing countries governments are well equipped with various policies and legal frameworks to guide and ensure local communities and affected people's views are well incorporated in DIDR processes, ranging from policies on sustainable hydropower development, EIA review, grievance and compensation, resettlement action plan, to land concession rules and procedures.

Nonetheless, these policies guidelines and legal frameworks continue to treat DIDR in a fragmented way, while neglecting the complex power interplays (including the issue of vested interests and power asymmetry) that shape DIDR processes and outcomes. For example, focusing mainly on procedures for information sharing and consultation, key policy actors fail to address the problem of power asymmetry hampering local community's ability to convey their voice and negotiate their needs (Price 2015). Similarly, viewing EIAs in isolation from institutional arrangements, policy actors fail to address the problem of vested interests in project approval processes (Campbell et al. 2015). Theoretically speaking, EIAs are meant to be an unbiased source of information presented by an impartial or neutral viewpoint. In reality, however, it is universal practice that the consultant is hired by the project developer, which introduces a source of bias into the EIA process. Both the lead agency and the consultant want projects to be approved so there is a strong incentive to ignore or downplay negative impacts and to exaggerate potential benefits. Further, the developer has the opportunity to review the EIA before it is made public and can therefore express their opinion regarding the description of impacts. In most cases, the consultant has an incentive to appease the developer in order to ensure that the consultant is hired to write EIAs for future projects. This problem of vested interests is also apparent in the way International Financial Institutions' (IFIs) efforts to promote more inclusive and participatory decision-making in DIDR are hampered by both IFIs' and national governments' development agenda to promote economic growth, regardless of how costs and benefits of such growth is distributed.

When aiming to convey displaced and resettled community voices, and to better represent local community's development needs and aspirations, scholars and practitioners need to also look at processes of institutional emergence across scales, and try to incorporate key elements of such emergence as an integral part of policy reform processes towards more inclusive and accountable policy guidelines, legal frameworks, procedures and mechanisms for DIDR. This includes better understanding of how affected communities can respond to and navigate through DIDR and potentially improve their circumstances, including through reterritorialization and strategic alliance formation.

From displacement to reterritorialization: strategic alliance formation

Current literature on socio-political production of space has highlighted the importance of power analysis surrounding the logic of inclusion and exclusion. Here, understanding of power dynamics and power relations that (re)shape displacement and resettlement processes and outcomes is closely interlinked with unpacking the creation, sustenance and reproduction of social relations, how these relations are shaped by diverse local perspectives, its implications for one's ability to convey their voice through processes of institutional emergence and the creation of new political spaces through, for instance, reterritorialization. As stated by Brun (2001, 19 and 23):

> *Reterritorialization in Malkki's (1995) understanding means to lose one's territory, and then construct a new community within a new area. Reterritorialization is therefore not only the process of moving from one location to another. It may be understood as the way displaced and local people establish new, or rather expand networks and cultural practices that define new spaces for daily life. This understanding of reterritorialization involves the emergence of otherwise marginalized voices and alternate representation.*

Reterritorialization involves the process of how the displaced and resettled (re)build their networks, improve their bargaining power, adjust their livelihoods and adapt their strategies, while making the decisions to control their own lives, in relation to other groups' (e.g. host community/original settlers in the case of internal resettlement) interests, views and power positions (Lestrelin 2011). Reterritorialization is not only about the reconstruction of social relationships between and among the displaced and resettled, but also about making new connections with other groups within the local community and the larger society, while shaping such connection as an integral part of (transnational) social movement.

Unpacking relations and interactions shaped and reshaped by power interplay between the displaced and resettled with their host community and the larger society is key to better understanding processes of institutional emergence. How the latter results in the creation of new political spaces, could serve as stimulus for the evolution of local institutional arrangements and existing rights systems (e.g. water and land rights) towards more inclusive and accountable natural resource governance. Lastly, placing DIDR within the context of territorialization and reterritorialization enables analysis of the entire spectrum of contestation and power struggles as dynamic, evolving, and continuing processes within an ever-changing geography.

Processes of institutional emergence and the creation of new political space

In order to shed light on how reterritorialization functions as a process, this section applies the previous conceptual insights to a case of Laos. Laos is at the forefront of rapid infrastructure development projects with hydropower development, plantation agriculture, mining industries, roads and railway construction contributing significantly to the country's economic development through the government's revenue generation. Consequently, issues of displacement and resettlement have received a lot of attention, including in relation to large-scale land concessions and land use planning processes.

Drawing on my own research in Namai village (hereafter Ban Namai), Nambak district, Luang Prabang province in Laos, I show how farm households' responses to a recent land use planning initiative – aiming to demarcate farmers' farmlands and forest areas (Kramp et al.

2020) – are derived from their respective status as original settlers (or villagers who always have lived in the village), recent and late comers (or villagers who have been resettled from their original village to Ban Namai) as shaped by the government's earlier resettlement policies. In Laos, internal territorialization through involuntary displacement and resettlement have been pursued since pre-colonial time. Only recently growing attention has been given to non-state actors in particular those who have historically faced difficulties in getting government institutions to recognize their land tenure, and their role in making territory (Corson 2011; Lukas and Peluso 2019).

Ban Namai comprises of an ethnic Tai-Lue community that, in 1986, was merged with three surrounding Khmu ethnic minority communities (Ban Luk, Ban Klok, and Ban Huana) as part of the Government of Laos's (GoL) widespread forced resettlement of minority groups living in remote upland areas. At this time, the Tai-Lue community was split between those who welcomed the idea of new settlers in order for Ban Namai to grow and develop, and those who opposed the idea of newcomers settling on their land. When the first Khmu households arrived, they were given land to build their houses on garden areas of the Tai-Lue. Prior to the resettlement, this land was being reserved for future generations. As more resettled Khmu households arrived, however, the Tai-Lue became unwilling to provide more land, instead insisting that already resettled Khmu households partition their land.

Khmu recent and late settlers struggled to get access to upland areas for their swidden cultivation practices because the Tai-Lue community predominantly controlled these areas, while referring to their land rights based on customary tenure. The latter lacks formal land title, as government authorities allowed these households to reserve land but did not provide them with a formal right to either use or own the land, but they nonetheless control land access. In response to the growing influx of Khmu migrants Tai-Lue farmers who cleared upland plots and paid land tax were allowed by the district authority to reserve the land but were not provided any legal document. Only in 2007 the first land records were issued in the form of family land books which delineated residential area and registered the measurements of low-lying paddy fields. This formalization process did not record upland use, however, rendering these areas as state land. In response to this lack of formal right to secure tenure and land access, Tai-Lue upland farmers have increasingly converted plots of reserved land into small cash crop plantations – mostly rubber. Through such conversions they have been able to cement their land access, while ensuring that their land use mimics, or is in line with, the government's policy to promote market integration through commercial farming and plantation agriculture. While this strategy has disadvantaged Khmu farmers' livelihood options by limiting access to Tai-Lue's reserved land, the latter have also engaged Khmu residents in their farming activities. For example, Tai-Lue farmers have hired Khmu farmers as labourers, in exchange for access to upland plots elsewhere. Moreover, Tai-Lue farmers have rented parts of their land to Khmu farmers, in exchange for money (depending on soil quality and size of the plot) or, alternatively, part of the rice yield (around 2–3 bags).

Common assumptions tend to suggest that customary land rights would get swept away or disregarded in the face of crop booms, in this case rubber plantations. Nonetheless, farm households' strategies to reterritorialize, by transforming their (reserved) upland rice fields into rubber gardens, shows how customary land rights could also serve as a tool for farmers to strengthen their land tenure while mimicking private sector actors' strategy to gain access to land. As stated in Kramp et al. (2020):

> When farmers changed their upland rice fields into a rubber plantation area to defend their access
> to and control over land from the government's policy interventions and private sector actors' land

concession – while at the same time benefitting from the rubber boom – they transformed their customary land rights from merely a local institutional arrangement into a territorial strategy.

As farm households view rubber planting both as a market but more importantly as a territorial strategy to secure their land tenure, it becomes a tool for farmers to reterritorialize and contest state's territorialization strategy, centred on land use planning. Moreover, as farm households mimic private sector actors' strategy to gain access to land through rubber planting, such strategy allows them to oppose state's territorialization project, albeit informally and indirectly, and within state spaces (Kenney-Lazar et al. 2018).

The case study of farm households' reterritorialization strategy in Ban Namai illustrates processes of *institutional emergence*, centered on farm households' ability to use their customary land rights as their key means to (re)arrange local institutional arrangements pertaining to access to land and land use. Land concessions have been a crucial component of the GoL's increasingly neoliberal economic agenda and, as Kramp et al. (2020) have shown, have provided a channel for smallholders to turn their land into capital and their customary rights into a territorial strategy to secure land tenure. What is interesting in this development is that farmers who have historically faced difficulties in getting government institutions to recognize their land tenure, were able to secure the latter through rubber cultivation. Local rent seeking of villagers in their scramble for land has in this sense worked together with a neoliberal mode of governance that has established a market for rubber and has allowed for an increasing influence of the rubber company over local resource use. The case study of Ban Namai thus demonstrates how displaced and resettled peoples can be active agents for change, who are capable to adapt and adjust their livelihood strategies, navigate through the power struggles and enact their collective interests and resistance. This analysis of displaced residents as change agents moves beyond current 'victimhood' framing of the displaced as merely passive development recipients and disaffected peoples.

Conclusion

DIDR is a global issue that pushes new people into poverty each year. There is an urgent need to link economic growth with inclusive development and the overall notion of social justice. Like Aiken and Leigh (2015, 80), I argue that DIDR is *'fundamentally a political phenomenon, involving the use of power by one party to relocate another'*. Development induced displacement and resettlement cannot be discussed and analysed in isolation from wider debates regarding people, place, mobility, and identity (Kibreab 1999), the overall (re)production of rights and authority (Lund and Rachman 2018), the (re)shaping of state spaces, and how these manifest in the overall voluntary and involuntary movement of people (Neef and Singer 2015). As stated by Kibreab (1999, 407):

> *The identity which people gain from their association with a particular place is not per se intrinsically fundamental. But in a world, in which many rights such as equal treatment, access to sources of livelihoods, access to land, rights of freedom of movement… are determined on the basis of territorially anchored identities, the identity people gain from their association with a particular country is an indispensable instrument to a socially and economically fulfilling end.*

This highlights the centrality of rights in DIDR discussion and its close interconnection with the economic and political (re)shaping of spatio-temporal territory. It also shows how the rights to make personal decisions are embedded in one's ability to create, sustain, contest, and reproduce various forms of territories upon which one's conception of political rights and authority are based. Building on earlier work and scholarship looking at space as processes of socio-political construction (Massey 2005), this chapter has looked at space as both physical and imaginary embodiments of people's rights and identities, how these evolve over time, how they are (re)shaped by actors' and institutions' interests, strategies and access to resources, and thus how they function as a terrain of contestation. Centring on the close interlinkages between territory, identity and rights, it has highlighted the importance of political economy and political ecology approaches in better understanding current discussions on DIDR.

What this chapter tells us is that we should not focus only on the harm wrought by DIDR. On the contrary, we need to also look at how communities respond to cope with the changes while trying to improve their livelihood. One such response is through reterritorialization. Future research on forced displacement and resettlement needs to put farm households' strategies central in the ongoing debates on natural resource governance and local community's livelihoods. Such research should place these struggles within the context of broader contemporary struggles over access to (natural) resources, how these contestations are closely interlinked with the (re)production of rights and political authority, as well as accompanying processes of institutional emergence and the creation of political spaces. Without this analysis, we lack understanding of how grass-roots forces can contribute to more inclusive and just development and natural resource governance.

Key sources

1. Crewe, E., E. Harrison (1998), *Whose development? An ethnography of aid.* London: Zed books.

Crewe and Harrison look at development as a political idea and undertaking, rather than a technical, managerial issue as often is presented in mainstream development debates. How is development linked with power relations? How does the overall idea of development change our understanding of society and its social complexity, and vice versa?

2. Evrard, O., Y. Goudineau (2004), Planned resettlement, unexpected migrations and cultural trauma in Laos. *Development and Change* 35(5): 937–962.

In this article Evrard and Goudineau describe the overall rationales behind the planned internal resettlement in Laos. What are the socio-economic and political reasonings driving internal resettlement? How does this reasoning impact people's views and perceptions of their identity and culture, and their (in)ability to claim their rights?

3. Kibreab, G. (1999), Revisiting the debate on people, place, identity and displacement. *Journal of Refugee Studies* 12(4), 384–410.

Kibreab brings to light the idea of 'territorially anchored identities,' or the identity people gain from their association with a certain place (e.g., country, region). How does the overall shaping of one's identity relate with access to resources, including in relation to people's livelihood options in particular places? How does displacement (re)shape people's identity especially in relation to their rights for freedom of movement?

4. Kramp, J., Suhardiman, D., Keovilignavong, O. (2020), Unmaking the upland: Resettlement, rubber and land use planning in Namai village, Laos. *Journal of Peasant Studies*.

Kramp et al. illustrate how farm households and local communities can mimic state's territorialization strategies and transform these as a means to reterritorialize, or territorialization from the ground up. What are key decisive factors enable farm households to do so? How does this strategy contribute to our understanding of the relations between state-society-private sector actors in the current development context and displacement?

5. Lestrelin, G. (2011), Rethinking state-ethnic minority relations in Laos: Internal resettlement, land reform and counter territorialization. *Political Geography* 30(6), 311–319.

Lestrelin brings to light a different way to look at state-ethnic minority relations in Laos. How do farm households' strategies to maintain their access to resources and agricultural practices represent the idea of counter territorialization vis-à-vis the state's internal resettlement policy? How does this counter territorialization enrich our understanding of power relations in (re) shaping state-society relations within the context of displacement?

References

Aiken, S. R. and Leigh, C. H. (2015), 'Dams and indigenous people in Malaysia: Development, displacement and resettlement', *Geografiska Annaler: Series B. Human Geography,* vol. 97, no.1, pp. 69–93.

Brun, C. (2001), 'Reterritorializing the relationship between people and place in refugee studies', *Geografiska Annaler: Series B. Human Geography,* vol. 83, no.1, pp. 15–25.

Bui, T.M.H., Schreinemachers, P. and Berger, T. (2013), 'Hydropower development in Vietnam: involuntary resettlement and factors enabling rehabilitation', *Land Use Policy,* vol. 31, pp. 536–544.

Campbell, L., Suhardiman, D., Giordano, M. and McCornick, P. (2015), 'Environmental Impact Assessment: Theory, practice and its implications for the Mekong hydropower debate', *International Journal of Water Governance,* vol. 3, no.4, pp. 93–116.

Cernea, M. (1997), 'The risks and reconstruction model for resettling displaced populations', *World Development,* vol. 25, no.10, pp. 1569–1587.

Corson, C. (2011), 'Territorialization, enclosure and neoliberalism: Non-state influence in struggles over Madagascar's forests', *Journal of Peasant Studies* vol. 38, no.4, pp. 703–726. DOI: 10.1080/03066150.2011.607696

Crewe, E. and Harrison, E. (1998), *Whose development? An ethnography of aid*, Zed books, London.

de Landa, M. (2006), *A new philosophy of society: Assemblage theory and social complexity*, Continuum, London.

Diepart, J. C., Sem, T. (2018), 'Fragmented territories: Incomplete enclosures and agrarian change on the agricultural frontier of Samlaut district, Northwest Cambodia', *Journal of Agrarian Change,* vol. 18, no.1, pp. 156–177.

Evrard, O. and Goudineau, Y. (2004), 'Planned resettlement, unexpected migrations and cultural trauma in Laos', *Development and Change,* vol. 35, no.5, pp. 937–962.

Gupte, J. and Mehta, L. (2007), 'Disjunctures in labelling refugees and oustees.' In: Moncrieffe, J., Eyben, R. (Eds), *The power of labeling: How people are categorized and why it matters,* pp. 64–79.

Hathaway, J. (2007), 'Forced migration studies: Could we agree just to date?', *Journal of Refugee Studies,* vol. 20, no. 3, pp. 349–369.

Hepner, T. R. and Tecle, S. (2013), 'New refugees, development-forced displacement, and transnational governance in Eritrea and exile', *Urban Anthropology,* vol. 42, no. 3/4, pp. 377–410.

Kenney-Lazar, M., Suhardiman, D. and Dwyer, M. (2018), 'State spaces of resistance: Industrial tree plantations and the struggle for land in Laos', *Antipode,* vol. 50, no. 5, pp. 1290–1310.

Kibreab, G. (1999), 'Revisiting the debate on people, place, identity and displacement', *Journal of Refugee Studies,* vol. 12, no. 4, pp. 384–410.

Kramp, J., Suhardiman, D. and Keovilignavong, O. (2020), 'Unmaking the upland: Resettlement, rubber and land use planning in Namai village, Laos', *Journal of Peasant Studies*. DOI: https://doi.org/10.108 0/03066150.2020.1762179

Lestrelin, G. (2011), 'Rethinking state-ethnic minority relations in Laos: Internal resettlement, land reform and counter territorialization', *Political Geography*, vol. 30, no. 6, pp. 311–319.

Lukas, M. C. and Peluso, N. L. (2019), 'Transforming the classic political forest: Contentious territories in Java', *Antipode*, 52(4), 971–995.

Lund, C. and Rachman, N. F. (2018), 'Indirect recognition: Frontiers and territorialization around Mount Halimun-Salak National Park, Indonesia', *World Development* 101: 417–428.

Massey, D. (2005), *For space*. London: Sage.

Neef, A. and Singer, J. (2015), 'Development induced displacement in Asia: Conflicts, risks and resilience', *Development in Practice*, vol. 25, no. 5, pp. 601–611.

Picciotto, R. (2013), *Involuntary resettlement in infrastructure projects: A development perspective In: Ingram, G. K., Brandt, K. L. (Eds)*, Infrastructure and Land Policies, Columbia University Press, New York, pp. 236–262.

Piquet, E., Pecoud, A. and De Guchteneire, P. (2011), *Migration and climate change*, Cambridge University Press, Cambridge.

Price, S. (2015), 'A no-displacement option? Rights, risks and negotiated settlement in development displacement', *Development in Practice*, vol. 25, no. 5, pp. 673–685.

Sadoff, C. and Grey, D. (2002), 'Beyond the river: The benefits of cooperation on international rivers', *Water Policy*, vol. 4, pp. 389–403.

Scudder, T. (2009), *Resettlement theory and the Kariba case. In: Oliver-Smith, A (Ed). Development and dispossession: The crisis of forced displacement and resettlement*, School of Advanced Research Press, New Mexico: pp. 25–47.

Scudder, T. and Colson, E. (1982), 'From welfare to development: A conceptual framework for the analysis of dislocated people.' In: Hansen, A., Oliver-Smith, A. (Eds). *Involuntary migration and resettlement*, Westview Press, Boulder, pp. 267–287.

UN Global Compact (2011), *'Guiding principles on business and human rights: Implementing the UN's 'Protect', Respect, Remedy' framework*. Rome: UN.

UNHCR (2004), *Guiding principles on internal displacement,* Geneva.

Utting, P. (2005), 'Corporate responsibility and the movement of business', *Development in Practice*, vol. 15, no. 3–4, pp. 375–388.

Vandergeest, P., Idahosa, P. and Bose, P. S. (2006), *Development's displacements: Economies, ecologies and cultures at risk*, UBC Press, Vancouver.

WCD (2000), *Dams and development: A new framework for decision making*, Earthscan, London.

Wilmsen, B. and Webber, M. (2015), 'What can we learn from the practice of development-forced displacement and resettlement for organized resettlements in response to climate change?', *Geoforum*, vol. 58, pp. 76–85.

33

HUMAN MOBILITY AND CLIMATE CHANGE

Andreas Neef and Lucy Benge

Introduction

The relationship between climate change and migration is highly complex. There is growing consensus that sea-level rise – coupled with the increasing frequency and intensity of weather-related disasters – is contributing to migration and displacement (IOM 2019). The World Bank estimates that, by 2050, climate change could force more than 143 million people to move within their countries in three major world regions, South Asia, sub-Saharan Africa and Latin America (Rigaud et al. 2018). Transnational climate migration is also expected to increase as sea-level rise is threatening the very existence of low-lying island countries in the South Pacific, such as Tuvalu and Kiribati (Farbotko et al. 2016). However, the difficulty of disentangling climate drivers from other more traditional drivers of migration – such as migration for employment or conflict-induced migration – raises questions of how human movement in the context of climate change is understood and governed. When people move, whether within or across borders, it is often the *cause* of their movement that determines both how they are defined and the solutions available to them. When an individual is defined as a 'migrant' this implies a greater degree of autonomy in their decision to move than if they were to be defined as a 'displaced person' or 'refugee.' Hence, the terms we choose to use say a lot about how we understand the triggers of movement, whether movement is seen to be voluntary or forced, and what policy solutions are available.

Contestation over how movement in the context of climate change is framed also poses significant issues regarding the presentation of accurate and objective data on this phenomenon. Less information exists regarding cases of 'voluntary' movement (i.e. when people move in anticipation of climate threats) than cases that are deemed 'forced' (i.e. evacuations). This may be because 'forced' movement is more visible – often in response to major disasters – than cases of so-called 'voluntary' movement, which occur in response to less visible, slower-onset triggers. Although it is widely acknowledged that mobility occurs along a continuum with 'forced' movement at one end and 'voluntary' movement at the other, policy approaches often contribute to the creation of a false dichotomy – conceptualizing climate-induced mobility as either predominantly 'forced' or 'voluntary.' In the case of Fiji's recently developed Planned Relocation Guidelines (2018), for example, the relocation of communities needs to be confirmed as community-driven and hence 'voluntary' in order to legitimize government

DOI: 10.4324/9781003017653-37

intervention. Similarly, policy emphasizing the economic development and livelihood benefits of migration tend to focus on its adaptive and voluntary nature, while narratives emphasizing the risks posed by migration generally focus on the forced nature of movement in the context of climate change and the responsibility of the international community to develop protection policies.

This chapter looks at how 'protection' narratives have been framed in opposition to 'adaptation' and attempts to understand why organizations such as the United Nations Refugee Agency (UNHCR) and international bodies such as the United Nations Framework Convention on Climate Change (UNFCCC) have come to prioritize and advocate for the use of 'adaptation' policies over those of 'protection.'

Approaches to addressing migration in the context of climate change

This section explores two narratives that have played a key role in determining how climate-induced human mobility is understood and governed. The first emphasizes the vulnerability of 'people on the move' and the responsibility of emitters in the Global North to *protect* them. This narrative was popular in the 1990s and early 2000s, but still persists today despite the shift towards a new narrative emphasizing the resilience of migrants and their ability to use migration to successfully *adapt* to risk (Methmann and Oels 2015).

The following comparison of 'protection' versus 'adaptation' approaches will demonstrate the different ways in which movement in response to climate change has been conceptualized and, in turn, how these conceptualizations contribute to the way people on the move are depicted and governed. Revealing how these narratives have been constructed and looking at how compatible they are with lived realities, beliefs and culture will help to demonstrate the power these narratives yield, the political purposes they serve and the voices they obscure.

Protection approaches

Responding to legal gaps

Protection approaches respond to gaps in international legal protection for people attempting to cross borders in response to slow-onset climate events. The protection narrative emphasizes the forced nature of movement in response to climate change and often advocates for the use of the term 'climate refugee' in response to the lack of protection offered by refugee law (Biermann and Boas 2010). Issues with human rights law have also led to calls for the development of new protection mechanisms that are more suited to movement in the context of climate change. While human rights law recognizes the right to migrate, and while states must protect the human rights of people who enter their country, the state-centric nature of human rights law means that there is no obligation for third-party states to actually allow migrants to enter their country (Thomas and Yarnell 2018). The lack of extra-territorial responsibility generated by human rights law has been identified as being at odds with efforts by the UNFCCC to recognize global accountability for climate change and as the biggest hindrance to the use of human rights law as a tool for protecting climate migrants (Gromilova 2014).

Efforts to address gaps through new protection mechanisms

The failure of current legal mechanisms to provide protection to people moving across borders due to climate triggers has led some authors to argue that new protection mechanisms are required – either within the UNFCCC in order to draw on normative established principles such as 'common but differentiated responsibility' (Biermann and Boas 2010); or independent of the UNFCCC in order to borrow useful concepts from human rights law, environmental law and refugee law and to tailor these concepts to suit migration in the context of climate change. Despite these efforts, there is currently little international political appetite to create protection policies for climate-related migration across borders (McAdam and Ferris 2015). The closest we get to a protection agenda comes from the Platform on Disaster Displacement (PDD), formerly The Nansen Initiative, which looks for ways to create a coherent approach for protecting people displaced due to disasters and climate change.

The PDD does not call for new legally binding conventions but instead works with states to create context-specific state-led approaches to protection. This is due to the lack of support UNHCR received in 2011, when it lobbied states to endorse its position as the facilitator of a new legal protection framework. UNHCR failed to receive support due to concern that such a mandate would compromise state sovereignty over borders (Hall 2016). Yet it is important to note that the PDD does not shy away from utilizing the term 'displacement' which recognizes the 'forced' nature of movement in response to disaster and the impacts of climate change.

Concerns with protection approaches

Numerous critiques have been brought against protection narratives, particularly those that utilize the term 'environmental refugee' (Biermann and Boas 2010; Gemenne 2015). Concerns with these narratives relate to their lack of acknowledgement of migrant autonomy and histories of migration, and their contribution to the creation of 'vulnerable' subjects (Farbotko and Lazrus 2012). The emphasis on vulnerability is problematic due to the risk of legitimizing external intervention in 'vulnerable' regions of the Global South (Boas and Rothe 2016) and the potential for creating dependency on countries of the Global North (Farbotko and Lazrus 2012). Many authors have argued that protection narratives fail to recognize the empowering potential of migration and the way people utilize migration to their advantage – i.e. for economic opportunity – as well as risk reduction (Barnett and Chamberlain 2010; Farbotko et al. 2016; McNamara and Gibson 2009). These authors tend to look at how migration can be used as a voluntary tool of adaptation, utilized by responsible and resilient actors to achieve their development aspirations.

Adaptation approaches

'Migration as adaptation' within the international climate change regime

The link between human movement and adaptation was first brought into the international climate policy space with the 2010 UNFCCC Cancún Adaptation Framework. This invited parties to undertake 'measures to enhance understanding, coordination and cooperation with regard to climate-induced displacement, migration and planned relocation, where appropriate, at national, regional and international levels' (UNFCCC 2010). The inclusion of migration (as distinct from displacement) within the Cancún Framework suggests that voluntary movement in response to climate change is possible. Nash (2018, 58) argues that this categorization is 'never explicitly questioned' creating an idea that this language is neutral, correct and uncontroversial.

The next major turning point for addressing climate-related migration within the UNFCCC came with the establishment of the Task Force on Displacement at the 21st Conference of Parties in Paris in 2015. Rather than looking at how to develop protection mechanisms for displaced people, the Task Force looked at how to avert and minimize displacement through sharing of information, tools and guidance among parties, the provision of financial, technological and capacity building support from UN agencies to affected community and through the facilitation of 'orderly, safe, regular and responsible migration and mobility of people [...] by enhancing opportunities for regular migration pathways, including through labour mobility' (UNFCCC 2018).

Linking planned migration to adaptation and development

Narratives emphasizing the adaptation and development benefits of migration focus on facilitating migration that is 'planned,' 'orderly' or 'regular.' The 'planned' component of this narrative is critical to understanding how migration in the context of climate change has been transformed (discursively at least) from a 'forced' process into a 'voluntary' and empowering one. Narratives that advocate for the use of migration as adaptation focus on several key factors: the voluntary nature of anticipatory, planned migrations; the ability to avoid creating trapped populations; and the livelihoods or development opportunities migration offers.

When migration is *planned* in anticipation of disasters or slow-onset climatic changes it is deemed to be more voluntary than movement that occurs without warning (UNHCR 2014). Making voluntary and planned migration pathways available through government support can help to avoid trapping populations who lack the capacity to move freely (Black et al. 2011), while also addressing underlying development problems related to 'scarce resources, overcrowding, rapid urbanization and environmental degradation' (McAdam 2011, 26). Significantly, planned migration has been framed as an opportunity to diversify incomes and create new livelihood and development opportunities, both for those who move and those who stay behind – through the receipt of remittances (Kälin 2015). The United Nations High Commissioner for Refugees, for instance, has suggested that by working within a sustainable development framework, planned relocation has the ability to improve livelihoods and standards of living while at the same time reducing physical exposure to environmental risks (UNHCR 2014).

Concerns with adaptation and development approaches

Issues raised against the 'migration as adaptation' narrative range from concerns that it fails to reflect migrant realities (Ransan-Cooper et al. 2015) and the subjective barriers to migration that people face (Adams 2016); that it is unable to acknowledge the potential for new vulnerabilities in the places migrants move to (Adger et al. 2015); and that it allows for responsibility to be diverted from international actors to local governments, communities and to migrants themselves (Bettini 2014; Methmann and Oels 2015).

This last point speaks to concerns relating to climate justice. By framing migration as a form of adaptation, there is less of an imperative for international actors to adopt mitigation measures or to protect people affected by climate change when they 'choose' to move (McNamara and Gibson 2009; Gemenne 2015). Furthermore, by focusing on 'pull factors' – such as better job opportunities – that contribute to decisions to migrate, the role climate change plays as a 'push factor' is concealed. Bettini (2014, 185) has described this as a 'political convenience', allowing the UNFCCC to frame climate migration as a 'low controversy' issue. Methmann and Oels (2015, 53) reiterate this point by looking at how depicting migrants as 'adaptive' and 'resilient' has contributed to the 'elimination of the political' by framing migration as a 'choice' rather than the result of a global political failure to reduce emissions. This allows for the loss and damage associated with climate-related migrations to be concealed. For example, focusing on the economic development benefits of migration risks obscuring what Adams (2016, 445) describes as 'the non-economic benefits of place and the attachment people form to them'. Similarly, by linking migration to development opportunities, it is easy to conceal the way in which patterns of development in the Global North (particularly those that have emphasized unfettered economic growth) are responsible for contributing to increased exposure to climate risks in the first place (Cannon and Müller-Mahn 2010).

Discussion: place-based implications, contesting dominant narratives

Understanding the link between migration and climate change, and how this phenomenon should be responded to, is complicated by a lack of knowledge around the extent to which climate change actually drives migration and thus ambiguity around whether migration, in this context, should be understood as 'voluntary' or 'forced.' Despite these challenges, policy makers responding to climate-related movement are careful not to draw attention to them. This contributes to the objectification of policy solutions by concealing the different historical, social, and political contexts in which migration in response to climate change occurs. Revealing these contexts is important as it helps to (1) challenge the idea that policy is responding to an objective (as opposed to a subjective and context specific) phenomenon, (2) reveal the power dynamics and political motivations behind seemingly 'neutral' policy solutions, and (3) demonstrate the existence of diverse alternative narratives.

In order to do this, we need to look at how dominant policy solutions have travelled and how comfortably or uncomfortably they sit within places of implementation. Here, case study analysis is a useful tool, and in the following we provide one such case study from our research in Fiji which looked at how international guidelines on planned relocation reflect place-based subjectivities and ideas of acceptable risk (Benge and Neef 2020).

Place-based contestations of dominant narratives

The work of Carol Farbotko is an excellent place to begin when aiming to understand place-based narratives and, in particular, how identity informs perceptions around appropriate responses to climate change. Farbotko and McMichael (2019) recently explored how voluntary immobility in the Pacific provides a counterargument to the notion of 'migration as adaptation' and how the decision to remain in place might act as an important adaptation strategy in and of itself – helping to address the threat to culture, identity and place-based connection that migration in response to climate change poses. Current narratives focusing on migration as an adaptive response to climate change have contributed to concealing experiences relating to both voluntary and involuntary immobility – ignoring the stories of those who choose to stay, as well as those who are unable to migrate because they lack the resources to do so (commonly referred to as 'trapped populations'). While trapped populations have received some mention in climate migration literature (Black and Collyer 2014), populations who *choose* to remain in place have received very little.

Looking at these populations, and importantly the reasons why they choose to remain in place, offers a strong counter to the narrative of 'migration as adaptation'. While the 'migration as adaptation' narrative has focused on promoting migration for its economic development and livelihood opportunities, Farbotko (2018) reveals the voices that are silenced by this narrative. These include those that are concerned about the risk of losing their identity, their communal unity and ties to their ancestors by leaving their land. Similarly, our own research in Fiji sheds light on the way in which policy discourse often places more emphasis on physical risk factors and economic opportunities propelling movement rather than the cultural or socio-psychological factors contributing to decisions to migrate or to remain in place.

In 2014 it was estimated by the Fijian Government that 676 villages would be likely to require relocation in the future due to the impacts of climate change (Leckie and Huggins 2016) and more recently it has been suggested that some 80 communities are in need of relocation now (Piggott-McKellar et al. 2019). This year the Internal Displacement Monitoring Centre risk profile for Fiji also estimated that 5,800 people could be displaced in any given year due to storm surges, earthquakes, cyclones and tsunami (IDMC 2020). Responding to these risks the Government of Fiji published its Planned Relocation Guidelines in 2018 to assist stakeholders in implementing coordinated, *voluntary* and participatory community relocations within Fiji. Interestingly, the guidelines reflect dominant international policy narratives around migration in the context of climate change by considering planned relocation as an 'adaptation strategy' and by framing relocation as a solution to address development pressures such as 'overcrowding, unemployment, infrastructure, pollution and environmental fragility' (Government of the Republic of Fiji 2018, 7). While Fiji's Planned Relocation Guidelines talk about relocation as a 'last resort', due in part to the recognized socio-cultural risks of moving, relocation can still be justified by emphasizing the opportunity for a 'progressive standard of living' (2018, 14).

Our interviews carried out in 2016 with practitioners involved in the development of Fiji's guidelines suggest a degree of complexity and stakeholder contention that is not reflected in the policy text. While practitioners recognize the opportunities that relocation provides for access to better land, livelihoods and employment, it is also understood that relocation is likely to be met with resistance given the attachment of people to land and burial sites and the risk of social and cultural disturbance that relocation poses (Benge and Neef 2020). It is for these reasons that the guidelines emphasize the need for community participation in decision-making, so that relocation can be carried out 'voluntarily.' However, while practitioners acknowledged that community participation would ensure relocation was not 'forced by government' they noted that relocation could still be 'forced' from the perspective of having no other choice once in-situ adaptation efforts had been exhausted (Benge and Neef 2020). This is a significant point as it introduces questions of responsibility and justice which climate change adaptation narratives have often been criticized for overlooking. It suggests also that while relocation might offer development opportunities, these opportunities should not conceal the social and cultural losses associated with migration or the international responsibility for contributing to this loss through their role in the production of global emissions.

These examples are not meant to suggest that the majority of Pacific Island communities would choose voluntary immobility, yet highlighting these perspectives is important as it demonstrates how easily some voices – in particular those that adopt alternative perspectives on risk and opportunity – can be concealed by dominant and seemingly 'neutral' policy narratives. This concealment is the result of policy narratives that have been constructed as real despite being a reflection of power, namely of who has the power to speak and be heard.

When reading policy or examining proposed policy solutions it is critical to look closely at whose voices are being prioritized, whose stories or perspectives are obscured and what the political implications of this might be. While this may not always be self-evident, we recommend starting by examining lived realities, indigenous beliefs, stories and values. Doing so will help to demonstrate what is missing from policy narratives and why. The following three-step pedagogical guide sets out a methodology for critically examining the migration and climate change nexus and for revealing the political intentions and relations of power that sit behind dominant policy solutions.

Understanding climate mobility narratives using a critical pedagogy

Begin with case studies

Beginning with case studies ensures students do not formulate their understanding of the subject through conceptual papers, theory, or policy alone. While policy concepts and theoretical frameworks are important, empirical case studies demonstrate how policies are translated at the local level and help students link theory with development praxis. Good case studies adopt a critical lens and offer a diversity of perspectives on how people understand, give meaning to, and respond to the impacts of climate change are particularly useful.

Farbotko, C. and Lazrus, H. (2012) 'The First Climate Refugees? Contesting Global Narratives of Climate Change in Tuvalu', *Global Environmental Change* 22, 382–390.

This article juxtaposes global climate refugee narratives with Tuvaluan conceptions of climate challenges and mobility practices.

Piggott-McKellar, A.E., McNamara, K.E., Nunn, P.D. and Sekinini, S.T. (2019) 'Moving People in a Changing Climate: Lessons from Two Case Studies in Fiji', *Social Sciences* 8(5), 133.

This article documents people's lived experiences in two relocated communities in Fiji and assesses the outcomes of the relocations on those directly affected.

Consider the following questions when reading these case studies:

- How is the relationship between human mobility and climate change depicted?
- How has this perspective been informed?
- Whose voices are being considered in these case studies? Whose might be obscured?
- What are the possible implications of obscuring these voices / narratives?
- What suggestions do these articles make about possible policy solutions?

Examine policy proposals and critical policy commentaries

Policy is not 'neutral, rather it is the cumulative output of a particular social, political and historical context (Nash 2018). Revealing this context is critical for understanding the intentions of policy and its likely implications. It also helps us to recognize how policy is built upon a knowledge base that is often fluid, subjective and contested – despite often being depicted as objective and stable. Reading policy proposals and critical policy commentaries alongside case studies allows students to reflect on how this knowledge is constructed and the perspectives that may be concealed when policy is presented as universally applicable.

1. Biermann, F. and Boas, I. (2010) 'Preparing for a Warmer World: Towards a Global Governance System to Protect Climate Refugees', *Global Environmental Politics*, 10(1), 60–88.

This article argues for a new protection regime for 'climate refugees' as opposed to an expansion of the Geneva Convention.

2. McAdam, J. (2011) 'Swimming against the Tide: Why a Climate Change Displacement Treaty is Not the Answer', *International Journal of Refugee Law* 23(1), 2–27.

This article presents a set of arguments against the benefits of a climate displacement treaty and maintains that the focus should be on bilateral or regional agreements instead.

You might ask the following questions when reading these policy proposals:

- What assumptions about climate mobilities are made in these articles?
- Is a consistent policy narrative presented? What anomalies exist within and between these narratives? What narratives might be concealed?
- What implementation challenges might be encountered when policy moves from institutions to places of implementation?

Look for critical deconstructions of dominant narratives

Critically examining the climate migration nexus requires asking questions that help us to understand how problems and their solutions have been framed, by whom and with what political effect. Papers that successfully address these questions help to pull climate migration narratives out of the depoliticized realm and into a space that reveals political motivation, power, inequality and hidden questions of justice.

1. Bettini, G. (2014) 'Climate Migration as an Adaptation Strategy: De-securitising Climate-Induced Migration or Making the Unruly Governable?', *Critical Studies on Security* 2(2), 180–195.

This article takes a critical perspective on the shift from the 'forced migration and climate refugee' narrative to a 'migration as adaptation' discourse.

2. Methmann, C. and Oels, A. (2015) 'From 'Fearing' to 'Empowering' Climate Refugees: Governing Climate-Induced Migration in the Name of Resilience', *Security Dialogue* 46(1), 51–68.

The authors critique how climate change-induced migration is increasingly presented as a rational strategy of adaptation to unavoidable levels of climate change, rendering the relocation of millions of people acceptable and rational.

Consider the following questions when reading these critical articles:

- What do these articles tell us about how migration in the context of climate change is governed?
- To what extent do the authors adopt a self-reflective approach?
- Do these authors offer alternatives to the narratives they critique? How do these alternatives avoid reproducing new forms of mobility governance?

Conclusion

As we have shown in this chapter, debate over how to address climate-related human mobility has depended on whether movement is understood principally as a 'voluntary adaptive oppor-tunity' adopted by 'resilient' migrants, or as a 'forced impact' experienced by 'vulnerable' people in need of protection. While serious problems have been identified with both narratives, per-haps the greatest issue lies in the inability to reconcile these perspectives – that is, to understand how human movement can be simultaneously both an opportunity *and* a loss. In Fiji, and other parts of the Pacific, it is easy to see how this is possible: mobility has the potential to *decrease* vulnerability caused by environmental degradation at the same time as *increasing* vulnerability to cultural harms – such as loss of attachment to place and identity. Which side of this narrative is emphasized will depend on who is telling the story and whose interests are being served. When the adaptive, voluntary quality of mobility is emphasized by international actors this may help to avoid an international responsibility to protect migrants, while local governments and communities may use this same narrative to promote self-determination and to protect against international interference. On the other hand, when the forced quality of mobility is emphasized this can help to create an imperative for ongoing emissions reduction as well as a case for international compensation.

Acknowledgement

Work on this chapter has been made possible by funding provided by the Worldwide Universities Network (WUN) for the project 'Climate-Induced Migration: Global Scope, Regional Impacts and National Policy Frameworks.'

References

Adams, H (2016), 'Why Populations Persist: Mobility, Place attachment and Climate Change', *Population and Environment*, vol. 37, no. 4, pp. 429–448.

Adger, W.N., Arnell, N.W., Black, R., Dercon, S., Geddes, A. and Thomas, D.S (2015), 'Focus on Environmental Risks and Migration: Causes and Consequences', *Environmental Research Letters*, vol. 10, no. 6, pp. 1–6.

Barnett, J. and Chamberlain, N (2010), *Migration as Climate Change Adaptation: Implications for the Pacific', in B. Burson (ed.), Climate Change and Migration: South Pacific Perspectives,* Milne Print, Wellington, pp. 51–60.

Benge, L. and Neef, A (2020), *Planned Relocation as a Contentious Strategy of Climate Change Adaptation in Fiji', in A. Neef and N. Pauli (eds) Climate-Induced Disasters in the Asia-Pacific Region,* Emerald Publishing Limited, Bingley, pp. 193–212.

Bettini, G (2014), 'Climate Migration as an Adaptation Strategy: De-securitizing Climate-Induced Migration or Making the Unruly Governable?', *Critical Studies on Security*, vol. 2, no. 2, pp. 180–195.

Biermann, F. and Boas, I (2010), 'Preparing for a Warmer World: Towards a Global Governance System to Protect Climate Refugees', *Global Environmental Politics*, vol. 10, no. 1, pp. 60–88.

Black, R., Bennett, S.R.G., Thomas, S.M. and Beddington, J.R (2011), 'Climate Change: Migration as Adaptation', *Nature*, vol. 478, pp. 447–449.

Black, R. and Collyer, M (2014), 'Trapped' Populations: Limits on Mobility at Time of Crisis', in S.F. Martin, S. Weerasinghe and A. Taylor (eds), *Humanitarian Crises and Migration*, Routledge, London and New York, pp. 287–305.

Boas, I. and Rothe, D (2016), 'From Conflict to Resilience? Explaining Recent Changes in Climate Security Discourse and Practice', *Environmental Politics*, vol. 25, no. 4, pp. 613–632.

Cannon, T. and Müller-Mahn, D (2010), 'Vulnerability, Resilience and Development Discourses in Context of Climate change', *Natural Hazards,* vol. 55, pp. 621–635.

Farbotko, C (2018), 'Voluntary Immobility: Indigenous Voices in the Pacific', *Forced Migration Review*, vol. 57, pp. 81–83.

Farbotko, C. and Lazrus, H (2012), 'The First Climate Refugees? Contesting Global Narratives of Climate Change in Tuvalu', *Global Environmental Change,* vol. 22, pp. 382–390.

Farbotko, C. and McMichael, C (2019), 'Voluntary Immobility and Existential Security in a Changing Climate in the Pacific', *Asia Pacific Viewpoint*, vol. 60, no. 2, pp. 148–162.

Farbotko, C., Stratford, E. and Lazrus, H (2016), 'Climate Migrants and New Identities? The Geopolitics of Embracing or Rejecting Mobility', *Social and Cultural Geography*, vol. 17, no. 4, pp. 533–552.

Gemenne, F (2015), 'One Good Reason to Speak of "Climate Refugees"', *Forced Migration Review*, vol. 49, pp. 70–71.

Government of the Republic of Fiji (2018), *Planned Relocation Guidelines. A Framework to Undertake Climate Change Related Relocation*, Suva: Ministry of the Economy.

Gromilova, M (2014), 'Revisiting Planned Relocation as a Climate Change Adaptation Strategy: The Added Value of a Human Rights-Based Approach', *Utrecht Law Review*, vol. 10, no. 1, pp. 76–95.

Hall, N (2016), *Displacement, Development, and Climate Change: International Organizations Moving Beyond their Mandates*, Routledge, London and New York.

IDMC (2020), *Fiji Disaster Displacement Risk Profile*, Internal Displacement Monitoring Centre (IDMC), Geneva.

IOM (2019), *World Migration Report 2020,* International Organization for Migration (IOM), Geneva.

Kälin, W (2015), 'The Nansen Initiative: Building Consensus on Displacement in Disaster Contexts', *Forced Migration Review*, vol. 49, pp. 5–7.

Leckie, S. and Huggins, C. (eds) (2016), *Repairing Domestic Climate Displacement: The Peninsula Principles*, Routledge, London and New York.

McAdam, J (2011), 'Swimming against the Tide: Why a Climate Change Displacement Treaty is Not the Answer', *International Journal of Refugee Law,* vol. 23, no. 1, pp. 2–27.

McAdam, J. and Ferris, E (2015), 'Planned Relocations in the Context of Climate Change: Unpacking the Legal and Conceptual Issues', *Cambridge Journal of International and Comparative Law*, vol. 4, no. 1, pp. 137–166.

McNamara, K.E. and Gibson, C (2009), '"We Do Not Want to Leave Our Land": Pacific Ambassadors at the United Nations Resist the Category of "Climate Refugees"', *Geoforum*, vol. 40, pp. 475–483.

Methmann, C. and Oels, A (2015), 'From "Fearing" to "Empowering" Climate Refugees: Governing Climate-Induced Migration in the Name of Resilience', *Security Dialogue*, vol. 46, no. 1, pp. 51–68.

Nash, S.L (2018), 'From Cancun to Paris: An Era of Policy Making on Climate Change and Migration', *Global Policy*, vol. 9, no. 1, pp. 53–63.

Piggott-McKellar, A.E., McNamara, K.E., Nunn, P.D. and Sekinini, S.T (2019), 'Moving People in a Changing Climate: Lessons from Two Case Studies in Fiji', *Social Sciences*, vol. 8, no. 5, pp. 133.

Ransan-Cooper, H., Farbotko, C., McNamara, K.E., Thornton, F. and Chevalier, E. (2015), 'Being(s) Framed: The Means and Ends of Framing Environmental Migrants', *Global Environmental Change*, vol. 35, pp. 106–115.

Rigaud, K., Kanta, de Sherbinin, A., Jones, B., Bergmann, J., Clement, V., Ober, K., Schewe, J., Adamo, S., McCusker, B., Heuser, S. and Midgley, A (2018), *Groundswell: Preparing for Internal Climate Migration*, The World Bank Washington.

Thomas, A. and Yarnell, M (2018), *Ensuring that the Global Compacts on Refugees and Migration Deliver,* Refugees International, Washington DC.

UNFCCC (2010), *Report of the Conference of the Parties on its Sixteenth Session* (held in Cancún from 29 November to 10 December 2010), United Nations Framework Convention on Climate Change (UNFCCC) Cancún.

UNFCCC (2018), *Report of the Conference of Parties on its Twenty-Fourth Session* (held in Katowice from 2 to 15 December 2018), United Nations Framework Convention on Climate Change (UNFCCC), Katowice.

UNHCR (2014). *Planned relocation, disasters and climate change: Consolidating good practices and preparing for the future.* United Nations High Commissioner for Refugees (UNHCR), Geneva. Retrieved from http://www.unhcr.org/54082cc69.pdf

34

EDUCATIONAL INEQUALITY AND DEVELOPMENT

Youyenn Teo

Introduction

Formal education is widely regarded and presented as an important goal in the quest for human wellbeing, as well as a crucial pathway for fostering greater equality in societies. Low rates of schooling, poorly resourced schools, and unequal access to schooling – particularly along gender or regional lines – are, in development discourse, unequivocally framed as challenges that should and must be overcome. Global governance institutions such as the United Nations (UN), the Organisation for Economic Co-operation and Development (OECD), and the World Economic Forum (WEF), place access to schooling at the forefront of developmental goals and on their institutional agendas of research and advocacy. The United Nations' Sustainable Development Goals (SDG), for example, includes 'Quality Education' among its 17 aspirations, explicitly articulating education as key to offering people routes for escaping poverty and upward socioeconomic mobility (United Nations 2015).

This global development agenda – its framing of education as having ameliorative effects on wellbeing and equality, as well as its tracking of indicators such as school-going rates and scores in standardized tests – is important as well as useful. Schools, and public education in general, have indisputably been sites of respite, safety, and possibility for children. The suspension of schools during the COVID-19 crisis has provided stark evidence of this (Sharma 2020). Ongoing research through assessments like the OECD's Programme for International Student Assessment (PISA) help us understand how countries fare relative to one another, frequently drawing attention to the contrasts between the wealthier Global North and poorer Global South.

That education can lead to better outcomes for individual persons, particularly those who live in conditions of high deprivation, is indisputable. The claim of education leading to upward socioeconomic mobility, however, is a far stronger claim and aspiration that we should scrutinize. It implicates education in another major developmental goal – that of reducing inequality (SDG10). Framing education as a solution to reducing inequality – either between countries or within national contexts – is a leap that perhaps relies more on faith than data. Accepting without question that better education will lead to greater equality requires sidestepping important empirical evidence that inequality finds fertile ground to grow in inventive ways precisely *through* education systems; it is here that inequality thickens beyond different material

DOI: 10.4324/9781003017653-38

outcomes into norms, beliefs, and habits that rationalize, naturalize, and *legitimize* unequal outcomes (Bourdieu 1989).

The case of contemporary Singapore is useful for examining what development discourse on education captures and misses, and particularly for interrogating the claim of education as a solution to inequality. Here is a high-income country that performs superbly on global indicators of school attendance and student performance, and not so well on measurements of equality and student wellbeing. The co-existence of these two results suggests that achieving some aspects of educational success does not automatically lead to it being a solution to inequality. Yet, this is also a case often lauded as model because the 'success' of its system is well known and the 'failure' is not, indicating that measurements and global rankings have themselves become significant legitimating instruments *and* that the slippery connections between education and equality have too frequently gone unexamined. When scrutinized, the Singapore case in fact offers cautionary lessons: its ostensible successes have come with particular costs to wellbeing and equality. Undoing these – precisely because of the legitimation of its 'success' – is a difficult challenge.

To better understand that which is taken for granted, we must scrutinize both that which is obvious and visible, and what is obscured and invisible. In what follows, I pay attention to two sets of visibilities and invisibilities. I alluded to the first earlier: development discourse and practices highlight aspects of education while obscuring others, rendering visible its potential for positive change and invisible how it can deepen inequality. Second, in discussions of inequality, we tend to focus on 'the poor' or deprived. I offer instead empirical contrasts of the visible practices of wealthier families to the less visible challenges of poorer ones, in order to demonstrate the relationality between wealth and poverty. This illustrates what education systems demand and reward, drawing attention to how systems that score extraordinarily well on global indicators may nonetheless embed in their designs mechanisms that generate rather than disrupt inequality.

Education in Singapore: visible successes, obscured limitations

The Singapore education system has been internationally lauded, particularly for its testable outcomes. In the 2018 Programme for International Student Assessment (PISA), for example, Singapore students scored higher than the OECD average for reading, mathematics, and science; a high proportion of its students are at the highest levels of proficiency in these subjects (OECD 2019b). Public education is widely valorised by the state and coordinated under the auspices of the Ministry of Education (MOE). Fees are heavily subsidized and kept relatively low, particularly at the primary and secondary school levels (Ng 2014). Teachers receive state-subsidized and regular training – at the National Institute of Education (NIE) – as well as respectable compensation.

While Singapore students score very well relative to those in other countries, within-country inequalities have been highlighted by OECD reports on the PISA (OECD 2019a, OECD 2018). On measurements the reports identify as problematic because they tend to intensify socioeconomic inequality – early tracking of students; focus on academic results; segregation of students into different schools by academic performance; segregation of socioeconomically advantaged and disadvantaged students into different schools – Singapore is at one end of the spectrum in having a high degree of all four. This puts it in stark contrast to other 'high-performing' places like Finland. Correspondingly, differences in scores between socioeconomically advantaged and disadvantaged Singapore students is higher than the OECD average (OECD 2019b). On measures that compare socioeconomically advantaged and disadvantaged students' wellbeing – their sense of self-efficacy and belonging – Singapore also fares more poorly compared to other countries.

Irene Ng argues that intergenerational mobility in education has been moderately low over the past decades, and that this is due to its design focus, including tracking students of different academic abilities into different schools (Ng 2014). Ong Xiang Ling and Cheung Hoi Shan found in a study that – as measured by parents' educational attainment, housing type, and monthly per capita household income – students in 'elite' schools are disproportionately from higher socioeconomic status households compared to students from 'neighbourhood' schools (Ong and Cheung 2016). On the other side of the spectrum, among students placed in lower educational tracks, there are disproportionate numbers of students from ethnic minority and lower-income households (Wang et al. 2014).

The highly competitive and stressful nature of schooling is a frequent lament of Singapore parents in multiple spheres – online, on parent chat groups, in national news media and pop culture, as well as in private conversations. Despite official rhetoric, often from top government officials, that there are multiple pathways to success, that academic achievements are not everything, and that all schools are good, no one is buying it. The reality everyone sees is that some schools have better outcomes than others in terms of exam results and, more importantly, that educational credentials are deeply correlated to job opportunities and wages. Figure 34.1 illustrates the consequences of educational qualifications on incomes. The average monthly income of households with a university graduate as main income earner (S$18,255) is almost three times that of households with a secondary school graduate as main income earner (S$6,580). The major jump in income happens for university graduates, highlighting the stakes involved in academic performance. University graduates are also well aware that their graduating grade point average and corresponding category of honours affect their job prospects and starting salaries, particularly for civil service jobs.

It is important to see these outcomes not as aberrations but as having systemic roots.

Historically, the Singapore education system has developed around the goal of identifying and nurturing 'talent' who can contribute to the nation's developmentalist goals. While

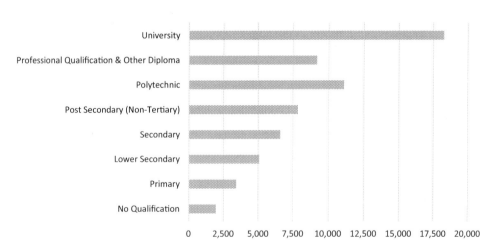

Figure 34.1 Average monthly household income among resident households by highest qualification attained of main income earner

Source: Department of Statistics Household Expenditure Survey 2017/18

educating the masses is important, embedded within the system is deep commitment to the idea that true talent is limited, partly because intelligence is biologically predetermined (Lee 1983). The school system thus has an important job of identifying and specially nurturing elites (Barr and Skrbis 2008). To this end, a national system of tracking was put in place beginning around 1979 and refined over the years (Deng and Gopinathan 2016, Ng 2014).

The state has been steadfast in pursuing its goal of identifying 'talent' through sorting and hierarchization over the past decades. Despite mounting international research critiquing the practice, tracking begins early, about 9 years old. Three major examinations – each focused on narrow academic skills – stand between a child and university. After each major examination, students are sorted into different schools and 'streams' based on their test results. Few students move between streams or schools once they are sorted (Ministry of Education 2012). Throughout the school years, English, Mathematics, and Science are the most highly emphasized subjects.

The proportions of students in the different tracks over time provides a glimpse of the durability of these hierarchical tiers. The tiers, as well as the proportions of students who fit in each, have been relatively stable. For each cohort that started primary school between 1996 and 2008, Figure 34.2 illustrates the proportions in the various streams and Figure 34.3 their educational pathway and attainment.

Over this 13-year period, there has been limited changed in broad structure. Where there can be said to be some upward mobility – fewer students who drop out of secondary school and more students who acquire university degrees – this reflects an overall shift upwards as development progressed, the country became wealthier, and people's demands for university education grew. The layers themselves – the categories, size, and relative position – have seen remarkably few shifts.

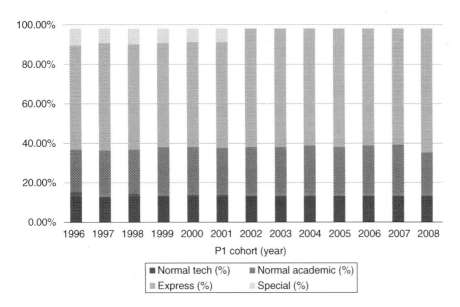

Figure 34.2   Proportion of students enrolled in Secondary School by stream

Note [1] The Special and Express streams were merged, in accounting though not in practice, from the 2008 Sec 1/2002 Primary 1 cohort onwards

Source: Ministry of Education. (2019). 'Enrolment – Secondary, By Level and Course'

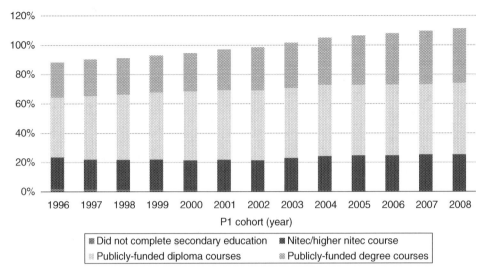

Figure 34.3 Level of participation at post-secondary level by cohort (year starting P1)

Note [1] Numbers add up to more than 100 per cent because some students completed Nitec/Higher Nitec courses and/or diplomas before going on to publicly-funded degree courses, and so are counted in more than one category. For the most part, however, this chart is indicative of highest levels of educational attainment

[2] Nitec/Higher Nitec courses are courses in vocational schools known as Institute of Technical Education (ITE)

Sources: Ministry of Education. 2019. 'Percentage of P1 Cohort Who Did Not Complete Secondary Education' Ministry of Education. 2019. 'Percentage of P1 Cohort that Progressed to Post-Secondary Education'

I was in elementary and high school in the 1980s and 1990s, and vividly recall principals and teachers unabashedly goading us to earn our place among the 'cream of the crop,' 'the top 10 per cent.' Openly elitist language has receded from public view in recent years, and tracking is now described as a pedagogical tool to identify different learning styles and provide multiple pathways in learning and achievement. Yet, looking at the structural organization of schooling, there has been remarkable consistency rather than change. Recalling where Singapore stands vis-à-vis global indicators, both its successes and failings start to make sense. The system rewards narrow forms of learning that translate well into standardized testing under time pressures; the 'top' segment of its student population is indeed identified precisely by their mastery in these types of assessments. As I shall elaborate, this mastery is acquired through private investments in tutoring. Importantly, each of the qualities identified by OECD reports as problematic from the perspective of reproducing inequality – early testing, strong focus on academic results, segregation by track and school – are neither negative externalities nor accidental features; they are purposeful design.

The pattern of hierarchized tiers, and the principles that have brought them about in the Singapore case, alert us to a key problem with applauding education as a route to upward social mobility: we can accept this as self-evident and as a worthy end in itself only if we avert our eyes from the hierarchies, how they are enacted, and their very real consequences for lives and wellbeing.

Education systems and social norms: entrenching inequality

Ignoring education as sites that reproduce and naturalize inequality is, as I argue earlier, problematic. When obscured, cases such as Singapore's come to be conceived as models. This affects other countries seeking to emulate models and also itself; learning and reforms cannot occur where there are deep misrecognitions. The patterns of inequality I describe in the previous section have not gone completely unnoticed within Singapore, but the belief, nationalist pride, as well as everyday habits invested in the Singapore model has made reckoning with its failings extremely difficult and complex.

In this section, I focus on the last of these – everyday habits – to illustrate the shape and texture of inequality and its embeddedness in ordinary lives within this highly 'successful' education system. I describe how parents and children across the class spectrum respond to the school system they are in. I hope this section serves as a reminder that trends of inequality map onto real human costs. Singapore is perhaps an unexpected case to use for speaking of human costs, particularly in the development context. It is so ostensibly successful that the pitfalls of its policy regime are obvious only to its own inhabitants in rare moments of honesty. This then is the specific provocation: here lies an important cautionary tale that if inequality is embedded in policy principles and design, then one should not be surprised when it is the outcome.

From 2013–2016, researching the everyday lives of low-income Singaporeans, I spoke with more than 200 people in several low-income neighbourhoods (Teo 2018). Between 2018 and 2020, I interviewed 92 parents from various socioeconomic backgrounds about how they manage work and care responsibilities.

Across the two projects, many conversations centred on how children do in school. This is a major part of parents' sense of themselves as parents. Parental involvement is framed as a necessity, particularly at the PSLE (Primary School Leaving Examination) stage – this is the exam they see having make-or-break effects on a child's path. Because 12 years old is an age where children themselves are often nonchalant, parents, and mothers in particular, express high levels of anxiety and a sense that they are personally accountable for their children's exam results. It is common in Singapore to hear of women quitting their jobs for the much-dreaded 'PSLE year.' These feelings – of education as parental responsibility and exams as high-stakes – hold true for parents across the class spectrum.

The difficulty of examinations; the deep belief that they have huge consequences for their kids' futures; and the difficulty of teaching one's own children, fuel the growth of the private tuition industry. Parents with ample means use these to help their children from the get-go (in some cases as young as preschool) and on a regular basis (i.e. throughout the school years); many hire tutors pre-emptively – before children ever struggle in school – because the aim for their children is not to pass exams but to ace them. Parents with moderate means forgo other household needs to pay for tutors in crucial exam years and/or on subjects especially tough for their kids. Tuition is a billion-dollar industry, with parents spending significant proportions of household income on it (Wise 2016). In 2017/18, monthly expenditure on private tuition and other educational courses varied widely by household income: households in the 1st to 20th percentiles spent S$45.30 per month. Relative to this, those in the 21st to 40th percentile spent 1.7 times the amount (S$75.80); those in the 41st to 60th percentiles 2.7 times (S$121.20); those in the 61st to 80th percentiles 3.4 times (S$152.60); and those in the 81st to 100th percentiles 3.7 times (S$167) (Department of Statistics 2018).

Tuition has become a social norm – parents think of it as necessary; children see that most of their friends have it, and school teachers themselves often recommend it. Social scripts have developed around it: parents lament that schoolwork has become more challenging and their kids have much busier lives now compared to their own childhoods. They express resignation and empathy for school teachers who have 'so many other kids' to teach that they cannot really give personal attention to their kid. University-educated parents who *should* be able to do primary school mathematics, and who therefore feel sheepish that they are not the ones teaching their children, talk about how hard it is to coach their own kids, how often they lose their tempers, and therefore how they prefer to 'outsource' this labour in order to maintain good parent-child relationships. In any case, they tell me, students are expected to show how they have arrived at answers in ways so different from when they were in school that they cannot teach methods that translate to the 'correct' answers in tests and exams.

The ubiquity of private tutoring, importantly, has shifted everyone's expectations of what school does and how teachers teach. Instead of making demands for teachers to slow down for their children, parents accept – since 'everyone has tuition' – that the pace at school will be fast and that their responsibility as parents is to pay for extra coaching outside of school in order for their children to 'catch up' or 'not fall behind' within it. Recognizing tuition as a norm, parents feel urgent pressure to 'give tuition' to *keep up*, indicating that the widespread use of private tutoring may also have distorting effects in influencing how school teachers operate. Parents speak often of standards being 'much higher than in our time.' Some even lament that their children are doing most of their learning at tuition and not in school.

The skills required for doing well in Singapore schools today, far from reflecting 'natural talents,' need to be purchased. A closer look at tuition services reveals a tiered market to serve customers with different budgets. In this massive business, there are branded chains as well as bespoke services. There are smaller and cheaper centres located in so-called heartland neighbourhoods as well as fancier chain stores located in major shopping malls. Many university students work part-time as tutors, but the industry is also heavily staffed by ex-school teachers. These teachers – and especially those who have worked in well-known schools – command the highest prices, indicating their value in being able to teach the official curriculum. Tuition centres unabashedly advertise their capacity to teach students test-taking skills that will translate into improved grades at major exams. So intense is the need for tuition that this has also become a major role played by the social service sector – volunteer-run and/or charity-based tuition services are in high demand, particularly in low-income neighbourhoods. The whole tuition landscape is well known to parents: they talk knowledgeably about different price points, tutors with varying teaching credentials and years of experience, variant levels of customized attention, and therefore to some extent different 'returns on investment.' On education as in many other things in life, people expect to get what they pay for (Teo 2017).

As one should expect when private resources vary and market choices abound, inequalities result. Kids from low-income families fall behind almost immediately as they enter primary school. They are less advanced than kids from wealthier families, who are often better prepared – through kindergarten, enrichment classes, and home learning – to read and write. Very quickly, many barely pass or completely fail English and Mathematics. In Primary 1 and 2, many of the kids from low-income families are identified as having problems and pulled out of class for extra coaching. While this can help, it is not easy for the kids to catch up, since the more advanced kids continue to move forward at a fast pace. By Primary 3, many kids from low-income families are tracked and banded into lower-performing classes. Although schools vary in how obvious the banding is at Primary 3 and 4, kids themselves are well aware of where they stand vis-à-vis others. By Primary 5 and 6, many of the kids do so poorly that they have to switch to what is known as Foundation level for some or all of their subjects. Many develop a sense of themselves as inferior to others and start to feel demoralized. In some cases, kids begin to resist going to school. If conditions are unstable at home – changing

Level	A-Level / Diploma Tutor	Undergrad Tutor	Graduate Tutor	MOE Teacher
Per Hour Rates				
Lower Pri	$15-$20	$18-$30	$20-$35	$45-$55
Upper Pri (PSLE)	$18-$25	$20-$35	$22-$38	$45-$55
Lower Sec	$20-$30	$25-$35	$30-$45	$45-$60
Upper Sec	$25-$$35	$25-$40	$30-$50	$50-$80
JC	$30-$35	$30-$50	$40-$60	$70-$100
Music	fr $30			

*This is just an estimate. Prices may be lower or higher in some cases.

Figure 34.4 Tuition rates. Screenshot from online tutoring agency taken by Youyenn Teo, March 2020

living arrangements, irregular job schedules, unstable income, marital conflict, need for children to take on responsibilities such as caring for younger siblings – parents also find it difficult to manage their children's school schedules. If they do indeed stop going to school regularly, their social connections at school suffer. When they return, the absence of friends, teasing or bullying from schoolmates, can make staying the course difficult.

The effects of early sorting and labelling are profound for parents as well as kids. Parents with more means also tend to be parents who have the confidence to think their kids are not stupid, and just need more help. How kids are branded by schools, the information parents get from schools about how they compare vis-à-vis other children – these shape parents' conceptions of their children's capabilities. Parents' opinions and actions feed into children's sense of themselves and their potential. I met many parents with high hopes that their children will do better than them and not end up with such hard lives, but they parent in a context where their children are already branded as weaker, as not as smart or capable. For many low-income parents, a sense of resignation sets in. An acceptance of a child's poor results and lacks come to define the dynamic within the family.

In recent years, as inequality in Singapore has increased and public attention on it intensified, political leaders have spoken of 'levelling the playing field' and 'doubling down on meritocracy' (Ong 2018). Accompanying this, they have directed funds towards 'uplifting' those at the bottom. Apart from the problem of inequality, some Singaporean educators and parents have spoken up about the perverse effects this high-pressure, exams-centric system has on students' creativity, on their attitudes toward learning and risk-taking, as well as on their mental well-being. Despite tweaks, the state has been dogged in its commitment to maintaining the multiple tracks and streaming process, the examinations-oriented and high precocity system. Little has happened, in other words, to disrupt the role the formal school system plays in fuelling the shadow education industry *and* the shadow education industry's effects on how schools, parents, and students relate to one another.

Invoking the motif of visibility and invisibility, the problems of 'the bottom' are connected to those of 'the top' and for that matter 'the middle.' Everyone living within a society is compelled to navigate the national school system and the norms around performance and competition it engenders. Kids' experiences with schooling – either as places that push them to struggle incessantly to stay on top, or as places where they are failures – are a direct consequence of the larger social realities of the system's demands for precocity and narrow capacities, as well as the social habits of meeting those demands through private investments. The education game in Singapore is that of hierarchical jostling, involving – at various levels of consciousness and with variant sentiments – everybody.

Habits are difficult to break, particularly when so many people reproduce them on a daily basis. These are habits that have been invested with meaning – where no matter where one stands, there is acceptance that the Singapore education system is 'rigorous' and 'merit-based.' Those who come out 'on top' and end up in positions of power and influence have particular reason to perpetuate this. The habits are also linked to material investments: from parents who have already put in money, to teachers who have quit jobs to become private tutors, to centres that need to generate income to pay rent – there are costs to playing the game but now also costs to not playing it.

The Singapore education system, with all of its inequalities, has been naturalized and legitimized as a good one. That which is strongly declared, regularly said, and matter-of-factly presented takes on the status of common sense – accepted as the way things are, beyond reproach, outside the sphere of claims that need to be tested. 'World-class education' and 'meritocracy' have this type of status in Singapore society. From time to time there are critical questions raised about its specifics, but the belief – on the part of political elites and ordinary citizens – that it is a success is extraordinarily difficult to challenge.

Here is where we can see most clearly what is at stake when we introduce into the conversation not just the invisible but the *relationality* of the visible and invisible. This inserts a very simple question that disrupts the view of Singapore as model: world-class for whom? With this question in mind, I argue that the narrative of a 'world-class' public school system, where some students test spectacularly well, obscures the processes of individualization and privatization that are key to these 'successes' as well as the attendant costs of this version of success.

The chapter interrogates the analytic slippage between quality education, upward mobility, and equality. The case of Singapore is useful for illustrating the significance of policy design and serves as a reminder that education systems have often been sites for the thickening of inequality. If reducing inequality is a serious goal, then it must be pursued seriously rather than merely hoped for. Witnessing the pitfalls of an ostensibly successful system is a place to begin rethinking education and its promise for human flourishing and wellbeing.

Knowledge-production as practice and pedagogy

My research on inequality has thrown up findings beyond what I have presented about educational inequalities. In shedding light on the everyday experiences of very low-income persons in contemporary Singapore, I have contrasted the conditions they face with higher income persons in areas of wage work conditions, work-life balance, and access to public goods such as housing (Teo 2018). While I have chosen to focus primarily on education in this chapter, the inequalities in this area of life are just one of many facets. I picked this to illustrate inequality not because I hold out some naïve conception of education as the site for reducing inequalities, but because it serves especially well to illustrate the variant conditions faced by people making their way in the world, the ways in which the poor and powerless often play by rules on which they cannot win, and the ways in which game-rules often serve to naturalize existing hierarchies and inequalities (Bourdieu 1989, Khan 2011).

I write this chapter in the thick of the COVID-19 crisis. Inequality is one of the most important conceptual tools we have for understanding these times as well as considering the aftermath. Yet, as in 'normal' times, we are faced with a situation where the inequality problem appears high on the minds of scholars and activists but is on the periphery of elites' political discourse. The realm of education has again been the site for a kind of neutralization: there is much talk about getting laptops into the homes of low-income families so that kids can continue with schooling, but no talk about how the balance of public investments and private resources requires recalibration. In Singapore, much has been made of extra lessons for students who are taking major exams in 2020, and not a peep about rethinking the value of examinations as a pedagogical tool, as a mode of deciding who is worthy and who is not, nor how this logic translates into deep unevenness in how people are rewarded for their labours. In these times of massive global existential crisis, political leaders remain extraordinarily nonplussed. The Singapore state's responses to the crisis thus far remain framed in terms of uplifting those who need help, rather than how to rethink the future of the planet, the unequal wellbeing of different groups within the city's inhabitants, and the social arrangements that need to change to face the long fallout ahead.

What is the purpose of scholarship in these unsettling times and this troubling context? There are at least two ways to approach this – one centred on knowledge-production, the other oriented towards public sociology.

In the first, critical scholarship has to analyse polity and society in ways that raise fresh questions and overlooked perspectives, helping us better understand the mechanics of how things work and where they fail. Looking at public policy, scholarship can interrogate their real effects, which are separate from their ostensible goals. Inserting social scientific tools for thinking about variation, we can identify how their effects vary across social groups. Ethnographic approaches, in particular, can locate and describe how policies are negotiated by people and how repeated negotiations lead to the production of not just policy rules and regulations but social conditions that everyone in a given society relates to and navigates (Teo 2010). State practices, and the social norms that form around them, have deep implications for the ability of people to meet needs and flourish. Critical analysis of these is thus a crucial aspect of any quest to improve the conditions for people to meet needs and flourish (Wright 2010). The role of scholarship is partly to insert into view questions that go beyond the minutiae or technical aspects of policy regimes and to focus attention on big-picture principles and overlooked issues.

This leads me to the second issue – whose views should scholars try to reorient? In order to motivate evaluations of existing policies and practices, however modestly, there has to be pressure from civil society. A critical mass of a critical public has to create the pressure for policy change. While knowledge produced through scholarship may not generate this directly, it can nudge it along by inserting into the public sphere questions, perspectives, and explanations about present circumstances as well as future possibilities. Through its insights, findings, vocabularies, frameworks, scholarship can contribute to urgent conversations about what is to be done. It can, sometimes, shame governments into action by exposing injustices.

Existing arrangements of scholarly knowledge and its distribution are inadequate for the tasks critical scholarship *could* do. As I elaborate elsewhere (Teo 2021), academics have to position ourselves in a larger ecology of knowledge-producers – we have to find and create communities and bring others in the academy along; we have to stretch across generational divides; we have to do collective knowledge-production not only at the point of knowledge-dissemination but also at the point of conceptualization and production. In a world where our expertise is suspect, we have to build our own communities of legitimacy-granters, create legibility for our work outside the usual anointers of legitimacy. The labour of doing public sociology is collective labour, entailing time to create knowledge *and* solidarity, involving bodies in and out of the academy (Teo et al. 2020).

Michael Burawoy, in championing 'public sociology,' spoke of students as an important constituent within this public, and a teacher's first and immediate public audience (Burawoy 2014, Burawoy 2005). Where inequalities are salient and yet not understood – experienced in the everyday but rendered natural in public discourse and placed beyond conscious reflection – teachers have major roles to play in bringing its contours, textures, and consequences into conscious view.

What might a curriculum on inequality entail? First, given worldviews dominated by neoliberal capitalism, teachers could begin with broad unpacking of questions of political economy. Students could read the works of scholars such as Nancy Fraser (2019), David Harvey (2005), and Ananya Roy et al. (2016). Providing students language for naming capitalism and its logic and contours, and lenses for seeing capitalism as historical phenomenon, will be a first step in allowing them to unlearn what they take for granted. Disrupting the deep-seated individualistic common sense of contemporary times will require showing how this has come about historically.

Second, given the disproportionate representation among university students of the relatively advantaged, teaching inequality has to entail drawing students into self-reflection. Ethnographic works, either of their home societies or of other societies, can draw them in (Khan 2011, Lareau 2011 [2003], Teo 2018). The process of looking at aspects of life they relate to can cut uncomfortably close to students' own identities and is tricky terrain to navigate. Teachers must remind students here that the exercise of reflexivity is not about the self as an individual but a self as it is embedded in a social body. Reflexivity can be cultivated and ought to be about one's social position and not about one's selfhood.

Finally, teaching about inequality, particularly insofar as it draws students in through personal stakes, will lead students to ask, what can we do? Introducing comparative perspectives of how different countries have slightly different configurations of political economy will allow students to see that there are indeed variations and that capitalist economies are results of social forces and political choices of agents rather than deterministic natural phenomena (Wright 2010, Gornick et al. 2009, Razavi 2007, O'Connor et al. 1999). In addition, exposing students to ongoing activism and social movements, in their own country as well as elsewhere, will be an important component of activating imaginations and motivating action.

The following list of readings provide starting points for unpacking questions of political economy; drawing students into self-reflection; and encouraging students to see variations and alternative modes of organizing social life.

Key sources

1. Harvey, D. (2005). *A brief history of neoliberalism,* Oxford, England; New York, Oxford University Press.

Harvey provides a concise history of the rise of neoliberalism as ideology and practice. What are the mechanisms and agents responsible for this? What consequences have followed?

2. Khan, S.R. (2011). *Privilege: The Making of an Adolescent Elite at St. Paul's School,* Princeton and Oxford, Princeton University Press.

This ethnography illustrates how contemporary characteristics of privilege are generated and naturalized in the U.S. What practices of schools contribute to the reproduction of inequality? How might the insights from this case be applied to other contexts?

3. Roy, A., Negrón-Gonzales, G., Opoku-Agyemang, K. and Talwalker, C.V. (2016). *Encountering poverty: thinking and acting in an unequal world,* Oakland, California, University of California Press.

This book reflects on key trends in anti-poverty movements. What are key blindspots and contradictions in contemporary social actions against poverty and inequality? How can they be overcome?

4. Teo, Y.Y. (2018). *This is What Inequality Looks Like,* Singapore, Ethos Books.

This ethnography contrasts everyday experiences of those living with less and those with more, to illustrate the logics and pitfalls in Singapore's social welfare regime. How do structure and culture intersect in the shaping of inequality? What will it take to alter systemic problems?

5. Wright, E.O. (2010). *Envisioning Real Utopias,* London, Verso.

This book, part of a larger series of works that 'envision real utopias,' maps out institutional mechanisms necessary for longer-term alleviation of inequality. What are the characteristics of utopias? What is the role of imagination in scholarly work, and in the lives of students?

References

Barr, M.D. and Skrbis, Z. (2008), *Constructing Singapore: Elitism, Ethnicity and the Nation-Building project,* NIAS Press, Copenhagen.

Bourdieu, P. (1989), *The State Nobility: Elite Schools in the Field of Power,* Stanford University Press, California.

Burawoy, M. (2005). For public sociology. *American Sociological Review,* vol. 70, pp. 4–28.

Burawoy, M. (ed.) (2014). *Current Sociology Special Issue--Precarious Engagements: Combat in the Realm of Public Sociology.*

Deng, Z. and Gopinathan, S. (2016), 'PISA and high-performing education systems: explaining Singapore's education success', *Comparative Education,* vol. 52, pp. 449–472.

Department of Statistics (2018), 'Household Expenditure Survey', Singapore.

Fraser, N. (2019), *The old is dying and the new cannot be born: From progressive neoliberalism to Trump and beyond,* Verso Books.

Gornick, J.C., Meyers, M.K. and Wright, E.O. (eds.) (2009), *Gender equality: Transforming family divisions of labor,* Verso, London.

Harvey, D. (2005), *A brief history of neoliberalism,* Oxford University Press, Oxford, England; New York.

Khan, S.R. (2011), *Privilege: The Making of an Adolescent Elite at St. Paul's School,* Princeton University Press, Princeton and Oxford.

Lareau, A. (2011 [2003]), *Unequal childhoods: class, race, and family life (2nd edition),* University of California Press, Berkeley.

Lee, K.Y. (1983), *Talent for the future: Speech by Prime Minister Lee Kuan Yew at the National Day Cultural Show,* National Archives of Singapore, Singapore.

Ministry of Education (2012), 'Progress of Normal (Technical) Students', Singapore.

Ng, I.Y.H. (2014), Education and intergenerational mobility in Singapore, *Educational Review,* vol. 66, pp. 362–376.

O'Connor, J.S., Orloff, A.S. and Shaver, S. (1999), *States, markets, families: gender, liberalism, and social policy in Australia, Canada, Great Britain, and the United States,* Cambridge University Press Cambridge; New York.

OECD (2018), *Equity in Education.*

OECD (2019a), *Balancing School Choice and Equity.*

OECD (2019b), Programme for International Student Assessment (PISA) Results from PISA 2018 – Country Note (Singapore).

Ong, X.L. and Cheung, H.S. (2016), *Schools and the Class Divide: An Examination of Children's Self-Concept and Aspirations in Singapore,* Singapore Children's Society.

Ong, Y.K. (2018), 'Meritocracy and the paradox of success', *The Straits Times,* October 25.

Razavi, S. (2007), 'The political and social economy of care in a development context', *Gender and Development Programme Paper Number 3,* United Nations Research Institute for Social Development (UNRISD), Switzerland.

Roy, A., Negrón-gonzales, G., Opoku-Agyemang, K. and Talwalker, C.V. (2016), *Encountering poverty: thinking and acting in an unequal world,* University of California Press, Oakland, CA.

Sharma, N. (2020), 'Torn safety nets: How COVID-19 has exposed huge inequalities in global education', Viewed 17 June 2020, www.weforum.org/agenda/2020/06/torn-safety-nets-shocks-to-schooling-in-developing-countries-during-coronavirus-crisis/.

Teo, Y. (2010), 'Shaping the Singapore Family, Producing the State and Society', *Economy and Society,* vol. 39, pp. 337–359.

Teo, Y. (2017), 'The Singaporean welfare state system: with special reference to public housing and the Central Provident Fund. *In:* ASPALTER, C. (ed.)', *The Routledge International Handbook to Welfare State Systems,* Routledge, London; New York.

Teo, Y. (2021), 'From public sociology to collective knowledge-production', *In:* Hossfeld, L., Kelly, B. and Hossfeld, C. (eds.), *Routledge International Handbook on Public Sociology,* Routledge, London; New York.

Teo, Y.Y. (2018), *This is What Inequality Looks Like,* Singapore, Ethos Books.

Teo, Y.Y., George, C., Lim, L. and Chong, J.I. (2020), Academia (and knowledge) in the age of pandemic, https://www.academia.sg/editorials/editorial-18april2020/.

United Nations, (2015), Sustainable Development Goals: Quality Education [Online], Viewed 16 June 2020, www.un.org/sustainabledevelopment/education/.

Wang, L.Y., Teng, S.S. and Tan, C.S. (2014), 'Levelling up academically low progress students (NIE Working Paper Series No. 3)', *National Institute of Education.*

Wise, A. (2016), 'Behind Singapore's PISA rankings success – and why other countries may not want to join the race.' *The Conversation,* Viewed 13 December 2016, http://theconversation.com/behind-singapores-pisa-rankings-success-and-why-other-countries-may-not-want-to-join-the-race-70057.

Wright, E.O. (2010), *Envisioning Real Utopias,* Verso, London.

35

GENDER INEQUALITY AND DEVELOPMENT

Archana Preeti Voola and Bina Fernandez

Introduction

The past half century has witnessed several significant shifts in research, policy, practice and pedagogy focused on redressing gender inequalities in development. This chapter provides a synoptic review of these trends, and a selected set of learning resources to help navigate a field that is now highly complex in both its depth and breadth. Drawing on this review, we argue that an intersectional feminist approach to challenging power inequalities must inform research, policy, practice, and pedagogy initiatives, if they are to bring about transformative changes towards a more gender equal world.

In the first section, we briefly trace conceptual shifts in the definition and usage of the term gender inequality in development, emphasizing the importance of an intersectional lens to uncover omissions and inclusions in the deployment of a gender lens. We then discuss forms of gender inequalities in the economic, political, and social domains. The second section offers a critical analysis of how gender inequalities have been addressed in policy and practice over the past five decades, paying particular attention to gender mainstreaming and the evolution of gender analysis tools and frameworks. Finally, the third section concludes by emphasizing the importance of a continued commitment to feminist values in pedagogies of training and education in the field of gender and development.

Conceptualizations of gender inequality in development

A milestone in making visible the inequalities women faced was the publication of Ester Boserup's *Women's Role in Economic Development* (1970). This early thinking relied heavily on assumptions about development as a project of modernization, and on liberal feminist ideas of achieving equality by integrating women in the development process through training, education, and increased labour force participation. Boserup (1970) argued that women's participation in agricultural work had been marginalized by the international development programmes of the times which offered agricultural capital, raw materials and market access to men while relegating women to subsistence farming with negligible profits. Boserup's work was a watershed moment in international development, and heralded international attention to equality for women. Subsequent decades saw the organization of a series of international events at the

DOI: 10.4324/9781003017653-39

United Nations such as the International Year for Women 1975, the UN Decade on Women 1976–1985 and four world conferences on women 1975–1995, culminating with the UN Fourth World Conference on Women held at Beijing in 1995. National policy responses to these international events were to target female populations in development practice and provide them with more equal access to resources.

Although the category 'women' was initially a powerful organizing principle for advocacy and social change, by the late 1990s consensus had shifted to the use of gender broadly defined as 'the socially constructed relationship between women and men.' This reframing urged closer attention to 'effective strategies for transforming the relations of power that sustain social injustice' (Cornwall and Rivas 2015, 401–2). The attention to identities as socially constructed allowed two important conceptual shifts: first, the recognition that 'women' was not a homogenous category, and that it was vital to incorporate intersectional analyses that considered how gender inequalities are intertwined with other axes of marginalization such as race, ethnicity, sexuality, age, physical ability, or nationality. The second was a shift away from the focus on 'women' as the protagonists of the international development story as well as the hetero-normative and binary gender categories of men-women to include an analysis of men and masculinities, and gender non-conforming groups.

The analysis of gender moved from gender as a noun (male/female) to gender as relational (verb) 'constantly being produced, renewed, and changed' (Connell 2005, 6). A sharper focus on existing unequal relations between men, women and gender non-conforming groups led to numerous interventions aimed not only at changing outcomes for these groups but transforming norms of gender and sexuality that perpetuate these inequalities (Fisher 2019; Wanner and Wadham 2015). The definition of gender expanded to mean 'the constellation of institutions, including policies, laws, and norms, that constitute the roles, relations, and identities of women and men, and the feminine and the masculine, in a given context' (Htun and Weldon 2010, 208). For instance, the present trend in international development practice is to include men in the promotion and advocacy of gender equality (Jewkes, Flood and Lang 2015; Wanner and Wadham 2015). But if inclusion happens at a superficial level, with a focus on men's roles and tasks rather than a substantive focus on the power relations, this could result in material privilege for some men and harms experienced by other non-gender conforming groups. As such, rather than re-casting male patriarchy by realigning opportunities, resources and positional power for women (Cornwall and Rivas 2015, 402), a rethinking of masculinities is needed in international development (Fisher 2019).

Some feminist scholars have also critiqued the way in which conceptualizations of gender inequality were deployed in development policy and practice. For instance, Maxine Molyneux suggested that the repetition of references to gender may, in fact, be causing 'gender fatigue' (2007, 227) within the field. A cautionary strand of literature emerged in the field of international development bemoaning a loss of critical reflection surrounding usage of the concept of gender (Cornwall et al. 2007; Prugl and Lustgarten 2007). There was often limited understanding of gender inequality and gender ordering in programmes and policies, where women's issues were viewed as separate and independent rather than a core element of development policy and practice. This was evident in The World Bank's (2006) report *Gender Equality as Smart Economics* that advocated for women's increased labour force participation and expanding their economic opportunities, but which feminists critiqued for not only instrumentalizing gender equality, but also for deepening gendered inequalities in order to sustain and strengthen processes of global capital accumulation (Wilson 2015).

A variety of reasons were proposed by scholars for the gender ennui. Firstly, a slippage emerged in mainstream development policy and practice, wherein gender began to be reductively understood as biological difference rather than a social construction. The incorporation of women in the development agenda carried a moral imperative based on biological universals, rather than sociological and anthropological specificities. Chandra Talpade Mohanty (1986) lamented the reductive descriptors of gender as men and women producing restrictive social change as 'sexual difference becomes coterminous with female subordination, and power is automatically defined in binary terms: people who have it (read: men), and people who do not (read: women)' (p. 344). Gender became conflated with women and gender issues became equated with women's issues which gave rise to the development recipe of 'add women and stir.' At the policy level, international organizations such as the World Bank and United Nations subsumed gender equality goals under organizational agendas and strategies which transformed 'a radical movement idea into a strategy of public management' (Prugl and Lustgarten 2007, 55).

Secondly, the representations of women within the field of development began to essentialize women through their portrayals as nurturers, carers, inherently peaceful, closer to nature, less corrupt than men, and so on. Such rhetoric in policy and practice instrumentalized women to achieve certain ends – such as poverty reduction, good governance, environmental sustainability, and the like. This provided ample justification for NGOs and donor agencies to shift the onus of development onto women. For instance, in environmental projects, Melissa Leach (2008) found that women were recruited to undertake activities such as tree planting, soil conservation, fruit tree agroforestry and the like without remuneration as part of their community development expectation. Nevertheless, when it came to work that involved remuneration, it was males who held control. Such reductive renderings of gender essentialized women and trivialized the complex social reality and structural constraints within which the lives of the poor and marginalized were enmeshed.

Thirdly, scholars observed policy makers and practitioners in the development arena were reluctant to leave the familiar territory of poverty and welfare, to tackle the critical issues of unequal power relations and social justice (Mukhopadhyay 2004; Molyneux 2007). This is evinced, for example, in the mainstreaming of gender in the early 2000s Millennium Development Goal number 3 to promote gender equality and empower women. This mainstreaming continued into the current Sustainable Development Goals number 5 to achieve gender equality and empower all women and girls. These goals were ratified by the United Nations member countries and often tracked via yearly country reports documenting progress on the goals. Nevertheless, these gender equality goals were seen as only advisory, with little if any penalties for noncompliance. Even 'compliance' could potentially be problematic, as it was often reduced to technical fixes that were cost effective, efficient and replicable on a mass scale, but were tokenistic and rendered the gender equality agenda 'ahistorical, apolitical, decontextualized and technical' (Mukhopadhyay 2004, 97).

Domains of gender inequalities

Gender inequalities are constituted in different domains through the roles, relations, identities and practices of men, women, and gender non-conforming groups. In this section we examine these inequalities and power asymmetries in the economic, political and social domains, employing purposive case study examples from international development practice.

In the economic domain, gender as a description of power relations is overwhelmingly evident at the interface of productive, reproductive and community labour. Globally, women perform

76.2 per cent of unpaid care work relative to men. In other words, women are likely to perform reproductive and community labour 3.2 times more than men (ILO 2018, xxix). As a corollary, men's participation in reproductive and community labour has tended to remain minimal. In 2018, out of the 3.5 billion people in the global labour force, three in five were men (ILO 2019, 1). The gendered division of labour and care is conspicuous in the economic domain where men are overrepresented in productive labour with monetary benefits, while women are hyper-represented in reproductive and community labour which is either low-wage, unpaid or voluntary. The pre-existing gender myths that women by their 'feminine' nature are 'more hardworking, more caring, more responsible…relative to men' (Cornwall and Rivas 2015, 399) are reproduced in economic life, resulting in a patriarchal dividend for a majority of men.

Box 35.1 Teaching resource case study 1: valuing women's productive, reproductive and community labour

International Food Policy Research Institute (IFPRI) designed and implemented gender-responsive agricultural development projects in Africa and Asia with an emphasis on access to agricultural assets to women. IFPRI took pride in its gender response research and practice, which was part of its organizational strategy. However, analysis of the Gender, Agriculture and Assets Project (GAAP) programme based in countries such as India, Bangladesh, Mozambique, Burkina Faso, Uganda, Kenya, and Tanzania found unintended consequences that exacerbated gender inequalities (Johnson et al. 2016). Many women who undertook productive labour opportunities from IFPRI based projects faced increased domestic violence. The increased agricultural opportunities for women undermined the masculine identity of 'breadwinner' for men, triggering feelings of displacement and exclusion in men. Tip toeing around gender power imbalances without directly addressing them led to short-term gains in women's productive labour, but long-term losses in their physical security and interpersonal relations with their male kin. Little or no transformative changes were documented in gender power relations, leaving intact norms that maintain inequalities. Moreover, the productive labour added to women' time burden as they were still engaged in reproductive labour or care work. While women's visibility and participation in productive labour increased through interventions from organizations such as IFPRI, their participation in reproductive and community labour remained unchanged. Additionally, some projects that focused on intensive productive labour, such as dairy farming, meant that women had to pull their children out of school to help with the increased workload (Johnson et al. 2016).

Resources: www.ifpri.org/strategic-research-area/cross-cutting-theme-gender, Johnson et al. 2016.

Within the political domain of everyday life, the unequal nature of gendered power relations can be identified and transformed by the political spaces that surround and imbue them. These spaces can take the shape of policy instruments, legal frameworks, political participation, or everyday democracy in public spaces. Feminist movements of the 1960s and 1970s highlighted that private spaces can also be political and within the field of international development, political spaces exist in the public as well as intimate lives of men, women and gender diverse groups. Prevention of violence against women and girls (VAWG) has become an established strategy to address gender inequality in international development and provides a useful context to explore the political sphere. Several shifts have occurred in the past 20 years in how to work with men and boys to transform gender power relations (Jewkes, Flood and Lang 2015). Initially, men were invisible in violence prevention efforts except in their roles as perpetuators of the violence. By the 1990s, men began to be included as partners and allies in prevention of violence efforts.

More recent interventions in VAWG seek to change the way men see themselves through critical reflections of their power and privilege as well as in collaboration with women and women's groups. An instructive example, VAWG in the public-private political spaces, is the Prevention of Violence against Women and Children programme in the slums of Dharavi, India. The programme engages men by forming men's groups and employs training manuals on gender, health, and violence. Access to private or semi-private space was essential for the success of the programme as men could 'engage in affect and emotions, to negotiate vulnerability, especially in social contexts where there are few avenues for them' (Chakraborty, Osrin and Daruwalla 2020, 763). While the need for private spaces was in part to allow for men's discretion, it was also a venue for critical reflection of gendered power, privilege, and complicity in VAWG. However, the arrangement of public-private spaces in the political domain is shaped by gender power relations and vice versa, and unless the structural nature of VAWG is recognized intervention outcomes may be limited to acquiring knowledge and awareness, rather than upending gender unequal power relations.

Healthcare is an integral component of the social domain and provides an instructive context to examine how gender power imbalances are constituted and negotiated in the health care systems within the field of international development. Recent disease outbreaks such as Middle East Respiratory Syndrome (MERS), Zika and Ebola as well as the current pandemic of COVID-19, highlight how gendered norms around work and care perpetuate gender inequalities. In situations of health crises, urgency often supersedes any responses to the redressal of gender power relations, in order to attend to immediate biomedical needs. During the South American Zika outbreak in 2014 the World Health Organization advised nations impacted by Zika to provide equitable access to reproductive and sexual health care (WHO 2016). This messaging was not in tune with gender norms in South American countries which encouraged women to avoid pregnancies rather than provide them with access to reproductive healthcare (Miller 2016). Similarly, country specific legislation and socio-religious norms also severely restricts women's access to contraception and abortion.

In order to move from the impasse of gender deployed as a descriptor of biological difference, gender scholars in global health propose employing gender as an explanatory tool of power relations by examining 'who has what (access to resources); who does what (the division of labour and everyday practices); how values are defined (social norms) and who decides (rules and decision-making)' (Morgan et al. 2016, 1069). Such an examination, analyses the inequalities faced by women as they are cast as the primary custodians of reproductive and family healthcare as mothers, daughters, community health workers (Witter et al. 2017). It also highlights how institutional laws, policies and norms obscure the roles and responsibilities of men in reproductive healthcare, the rights of gender non-conforming groups to sexual health or the choices of those who do not want children. In the social domain of health, the gendered division of work and care dictates an exclusive focus on women as 'shock absorbers' (Harman 2016, 525) leaving intact the structures of exclusion and discrimination that marginalize them.

Box 35.2 Teaching resource case study 2: critiquing the 'Girl Effect'

In the early 2000s, several development initiatives increasingly focused on educating girls through a discourse of 'girl power,' as exemplified by the Girl Effect, a prominent campaign sponsored by the Nike Foundation. Such initiatives present 'putting girls in control' as creating positive change not only for the girls, but also their families, communities, and countries. Feminist critics of the 'Girl Effect' view it as an extension of the instrumentalist approaches to women as 'good investments.' They further argue that the gendered Orientalisms noted by Mohanty (1986) are currently being reworked through the subject position of the Third World girl as simultaneously oppressed and full of 'Third World potential' (Moeller 2018). However, as Switzer observes, the Girl Effect serves as 'a regulatory representational regime that works to explicitly racialize, depoliticise, ahistoricise and naturalise global structural inequalities that legitimise neoliberal interventions in the name of girls' empowerment.'

See: www.girleffect.org, Switzer 2013, Moeller 2018, Wilson 2015.

Addressing gender inequalities through development policy and practice

As previously noted, gender inequalities in development policy began to be substantively recognized in the early 1970s, with the articulation of key shifts in thinking and practice in the now well-known WID, WAD and GAD paradigms (Cornwall et al. 2007). Adopting a feminist standpoint that emerged from grassroots consciousness-raising activism, the GAD approach that emerged in the 1980s was based on a recognition of difference, and a shift away from the 'efficiency' based arguments of WID approaches to 'equality' and 'empowerment' based arguments for redressing gender inequalities. Central to the GAD approach is an intersectional perspective on gender, and a transformative agenda to redress the structural sources of the unequal distribution of power. The scaling up of the GAD approach to wider policies and institutions saw the emergence of 'gender mainstreaming.'

Gender mainstreaming

From the mid-1990s gender mainstreaming became a major global undertaking, involving states, multilateral institutions, domestic and international NGO's institutions, and other actors working across multiple scales. As defined in the much-cited UN Economic and Social Council definition, gender mainstreaming refers to:

> *the process of assessing the implications for women and men of any planned action, including legislation, policies or programmes, in any area and at all levels... The ultimate goal of mainstreaming is to achieve gender equality.*
>
> *(See: www.un.org/womenwatch/daw/csw/GMS.PDF)*

This impetus to 'mainstream gender' was first proposed at the United Nations Fourth World Conference on Women at Beijing in 1995, and was contentious right from the start (Baden and Goetz 1997). Conservatives opposed the idea of a socially constructed, mutable gender identity because they were committed to the idea of women's 'natural' domestic roles; while some progressives argued that the shift to 'gender' would detract attention from the continued need to focus on the inequality faced by women (ibid.). Later critiques of gender mainstreaming question the 'top-down' nature of the agenda, pointing out the loss of accountability to grassroots constituencies. Others critiqued gender mainstreaming for de-politicizing the explicitly feminist agenda that had animated the GAD approach, since gender mainstreaming approaches usually fail to specify or address the power relations which maintain gender inequalities and feminist goals were often co-opted by public and private institutions (Mukhopadhyay 2004) and gender mainstreaming morphed from 'a model of resistance to an instrument of power' (Caglar et al. 2013, S-6).

Notwithstanding these continuing feminist reservations, gender mainstreaming is now widely implemented across institutions in both the Global North and South, which has resulted in the creation of a plethora of guidelines, inter-ministerial committees, gender focal points, national policies and programmes on gender. It has also resulted in the production of gender analysis tools and frameworks, as discussed below.

Gender analysis tools and frameworks

An important first step towards gender-responsive policies and programmes is to undertake gender analysis, or the systematic collection and analysis of gender disaggregated data. Such gender analysis usually draws on two kinds of data: quantitative/statistical data at macro or national levels, and qualitative data at micro levels, each of which we discuss below.

The initial impetus for the gender disaggregation of macro level, national statistical data came from the UNDP in 1995, through the introduction of two analytical tools – the Gender Development Index (GDI) and the Gender Empowerment Measure (GEM). The GDI was a 'gender-sensitive' measure to account for disparities between women and men, using the same three basic component indicators as in the Human Development Index (HDI) – health, knowledge and living standards. The GEM was designed to capture dimensions of gender inequality not considered in the GDI, which were proxy indicators for women's empowerment: the proportion of seats held by women in national parliaments, the percentage of women in economic decision-making positions and women's share of income.

Although both measures allow the ranking of countries on a global scale, they have been widely critiqued for over-simplification of complex phenomena such as empowerment to a few reductionist variables, as well as, paradoxically, for the complexity of measures that are highly specialized and difficult to capture across widely varying national capacity for data collection (Klasen 2006). To ameliorate some of the problems with the GDI, in 2010, the UNDP introduced the Gender Inequality Index, which considers three dimensions: reproductive health, empowerment, and labour market participation. And to ameliorate some of the problems associated with the GEM, a larger set of 13 indicators of empowerment has been incorporated into a data visualization dashboard on Women's Empowerment;[1] importantly, violence against women and girls is now included in this measure. Nevertheless, persistent questions remain on about whether it is possible to meaningfully quantify and compare culturally loaded concepts such as 'empowerment.'

In contrast to the technocratic and top-down formulation of the UNDP's macro level statistical tools, a consortium of NGOs, academics and civil society organizations have collaborated over the past decade to create a new gender-sensitive measure of multi-dimensional poverty – the Individual Deprivation measure (IDM). Drawing on insights from participatory and feminist research across 18 sites in six countries, the IDM was developed to assess 15 dimensions of life identified by women and men experiencing poverty.[2] While not (yet) endorsed by the

UN or other national and international agencies, this tool offers the scope for a more nuanced calibration of the analysis of gender inequality, even when collecting macro, statistical data.

Turning to qualitative, micro-level gender analysis frameworks (GAFs), it is important to note that the purpose of such frameworks is to enable a contextually specific understanding of gender inequality to inform the implementation of development programmes. When undertaken well, GAFs can make visible the ways in which unequal gender relations in a given context are likely to influence the outcomes of any intervention and identify how interventions may affect women and men differently. GAFs can alert planners to the possibility that interventions that profess to be gender-neutral can in fact reflect, and potentially reinforce, existing gender inequalities. Gender analysis is therefore required to ensure that policy and programme intervention decisions can lead to the transformation of pre-existing gender inequalities, without producing new ones.

The five main GAFs include the Harvard Analytical Framework, the Gender Analysis Matrix, Moser's Framework, Longwe's (Empowerment) Framework and the Social-relations Framework (March et al. 1999). Central to debates about GAFs is the conflict between their underlying assumptions and ideologies. The early GAFs such as the Harvard Analytical Framework and the Gender Analysis Matrix tended to adopt an *integrationist approach* that focused to a great extent on gender roles and dynamics at the household and community level. They provided useful analytical tools for practitioners to adopt and use, especially at the level of micro-plans for projects. In contrast, the Moser, Empowerment and Social Relations Frameworks take a transformative approach in which demands for equity, empowerment and rights form the basis of women's claims. In these frameworks, substantive equality is the goal, since formal equality is meaningless when the institutional culture is discriminatory. This perspective implicitly recognizes the embeddedness of structures of patriarchy in the state and society, and examines the complex ways in which male/female hierarchies are reproduced through development interventions. Therefore, the focus of these frameworks is on rethinking institutional rules, agendas, priorities, goals and practices, and the ways they embody male agency and interests.

Institutional preferences for GAFs vary: those adopting integrationist frameworks tend to regard feminist politics as redundant (without explicitly saying so). The emphasis is on gender disaggregation (of data, for example) rather than on gender relations and conflict; on how gender can constrain or facilitate pre-determined development goals (growth, human resource development and fertility reduction) rather on how gender can re-define development goals or set alternative development agendas. Planning is seen as a top-down technical exercise (ignoring participatory methods of data gathering and development planning) rather than a political process where the marginalized can have a voice.

In contrast, institutions adopting transformative frameworks attach considerable importance to the role of feminist advocacy and politics in challenging the gender biases of state, market and community institutions. They draw attention to women's subordination and the need to address power inequalities through development interventions (the practical/strategic distinction pointing to a more transformatory planning agenda), and offer concrete examples of participatory, bottom-up development planning.

However, it is important not to exaggerate the degree to which GAFs correspond to organizational mandates and cultures. Development institutions are also eclectic in what they adopt from the different GAFs. Some analytical tools (such as the Harvard Analytical Framework) are widely adopted because they provide a 'neat mental grid' or 'user-friendly handle' on certain complex issues, even if their analytical underpinnings (and political implications) are not necessarily endorsed by the organizations that adopt them.

Finally, translating GAFs into practical tools to enable gender redistributive responses and strategies is challenging in practice, particularly when adopting complex intersectional analysis and strategies. The search for tools and frameworks to integrate gender-sensitive data and practices to projects and policies implies a faith that technique can override forms of prejudice embedded in organizational systems, work cultures and individual attitudes. Framing gender as a technical issue underestimates the role of discriminatory gendered patterns in incentive systems, accountability structures and the bureaucratic procedures and institutional practices of development organizations. As a result, change towards more equal gender relations often remains an elusive goal in spite of the incorporation of gender analysis frameworks into many projects and programmes.

Feminist pedagogies, gender training and education

The antecedents of gender training and education practices can be traced to reflexive pedagogies that emerged from feminist conscious-raising groups of the 1970s and 1980s, in both the Global North and South. In these groups, the process of women's collective reflection on, and sharing of, their experiences of gender inequalities was essential for political mobilization towards the goals of gender equitable social transformation. Vital to the creation of these feminist pedagogies and practices was first, a commitment to *feminist principles* and a *feminist political analysis* (see Box 35.2).

Box 35.3 Questions to consider: how feminist is your pedagogy, training, or programme?

1.   Does it incorporate **feminist principles**?
     -   Mutual respect
     -   An ethics of care
     -   Non-hierarchical modes of organization
     -   Reflexivity
2.   Does it incorporate a **feminist political analysis**?
     -   Connecting women's individual experiences of inequalities to the collective
     -   Recognizing the complexity of intersectional inequalities
     -   Seeking transformative change of the underlying causes of gender inequality

---

As feminists sought to transform oppressive institutions, 'gender training' became the primary tool through which gender mainstreaming initiatives were delivered. In the wake of the 1995 Beijing Conference on Women, by the early 21st century, standardized gender training workshops had become a widely used mechanism, often compulsory, to achieve gender mainstreaming. UN Women defined training for gender equality as:

> *A transformative process that aims to provide knowledge, techniques and tools to develop skills and changes in attitudes and behaviours. It is a long-term continuous process that requires political will and commitment from all parties involved (both decision makers and trainees) with the objective of creating an aware, competent and gender equitable society.*
>
> (UN Women Training Centre, www.unwomen.org/en/how-we-work/
> capacity-development-and-training.)

While the overarching goal of gender equitable social transformation may still be the same, the process through which this is assumed to happen is much more structured, and less explicitly feminist. For instance, gender training manuals produced by NGOs like Oxfam were more explicitly aligned with Paulo Freire's pedagogies of conscientization (Williams, Seed and Mwau 1994) than they were with explicitly feminist pedagogies and practices. Moreover, regardless of whether trainings were feminist or Freirean, they often reduced challenging and often contested processes of social change in gender relations to 'frameworks and checklists that can be easily applied, and … push for the condensation of complex theories and context-specific analysis into easy and digestible terms' (Davids and van Eerdewijk 2017, 87).

The need for 'gender training' has, inevitably produced a cadre of 'gender experts' who provide such training at global and local levels, however, a survey of 188 gender experts worldwide conducted by the Graduate Institute in Geneva found that gender expertise appears to be primarily obtained through work experience, rather than formal degrees. Although 92 per cent of those surveyed had graduate degrees and 72 per cent had PhDs (primarily in the social sciences) only a tiny minority held educational qualifications in gender and women's studies, and notably, 40 per cent did not identify as 'feminist' (Thompson and Prügl 2015). While gender expertise is valued within the development sector, globally, there are few specialist higher degrees in gender and development. Among the top 20 ranked Development Studies programmes,[3] there are only a handful of Development Studies master's courses with 'gender' in the title of the degree/course offering: the MA in Gender and Development at the Institute of Development Studies at the University of Sussex; the MSc Gender, Development and Globalization at the London School of Economics, the MA Gender Analysis in International Development at the University of East Anglia, and the MA in Development Studies (Gender and Development Specialization) at the University of Melbourne.

To conclude then, we can observe that this arena of gender training and expertise contains a similar underlying tension between the move to institutionalization and professionalization and the adherence to feminist principles and politics that we observed in the related arenas of gender mainstreaming and gender analysis frameworks. Some feminists continue to be optimistic about the potential to 'harness gender training as a catalyst for disjuncture, rupture and change in institutions' (Ferguson 2019, 20), while others are more sceptical, and urge caution. Clearly, in either case, the imperative would be to keep a feminist perspective in focus, if the goal of transformative gender equality is to be achieved.

Key sources

1. Cornwall, A., Harrison. E., and Whitehead. A. (Eds.) (2007). *Feminisms in development: Contradictions, contestations and challenges.* New York: Palgrave Macmillan.

The contributions to this edited volume by a broadly feminist, heterogeneous mix of distinguished academics, bureaucrats, activists, and practitioners explores the neutralization of the feminist impetus behind GAD initiatives. Collectively, the volume analyses the struggles over interpretation, interrogates the multiple ways in which the development institutions work to undermine feminism, and explores the challenges of re-politicizing feminist notions of rights, citizenship, and solidarity across difference.

2. March, C., I. Smyth, and M. Mukhopadhyay. (1999). *A Guide to Gender Analysis Frameworks.* Oxford: Oxfam Practical Action Publishing.

A valuable practical guide to the range of frameworks to analyse gender relations, plan gender-sensitive research projects, and design development interventions to address gender inequalities. The introduction provides a comparative summary of the advantages and disadvantages of the frameworks, which is followed by specific toolkits.

3. Mohanty, C.T. (1986). Under Western Eyes. Feminist Scholarship and Colonial Discourses. *Boundary 2, 12*(3), 333–358.

Mohanty critiques Western feminist scholars' characterization of the 'Third World Woman' as a homogenous, monolithic category. She challenges the ahistorical and implicitly colonialist frameworks employed by such scholars and problematizes their corollary assumptions about Western women as secular, liberated, progressive.

4. Wanner, T., and Wadham, B. (2015). Men and Masculinities in International Development:'Men-streaming' Gender and Development? *Development Policy Review, 33*(1), 15–32.

This paper documents the growing interest within international development policy and practice on the role of men and masculinities, referred to as 'men-streaming,' in addressing gender inequalities. While critiquing the current trend as mostly rhetoric with limited practical application, the paper suggests employing a critical feminist lens to examine the hegemonic masculinities and patriarchal dividends intrinsic to the gender status quo in development.

5. Wilson, K. (2015). Towards a radical re-appropriation: Gender, development and neoliberal feminism. *Development and Change, 46*(4), 803–832.

This paper offers an incisive critique of 21st century approaches to gender promoted by neo-liberal development frameworks that assume 'Gender Equality as Smart Economics.' Wilson argues that such approaches deepen gendered inequalities and strengthen processes of global capital accumulation. Drawing on Marxist, Black, Post-colonial and Queer feminisms she advocates for a radical re-appropriation of interventions.

Notes

1 See http://hdr.undp.org/en/content/dashboard-3-women's-empowerment
2 See www.individualdeprivationmeasure.org
3 According to the QS World University Rankings www.topuniversities.com/university-rankings/world-university-rankings/2020

References

Baden, S. and Goetz, A.-M. (1997), 'Who needs (Sex) when you can have (Gender)? Conflicting Discourses on Gender at Beijing', *Feminist Review*, vol. 56, pp.3–23.
Boserup, E. (1970), *Women's Role in Economic Development,* UK: EarthScan, London.
Caglar, Giilay, Elisabeth Priigl, and Susanne Zwingel (eds.) (2013), *Feminist Strategies in International Governance*, Routledge, Abingdon and New York.
Chakraborty, P., Osrin, D., and Daruwalla, N (2020), 'We Learn How to Become Good Men: Working with Male Allies to Prevent Violence against Women and Girls in Urban Informal Settlements in Mumbai, India', *Men and masculinities*, vol. 23, no. 3–4, pp. 749–771.

Connell, R. (2005), 'Advancing gender reform in large-scale organisations: A new approach for practitioners and researchers', *Policy and Society,* vol. 24, no. 4, pp. 5–24.

Cornwall, A., and Rivas, A.M. (2015), 'From 'gender equality and 'women's empowerment' to global justice: reclaiming a transformative agenda for gender and development', *Third World Quarterly,* vol. 36, no. 2, pp. 396–415.

Cornwall, A., Harrison. E., and Whitehead. A. (2007), 'Introduction: Feminisms in development: contradictions, contestations and challenges. In Cornwall, A.E. Harrison and A. Whitehead (Eds.)', *Feminisms in development: Contradictions, contestations and challenges,* pp. 1–20.

Davids, T. and van Eerdewijk, A. (2017), 'The Smothering of Feminist Knowledge: Gender Mainstreaming Articulated through Neoliberal Governmentalities. In: Bustelo, M., Ferguson, L., and Forest, M'. *The politics of feminist knowledge transfer: gender training and gender expertise.* New York: Palgrave Macmillan.

Fisher, S. (2019), 'Rethinking conceptual representations of masculinity in international development gender equality training', *NORMA,* vol. 14, no. 1, pp. 50–65.

Ferguson, L. (2019), *Gender training: a transformative tool for gender equality*, Palgrave Macmillan, Cham, Switzerland.

Harman, S. (2016), 'Ebola, gender and conspicuously invisible women in global health governance', *Third World Quarterly,* vol. 37, no. 3, pp. 524–541.

Htun, M., and Weldon, S.L. (2010), 'When do governments promote women's rights? A framework for the comparative analysis of sex equality policy', *Perspectives on Politics,* vol. 8, no. 1, pp. 207–216.

International Labour Organisation (ILO) (2019), World Employment Social Outlook: Trends 2019, Viewed 15 December 2021, www.ilo.org/wcmsp5/groups/public/---dgreports/---dcomm/---publ/documents/publication/wcms_670542.pdf

International Labour Organisation (ILO) (2018), Care work and care jobs for the future of decent work, Viewed 15 December 2021, www.ilo.org/wcmsp5/groups/public/---dgreports/---dcomm/---publ/documents/publication/wcms_633135.pdf

Jewkes, R., Flood, M., and Lang, J. (2015), 'From work with men and boys to changes of social norms and reduction of inequities in gender relations: a conceptual shift in prevention of violence against women and girls', *The Lancet,* vol. 385, no. 9977, pp. 1580–1589.

Johnson, N.L., Kovarik, C., Meinzen-Dick, R., Njuki, J., and Quisumbing, A. (2016), 'Gender, assets, and agricultural development: Lessons from eight projects', *World Development,* vol. 83, pp. 295–311.

Klasen, S. (2006), 'UNDP's Gender-Related Measures: Some Conceptual Problems and Possible Solutions', *Journal of Human Development,* vol. 7, no. 2, pp. 243–274.

Leach, M. (2008), 'Earth mother myths and other ecofeminist fables: How a strategic notion rose and fell. In A. Cornwall, E. Harrison and A. Whitehead (Eds.)', *Gender myths and feminist fables: The struggle for interpretive power in Gender and Development* (pp. 67–84). Malden, MA: Blackwell Publishing.

March, C., I. Smyth, and Mukhopadhyay M. (1999), *A Guide to Gender Analysis Frameworks*, Oxford: Oxfam Practical Action Publishing .

Miller, M. (2016), 'Infected with dogma: how South America's response to the Zika virus fails women', *The Humanist,* vol. 76, no. 2, pp. 9.

Moeller, K. (2014), 'Searching for Adolescent Girls in Brazil: The Transnational Politics of Poverty in 'The Girl Effect'', *Feminist Studies,* vol. 40, no. 3, pp. 575–601.

Mohanty, C.T. (1986), 'Under Western Eyes. Feminist Scholarship and Colonial Discourses', *Boundary 2,* vol. 12, no. 3, pp. 333–358.

Molyneux, M. (2007), 'The Chimera of Success: Gender Ennui and the Changed International Policy Environment'. In A, Cornwall., E. Harrison and A. Whitehead (Eds.), *Feminisms in development: Contradictions, contestations and challenges* (pp. 227–240). New York: Palgrave Macmillan.

Morgan, R., George, A., Ssali, S., Hawkins, K., Molyneux, S., and Theobald, S. (2016), 'How to do (or not to do)… gender analysis in health systems research', *Health policy and planning,* vol. 31, no. 8, pp. 1069–1078.

Mukhopadhyay, M. (2004), 'Mainstreaming Gender or "Streaming" Gender Away: Feminist Marooned in the Development Business', *IDS Bulletin,* vol. 35, no. 4, pp. 95–103.

Prugl, E., and Lustgarten, A. (2007), 'Mainstreaming gender in international organizations. In J. S. Jaquette and G. Summerfield (Eds.), Women and gender equity in development theory and practice: Institutions, resources, and mobilization (pp. 53–70). London: Duke University Press.

Switzer, H. (2013), '(Post)Feminist Development Fables: The *Girl Effect* and the Production of Sexual Subjects', *Feminist Theory,* vol. 14, no. 3, pp. 345–360.

Thompson, H., and Prügl, E. (2015), Gender Experts and Gender Expertise: Results of a Survey. Geneva: Graduate Institute of Geneva, Viewed 15 December 2021, www.graduateinstitute.ch/library/publications-institute/gender-experts-and-gender-expertise-results-survey

Wanner, T., and Wadham, B. (2015), 'Men and Masculinities in International Development: 'Men-streaming' Gender and Development?', *Development Policy Review,* vol. 33, no. 1, pp. 15–32.

Wilson, K. (2015), 'Towards a radical re-appropriation: Gender, development and neoliberal feminism', *Development and Change,* vol. 46, no. 4, pp. 803–832.

Witter, S., Namakula, J., Wurie, H., Chirwa, Y., So, S., Vong, S., Ros, B., Buzuzi, S. and Theobald, S. (2017), 'The gendered health workforce: mixed methods analysis from four fragile and post-conflict contexts', *Health policy and planning,* vol. 32, no.5, pp. 52–62.

Williams, S., Seed, J., and Mwau, A. (1994), *The Oxfam gender training manual,* Oxfam Practical Action Publishing, Oxford.

World Bank Group (2006), Gender Equality as Smart Economics: A World Bank Group Gender Action Plan (Fiscal years 2007–10), Viewed 20 July 2010, http://siteresources.worldbank.org/INTGENDER/Resources/GAPNov2.pdf

World Health Organization (WHO) (2016), Zika Strategic Response Plan, Geneva: World Health Organization, Viewed 15 December 2021, https://apps.who.int/iris/bitstream/handle/10665/246091/WHO-ZIKV-SRF-16.3-eng.pdf?sequence=1

36

GENDER INEQUALITY AND DEVELOPMENT PEDAGOGY

Sara N. Amin, Christian Girard, and Domenica Gisella Calabrò

Introduction

In 2015, the launch of the Sustainable Development Goals (SDGs) saw gender equality enshrined in a stand-alone goal (Goal 5), and as an underpinning goal of *all* other 16 goals. Considering it was only in the 1970s that development policies started to explicitly engage with gender inequality, this reflects the significant successes of feminist movements within development (Sen 2019). However, much remains to be achieved in pursuing gender equity. Four key domains that have become central – and contested – for how to change the status of women and produce gender equality more broadly are: education, employment, political representation, and gender-based violence. In this chapter, we show that while these policies and pathways have been important in changing girls' and women's experiences and life chances, many continue to be inadequate in changing the power structures that (re)produce gender inequality. We suggest that changing power structures of gender inequality requires taking seriously relationality, context, and the multidimensionality of gender. We then conclude by reflecting on our pedagogical practices in two universities in the Global South which attempt to apply these insights.

Education

Formal education is widely understood as a means to address gender inequality by providing girls and women with the necessary skills and resources to access economic opportunities, as well as to gain voice and confidence to enact change. The dominant policy focus here has been on closing gender disparities in enrolment rates, completion rates, years of schooling, literacy levels, and more recently, areas of study (Stromquist 2015). For example, the Education 2030 Framework for Action (EFA), established to facilitate the achievement of Sustainable Development Goal 4 on education, directly identifies gender equality as key to the right to education (UNESCO 2018). Relatedly, the 'Girl Effect' movement – the idea that investing in girls' education offers the greatest returns for ending poverty more widely – has become a major part of the international development agenda, including through the World Bank, the IMF, various INGOs, and corporations such as Nike and Goldman Sachs (Boyd 2016).

DOI: 10.4324/9781003017653-40

The above campaigns and associated policies and resources invested at global (e.g. MDGs and SDGs) and local levels (national level policies on education) have had an impact on increasing gender parity in primary and secondary education (UNESCO 2018). Differences between girls' and boys' out-of-school rates have narrowed significantly across the world. These numbers are promising and important, with numerous studies showing that formal education has an important impact on self-esteem, awareness of rights, and decision-making about marriage, sex and reproduction, as well as on increasing access to income-generating activities and contributing to shifts in access to economic and political power (Stromquist 2015).

Nevertheless, gender disparities continue for secondary and higher education (UNESCO 2018). There are persistent and context-specific gaps in global trends, including girls from the poorest socioeconomic background being less likely than boys to attend primary school, and girls from Arab and sub-Saharan African regions faring the worst in enrolment and completion rates, particularly for secondary schooling. While gender gaps in youth literacy are decreasing, the gender gap in adult literacy has not changed since 2000, with 63 per cent of adult women unable to read or write (UNESCO 2018). Moreover, while gender disparities in education are narrowing overall, disparities widen quite significantly as one moves up the education system (UNESCO 2018).

In addition, while the EFA campaign and goals were premised on the understanding that *quality* of education and changing values, norms and practices are central in changing inequities in education, including gendered inequities, both actual efforts and the measurement of progress has been limited to quantitative indicators and outcomes. This is an important issue because it overlooks the mechanisms by which gender inequality continues to be reproduced through education. Gender inequality in education is produced through biased curricula and pedagogy, gender inequalities in staffing within institutions, gendered inequalities in the infrastructure of schools (e.g. toilets), and gendered violence within schools. It is also produced by larger societal inequalities that produce barriers, challenges and disadvantaged outcomes for girls and women in formal educational settings – such as entrenched gender norms about (child) marriage and motherhood, gendered division of labour at home, gendered access to financial resources required for education, gendered violence in public spaces and transportation, societal menstruation taboos and lack of resources related to menstruation at schools (see Shel 2007 for a comprehensive review on these elements).

Policy wise, including in the EFA framework, there is recognition that addressing gender inequality in education will require addressing gender inequality in other spheres, including in the political, economic and home spheres (UNESCO 2018). While some of these efforts have seen substantive resources devoted to changing the gendered organization of schools, many policy efforts remain focused on changing attitudes and practices via gender sensitivity training of teachers and integrating elements of gender, sexuality and gender equality in curricula, without addressing the structural issues. For example, policies and indicators ignore or miss out on issues related to the large-scale trends in divestments in formal education that have taken place due to structural adjustment policies in the 1980s/1990s, as well as increased privatization of education under neoliberal agendas in the early 21st century (Stromquist 2015). Relatedly, there are critiques of the instrumentalization of gender equality in education policies for the purposes of 'neoliberal' development, as opposed to shifting power relations in gendered structures (Boyd 2016).

When interrogating and addressing gender equality in education, attention must also be given to challenges faced by men and boys. For example, completion rates for boys are *lower* than girls in secondary education and in upper middle and high income countries (UNESCO 2018), while numerous studies have highlighted how pressures on boys to perform hegemonic masculinities in school contexts continue to negatively impact both boys' own educational performance, as well as the performance of girls (Morojele 2011; UNESCO 2017). Studies in migration and gender have also shown that migration for development policies have led to unintended negative gendered outcomes in many places in the Global South, including pulling boys out of school and into the remittance economy, as well as sometimes leading to early marriages of girls or increased polygamy in communities with large international migrant labour (Rafique Wassan et al. 2017). In addition, while qualitative studies on LGBQT+ students in school settings have clearly shown the prejudice, marginalization and violence they experience, there are no comparative systematic monitoring systems in place to track educational outcomes of sexual and gender minorities in formal schooling (c.f. UNESCO 2017).

Employment and wealth

Despite some progress in terms of reducing economic inequality between men and women in the recent decades (from pay gap to assets and inheritance, Ortiz-Espinosa and Roser 2018), there are still substantial global gender disparities regarding employment and wealth. Women still face less employment, less quality jobs, less security, lower pay and higher participation in unpaid work. Men own 50 per cent more wealth than women (Oxfam 2020), women are still more likely to be unemployed than men (ILO 2018), and women are generally paid less (Oxfam 2020; ILO 2018; Ortiz-Espinosa and Roser 2018).

In addition, there are more women in unpaid work (e.g. contributing family workers – ILO 2018), care work (which is almost always undervalued and underpaid), and domestic work (Oxfam 2020). Women tend to be over-represented in the informal economy in low and lower-middle income countries, both including and excluding agriculture (ILO, 2018). This is particularly so in Sub-Saharan Africa (90%), South Asia (89%) and Latin America (75%) (ibid.). Furthermore, when women are in formal work, they are overwhelmingly in more precarious and more vulnerable working conditions than men all across the Global South (Oxfam 2020; ILO 2018; Ortiz-Espinosa and Roser 2018; Beazley and Desai 2014).

Multiple factors contribute to the above imbalances, including broader social and cultural norms that dictate women's 'place' in society, the triple productive, reproductive and community roles of women that include domestic chores and child-rearing and other care responsibilities (and resulting in limited mobility due to these responsibilities), as well as spaces and transportation services that remain unsafe for women (Beazley and Desai 2014). Finally, the lack or limited control of women over decision-making in the household, (particularly regarding financial decisions, including their own spending), also affects women's ownership of and access to wealth, with men often more likely to own land and control productive assets (Ortiz-Espinosa and Roser 2018). This is compounded by systems and laws that tend to largely favour men over women in terms of ownership and inheritance, making it even harder for women to make a claim or assert control of belongings or assets. Notably, land rights pose a particular challenge, particularly in countries and regions with strong patriarchal institutions in place.

Key areas and policies that can tackle these gaps and reduce gender inequalities in income and employment include the following (Oxfam 2020; and Ortiz-Espinosa and Roser 2018): equal pay for work of equal value; ending gender-based occupational and sectoral segregation through both expanding relevant educational pathways and addressing gender-biased work cultures and insecurities; addressing structural conditions that promote discrimination and violence against women in the workplace, at home, and within public spaces and transportation. The persistence of gendered divisions of labour in the family and women's disproportionate reproductive labour burden also need to be addressed through a range of practices, including (better) paid care work, childcare, maternity and parental leave, as well as other social policies and protection measures to facilitate the ending of the triple burden that women face. Finally, the specific needs and challenges faced in the informal economy require large-scale social protection measures.

In outlining these priorities, it bears noting that multiple approaches and dynamics have contributed to reducing gender gaps in employment and wealth, including: microcredit schemes and other forms of support for income-generating activities, migration (international and internal), the feminization of labour in transnational companies and export manufacturing, and changing gender-related laws and practices related to land, property, and inheritance rights. Such efforts have seen mixed results and have faced various lines of critique.

Regarding microcredit, for example, although many point to successes of these initiatives in terms of poverty reduction and women's empowerment, others raise concerns over the risk and burden of over-indebtedness, the lack of effective control over financial resources and decisions by women within households, and the making of women legally responsible for loans that, in some cases, are spent or controlled by male partners or family members (e.g. see Kabeer 2005). In addition, microcredit may worsen women's over-representation in what Rogerson (1997) calls 'over-traded income opportunities,' where competition is high, markets are saturated, and profit-margins are extremely low. This is further compounded by the restrained mobility faced by many women, as home-based enterprises often cater only to neighbourhood populations with limited purchasing power.

Regarding migration – particularly international and/or towards urban centres and special economic zones – new opportunities for women to access employment have been accompanied by countless 'stories of exploitation and abuse' (Beazley and Desai 2014, 594). Opportunities can, and often do, benefit households through financial remittances and the opening up of decision-making power for 'left-behind' wives (Rashid 2013), however, they can also negatively impact family relations, including leaving many women with increased domestic pressures and responsibilities (Beazley and Desai 2014). Expectedly, women are affected differently by migration in different contexts, with men being more likely to travel abroad in South Asia, while women in Southeast Asia (notably the Philippines) often migrate to undertake domestic work or health sector work in North America and Europe, as well as in the Middle East and the Pacific.

The role of manufacturing, export-oriented industrialization (EOI) and transnational corporations has also had a major impact on expanding women's opportunities and employment in the Global South, from factory garment workers to electronics and call centres (Beazley and Desai 2014). This work can bring extra income and freedoms and change the value of daughters (e.g. see Ahmed and Bould 2004), but also often comes with precarious (and even sometimes dangerous) work conditions. They are frequently low paid, and many target young, unmarried women, and/or women without children, who are too often fired the moment their status changes (as it is perceived to reduce productivity) (Beazley and Desai 2014).

Gender-based violence

Violence against women (VAW) is often used interchangeably with gender-based violence (GBV), for this disproportionately affects women and girls in the Global North and South alike. To identify the multiple facets of VAW, the World Health Organization has distinguished between intimate partner violence (IPV), including forms of physical, sexual, and psychological violence, and non-partner sexual violence, comprising sexual threat, sexual harassment, sexual abuse, sexual assault and rape (WHO 2013). Other forms include trafficking, sexual and labour exploitation, femicide, female genital mutilation, and child marriage. SDG 5 identifies the elimination of violence against women and girls in all spheres of life as its second target, following the more general target of eliminating all discrimination against them. Survivors of violence or women experiencing violence become more vulnerable to existing inequalities across different spheres of society. Furthermore, violence, particularly by an intimate partner, has profound repercussions on the wellbeing of families, communities, and society, making this target necessary to achieve other SDGs.

The indicators behind the SDG5 second target reflect the WHO differentiation between intimate and non-intimate partner violence. The first is commonly addressed as domestic violence (DV), the most frequent form of violence within the home or family being men – husbands, intimate partners, and ex-partners – attacking women. Detrimental to the physical, mental and social wellbeing of the abused person, DV has also profound negative repercussions on the economy and at societal level, including inter-generational trauma (Duvvury et al. 2013). In terms of sexual violence (SV), rape against women (and men) is often a 'weapon' in times of conflict. The causes of both forms of violence include gender inequalities, as much as structures of social and political inequalities (Merry 2006), histories of colonial violence (Atkinson 2002; Hunt 2008), and large-scale processes, including globalization and its components (Fulu and Miedema 2015). Different categories of identity like class, race, nationality, religion, indigeneity and others may complicate the experience of these forms of violence. Significantly, discussions often leave out the category of disability, silencing its added vulnerability.

In 1993, the UN Declaration on the Elimination of Violence against Women enshrined VAW as a violation of human rights. Although development entities and actors have been increasingly framing advocacy, projects and policies within the international discourse of human rights, the concept continues to meet forms of opposition at the local level, in that many perceive it as alien to their worldview, if not neo-colonialist (Merry 2006). Increasingly, initiatives aim to acknowledge and understand context, promote bottom up initiatives, and train local capacity in measuring, investigating, and responding to violence against women. For instance, following the 2030 Agenda for sustainable development, the UNDP, in collaboration with Australian Aid, launched the kNowVAWdata initiative to provide regional and national support to measure VAW in the Asia-Pacific. At the community level, initiatives include 'translating' notions of women's rights and gender equality within local cultures[1] and relevant training for police forces. More recently, EU and UN have partnered in the SPOTLIGHT initiative spanning across Africa, Asia, the Pacific, the Caribbean region, and Latin America, where specific countries are targeted over multiple years. Acknowledging the context-specific forms of violence predominant in those areas, it aims to generate awareness around those issues and identify contextual preventive and responsive strategies.

Progress for women in access to education and economic empowerment has so far had a limited effect on making women less vulnerable to violence. In fact, their economic empowerment may sometimes even increase violence from intimate partners, as women change roles or become breadwinners. This means that 'interventions aimed at economically empowering women must incorporate strategies to minimize unintended negative impacts on women, including risks of increased DV, and promote the empowerment of women from a holistic perspective' (Hughes et al. 2015, 281). Such promotion should come with the acknowledgement and understanding of the many intersectionalities that may make some women more vulnerable to violence. Interventions, as well as pedagogy, require an increasingly sustained analysis of the multiple facets of empowerment – what it means to women in the context of the social relations they are embedded in and what it does to them in the circumstances their lives happen to unfold. For many, empowerment will actually mean to be able to overcome attached stigmas.

These prevent women from reporting or legally pursuing perpetrators of violence, accessing support, or moving on with their lives.

At the same time, fighting stigma and addressing the understandings that lead individuals, women included, to condone or ignore violence against women require approaches that are mindful of the relational dimension of the gendered issue. Men, and understanding of masculinities, overall remain the big absentees in relevant advocacy, initiatives, debates, analysis and research. Only limited research is available on men as perpetrators (Fulu et al. 2013), and men and masculinities remain a novel topic in the development arena and pedagogies globally. Particularly, the topic of trans-generational trauma should be dislodged from the niche it still occupies, to consider both how boys and men re-enact their own trauma and how colonial violence has been internalized and absorbed in the Global South. The impact of the interactions between masculinities and large-scale processes like globalization equally deserves more attention. Finally, it is important to remember that boys and men may also be victims of forms of GBV and that gender and sexual minorities are equally targets of varied forms of GBV (Fulu et al. 2013).

Political voice and representation

Expanding and increasing women's voices, rights and representation in political spaces is essential to gender equality. The political landscape for women has changed due to women's and feminist movements, leading to a focus by national and international policy actors on increasing the number of women in formal leadership positions in both public and private spheres. Gender quota-based or affirmative action-based policies have been the major macro-policy focus to correct gender gaps in representation (Tadros 2014). While the gender gap in political representation has reduced over time, gender-based economic and political inequalities remain the most difficult to change (Sen 2019). Based on the data available from August 2020, globally, the proportion of women in all chambers of government is 25 per cent, with the Americas having 31.8 per cent, Europe 30 per cent, Sub-Saharan Africa 34.7 per cent, Asia 20.4 per cent, Middle East and North Africa (MENA) 16.6 per cent and the Pacific 19.6 per cent (Inter-Parliamentary Union [IPU] 2020). However, only five countries have 50 per cent or more women in one of the chambers of governments: Rwanda, Cuba, Bolivia, the UAE and Canada (IPU 2020).

Research shows that a combination of political, socioeconomic and cultural factors increases the proportion of women present in electoral political bodies (also known as descriptive representation). The structure of the electoral system, the presence of leftist parties in governments, political ideology, the time since women's suffrage, democratization processes, the proportion of women in professional occupations, the level of national economic development, secondary-school enrolment rates of girls, attitudes about women's place in politics, and the interaction of pressures between local and international women's movements are all important in production of gender (in)equalities of political voice and representation (Paxton, Hughes and Green 2006; Swiss, Fallon and Burgos 2012). These larger patterns need to be contextualized as evidenced in outlier cases such as Kerala (India), where the state's significant achievement of women's socioeconomic rights co-exist with lack of political gains for women and persistently high levels of VAW. Moreover, in the increased numbers of women legislators, women from ethnically, socio-economically or other marginalized communities continue to be underrepresented, underscoring the importance of integrating intersectionality in policy-making on women's political representation.

Moreover, it seems that at early stages of expansion of women legislators, there is a symbolic modelling effect which changes both attitudes towards women in politics and increases women's political participation (Liu and Banaszak 2017). However, these types of positive effects may be mediated by the specific forms gender inequality takes, i.e. what kinds of gaps exist in women's social vs. political rights. For example, in East and Southeast Asia, increases in women legislature seems to actually be followed by backlash against women's political engagement (Liu 2018). Liu explains that 'the link between women's political representation and political engagement becomes more complex in situations where a huge gap exists between power obtained by female leaders in the political structure and (the lack of) power legitimated by ordinary women in the social structure' (2018, 256). Effects of symbolic representation may also be more a product of substantive representation and not the number of women politicians – i.e. attitudes towards women in politics and women's political engagement may increase when women legislators pursue policies that produce change (Tadros 2014).

Another question about women's political representation relates to substantive representation, i.e. what impact has women's formal political representation had on pursuing policies that are aligned to women's interests. This is controversial as a question since women are not a monolithic homogenous group and women's interests and feminist interests are not necessarily always congruent. Nevertheless, research indicates that increases (even incremental ones) in the number of women in parliaments lead to policy actions that prioritize health and especially child's health, social policy spending, and broader 'women-friendly' policies (Swiss, Fallon and Burgos 2012). The degree to which women in politics are able to make changes to the way gendered power works depends on multiple things, including a critical mass of women in legislative bodies, but also and perhaps more importantly, how women in politics are able to enter the informal political spaces and networks that shadow formal political processes, as well as mobilize their own constituencies and support networks (Tadros 2014).

In their most promising (radical) forms, education, employment, expanded ownership of wealth, freedom from violence, and political power can expand women's options and their ability to influence, inform and change the way power is distributed – including how decisions are made at household, economic and political levels. As seen above, the focus on national and international policies has been on expanding girls' and women's access to *formal* institutions of education, employment/income, politics and developing laws, policing practices, and awareness that reduces violence against women. However, as the *Pathways of Women's Empowerment* project demonstrated through studies across 55 projects in 15 countries, informal (political) spaces which helped women to organize collectively were the strongest predictors of women's empowerment across multiple dimensions (Cornwall 2016). Drawing on radical feminist practices of consciousness raising, such spaces which facilitate cognitive empowerment and critical awareness about gendered power, produce a network of support and constituency and thus provide a basis for collective agency and action (Stromquist 2015). Tadros (2014) argues that since women's engagement in politics is made possible through the work of feminist and women's organizations at grassroots level, support for these organizations and their work is crucial in changing women's place in politics. Moreover, the creation of spaces in which women can come together and talk freely about their experiences, aspirations and how things can be changed is crucial. What feminist research argues is that these spaces are political spaces and that they are often missed in quantitative analyses and national/international policymaking when it comes to understanding how women's political voice gets amplified (Cornwall 2016; Tadros 2014).

These spaces, though, can also be problematic when their practices and approaches are top-down and decontextualized and/or when they are led by donor-dependent or foreign-led civil society organizations. They can also be spaces impacted by politics of race, class, sexuality, religion. However, researchers of women's advocacy and feminism in Muslim societies (e.g.

Saba Mahmood's work) or of women's ability to mobilize *kastom* (custom) to do peacebuilding work in Papua New Guinea and the Solomon Islands (e.g. see Nicole George's work) speak to the importance of not seeing these conflicting politics in 'informal' and 'customary' spaces as problems or anti-women's rights and gender equality.

Moreover, Murphy-Graham (2010), in her work with a rural Honduran women's empowerment programme, points to the importance of paying explicit attention to supporting women to develop relational skills and resources (e.g. negotiation, conflict resolution) in such spaces to enable women (and men) to transform the dynamics of women's relationships with those that matter to them and/or those who can exercise power over them (at home, work, and other spaces). These types of relational skills are necessary in transforming gender relations of power on an everyday basis. However, policymaking and action related to this has been missing in many efforts at using education, employment, political representation, and laws on gender-based violence. In the next section, we provide some examples of how we can do this.

Pedagogies and practice

In this final section, we reflect on our pedagogical practices in two unique higher education institutions in the Global South – The Asian University for Women (AUW) and The University of the South Pacific (USP). We argue, based on these two examples (institutional and programme-based), that if higher education institutions are going to contribute meaningfully to gender equality, there is a need for institutional and pedagogical practices to actively link students and their communities since education-based change regarding gender dynamics requires collective learning and creation of relationships of support. This also helps address the alienation that higher education and experiences can produce.

One example of a university which has attempted to address the multiple structural dimensions of gender inequality discussed above in its mission and practices is AUW in Bangladesh. AUW was established in 2008 as a liberal arts international residential university, currently catering to women from 15 different countries in Asia.[2] AUW was purposefully founded to address the gender gap in tertiary education in the Global South and in Asia in particular, with a mission to contribute to women's empowerment and foster exceptional women leaders in the region. The decision to create an all-women's university drew on the radical feminist understanding that empowerment requires the production of safe spaces for women to come together, critically question the sources of their oppression, as well as build solidarity networks across multiple differences (Amin 2018). Key curricular practices to facilitate this were that the university was a residential university, had gender-related courses across disciplines (including the sciences), that women learned martial arts, and that students did community-based, service-learning, and action-research projects throughout the four years of their studies. In addition, to ensure a plurality of voices and because women who access tertiary education in Asia tend to overwhelmingly come from privileged backgrounds, AUW had two important specific practices in terms of recruitment (Phillott 2019): (1) a needs-blind admission policy based on community activism and academic excellence that resulted in over 90 per cent of the students at the university to be on full or partial scholarship; and (2) building long-term relationships with schools and organizations in the 15 countries to actively identify and recruit women from socio-economically, culturally, or politically marginalized areas. Together, these practices ensure that women from different positions of privilege and disadvantage were building relationships and sisterhood, sharing resources, strategies, and experiences and that they were working together to tackle broad and fundamental structural challenges affecting women in different ways.

While the international, residential and transformative aspects of the education experience are seen by both the women and the university as empowering in many ways, these processes also produced anxieties, displacements and challenges for the women, both during and after their studies, which sometimes led to experiences of alienation, marginalization and even violence (Amin 2018). Key pedagogical practices that mitigated these negative outcomes related to embedding women's educational journeys in their communities through field research, service-learning projects and professional internships in their communities, development of explicit workshops during their studies which focused on how to transform relationships of power in the family, workplace and community, as well as story-sharing/telling spaces in which diverse social actors shared journeys of change, advocating for change, and the personal challenges and strategies used in these efforts (Amin 2018; Amin et al. 2015). Field research and service-learning projects ranged in disciplinary and issue focus (e.g. peacebuilding with communities in Sri Lanka, challenging menstruation taboos in India and Afghanistan) and were powerful in both making communities sites of learning, as well as creating spaces in which women became visible in ways (e.g. as researchers) and in spaces (e.g. religious institutions, public market places) that are less common. Internships were strategically developed to create pathways for women into areas where women are often absent in decision-making processes, e.g. internships of women in government ministries such as the strategic Ministry of Counter Narcotics in Afghanistan. These types of practices created opportunities of dialogue and conversation between the women and their communities, as well as were sites in which women gained experience and practice in challenging gender inequality, negotiating resistance to their work and presence, and building lasting relationships with their communities (Amin 2018; Amin et al. 2015; Phillott 2019).

Founded in Fiji in 1968, USP today spreads across 12 Pacific Island countries.[3] Owned by their respective governments, it aspires to be an internationally reputed teaching, learning and research centre serving the region's aspirations and needs. Its latest strategic plan 'Shaping Pacific Futures'[4] identifies gender equality as a priority for the institution. As the Priority Area of Research, Innovation and Internationalization develops around the SDGs, gender equality also features as a goal of its own within the thematic area 'governance, justice and equalities.' In 2012, USP equally established a post-graduate certificate programme in Gender Studies,[5] which is attracting increasing numbers of students, including cis-men. This was a direct response to civil society and government stakeholders' calls to create an academic programme supporting the development of feminist research and theory-based policies across different sectors, which would also be culturally and politically sensitive to the needs of Pacific Island communities.

Despite the increasing presence of 'gender' in the region, many people continue to dismiss feminism as 'a foreign flower' (Underhill-Sem 2010, 13), which may contribute to stripping Pacific Island women of their agency. Hence, many pit 'gender' against 'culture' rather than placing it within cultural processes (ibid.). Interestingly, other elements introduced during the colonial era, like rugby and Christianity, have become part of Pacific Islands' lived realities (ibid.), to the extent of being used to challenge gender equality's arguments. Attempts to counterbalance this hostility have included identifying cultural forms of feminism and detecting values associated to gender equality and human rights in Indigenous cultures. Additionally, many perceive gender as a proxy for women, labelling it as anti-men, or barely seeing space for discussions beyond the binary men-women, the latter being also ostracized by perceptions of culture and religion where gender and sexual diversity seem to hold no place (ibid., 11–12). Another issue is that development labels like 'Pacific women' and the 'Pacific' may respectively obliterate the diversity of women in the region and their intersectionalities and reproduce the colonially constructed insularity of Pacific Islands.

The challenge has then been to generate a learning experience conducive to ownership of values and discourses of gender equality, decoloniality, and inclusion of all women, men, and gender and sexual minorities. The programme has encouraged a process where students and lecturer constantly learn from one another and where students' cultures, languages, relations, lived experiences, and their pre-existent knowledge of the region's realities become learning material to analyse gender dynamics and actually 'translate' gender. Acknowledging the place arts occupy in the region's identity, as well as ancestral practices of storytelling, the programme has encouraged students' creativity in articulating and framing the resulting discussions. Students thus embrace a 'logic of the gift,' conciliating feminist epistemologies and Indigenous epistemologies' emphasis on (and value of) contextualized experience and sharing (Kuokkanen 2006). This contributes to creating a safe and inclusive space to look critically into the larger Pacific colonial and postcolonial experience and its role in producing gender.

For instance, discussing women in the region, students realize the impact of colonial categories like Melanesia, Polynesia, and Micronesia on representation and inclusion in the overall picture of Pacific women; the forms of separation that different colonial experiences (including distinct colonial empires) have translated into and the space for non-Indigenous Pacific women and 'overseas' Pacific women. They consider how rank can overtake gender, how other categories of identity, particularly disability and location, can create power differentials among women, and they look in to forms of agency and empowerment within their cultures. Analysing violence against women, students get used to include men and the analysis of how masculine norms may shape men's existences, but also start approaching Pacific masculinities as postcolonial masculinities, reflecting on the possible internalization of forms of colonial violence and the impact of continuous local and global socioeconomic changes. Likewise, students approach gender and sexual diversity issues as the interaction of global LGBTQI+ activism and discourse, cultural histories of gender and sexual diversity, and colonialism. In addition to these discussions, the programme has introduced working sessions with gender and development practitioners. Further ways to connect students with their communities, societies, and context could include fieldtrips to villages to dialogue with women's groups and to urban feminist associations, as well as identifying alternative locally produced learning sources.

In conclusion, based on our experiences of teaching at institutions and programmes that aim at contributing to gender equality, three practices are necessary: 1) policies and interventions ensuring that women from marginalized groups and gender and sexual minorities are proactively recruited and supported into relevant programmes and that men and analysis of masculinities are involved in discussions and initiatives; 2) curricular and extracurricular content and practices developed in dialogue with grassroots organizations, civil society actors, governments and private sector actors advocating for gender equality; and 3) context-responsive pedagogical practices where students and communities are linked and put into conversation with each other.

Key sources

1. Cornwall, A. (2016). Women's empowerment: What works? *Journal of International Development*, 28(3), 342–359.

This article brings together the most important findings from the interdisciplinary *Pathways of Women's Empowerment* project based on studies across 55 projects in 15 countries in the Global South. It provides key insights on policies that work to address gender inequality across different contexts, and where there continue to be important policy gaps. The details and additional outputs from the project can be found through: www.ids.ac.uk/projects/pathways-of-womens-empowerment-research-programme-consortium/

2. Fulu, E., Warner, X., Miedema, S., Jewkes, R., Roselli, T. and Lang, J. (2013). *Why do some men use violence against women and how can we prevent it? Quantitative Findings from the United Nations Multi-Country Study on Men and Violence in Asia and the Pacific.* Bangkok: UNDP, UNFPA, UN Women and UNV.

This UN multi-country study addresses violence against women by focusing on the male perpetrators. It examines and contextualizes their perceptions of their own actions, analysing the relationship between any experience as victims or witness of violence during their childhood and adolescence and their violence against women as adults. Importantly, this work fills a gap, as little research is available on the role of perpetrators in the understanding and prevention of the phenomenon.

3. Oxfam (2020). *Time to care: Unpaid and underpaid care work and the global inequality crisis.* Oxfam Briefing Paper – January 2020. Oxfam International, Oxford.

This document gives an overview of the state of gender inequality in relation to employment and wealth around the globe and underlines the over-representation of women in care work, an often unpaid or underpaid sector of human societies. The document also summarizes the core dynamics and challenges that contribute to gender (in)equality and highlights key areas for policy and action.

4. Stromquist, N. P. (2015). Women's Empowerment and Education: linking knowledge to transformative action. *European Journal of Education*, 50(3), 307–324.

This article both summarizes the research and policy outcomes related to education as a pathway for women's empowerment in the Global South, as well as the key processes that continue to impact on gender inequality in education. Importantly, like most of Stromquist's work, the article provides insights and recommendations on how research and education can be used to produce transformative change and action.

5. Tadros, M. (Ed.). (2014). *Women in Politics: Gender, Power and Development.* Zed Books Ltd.

Part of the Pathways of Women's Empowerment project, this edited volume provides an excellent review of the research on women in politics, what policies have worked and what does not work. Most importantly, the book provides life-stories of women in politics across different countries in the Global South and the environments that enable women's participation in politics. The collection provides new insights into the very specific mechanisms that play out in gender and politics.

Notes

1 See for instance SPC's initiatives https://rrrt.spc.int/resources/videos/cultural-mapping-for-social-inclusion-in-oceania

2 Afghanistan, Bangladesh, Bhutan, Cambodia, China (Tibet), India, Indonesia, Malaysia, Myanmar, Nepal, Pakistan, Palestine, Sri Lanka, Syria, Vietnam
3 Cook Islands, Fiji, Kiribati, Marshall Islands, Niue, Nauru, Samoa, Solomon Islands, Tokelau, Tonga, Tuvalu, Vanuatu
4 www.usp.ac.fj/fileadmin/files/SP-2019–2021/USP-Strategic-Plan-2019–2021.pdf
5 The programme currently offers a postgraduate certificate in the discipline, including a course on feminist theory, debates and research, and another one on gender and development. The proposal has been presented to offer a Postgraduate Diploma in Gender Studies, including a new course on gender and environment.

References

Ahmed, S., and Bould, S. 2004, 'One able daughter is worth 10 illiterate sons': Reframing the patriarchal family, *Journal of Marriage and Family*, vol. 66, no. 5, pp. 1332–1341.

Amin, S.N. (2018), 'Somehow, I will convince my people,… . I will have to follow but I won't accept everything: Asian women's empowerment via higher education', *Asian Journal of Women's Studies*, vol. 24, no. 2, pp. 183–204.

Amin, S.N., Mostafa, M., Kaiser, M. S., Hussain, F., and Ganepola, V. (2015), 'Beyond classroom knowledge and experience: How can fieldwork enrich students' learning and perception on gender', *At the center: Feminism, social science and knowledge*, Advances in Gender Research, Edited by Marcia Segal and Vasilike Demos, 20, 199–222.

Atkinson, Jane (2002), *Trauma trails, recreating song lines*, Spinifex Press, North Melbourne, Victoria.

Beazley, Harriot and Desai, Vandana (2014), 'Gender and globalization'. In Desai, V. and Potter, R.B. (Eds), *The Companion to Development Studies, Third Edition*. Routledge.

Boyd, G.G.D. (2016), 'The girl effect: A neoliberal instrumentalization of gender equality', *Consilience*, vol.15), pp. 146–180.

Cornwall, A. (2016), 'Women's empowerment: What works?', *Journal of International Development*, vol. 28, no. 3, pp. 342–359.

Duvvury, Nata, Aoife Callan, Patricia Carney, and Srinivas Raghavendra (2013), 'Intimate Partner Violence: Economic Costs and Implications for Growth and Development', *Women's Voice, Agency & Participation Research Series,* no. 3. *World Bank, Washington, DC. © World Bank.* https://openknowledge.worldbank.org/handle/10986/16697 License: CC BY 3.0 IGO.

Fulu E., and Miedema S. (2015), 'Violence Against Women: Globalizing the Integrated Ecological Model', *Violence Against Women* vol. 21, no. 12, pp. 1431–1455.

Fulu, E., Warner, X., Miedema, S., Jewkes, R., Roselli, T. and Lang, J. (2013), *Why do some men use violence against women and how can we prevent it? Quantitative Findings from the United Nations Multi-Country Study on Men and Violence in Asia and the Pacific*, UNFPA, Bangkok.

Hughes, C., Bolis, M., Fries, R., and Finigan, S. (2015), 'Women's economic inequality and domestic violence: exploring the links and empowering women', *Gender & Development*, vol. 23, no.2, pp. 279–297.

Hunt, Nancy Rose (2008), 'An acoustic register, tenacious images, and Congolese scenes of rape and repetition,' *Cultural Anthropology,* vol. 23, no. 2, pp. 220–253.

ILO (2018), *Women and Men in the Informal Economy: A Statistical Picture (Third Edition)*, International Labour Office, Geneva.

Inter-Parliamentary Union (IPU) (2020), Monthly ranking of women in national parliaments, Viewed 15th August 2020, https://data.ipu.org/women-ranking?month=8&year=2020.

Kabeer, Naila (2005), 'Is Microfinance a 'Magic Bullet' for Women's Empowerment? Analysis of Findings from South Asia', *Economic and Political Weekly*, vol. 40, no. 44/45, pp. 4709–4718.

Kuokkanen, Rauna Johanna (2006), 'The Logic of the Gift: Reclaiming the Indigenous People's Knowledge. In Thorsten Botz-Bornstein, Jürgen Hengelbrock (eds)', *Re-ethnicizing the minds? Cultural revival in contemporary thought,* Amsterdam: Rodopi

Liu, S.J.S. (2018), 'Are female political leaders role models? Lessons from Asia, *Political Research Quarterly*, vol. 71, no. 2, pp.255–269.

Liu, Shan-Jan Sarah, and Banaszak, Lee Ann (2017), 'Do Government Positions Held by Women Matter? A Cross-National Examination of Female Ministers' Impacts on Women's Political Participation', *Politics & Gender,* vol. 13, no. 1, pp. 132–62.

Merry, Sally Engle (2006), *Human rights and gender violence: Translating international law into local justice*, The University of Chicago Press, Chicago.

Morojele, P. (2011), 'What does it mean to be a boy? Implications for girls' and boys' schooling experiences in Lesotho rural schools', *Gender and Education*, vol. 23, no.6, pp. 677–693.

Murphy-Graham, E. (2010), 'And when she comes home? Education and women's empowerment in intimate relationships', *International Journal of Educational Development*, vol. 30, no. 3, pp. 320–331.

Ortiz-Ospina, Esteban and Roser, Max (2018), 'Economic inequality by gender', Viewed 26th September 2020, https://ourworldindata.org/economic-inequality-by-gender.

Oxfam (2020), *Time to care: Unpaid and underpaid care work and the global inequality crisis*, Oxford, Oxfam Briefing Paper – January 2020.

Paxton, P., Hughes, M.M., and Green, J. L (2006), 'The international women's movement and women's political representation', *American Sociological Review*, vol. 71, no. 6, pp. 898–920.

Phillott, A.D. (2019), 'Meeting Strategic Gender Needs: The Case of Asian University for Women, Bangladesh', *Doing Liberal Arts Education*, pp. 135–145.

Rafique Wassan, M., Hussain, Z., Ali Shah, M., and Amin, S.N. (2017), 'International labor migration and social change in rural Sindh, Pakistan', *Asian and Pacific Migration Journal*, vol. 26, no. 3, pp. 381–402.

Rashid, S.R. (2013), 'Bangladeshi women's experiences of their men's migration: Rethinking power, agency, and subordination', *Asian Survey*, vol. 53, no, 5, pp. 883–908.

Rogerson, C.M. (1997), *SMMEs and Poverty in South Africa*. Input Report for the National Project on Poverty and Inequality. Government of South Africa.

Sen, G. (2019), 'Gender Equality and Women's Empowerment: Feminist Mobilization for the SDGs', *Global Policy*, vol. 10, pp. 28–38.

Shel, T.A. (2008), Gender and Inequity in Education. Gender and inequity in education: literature review, Viewed 17 September 2020, https://unesdoc.unesco.org/ark:/48223/pf0000155580

Stromquist, N.P. (2015), 'Women's Empowerment and Education: linking knowledge to transformative action', *European Journal of Eudcation*, vol. 50, no. 3, pp. 307–324.

Swiss, L., Fallon, K.M., and Burgos, G. (2012), 'Does critical mass matter? Women's political representation and child health in developing countries', *Social Forces*, vol. 91, no. 2, pp. 531–558.

Tadros, M. (Ed.) (2014), *Women in Politics: Gender, Power and Development*. Zed Books Ltd.

Underhill-Sem, Yvonne (2010), *Gender Culture and the Pacific. Asia Pacific Human Development Report Paper Series 2010/05*, UNDP Asia Pacific Office, Bangkok.

UNESCO (2017), School violence and bullying: global status report. Presented at the International Symposium on School Violence and Bullying: from evidence to action. Seoul., Viewed 17 September 2020, https://unesdoc.unesco.org/ark:/48223/pf0000246970

UNESCO (2018), Gender Review: Meeting our commitments to gender equality in education, Viewed 17 September 2020, http://gem-report-2017.unesco.org/en/2018_gender_review/

WHO (2013), *Global and regional estimates of violence against women: prevalence and health effects of intimate partner violence and non-partner sexual violence WHO*, London School of Hygiene and Tropical Medicine and South African Medical Research Council, Geneva.

37

VIOLENT DEVELOPMENT IN XINJIANG UYGHUR AUTONOMOUS REGION

Anna Hayes and Kearrin Sims

Introduction

The violent effects of 'development' on ethnic minorities and other marginalized groups have been widely documented (see for example Scott 2009). Key concerns raised within extant literature include the relationships between development, state expansion, and the loss of minority livelihoods, autonomy and ways of life. Another prominent concern has been the Eurocentrism that underpins much development thinking and practice, and its devaluing of ethnic minority knowledge and cultural practices. Through each of the above forces, ethnic minorities are positioned as backward, uncivilized and in need of 'development' and modernization.

Critiques of Eurocentric development and the culturally-homogenizing effects of state expansion are part of a broader turn within development studies that calls for greater attentiveness to (and appreciation of) the importance of culture within development processes (see Escobar 2012). Culture has become a central thematic within development debates, shifting from a peripheral to a mainstream concern. This has not, however, prevented the perpetuation of development-induced cultural violence, or led to the widespread adoption of more culturally-sensitive development practices. In short, shifts in academic and activist discourses have not been accompanied by equal shifts in development policy and practice (Sims 2015).

With the continued expansion of South-South cooperation over recent decades, development actors from beyond the 'West' have become increasingly influential (Mawdsley 2012). Such actors offer alternative ways of thinking about and doing development, bringing new potentialities for positive social and environmental change and new forms of social, cultural, economic and environmental violence. Empirical shifts in development norms, financing, ideas and practice require new analysis to examine how, and in what ways, new development actors are reworking the global development sector. In addition, continued research is also needed into how new development interventions are reworking the lives of marginalized and disadvantaged communities.

No country has been more influential in reworking the global development arena than China. Through its ambitious Belt and Road Initiative (BRI), China is financing and/or facilitating projects across some 130 countries. The Chinese Communist Party (CCP) has established new multilateral development banks, and resisted participation in prominent existing architectures of multilateral development cooperation such as the Organisation for Economic Cooperation

DOI: 10.4324/9781003017653-41

and Development (OECD) Development Assistance Committee (DAC). In official speeches, the CCP advocates for 'mutual-benefits' and 'win-win' exchanges of South-South cooperation, frequently dichotomizing its development cooperation against what it identifies as traditional, Western imperialist, development cooperation. However, this raises the question of what is the nature of Chinese development at home?

As this chapter interrogates, while China's development cooperation may push back against the norms and assumptions embedded in DAC approaches to development – sometimes in a manner than promotes alternative visions and a greater plurality of voices – this does not also mean that the cultural violence of development necessarily recedes. Such debates and concerns are evident in Chinese development interventions in the Xinjiang Uyghur Autonomous Region (XUAR).

Due to its geographic location in China's north-western border region, XUAR is widely perceived as of critical importance to the BRI and the CCP's efforts to establish new overland connectivity routes into Central Asia and onwards to Europe. As a region where (national) ethnic minorities have long made up the majority of the population, XUAR provides a useful case study for examining how the CCP has wedded development to state-expansionism and efforts, critics maintain, to erase cultural diversity. While there are differences between the CCP's domestic and international development interventions, examination of development efforts in XUAR reveal that there are also continuities – including high-modernist forms of violence against ethnic minorities. In the next section we provide a brief summary of the relationships between development and violence, followed by the argument that Chinese development cooperation is undergirded by technocratic high-modernist ideologies. This is then followed by an examination of such high-modernist development through the XUAR case study, before turning to consider some of the implications of new forms of violent development for development studies students, educators, and practitioners.

The violence of development

The violent and harmful consequences of development have been widely interrogated. To provide a non-exhaustive summary, extant research has examined structural violence, infrastructural violence, epistemic violence, slow violence, spatial violence and cultural violence. Recently, a growing body of research has also examined how development produces intersectional forms of violence that bring harm to disadvantaged communities in multi-layered ways (Davies 2019; Sims forthcoming).

As previously noted, regarding ethnic minority communities, two prominent themes of inquiry have been the cultural violence of state-building/expansion and the violent effects of Eurocentric development. Beginning with state expansion, historic and ongoing processes of state territorialization in areas populated by ethnic minorities have been accompanied by the acquisition of land and other resources, the compromising or erasure of ethnic minority livelihoods, prohibitive legislation and other coercive constraints on cultural practices, and efforts to promote cultural homogenization through practices such as standardized national curricula in schooling. In most cases, state territorialization efforts have also pursued growth-centric and technocratic development agendas that prioritize economic growth, infrastructure expansion, urbanization, industrialization and marketization, and top-down national-scale planning (Scott 1998).

Turning to Eurocentric development, extensive post-colonial, decolonial and post development critiques have revealed how prevailing development discourses position Europe's (and post-colonial North America's) social, political, economic and cultural systems as universally valid frameworks for the socially and intellectually 'inferior' peoples of the 'non-Western' world.

Violent and culturally-homogenizing discourses of Eurocentric development perceive peoples as living in different historical stages of an imagined, and universally pre-determined, Western chronology of development, whereby non-Western countries are required to 'modernize' in order to 'catch up' to the West. Interwoven with high-modernist development approaches, such Eurocentrism has contributed to the loss of ethnic minority knowledge, languages, and cultural practices, as minority groups become positioned as 'backward' peoples living 'traditional' (read: historic and outdated) ways of life (Escobar 2012; Kapoor 2008).

While China cannot be accused of advancing Eurocentrism, scholars have argued that its development efforts further expand high-modernist logics of development typically ascribed as being Eurocentric (Sims 2020). Indeed, in many respects, Chinese development narratives are more high-modernist than most contemporary Western development discourses, which have shifted away from the 'the idea of the continual improvability of the human condition' that China continues to promulgate (Nyiri 2012, 557). As the following two sections demonstrate, China's commitment to high-modernist development carries the potential for severe cultural violence against ethnic minority communities.

Violent development in the Xinjiang Uyghur Autonomous Region

Since 1949, the CCP's approach to development in XUAR, and elsewhere, can be described as high-modernist: holding faith in continued linear progress via the expansion of scientific and technical knowledge, economic production, and 'the rational design of social order' (Scott 2009, 89). XUAR has been a priority site for development interventions since its annexation into the People's Republic of China (PRC) in 1949, constituting a substantial part of Han-China's 'backward' western regions. Upon annexation, the Uyghur ethnic group constituted approximately 76 per cent of the total population, with Han Chinese representing just 6.7 per cent, and other minority nationalities comprising the remainder. After years of government programmes to stimulate Han migration into this frontier region, later followed by private economically motivated migration, the region's ethnic ratio has significantly altered. Uyghurs now account for approximately 42 per cent of the overall population with Han Chinese accounting for 40 per cent (Clarke and Hayes 2016, 4). At least partially legitimized via discourses of development, Han migration to the region has served two main purposes. First, it has enabled the region to become Sinicized, ensuring that it could not separate from the PRC, and second, is has assisted modernization efforts through urban–industrial expansion, increased marketization, the promulgation of nationalist narratives and greater state territorialization.

One key organization responsible for undertaking modernization and economic development efforts, as well as frontier maintenance and stability, has been the Xinjiang Production and Construction Corps (XPCC) (*Xinjiang Shengchan Jianshe Bingtuan*). Established in 1952, the XPCC is a paramilitary force tasked with land reclamation for farming and industrialization and is hugely significant in terms of its economic power (O'Brien 2016). The XPCC has been a key driver of the modernization agenda in XUAR, operating large and small-scale farms, factories, and mining operations, as well as tourism, insurance, trade, real estate and construction ventures. Together with the People's Liberation Army (PLA), the XPCC has also been a key force in the securitization of the region – providing modernization and development in peaceful times and ensuring/enforcing regional security and stability during periods of heightened tensions or conflict (Smith Finley 2016: 92). Finally, the XPCC has also been instrumental in the establishment of 'Han cities,' which are designed to further attract Han migration from other parts of China.

Through all of the above activities, the XPCC has contributed to deep ethnic segregation. The economic benefits of the XPCC's work has largely flowed to Han migrants, both long-term and more recent migrants, while Uyghurs and other Muslim minorities have had minimal benefits or been further marginalized within the economic structures (Leibold and Deng 2016). Wealth disparity and inequalities have increased alongside modernization efforts, as has environmental degradation resulting from large-scale agribusiness and mining enterprises. Key environmental impacts have included the unsustainable overuse of underground water supplies, afforestation, and emerging desertification (Feng et al. 2015).

Alongside economic and industrial modernization, Beijing's development efforts in XUAR have also sought to devalue, and in some cases erase, minority knowledge systems, cultural values, and cultural practices. In official and academic rhetoric, Uyghurs and other minority nationalities are framed as being 'backwards' and in need of modernization and development. Such positioning of minorities has been referred to as 'Han Chauvinism,' with both O'Brien (2016) and Xu (2007) arguing that these chauvinistic views still widely permeate how XUAR and its non-Han peoples are regarded – and underpin the 'civilizing mission' that undergirds Beijing's development interventions within the region. For O'Brien, the colonialist mindset of Beijing is poignantly revealed within the statements of a government official that he interviewed during field research in XUAR, who stated that he believed the Han to be:

> the most advanced of the Chinese people and we are here [in Xinjiang] to help Uyghurs to develop.... . We are a Chinese family and we are here to help our younger brothers.
>
> (2016, 34.)

In some cases, such perceptions of 'backwardness' or 'behindness' are even extended to Han migrants to the region, who are seen to have entered into, and embodied, a place of 'behindness,' even if their relative social status increased by virtue of their migration (2016, 5).

The self-perceived superiority of the Han, and the resultant perceived inferiority of the Uyghurs, sets up a significant power imbalance, which in the past has led to deliberate acts of targeted violence against Muslim minorities. During China's Great Leap Forward – a mass campaign aimed at spurring rapid development and modernization across China – ethnicity and religion were singled out for eradication because they were perceived to be 'obstacles to progress' and 'backwards custom' (Bovingdon 2004, 19). Later, during the Cultural Revolution, any practices or traditions that did not meet the CCP's parameters for correct and progressive collective culture (in the sense of being entirely of Communist virtue) were again identified as being backward and a danger to the goals of the state. During both mass campaigns, minority nationalities were subjected to widespread cultural and religious insults, violence, and human rights abuses (Millward 2007). In XUAR, targeted ethnic violence was carried out by predominantly Han cadres and activists against the minority nationalities. Violent acts were particularly extreme during the Cultural Revolution and included the large-scale destruction and desecration of mosques, Qur'ans were burned, ethnic dress and practices were banned, and Imams and Muslim intellectuals were publicly humiliated with some forced to engage in religiously prohibited activities such as raising pigs (Bovingdon 2004; Millward 2007). These mass campaigns were aimed at the rapid transformation and assimilation of the Muslim minorities of XUAR in order to hasten the modernization and development of this region. In XUAR then, state violence against ethnic minorities results from both direct state persecution and through more indirect forces of modernization, both of which are legitimized through the need to bring 'development' to this 'uncivilized' region.

Another notable development campaign rolled out by Beijing was the Great Western Development (GWD) campaign of 2000. The GWD campaign was another attempt by the Chinese government to rapidly improve living standards, infrastructure and ethnic relations in frontier regions like XUAR, but also included the Tibet Autonomous Region (TAR), the Inner Mongolia Autonomous Region and other inland regions of China. Despite the regional minority nationalities being a target group for development, the hegemonic nature of the development strategy ultimately led to what Fischer (2013) called 'disempowered development.' The strategy disempowered its intended targets, the minority nationalities, because it was undergirded by a form of 'flow on' or 'trickle-down' economic development. Beijing proposed that initially some members of the regional populations would fare better than others, but economic benefits would eventually trickle out to all members of the community. The key problem with the strategy was its intended beneficiaries were largely Han Chinese populations living in these regions, as well as new migrants pursuing greater economic opportunities. The strategy was detrimental to most members of the regional minority nationalities who experienced deeper economic and societal marginalization. Therefore, while the GWD campaign was intended to improve living conditions and opportunities for the Muslim minority populations of XUAR it did not achieve these goals. By primarily focusing on the development of mostly Han urban cities in the north, particularly the region's capital Urumqi, rather than the mostly Uyghur poorer rural areas in southern XUAR, like Kashgar, it also escalated ethnic tensions and contributed to periodic outbreaks of localized violence across the region. The campaign was successful however in continuing the Sinification of XUAR, and cemented modernization, industrialization and urbanization as the primary goals of CCP development strategies in the region.

Violence expanded: the belt and road initiative in XUAR

The BRI is the latest grand development and modernization strategy rolled out by Beijing. It was initially launched in September 2013 when Chinese President Xi Jinping announced plans for a new Silk Road Economic Belt during his 10-day tour of the Central Asian Republics. In official discourse, the BRI is presented as a development and investment initiative that seeks to open up the Chinese hinterland as well as promoting Chinese development and economic partnerships with other states. It also purports to offer a new development model to states largely located in the Global South, with promises of 'win-win' outcomes for participating states (National Development and Reform Commission 2015). XUAR holds particular significance to the BRI due to its geographic location, serving as a gateway for China to Central Asia and the Middle East. Accordingly, the BRI has subsumed ongoing and planned projects started under the GWD campaign. One key difference between the BRI and the GWD campaign, however, has been that the BRI has more aggressively targeted Uyghur areas. This has resulted in XUAR's Muslim minorities experiencing increasing levels of control and oversight by regional authorities who wish to successfully implement this nationally significant development strategy. It has also led to widespread human rights abuses, demonstrating that the BRI, as it is being applied in XUAR, is an extremely violent development strategy.

To achieve the development goals of the BRI, and to force a social re-engineering of XUAR's population, Beijing has implemented heightened surveillance and digital authoritarianism across XUAR. Early indicators of the increasing surveillance of Uyghurs came in 2014, when local residents in Peyziwat county in Kashgar prefecture were instructed to spy on their neighbours. Residents were required to sign a 'Joint Responsibility Contract,' by which they agreed to report prohibited activities such as 'unusual travel,' the transmission of information considered to be 'politically sensitive,' the sale of land, quitting smoking or alcohol consumption, 'unusual purchases,' the promotion or teaching of Islam, and even the refusal 'to watch or read official news media' (Niyaz and Finney 2016). They were informed that if they failed to carry out their surveillance duties, all residents would face a collective punishment. Under duress, villagers did comply, thereby dividing the Muslim minority population and turning them against one another. Their compliance was confirmed by a deputy Party secretary who in 2016 reported the policy was having 'great success' and that 'the few who have failed to uphold it have been punished according to the law' (Niyaz and Finney 2016).

In addition to such 'divide and conquer' governance, the CCP's digital authoritarianism in XUAR involves high-tech strategies including face and voice recognition, iris scanners, DNA sampling and 3D identification imagery of the Muslim minorities. Security checkpoints have proliferated, facial scanners have been introduced, and smartphone searches are now routinely conducted by authorities looking for encrypted apps and suspicious content (Millward 2018). According to Chin and Burge, surveillance spending in XUAR for 2017 was US$9.1 billion (mainly on technology-based surveillance), an increase of 92 per cent from the previous year's total (Chin and Burge 2017). This spending has included mosque surveillance, leading to a further erosion of religious freedom for the Muslim minorities. What we see in XUAR is the coming together of longstanding forms of cultural violence against ethnic minorities that is enacted through state modernization programmes with new technologies of digital violence, which are also legitimized through discourses of security and development.

The mass detention of XUAR's Muslim minorities provides further evidence of the regional crackdown occurring alongside the BRI. Towards the end of 2016, members of the Uyghur diasporic community became increasingly concerned that their relatives and friends still living in XUAR were disappearing. By mid-2017, communications into the region became even more fragmented. Twitter became a site for growing transnational awareness of the disappearances with the #MeTooUyghur campaign generating lists of the missing. A separate list compiled in January 2019 identified more than 221 academics, University heads, cultural figures, musicians, journalists, authors, folklorists and other custodians of Uyghur culture who had disappeared (Uyghuryol 2019), while further additional lists include prominent businesspeople, pregnant women and mothers of young children, housewives, and even retired civil servants who spent years working for the CCP (Wong 2019).

In mid-2018, the Congressional-Executive Commission on China (2018) reported over 1 million Uyghurs and other Muslim minority nationalities had been imprisoned in detention camps across XUAR. Beijing initially denied the existence of the camps, but under increased pressure, including satellite imagery, they were forced to concede. Although the CCP calls the camps 'vocational training' centres, given that the disappeared includes highly skilled professionals who were in employment as well as retired elderly people, this explanation has largely been discredited. The Congressional-Executive Commission Report on China (2018) identified their purpose as being 'political re-education' camps, part of which is centred on development goals. The CCP is attempting to force the Muslim minorities to undergo modernization. They are instructed to overcome the 'backwardness' of their religious and cultural identities by embracing modernity and development. They are also taught Chinese Mandarin so they can be employed in the factories that are being established across XUAR, many of which manufacture products for multinational corporations.

The linkages between the detention camps and development goals are apparent in what has occurred in Kashgar, formerly a majority-Uyghur city in southern XUAR. In 2009, Beijing began to redevelop this Old City, which then mostly comprised traditional Islamic mud-brick architecture. This transformation resulted in the relocation of 220,000 Uyghurs from their homes, which were demolished to make way for reconstructed parts of the Old City designed for tourism and new high-rise towers resonant of other parts of China. The forced resettlement and displacement of ethnic minorities under the guise of development has been common across much of the Global South, but has been particularly prominent within China, as is also demonstrated in work by scholars writing on the border regions of Laos, Thailand and Myanmar (Stuart-Fox 2003; Sturgeon 2010). In Kashgar, the displaced Uyghurs were relocated to apartment buildings on the outskirts of the city, causing much social, cultural and religious fragmentation, as well as increasing the Sinicization of the city core (Hayes 2020).

The rapid transformation of Kashgar was one of the first stages in achieving what the CCP has called the Great Kashgar Dream (Liu 2010). This Dream, which is linked to both Xi Jinping's China Dream concept and the BRI, involves the rapid economic development and modernization of Kashgar so it can become one of the pivot points in the economic arteries of the BRI.[1] Beijing hopes to more than double the population of Kashgar and has planned more than 100 square kilometres of construction in its long-term development planning (ibid.). The long-term aim is that Kashgar will eventually become a central city surrounded by nine major bases including: textiles; a large-scale metallurgical industrial base; a petrochemical base; a processing base for agricultural and sideline products; an export commodity processing and manufacturing base for neighbouring countries; halal food production and supply base for Muslim countries; a building materials base for neighbouring countries; a trade logistics base; as well as Kashgar being an international tourism destination (ibid.). Already there is evidence that some detention camps across XUAR house the workers for co-located factories and enterprises with other detainees being sent as labour to factories in other parts of China (Xu et al. 2020).

There is mounting evidence of violence taking place in the detention camps. Documented violence includes multiple human rights violations such as sleep deprivation, inadequate clothing for conditions, physical and sexual assault, and other forms of abuse (CECC 2018; Roberts 2020; Smith Finley 2021; Turdush & Fiskesjo 2021). The large-scale incarceration of adult Uyghurs has also seen Uyghur children being sent to orphanages, not only in XUAR but also to eastern parts of China (Millward 2018; Zenz 2019). The violence extends to limiting reproduction among the Muslim minorities. Zenz (2020) has reported that between 2015 and 2018 the birth rates in the largest two Uyghur prefectures in XUAR fell by 85 per cent. In addition, as many as 80 per cent of Uyghur women in some areas have undergone invasive mandatory birth control measures including forced sterilization and IUD insertion, with one Uyghur region recorded a near-zero birth rate (Zenz 2020).

The violence unfolding in XUAR is directly linked to the development goals of the BRI, and the CCP's attempts to bring modernization and security to XUAR – its gateway to Central Asia and beyond. As Brown succinctly states, for the BRI 'to work, Xinjiang has to work' 2018). Muslim minorities are viewed by the CCP as backwards peoples who present obstacles to development and modernization, and accordingly the 'development' of XUAR has been accompanied by intense state violence and policing. This deep suppression of Muslim minorities also exposes an acute sensitivity Beijing harbours in relation to the development and opening up XUAR to neighbouring partner states, many of whom share kinship and religious links to the Muslim minorities. XUAR is a site of acute anxiety for Beijing, and its 'brutal suppression conforms to [CCP fears over] what may happen in the future if tough measures are not taken at present' (ibid.). This is violent development on a massive scale.

Lessons for pedagogy and practice

While it has long been known that development has multiple violent effects, empirical shifts within the global development arena mean that the violence of development requires continual (re)interrogation. New development actors, modalities, materialities, and ontologies bring new forms of violence, articulated in new ways. This chapter has focused on XUAR because it represents an extreme case – where many forms of development-induced violence that occur elsewhere are more pronounced. It also focuses on XUAR because of the CCP's attempts to reconfigure the global development arena, and because of the region's critical importance to the success of Beijing's grand BRI. In XUAR we see the perpetuation of many well-known violent effects of high-modernist development, coupled with violent new digital surveillance technologies, and colonialist Han Chinese perceptions of ethnic minority communities that are also embedded within the CCP's international development cooperation.

We see three primary 'lessons' from this chapter for development pedagogy and practice. The first is the need for greater attentiveness to the violent effects of development, as well as how different forms of violence – cultural, structural, material, epistemic, etc. – intersect and reinforce one another. The violent effects of development have been widely studied, but this chapter seeks to serve as a reminder that violence is a constitutive feature of development. It is a not a 'side-effect,' but a central component of development and modernization.

The second key point for further consideration is how growing South-South cooperation is laden with new forms of violence, and particularly violence against already marginalized or disadvantaged communities. This requires new empirical research and new (and increased) engagement with 'Southern' actors. The point here is not to imply that 'traditional,' Western, development actors offer less violent or more inclusive development. Rather, it is simply to be more attentive to shifting landscapes of development and the myriad of new opportunities and challenges that are emerging as a result of an increasingly multipolar development arena. This chapter focuses on violence, but this is just one of many themes regarding South-South cooperation and China's BRI that requires more research and analysis. As Waisbich and Mawdsley note (this volume): 'One task for those interested in critically researching South-South (Development) Cooperation in the decade ahead resides therefore in continued questioning of the gaps between the transformative claims by Southern partners and their uneven materializations on the ground…' This chapter suggests that violence is one entry-point for such questioning.

The third, less explicit, point of significance within this chapter is the need to take culture more seriously when thinking about development and its effects. To take culture seriously within development requires a contextually based and locally informed re-evaluation of development objectives in accordance with disparate socio-cultural norms and aspirations. It requires recognizing that culture represents a fluid arena of struggle that is central to the politics of development, and that performing culturally-sensitive development is 'not simply a matter of including culture but also of interrogating culture as a terrain of power' (Pieterse 2010, 77). Regarding violence, this requires attention to both the many material forms of violent development discussed throughout this chapter, as well as epistemic violence. By epistemic violence, we refer here to practices of silencing that police ideas and establish who can speak and with what authority (Dotson 2011).

Questions of culture and epistemic violence are highly important when attempting to make sense of the BRI and its effects for, as Winter explains, the BRI represents 'a site of cultural production and cultural politics' that establishes particular 'political project[s] of historical imagination' and attempts to engineer particular futures into existence (Winter 2019, xiii). In the above example of XUAR we see how BRI reinforces state-centric narratives of development that position ethnic (and other) minorities as backward, peripheralized, peoples. Through the cultural politics of BRI, the CCP seeks to maintain 'depoliticized aesthetics of ethno-cultural diversity' that gloss over power imbalances between different ethnic groups, and which seek to mask violence against ethnic minorities through discourses of development and representations of shared national values (Winter 2019, 117).

While Beijing's development cooperation abroad differs from its interventions in XUAR, the CCP's perceptions of ethnic minorities as backward also extend beyond its territorial borders into its state-centric and high-modernist development cooperation. As Stuart-Fox has argued in his work on Laos, Chinese development discourse is underpinned by a Middle Kingdom mentality that perceives China as being surrounded by 'less advanced' nations [and peoples] in need of China's guidance and assistance (Stuart-Fox 2003, 9). Where much development studies literature calls for the greater empowerment of minority groups in identifying their own needs and having the means to shape their own development futures, BRI's high-modernist development discourse calls for rapid modernization driven by Chinese technical expertise and investment. Recognizing this, development pedagogy and practice should be attentive to the ways in which BRI may advance a new 'cognitive empire,' whereby what are frequently identified as Eurocentric tropes of development are rethought and contested, but may also be replaced by new 'failures in the domain of knowledge to recognize the different ways of knowing by which diverse people across the human globe make sense of the world and provide meaning to their existence' (Ndlovu-Gatsheni 2020, 6). The following questions and activities seek to provoke further thinking on some of these issues:

1. How can the perspectives of vulnerable and marginalized people be given primacy within the development process?
2. If the development process requires a reordering of society, how can the impacts be minimized to reduce harm and violence to affected communities?
4. In what ways have development and modernity become intertwined? How can we rethink development within ethnic minority contexts beyond attempts to modernize peoples and societies in rapid and violent ways?
5. Assess the domestic rollout of the Belt and Road Initiative. What are the lessons from XUAR? Does the domestic rollout provide a cautionary tale to BRI partner states?
6. How can we take culture more seriously in development practice? What does it mean to engage in culturally-sensitive development and what are some useful modes of practice for culturally-sensitive development?
7. How might the BRI and the wider expansion of South-South cooperation (SSC) lead to new ways of thinking about development? Can you provide any examples of how SSC has challenged Eurocentric development thinking/practice? And how might new forms of SSC produce new 'failures in the domain of knowledge to recognize the different ways of knowing by which diverse people across the human globe make sense of the world and provide meaning to their existence'? Can you provide any examples here?

Key sources

1. Gu, Jing, Alex Shankland, and Anuradha M. Chenoy. (2016). *The BRICS in International Development*. Basingstoke: Palgrave Macmillan.

This edited volume includes a collection of insightful pieces on the actors, institutions, and ideas shaping the first years of the re-emergence of BRICS countries as 'rising powers in international development.' The authors are Northern and Southern-based scholars with sound understanding and empirical knowledge of national and global development dynamics shaping BRICS countries rising roles in the field.

2.   Hillman, Jonathan. (2020). *The Emperor's New Road: China and the Project of the Century*. New Haven: Yale University Press.

This book provides an excellent examination of the BRI. Hillman examines views of the BRI, both inside and outside of China, the challenges and promised opportunities of the BRI, as well as documenting some of the implementation challenges that are complicating the rollout of the strategy. The book is rich with personal observation, interview data and in-depth knowledge and analysis of both the policy and academic literature on the BRI.

3.   Mawdsley, Emma, Elsje Fourie, and Wiebe Nauta, eds. (2019). *Researching South–South Development Cooperation: The Politics of Knowledge Production*. 1st ed. Routledge. https://doi.org/10.4324/9780429459146

This recent edited volume reflects on politics of knowledge production in the field of South-South Cooperation for development. It gathers a collection of conceptual, methodological, and reflexive first-hand accounts by both Northern and Southern, senior and junior scholars and offers honest accounts on the promises, opportunities and impossibilities in co-producing critical knowledge on SSC.

4.   Mohan, G., and B. Lampert. (2013). 'Negotiating China: Reinserting African Agency into China-Africa Relations.' *African Affairs* 112 (446): 92–110. https://doi.org/10.1093/afraf/ads065

One of the first and most accomplished papers on African agency in China-Africa relations, offering a complex account of how different actors within African countries negotiate the Chinese partner. The paper highlights the relationships between African and Chinese political and business elites and indicates emerging forms of contestation of China presence within African societies.

5.   Roberts, Sean R. (2020), *The War on the Uyghurs: China's campaign against Xinjiang's Muslims*. Manchester: Manchester University Press.

This book documents the modern history of Chinese colonialism in the homeland of the Uyghurs. Roberts identifies how the CCP recast Uyghur opposition to domestic policies and practices as acts of terrorism. These efforts were emboldened by the Global War on Terror, by which time the Uyghurs had become collectively framed as a terrorist threat against the Chinese state. The book also identifies how China's actions in Xinjiang have influenced other states to increase their own persecution of ethnic minority populations. Following China's lead, these states are also recasting acts of domestic opposition as terrorism as they too attempt to legitimize crackdowns on ethnic minority populations.

6.   Shankland, Alex, and Euclides Gonçalves. (2016). 'Imagining Agricultural Development in South–South Cooperation: The Contestation and Transformation of ProSAVANA.' *World Development* 81 (May): 35–46. https://doi.org/10.1016/j.worlddev.2016.01.002

An empirically rich contribution on one particular triangular cooperation initiative, the Brazil, Japan, Mozambique ProSAVANA agricultural development programme, and its multiple negotiations on the ground. Based on a multi-sited ethnographic work, this paper offers a powerful illustration of how South-South affinities and imaginaries are built and contested in the course of one initiative.

7. Westhuizen, Janis van der, and Carlos R.S. Milani. (2019). 'Development Cooperation, the International–Domestic Nexus and the Graduation Dilemma: Comparing South Africa and Brazil.' *Cambridge Review of International Affairs* 32 (1): 22–42. https://doi.org/10.1080/09557571.2018.1554622

An important contribution to the conceptualization of the domestic politics of South-South Cooperation in two emblematic Southern partners: Brazil and South Africa. Here the authors emphasize the multiple constituencies bargaining around SSC at home and abroad, and the tensions and dilemmas Southern providers face while promoting development beyond borders.

8. Zoccal, Geovana, and Paulo Esteves. (2018). 'The BRICS Effect: Impacts of South–South Cooperation in the Social Field of International Development Cooperation.' *IDS Bulletin* 49 (3). https://doi.org/10.19088/1968–2018.152

A good representative of 'second generation' South-South Cooperation studies, which aims to unpack the effects of the political and material rise of Southern development actors in the norms, governance, and practice of global development.

Note

1 For further information on these connections and an overview of the China Dream concept see Hayes (2020).

References

Bovingdon, G. (2004) 'Autonomy in Xinjiang Uyghur Autonomous Region: Han Nationalist Imperatives and Uyghur Discontent,' *Policy Studies* 11, East-West Center, Washington.

Chin, J., and Burge C. (2017) 'Twelve days in Xinjiang: How China's Surveillance State Overwhelms Daily Life,' *Wall Street Journal* December 19, accessed March 11, 2018, www.wsj.com/articles/twelve-days-in-xinjiang-how-chinas-surveillance-state-overwhelms-daily-life1513700355

Congressional-Executive Commission on China (CECC), (2018), Annual Report 2018–115 Congress, Second Session, October 10, accessed November 9, 2018, www.cecc.gov

Davies, T., (2019), Slow Violence and toxic geographies: 'Out of sight' to whom? *Environment and Planning C: Politics and Space*, 0(0), 1–19.

Dotson, K., (2011), Tracking Epistemic Violence, Tracking Practices of Silencing, *Hypatia*, 26(2), 236–257.

Escobar, A. (2012), *Encountering development: The making and unmaking of the Third World*, Princeton University Press, New Jersey.

Feng, Q., Ma, H., Jiang, X, Wang, X. and Cao, S. (2015), What has caused desertification in China? *Scientific Reports* 5(15998), 1–8.

Fischer, A. (2013), *The Disempowered Development of Tibet in China: A Study in the Economics of Marginalization*, Lanham: Lexington Books.

Hayes, A. (2020), 'Interwoven Destinies': The significance of Xinjiang to the China Dream, the Belt and Road Initiative, and the Xi Jinping Legacy, *Journal of Contemporary China* 29(121), 31–45.

Hayes, A. and Clarke, M. (eds) (2016), *Inside Xinjiang: Space, place and power in China's Muslim Far Northwest*, Oxon: Routledge, 32–51.

Kapoor, I. (2008), *The Postcolonial Politics of Development*, Routledge, London.

Leibold, J. and Deng, DX (2016), Segregated Diversity: Uyghur Residential Patterns in Xinjiang, China (in) A. Hayes and M. Clarke (eds) *Inside Xinjiang: Space, place and power in China's Muslim Far Northwest*, Oxon: Routledge, 122–48.

Liu, H. (2010) 'Special Zone: Kashgar,' *National Financial Weekly* August 2, accessed July 25, 2019.

National Development and Reform Commission, Ministry of Foreign Affairs, and Ministry of Commerce of People's Republic of China, with State Council authorization (2015), *'Vision and Actions on Jointly Building Silk Road Economic Belt and 21st Century Maritime Silk Road, Beijing'*, National Development and Reform Commission, accessed 18 April 2016, http://en.ndrc.gov.cn/newsrelease/201503/t20150330_669367.html

Mawdsley, E. (2012), *From Recipients to Donors: Emerging Powers and the Changing Development Landscape*, Zed Books, London.

Mawdsley, E. (2019), South–South Cooperation 3.0? Managing the Consequences of Success in the Decade Ahead, *Oxford Development Studies*.

Millward, J.A. (2007), *Eurasian Crossroads: A History of Xinjiang*, Columbia University Press, New York.

Ndlovu-Gatsheni, S.J. (2020), *Decolonization, Development and Knowledge in Africa: Turning over a New Leaf*, Routledge, London.

Niyaz, K. and Finney, R. (2016) *China: Uyghur villagers forced by 'contract' to spy on neighbor,'* RefWorld April 229, accessed March 16, 2018, www.refworld.org/docid/5760fbbc15.html

Nyiri, P. (2021), Enclaves of Improvement: Sovereignty and Developmentalism in the Special Zones of the China-Lao Borderlands, *Comparative Studies in Society and History,* 54(3), 533–562. Doi:10.1017/S0010417512000229

O'Brien, D. (2016), 'If there is harmony in the house there will be order in the nation': an exploration of the Han Chinese as political actors in Xinjiang, in A. Hayes and M. Clarke (eds) *Inside Xinjiang: Space, place and power in China's Muslim Far Northwest*, Oxon: Routledge, 32–51.

Pieterse, J.N. (2010), *Development theory: Deconstructions/reconstructions*, SAGE, London.

Roberts, Sean R. (2020), *The War on the Uyghurs: China's Campaign against Xinjiang's Muslims.* Manchester: Manchester University Press.

Scott, J.C. (2009), *The Art of Not Being Governed: An Anarchist History of Upland Southeast Asia*, Yale University Press, New Haven.

Scott, J.C. (1998), *Seeing Like a State: How Certain Schemes to Improve the Human Condition Have Failed*, Yale University Press, New Haven.

Sims, K. (2020), China's Development Contribution to Laos: Beyond Connectivity and Growth, in Don Emmerson (ed.), *The Deer and the Dragon: Southeast Asia and China in the 21st Century*, Stanford University Press, California, pp. 271–298.

Sims, K. (2021), *Infrastructure violence and retroliberal development: connectivity and dispossession in Laos, Third World Quarterly,* 42(8), 1788–1808.

Sims, K., (2015), The Asian Development Bank and the production of poverty: Neoliberalism, technocratic modernization and land dispossession in the Greater Mekong Subregion, *Singapore Journal of Tropical Geography* 36(1), 112–126.

Smith Finley, J. (2016), Whose Xinjiang? Space, place and power in the rock fusion of Xin Xinjiangren, Dao Lang, in A. Hayes and M. Clarke (eds) *Inside Xinjiang: Space, place and power in China's Muslim Far Northwest*, Oxon: Routledge, pp. 75–99.

Smith Finley, J. (2021), Why Scholars and Activists Increasingly Fear a Uyghur Genocide in Xinjiang, *Journal of Genocide Research*, 23(3), 348–370.

Stuart-Fox, M., (2003), *A Short History of China and Southeast Asia: Tribute, Trade and Influence*, Crows Nest, Allen & Unwin.

Sturgeon, J.C., (2010), Governing Minorities and Development in Xishuangbanna, China: Akha and Dai Rubber Farmers as Entrepreneurs, *Geoforum*, 41(2), 318–328.

Turdush, R. and Fiskesjo, M. (2021), Dossier: Uyghur Women in China's Genocide, *Genocide Studies and Prevention: An International Journal*, 15(1), 22–43.

Uyghuryol (2019), List of Uyghur Intellectuals imprisoned in China from 2016 to the present, January, accessed October 1, 2019, https://uyghuryol.com/2019/01/1003/

Winter, T., (2019), *Geocultural Power: China's Quest to Revive the Silk Roads for the Twenty-First Century*, Chicago, University of Chicago Press.

Wong, C.H. (2019), China Says Majority of Xinjiang Detainees Released, but Activists Question Claim, *The Wall Street Journal* July 30, accessed September 25, 2019, www.wsj.com/articles/china-offers-rare-update-on-detainees-in-xinjiang-detention-camps-11564490302

Xu, J. (2007), Community Participation in Ethnic Minority Cultural Heritage Management in China: A Case Study of Xianrendong Ethnic Cultural and Ecological Village, *Papers from the Institute of Archaeology* 18, 148–160.

Xu, V.X. with Cave, D., Leibold, J., Munro, K. and Ruser, N. (2020), Uyghurs for Sale: 'Re-Education', forced labour and surveillance beyond Xinjiang, *Australian Strategic Policy Institute* March 1, accessed March 5, 2020, www.aspi.org.au/report/uyghurs-sale 29–30.

Zenz, A. (2019), Break Their Roots: Evidence for China's Parent-Child Separation Campaign in Xinjiang, *Journal of Political Risk* vol. 7, no. 7, accessed October 1, 2019, www.jpolrisk.com/break-their-roots-evidence-for-chinas-parent-child-separation-campaign-in-xinjiang/

Zenz, A. (2020), *Sterilizations, IUDs and Mandatory Birth Control: The CCP's Campaign to Suppress Uyghur Birthrates in Xinjiang,* Updated July 21, Washington DC: The Jamestown Federation, viewed August 23, 2020, https://jamestown.org/wpsterilizations-and-IUDs-UPDATED-July-21-Rev2.pdf?x70493-content/uploads/2020/06/Zenz-Internment

Zoccal, G. and Esteves, P. (2018), The BRICS Effect: Impacts of South–South Cooperation in the Social Field of International Development Cooperation, *IDS Bulletin*, 49(3).

PART 4

Game changers of Global Development?

38

INTRODUCTION

Game changers of Global Development?

Nicola Banks and Jonathan Makuwira

When this section was initially envisioned, we could never have predicted the trajectory that 2020 and 2021 would follow. The emergence and spread of the COVID-19 virus prove to be the ultimate 'game changer' globally, as economies and borders closed and governments around the world scrambled to deal with the tri-fold health, economic and social crises that accompanied its spread. Through the failure of public health as a global public good, the virus poses a significant development challenge for all countries, accentuating the case for a global, rather than international development paradigm (Oldekop et al. 2020).

Situating the COVID-19 pandemic and responses to it within global public health systems, **Stephanie Topp** illustrates the game-changing nature of the global pandemic. As a mixture of public health and social measures were implemented globally to slow the spread of the virus, economies and social interactions were curtailed to huge social and economic cost. Reinforcing existing and deeply entrenched social inequalities, the crisis highlighted the incapacity of health systems around the world to address inequality and highlighted the need for a radical rethink of the neoliberal institutions – including health systems – that underpin modern life. Building systems through a lens of social justice requires greater focus on local, national, and global coordination and an explicit focus on addressing social and health inequalities. The speed of change occurring in early 2021 means her chapter misses out breakthroughs that took place in the approval and rolling out of vaccines. These too add strength to her arguments around neoliberal institutions and global cooperation, with the political wranglings that accompanied these successes – including protectionist export controls to keep vaccines 'at home' and important debates over access and distribution of vaccines across the Global North and South – highlighting the ways in which inequality are structured into health systems and outcomes globally.

Looking beyond COVID-19, **Pranee Liamputtong** and **Zoe Sanipreeya Rice** look more broadly at issues of health and illness. They highlight the imperative of a broader pursuit of social justice in order to transform and reduce health inequalities globally. While life expectancy has increased around the world, inequalities in life expectancies have widened within and across countries. Despite a multitude of determinants of health that influence health and illness, they argue strongly for the need to prioritize the social determinants of health in our understanding and analysis. Inequalities in health reflect the inequalities that underpin our societies globally,

DOI: 10.4324/9781003017653-43

making a broader pursuit of social justice critical to overcoming health inequalities and assisting human development in more equitable fashion within and across countries globally. In particular, they highlight the need to explore how gender, ethnicity and social class intersect to influence health and illness; indeed – and very strikingly – COVID-19 deaths were particularly pronounced along these dimensions in many countries around the world. Broader inequalities and exclusion also mean that migrants should be a particular focus of global health policy, and that policies must move beyond 'protectionist' policies to enhance their health outcomes.

Moving away from a medical-focused approach to a social one is also an important thread in **Jonathan Makuwira's** chapter on disability-inclusive development. Recognizing the 'social' is critical in addressing the challenges of the nearly one billion people with disabilities globally (80 per cent of whom live in developing countries); this requires seeing disability not as impairment but a failure in social institutions and physical environments that prevent people with disabilities from fully participating in development processes. He highlights the failures of mainstream development (and theories of development) in this respect, and advocates Critical Disability Theory as a new mode of enquiry that resituates the debate and produces new forms of knowledge that recognize the social and political constructions of disability, thus enabling research and policy to recommit to more inclusive development moving forwards. He highlights the importance of a 'twin-track' approach to mainstreaming disabilities in development processes. First, is the removal of these social barriers to the participation of people with disabilities. Second is an empowerment-based approach through which disability-specific initiatives are undertaken to ensure that they have equal access to services, livelihoods, and social and political participation. Makuwira also thoughtfully highlights the importance of confronting these deep-rooted discriminatory social issues in our teaching; these are complex and require critical pedagogies, must recognize the roles of language, stereotypes and other non-physical obstacles, and push students and educators out of their comfort zone, to 'unlearn' knowledge and behaviours and to destabilize and challenge their ideas about development.

The game changers that we look at in this section are not all making or breaking development outcomes at COVID-19's speed or scale. We look at several significant global shifts that are taking place across the world and the implications of these in presenting opportunities and challenges for the global population national and internationally.

Increasing urbanization around the world has led to the world's population being predominantly urban for more than a decade. 'Urban living' is not synonymous with living in big cities, but this is a trend that is set to accelerate (Hoornweg and Pope 2017). Urbanization – particularly through big cities – is a key driver of economic growth, yet one that exacerbates pressure on urban jobs, housing, services, infrastructure and environments (among others). Meeting these challenges is heavily dependent on strong urban governance and poses urgent questions for urban planning, policies, and programmes. Widespread urban informality deepens these challenges, making traditional planning systems (and educations) unable to face up to the realities of urban development or to meet the needs of city residents that access housing, services and infrastructure informally (Banks, Lombard and Mitlin 2020). Even where informality may be less prevalent, urban development is often exclusionary of low-income or marginalized groups; inclusion in urban planning processes is critical.

Diana Mitlin, Jack Makau, Sophie King and **Tom Gillespie** highlight the strong parallels in exclusionary practices across cities of the Global North and South. In Manchester, UK and Nairobi, Kenya, the authors highlight the challenges facing low-income households: many are displaced through gentrification and poverty programmes that are designed by professionals without input from low-income people and communities. This makes strong and mobilized local communities (with strong roles for women) critical to overcome exclusionary practices and secure better access to essential public services. They highlight the ways in which moving towards a 'global' understanding of poverty and exclusion can promote solidarity between North and South, but also highlight that a more 'global' agenda must not overshadow the experiences of groups at the most local levels. We must continue to focus on how these trends, systems and practices manifest themselves locally to see how local groups of mobilized residents pursue solutions to local deprivations. Transnational solidarity, through community exchanges across the North and South, can support local work and progress in these areas.

Financial crises and urban politics have left millions homeless and in financial desperation across the world's cities (Rolnik 2019). Looking in more detail at the critical issue of housing, **Poonam Devi and Naohiro Nakamura** investigate the causes and nature of the global housing crisis, a situation in which – across North and South – a severe mismatch between demand and supply for affordable housing has led to a severe shortage of decent houses, an increasing prevalence of informal housing, and evictions and displacements. Despite the global nature of these problems, they argue for a deeper understanding of housing issues and how they can be addressed within national contexts, rather than seeking to address them through singular or monolithic, neoliberal approaches.

These chapters both highlight the importance of recognizing and integrating informality into teaching. Future planners and professionals cannot 'learn' and address the manifestations of poverty, disadvantage and the urban experience without understanding the experiences of urban residents in accessing housing and public services informally; perhaps more importantly, they will continue the tendency for professionals to design and implement programmes while excluding the solutions devised by the low-income urban groups who are meeting these challenges on the ground. In Manchester, groups of low-income urban residents were brought into the classroom to teach masters students; in Fiji, field labs for undergraduate students allowed them to experience informal settlements. Place-based and people-based learning may well stay in the minds and hearts of students for longer than theory-based learning from textbooks and 'experts.'

Questions of employment are increasingly prominent in a changing global context. That the world is increasingly urban and increasingly young increases the scale and urgency with which more and better jobs are required. Both these changes result in higher numbers of entrants to the labour force: urbanization represents a move from agricultural-based livelihoods towards services and industry; growing youth populations means large numbers of young people entering the labour market. Sustainable Development Goal 8 highlights the need for inclusive, sustained, and sustainable economic growth that supports full, productive, and decent work for all.

While growth is a prerequisite for job creation, growth is not naturally inclined towards 'inclusivity' or 'decent work.' This is evidenced by a rise in the number of 'precarious' workers, working outside legal and social protections and lacking decent working conditions or sufficient wage rates across both North and South. Informality in jobs is widespread across the Global South (c.f. Vanek et al. 2014), and there is a strong gendered aspect to this that disadvantages women. Across the Global North, too, there are 'non-standard' employment arrangements that would be characterized as 'informal' in developing countries. We have seen an increase in the global labour force working under arrangements that offer limited benefits and social protection.

This makes more *and* better jobs critical globally. In her contribution, **Aarti Krishnan** explores efforts to improve working conditions and wages along global value chains in three key sectors: apparel and textiles; agriculture and agro-processing; and tourism. Power inequalities across these chains are unbalanced towards lead firms. The motivations of lead firms are thus pushed downwards to all actors in the chain, shaping working conditions and wages at the bottom of the chain. All too often, these motivations have been rooted in low costs and high profits. Krishnan focuses on the strategies that have tried to improve how these working conditions are governed and analyses the impact(s) that these have had on the number and quality of employment opportunities worldwide. Governance has largely been through *compliance*-based strategies, in which firms within GVCs have to comply with standards dictated by the lead firm if they want to access a supply chain. Other actors have also been able to influence these standards; this includes closer partnerships with supply firms (mentor-driven strategies), broader coalitions of actors (associative strategies e.g. Fairtrade), or pressure from workers themselves (bottom-up strategies). While higher wages or better working conditions are positive changes often associated with these initiatives, Krishnan illustrates that there are rarely 'win–win' strategies that promote all desired improvements. Change is often partial and accompanied by social downgrading (i.e. negative changes). Public governance is critical, with important roles for governments in creating and implementing labour regulations and engaging with powerful global lead firms.

Migration is another movement changing the shape and dynamics of global populations. **Tanja Bastia and Ron Skeldon** detail how current estimates of internal and international migration downplay their importance to global, regional, and national dynamics, to migrants themselves, and to the households and communities that migrants support. They argue that research and teaching must better recognize the complexity of migration processes, including delving beneath headline global or national figures to explore how these play out at regional or local levels. Debunking migration myths must be a priority, as well as using case studies to highlight the depth and complexity of migration issues. For example, figures for the proportion of a population who are 'foreign born' may look drastically different in cities (and in particular, capital cities) and in carefully breaking down international migration we see that most migrants migrate regionally (e.g. between countries in Africa or Asia) rather than, as commonly perceived, from Global North to Global South. We must also understand *who* is migrating, and *why* – and contextualize these movements against the relevant developmental, demographic, societal and political transitions, to see how these shape migration trends at national, regional and global levels. The link between migration and development is another key question, and they look particularly at the role of remittances, skilled migration and diaspora in contributing to desired development goals and objectives. The flows of people and resources between sender and recipient countries highlights the importance and influence of global interconnectedness both at 'home' and 'abroad.'

Joseph Besigye Bazirake and Carolina Suransky look in detail at the issue of forced migration, which has fundamentally changed globally; not only on the rise, it is often no longer a short-term or temporary phenomenon but a protracted one with roots in conflict, persecution, displacement and inequality. The vast majority of forced migrants are *internally*-displaced migrants and most of those fleeing *externally* are accommodated in developing countries. Only a small proportion of externally-displaced refugees are afforded the status of asylum-seekers, an important legal distinction that grants legal protection in refugees' new host countries in recognition of the persecution or human rights violations they would face at home. Asylum-seeking systems have been slow to recognize and accommodate the increasing diversity and complexity of factors that force migrants to leave their home country. Instead they have moved towards stricter border control policies and an increasing criminalization of migration that delegitimizes forced migration. Improvement in the key 'durable' solutions of voluntary repatriation, resettlement and local integration are also necessary; key here is bridging the gap between short-term humanitarian response and longer-term developmental goals.

With the number of state-based conflicts rising, and its heavy impacts on civilians and institutions in conflict-affected (and their neighbouring) countries, **Jessica Hawkins** argues that conflict must be brought central to global development agendas. While development studies has been criticized for keeping conflict out of development, enduring development is dependent on understanding the causes and nature of conflict and the ways in which conflict affects development. Her chapter outlines a pedagogy of conflict and development, starting with an outline of the key (intersecting) theoretical approaches that have emerged for understanding the causes of conflict and going on to explore the multiple impacts of conflict on development. She looks, in particular, at three areas in which conflict impacts strongly on human and social development: displacement, gender, and peacebuilding and security. Following on discussions of forced migration in the previous chapter, we see in the contribution of conflict and displacement to patterns of human movement and broader developmental outcomes. Issues of shelter, food, healthcare, education, and livelihoods are all critical to displaced communities. Host communities also face development challenges stemming from increased pressures on food and health systems. Any study of conflict and its aftermath must be gendered, looking with nuance at the ways in which conflict permeates throughout affected societies to affect men, women, and children differently. Yet research and teaching agendas have been slow to look at issues of gender beyond sexual violence. In terms of security and peacebuilding too, she highlights the need to expand our understandings and agendas to move beyond traditional neoliberal assumptions that peace is 'built' towards locally-rooted and more sustainable forms of peace formation that promote and prioritize local agency and empowerment, so that peacebuilding efforts are less likely to institutionalize hierarchies of injustice and inequality.

Hawkins emphasizes the urgent need for theories, policy and practice to do things differently, starting from a more nuanced understanding of the many aspects of and interconnections between the concepts of conflict and development; these two distinct fields must no longer be siloed in research, policy and practice, and teaching. Hawkins outlines a thoughtful and comprehensive pedagogy of conflict and development that is interdisciplinary and rooted in case studies. Focusing on case studies in depth allows students to delve deeply into multiple theories of conflict to investigate how these play out simultaneously. Depth is also critical to do justice to the historical and context-specific nature of conflict and violence. Importantly, she also emphasizes that teaching must look beyond lessons and learnings delivered, to ensure that the trauma-inducing aspects of learning about conflict and violence does not become a mental health burden for students.

Changes in population dynamics can also have a major impact on development outcomes. The impact of ageing populations or large populations of children and youth on national developmental outcomes are two key policy issues here. In 2019, nearly 2 billion children (0–14) made up just under 26 per cent of the world's population of 7.674 billion (World Bank 2021). An estimated 1.2 billion young people (aged 15–24 years) in 2020 made up around 15.5 per cent of the world's population (UN 2020). Across sub-Saharan Africa the youth population will continue to grow, where it estimated to reach 30 per cent of the world's youth population by 2050 (up from 18 per cent in 2020) (ibid.). At the other end of the life cycle, nearly 700 million people over the age of 65 made up 9 per cent of the world's population in 2019 (World Bank 2021); this proportion has been increasing continually, leading to strong policy concerns around the implications of this on national outcomes. Children and young people are recognized as key development stakeholder groups, as demonstrated by a UN body designated especially for children (UNICEF) and the integration of children and youth into the Sustainable Development Goals, among others. Despite holding well-defined spaces in development, the two contributions here highlight that much remains to be done to understand and integrate children, youth and ageing populations into development thinking and practice more effectively.

Vandra Harris Agisilaou argues that despite increased attention and 'space' in global development agendas, children and young people remain peripheral. Theories and tools for understanding these developmental phases have been insufficient in understanding the developmental processes occurring throughout this period and therefore fails to attribute them with the agency they deserve. The disaggregation of data to such broad categories (0–14 years; 14–24 years; 25–64 years; 65 years +) overlooks the social and economic importance of these categories. It brings focus to what these groups *need* rather than contribute and overshadows the changes and complexities that occur within each category. Globally, effective and high-quality investments in early childhood development (that take culture, context and family priorities into consideration as leading influences) have shown to be positive and enduring; the period of childhood sets the foundations for all future pathways of human development. Processes of child development cannot be viewed in isolation of biological and brain developments, or out with the social contexts in which these processes take place. The interplay of biological, psychological and environmental processes is also necessary in our understanding of for young people as well as children, as Banks (2021) highlights. She brings in Developmental Psychology to her analysis of young lives and experiences, a discipline which Development Studies has overlooked. Yet Development Studies has much to offer the field in doing so; such analysis must be done in a socially-, economically- and politically-contextualized way if we are to understand the ways in which space and place, social norms and practices, economic realities and political landscapes come together to enhance or undermine critical developmental processes in children and young people.

In her contribution, **Penny Vera-Sanso** highlights passionately that the game-changing nature of population ageing dynamics is not in the demographic theories, models and estimates that predict doomsday scenarios of daunting financial and care burdens. Such theories are based on Western and neoliberal theories and experiences, and move counter to a more effective global policy framework. The real game changer, she argues, will be moving away from these tropes towards a more critical, decolonial and less 'ageist' sociology of ageing that recognizes diversity in how ageing is produced globally. Key here will be eradicating assumptions that the experience of developed countries is universally desired or feasible (e.g. retirement ages), better situating population growth rates and predictions within the local, national and global contexts that influence mortality rates and labour market outcomes for different generations, and moving away from the ageist assumption that young people are productive, but the elderly are not. Assumptions in which the elderly require the care burden sit uneasily alongside many global contexts in which it is the 'productive' youth labour force that finds themselves struggling to support themselves given a lack of regular, secure and well-paid employment opportunities. Understanding the local labour market realities – in which employment patterns are not shaped by age, as currently assumed, but by other social, economic and political factors – is critical, before we can truly understand household wellbeing, dynamics and relative care burdens, or before we can determine all household members' (including elderly members) opportunities for independence or need for support. Critical here is whether families can support their elderly members and whether individuals can save for their own future life. These questions become not ones of demography but of the local and national labour markets and social provision(s) which may, or may not, enable this. A much broader – and gendered – analysis of life trajectories across the whole life course is critical to a better understanding of the needs and vulnerabilities of ageing.

These chapters are particularly powerful together. Shedding more light onto the production of *social* concepts of 'youthhood' and ageing is critical; we must move beyond simplistic categories demarcated by age and beyond theories that do not recognize the situated complexity of contributions from children, youth and ageing populations. Two issues are critical here. First is recognizing and valuing the agency and capability of children, young people and ageing populations rather than viewing them through a lens of dependency or passivity. Second is that 'zooming in' to look at children, young people or ageing populations is too narrow a scope. A situated understanding of these categories requires a much deeper and broader analysis; we must look at these groups within the complex intergenerational relationships in which they are situated, as well as their positioning within the realities of the local labour markets, social and cultural systems, and government support (among others) that provide the backdrop to their lives and life-courses. It requires an intersectional approach that recognizes how experiences are influenced not only by age, but are also socially-produced, along lines that include gender, class, ethnicity or race. It also requires incorporating insights from new disciplines to those with greater influence and experience in development studies and new critical theories and methodologies to displace the predominantly Western and neoliberal ones that dominate. As Vera-Sanso herself argues, a spirit of 'unlearning and a conviviality that recognizes the incompleteness of knowledge' is necessary. Only then can we begin to give equal space, priority and respect to different viewpoints, perspectives and knowledge to create a new "geopolitics of knowledge."

In sum

Any headline figure can suggest that the world's problems are increasing – and increasingly complex. The 11 insightful chapters here evidence this and provide an overview of some of the changes happening at the global level that can have a huge impact on local, national, and global development outcomes. As well as giving an overview of key issues and agendas in their respective areas of interest, they also highlight the way forward to more effective research and teaching agendas and policy responses. In fact, the game-changing nature of all of these issues is not just in their role and contributions to developmental outcomes; new critical and multi-disciplinary studies cognizant of lives and livelihoods around the world will play an important role in advancements in teaching, policy and practice.

As Horner and Hulme (2019) point out, the increasing prevalence of shared global problems has begun to question the reality of the traditional North-South binary that has underpinned the concept of international development; the shift towards 'global' development is important to reposition power and knowledge dynamics. Yet as the diverse chapters highlight here, this doesn't overshadow the need for heavily contextualized studies of and responses to each of these issues. They call for an end to neoliberal frameworks and solutions and to greater space for learning from local and national spaces globally. Deeper understanding based upon broader and multi-disciplinary research is critical, and this highlights an important role for Development Studies in bringing these together.

That within-country and between-country inequality are now firmly on the research and policy agenda is unquestionably a hugely important advancement that must continue. But the chapters here also highlight that this must not prevent us from keeping poverty – and what to do about it – firmly on the global agenda. The same can be said for inclusivity, and taking an intersectional approach to this to understand the experiences of all groups and classes, whether this be women, ethnic minorities, people with disabilities, children or youth, among others. Despite decades of work and recognition of the central role gender plays in poverty, disadvantage, and discrimination, disappointingly the chapters here still highlight a lack of gendered analysis and understanding in their respective fields. How can we move beyond having separate 'gender experts' in different fields to a more mainstreamed approach in which all research, policy and practice takes questions of gender seriously (see Ferguson 2015 for important insight here)? Is this really too much to ask?

The chapters here also all outline clearly how these discussions feed through into important teaching agendas. Students must be taught to think critically, beyond existing disciplinary boundaries and traditional (predominantly Western) theories and models. It must start deep, engaged, and uncomfortable discussions that challenge students to question and 'unlearn' existing knowledge and to confront their own thoughts and beliefs in more inclusive, non-discriminatory ways. We must consider support for students where the content of this teaching can evoke a mental health burden (e.g. conflict and humanitarianism). Effective case studies that prioritize depth are important so students can trace the influence of competing and intersecting theories, recognize nuance, and do sufficient justice to historical economic, social, political and cultural processes in understanding change or outcomes. Critical in these discussions is how important it is that the next generation of professionals are taught in ways that reflect the global realities of how millions of people live, work and access services; without this, problems of poverty, exclusion and injustice will continue to be reproduced and solutions will continue to be insufficient. This is not only in *what* we teach, but also *how* we teach it. This section outlines various innovations through which students learn not only through academic educators or professionals, but also from the real experts – those who have lived experience of the subjects being taught. This may be through bringing marginalized groups into the classroom as teachers or through carefully thought out fieldwork trips or 'living laboratories'; the ethical implications of these innovations must be carefully thought through. Through this the role of listening becomes an incredibly important part of the learning journey.

References

Banks, N. (2021) 'Developmental spaces? Developmental Psychology and urban geographies of youth in sub-Saharan Africa', *Children's Geographies*, vol. 19, iss. 2, pp. 210–224.

Banks, N., M. Lombard and D. Mitlin, (2020), 'Urban informality as a site of critical analysis', *Journal of Development Studies*, vol. 56, iss. 2, pp. 223–238.

Ferguson, L., (2015), '"This is Our Gender Person" The Messy Business of Working as a Gender Expert in International Development' *International Feminist Journal of Politics*, vol. 17, iss. 3, pp. 380–397.

Hoornweg, D. and K. Pope, (2017), 'Population predictions for the world's largest cities in the 21st Century' *Environment and Urbanization*, vol. 29, iss. 1, pp. 195–216.

Horner, R. and D. Hulme, (2019), 'From International to Global Development: New Geographies of 21st Century Development', *Development and Change*, vol. 50, iss. 2, pp. 347–378.

Oldekop and the whole GDI universe, (2020) 'Covid-19 and the case for global development' *World Development*, vol. 134, pp. 1–3.

Rolnik, R., (2019), *Urban Warfare: Housing Under the Empire of Finance* London, Verso.

Vanek. J., M. A. Chen, F. Carre, J. Heintz and R. Hussmans, (2014), 'Statistics on the Informal Economy: Definitions, Regional Estimates and Challenges', *WIEGO Working Paper (Statistics) No. 2*, Cambridge, MA and Manchester.

United Nations (UN), (2020), *World Youth Report: Youth, Social Entrepreneurship and the 2020 Agenda*.

World Bank (2021), *World Bank Data*, available at: https://data.worldbank.org/indicator/SP.POP.0014. TO (accessed 9 March 2021).

39

COVID-19 AND GLOBAL HEALTH SYSTEMS

Stephanie M. Topp

Background

COVID-19 is an infectious disease that causes respiratory illness, with a range of symptoms including fever, cough, and in more acute cases, difficulty breathing, pneumonia and death. As of October 1–2020, there were more than 35 million cases globally, over 1 million deaths, with almost every country affected. Although at the time of writing there is still much unknown about the disease and its long-term consequences for individual and population level health, experience suggests approximately one in five people infected require hospitalization, with high rates of more severe illness in people over 50, as well as those with underlying chronic conditions. Thought to have originally made the jump from bats to humans, COVID-19 is transmitted human to human via airborne droplets. Prior to the development of a vaccine, the slowing of COVID-19 infection required physical distancing between individuals, improved hygiene practices and effective public health measures including testing, contact tracing and quarantining of exposed individuals. Together these are collectively referred to as 'public health and social measures' and aggressive application of such measures successfully slowed the spread of COVID-19 in a number of countries at different times throughout 2020 (Hale et al. 2020). Yet the introduction of such measures also had devastating social and economic consequences, simultaneously laying bare the inequities on which global trade relations are based and exacerbating existing fault lines within and between societies.

The focus of this chapter lies at the intersection of a long-standing debate in global health and development regarding how best to design, finance and govern health systems to support human development, and the very contemporary question of how we should respond to COVID-19 in the short- mid- and longer-term. The chapter starts with an overview of the health system debates between the 1950s and 60s and the current day. It tracks the ebb and flow of arguments that have positioned health systems on the one hand as levers for broader social equity, versus those that invoke neoliberal thinking and position health systems as more or less exclusively responsible for the delivery of efficient and effective health care. To the extent possible, given the fast-moving nature of the COVID-19 pandemic and responses to it, the chapter then seeks to map and interrogate several emerging threads of the health system debate demonstrating significant continuity between past and present arguments in spite of the political and public health upheavals of recent times.

DOI: 10.4324/9781003017653-44

On the one hand, the chapter observes policy makers, media pundits and academics making the case for investment in better technologies and health care systems, based on an instrumental frame in which health systems are viewed primarily as vehicles to enable access to medical technologies. A counter narrative, linked to a 'population health' lens and the need to address structural determinants of health, argues that health systems should be viewed as complex social systems and consciously designed to mitigate health inequities. This view is anchored by recognition of structural inequities which, in crises such as the COVID-19 pandemic, heighten certain populations' vulnerability. The chapter ends by considering the way economic consequences of COVID-19 policy responses are playing into health system debates and the different interpretations of the phrase 'build back better' across the political and ideological spectrum.

Box 39.1 What is a health system?

The World Health Organization (WHO) defines health systems as comprising 'all organizations, institutions and resources devoted to producing actions whose primary intent is to improve health' (WHO 2000). Arguably the central task of a health system is to facilitate and/or organize the delivery of health care services. But many, including WHO, argue that health systems have wider goals beyond that of delivering medical care and improving population health; goals such as health equity and fairness, the (re)distribution of health care costs, and protection for households from catastrophic out-of-pocket expenditure. Further, while historically health systems have been viewed as mechanical or technical in nature, a growing body of scholars, policy makers and practitioners now characterize health systems as complex social systems; that is, systems that reflect the values (and biases), norms and power dynamics of the societies in which they are located (Sheikh et al. 2011, Whyle and Olivier 2020). While recognized for delivering services for the prevention and treatment of ill-health (a medical model), health systems are thus also understood to be central to government and civil society attempts to address inequities in health and development indicators and wider social injustice (Commission on Social Determinants of Health 2008).

For the past half century robust debates regarding *how* health systems should be designed, financed, and governed to address either the narrow or wider-reaching goals outlined in Box 1 have formed the backdrop for national and global health policy. These debates have been shaped by varied ideological, disciplinary, political, and experiential backgrounds.

The rise, fall, and rise again of comprehensive primary health care

Thinking about health systems has oscillated and evolved over the past 75 years. In the decades following the Second World War the emergence of new medical technologies such as new antibiotic and antimalarial drugs combined with growing momentum for decolonization, led to a focus in both international and national arenas on health systems designed for high quality and technologically advanced medical care (Hall and Taylor 2003). The establishment or scaling-up of teaching hospitals, nursing and medical schools modelled on those in highly industrial nations was popular during this period among national health policy makers in the Global South. The World Health Organization (WHO) and United Nations Children's Fund

(UNICEF) too, were launching ambitious disease elimination campaigns largely premised on new vaccine and vector control technologies.

By the late 1960s and early 1970s, however, there was a shift in thinking about health systems both within international organizations and among national policy makers. Investment in high-tech and hospital-focused systems had not produced the rapid improvements in population health many had expected. Such systems tended to concentrate investment in urban-based hospitals while still largely rurally based populations lacked access to basic services (Hall and Taylor 2003). A series of technical reports published by WHO in the early 1970s were openly critical of an approach that had pushed expensive curative-models of care, advocating instead for a shift towards strategies that ensured equity of access, and renewed focus on upstream or social determinants of health such as social exclusion, poverty, and structural inequities (Newell and World Health 1975).

In 1978, WHO and UNICEF co-convened the first International Conference on Primary Health Care in the then Soviet city of Alma Ata (modern-day Almaty in Kazakhstan), culminating in the **Declaration of Alma Ata** (the Declaration). The Declaration was seen as a watershed event, capturing a rare moment of international consensus regarding the need to shift away from the dominant medical model, towards an approach that embedded community-empowerment and social justice; enabled action on social determinants of health; involved all sectors in the promotion of health; ensured community participation in planning, implementation and regulation of health care; and focused on equity in health status as well as high quality care.

The Declaration has served as a touchstone for advocates of comprehensive primary health care and socially inclusive health systems in the decades since. But notwithstanding its bold aspirations, the consensus that underpinned the Declaration's formulation was short-lived. Almost immediately following the conference, critics began to voice concerns over the values and recommendations it encapsulated. Some interpreted the vision of community-oriented services as an attack on the medical establishment generally, and specialized health care in particular; others saw the concept of comprehensive primary health care as lacking pragmatism, being an unrealistic attempt to be everything to everyone; and still others saw the Declaration through the lens of geo-politics, as an ideological push by the Soviet states for so-called 'socialized' medicine (Birn and Krementsov 2018).

Barely one year after the signing of the Declaration of Alma Ata with its vision for comprehensive primary health care, an alternative concept of 'selective' primary health care was floated as a more pragmatic guide to international and national health service design. Contra the vision outlined in the Declaration, *selective* primary health care was characterized by a focus on targeted actions for specific diseases or health issues, chosen for a combination of: i) high need and ii) availability of technologies that could be delivered via standalone services or 'vertical' interventions (i.e. without being reliant on government run health services). Selective primary health care was first operationalized by UNICEF in 1979 and was the template for operations throughout the 1980s and 1990s with a rolling series of highly targeted programmes such as those for Oral Rehydration Therapy, Breastfeeding, Immunization and Growth Monitoring.

Why were these trends within WHO and UNICEF so important for health services in low- and middle-income countries during this period? The answer lies in part with to the global economic backdrop. The oil and commodity slump of the mid-1970s and resultant debt crisis were compounded by the Structural Adjustment Programmes (SAPs) introduced by the International Monetary Fund and World Bank as a condition of continued loans. One of the much criticized features of SAPs was the requirement that governments make large cuts to public spending, including the wage bills for the health and education sectors. During this period also, the World Bank became a more active participant in global health policy proscriptions, with the 1993 World Development Report themed *Investing in Health* and outlining proscriptions for neoliberal policy reforms including for the introduction of fee-for-service health care, and privatization of both health service delivery and health insurance.

Over the course of the 1980s and 1990s, this confluence of debt and other macro-economic policy pressures saw a precipitous decline in the coverage and quality of public sector health care and concurrent growth of private (largely unregulated) health markets that were expensive to access. In this context many countries became increasingly dependent on health financing from international organizations. Consequentially, the type and coverage of health services was often defined by the targeted (disease- or condition-specific) priorities of donors themselves. Public primary care services, national health information systems and human resourcing and logistics that underpinned basic services or action on upstream social determinants, saw little investment.

An era of verticalization: MDGs and the global health initiatives

By the early 2000s, and with the rise of the HIV and AIDS epidemic, a new wave of vertical, disease-specific programmes were being advocated for in international circles, spurred in part by the formation of the Millennium Development Goals (MDGs) and the emergence of several high-profile global health initiatives. These included the Global Fund for AIDS, Tuberculosis and Malaria (Global Fund); the Global Alliance for Vaccines and Immunization (GAVI) and the United States' President's Emergency Plan for AIDS Relief (PEPFAR) each of which could be viewed as successors to the selective primary health care approach.

While varied in modality (e.g. GAVI was a funding mechanism; PEPFAR engaged in service programming and service delivery too), all brought a medical-technical and disease-specific focus. Priority lay with the rapid establishment or scaling-up of medical technologies (vaccines or diagnosis and treatment) in response to a specific disease or group of diseases. Services funded through these global health initiatives were commonly framed by a moral imperative to 'save lives,' simultaneously invoked as a rationale for bypassing – where possible – weak local public sector health services.

Rather than slower consultative or country-led processes, it was argued that parallel logistics and standalone services utilizing medical (often Global North) experts would be capable of delivering care more quickly (WHO 2006). Yet in many places what evolved was an uneasy hybrid system, still utilizing elements of local health infrastructure and health workforce alongside donor-funded human resources, supply chains and information systems (Schneider et al. 2006). Ambitious health-related targets – both specific to each global health initiative as well as those linked to the MDGs were thus still reliant on national health systems. And in fact, by the late 2000s, delayed progress in achieving disease-specific targets had once again forced a sector-wide re-evaluation. Notwithstanding some important gains – particularly in relation to HIV treatment access – there was growing recognition that to sustain and make further progress on key disease outcomes, investment in the national health systems that continued to underpin such endeavours would be essential.

Primary health care: now more than ever?

In 2008 WHO's Annual Report celebrated the 30-year anniversary of the Declaration of Alma Ata entitled *Primary Health Care – Now More than Ever.* Published against a backdrop of increased health spending globally, but also accelerating rates of health inequity and the rapid rise of disease-specific global health initiatives placing enormous pressure on weak health systems, the report flagged the need to re-orient efforts towards strengthening health systems through a focus on primary health care. In the same year WHO launched its 'Maximizing Positive Synergies' initiative, which similarly highlighted the importance of designing disease-specific programmes and initiatives with a view to strengthening national health systems, rather than hollowing them out or disproportionately skewing their focus to select diseases.

A decade of widespread rhetorical support for the principles of primary health care followed, culminating in a new Declaration of Astana, at the 2018 Global Conference on Primary Health Care held in Kazakhstan to celebrate the 40th anniversary of the Alma Ata Declaration. Yet many questions have been raised about the degree to which any substantive shift occurred in the international sphere, as disease-specific targets and funding streams driving disease-specific programming and service models proved difficult to displace (Rifkin 2020).

A new phase in the health systems debates

The impact of the COVID-19 pandemic has underscored the value of health in our society but also highlighted the variance and continued dissonance in public, academic and political views on the role of health systems in the 21st century. This chapter argues that with the COVID-19 pandemic, we entered a new phase in the health systems debate as laid out above, regarding the key purpose and design features of health systems and their relationship to human development. It argues that the simultaneous health, economic, and political crises wrought by the pandemic on an almost universal scale created a new space, larger than in previous times, in which to consider the health systems implications of different values and ideologies. Although framed by the immediate and ongoing experiences of COVID-19, the chapter describes how such discussions draw on long-standing public health paradigms. These include on the one hand, a view of health systems primarily as vehicles for the delivery of modern medical services; and on the other hand, a much broader conceptualization of health systems as social institutions, shaped by societal values and, themselves capable of social transformation (Van Olmen et al. 2012, Freeman 2005). In the remainder of the chapter I consider how the experience of COVID-19 has interacted with this paradigmatic thinking and discuss some of the implications for health systems globally moving forward.

Ventilators, vaccines, and the threat of 'Covid-ization'

With the onset of the COVID-19 pandemic and the experiences of Wuhan Province, Italy, and Spain being broadcast globally in early 2020, policy makers all round the world were forced to rapidly evaluate the preparedness of their national health systems. Early evaluations tended to focus on the number of intensive care unit (ICU) beds and ventilators, and secondarily on availability of trained clinical staff and personal protective equipment (PPE). Academic modelling exercises, policy debates and more local planning zeroed in on hospital capacity and the question of whether ICUs could cope with a surge of acutely symptomatic COVID-19 infected individuals. In some countries (including the United States and United Kingdom) this resulted in large one-off investments in ventilators (Davies 2020). And as the pandemic evolved, the conversation shifted to pay increasing attention to vaccine development and associated health system capacity for universal distribution (Kupferschmidt 2020).

Concurrent to this focus on health system capacity to deliver COVID-19 specific medical care, were concerns regarding the impacts of that same focus on other medical priorities and services. In particular, concern for the implications of diverting funds or human resource capacity for other disease-specific services such as HIV, tuberculosis (TB) and malaria came to the fore. Citing modelling from the UN-housed STOP TB programme (STOP TB Partnership et al. 2020), for example, commentators warned of the need for 'damage control' in relation to the impact of COVID-19 on TB care and treatment (Bigio and Pai 2020). The World Malaria Program issued an analysis which framed the pandemic as a 'threat' to malaria service delivery particularly in countries with fragile health systems and called for responses to protect malaria services in this new context. The Lancet published modelling that predicted disruption to HIV services, with the researchers concluding that '[m]aintaining the most critical prevention activities and health-care services for HIV, tuberculosis, and malaria could substantially reduce the overall impact of the COVID-19 pandemic' (Hogan et al. 2020). A report from the Global Fund for AIDS Tuberculosis and Malaria similarly warned of the need to 'protect extraordinary progress' in HIV, TB and malaria outcomes 'as health and community systems are overwhelmed, treatment and prevention programmes are disrupted, and resources are diverted' to COVID-19 activities.

Undoubtedly figures have shown precipitous drop-offs in the provision of non-COVID-19 services. In a WHO-run survey of COVID-19 impact on non-communicable disease services in 155 countries, more than half (53 per cent) reported partially or completely disrupting hypertension treatment services; 49 per cent disrupting treatment for diabetes and associated complications, and 42 per cent disrupting cancer treatment. Data from India, China and Pakistan from the three months preceding September 2020, pointed to a daily decline of TB case notifications of between 75–80 per cent and 66.8 per cent decrease in Bacillus Calmette-Guerin (BCG) vaccination coverage for newborns, which protects against TB in early years of life (Malik et al. 2020). TB services were affected in high-income settings too with the United States' Centres for Disease Control reporting 60–72 per cent of contactable domestic TB programmes reporting weakened staff capacity; 52 per cent diagnosing and treating fewer patients with TB; and 64 per cent reporting fewer resources for contact tracing (Kuehn 2020).

Notwithstanding the public health concerns for the contraction in non-COVID-19 services, there was some irony to be observed in these concerns being expressed most frequently and most loudly by those working on the 'big three' diseases – HIV, TB and malaria, erstwhile beneficiaries of unparalleled levels of disease-specific financing and alongside that, criticism for skewing the priorities of national health programmes. Indeed, it may almost be possible to characterize the attention paid to COVID-19 as the most recent manifestation of a well-established pattern of disease 'exceptionalism' of which HIV, TB and malaria were simply earlier examples.

Whether acknowledged or not, however, these arguments were also underpinned by the same logic invoked during the late 1980s and 1990s in support of 'selective' primary health care and other targeted approaches to global health programming. The point of departure for such arguments was a high burden of disease and the concurrent availability of medical treatment, together which presented an opportunity for targeted and thus highly efficient interventions. In the context of COVID-19, predictions of increased morbidity and mortality and associated costs were linked to the combination of weakened service capability and/or access to these specific services caused by diversion of resources to other priorities. Acknowledgement or even consideration of the *broader* inequities in pre- or post-COVID-19 resource distribution or health care access – discussed further below – found little place in these submissions.

As observed by Halpern et al. (2020) both historically and in the COVID-19 era, these types of responses which focus on highly targeted medical intervention are in part linked to the widespread 'cognitive biases' which lead health policy makers and citizens alike to prioritize investments in *readily imaginable threats* (e.g. patients in ICU) over the statistical ones (e.g. population rates of COVID-19 infection); the needs of the *present* (someone sick now) over the future (preventing someone getting sick), and *direct interventions* (drug or device-based medical care) over indirect ones (public health and preventive measures) (Halpern et al. 2020). Such biases – which are pervasive and have psychological and socio-cultural elements, are key drivers of targeted and medicalized approaches to health crises and to community responses to them. Such biases are also well supported by public health and global health discourses that continue to, both consciously and unconsciously, depoliticize health and health care through the use of terms such as 'evidence-based,' 'data-driven' which imply health policy responses can be held separate from the social, cultural and political contexts in which they are operationalized.

So too do these biases intermingle with ideological influences. As described earlier, neo-liberalism is reflected economically in its valorisation of market efficiency and culturally in the promotion of individualism and competitiveness (Nunes 2020). These traits align comfortably with an understanding of health systems as primarily committed to highly efficient and individualized access to medical technologies and have underpinned a trend towards disease 'exceptionalism' observed in global health financing over the past several decades. Despite recognition of the way health systems underpin disease-specific programmes, analysis from the Institute for Health Metrics and Evaluation demonstrates global investment in health system strengthening (e.g. health worker training; cold chain equipment or health information systems) as a proportion of total development assistance for health (DAH) shrank from 19 per cent to 14 per cent between 1990 and 2019. Of the 14 per cent allocated to health system strengthening in 2019, moreover, the majority was focused on systems for specific diseases, predominantly HIV/AIDS. Whether conscious or not, the thread of conversation focusing on the negative impacts of COVID-19 on specific health services and programmes thus supported a view of health systems primarily as a vehicle for delivering proven medical technologies, and particularly those amenable to individualized intervention.

COVID-19, inequities and the population health perspective

Contrasting the medically focused conversation outlined above, a different thread of the health system-related COVID-19 conversation focused on broader pre-existing health inequities and their intersection with the pandemic. Before COVID-19, there were more than 700 million people around the world living in extreme poverty; one in five women and girls between ages 15–49 had experienced physical or sexual violence by an intimate partner in the previous 12 months; and around 70 million people had been forcibly displaced from their homes. Nearly one billion lived in slums, with unreliable access to running water. Millions, mostly people of colour, had precarious jobs. All over the world, billions lacked access to the basic necessities that made good health possible.

With the onset of COVID-19, these social, economic and health inequities were highlighted and in many cases exacerbated. In the United States, where much of the initial equity-focused analysis was conducted (Brown et al. 2020), early containment efforts, mitigation policies and subsequent re-openings all brought evidence of heightened vulnerability among African American, First Nations and Latina populations, intersecting with socio-demographic differences. As the pandemic spread, concerns about the profound and differential impact of government response measures – made in very specific political contexts – emerged in a number of countries (Hale et al. 2020).

Following the decision by India's Modi government to order a nationwide lockdown – with little consideration for the livelihood needs and social security of millions – the International Labour Organization projected that 400 million Indians were at risk of falling into poverty. In the United States, critics of the Trump administration pointed to delays in making sufficient numbers of tests available; mixed messaging regarding physical distancing and masks; and the way long-standing structural inequities (unaddressed in the COVID-19 context) were resulting in dramatically higher mortality rates among racial and ethnic minorities. In a best-case scenario across the African continent, it was estimated that between 1.8 to 5.3 million direct and indirect COVID-19 associated deaths would occur by 2030, in part due to a 35 per cent increase (603 million) in the number of extremely poor people linked to containment measures (Cilliers et al. 2020).

In the sense that socially and economically marginalized groups were experiencing unequally higher rates of infection, ill-health and death, COVID-19 was not dramatically different to other pandemic events. Health crises such as pandemics are well recognized among development and social theorists for highlighting aspects of societies that might otherwise be taken-for-granted or hidden, such as entrenched social inequalities and social marginalization. Commentary regarding the incapacity of health systems to address such inequities built on these observations, experiences and analyses. Pointing to the United States once more, for example, Metzl et al. (2020) observed that the US health system would struggle to fulfil its mission to promote health and wellbeing as well as treat disease, as long as it was structured to focus on individualized medical care, with the pandemic daily highlighting the extent to which illness for many people was the product of larger structures, systems and economies.

The COVID-19 era health system conversation thus came to echo debates from earlier decades regarding the trade-off between health systems focused on population health (see Box 39.2) versus medical care; and a need for greater recognition of a disconnect between an understanding of health as fundamentally influenced by social determinants and the enduring medical orientation of most modern health systems. In both contemporary and historical debates, recognition of the underfunded nature of population health and public health both in absolute terms and as a proportion of total health spending is well recognized, as health spending in both national and international spheres remains dominated by medical services (Hemenway 2010).

Box 39.2

Population health is concerned with health trends and patterns at the population (versus individual) level, and gives particular recognition to the importance of public health measures in addressing these. Public health as famously defined by Winslow, and later Acheson (1988) as 'the art and science of preventing disease, prolonging life and promoting health through the organized efforts of society.' Although sometimes linked to medical technologies such as vaccines, public health measures often lie beyond the medical field and include provision of improved water sources, food fortification, sewerage systems, vector control, and community education. Thus, while medicine focuses on individual health outcomes, population health and public health within that, are much more concerned with equity of health outcomes across a larger population. It is this (largely public health enabled) potential to actively combat systemic inequity that leads proponents of population health thinking to describe health systems in their ideal form, as 'determinants of health.'

Systemic disruption and 'building back better'

As should be clear, cyclical debate over the values and core role of health systems has been a feature of the global and public health conversation for many decades. Periods of health crisis tend to resurface and intensify the attention paid to such issues, and in this the COVID-19 pandemic was no different. It brought into sharp focus the importance of health systems, the interconnections between those systems and broader social and economic policy, and raised questions about different countries' capacity to respond effectively to a range of challenges (Ludovic 2020). But in one respect COVID-19 was different. Unlike previous pandemics of the past half century, COVID-19 produced not only a health and health systems crisis, but a con-joint economic crisis.

Since the late 20th and into the 21st century, most economies worldwide have been organized around a neoliberal logic which assumes the primacy of the market, minimal state intervention and private sector efficiency. Such economies are fundamentally ill-suited to even a temporary 'suspension of circulation' (Nunes 2020). The COVID-19 pandemic revealed not only the inherent vulnerability of such markets to large scale (non-economic, non-financial) shocks, but a contradiction in their demand for continued 'circulation' in a situation where such activity was leading to sickness and death of large numbers of people. Policy makers globally were thus faced with difficult – arguably impossible – decisions that weighed public health and biosecurity concerns against the social, economic, and indirect health impacts of economic shut down. And in such choices, the vulnerability of the neoliberal model as a basis for health system function was revealed. As Nunes (2020) observes:

> *Part of the force of neoliberalism stems from its façade of inevitability, since it draws on the idea that there is no alternative to the austerity that aims to deflate the State as a guarantor of the common good. The COVID-19 pandemic shows that true resilience lies neither in the market (which is usually the first thing to collapse when a large-scale shock occurs) nor in privatized health. Resilience comes from strengthening a public and universal system, based on the premise of health as a common good and on social participation as an essential democratic mechanism for the health system's definition and implementation.*

In a collection published early in the epidemic, cultural theorist and philosopher Slavoj Žižek insisted that '*we should resist the temptation to treat the [...] epidemic as something that has a deeper meaning*' (Žižek 2020). Yet others including the Editor in Chief of the prestigious medical journal Lancet challenged this view by asking: '*What is wrong with our system[s] that we were caught unprepared by the catastrophe despite scientists warning us about it for years?*' (Horton 2020). And in part responding to this question, a distinctive thread in the post-COVID-19 health systems conversation focused on how pandemic related economic and social policy disruption had created a space for '*building back better.*' While taken up by individuals and organizations of varied backgrounds, the phrase was reflective of an idea that there was an opportunity, in the COVID-19 era, to do more than simply get economies and livelihoods back to a pre-pandemic normal. It hinted at making changes that would alter the social and economic landscape through renewed investment in systems that were more sensitive to, and capable of mitigating structural inequity and in turn, more likely to promote the resilience of individuals, communities, and nation states.

Among health system commentators the phrase 'building back better' in part reflected an attempt (perhaps pre-emptively) to direct post-COVID-19 conversations towards recognition of and action against the creep of medicalization in the health systems of most countries, through a conscious return to policies guided by the concept of equity and the principles of (comprehensive) primary health care. By late 2020 even organizations considered erstwhile champions of neoliberal dogma such as the World Bank were engaging with this narrative, admitting for example, the critical importance of state-based reforms that would enhance social safety nets and strengthen scaled-up service delivery systems (Bodewig and Hallegatte 2020).

Others however suggested that such iterative action was not enough, with Baum and Friel advocating a 'social vaccine' (Baum and Friel 2020) which targeted the conditions that underpin four basic requirements for global health and equity, including: i) a life with security; ii) opportunities that are fair; iii) a planet that is habitable and supports biodiversity; and iv) governance that is just. Such proscriptions clearly go beyond 'building back better,' implying the need for a radical rethink of the institutions – including our health systems – that underpin modern life. Such a call speaks to the timeless advice of Lynn Freeman who, writing in 2005 noted (Freeman 2005): *'the answer is not just money; it is the entire way in which we think about the connection between health and development – and the priority actions that result'*.

Key sources

1. J. Ludovic, S. Bourdin, F. Nadou, G. Noiret (2020) Economic globalization and the COVID-19 pandemic: global spread and inequalities. *Bulletin of WHO*. E-pub; 23 April 2020.

Published still near to the beginning of the COVID-19 pandemic, this article models the link between the pattern and speed of transmission during the first three months of the pandemic and the forms and characteristics of economic globalization. Using public data on the global spread of the virus and commonly used socio-economic variables, the authors identify the impact of socio-economic factors on the number of cases and the patterns of transmission around the world. The paper is one of the first analyses to demonstrate empirically the (now intuitive) link between the globalization of economic relations and the structure and speed of the international spread of COVID-19, showing how inter-dependencies and reciprocities of the world's mature market economies rendered them at greater risk of early spread of the virus. Furthermore, the authors characterize COVID-19 as the 'first real pandemic of the age of globalization' due to the profound inter-relationship between *health* and *economic* risks and their major impacts on each other in terms of public decision-making processes. The article closes by posing two important questions at the intersection of post-COVID-19 public health and development studies. First, whether the international spatial mobility of people and goods – a characteristic feature of modern market economies – should continue given the inherent health and economic risks? And relatedly, how in the then absence of a vaccine, population versus individual rights should be accounted for in public policy decision making designed to mitigate the pandemic?

2. Lynn Freedman (2005) Achieving the MDGs: Health systems as Core Social Institutions. *Development*. March 2005. DOI: 10.1057/palgrave.development.1100107

This older, but seminal article is one of the first to clearly articulate the case that health systems should be understood as much more than just 'cost-effective delivery systems' oriented to target-based achievements, and should be understood as 'core social institutions.' Freedman argues that recognizing health as a central part of the wider development agenda provides an opportunity to 're-ground' our understanding of health systems in the policy debates of the day – globalization, human security, equity, human rights and poverty reduction. She argues that viewed from a developmental perspective, health is not a 'residual category' or side effect of development policy, but rather a core concern. The author outlines two common rationales used to link health and development or poverty reduction. The first is that health is intrinsically valuable – an argument most famously invoked by Amartya Sen using the 'capabilities' approach. The second rationale is that health is instrumentally valuable – an approach most commonly invoked by economists who note population health to be a pre-condition for economic growth. To these rationales Freedman adds a third. The idea of ill-health and poverty linked through individual or community interactions with, and experiences of, broader structures of power. Freedman notes that health systems:

> *function at the intersection between people and the structures of power that shape their broader society. Neglect, abuse and exclusion by the health system is part of the very experience of being poor. Conversely [...] legitimate claims of entitlement to the services and other conditions necessary to promote health are assets of citizens in a democratic society.*

Health systems are thus not only producers of health and health care, but 'purveyors of a wider set of societal norms and values.' While dated in its references to the long-past Millennium Development Goals, this article is striking for its continued relevance to health system debates, including in the face of intensified ethical and philosophical challenges being faced by health policy makers in a post-COVID-19 world.

3. J. van Olmen, B. Marchal, W. van Damme, G. Kegels, P.S. Hill (2012) Health system frameworks in their political context: framing divergent agendas. *BMC Public Health.* 12:774.

This article provides a critical grounding in the political context of some of the most popular health systems frameworks, demonstrating the way they are shaped by political agendas and paradigms of the day. Consciously engaging with the political history and political economy of health system debates, the authors document landmarks in the development of health systems thinking, and (complementing this textbook chapter) link this backdrop to the development and application of certain health system frameworks. Grouping frameworks into 'reform-focused' (considering health systems as projects to be engineered) and more 'organic' (characterized by their view of health systems as mirrors of society) the authors observe that such frameworks have not reflected a 'progressive accumulation of insights' but rather co-exist and have emerged partly in line with prevailing political paradigms, and partly in response to different needs. Nor therefore, the article concludes, are the frameworks neutral. Rather they 'frame health, health systems and politics in particular political and public health paradigms, although these underlying assumptions are virtually never specified by their authors or proponents.' This article is valuable for its contextualization of health system frameworks and for drawing attention to the profoundly political nature of health policy and health system function – a feature often overlooked in biomedically-oriented descriptions and analyses of the same.

4. A.E. Birn, N. Kremenstov (2018) 'Socialising' primary care? The Soviet Union, WHO and the 1978 Alma Ata Conference. *BMJ Global Health.* 24 Oct. 2018, 3:e000992.

Complementing the far more common, and thus accessible, Western-dominated accounts of the events and decisions leading to the watershed Alma Ata Conference in 1978, this historical analysis draws on archival analysis and interviews to develop the Soviet side of the Alma Ata story. The article highlights how throughout the decade-long lead up to the event, there remained contention among key planners within and outside World Health Organization (WHO) regarding the relative emphasis that should be placed on medical expertise and infrastructure versus social platforms – such as community engagement and participation – as the foundations for future health system design. Birn and Kremenstov demonstrate how seemingly limited Soviet interest in the conference beyond the opportunity it presented to showcase Soviet achievements, contrasted with the global resonance it achieved during and afterwards, pointing to the different meanings and significance given to the idea of 'primary health care' by different State actors. While ensuring a more nuanced and comprehensive understanding of the Alma Ata Conference is useful, the article is also an excellent resource for students / readers seeking to better understand the role of health-specific donor aid (development assistance for health) in broader political and diplomatic machinations in the post-war period and up until comparatively recently. Although not commonly at the centre of development or international relations narratives, this article demonstrates the often important role of health diplomacy in strategic and ideological agendas, and the implications for health system development in low- and middle-income (particularly post-colonial) countries. The article is an excellent example of the role of cross-over social science and humanities-informed research examining the role and function of health systems in society.

5. E.B. Whyle, J. Olivier (2020) Towards and Explanation of the Social Value of Health Systems: An Interpretative Synthesis. *Int J Health Policy Manag.* 2020 epub 26 Aug.

The point of departure for this interpretative synthesis is an understanding that health systems are not only *influenced by* social values, but also capable of *influencing* and *generating* social values in the societies they serve. This understanding, the authors argue, necessitates an exploration of the mechanisms that underlie it, in order to examine whether and how values-driven health system reform can be harnessed to bring about positive social change. Adapting the methods of meta-ethnography, the authors synthesize claims from the literature about the relationships between health systems and social values within a unifying frame; and subsequently consider the implications of an emergent explanatory theory for promoting 'values-based systems reform in complex social systems.' Key lessons emerging include first, that health systems are change-resistant, in part, because values become institutionalized and legitimized over time, meaning attempts to influence health system status via the introduction of progressive values in a single programme or policy are unlikely to substantially effect the system as a whole. And second, that the policy-making processes matter as much as the policies themselves, since health policy decisions most often require trade-offs between competing values and thus a process of deciding which value to prioritize. The authors conclude 'policy processes should be a dialogic site for deliberations and consensus-building' involving policy makers in partnership with an informed public. Extending a key theme in the accompanying chapter, the authors highlight too the central role that language plays in articulating and embedding values in health systems, noting that 'pernicious ideologies in policy discourse can become popularly accepted values [since] policy frames incorporate particular norms of fairness.' For example, 'when goods and services are portrayed as marketable commodities, fairness is defined primarily in terms of individual choice and personal derservingness and these notions of fairness [can] become the primary way of judging equity.' The article provides the reader/student with a deeper appreciation of the socially-embedded nature of health systems and the mechanisms by which different values come to shape their operation.

References

Acheson, D. (1988), *Acheson Report: Independent Inquiry into Inequalities in Health Report*, The Stationery Office, London.

Baum, F. and Friel, S. (2020), COVID-19: the need for a social vaccine, viewed 6th October 2020, https://insightplus.mja.com.au/2020/36/covid-19-the-need-for-a-social-vaccine/

Bigio, J. and Pai, M. (2020), 'How Covid is making it tougher to tackle TB, AIDS, malaria and child health', *ThePrint*, 18 June, 2020.

Birn, A. E. and Krementsov, N. (2018), 'Socialising' primary care? The Soviet Union, WHO and the 1978 Alma-Ata Conference', *BMJ Glob Health*, vol. 3.

Bodewig, C. and Hallegatte, S. (2020), 'Building back better after COVID-19: How social protection can help countries prepare for the impacts of climate change', *World Bank Blogs*, 14 July, 2020.

Brown, I. M., Khan, A., Slocum, J., Campbell, L. F., Lacey, J. R. and Landry, A. M. (2020), 'COVID-19 Disparities and the Black Community: A Health Equity-Informed Rapid Response Is Needed,' *Am J Public Health*, vol. 110, pp. 1350–1351.

Cilliers, J., Oosthuizen, M., Kwasi, S., Alexander, K., Pooe, T. K., Yeboua, K. and Moyer, J. D. (2020), 'Exploring the impact of COVID-19 in Africa: a scenario analysis to 2030,' *Institute for Security Studies*. https://reliefweb.int/report/world/exploring-impact-covid-19-africa-scenario-analysis-2030

Commission on social determinants of health (2008), 'Final report: Closing the Gap in a Generation – Health Equity Through Action on the Social Determinants of Health.' World Health Organisation (WHO) (ed.) Geneva.

Davies, R. (2020). UK scraps plans to buy thousands of ventilators from Formula One group, *The Guardian*, 14 April 2020.

Freeman, L. P. (2005), 'Achieving the MDGs: Health systems as core social institutions', *Development*, 48.

Hale, T., Webster, S., Petherick, A., Phillips, T. and Kira, B. (2020), Oxford COVID-19 Government Response Tracker, viewed, www.bsg.ox.ac.uk/covidtracker.

Hall, J. J. and Taylor, R. (2003), 'Health for all beyond 2000: the demise of the Alma-Ata Declaration and primary health care in developing countries', *Med J Aust*, vol. 178, pp. 17–20.

Halpern, S. D., Truog, R. D. and Miller, F. G. (2020), 'Cognitive Bias and Public Health Policy During the COVID-19 Pandemic', *JAMA*, vol. 324, pp. 337–338.

Hemenway, D. (2010), 'Why we don't spend enough on public health'. *N Engl J Med*, vol. 362, pp. 1657–8.

Hogan, A. B., Jewell, B. L., Sherrard-smith, E., Vesga, J. F., Watson, O. J., Whittaker, C., Hamlet, A., Smith, J. A., Winskill, P., Verity, R., Baguelin, M., Lees, J. A., Whittles, L. K., Ainslie, K. E. C., Bhatt, S., Boonyasiri, A., Brazeau, N. F., Cattarino, L., Cooper, L. V., Coupland, H., Cuomo-dannenburg, G., Dighe, A., Djaafara, B. A., Donnelly, C. A., Eaton, J. W., Van Elsland, S. L., Fitzjohn, R. G., Fu, H., Gaythorpe, K. A. M., Green, W., Haw, D. J., Hayes, S., Hinsley, W., Imai, N., Laydon, D. J., Mangal, T. D., Mellan, T. A., Mishra, S., Nedjati-gilani, G., Parag, K. V., Thompson, H. A., Unwin, H. J. T., Vollmer, M. A. C., Walters, C. E., Wang, H., Wang, Y., Xi, X., Ferguson, N. M., Okell, L. C., Churcher, T. S., Arinaminpathy, N., Ghani, A. C., Walker, P. G. T. and Hallett, T. B. (2020), 'Potential impact of the COVID-19 pandemic on HIV, tuberculosis, and malaria in low-income and middle-income countries: a modelling study,' *Lancet Glob Health*, vol. 8, no. 9, E1132–E1141.

Horton, R. (2020), 'After COVID-19-is an "alternate society" possible?', *Lancet*, England.

Kuehn, B. M (2020), 'News From the Centers for Disease Control and Prevention', *JamA*, 324, 929.

Kupferschmidt, K (2020), 'WHO unveils global plan to fairly distribute COVID-19 vaccine, but challenges await', *ScienceMag*, 21 September 2020.

Ludovic, J., S. Bourdin, F. Nadou, G Noiret (2020), 'Economic globalization and the COVID-19 pandemic: global spread and inequalities', *Bulletin of WHO*, 23 April 2020.

Malik, A. A., Safdar, N., Chandir, S., Khan, U., Khowaja, S., Riaz, N., Maniar, R., Jaswal, M., Khan, A. J. and Hussain, H. (2020), 'Tuberculosis control and care in the era of COVID-19', *Health Policy Plan*, vol. 1; 35, no. 8, pp. 1130–1132.

Metzl, J. M., Maybank, A. and De Maio, F. (2020), 'Responding to the COVID-19 Pandemic: The Need for a Structurally Competent Health Care System', *JAMA,* vol. 324, pp. 231–232.

Newell, K. W. and WHO (1975), 'Health by the people', eds by Kenneth W. Newell. Geneva, World Health Organization.

Nunes, J. (2020), The COVID-19 pandemic: securitization, neoliberal crisis, and global vulnerabilization, *Cad Saude Publica,* vol. 36, no. 5, pp. e00063120. doi: 10.1590/0102-311x00063120

Rifkin, S. B. (2020), 'Paradigms, policies and people: the future of primary health care', *BMJ Glob Health,* vol. 5, pp. e002254

Schneider, H., Blaauw, D., Gilson, L., Chabikuli, N. and Goudge, J. (2006), 'Health systems and access to antiretroviral drugs for HIV in Southern Africa: service delivery and human resources challenges', *Reprod Health Matters,* vol. 14, pp. 12–23.

Sheikh, K., Gilson, L., Agyepong, I. A., Hanson, K., Ssengooba, F. and Bennett, S. (2011), 'Building the field of health policy and systems research: framing the questions,' *PLoS Med*, vol. 8, no. 8, pp. e1001073. https://doi.org/10.1371/journal.pmed.1001073

Stop TB Partnership, Imperial College, Avenir Health, Johns Hopkins University and USAID (2020), 'The potential impact of the COVID-19 response on tuberculosis in high-burden countries: a modelling analysis', *STOP TB Partnership.*

Van Olmen, J., Marchal, B., Van Damme, W., Kegels, G. and Hill, P.S. (2012), 'Health systems frameworks in their political context: framing divergent agendas,' *BMC Public Health,* vol. 12, pp. 774.

WHO (2000), 'World Health Report 2000: Health Systems – Improving Performance,' *Geneva.*

WHO (2006), 'Scaling up HIV/AIDS prevention, treatment and care: a report on WHO's support to countries in implementing the "3 by 5" Initiative, 2004–2005,' Geneva.

Whyle, E.B. and Olivier, J. (2020), 'Towards an Explanation of the Social Value of Health Systems: An Interpretive Synthesis', *Int J Health Policy Manag*, doi: 10.34172/ijhpm.2020.159. Epub ahead of print. PMID: 32861236.

Žižek, S. (2020), *Pandemic!: COVID-19 Shakes the World*, OR Books, New York.

40

HEALTH AND ILLNESS

Pranee Liamputtong and Zoe Sanipreeya Rice

Introduction

Globally, people are now living longer than in the past. However, life expectancy between nations in the world is unequal (de Souza and Rêgo, 2018). Life expectancy refers to a principle measure of the health and wellbeing of people in the nation. It mirrors the social and economic conditions of the country, as well as the quality of health care and its infrastructure. Individuals living in more affluent nations tend to live longer than those in the poorer nations. In the developed nations, except during war, infection outbreaks, and famine, life expectancy has continuously improved for decades. However, within these nations, we still witness the deterioration of life expectancy among poor and marginalized groups. When comparing life expectancy across the globe, inequalities are significantly marked.

People experience health and illness in different ways. Some of this is due to cultural differences and location divergences, but more often, it is due to marked inequalities that influence the life expectancy of individuals, causations of illness, and availability of effective health care (Broom et al. 2019). For instance, in Africa, one of the main causes of health inequality across nations was due to the epidemic of human immunodeficiency virus (HIV) in the 1990s (de Souza and Rêgo, 2018). Many infectious diseases that have been eradicated in richer countries, such as polio, malaria, typhoid, and dengue fever, continue to afflict poor nations. Poor nations continuously endure a dual burden of these diseases besides emerging non-communicable diseases such as diabetes, cardiovascular disease, and cancer (Broom et al. 2019).

People's health, illness, and wellbeing can be determined by many factors. These range from societal influences including health care provision and access, to individual aspects such as genetics. These are referred to as 'determinants of health,' which 'influence how likely we are to stay healthy or to become ill or injured' (AIHW 2016, 128; Gleeson and Chong 2019). In this chapter, we discuss key determinants of health and the social determinants of health framework. Within this framework, the intersection of gender, ethnicity and social class is elaborated. We also look at health inequalities and inequities around the globe. We summarize key global health issues that have created ill health and suffering. In this chapter, we propose social justice as an important means that can address and respond to health inequalities in the world, and certainly to the health inequality among the migrant population. Migration has become a global

DOI: 10.4324/9781003017653-45

phenomenon in recent years. Migration has a great impact on the health and wellbeing of migrants and has influenced human development globally.

Determinants of health

Health, illness, and the wellbeing of individuals, communities, nations are determined by a diverse range of complex individual, social, cultural, environmental and economic factors, as well as by health care systems (Germov 2019; Liamputtong 2019). This multi-dimensional approach is referred to as 'determinants of health' (Marmot 2015; Oldroyd 2019). The focus of this perspective is on factors which could influence and determine the health of people, instead of on the state and outcomes of their health. It also emphasizes the prevention of ill health, rather than the measurement of illness (Oldroyd 2019).

The determinants of health are characteristics or factors which can bring about a change in the health and illness of individuals and populations, either for better or worse (Oldroyd 2019). Determinants of health include biological and genetic factors; health behaviours (such as risky lifestyles, abuse of alcohol and cigarette smoking); environmental factors (including housing, social support, social connection, geographical position, and climate); and socio-cultural factors (such as gender, ethnicity, education, income, and occupation) (Liamputtong 2019) (see Figure 40.1). Resources and systems also have effects on the health and wellbeing of individuals and populations. These include access to health services, health care policy and the health care system.

Determinants of health outcomes are interconnected with conditions that can either hinder or improve individuals' possibilities of having and sustaining good health. Some conditions have a direct impact on the health and illness of individuals, for example, direct contact with

Figure 40.1   Determinants of health of an individual

heat or asbestos in their environment, cigarette smoking or lack of physical activities. Other conditions have an indirect impact on individuals that can increase or reduce the influences of other factors, for example, when individuals cannot afford or access suitable health care (Oldroyd 2019). These conditions interact and function in complex ways. For instance, when people do not have good health, they may not be able to participate in employment or physical activities. This, in turn, will have a further impact on their health (Oldroyd 2019). Within the determinants of health, social determinants of health play a significant role in determining the health and wellbeing of individuals.

Social determinants of health and health inequality

Social determinants of health are described by the World Health Organization (2017) as 'the circumstances in which people are born, grow, work, live, and age, and the wider set of forces and systems shaping the conditions of daily life. These forces and systems include economic policies and systems, development agendas, social norms, social policies, and political systems.' Social determinants of health are created by the unequal distribution of power, money, and resources.

Social conditions are the most influential foundation of good health or illness (Marmot 2015; Marmot and Bell 2016). They are the root cause of health inequities, the unjust and preventable discrepancies in health status that we have witnessed within and between nations (Gleeson and Chong 2019). Thus, social determinants can be perceived as 'causes of the causes' – the living and working conditions that shape the social patterning of health inequality (Williams and Germov 2019, 254). Marmot (2015, 259) puts it clearly:

> Societies have cultures, values and economic arrangements that set the context for conditions through the life course that influence health…inequalities in power, money and resources give rise to inequities in the conditions of daily life, which in turn lead to inequities in health.

Important social determinants of health are related to positions of social life including gender, ethnicity and social class (Schofield 2015). Each of these social determinants intersect in ways that can create inequalities in health among individuals (Schofield 2015). The term health inequality signifies the disparities in health of individuals and groups.

Health inequalities also exist globally (Commission on Social Determinants of Health 2014; Broom et al. 2019; Gleeson and Chong 2019). In the past three decades, we have witnessed a steady growth of economic inequality in the world (Gleeson and Chong, 2019; see also an earlier section on poverty and food insecurity). This has crucial ramifications for global health. There is clear evidence of the negative impacts of income inequality on the morbidity and mortality of people (Commission on Social Determinants of Health 2014). Such inequality disproportionately affects the health of the most disadvantaged members of society, but it is detrimental to everyone (Turnock 2016). Notably, a growing body of research has revealed that polluted and degraded environments tend to occur more often in unequal societies, and this helps to explain why unequal societies tend to be less healthy than equal ones.

Social inequalities also intersect in ways that can create further inequalities in health among some individuals and groups. For example, men from ethnic minority groups and lower social classes are also likely to be disadvantaged in terms of health and wellbeing in comparison to white Anglo-Celtic men with higher incomes (Germov 2019). Among migrant populations, those who are from refugee backgrounds are more likely to fare worse in all aspects (including health status and health needs) than those economic migrants who come from higher social classes and have higher educational attainments. Among these populations, too, women are more disadvantaged than men (Germov 2019). Within the Australian context, we have witnessed ample examples of these interrelationships within the Indigenous population. Indigenous Australians fare worse in health issues than non-Indigenous Australians, and often women and children and those who live in remote areas fare worse than Indigenous men who live in more urban settings (Jackson Pulver et al. 2019).

Key global health issues
Poverty and food insecurity

According to the World Bank (2018), prior to the outbreak of COVID-19, 767 million people are living below the international poverty line of US$1.90 per day. Although intense poverty has been curtailed around the globe, the divergence between 'the extremely poor' and 'the extremely rich' has reached new heights (Broom et al. 2019, 75). The richest 1 per cent of individuals possess more wealth than the rest of the global population. Poverty has become more entrenched and difficult to eradicate, particularly in remote areas and in nations experiencing violent conflict and weak policies (World Bank 2018). Most of the world's poor are in Sub-Saharan Africa, and the total number of poor people in this region is increasing. Indeed, of the world's 28 poorest countries in the world, 27 are located in Sub-Saharan Africa, with poverty rates above 30 per cent (World Bank 2018).

Poverty threatens food security around the globe (Castleman and Bergeron 2015), and food security is central to both the reduction of hunger and malnutrition, and to the achievement of other national development goals (Castleman and Bergeron 2015). Food insecurity refers to 'the limited or uncertain availability of nutritionally adequate or safe foods or limited or uncertain ability to acquire foods in socially acceptable ways' (Pollard and Booth 2019, 2). Although food insecurity occurs mainly in poorer nations, it is also prevalent among vulnerable groups in affluent countries (Mansour et al. 2021).

Due to household food insecurity, hunger and malnutrition are leading causes of morbidity and mortality in poor nations (Pollard and Booth 2019). Worldwide, about 795 million people are undernourished, of whom 780 million are in developing countries (FAO 2019). Similarly, according to FAO (2019, 3), over 820 million people around the globe are 'still hungry today.' This is particularly so in all subregions in Africa, although hunger is also rising in Asia, Latin America, and the Caribbean (FAO 2019).

Food insecurity has serious health ramifications for both children and adults. For children, food insecurity can lead to poor general health, slow progress, and unsettled behaviour (McKay et al. 2019). The majority of stunted children in the world reside in the 36 countries that have the highest level of lifelong undernutrition (Pollard and Booth 2019). Children experiencing hunger and food insecurity also have a high chance of having weight problems, being obese and susceptible to non-communicable diseases later in life (FAO 2019). For adults, food insecurity is linked with a high risk of mental health problems including depression, anxiety and mood disorders, as well as chronic conditions such as diabetes and hypertension (McKay et al. 2019). Importantly, food insecurity may aggravate existing health inequalities, within and across nations.

Overweight and obesity

Overweight and obesity have now become a major part of the global disease burden (FAO 2019). The fast increase in obesity is particularly disturbing. Since 1980, the prevalence of obesity has doubled in more than 70 countries, and it continues to increase in most other countries. In 2015, a total of 107.7 million children and 603.7 million adults were obese. In 2016, the number of obese individuals around the globe exceeded the number of undernourished people (FAO 2019).

Overweight and obesity are among the main risk factors for non-communicable diseases, including type 2 diabetes, circulatory disease and musculoskeletal problems (Schneider 2014). In many developing nations, such as the USA, the UK, and Australia, a high prevalence of obesity is caused by the combination of eating too much and exercising too little (Schneider 2014; Williams and Germov 2019). People from low socio-economic backgrounds and Indigenous communities tend to be more overweight and obese than better-off groups (Williams and Germov 2019). Growing rural-urban migration has also contributed to increasing overweight and obesity in many countries (Schneider 2014).

The obesity pandemic has become a challenge for both developed and developing nations (White et al. 2014). For poor countries with overstretched health care systems, this is a major concern. They have to deal with a 'double burden' of an 'unfinished agenda' of widespread undernutrition and infectious diseases, as well as an emerging burden of diseases linked with over-nutrition (White et al. 2014, 292). Prevalence rates for overweight and obesity vary considerably across different regions – with higher rates in Central and Eastern Europe, North America and the Middle East than in northern Europe and Asia – but it is estimated that rapidly increasing rates of overweight and obesity in developing nations could see the number of obese people double by 2025 (White et al. 2014).

HIV/AIDS epidemic

The HIV/AIDS epidemic has entered its fourth decade and continues to pose a major public health problem worldwide (AIDSInfo 2019; Carlson 2019). While HIV initially, primarily, affected intravenous drug users, sex workers, and homosexual men, it now affects a large number of women around the globe (UNAIDS 2017). Many of these women are also mothers with young infants. According to Carlson (2019), in 2018, there were 37.9 million people worldwide living with HIV/ AIDS. In 2018, 770,000 people died of AIDS-related illnesses (Carlson 2019). Among young women in developing countries, in particular, the rates of infection are increasing rapidly. The majority of people living with HIV reside in low- and middle-income countries; about 68 per cent live in Sub-Saharan Africa. Among this group 20.6 million are located in East and Southern Africa, and in 2018, there were 800,000 new HIV infections within this subregion (Carlson 2019; see also AIDSInfo 2019).

Globally, women comprised more than half of the total number of HIV/AIDS infections (UNAIDS 2017). Overwhelmingly, the HIV/AIDS pandemic disproportionately affects women of reproductive age, with young women (aged 15–24) particularly at risk. Among young people aged 15–24, there are approximately 6,200 new infections each week, with young women being twice as likely to be HIV-positive as young men (Carlson 2019; UNAIDS 2019). Children are also affected by the HIV/AIDS epidemic, with 1.7 million people aged 10–24 living with HIV/AIDS in 2017 (UNAIDS 2017). While the number of children living with HIV is declining, in 2018, there were still 160,000 new infections among children globally (Carlson 2019). Again, young females (girls) are disproportionately affected relative to boys.

Mental health issues

Mental health issues exist in all nations. Globally, they are the most expensive global health problem, and are experienced across all age groups (Ngubane et al. 2019). Worldwide, twenty-four million people live with schizophrenia and 90 per cent of these reside in developing countries (Ngubane et al. 2019). Globally, official statistics show that over 300 million people (4.7 per cent of the world's population) suffer from depression. According to the World Health Organization (2019a, b), depression is the largest contributor to global disability and suicide deaths.

In developed nations, mental health issues constitute more health burdens than other illnesses, with wide-ranging symptoms including depression, changes in sleep patterns, weight loss/gain, irritability and violence (Schneider 2014). Further, clear links have been demonstrated between mental health issues and chronic diseases (for example, asthma, diabetes, epilepsy, cardiovascular disease, and cancer). People with mental illnesses are also at increased risk of injuries (both intentional and unintentional), and have higher rates of substance abuse (Schneider 2014; Sawyer 2019).

Many mental health issues emerge from social and environmental circumstances, including conflict, disasters such as bushfires and earthquakes, or other stress-inducing events (Schneider 2014). However, the anxieties and depressive disorders that are becoming increasingly common in Western nations are mainly due to the commonplace stressors such as financial strain, unemployment, economic hardship, overwork, the pressure to achieve, relationship breakdown and drought (Schneider 2014; Sawyer 2019). These stresses are intrinsically connected with salient social and environmental factors such as an increased sense of 'individualism,' lack of social support, and anxieties about environmental threats (Sawyer 2019). Conversely, for many marginalized groups – such as refugees, immigrants, Indigenous people and people with disabilities, – common stressors include forms of discrimination such as racism and ableism.

Migration and health

The health of migrants is now acknowledged as a 'global public health priority' (McMichael 2019; Wickramage et al. 2018). Currently, there are approximately 258 million international migrants living outside of their countries of origin, and 763 million internal migrants and displaced people live within their countries of birth (WHO 2019c). This includes 20.4 million refugees and 3.5 million asylum seekers (IOM 2018). It is projected that by 2050, there will be 405 million international migrants in the world.

Migration affects individuals and communities in many ways – often requiring a complete change of daily life practices and producing complex social, economic and health consequences (Castaneda et al. 2015). Migration is a result of a number of social determinants that 'push' people away from their current locations and 'pull' towards new ones (Castaneda et al. 2015; McMichael 2019). Many people flee armed conflict, persecution, and natural disaster. Movements can also be associated with human rights issues in the receiving countries. This can be witnessed clearly in the recent flood of Rohingya refugees who travelled by boats from Myanmar and were denied entry to Thailand, Malaysia and Indonesia, and Syrian refugees who attempted to enter European nations. In contrast, economic or educational opportunities may 'pull' people to new destinations. Economic migrants are the largest growing portion of the migrating population globally, and with worsening global economic problems, this is likely to continue (Schneider 2014; McMichael 2019).

An important migration issue that has ramifications for global health practice is the increase in the number of female migrants (UN Women 2018). Women generally make up half of the migrating population, and in some countries, it is 70–80 per cent (UN Women 2018). Often, migrant women are put in low paid jobs that are unregulated, such as domestic work. They are at high risk of exploitation, violence, and abuse, such as human trafficking. These lead to long-term health and social problems for women, including sexually transmitted diseases and increased numbers of unplanned pregnancies. Often too, they are rejected by their own families when they return home (Schneider 2014).

Migration itself is not 'intrinsically unhealthy' (McMichael 2019). Migration can bring many benefits such as increased educational and economic educational opportunities in a new country which leads to better health and wellbeing. However, the circumstances that many people travel, live and work often endanger health. This is particularly so for refugees, asylum seekers, and irregular migrants, who frequently experience discrimination, exploitation, violence, poor living and work conditions, and lack access to health care. These are social determinants that impact their health and wellbeing, physically and mentally (McMichael 2019).

Social justice as a means to combat health inequality: the case of migrants

One way to approach the health of migrants is through a social justice lens. As outlined above, health inequalities are the consequence of inequitable societies. Framed within the social determinants lens, social inequalities can be addressed through the concept of social justice (Miles 2019). It has been suggested that social justice approaches are crucial for global health; individuals have the right to achieve good health outcomes (CSDH 2014; Miles 2019). Social justice is 'the foundation of public health' (Turnock 2016, 19). The concept emerged around 1848, which was a time that has been referred to as the 'birth of modern public health' (Turnock 2016, 19). Social justice suggests that public health is 'a public matter.' The consequences of social injustice by means of disease, ill health and death are the reflection of 'the decisions and actions that a society makes, for good or for ill' (Turnock 2016, 19). Justice means that there is an equitable distribution of benefits and burdens among populations. Injustices happen when some burden is unwarrantedly placed on some individuals and groups and they lack access to some benefit to which they are entitled. We contend that social justice can easily be adopted for global health issues of migrants. When health and access to health care are perceived as a societal benefit (or if poor health is seen as a burden), the connections between the concepts of justice and migrants' health in the globe are very clear (Turnock 2016).

Globalization is inevitably linked to population mobility and migration. Ensuring safe and healthy migration is crucial for the health and wellbeing of mobilized populations. Innovative approaches in migrant health policymaking and practice should endorse the social determinants of migrant health and emphasize the reduction of health inequalities, social inclusion, rights-based approaches, and universal health coverage. They should move beyond 'protectionist approaches,' such as screening of infectious diseases among migrants, to pursue actions that can address and react to the vulnerability and health rights of migrants (McMichael 2019).

Below we provide examples of how social justice – framed within the social determinants of health framework – can enhance the health of immigrants. Health inequities are common among vulnerable and marginalized groups, including migrants; a group that frequently experience various forms of deprivation. The aims of the Sustainable Development Goals (SDGs) may provide a means to reduce health inequalities of this population. The Sustainable Development Goals (SDGs) determine migration as both an impetus and an agent for sustainable development.

Within the SDGs, Goal 3 designs to 'ensure healthy lives and promote wellbeing' for all individuals at all ages (p. 16) and Goal 10 aims to 'reduce inequalities within and among countries' (United Nations General Assembly 2015, 21). Health equity can be improved by improving the quality and accessibility of health care. The SDGs calls for an attempt to 'leave no one behind,' regardless of their migration status, so that Universal Health Coverage (UHC) for all can be achieved (United Nation General Assembly 2015). This is stated in SDG target 3.8 which aspires to 'achieve universal health coverage (UHC), including financial risk protection, access to quality essential health care services, and access to safe, effective, quality, and affordable essential medicines and vaccines for all' (United Nations General Assembly 2015, 16). The term 'all' includes migrants and refugees irrespective of their migratory status (World Health Organization 2019c, d). To achieve UHC, it will require the nation to pay more attention to the broader aim of health equity and to strengthen their health care systems. Thus, no one will be left behind.

The Rio Political Declaration (WHO 2011) passed at the World Conference on Social Determinants of Health in 2011 and the 2008 World Health Assembly Resolution on Health of Migrants (WHO 2008) acknowledged the need of migrant-inclusive health policies as well as coherent inter-sectoral policy to address health inequalities and vulnerabilities impacting migrants and to ensure that migrants have healthy lives and can achieve as productive members of society. In order to promote migrant-sensitive health policies, five main action areas were pinpointed:

1. Embrace better governance for health and development
2. Endorse participation in policymaking and implementation
3. Overhaul the health sector towards reducing health inequities
4. Enhance global governance and collaboration
5. Oversee progress and increase accountability (IOM 2020)

It is argued that attempts to improve the health needs of migrants not only enhance their health outcomes, but can also assist migrants with settlement and integration, curtail health expenses, and enrich the socio-economic development of the nation. This will ultimately lead to the protection of human rights and global health. As migration is a fundamental element of the global world, it is crucial to develop inclusive approaches to global health practice and policy that will safeguard healthier and safer migration for all. This will ultimately lead to the achievement of social justice for this population group.

Conclusion

Health, illness, and wellbeing are of fundamental importance to all human beings. There are diverse factors which can have an impact on the health, illness, and wellbeing of individuals, groups and populations. These are known as determinants of health and include biological, environmental, social, cultural and economic determinants. These determinants can affect how healthy or sick an individual can be. Additionally, within the social determinants of health, there are three crucial social structures that can affect the health and wellbeing of people. These include gender, ethnicity, and social class. These three factors do not operate in isolation, but rather interrelate to the extent that they create inequality in health among people. We have argued that health inequality and inequity are the root cause of differences in health and wellbeing among people around the globe, particularly among migrant populations.

Although health care delivery is important, Broom and colleagues (2019, 68) contend that 'solutions to global health inequalities lie with interventions that address the broader social determinants of health.' We totally agree with their position. We thus propose social justice as an important means that can address and respond to health inequality in the world, and certainly to the health inequality among the migrant population. This, we argue, is a vital means to witness a rapid positive transformation of human development outcomes globally.

Key sources

1. Germov, J. (ed.) (2019). *Second opinion: An introduction to health sociology*, 6th edn. Melbourne: Oxford University Press.

This text provides important sociological knowledge about the social context of health and illness. It is essential for readers who are interested in the social causes of ill health.

2. Liamputtong, P. (2019). *Social determinants of health*. Melbourne: Oxford University Press.

This book focuses solely on the social determinants of health. It provides a good foundation for an in-depth understanding of the social aspects of health and illness.

3. Marmot, M. (2015). *The health gap: The challenge of an unequal world*. London: Bloomsbury.

The book underscores health inequality around the globe. Readers will have a good understanding about the health gap and its challenges in the world.

4. Schofield, T. (2015). *A sociological approach to health determinants*. Port Melbourne: Cambridge University Press.

It is a good text for a good understanding about the social approach of health. It discusses several sociological issues that influence the health and wellbeing of individuals and societies.

5. Turnock, B. J. (2016). *Public health: What it is and how it works* (6th ed.). Burlington, MA: Jones & Bartlett Learning.

This book discusses the essence of public health. It provides basic understanding about public health and how it contributes to the health of people in the world.

References

AIDSInfo (2019). *HIV/AIDS: The basics.* Retrieved 0n 17 June 2018 from https://aidsinfo.nih.gov/understanding-hiv-aids/fact-sheets/19/45/hiv-aids--the-basics

Australian Institute of Health and Welfare (AIHW) (2016). *Australia's health 2016.* Canberra: Australian Institute of Health and Welfare. Accessed 15 November 2017 from: www.aihw.gov.au

Broom, A., Kenny, K., and Germov, J. (2019). Global public health. In J. Germov (ed.), *Second opinion: An introduction to health sociology*, 6th edn (pp. 66–87). Oxford University Press: Melbourne.

Carlson, G. (2019). *Global HIV/AIDS statistics.* Retrieved www.avert.org/global-hiv-and-aids-statistics

Castañeda, H., Holmes, S.M., Madrigal, D.S., De Trinidad Young, M-E., Beyeler, N., and Quesada, J. (2015). Immigration as a social determinant of health. *Annual Review of Public Health*, 36, 375–392.

Castleman, T., and Bergeron, G. (2015). Food security and program integration: An overview. In L. Ivers (ed.), *Food insecurity and public health* (pp. 1–22). CRC Press: New York.

Commission on Social Determinants of Health (CSDH) (2014). *Closing the gap in a generation: Health equity through action on the social determinants of health. Final report of the Commission on Social Determinants of Health.* Geneva: World Health Organisation.

de Souza, F.C., and Rêgo, L.C. (2018). Life expectancy and healthy life expectancy changes between 2000 and 2015: An analysis of 183 World Health Organization member states. *Journal of Public Health: From Theory to Practice*, 26, 261–269.

FAO, IFAD, UNICEF, WFP and WHO. (2019). *The state of food security and nutrition in the world 2019: Safeguarding against economic slowdowns and downturns.* FAO: Rome.

Germov, J. (ed.) (2019). *Second opinion: An introduction to health sociology*, 6th edn. Melbourne: Oxford University Press.

Gleeson, D., and Chong, S. (2019). Social determinants of health on a global scale. In P. Liamputtong (ed.), *Social determinants of health* (pp. 353–381). Melbourne: Oxford University Press.

International Organization for Migration (IOM) (2018). *Data bulletin series: Informing the implementation of the global compact for migration.* Geneva: IOM.

International Organization for Migration (IOM) (2020). *Social determinants of migrant health.* Retrieved from www.iom.int/social-determinants-migrant-health

Jackson Pulver, L., Williams, M., and Fitzpatrick, S. (2019). Social determinants of Australia's First People's health: A multi-level empowerment perspective. In P. Liamputtong (ed.), *Social determinants of health* (pp. 175–214) Melbourne: Oxford University Press.

Liamputtong, P. (2019). *Social determinants of health.* Melbourne: Oxford University Press.

Mansour, R., John, J.R., Liamputtong, P., and Arora, A. (2021). Food insecurity and food label comprehension among Libyan migrants in Australia. *Nutrients*, 13, 2433. https://doi.org/10.3390/nu13072433

Marmot, M. (2015). *The health gap: The challenge of an unequal world.* London: Bloomsbury.

Marmot, M., and Bell, R. (2016). Social inequalities in health: A proper concern of epidemiology. *Annals of Epidemiology*, 26, 238–240.

McKay, F.H., Haines, B.C., and Dunn, M. (2019). Measuring and understanding food insecurity in Australia: A systematic review. *International Journal of Environmental Research in Public Health*, 16, 476. Doi:10.3390/ijerph16030476

McMichael, C. (2019). The health of migrants and refugees. In P. Liamputtong (ed.), *Public health: Local and global perspectives*, 2nd edn (352–370). Port Melbourne: Cambridge University Press.

Miles, D. (2019). Social justice, human rights and social determinants of health. In P. Liamputtong (ed.), *Social determinants of health* (pp. 107–130). Melbourne: Oxford University Press.

Ngubane, S.N., McAndrew, S., and Collier, E. (2019). The experiences and meanings of recovery for Swazi women living with "schizophrenia". *Journal of Psychiatric and Mental Health Nursing*, 26, 153–162.

Oldroyd, J. (2019). Social determinants of public health. In P. Liamputtong (ed.), *Public health: Local and global perspectives*, 2nd edn (105–123). Port Melbourne: Cambridge University Press.

Pollard, C.M., and Booth, S. (2019). Food insecurity and hunger in rich countries – It is time for action against inequality. *International Journal of Environmental Research and Public Health*, 16, 1804. Doi:10.3390/ijerph16101804

Sawyer, A-M. (2019). Mental illness: Understanding, experience, and service provision. In J. Germov (ed.), *Second opinion: An introduction to health sociology*, 6th edn (pp. 307–342). Melbourne: Oxford University Press.

Schneider, M-J. (2014). *Introduction to public health* (4th ed.). Jones and Bartlett: Sudbury. Taylor, S. (2008). The concept of health. In S. Taylor, M. Foster and J. Fleming (Eds), *Health care practice in Australia* (pp. 3–21). Melbourne: Oxford University Press.

Schofield, T. (2015). *A sociological approach to health determinants.* Port Melbourne: Cambridge University Press.

Turnock, B. J. (2016). *Public health: What it is and how it works* (6th ed.). Burlington, MA: Jones & Bartlett Learning.

UNAIDS (2017). The global HIV/AIDS epidemic. Retrieved www.hiv.gov/hiv-basics/overview/data-and-trends/global-statistics

UNAIDS (2019). *UNAIDS data 2019.* Retrieved from www.unaids.org/sites/default/files/media_asset/2019-UNAIDS-data_en.pdf

United Nations General Assembly (2015). *Transforming our world: the 2030 agenda for sustainable development, Resolution RES/70/1.* New York: UN.

UN Women (2018). *Women refugees and migrants.* Retrieved from

White, F. M. M., Stallones, L. and Last, J. M. (2014). *Global public health: Ecological foundations.* New York: Oxford University Press.

Wickramage, K., Vearsey, J., Zwi, A.B., Robinson, C., Knipper, M., (2018). Migration and health: a global public health research priority. *BMC Public Health*, 18, 987. https://doi.org/10.1186/s12889-018-5932-5

Williams, L., and Germov, J. (2019). The social determinants of obesity. In J. Germov (ed.), *Second opinion: An introduction to health sociology*, 6th edn (pp. 252–279). Melbourne: Oxford University Press.

World Bank (2018). *Poverty and shared prosperity 2018: Piecing together the poverty puzzle*. Overview booklet. Washington, DC: World Bank.

World Health Organization (2008). *Sixty-first world health assembly*. Retrieved from http://apps.who.int/gb/ebwha/pdf_files/WHA61-REC1/A61_REC1-en.pdf

World Health Organization (2011). *Rio Political Declaration on Social Determinants of Health*. Retrieved from www.who.int/sdhconference/declaration/Rio_political_declaration.pdf?ua=1

World Health Organization (2017). Social determinants of health. Geneva: WHO. Retrieved on 10 November 2017 from www.who.int/social_determinants/sdh_definition/en/

World Health Organization (2019a). *Mental disorders*. Retrieved from www.who.int/news-room/fact-sheets/detail/mental-disorders

World Health Organization (2019b). The WHO Special Initiative for Mental Health (2019–2023): Universal Health Coverage for mental health. Retrieved from https://apps.who.int/iris/bitstream/handle/10665/310981/WHO-MSD-19.1-eng.pdf?ua=1

World Health Organization (2019c). *Refugee and migrant health*. Retrieved from www.who.int/migrants/en/

World Health Organization (2019d). Draft global action plan: 'Promoting the health of refugees and migrants' (2019–2023). Retrieved from www.who.int/migrants/GlobalActionPlan.pdf?ua=1 https://apps.who.int/gb/ebwha/pdf_files/WHA72/A72_25-en.pdf

41

DISABILITY-INCLUSIVE DEVELOPMENT

Jonathan Makuwira

Introduction

On the cover page of the Department for International Development's (DfID's) *Strategy for Disability Inclusive Development 2018–2023* stands a man holding a placard which reads 'WE ARE ALL EQUAL.' Below him are bold words 'NOW IS THE TIME.' Equally striking in this document is the foreword by the Secretary of State for International Development, The Rt Hon. Penny Mordaunt, MP, who emphatically states: 'We will not eradicate poverty, deliver Sustainable Development Goals (SDGs) or implement the UN Convention on the Rights of Persons with Disabilities (UN-CRPD), without including people with disabilities in all our work' (DfID 2018, n. p.). These sentiments echo a chorus that has now become a central feature of Global Development discourse – 'Leave no one behind.' Furthermore, the sentiments, which are also fully expressed in SDGs and other development instruments and conventions, underscore the core purpose of development – a development seen from a point of social justice. In other words, I contend, in this chapter that development should, first and foremost, not only be informed by social justice, but that development is a central tenet of justice. This reflects both principles of human rights, and that development is a human right; 'Leaving No One Behind' and being 'ALL EQUAL,' is to declare development as all-encompassing and inclusive.

The purpose of this chapter is to critically examine disability in the context of development. It teases out various perspectives on how 'inclusivity' is theorized in development lexicon and applied in practice, with a particular focus on disability and development. The chapter starts with a reflection on historical discourse and some of the shifts that have occurred regarding conceptualizations of disability from a medical lens to a social lens. It also briefly takes a panoramic view of other models of disability that emerge from a social lens. In doing so, the chapter locates the underpinning theoretical ideas informing contemporary thinking on disability-inclusive development. An in-depth analysis on various conventions is undertaken to highlight tensions and contradictions. Finally, the chapter dissects prevailing current approaches to disability-inclusive development and examines them in light of the current SDGs. It concludes by suggesting ways in which disability-inclusive development can better be theorized and implemented in practice.

DOI: 10.4324/9781003017653-46

Philosophical foundations of disability-inclusive development

The need to be 'inclusive,' as far as development discourse is concerned, is an implicit acknowledgement of a deficit in a development system. It is a recommitment to achieving social justice. It is about liberation and instilling a sense of belonging. It is from this position that I engage 'Critical Disability Theory' (CDT) (Minich 2016), in order to scrutinize the normative ideologies that should help unpack the disability-inclusive development debate and the kind of knowledge produced in this relatively new domain of academic inquiry. CDT not only enables us to produce knowledge that supports justice for people with disabilities, but also to (re)consider how we can liberate the field of development studies which, from an academic perspective, requires re-centring to better recognize that pursuing justice requires being active in bringing to the fore issues that are silent(ced) within public mainstream debates. Inevitably, CDT is about emancipation from the bondage of pathologizing disability in order to recognize the social-political constructions of disability and, therefore, to track the impact such representations have on what Goodley, Liddlard and Ruuswick-Cole (2019) call 'oppressed persons.' In the context of this discussion, 'oppression' is not used to synonymize colonial bondage but is understood in the context where society's construction of disability is isolationist in nature and excludes them based on either their physical appearance or 'deformity,' or the 'dysfunctionality' of their bodies. Of significance to this perception is the role of CDT in demystifying how disability is seen as something static or something with a unitary understanding. But looked at from a wide context, disability is NOT 'inability.' Over the past two decades, the slogan 'disability is not inability,' has widely been used to quell poor framing in disability studies. Recognizing this, the following section(s) critically examine the predominant models that have broadly informed understandings of disability and development. While the focus is on models, it has to be understood that there seems to be no fine line between models and theoretical underpinnings.

Models of disability and their implications on development

Critical reflections on disability-inclusive development begin with an analysis of the various models of disability (McEwan, 2007). While the use of 'models' has been a popular approach to understand disability studies, there have been critics to the debate. In following the trajectory in the discourse of disability-inclusive development and the models that have informed thinking, the work of Retief and Letšosa (2018) is very significant. Acknowledging the nascence of disability studies broadly speaking, and the role of disability studies in informing the current thinking in disability-inclusive development, Retief and Letšosa (2018) demonstrate how the influence of biblical studies, systematic theology, moral theology and, more broadly, theological studies, have shaped the links between development and disability. However, outside theological studies, the current thinking and critique has been informed by sociology, ethics, education, psychology, and philosophy (see Table 41.1).

While an interrogation of each of the models presented in Table 41.1 is beyond the scope of this chapter, those canvassed provide a glimpse of the critical issues that the chapter seeks to address, especially now when the current SDGs have come out clear to emphasize the need for inclusion of people with disabilities. The shift from clinically observable impairment in bodily structure or function, to understanding social isolation, oppression, and exclusion from participating in social life, is fundamental in theorizing disability-inclusive development. The mix between impairment and disability needs clarity right at the outset. Talking

Table 41.1 Disciplinary influences on disability-inclusive development debates

Model	Meaning and/or major assumption	Origin	Goal of intervention
Moral and/or religious	Punishment from God Lack of adherence to social morality and /or religious proclamation or edicts, failure of faith.	Traditionally the oldest. Common in a number of religious tradition.	Spiritual or more so divine acceptance.
Medical or biological	Disability is a disease or a defect as a result of a failure of bodily system (pathological), or abnormality due to genetics.	Mid – 19th century	Being 'sick' requires exemption or a cure.
Social	Disability as socially constructed phenomenon. Problem resides in the environment that fails to accommodate people considered 'disabled'.	1960s and 1970s	Political, economic, social and policy systems, increased access, and inclusion.
Identity	Disability as an identity issue. Closely related to social model. None interested in the experiences and circumstances that have created a 'minority' group of 'people with disabilities.'	1980s onwards	Revolutionary vision and desire for 'equal opportunities.'
Human righ; ts	Disability as a human right issue differs from social and identity in that it focuses on 'human dignity.'	1990s	Social justice Improved life situation human rights protection Minority and cultural identification.
Economic model	Disability as a challenge to productivity.	Recent	Capability, respect, accommodation, civil rights.
Charity	Disability as victimhood. PWD are to be pitied. Victims of their own impairment and thus, 'different.'	One of the oldest	Able-bodied people should assist special services and /or institutions. Humane treatment.
Limits	All human beings experience some level of limitations in their daily lives. This is embodied in humans. PWD's are 'categorized' by 'disabled,' 'able-bodied,' 'normal,' 'abnormal.'	Not known	Capabilities

Source: Adapted from Retief and Letšosa 2018.

of impairment – where impairment is understood, at least from a medical model, as a natural and biological fact, versus disability – where disability is understood as an artificial social classification, propagated by factors outside individual rather than the impairment itself, is equally critical in understanding the current disability-inclusive development. In other words, the surrounding social institutions and the physical environment within which persons with disabilities must deal with, are a major impediment to their inclusion in development processes (Makuwira 2013; Bennett and Volpe 2018; Oliver 1996; Nguyen 2018; Toro, Kiverstein, and Rietveld, 2020). The models that have been examined so far – those gravitating around the 'social model,' attest to one fact, that is, disability is not inability. Hence the current narrative on The 2030 Agenda for Sustainable Development, where it is made clear that participation of people with disabilities (PWD) is essential and critical if we are to address the challenge of almost one billion people who live with a disability of some kind, 80 per cent of whom live in the developing world.

The mainstreaming debate: a historical overview

The discussion so far has established the distinction between a causal relationship between individual impairment, and disability as 'restriction in abilities to perform tasks' (Terzi 2004, 142). It is this kind of theorizing that, in the early 1980s, saw a global movement of disability activists pushing the frontiers of the debate to allow further interrogation on how such very important issues can find their way into mainstream debates, particularly in development policies. The UN Declaration on the Rights of Mentally Retarded Persons and the Declaration on the Rights of Disabled Persons adopted in 1970, were the first international instruments to explicitly set human rights guidelines that specifically targeted people with disabilities together. Although it was hoped that development of these instruments would situate disability rights on the international agenda, it soon became apparent that the philosophical tone was guided by charity and medical models. Furthermore, it was also obvious that such instruments were top-down and paternalistic. The push, which emanated from the formation of Disabled People's International (DPI) in 1980–81, pushed the agenda for full participation, equalization of opportunities and development. Subsequent meetings of DPI and other human rights organizations focused on the shift from medical models of disability to the adoption of the interactive model and the impacts of environmental factors associated with all aspects of health and functioning.

Among the pioneers in pushing the agenda for mainstream were the Nordic countries who, in the process, took the lead in the drafting of the Convention on the Rights of Persons with Disabilities (CRPD) which became effective on 3 May 2008 with 168 signatories. Also significant was the 1994 Committee on Economics, Social and Cultural Rights, which voiced against any denial to provide reasonable accommodation to people with disabilities, and strongly observed that such an act runs counter to the convention on Economic, Social and Cultural Rights. A further landmark development leading to mainstream disability in development was the declaration of the period between 1983 and 1992 as the UN Decade of Disabled Persons. The major thrust of those years was to ensure that there were global efforts to create an enabling legal environment and architecture to facilitate societal integration of persons with disabilities. This initiative resulted in the development of such instruments as:

- Conventions on the Rights of the child.
- The African Charter on Human and People's Rights.
- Standard Rules on the Equalization of opportunities for Persons with Disabilities.
- Biwako Millennium Framework for Action.

Not only did the initiative raise the profile of disability issues and help to foster the burgeoning of a global disability community but it also emphasized the importance of including disability issues in development policies and programming in the international dialogue. These efforts further necessitated organizations within the UN system to urge governments to act swiftly on the critical issues. First, to enhance disability prevention; second, to fully engage disability rehabilitation and, third, to equalize opportunities in national development – as highlighted in the UN Standard Rules on the Equalization of Opportunities for Persons with Disabilities, Rule 21 (UN 1993).

As stated earlier, the 1980s saw a global human rights movement which mustered support for inclusion and full participation of people with disabilities. In search of this position, the discourses on gender and non-discrimination ensued. Although the Convention on the Elimination of all Forms of Discrimination Against Women (CEDAW) was adopted in 1979, and became effective in 1981, it soon entered disability debates given that women with disabilities were also likely to be discriminated against in these international development endeavours. It was also 1981 that saw the declaration of the International Year of Disabled Persons (IYPD).

The 1990s

While efforts were underway to improve policy and international instruments, the UN was equally busy raising awareness of including disability issues on the agenda through various forums. Table 41.2 provides a summary of the key events of note in the 1990s.

The mounting pressure through various proposals submitted by human rights actors and other governmental and non-governmental agencies culminated in the establishment of an Ad Hoc Committee in 2001. In December 2006, the final negotiation on the final drafts were concluded and, in May 2008, the birth of the UN Convention on the Rights of Persons with Disabilities (CRPD), was born with a focus not only on advancing the universal human rights for persons with disabilities but also on shaping the way development was inclusive of persons with disabilities.

Table 41.2 Key major events in the 1990s

Name of event	Major emphases and linkage to disability
UN Conference on Sustainable Development (Rio 1992)	Pursuit of inclusive, equitable, and sustainable development can only be attained if human beings become a major concern and human rights principles are internalized in principle and practice.
World Conference on Human Rights (Vienna 1993)	Recognition of the Universality of Human Rights and Fundamental Freedoms that are inclusive of people with disability.
International Conference on Population and Development (Cairo 1994)	Equalization of opportunities for people with disabilities through realization of rights and full participation.
World Summit for Social Development (Copenhagen 1995)	Declaration that clearly stipulated the need to consider disadvantaged groups such as persons with disabilities as deserving special attention in attaining social development.
Fourth World Conference on Women. Being Platform for Action. (1995)	Considered women, especially those with disabilities, intensifying efforts for equal opportunity, empowerment, employment, and active participation.

Source: Adapted from UN 2018, 10–11.

Article 3 of the convention is of significance for our understanding of the foundations of disability-inclusive development. The following are the principles as summarized by Christian Blind Mission (CBM) (2017, 16):

- Respect for inherent dignity, individual autonomy, including the freedom to make one's choices, and independence of persons.
- Non-discrimination.
- Full and effective participation and inclusion in society.
- Respect for difference and acceptance of persons with disabilities as part of human diversity and humanity.
- Equality of opportunity.
- Accessibility.
- Equality between men and woman.
- Respect for evolving capacities of children with disabilities and respect for the right of children with disabilities to preserve their identities.

Development that is 'disability-inclusive'

For a long time, people with disabilities have been treated as 'objects' of development. This has been the theme where charity-based and medical models have been prevalent. As discussed so far, the major thrust in the fundamental issue gravitates around 'inclusion' which only addresses the issue partially. From a development perspective, inclusion does not equal having 'control' over one's destiny. It can still be tokenistic. To understand 'inclusive development,' it is important that we start by analysing 'disability mainstreaming.' According to Albert, Dube and Riis-Hansen (2005, 6), mainstreaming disability in development simply means:

> *The process of assessing the implications for disabled people of any planned actions, including legislation, policies and programmes, in all area and at all levels. It is a strategy for making disabled people's concerns and experiences an integral dimension of the designs, implementation, monitoring and evaluation of policies and programmes in all political, economic and societal spheres so that disabled people benefit equally and inequality is not perpetuated. The ultimate goal is to achieve equality.*

The definition, as comprehensive as it is, can best be simplified by asking the question: 'to what extent do the actions we are undertaking, address and/or respond to the needs and aspirations of persons with disabilities?' In attempting to respond to this question, we must attend to five principles that define disability-inclusive development; equality, non-discrimination, accessibility, full participation, and human diversity (CBM 2017). Because development is a human right, it is essential to engage persons with disabilities at every stage of development processes. In other words, approaches to any development intervention must be disability-sensitive.

The approach

So far I have made reference to mainstreaming disability in development processes. However, 'inclusion,' as the core aspect of mainstreaming, is not sufficient by itself. Neither is participation. Inclusion is a form of participation. However, 'participation' has different levels, spanning from tokenism to empowerment. To quote Arnstein (1969, 216), meaningful participation is a 'categorical term for citizen power.' Arnstein continues:

> *It [participation] is the redistribution of power that enables the have-not citizens, presently excluded from the political and economic processes to be deliberately included in the future…*
> *There is a critical difference going through the empty ritual of participation and having the real power needed to affect the outcomes of the process.*

These sentiments are very critical in that they highlight that mainstreaming alone is but one component of inclusion. Citizen power, which contains three elements – partnerships, delegated power, and citizen control, paves the way to our understanding of why one of the approaches for disability-inclusive development focuses on the 'Twin-Track' approach, to which I now turn.

Twin-Track

I have already explained part of the definition of 'twin-track.' But to elaborate further, a 'Twin-Track' approach involves mainstreaming disability in all development processes to ensure all barriers to the participation of PWD, often perpetuated by society, are removed. The second component of the Twin-Track is an empowerment process whereby disability-specific initiatives are undertaken to ensure that their families, representative organizations, and social networks have equal access to support services, health care, education, livelihoods, and political empowerment at their disposal. Together they constitute what the UN has emphasized as equalization of opportunities for people with disabilities. A society for all cannot be possible if this approach is not adopted in its entirety. It is important to note that critical to the success of a twin-track approach is to acknowledge where power is and how it is used to include and exclude. Twin-track approaches understand that legislation alone is no panacea to inequality; rather what is needed is to deal with societal attitudes in order to build an enabling environment where empowerment and mainstream will equal full participation (real power) and barrier-free societies.

The desire for a barrier-free society, however, hinges on individuals and development institutions. This is particularly so in respect to disability-specific initiatives, which must consider engagement through disability people's organizations (DPOs).

DPOs and disability-inclusive development

Disability People's Organizations, as a wider part of civil society to which all non-governmental organizations (NGOs) belong, have a critical role in enhancing disability-inclusive development. One of the major advantages is the fact that most DPOs are often established and administered or managed by people with disabilities themselves. This is often through membership organizations. Table 41.3 provides a summary of some International, Regional, National and Local DPOs.

Table 41.3 highlights not only the diversity of the organizations, but also a proposed scope such organizations intend to cover. Issues addressed at the grassroots levels intertwine with those international DPOs address – an indication of the thrust being pushed from below but resonating with international instruments such as CRPD.

While DPOs are specific in dealing with a particular development issue – that of inclusion, their relationships with other NGOs matter most. As argued by Kett, Carew, Asiimwe, Bwalya et al. (2019), in their reflection about partnerships, one of the key issues they have observed is that when DPOs partner with other organizations, they learn new things including an improved capacity to appreciate the intricacies of joint evidence-based

Table 41.3 Examples of international, regional, national and local DPOs

Name of DPO	Year formed	Mission/mandate	Reach
International Disability and Development Consortium	1994	Promote disability-inclusive development and human rights.	International
Southern Africa Federation of the Disabled (SAFOD)	1986	Advocates for the rights of persons with disabilities and nurturing and strengthening of its members in Southern Africa.	Southern Africa
Australian Federation of Disability Organization	2004	National Voice for people with disabilities through policy advice and representation to government.	Australia-wide
Malawi Union of the Blind	1994	Bring together blind and partially sighted to participate in development.	Malawi but only targeting the visually impaired.

Source: www.iddcconsortium.net/; www.safod.net/; www.afdo.org.au/

research and capacity development. To effectively combat poverty among people with disabilities, not only is it important to increase DPOs' networks – donors, private sector, government departments, other DPOs and NGOs – it is also important that such partnerships are founded on a common purpose with a high level of mutual understanding and respect (Cleaver, Magalhaes, Bond and Nixon, 2017). By the mere fact the majority of DPOs tend to be consortia, federations, associations, or networks, explain the centrality of the 'organizing' element. In other words, it is commonality of the central issue which Cleaver calls 'Pan-disability or cross – disability approach to the organizing of persons with disabilities' (2016, 11). In essence, the style of partnership often starts with national membership and subsequently expands to include members outside national boundaries and regions. In all, disability–inclusive development requires approaches that transcend tokenism to embrace a multiplicity of ideological positions in order to understand the complex landscape navigated by various persons and actors in the process.

Pedagogical implications

The discussion on disability-inclusive development warrants new thinking and new pedagogies. The issues people with disabilities are going through gravitate around poverty and societal issues that are complex and messy. The literature is replete with definitions of poverty (Banks, Kuper and Polack, 2017), but to unpack the subtleties of poverty and disability in an international development studies classroom requires engaging with pedagogical approaches that disrupt power and domination. As such the first point of entry is to engage critical theories (Calhoun 1995). Because development is initially framed and conceptualized as a process designed to enable people to free themselves from entrenched structural disadvantage, the deployment of critical pedagogies is to allow 'agency'– and particularly the agency of persons with disabilities themselves – to not only question but also, where possible, reject unjust social norms. A further advantage of critical theory is also its ability to offer a practical means to undertake the consciousness and/or awareness-raising required to effectively engage in the political economy of development. Finally, the emphasis on policy engagement and participation cannot be operationalized without understanding how language works and why certain words and structures operate the way they do. Thus, structuralism and critical discourse analysis must also form part of the peda-gogical package in the teaching of disability-inclusive development (Pieterse 2011).

Pedagogy of discomfort

The pedagogy of discomfort, as first used by Megan Boler (1999) in her seminal work *Feeling Power*, is best suited in teaching disability-inclusive development because of three main reasons. First, the pedagogy of discomfort allows both students and teachers to engage in critical conversations that destabilize and challenge their preconceived ideas about development so that eventually they can come out of their 'comfort zones' and embark on a journey of self-reflec-tion and transformation. Second, given that development processes thrive on power dynamics, the use of pedagogy of discomfort helps to question cherished beliefs which in development processes can be counterproductive as often development practitioners tend not to listen. Third, pedagogy of discomfort helps to transform teachers and students to learn to unlearn certain behaviours that are counterproductive to social transformation. Fourth, the pedagogy of discomfort is essential in development studies not just because it helps people to get out of their comfort zones, but it can also be used to get those in 'discomforting zones' to believe in their capabilities; to believe 'they can' do things; they can change the course of their situations. This is very important in disability-inclusive development where people with disabilities have to emancipate themselves from societal norms and power structures. Therefore, pedagogy of discomfort is not only suited to a classroom situation, it is also an approach that can be used in participatory decision-making processes.

Critical Discourse Analysis

I have already alluded to Critical Discourse Analysis (CDA) earlier in the chapter but, to reiterate, CDA is rarely seen as a pedagogical tool, especially in development studies. Because of the widespread criticism of development, especially by post-development thinkers, one of whom is Escobar (1995), and post-structuralist thinkers, especially Said (1978), there is more realization of the role of CDA and how its related concepts can play in shaping realities and relationships. Therefore, CDA as a teaching strategy would aim to delve into understanding how social relations, identities, knowledge and power are constructed through written, spoken and symbolic texts in understanding disability-inclusive development (Foucault 1972). It is, in part, for this reason that CDA is so well suited to teaching on disability-inclusive development.

First, one can adopt a constructivist approach where power issues are examined by looking at who or what is exerting power. This calls for understanding not only 'actors' and their intentions but to also better understand for whose or what purpose it is exerted. This is not an easy process but, as a matter of illustration, in teaching disability-inclusive development, we can look at how the textual mechanisms used by donors, government agencies, international NGOs, construct a reality by constituting subjects with particular attributes, for example those with impairment. This kind of analysis helps to understand how social hierarchy is recreated through a discourse of pity (Naylor 2011). Importantly, by engaging CDA we are able to navigate how inter-sub-jective meanings or understandings are constructed not in an individual's mind but through overt social practices in which people engage. This is where the global actors, such as donors take advantage and construct dominant narratives, texts, symbols, and representations of people and come up with policies which they (global actors) come to agreement with one another.

In the words of Naylor, 'dominance requires not just the texts of multiple actors to be in agreement but also that the texts come from actors at each conceptual level of global activity'

(Naylor 2011, 181). Critical Discourse Analysis further helps, in the case of understanding disability-inclusive development, to identify not only the dominant narratives, but narratives that are successful in their dominance. According to Doty (1993, 302), a successful narrative is 'a system of statements in which each individual statement makes sense, produces interpretive possibilities by making it virtually impossible to think outside it.' Good examples abound: 'people with disabilities,' 'children with special needs,' 'mainstreamed children,' 'exceptional learners,' 'differently abled' are but a few labels and narratives that are impossible for one to find anything wrong in them, yet, they are powerful enough to trigger a special approach to policy and practice. In dealing with such labels, we are forced to produce a kind of knowledge which Doty argues, may be taken to be true. To consider examples beyond disability, we may also consider discourses of the 'Third World' or 'Global South.' What do these terms mean? Does the Global South exist? How and in what ways?

Not only does such labelling construct a particular kind of subject with unique qualities, it also determines what such a person *is* and what he/she can or cannot do. This, ultimately, determines hierarchical binaries rich/poor; inferiority/superiority; similarity/difference; developed/developing. It is therefore suited to not only deploy CDA as a pedagogical tool in order to disrupt and uncover some of the subtleties in development policies but also to question the language used; the various assumptions embedded in such a language; and develop a deeper understanding of what shapes the thinking behind the language.

The pedagogy of listening

When Paulo Freire (1996a) wrote his seminal work on Pedagogy of the Oppressed, it was because he had seen and, more importantly, 'listened' to peoples' stories about oppression. To many of us in the development field, we sometimes struggle to make sense of the richness of the spaces we work in and deliberately pay attention to emerging issues in those spaces. In development studies, many key pedagogical approaches are rooted in participatory development. While participatory approaches to development allow people to get together and decide on their preferred destinies, the implicit driver of such processes is imbued in the skills of listening to one another. However, as pedagogy, there is much more to listening than we can imagine.

Listening, as an idea, comes in different forms. By definition, according to Freire, listening is much more than remaining silent when someone else is speaking. Rather it is a conscious decision one makes as they enter into a conversation with others. In the words of Freire himself, listening allows us to 'discover the rich possibility of doing things and learning things with different people' (Freire, 1996a, quoted in Manyozo 2016, 958). From a pedagogical perspective, this implies exercising tolerance, being accommodating and flexible to differing views that we may encounter in our conversation with others. If we are to enhance disability-inclusive development, this form of pedagogy cannot be over-emphasized.

Manyozo (2016, 958) offers another view to listening which is very relevant to the teaching of disability-inclusive development. Listening is also an act of war. He goes on to point out that when we listen, we are bound to not only consolidate our arguments or reaffirm our solidarity with ideologies we often espouse but also to develop an appreciation for difference, develop empathy and, in most cases, feel sympathetic to others. In other words, in being 'in their shoes,' we develop a feeling and sense of humanism and respect and tolerate others even if we may not be equals. Listening, Manyozo (2016) argues, offers us an opportunity to become students of society. It helps us to nurture the knowledge we have by accommodating other knowledges – those from disability studies.

Conclusion

This chapter has attempted to navigate the complex terrain of development in relation to disability studies. In particular, I have tried to bring to the fore why development has to be inclusive of people with disabilities. While it has been challenging to bring some of the emerging issues within the disability-inclusive development together, I hold the thesis that given that people with disabilities suffer all sorts of discrimination which results in their impoverishment, the time is overdue for bringing a narrative of justice, that reflects PWD's aspirations, into the current development agenda.

I am deeply aware that the enactment of various conventions and development instruments is meant to enhance unconditional inclusion of disability issues at every level of development process. However, the key challenges rest with society – and most prominently the perception society has towards people with disabilities. Their powerlessness to be part of mainstream development debates is, in actual fact, the powerlessness of society – the failure of governments, donors, development agencies to move boldly and engage, without any form of discrimination, with PWD. As I have argued, perhaps the time has come to take disability studies further by engaging the kind of pedagogies highlighted in the chapter. In all, there remains a lot of work to be done.

Key sources

1. Oliver, M. (1996). *Understanding disability. From theory to practice.* London: Palgrave Macmillan.

The book emphasizes that disability is a societal rather than an individual problem. It challenges disempowering approaches in dealing with disability issues and makes it clear that disability is not inability.

2. Boorse, C. (2010). Disability and medical theory. In: *Philosophical reflections on disability.* (eds.) D. C. Ralston and J. Ho. Dordrecht, NL: Springer, pp. 55–88. doi:10.1007/978–90–481–2477–0_4

This chapter challenges the medical theory and/or framing of disability and argues for a shift from medical to social and human rights models of disability.

3. Jampel, C. and Bebbington, A. (2018) Disability studies and development geography: Empirical connections, theoretical resonances, and future directions. *Geography Compass,* 12(12): e12414. https://doi.org/10.1111/gec3.12414

This paper reviews current scholarship on the nexus between disability and development and seeks to bring to the fore under-explored empirical connections and theoretical resonances between disability and development geography.

4. Naylor, T. (2011) Deconstructing development: The use of power and pity in the international development discourse. *International Studies Quarterly,* 55(1): 177–197.

The article explores how power and pity is used in advancing development agendas.

5. Escobar, A (1995). *Encountering development: The making and unmaking of the Third World.* New Jersey: Princeton University Press.

The book shows how development policies have, over the years, become mechanisms of control that were just as pervasive and effective as colonialism itself.

References

Albert, B., Dube, A. K. and Riis-Hansen, C. (2005). Has disability been mainstreamed into development cooperation? https://hpod.law.harvard.edu/pdf/Mainstreamed.pdf

Arnstein, S. R. (1969). A ladder of citizen participation. *Journal of the American Institute of Planners,* 35 (4), 216–224.

Banks, L.M., Kuper, H. and Polack, S. (2017) Poverty and disability in low- and middle-income countries: A systematic review. PLoS ONE 12(12), e0189996. https://doi.org/10.1371/journal.pone.0189996

Bennett, J. M. and Volpe, M. A. (2018) Models of Disability from Religious Tradition: Introductory Editorial, *Journal of Disability & Religion,* 22(2), 121–129, DOI: 10.1080/23312521.2018.1482134

Boler, M. (1999). *Feeling power: Emotions and education.* New York: Routledge.

Boorse, C. (2010). Disability and medical theory. In: *Philosophical reflections on disability.* (eds.) D. C. Ralston and J. Ho. Dordrecht (55–88), NL: Springer, doi:10.1007/978–90–481–2477–0_4

Calhoun, C. (1995). *Critical social theory: Culture, history, and the challenge of difference.* Oxford: Blackwell.

Christian Blind Mission (CBM) (2017) Disability-Inclusive Development Toolkit. www.cbm.org/fileadmin/user_upload/Publications/CBM-DID-TOOLKIT-accessible.pdf

Cleaver, S. (2016). Postcolonial encounters with disability: Exploring disability and ways forward together with persons with disabilities in Western Zambia. Doctoral dissertation. University of Toronto, Toronto, Canada.

Cleaver, S. R., Magalhaes, L., Bond, V. and Nixon, S. A. (2017). Exclusion through attempted inclusion: Research experiences with disabled persons' organisations (DPOs) in Western Zambia. *Disability, CBR & Inclusive Development,* 28(4), 110–117, doi:https://doi.org/10.5463/dcid.v28i4.670

Department of International Development (DfID) (2018). DfID Strategy for Disability Inclusive Development 2018–2023. https://assets.publishing.service.gov.uk/government/uploads/system/uploads/attachment_data/file/760997/Disability-Inclusion-Strategy.pdf

Doty, R. (1993). Foreign policy as social construction: A post-positivist analysis of U.S. counterinsurgency policy in the Philippines. *International Studies Quarterly,* 37(3), 297–320.

Escobar, A. (1995). *Encountering development: The making and unmaking of the Third World.* New Jersey: Princeton University Press.

Foucault, M. (1972). *The archaeology of knowledge.* New York: Pantheon Books.

Freire, P. (1996a). *Letters to Christina.* New York: Continuum.

Goodley, D., Lawthom, R., Liddlard, C. and Runswick-Cole, K. (2019). Provocations for critical disability studies. *Disability & Society,* 34(6), 972–997, DOI: 10.1080/09687599.2019.1566889

Jampel, C. and Bebbington, A. (2018). Disability studies and development geography: Empirical connections, theoretical resonances, and future directions. *Geography Compass,* 12(12), e12414. https://doi.org/10.1111/gec3.12414

Kett, M., Carew, M. T., Asiimwe, J-B, Bwalya, R. et al. (2019). Exploring partnerships between academia and disabled persons' organisations: Lessons learned from collaborative research in Africa. IDS Bulletin, 50(1), 65–78 https://opendocs.ids.ac.uk/opendocs/bitstream/handle/20.500.12413/14520/IDSB50.1_10.190881968–2019.106.pdf?sequence=1&isAllowed=y

Makuwira, J. J. (2013). People with Disabilities and Civic Engagement in Malawi, *Development Bulletin,* 75, 66–71. https://crawford.anu.edu.au/rmap/devnet/dev-bulletin.php

Manyozo, L. (2016). The pedagogy of listening, *Development in Practice,* 26(7), 954–959, DOI: 10.1080/09614524.2016.1210091

McEwan, C. (2007). Disability and development: Different models, different places. *Geography Compass,* 1(3), 448–466. https://doi.org/10.1111/j.1749–8198.2007.00023.x

Minich, J. A. (2016). Enabling Whom? Critical Disability Studies Now, *Lateral,* 5(1), doi:10.25158/L5.1.9

Naylor, T. (2011). Deconstructing development: The use of power and pity in the international development discourse. *International Studies Quarterly,* 55(1), 177–197.

Nguyen, X. T. (2018). Critical disability studies at the edge of global development: Why do we need to engage with Southern Theory? *Canadian Journal of Disability Studies,* 7(1), 1–25 DOI: https://doi.org/10.15353/cjds.v7i1.400

Oliver, M. (1996). *Understanding disability. From theory to practice.* London: Palgrave Macmillan.

Pieterse, N.-J. (2011). Discourse analysis in international development studies, *Journal of Multicultural Discourses*, 6(3), 237–240.

Retief, M. and Letšosa, R. (2018). 'Models of disability: A brief overview', HTS *Teologiese Studies/ Theological Studies* 74(1), a4738. https://doi.org

Said, E. (1978). *Orientalism: Western Conceptions of the Orient.* London: Penguin.

Terzi, L. (2004) The social model of disability: A philosophical critique. *Journal of Applied Philosophy*, 21(2), 141–157.

Toro, J., Kiverstein, J., and Rietveld, E. (2020). The ecological-enactive model of disability: Why disability does not entail pathological embodiment. *Frontiers in Psychology*, 11(1162). https://doi.org/10.3389/fpsyg.2020.01162

UN (1993). UN Standard Rules for the Equalization of Opportunities for Persons with Disabilities'. 85th Plenary Meeting, para. 1, Part IV, A/Res/48/96. December 20. www.ohchr.org/english/law/opportunities.htm

42

CITIZENSHIP, RIGHTS, AND GLOBAL DEVELOPMENT

Diana Mitlin, Jack Makau, Sophie King and Tom Gillespie

Introduction

This chapter discusses a three-year initiative in citizen-led development undertaken by social movement activists to share learning between the Global North and South. This initiative, located in Manchester and Nairobi, offers comparative grassroots perspectives from participants in order to foster new insights into global development challenges.

The experiences and perspectives we share here highlight the potential of community-to-community engagement for grassroots learning about processes of poverty, marginalization and disadvantage, social relations in diverse capitalist economies, forms of capital accumulation that advantage elites, and associated processes of social economic and political discrimination. Both (urban) social movements are struggling to manage the consequences of current trends including inadequate access to essential public services and insufficient attention to the voice and rights of women to secure better development options for themselves and their families.

Through this case study we identify reasons to enable grassroots communities to think and act globally. Our objective is to share the perspectives of social movement activists and facilitate their contribution to debates about development that are too often restricted to academics and professionals. In doing this, we seek to expand current understandings of global development, and test the value of the concept of transition towards a more socially just world.

Academic literature analysing global development has often focused on macro-scale trends and industrial processes, and on the consequences of movements of people, and ideas (both positive and negative). These debates give insufficient attention to the localized manifestations of emergent global trends. Moreover, these efforts give insufficient attention to how those experiencing disadvantage understand under-pinning structural relations and consequential patterns of exploitation and marginalization. Nor do they explore how those people can themselves be a part of learning and knowledge creation to address the problems identified by the critical analysis of on-the-ground conditions. In recognition of this gap, this chapter engages with the realities of global poverty and exclusion as seen by those living in low-income urban neighbourhoods, sharing their understanding of their situation and what they think will address their needs and interests. Following this, the chapter highlights the potential of experiential learning for grassroots communities, as well as the potential of engagements between academic

DOI: 10.4324/9781003017653-47

teaching programmes and grassroots communities to provide students with new insights and activist empowerment via recognition of their experiences.

The chapter is structured as follows. The following section discusses relevant literature including previous experiences of learning and solidarity between communities in the Global North and South to understand the specificity of the experiences shared here. We also consider literature on global development and North-South policy mobility to understand the potential value of this paper to current academic debate. The third section describes the two locations and summarizes our methodology. The fourth section introduces the learning exchanges. The fifth and sixth section share the perspectives of the Kenyan team on Manchester and the perspectives from Manchester activists on the experiences in Kenya. The final section concludes.

Global development, civil society, and poverty reduction

This section places the exchanges between community organizations based in low-income neighbourhoods in Manchester and Nairobi within a broader development literature. We consider grassroots exchange, and particularly labour movement organizing as one long-standing effort to build solidarity between North and South, policy mobilities and the significance of learning between North and South for new approaches to poverty reduction, and discussions of relational poverty within global development debates.

Within broadly-based development scholarship and practice, relatively little attention has been given to exchanges between low-income urban or rural communities between the Global South and Global North. This reflects the common emphasis within development circles on charity and empathy together with professionally-designed programming rather than self-empowerment, redistribution and social justice. While an extensive body of literature has discussed transnational engagement for policy change (Batliwala and Brown 2006), there have been fewer opportunities for citizen-to-citizen engagement. For example, when Oxfam sought to learn from the Global South to inform its poverty reduction programming in the Global North this focused on professional inputs rather than community exchanges (Lewis 2017). Herrle, Fokdal and Ley (2015) do discuss citizen collaboration, but their examples do not consider exchange between North and South.

One exception has been learning between movements of the Global North and South grounded in values of internationalism and solidarity between those who are oppressed. This offers an important legacy in terms of understanding the challenges in global alliance building as well as recognizing the ways in which colonial legacies of difference have been transformed into relations of solidarity. Many of these efforts have been rooted in the peace movement and labour unions, reflecting their political origins and efforts to protect the working class from the horrors of war and capitalist exploitation. An illustration of the challenges and potential of organizing around employment issues is offered by Evans (2014) who engages with the realities of globalized production when he analyses labour interests in international supply chains in the US, Latin America, Liberia and Europe. Evans argues that globalization incentivizes stronger links and strategic engagement between workers, and suggests that, despite neoliberalism, there is evidence that unions can still act internationally to protect workers' rights and interests. However, notable challenges are also evident, including informalization in the labour markets in both North and South that have weakened union activism.

In addition, a growing literature about policy mobilities has highlighted the sharing of poverty reduction strategies at a governmental level (Roy 2015; Lewis 2017). This issue is showcased by the COVID-19 pandemic, as measures to promote and enforce social distancing are being rolled out across the Global North and South (Wilkinson 2020).

Recent scholarship in development geography has called for a shift away from the 20th century paradigm of 'international development,' premised on an outdated distinction between a rich North and poor South, towards a 21st century paradigm of 'global development' that recognizes the universal character of development challenges such as poverty and inequality (Horner 2020; Horner and Hulme 2019). However, much of this discussion has focused on global data and its trends without critically analysing this data. As Ziai (2017) argues, dealing with large data sets risks ignoring the lived realities that people experience, and what those people do to act on the injustice that they experience (Ziai 2019). It fails to problematize the question about what is 'development' and who defines its meaning (Fischer 2020). Alemany, Slatter and Rodríguez-Enríquez (2020), for example, have argued that there is a need to consider a gendered analysis of global development trends, and this is borne out by the discussion below. What is also highlighted below is the need for an intersectional approach to disadvantage with class and place being rationales used for 'othering' (Castán Broto and Neves Alves 2018).

To advance debates around global development, we adopt a relational approach to understand urban poverty, social movements and citizenship in Manchester and Nairobi. Building on international development research that understands poverty in the Global South as a product of adverse incorporation into unequal social relations (Green and Hulme 2005; Hickey and du Toit 2007; Mosse 2010), we consider 'relational poverty politics' in the Global North and South (Elwood and Lawson 2018). Relational analysis challenges us to ask questions about the production and contestation of poverty that cuts across traditional North-South distinctions (Crane et al. 2020; Elwood and Lawson 2018).

Contexts and methodology

This chapter is authored by four professionals and academics who have all been engaged in grassroots organizational learning between North and South.

The organizing methodology that is being shared has emerged from Slum/Shack Dwellers International (SDI). SDI is a transnational movement of shack and slum dweller federations. Its membership is made up from women's led savings schemes, located for the most part in informal settlements. Launched in 1996, SDI has 32 affiliates across the Global South with a concentration in sub-Saharan Africa and South Asia.[1] SDI affiliates practice savings-based organizing, building strong locally grounded resident associations that understand local needs, define collective priorities, amplify women's voices, and address those priorities through engaging with local and national governments. The Muungano Alliance is SDI's affiliate in Kenya, and is a collaboration between Muungano waWanavijiji (the federation of slum dwellers), SDI Kenya (a technical assistance agency) and the Akiba Mashinani Trust (a financing facility).

The chapter has been written through a process of immersion in the lives, hopes and aspirations of grassroots leaders across the Global South and UK. We have drawn on multiple engagements with grassroots activists as they share their reflections on exchanges between Manchester and Nairobi. These engagements include formal meetings and workshops that have discussed learning, reflection sessions among activists, and less formal conversations with individuals and groups of residents.[2] This methodology is rooted in political ethnography (Auyero 2006). As scholars we believe in '… profound, engaged, and immersive ethnographic research that facilitates a deeper understanding of how communities work; how social relations are created, maintained, restored, and eroded; and how institutions evolve' (Pacheco-Vega and Parizeau 2018, 3). As activists we recognize the importance of a shared commitment to values of advancing social justice and reducing inequalities and action to achieve these values. We recognize problems of representation and attribution and believe that they are only partially overcome by efforts to improve accountabilities between academics and professionals and organized communities. We acknowledge that this text is made possible through the knowledge that activists have shared.

Nairobi is a city of 4.4 million in 2018, and home to just under 10 per cent of the Kenyan population. Economic growth in Kenya is currently estimated at 6 per cent, relatively strong for sub-Saharan Africa, but despite this growth there is considerable poverty. Over 90 per cent of households rent (Gulyani et al. 2018) with an estimated half of Nairobi's population living in 'slums' on 2 per cent of the land (Lines and Makau 2018). Very low wages mean that 69 per cent of residents are in household that live in one room (KNBS 2019b quoted in Mwau et al. 2020, 7). Contested land ownership and corrupt practices have reduce government's use of public land to address housing needs (Mwau et al. 2020). Decentralization in 2010 led to responsibilities being devolved to local authorities but only 6 per cent of county budgets are spent servicing urban areas (ibid.). In 2017, the president unexpectedly announced that affordable housing would be one of his four 'pillars' with a promised 500,000 dwellings but there is every indication that these units will be unaffordable to the majority of those in need.[3] Unequal access to land has been contested by civil society organizations (Klopp 2000; Lines and Makau 2018), with Nairobi County's declaration of a Special Planning Area (SPA) in Mukuru following years of campaigning and extensive housing innovations by the Muungano Alliance, as well as their engagement with utilities to (re)design options for improving access to basic services (Lines and Makau 2018).

Nineteenth century Manchester was one of the first global cities and the pioneer of industrial capitalism. Just as the challenges facing Nairobi's low-income residents reflect its colonial history, so Manchester's low-income communities have a history of displacement, declining job opportunities with deindustrialization and the rise of the welfare state with associated stigma and state management of poverty. In seeking to navigate a way through a conservative neoliberal agenda during the 1980s, the Labour dominated council embraced entrepreneurialism, and partnered with business elites (Cochrane, Peck and Tickell 1996; Ward 2003). Public housing estates were stigmatized as 'sink estates' to justify their redevelopment, leading to state-led gentrification in central neighbourhoods (de Noronha and Silver forthcoming). Employment is highly dependent on low-paid, insecure retail and hospitality jobs. In 2019, there were 98,898 Greater Manchester households on the housing waiting list (Bower 2020). Manchester city has the 17th highest child poverty rate in the UK, with 40.6 per cent of children falling below 60 per cent of the median income after housing costs (Hirsch and Stone 2020). In terms of multidimensional poverty, Rubery et al. (2017) find that Greater Manchester is below the national average for all dimensions of human development.

The development of Muungano's links with Manchester

The Muungano Alliance began to work with the University of Manchester in 2017 when it was asked to support student education. The Global Development Institute (GDI) at the university had been facilitating a course known as Citizen-led Development since 2011, with community members and SDI activists from different SDI affiliates being actively engaged in teaching. This pedagogical reform challenged traditional knowledge hierarchies. The academic engagement provided an opportunity for SDI community activists to meet and build a relationship with activists in Manchester. These visits built on exchanges between SDI and community activists in Manchester that had taken place from 2005. The groups visited included refugee and migrant associations, income-generation and micro-finance groups, credit unions and neighbourhood organizations. These exchanges had highlighted that the resonance between activists was strongest between women working in their own neighbourhoods to make lives better for their families. The consistency in their gendered role in social reproduction provides an important bond.

In June 2017, following an earlier visit of the South African SDI affiliate to Manchester, representatives of Mums Mart (from Wythenshawe) and community groups in Lower Broughton visited SDI groups in South Africa. This international exchange transformed Mums Mart's understanding of SDI approaches, inspiring them to adopt, and begin to adapt to their neighbourhood contexts, the broader set of practices that make up the SDI model of community organizing based on women-led community savings.

The first visit by Muungano to Manchester was in October 2017. Alongside teaching the Citizen-led Development class, the Muungano team met with community groups across Greater Manchester. Academics at the University of Manchester, and subsequently in Sheffield University (with long-standing associations with SDI) provided the vital link between the Kenyans class and the Manchester groups.[4] In June 2018, Mums Mart and the Lower Broughton communities visited Nairobi together with activists from two new areas, Brinnington in Stockport and Collyhurst. Exchanges intensified with further visits from Muungano to Manchester in October 2018 and then in 2019, and an exchange from Manchester to Nairobi in May 2019. In June 2019, the Greater Manchester Savers network invited Muungano and activists from Zimbabwe and South Africa to a two-day resident-led conference in Manchester. One of the key aims was to explore the potential of forming a city-wide grassroots network around issues affecting low-income neighbourhoods.

Muungano's perspectives on Manchester

With community groups spread across 400 informal settlements, all of which are seeking solutions to local deprivations, Muungano is in a constant state of change. The groups have a basic toolkit of practices that are common across SDI. As a matter of course, the groups form specific relationships with their respective local governments and a broad range of other development partners. At the same time, groups negotiate with other local associations. Through peer exchanges successful practices spread and are amended.

The initial meeting between Muungano and the community groups in Manchester took place in Lower Broughton, Salford. The Muungano activists walked through the social housing neighbourhood, with the Broughton members talking about the challenges they faced including the displacement of low-income residents, the patronizing attitude of professionals, hostile government policies, mental health challenges, and seniors living and dying alone. Though the problems sounded different to those that the Kenyan faced, Muungano felt a kinship stemming from these stories of marginalization. At the same time, the physical circumstances of Lower Broughton caused confusion. The homes that the team visited would qualify as middle-income homes in Kenya. One Muungano leader remarked, 'When Joanne (from Lower Broughton) speaks, I can hear Emily back at home (a community activist in Kenya). But Joanne's house is bigger than the house of the director of the NGO that supports us.'

A community savings group meeting with Mums' Mart at a gym in Wythenshawe deepened Muungano's consternation. The differences, from the mundane to the fundamental, appeared thoroughly stark. For instance, many group members brought along a dish to share and a heady buffet was laid out. For the Kenyan slum dweller, who does not have electricity, let alone an oven, cakes and pies come from the shop on birthdays and Christmas day. It was unimaginable that this shared meal happened every Friday. Then the local community answered the question, 'what will your savings be used for?' Mums Mart members explain this was so families could go to Blackpool on vacation during the December holidays. In the words of the Muungano leader in the visit, 'If the Mums Mart savers had not polished off the food, packed and carried home everything that remained, exactly as we do at home, I would have said that we were too different to be part of one movement.' However, what was clear for Muungano was the limited access that low-income people had to safe and secure accommodation and affordable housing; also recognizable were patterns of urban exclusion.

Interaction with academia has provided to Muungano critical moments of reflection and led to the articulation of its evolving practice. The Citizen-led Development class at Manchester was one of these moments of reflection. For Kenyan slum dwellers and activists, the opportunity to stand before and teach a university class, somewhere in the Global North, invoked great pride. This affirmation evidenced and reinforced the consciousness of Kenyan movement activities that their work was of value to development processes. For University of Manchester staff and students, the learning experience was equally treasured because it challenged the lack of expertise and knowledge that is frequently ascribed to slumdwellers.

In addition to solidarity, Muungano gained new knowledge. Slum upgrading is a central goal of Muungano. After decades of calling for recognition of informal settlements as human settlements and investment in housing and infrastructure, the government of Kenya has committed to provide 200,000 social houses before 2022. However, since the turn of the millennium, the government had constructed less than 1,000 low-income houses. For the activists visiting Manchester, the administration of council and housing association accommodation, and the outcomes that this had produced seemed to offer important learning.

Moreover, as the visit of the Manchester groups to Kenya took place, Muungano realized it was producing new relations and positive news stories. Expectations related to Kenya's colonial history could be disrupted by these white working-class women. Quotes from three participants demonstrate this. Anastasia (from Nakuru) explained that:

> *It never occurred to me that there are poor people in Manchester. I could not think that there are people who need food. Never occurred to me that there were people from the lower class. I thought they were rich… We prepared the group in Korochoco – there are people from the UK they are not rich – they are not donors. They are just people. …Then in the meeting – they began to understand – they were so excited.*[5]

Sharon (from Manchester) explained that Nairobi's political elite were anxious to benefit from the apparent reversal of roles: 'Local politicians [were] taking all the credit for the white people coming to see how well they run things… The Governor of Nairobi put us on Facebook.' And Kimani (a staff member at SDI Kenya) recognized that: ' [the government of Kenya] came to see great restoration that we have been doing… '[6] Informal settlement residents playing host to this group, accommodating them, caring for them and teaching them about community organizing had the potential to make senior policy makers in the Kenyan government look at Muungano differently.

Reflections and learning from Manchester

For community leaders in Greater Manchester, these exchange visits with Muungano represented refreshing new spaces for thinking and relationship building, leading to the overturning of prior expectations about each other's neighbourhoods and poverty and citizenship. For these women, juggling the pressures of living (and often parenting) on low incomes, while also volunteering within their communities, these trips were an oasis from the multiple demands of daily life. For five days, they were able to be valued as someone with important expertise to share and learning to participate in.

The exchanges created new platforms through which people from different areas of the city-region began to come together. This process, as well as the exchanges, brought the commonalities in experiences of poverty and exclusion into clear focus. Sharon reflected, after multiple engagements with Muungano at home and in Nairobi, how: 'we're all in the same boat. We've got the same issues as Africa's got. None of us have got the answers but we can all struggle through to try and get them.' This solidarity and increased awareness of commonality creates internal strength and determination. Tina, a community leader in Hulme reflected that 'I don't feel so alone.' Feeling alone in the struggle against poverty can be exhausting and leads to burn out, she explained, elaborating that 'I feel like I am at war.'

Returning to the story above of Muungano leaders' initial experiences of a housing estate in Salford, one of the women hosts that had previously participated in a visit to South Africa, reflected on their confusion about the good quality of housing. She observed that 'in Africa the poverty is visible, you can see it all around you, but here in the UK, our poverty is invisible.' She went on to explain that to understand someone's experience of poverty in an estate like hers, you would need to go in and talk to someone about their life, which is when you might learn about how they are struggling to pay their bills, how they are relying on a local food bank to feed their children, or the mental health impacts they are experiencing from living in poverty.

While lived realities differ, when exchange participants were invited to reflect on the way that Muungano organizes in their communities, leaders all recognized the importance of women's leadership. '[O]ne similarity is we're all women, we are all in it for our communities. And just to help each other,' and 'One similar thing I've seen is the women doing everything.' The women in Manchester were delighted to see women's importance being recognized in a way that perhaps wasn't always recognized in their own communities.

Another common theme in relation to discussions of 'community' and 'community spirit' was neighbourhood change. Reflections on the importance of organizing led to discussions about the effects of austerity policies and state welfare on people's openness and ability to organize. Tina in Hulme talked about how people in her community have been 'terrorized with austerity':

> One of the biggest things why people don't come is because of fear, and it's not because of fear of the police, it's this fear that whatever meagre benefits they're on, it'll be taken off them. Or because the discourse around people on benefits is that you're scammers, you're workless and all this lot is they don't share that information with each other. So, they're very private about stuff. And that was very different, when I was growing up, my mum was on benefits, everyone was on [benefits]. We knew. And you shared... .

In this context, hearing about how door-to-door savings collections had been used as a strategy for information sharing during authoritarian periods in the Kenyan movement's history resonated strongly and gave a new perspective on the citizenship potential of the savings approach for some of the women who until then had only thought of financial inclusion. The movements in South Africa and Kenya were perceived by women in Greater Manchester to have 'more community' and 'togetherness.' The issues that women from Wythenshawe and Lower Broughton were keen to learn about from their first international exchange were indicative of the challenges that they had experienced. They asked: 'how do they keep it all going?' 'what do they do about "people coming along to things as individuals" and creating "unity"'; and 'how do they stop their community activity being taken over by their local council or other agencies?' The question of 'how do we pull people together and develop a collective voice?' came up repeatedly. From the perspective of Greater Manchester this is indicative of the extent to which working class communities have become fragmented in Britain.[7]

Community leaders in Greater Manchester perceive they are not treated as equal partners by government and that their expertise and knowledge is unrecognized and/or under-valued. Many felt that they were poorly represented by their local councillors and that councillors turned up for things when it was of benefit to them politically but were not interested in working with the community on the issues that mattered to them.

> *I do think that councillors and the housing group really couldn't give a toss so long as they're making their money and they're voted in. I do think that anything they do, not just here, anything that they do is for them. And as quick as they help these groups set up, they whip it away when they're doing good.*

> *(activist)*

At the same time, when communities develop their own solutions, they often find that these are then claimed or taken over by other agencies in ways which undermine the community initiative or introduce so much formality it becomes too difficult for the community solution to work. One example given was an informal Mums and Tots group that had been running for many years with many positive ripple effects in the local community which closed due to too many regulatory requirements when the more formal UK children's centre service SureStart was introduced.

Looking at how the movements in South Africa and Kenya have built critical mass and developed effective strategies for influence gave women in Greater Manchester a renewed belief that change through collective action is possible. Muungano activists were able to share strategies for engagement with the state – moving between cooperation and confrontation – characterized by one male leader as: 'sometimes you have to offer the hand, and sometimes you must show the fist.'

Community leaders in Greater Manchester were also inspired by the partnerships that Muungano has been able to cultivate with local and international universities. They have built on this experience to change the character of their own relationship with the academy. As more GDI staff learned of the exchanges between communities in Nairobi and Manchester, they were curious to explore potential learnings for students. Leaders from Mums Mart came to lecture GDI students at the University and GDI students came out to Wythenshawe to meet with leaders from four savings groups across the newly established network. Community leaders were delighted. After many years of not having their voices heard by middle class professionals in Manchester, and/or having academics extract knowledge from communities, they were being recognized and rewarded for sharing their expertise with academics, students and international audiences of policy makers and practitioners.

Building on these experiences and learning, in 2020 leaders from across Manchester entered into a knowledge exchange partnership with housing providers and the local government neighbourhoods' team to understand how neighbourhood-level decision-making could be more inclusive. Out of these interactions, new relationships have emerged between community leaders and service providers which are leading to partnerships to explore options for supported accommodation for older people within inner city areas.

Conclusion

What emerges from these exchanges is that while some academics have discussed global development as an emerging new phenomenon identified by the convergence of trends such as national inequalities, grassroots activists recognize that the commonality of their experience is long-standing. While poverty in the Global North and South may appear very different because of access to material assets, and the relative wealth of countries, the experience of marginalization and disadvantage is very similar. Low-income residents face physical displacement as neighbourhoods are 'improved' by state interventions. Poverty programmes are designed without the involvement of low-income groups or through very superficial participatory processes. Low-income households and community groups must adjust to the rules set by others including charities such as housing associations and NGOs, and local government. And a democratic deficit means that politicians at all levels show little interest in the needs and interests of low-income groups.

The experiences of grassroots women demonstrate the continuing benefits of transnational learning and solidarity in a way that is relevant to current realities. While Evans (2014) rightly recognizes the challenges that growing informalization in labour markets creates for union-based organizing, we find that the gendered patterns of social reproduction combined with patterns of marginalization imposed by the state results in a link between women neighbourhood activists. Here we reflect on the consequences that this group of activists draw from this realization and how it may influence their work going forward.

In terms of immediate benefits, Muungano activists recognize the value in engaging local grassroots organizations in the Global North so as to be informed about potential policy and programming solutions. In Nairobi, for example, higher-density housing is needed, and the exchanges provided an opportunity to analyse approaches used in Manchester.

Also relevant is learning about approaches to poverty reduction and possible consequences. Increasingly policies designed in the Global North are being adopted by governments in the Global South. Social movements in the Global South need to understand better what might be coming their way. For example, the exchanges led women in Greater Manchester to reflect on social welfare provision and the social isolation people experienced in part because of the stigma associated with the benefit system. Muungano activists were struck by stories which were hard for them to comprehend. Kimani was surprised when he was told by women in Manchester that: 'I have lived here for 15 years – I don't know people. I go to the nursery school; we don't talk.'[8] As highlighted by the literature on policy mobilities, ideas and prescriptions are shared trans-nationally among professionals and government officials. In this context, grassroot understandings of the consequences of such policies are clearly of significance.

Both men and women take part in exchanges. However, international solidarity is rooted in women's experiences of discrimination within and beyond movements, just as savings-based organizing builds on gendered responsibilities and disadvantage within neighbourhoods. Women activists shared their frustration with not being recognized in male-dominated public spaces and processes at the same time as they celebrated their own and each other's contribution to neighbourhood development. The day-to-day work of women in addressing their extended family's needs, reaching out to friends and neighbours in trouble, anticipating – and mitigating – potentially negative impacts of elite actions, and working to provide protection and comfort was recognizable across international boundaries. Such realities, frequently invisible to men, were instantly acknowledged and celebrated.

This exploration reveals how transnational solidarity supports local work. When Muungano activists saw the scale of poverty and developmental problems in the North, they were surprised.

This was not a vision common in Kenya. The structuring of the world into North and South, together with the value-laden terms of 'developed' and 'developing' had led Kenyan community activists to assume that development had been achieved in the UK. Then, as the Kenyans became familiar with Greater Manchester activists, they saw the exchange offered them opportunities to challenge adverse relations. Muungano used their relations with Greater Manchester's activists to 'play' with racial stereotypes, revelling in their ability to disrupt expectations. Just as the slum dwellers had gained relational capital from teaching students at the University of Manchester so they gained further status from demonstrating that they were sharing their organizational capabilities with low-income communities living in Greater Manchester.

Observing what the Muungano Alliance had been able to achieve, the Greater Manchester groups gained the skills and confidence they needed to change relationships in their own context. For the Greater Manchester activists, stories of Kenyan communities fighting against invisibility, struggling for recognition from their government and against evictions gave them a sense of a global struggle and hope that such change might be possible at home. New organizing methodologies gave them the confidence to challenge the scepticism they faced from other residents who did not believe they could change things.

In terms of broader debates about global development, what is evident is that if the purpose of academic debates about globalization and its consequences is to recognize common challenges and advance social justice then the perspectives of grassroots communities need to be brought in. Our findings demonstrate how transnational learning between North and South aids efforts to address adverse processes and their local consequences. Exposing community activists to the realities in other regions of the world is empowering, enabling them to understand their own realities anew and engage the state more effectively. Our analysis also highlights the significance of intersectionality through the multiple forms of disadvantage faced by low-income communities.

Unfortunately, education has sometimes exacerbated rather than addressed social inequalities, as education-based differences in social status continue to disadvantage the working class. In our example, education offers a platform for empowerment. Recognition of the capabilities of grassroots activists through their invited contribution to teaching programmes is empowering for activists. And the stronger voices that emerge through local organizing enable them to challenge relational disadvantage in academic institutions and beyond. If global development is about recognizing commonalities between countries, then this chapter highlights that this recognition should not be left to data analysts. New insights are generated when local experiences co-produce knowledge between urban social movements.

Key sources

1. Elwood, S. and Lawson, V. (2018). (Un)Thinkable Poverty Politics. In Lawson, V. and Elwood, S. (eds.) *Relational Poverty Politics: Forms, Struggles, and Possibilities*, University of Georgia Press, pp. 1–24.

This volume (and the introduction) discusses the significance of social relations in creating and maintaining disadvantage. The authors show how uneven power – uncovered through a relational analysis – enables disadvantage to be exacerbated. And they suggest ways to undermine such processes.

2. Horner, R. (2020). Towards a new paradigm of global development? Beyond the limits of international development. *Progress in Human Geography*, 44(3), 415–436.

This article is important to read to understand the constraints that development studies scholars face because changes between the Global South and Global North, and changes within the Global South and North, challenge this binary categorization.

3. Lines, K. and Makau, J. (2018). Taking the long view: 20 years of Muungano wa Wanavijiji, the Kenyan federation of slum dwellers. *Environment and Urbanization*, 30(2): 407–424.

In Nairobi, the strategies of the political elite to maintain control led to very large numbers of tenants who secure shelter in a highly commodified urban economy. The experiences of Muungano across Kenya demonstrate the potential of an urban social movement to protect their members and identify and realize development alternatives despite this adverse context and considerable poverty.

4. Roy, A. (2015). Introduction: The Aporias of poverty. In A. Roy & E. S. Crane (Eds.) (2015). *Territories of poverty: Rethinking North and South*. Athens (Georgia): University of Georgia Press.

Roy begins this volume by reflecting on historical and current injustices within Chicago and an anti-eviction movement located in the city that drew on mobilization practices from Cape Town. Roy articulates the value in reflecting across urban territories in the Global North and Global South to understand the processes causing and perpetuating disadvantage and injustice.

Notes

1 www.sdinet.org
2 Between 2017 and 2019, some of these interactions were supported through a project called 'Seeing the Inner City from the South', within the Realising Just Cities programme at the University of Sheffield. https://gmsavers.org.uk/
3 www.standardmedia.co.ke/business/article/2001327402/why-state-affordable-housing-plan-is-built-on-shaky-ground. www.the-star.co.ke/news/big-read/2020–02–14-who-stands-to-benefit-from-uhurus-affordable-housing-project/ Accessed 9 August 2020.
4 By 2017 a professional group had emerged in Manchester to support this process.
5 Discussions on 1 November 2019, Chorlton (Manchester).
6 10 June 2019, inner-city exchange, Greater Manchester.
7 Although SDI federations seem well established in many African locations, when women first begin to meet in saving groups, they were often isolated and felt uncertain about the process. Existing community leaders and councillors dismissed their efforts.
8 1 November 2019, Chorlton (Manchester).

References

Alemany, C., Slatter, C. and Rodríguez-Enríquez, C. (2020). Gender blindness and the annulment of the development contract. *Development and Change*, vol. 50, no. 2, pp. 468–483.
Auyero, J. (2006). Introductory note to politics under the microscope: Special Issue on Political Ethnography I. *Qualitative Sociology*, vol. 29(Fall), pp. 257–259.
Batliwala, S. and Brown, D. L. (Eds.) (2006). *Transnational Civil Society: An Introduction*. Bloomfield: Kumarian Press, Inc.
Bower, C. (2020). Social housing is key to alleviating the housing crisis, so why are we not building it? *The Meteor*. August 13–2020: https://themeteor.org/2020/08/13/social-housing-is-key-to-alleviating-the-housing-crisis-so-why-are-we-not-building-it/ (accessed 20 October 2020).

Castán Broto, V. and Neves Alves, S. (2018). Intersectionality challenges for the co-production of urban services: notes for a theoretical and methodological agenda. *Environment and Urbanization*, vol. 30, no. 2, pp. 367–386.

Cochrane, A., Peck, J., and Tickell, A. (1996). Manchester plays games: exploring the local politics of globalisation. *Urban Studies*, vol. 33, no. 8, pp. 1319–1336.

Crane, A., Elwood, S., and Lawson, V. (2020). Re-politicising poverty: Relational re-conceptualisations of impoverishment. *Antipode*, vol. 52, no. 2, pp. 339–351.

de Noronha, N. and Silver, J. (forthcoming). The housing question in the district of Ancoats. In Burgum, S. and Higgins, K. (eds.) *How the Other Half Lives*. Manchester: Manchester University Press.

Elwood, S. and Lawson, V. (2018). (Un)Thinkable Poverty Politics. In Lawson, V. and Elwood, S. (eds.) *Relational Poverty Politics: Forms, Struggles, and Possibilities*, Athens (Georgia): University of Georgia Press, pp. 1–24.

Evans, P. (2014). National labour movements and transnational connections: Global labor's evolving architecture under neoliberalism1. *Global Labour Journal*, vol. 5, no. 3, 258–282.

Fischer, A. M. (2020). Bringing development back into development studies. *Development and Change*, vol. 50, no. 2, pp. 426–444.

Green, M. and Hulme, D. (2005). From correlates and characteristics to causes: Thinking about poverty from a chronic poverty perspective. *World Development*, vol. 33, no. 6, pp. 867–879.

Gulyani, S., Talukdar, D. and Bassett, E. M. (2018). A sharing economy? Unpacking demand and living conditions in the urban housing market in Kenya. *World Development*, vol. 109, pp. 57–72.

Herrle, P., Ley, A. and Fokdal, J. (Eds.) (2015). *From local action to global networks: Housing the urban poor*. Farnham: Ashgate Publishing Ltd.

Hickey, S. and du Toit, A. (2007) Adverse incorporation, social exclusion and chronic poverty. CPRC Working Paper 81.

Hirsch, D. and Stone, J. (2020) *Local indicators of child poverty after housing costs, 2018/19: Summary of estimates of child poverty after housing costs in local authorities and parliamentary constituencies, 2014/15 – 2018/19* http://www.endchildpoverty.org.uk/child-poverty-in-your-area-201415–201819/ (accessed 20 October).

Horner, R. (2020). Towards a new paradigm of global development? Beyond the limits of international development. *Progress in Human Geography*, vol. 44, no. 3, pp. 415–436.

Horner, R. and Hulme, D. (2019). From international to global development: new geographies of 21st century development. *Development and Change*, vol. 50, no. 2, pp. 347–378.

Klopp, J. M. (2000). Pilfering the public: The problem of land grabbing in contemporary Kenya. *Africa Today*, vol. 47, no. 1, pp. 7–26.

Lewis, D. (2017). Should we pay more attention to South-North learning? *Human Service Organizations: Management, Leadership & Governance*, vol. 41, no. 4, pp. 327–331.

Lines, K. and Makau, J. (2018). Taking the long view: 20 years of Muungano wa Wanavijiji, the Kenyan federation of slum dwellers. *Environment and Urbanization*, vol. 30, no. 2, pp. 407–424.

Mwau, B., Sverdlik, A. and Makau, J. (2020). *Urban transformation and the politics of shelter: Understanding Nairobi's housing markets*. London: International Institute for Environment and Development.

Mosse, D. (2010). A relational approach to durable poverty, inequality and power. *Journal of Development Studies*, vol. 46, no. 7, pp. 1156–1178.

Pacheco-Vega, R. and Parizeau, K. (2018). Doubly engaged ethnography: Opportunities and challenges when working with vulnerable communities. *International Journal of Qualitative Methods*, vol. 17, pp. 1–13.

Roy, A. (2015). Introduction: The Aporias of poverty. In A. Roy & Crane E.S. (Eds.) (2015). *Territories of poverty: Rethinking North and South*. Athens (Georgia): University of Georgia Press.

Rubery, J., Johnson, M., Lupton, R., and Roman, G. Z. (2017). Human Development Report for Greater Manchester Human Development Across the Life Course. Manchester: European Work and Employment Research Centre, Alliance Manchester Business School.

Ward, K. (2003). Entrepreneurial urbanism, state restructuring and civilizing 'New' East Manchester. *Area*, vol. 35, no. 2, pp. 116–127.

Wilkinson, A. (2020). Local response in health emergencies: Key considerations for addressing the COVID-19 pandemic in informal urban settlements. *Environment and Urbanization*, vol. 32, no. 2, pp. 503–522.

Ziai, A. (2018). *Development Discourse and Global History: From Colonialism to the Sustainable Development Goals*, London and New York: Routledge, viii+244 pp., UK £38.99 (hbk), ISBN 9781138735132.

Ziai, A. (2019). Towards a more critical theory of 'Development' in the 21st Century. *Development and Change*, vol. 50, no. 2, pp. 548–467.

43

HOUSING AND DEVELOPMENT

Poonam Pritika Devi and Naohiro Nakamura

Global housing crisis and informal housing

It is widely argued that neoliberal trends in housing policies – the privatization of housing production and promoting 'solutions' from the market (Gilbert 2004) – have substantially increased housing cost (as developers and property owners look for the highest returns from land), while the supply of 'unprofitable' low-cost housing has decreased (Santoro 2019). Neoliberal governments have also 'sought to protect the value of property above many other concerns' (Jacobs 2019, 18). This trend has hit hard low-income earners. Renters in private or public housing, squatters, and even homeowners have experienced evictions, displacements and banishment (Molina et al. 2019; Rolnik 2019). As such, the currently ongoing housing crisis does not necessarily indicate housing shortage. Rather, as Potts (2020) states, the current housing crisis is happening because of the mismatch between demand and supply. Only a handful of top-income earners can access the housing market, while the rest simply cannot afford even so-called 'decent' or 'low-cost' housing. To find a (often tiny) space to live, those who cannot afford formal housing have been looking for informal housing as their solution.

Informal housing is an 'accommodation provided beyond the "formal" regulations governing residential production (e.g. planning/zoning and building controls) and the housing market (such as property or tenure laws)' (Gurran et al. 2020, 2). Traditional perspectives on informal housing perceive informality either as a failure (Harris 2018), or a temporary characteristic that will eventually disappear. In reality, however, informal housing (or informal settlements) have long functioned as an alternative to formal housing within the Global South. This is due to both the large and swelling demographic size of urban populations and the failure of urban planning to reach disadvantaged residents and, as such, informal housing is a practical solution to many – including residents, planners, and even the government (Neuwirth 2007). Furthermore, not only is the existence of informal housing in the Global South near permanent, but the Global North is also observing an increase in informal housing.

With these trends, the perspectives of scholars, policy planners, and developers towards informal housing have shifted to better recognize informal housing/settlements as permanent features of cities, rather than a temporary aberration. Similarly, a substantial portion of housing policies now consists of informal settlement upgrading plans, particularly in the cities of the

DOI: 10.4324/9781003017653-48

505

Global South. In other words, the upgrading of informal settlements has become an accepted mode of urbanization (Mitlin and Walnycki 2020).

In this chapter, we look at informal housing, associated policies and the experiences of informal residents in three countries located in the southern hemisphere (not the Global South): Chile, Fiji, and Australia. Chile is considered a success case to promote homeownership and urge informal residents to become property owners. Fiji is implementing tenure security for informal residents. Australia is increasingly observing informal housing in a different feature from those in developing countries. Through comparative analysis, we seek to deepen understandings on housing issues and how they can be dealt with in different national contexts, rather than through singular, and monolithic, neoliberal approaches to the Global housing crisis (cf. Beswick et al. 2019).

Informal housing case studies

Case 1: Chile

In many contexts, homeownership is seen as a privilege of the wealthy. However, housing policies implemented by Latin American governments have, broadly speaking, aimed to distribute subsidies to low-income households that will enable them to access homeownership. This has resulted in Latin American countries becoming 'nations of homeowners,' with homeownership becoming the norm for both low and middle-class families (Contreras et al. 2019; Gilbert 2014). Chile, in particular, is considered a success case in terms of the number of citizens who have acquired homeownership.

In Chile, it was estimated that around 1.2 million people (near 20 per cent of the country's population) were living in informal settlements in 1990. Residents of these settlements lacked access to electricity, sewage, and drinking water (Salcedo 2010). Thanks to a state-led national program to invest in subsidized housing by the new democratic governments from 1990 onwards, between 1990 and 2010 more than a million Chileans who were formally residents of slums or shantytowns became homeowners (Salcedo 2010). By 1996, the number of informal residents decreased by about 500,000, totalling just 120,000 in 2006 (Salcedo 2010). By the end of 2006, La Toma de Peñalolén, the largest remaining slum in Santiago, was eliminated.

In addition to housing provision, investment into urban infrastructure by the government was substantial, and contributed to significant improvements in the living conditions of the urban poor. Although there were still around 60,000 homeless people in 2010, Salcedo (2010, 91) has described this transformation as Chile's success story: 'No other country in the world can show a housing policy with such a large number of built housing units relative to the country's population.' Similarly, Brain et al's descriptive study (2003, cited in Salcedo 2010) shows that approximately two-thirds of former informal residents were satisfied with the new housing units, infrastructure, and family relationships.

Of course, Chile's housing programs have also faced some critique. According to Contreras et al. (2019, 418), for example, migrant populations have been excluded from Chile's housing tenure model, with 21 per cent of the migrant population still renting (without formal contracts) due to the difficulties they face meeting the requirements for access to formal rent – such as a residence permit, the signature of a guarantor, a work contract, and a regular and sufficient income. Hence, many migrant populations continue living in informal settlements and self-built housing.

Another negative consequence of Chile's housing program has been an increase in crime, violence, drug trafficking, and noise (Brain et al. 2003) – all of which are higher, even in formal residential areas (Salcedo 2010). Restricted outsider access to privatized spaces and resulting social segregation has seen a decline in social support networks that have contributed to rising social issues including deviant and criminal activity (Celhay and Gil 2020). Indeed, some residents of low-income subsidized housing prefer to stay in an informal settlement because of its functionality – lower rates of vandalism, greater community support, and proximity to their workplace (Celhay and Gil 2020).

In summary, while the government of Chile has supported private property ownership and subsidized affordable housing as an upgrading of informal life, this 'has not been enough to overcome marginality and disintegration' (Beswick et al. 2019). Chile's case also shows that informal housing has not lost its functionality, even after housing policies have strongly (and successfully, to some extent) promoted homeownership. Informal settlements, that were expected to disappear in this transition, are still serving the needs of those who cannot, or who do not want to live in formal subsidized housing (Celhay and Gil 2020). As such, in the context of Chile, 'acknowledging self-built housing as another form of urban space production' for migrants and other disadvantaged populations provides additional benefits for an inclusive housing policy, alongside the state efforts to increase homeownership (Contreras et al. 2019, 284).

Case 2: Fiji

A major restriction on urban development and housing construction in Pacific Island countries (PICs) is the limited access to a large portion of customary land. For instance, in Fiji, indigenous Fijians communally own approximately 85 per cent of the country's land (Kiddle 2010), and the constitution protects their land rights. Transferring the ownership of customary land is restricted, which hinders formal urban expansion and development. Indigenous Fijians have also rejected neoliberal approaches to urban development that prioritize housing development by the private sector for economic profit, as such approaches conflict with indigenous values of communal share (Ben and Gounder 2019).

Although the population sizes of PICs are small, due to limited land size their major urban areas have faced formal housing shortages as a result of natural population increase and rural-urban migration. To accommodate Fiji's increasing urban population, the government has constructed 'affordable housing' (such as the Public Rental Board) and private sectors have developed housing on freehold land. However, the pace of population increase has been much faster than the infrastructural establishment, and this has resulted in a rapid growth of informal settlements in the city (Government of Fiji 2010). For example, the Greater Suva Urban Area (comprising the capital city of Suva and the municipalities of Lami, Nasinu and Nausori) has the greatest proportion of informal settlements in Fiji, housing 71 per cent of nation's informal residents (54,200) across 117 settlements (Kiddle and Hay 2017). These informal settlement numbers are increasing every year (120 in 2020). Similarly, in Nasinu, the municipality with the second largest population in the country, 60 per cent of households are believed to be squatting (Jones 2012).

Informal settlements in Fiji have been a major issue for national housing policy due to restrictions on land access and formal urban development. In fact, with a few exceptions informal settlement residents have hardly been the target of eviction by the authority. Instead, the government has planned to improve living conditions of informal settlements. In 2004, for example, the Urban Policy Action Plan for Fiji was formulated, which led to the implementation of an informal settlement scoping study in 2007 (Asian Development Bank 2012). In 2011, the government of Fiji introduced the 'Republic of Fiji National Housing Policy' to facilitate provisions of affordable and decent housing for all residents and implemented informal settlement upgrading plans – including infrastructure upgrading, formalization and relocation.

In Fiji, formalization is conducted via *in-situ* upgrading, which functions by two main approaches: tenure legalization to provide legal security through the issue of lease titles; and enhancing perceived and *de-facto* tenure security through provisions of infrastructural services (water supply, electricity and sewage) (Mohanty 2006). *In-situ* upgrading allows existing squatters in the selected settlements to remain in their current dwellings while upgrading program unfolds in the settlement. By 2015, *in-situ* upgrading had delivered 453 serviced housing lots to squatters while another 920 lots were in process (Kiddle and Hay 2017). Tenure formalization under *in-situ* upgrading is in the form of 'occupancy/development lease' – a provisional lease document providing households with land rights for 99 years. Under this arrangement, resident's occupancy status is formalized and the rights to enjoy infrastructure serviced by the local government is granted; yet, the ownership of land is not confirmed.

Granting formal tenure to residents is a cost-effective approach, however, residents do not experience substantial improvement in the physical condition of their housing under tenure legalization. In many settlements, most houses are constructed with tin boards and are prone to cyclones. Access to water, electricity and other services is limited. Many settlements are located on vulnerable lands – mangrove swamps, hillside, or valley – and such physical conditions do not change after tenure legalization. Nevertheless, our research shows that this approach has been positively perceived by informal residents.[1] The most positive outcome is that granting tenure has eradicated fears of eviction. Also, with legal tenure rights, the residents have become able to access credit from financial institutions to improve housing and fence their occupied area. Some households of Nauluvatu settlement (Suva) and Field 40 settlement (western city Lautoka), have informally occupied land over 50 years (3 generations) and retain de-facto tenure security. Still, tenure formalization was perceived as substantial progress. For instance, in Nauluvatu settlement, the residents asserted that the delivery of community lease would increase the sense of community-ship and aid in establishing firm protocols to represent the settlement as a village and necessitating all residents to adhere to it.

With restrictions to land access for formal housing development and the country's limited budget, Fiji has adopted a practical housing policy: let people construct housing by themselves without authorization, recognize informal settlements for their near-permanent existence, and grant them security. A problem with this approach is that the government cannot regulate 'bamboo-shoots-after-a-rain' like informal housing construction, which often causes conflicts with landowners or formal housing residents (Figure 43.1). For instance, landowners have to go through a long legal process to evict informal residents for development purposes. Our research also finds that formal residents often complain about informal residents leaving rubbish for collection even though they do not pay the service fee. In addition, in informal settlements where formalization is in progress, residents have become exclusive against newcomers. That is, existing residents claim that newcomers do not always follow protocols such as paying land rent to claim occupancy or that housing by newcomers occurs without proper procedures, which may negatively influence the formalization process. Newcomers are often merely perceived as free-riders of the benefits associated with formalization and the tenure security process has caused a disparity between newcomers and the existing residents, who appear to have forgotten that they, or their ancestors, were newcomers at some point and constructed housing without proper procedures.

Figure 43.1 A new house under construction in Nauluvatu settlement (Photo: P. P. Devi)

Case 3: Australia

Recent literature on housing has increasingly looked at informal housing in the Global North. Some pertinent case studies include North America (Durst and Wegmann 2017; Kinsella 2017), Europe (Esposito and Chiodelli 2020; Lombard 2019), and Australia (Gurran et al. 2020; Iveson et al. 2019; Nasreen and Ruming 2021). Squatting has existed in the Global North since at least the nineteenth century, often to claim a right to the city (Iveson et al. 2019). However, the construction of informal housing has been rare and informal settlements with a near-permanent style usually do not form (Lombard 2019; Nasreen and Ruming 2021). Instead, informal housing takes the form of secondary units (informal subdivisions), room rentals, unpermitted dwellings, 'beds in sheds,' or boarding house accommodation, 'with some lower level of regulatory compliance (by owners, landlords or builders) or protection (for residents), which in turn reduces the costs of construction and or rent relative to the formal market' (Gurran et al. 2020, 2). Because of the concealed nature of its existence, and its scattered locations, investigating informal housing is, arguably, more difficult and complicated in the Global North (Gurran et al. 2020; Nasreen and Ruming 2021).

Residents of informal housing in cities of the Global North include both those who cannot access formal housing because of their political status (e.g., illegal immigrants) and low-income earners who cannot afford urban housing – students, early-career workers, and recent immigrants (Lombard 2019). In respect to the latter group, the city with the world's second most expensive housing market, Sydney, offers a pertinent case study. Although Australia has long been a nation of homeownership, it began declining in the 2010s, because of its increasing cost and planning system constraints (Gurran et al. 2020). To improve housing affordability, the State of New South Wales permitted secondary dwellings and boarding houses under the Environmental Planning Policy (Affordable Rental Housing) 2009 (ARHSEPP). This deregulation was a strategy to stimulate a new rental market at a lower cost (Gurran et al. 2020). As a result, Sydney's housing supply more than doubled over the decade 2007/08–2015/16, with the increase of secondary dwellings per new dwellings from 4 per cent to 13 per cent. However, the rent, determined by the market, remained high and 'the supply of rental units affordable to low-income earners in Sydney actually declined' (Gurran et al. 2020, 10). As such, Sydney appears to be a textbook case of what Potts (2020) describes as the Global housing crisis: low-income earners just do not earn enough to secure a decent legal home.

Where then, do low-income earners who are experiencing difficulties accessing formal rental markets, go? Gurran et al. (2019) examined low-cost rental advertisement and found that in 2018 alone, there were 960 notifications of dwelling units that do not comply with regulations in just one local municipality of Sydney. Such units lack natural light or ventilation, for instance, or the buildings are old and may not meet the current standards. Some dwellings are 'created within outbuildings, garages, or garden sheds, and [are] lacking appropriate insulation, electricity, stormwater, or access provisions' (Gurran et al. 2020, 11). Often a boarding house owner multiplies a room into five or six. Despite such conditions, some of them are still expensive.

Relatedly, Nasreen and Ruming's (2021) interviews with residents of shared room housing in Sydney, the majority of whom are international students or low-income earners, found that tenants are unaware of their rights and obligations; hence, they are vulnerable to abusive landlords and often have to endure poor and unhygienic housing conditions or issues like overcrowding. A written agreement between the landlord and tenants often does not exist. Local authorities 'often lack resources to collect data on overcrowded and unauthorised properties, so enforcement [of regulations] is limited' (Nasreen and Ruming 2021, 19). A dilemma is that the strict enforcement may end up in shutting down such illegal dwellings and render vulnerable populations homeless, without alternative affordable housing (Nasreen and Ruming 2021). Gurran et al. (2020, 1) conclude that the market-led deregulation initiative failed to accommodate low-income renters and call for 'more systemic reforms beyond the planning system.'

Informal experiences

Cociña (2018, 1) states that 'housing policies play a central role as drivers of urban development, affecting many of the challenges that define the current urban condition.' Housing policies have the potential to reduce economic, social, and political inequalities. This is because housing does not solely denote the material construction or structure of walls and a roof, but housing policies have various non-housing effects, such as 'labour markets, education, health, community viability, violence and income distribution' (Cociña 2018, 1). The three selected countries looked at here are somewhat random and never representative. Nevertheless, we attempted to look at relatively under-represented countries in the literature of housing, given that North America and Europe have dominated discussions (see for example, Clapham et al. 2012). Over the past two to three decades, both Chile and Fiji have implemented housing policies to support disadvantaged groups. Although a few downsides have been observed, citizens appear to have positively perceived the attempts to maximize beneficiaries from the formalization of housing policies. However, the formalization process never means that informality has lost its function or informality is negatively perceived in these countries. Informal housing does serve those who are left out from the policies or those who do not want to live in formality. In Sydney, too, informal housing should play an important role for disadvantaged groups; however, informality is much associated with negative experiences and the policy reform has not been functioning well. Led by the market, the new deregulated housing policy appears to have increased economic inequality. Nevertheless, all the three case studies affirm the significance of government's initiatives in housing policies, despite the neoliberal trend. According to Jacobs (2019), so-called neoliberal governments are still intervening into housing markets and influencing social discourses on housing. Regarding feasible context-based housing policies, there appears much that the Global North can comprehend and adopt from practices in the Global South, in terms of dealing with informality well.

Learning informality in the city

Informal settlements are a substantial component of urban geography in the Global South that cross multiple sectors, including urban structure, urban planning, landscape, housing policies, economic activities, politics, and social inequality. Since 2014, we have delivered a university second-year urban geography course titled 'Urban Well-Being' at the University of the South Pacific (USP), Fiji. Naturally, course activities and student learning outcomes reflect discussions on informal settlements. The course involves a field lab component – students are required to choose any topic related to informal settlements, make a research plan, visit a few informal settlements in the Suva area (chaperoned by the teaching team), and conduct observations and interviews with residents. Although fieldwork is used intensively in geographical research, field labs hardly take place in university undergraduate teaching, because of their complexity in procedures and some associated risks. However, at USP, all geography courses involve a field lab component (see Muliaina, Chapter 52 in *this volume*).

In our urban geography course, informal settlements have been functioning as a teaching material, and the experiences of informal residents have been contributing to student learning experiences. In the field lab, students can learn what research articles on informal settlements argue through their own first-hand experiences. For instance, expenses for formal housing substantially influence one's decision to continue squatting, even if an opportunity to move elsewhere is provided (e.g. Celhay and Gil 2020; Cociña 2018; Erman 2019). Informal residents tend to prefer an informal lifestyle of proximity and close interaction with others to a life in demarcated modern, concrete built housing, even when infrastructure is less established (Celhay and Gil 2020; Erman 2019). Another significant insight demonstrated through field lab activities is that not all residents are economically disadvantaged. Informal residents are fighting against various stigmas attached to informal settlements, and students can see first-hand that informality is a driving force of urbanization. As Neuwirth (2007, 72) states: 'the squatters build and rebuild and build again, in a ceaseless drive to make their settlements permanent' (see Figure 43.1).

For this pedagogy, we have been able to appropriate our advantage – belonging to an institution in the Global South. Imagine if you plan an undergraduate student visit to informal housing in Sydney, Australia, for example. As seen, informal housing in cities of the Global North are sporadically located and do not have unique settlement features. Even if you were able to spot informal housing, the way to successfully let students observe and interview the vulnerable residents would have to go through a very complicated process, including obtaining approval from so many individuals and organizations. The context of informal housing in Australia raises difficult ethical questions that are less pronounced in Fiji.

However, in our case, the process is much simpler. Informal settlements are literally informal and, consequently, no individuals or institutions are in a position to grant formal approval.[2] On the day of the visit, we simply take students to a settlement and let them work there. Most informal residents are open and pleased to participate in student interviews. The openness of residents derives from several reasons. First, those who are unemployed are often keen to participate in the learning process. Second, some residents are eager to share their experiences of

Figure 43.2 A group of local and exchange students interviewing an informal resident (Photo: N. Nakamura)

informal life with students – and by contrast they are rather defensive against visits by government officials, who may distribute an eviction notice. Third, residents are also excited to see students from another Pacific Island country – as a regional university, USP has students from 12 different Pacific nations. Interaction with exchange students from the Global North is a particular excitement for informal residents (Figure 43.2). Some residents even serve a meal. Such informal interactions gradually deconstruct students' stereotypes against informal settlements being dangerous and associated with crimes.

Surprisingly and interestingly, for most students, learning about informal settlements is a totally new experience. Despite the large number of informal settlements in the Greater Suva Urban Area, we have not had any student in the course who openly stated that they are living in an informal settlement, while only a few have stated that their relatives or acquaintances live there. The chances are that there may have been a few such students but they did not openly say so for some reasons, perhaps because of stigma attached to informal settlements. In any case, most students have not visited an informal settlement until they take this course (and indeed, Poonam was one of such students). According to their feedback, some students were not even aware of the existence of informal settlements in close proximity to their homes.

Unfortunately, as a mere undergraduate student project, informal residents hardly benefit from our visit. While at the end of the semester, a few students state that visiting informal settlements in field lab was the best highlight in the course, many of them, once having completed the project, forget about the visit. However, involving 70 to 100 students from at least 12 different nations every year will arguably have an impact. Our hope is that some of them talk about their experience to friends or family members, which may contribute to deconstructing negative images attached to informal settlements, or that a few of them consider working with an NGO which supports informal residents. At least, with this pedagogy, students form place-based knowledge, which is much more practical over theory-based learning in the context of the Global South (see Brugman and Kol, Chapter 58 in *this volume*).

Conclusion: seeing housing policies from the South

Recently, scholars have attempted bridging the gap in the studies of urban housing between the North and the South through the concept informality (Beswick et al. 2019; Esposito and Chiodelli 2020; Harris 2018; Lombard 2019). The debate on urban informality largely focuses on growth in the South, while the focus of informality in the North is on a much smaller scale (Chiodelli 2019; Harris 2018). Nonetheless, as we have sought to demonstrate through our three country cases, informality and informal housing is no longer unique to the Global South. Recent research has shown that countries in the North do experience housing issues despite advanced technologies in housing construction (Banks et al. 2020; Chiodelli 2019). What is missing from the context of the Global North is the accumulation of empirical studies on informal housing (Nasreen and Ruming 2021). If you are teaching at a university in the Global North, your students may be renting a non-complying dwelling unit. Involving those students (or seeking help from student's network) to collect the voices of informal residents, instead of planning a student visit to informal housing in class, may help grasp a picture of informal housing and identify needs. This may eventually contribute to housing policy reforms for inclusiveness.

Key sources

1. Beswick, J., Imilan, W. and Olivera, P. (2019) 'Access to housing in the neoliberal era: a new comparativist analysis of the neoliberalisation of access to housing in Santiago and London,' *International Journal of Housing Policy,* 19(3), 288–310.

This article provides a rich comparison of two significantly different housing systems, Santiago, Chile, and London, UK. Instead of applying single and monolithic Western-dominated theories of neoliberalism to housing policy, this research adopts a context-based approach, i.e. how neoliberalism affects both cases differently. The article also attempts a cosmopolitan comparison between the Global North and South.

2. Harris, R. (2018) 'Modes of informal urban development: A global phenomenon.' *Journal of Planning Literature*, 33(3), 267–286.

This review article clarifies the urban character of informality, articulates its global relevance, and proposes a framework for understanding its major modes. Together with examples of urban informal development from cities in the Global North, the article demonstrates the applicability of concept informality across the world.

3. Kiddle, L., and Hay, I. (2017) 'Informal settlement upgrading: lessons from Suva and Honiara,' *Development Bulletin*, 78, 25–29.

This article explains the process of informal settlement upgrading in Fiji and the Solomon Islands and reviews the housing policies of each country. Fiji's strategies such as early government recognition, the right to non-eviction of informal residents, and policy frameworks are praised, compared to the lack of a comprehensive plan in the Solomon Islands.

4. Nasreen, Z. and Ruming, K. J. (2021) 'Informality, the marginalised and regulatory inadequacies: a case study of tenants' experiences of shared room housing in Sydney, Australia,' *International Journal of Housing Policy*, 21(2), 220–246.

This article is on the experiences of shared housing (as an example of informal housing) in Sydney, Australia. Readers can see the similar functions and characteristics of informal housing in the Global North to those of the Global South, i.e. affordability, flexibility, insecurity, exploitation, and risks.

    5.   Potts, D. (2020) *Reading Broken Cities: Inside the Global Housing Crisis.* London: Zed Books.

The author discusses how the current global housing crisis is occurring under neoliberalism. It is shown that the cause of crisis is not because of the shortage of the supply of housing, but the mismatch between supply and demand. Only the handful have access to the so-called 'decent affordable' housing, while many cannot even afford those housing.

Notes

1 This is based on Poonam's doctoral project: A critical examination of squatter upgrading plans in Fiji.
2 We note that the procedure is different in indigenous Fijian villages, where visitors must seek approval of village headman through *sevusevu* ceremony.

References

Asian Development Bank. (2012). *The State of Pacific Towns and Cities: Urbanization in ADB's Pacific Developing Member Countries*, Manila: Pacific Studies Series.

Banks, N., Lombard, M. and Mitlin, D. (2020). Urban informality as a site of critical analysis. *The Journal of Development Studies*, 56(2), 223–238.

Ben, C. and Gounder, N. (2019). Property rights: Principles of customary land and urban development in Fiji. *Land Use Policy*, 87, 104089. https://doi.org/10.1016/j.landusepol.2019.104089.

Beswick, J., Imilan, W. and Olivera, P. (2019). Access to housing in the neoliberal era: A new comparativist analysis of the neoliberalisation of access to housing in Santiago and London. *International Journal of Housing Policy*, 19(3), 288–310.

Brain, I., Concha, M. J. and del Campo, P. (2003). Estudio descriptivo de la situación post erradicación de las familias de campamentos de la RM. *CIS–Un techo para Chile*, 2(2), 14–20.

Celhay, P. and Gil, D. (2020). The function and credibility of urban slums: Evidence on informal settlements and affordable housing in Chile. *Cities*, 99. https://doi.org/10.1016/j.cities.2020.102605

Chiodelli, F. (2019). The dark side of urban informality in the global north: housing illegality and organized crime in northern Italy. *International Journal of Urban and Regional Research*, 43, 497–516.

Clapham, D.F., Clark, W.A.V. and Gibb, K. (2012). *The SAGE Handbook of Housing Studies*. London: Sage.

Cociña, C. (2018). Housing as urbanism: The role of housing policies in reducing inequalities. lessons from Puente Alto, Chile, *Housing Studies*, 1–23. https://doi.org/10.1080/02673037.2018.1543797

Contreras, Y., Neville, L. and González, R. (2019). In-formality in access to housing for Latin American migrants: A case study of an intermediate Chilean city. *International Journal of Housing Policy*, 19(3), 411–435.

Durst, N. and Wegmann, J. (2017). Informal housing in the United States. *International Journal of Urban and Regional Research*, 41, 282–297.

Erman, T. (2019). From informal housing to apartment housing: Exploring the 'new social' in a gecekondu rehousing project, Turkey. *Housing Studies*, 34(3), 519–537.

Esposito, E. and Chiodelli, F. (2020). Juggling the formal and the informal: The regulatory environment of the illegal access to public housing in Naples. *Geoforum* 113, 50–59.

Gilbert, A. (2004). Helping the poor through housing subsidies: Lessons from Chile, Colombia and South Africa. *Habitat International*, 28(1), 13–40.

Gilbert, A. (2014). Free housing for the poor: An effective way to address poverty? *Habitat International*, 41, 253–261.

Government of Fiji. (2010). *Fiji National Assessment Report*. Suva: Government of Fiji.

Gurran, N., Pill, M. and Maalsen, S. (2019). *Informal accommodation and vulnerable households in metropolitan Sydney: Scale, drivers and policy responses*. Sydney: Policy Lab.

Gurran, N., Maalsen, S. and Shrestha, P. (2020). Is 'informal' housing an affordability solution for expensive cities? Evidence from Sydney, Australia. *International Journal of Housing Policy*, 1–23. https://doi.org/10.1080/19491247.2020.1805147

Harris, R. (2018) Modes of informal urban development: A global phenomenon. *Journal of Planning Literature*, 33(3), 267–286.

Iveson, K., Lyons, C., Clark, S. and Weir, S. (2019). The informal Australian city. *Australian Geographer*, 50(1), 11–27.

Jacobs, K. (2019). *Neoliberal Housing Policy: an international perspective*. London: Routledge.

Jones, P. (2012). Searching for a little bit of utopia – understanding the growth of squatter and informal settlements in Pacific towns and cities. *Australian Planner*, 49(4), 327–338.

Kiddle, G. (2010). Perceived security of tenure and housing consolation in informal settlements: Case studies from urban Fiji. *Pacific Economic Bulletin*, 25(3), 193–214.

Kiddle, L. and Hay, I. (2017). Informal settlement upgrading: Lessons from Suva and Honiara. *Development Bulletin*, 78, 25–29.

Kinsella, K. (2017). Enumerating informal housing: A field method for identifying secondary units. *The Canadian Geographer*, 61(4), 510–524.

Lombard, M. (2019). Informality as structure or agency? Exploring shed housing in the UK as informal practice. *International Journal of Urban and Regional Research*, 43(3), 569–575.

Mitlin, D. and Walnycki, A. (2020). Informality as experimentation: Water utilities strategies for cost recovery and their consequences for universal access. *The Journal of Development Studies*, 56(2), 259–277.

Mohanty, M. (2006). Urban squatters, the informal sector and livelihood strategies of the poor in the Fiji Islands. *Development Bulletin*, 70, 65–68.

Molina, I., Czischke, D. and Rolnik, R. (2019). Housing policy issues in contemporary South America: an introduction. *International Journal of Housing Policy*, 19(3), 277–287.

Nasreen, Z. and Ruming, K.J. (2021). Informality, the marginalised and regulatory inadequacies: A case study of tenants' experiences of shared room housing in Sydney, Australia. *International Journal of Housing Policy*, 21(2), 220–246.

Neuwirth, R. (2007). Squatters and the cities of tomorrow. *City*, 11(1), 71–80.

Potts, D. (2020). *Reading Broken Cities: Inside the Global Housing Crisis*. London: Zed Books.

Rolnik, R. (2019). *Urban warfare: Housing under the empire of finance*. London: Verso.

Salcedo, R. (2010). The last slum: Moving from illegal settlements to subsidized home ownership in Chile. *Urban Affairs Review*, 46(1), 90–118.

Santoro, P.F. (2019). Inclusionary housing policies in Latin America: São Paulo, Brazil in dialogue with Bogotá, Colombia. *International Journal of Housing Policy*, 19(3), 385–410.

44

GLOBAL VALUE CHAINS AND DEVELOPMENT

Aarti Krishnan

Introduction: what are global value chains

Accounting for almost 50 per cent of global trade and generating over 70 per cent of employment opportunities around the world, global value chains (GVCs) are a critical player in labour
market outcomes globally (Kowalski et al. 2015). GVCs are especially important for national
income and employment creation in low- and lower-middle income countries (World Bank,
2019). GVCs emerge due to the fragmentation of trading structures and explicate how goods
and services are produced and flow from stages of production to consumption. Research
into GVCs primarily focuses on the importance of global lead firms (such as multi-national
corporations), and how they exert power over other actors (suppliers such as southern firms,
sub-contractors, and labour) within the chain (Gereffi 1999). Since the mid-2000s, the importance of the role of labour has increasingly become integral to GVC analysis. While significant research has seen a sharp increase in the quantity of employment in GVCs (UNCTAD
2017), this has been accompanied with a rise in labour force flexibility, precarity and worsening
work conditions. This is particularly the case in low-income countries where engagement
with GVCs occurs primarily in low-value-added or low skill sectors (Barrientos et al. 2011).
A critical question arises as to how GVCs should be governed to ensure better labour working
conditions, rights, and entitlements.

This chapter unpacks this question in relation to low-income countries, focusing on three
key sectors of agriculture, apparel and tourism. This is achieved by looking at two key tenants of
GVCs, namely governance and social upgrading. Governance explores how lead firms (multinationals) exert power over other suppliers in the chain. Social upgrading is understood as the
outcomes of governance as experienced by workers. These outcomes may be measurable (e.g.
wages, working conditions, social security) and/or immeasurable (e.g. labour rights, entitlements) (Barrientos et al. 2011, Barrientos 2019).

Conducting a meta-analysis of 35 studies,[1] this chapter demonstrates emerging patterns of
labour governance strategies in low/low-middle income countries and links these to social
upgrading. Four core governance strategies (plans of action) are identified to be frequently
employed by GVC actors (e.g. lead firms, governments, civil society) to govern labour, each
which creates varied social upgrading outcomes. These include: compliance, mentor, associative and bottom-up strategies. Some strategies facilitate improvement in worker conditions in

DOI: 10.4324/9781003017653-49

measurable *and* immeasurable aspects, while others lead to an improvement in one and a fall in the other.

The chapter is structured as follows. The following section discusses how governing labour is a complex multi-layered process across GVCs and describes the different forms of governance that exist in relation to labour in GVCs. We combine these different forms of governance to develop four key labour governing strategies – compliance, mentor, associative, bottom-up – before linking these to social upgrading. The third section presents the results of a meta-analysis of 35 case studies. It disusses the predominant labour governing strategies, key actors involved, and the implications of these on workers' social upgrading possibilities. The final section offers a pedagogy of labour and GVCs and a conclusion to the chapter.

How is labour governed in GVCs?

Governance of a GVC is complex, as GVCs include a range of actors from lead firms to national and subnational governments, workers (labour) and civil society. 'Lead' firms are becoming increasingly more powerful actors in GVCs, controlling how transactions within these chains are governed (Gereffi 1994; 1999; Gereffi et al. 2005). The power lead firms have over other actors is predominantly referred to as 'private governance' in GVCs. If all tasks from production to consumption are coordinated by a lead firm, private governance in a GVC is said to be 'vertical' or hierarchical, and if there are multiple intermediaries throughout the GVC, governance is said to be 'arms-length'(Gereffi et al. 2005).[2]

How does private governance work? To participate in GVCs, supplier firms and their labour, usually based in the Global South must comply with various demands set by lead firms to deliver high-quality products with low production costs (Barrientos 2008). These demands can vary from complying to standards or certifications, to using certain types of machinery. For example, agricultural GVCs are buyer-driven and governed primarily by large supermarkets from the Global North (Evers et al. 2014). Low-income countries tend to have a comparative advantage when it comes to the export of agro-related products. While trade preferences (e.g. Generalised Scheme of Preferences) and several bilateral agreements like the China-Kenya avocado deal exist, standards are widely considered the key governing instruments for GVCs. For instance, to participate in European markets, low-income countries need to adhere to international food safety standards (e.g. sanitary and phyto-sanitary measures) and private voluntary standards (e.g. ISO 22000, which covers business-to-business trade, or GlobalGAP and Fairtrade which cover business-to-consumer trade). Lead firms often push suppliers (who can be export companies or farmers) to comply with standards. In many cases this leads to marginalization from participation in the chain, as these standards have a high level of complexity and require significant paperwork to comply with traceability requirements (knowing the origin and the journey of the product from cradle to grave).

The apparel sector is also lead firm dominated (especially by MNCs headquartered in the US, Europe, and China). In general, the apparel value chain is divided into (1) raw natural and man-made fibres supply; (2) yarn and fabric production; (3) product development and design; (4) apparel assembly; and (5) post-production activities, including branding, distribution, and retail. The work outsourced to low-income countries (LICs) is largely labour-intensive activities that require low technological and capital inputs. Some common examples here include 'cut-make-trim' (CMT) for fast fashion and sports clothing (in which suppliers' activities are limited to sewing the apparel and cutting the fabric under a processing fee) and 'free-on-board' (FOB) (in which suppliers are also responsible for sourcing or producing fabrics, as well as finishing and packaging). This usually involves significant number of regular and irregular workers, who need to meet the fast-changing demands of lead firms in terms of product type, quantity, and processes of manufacturing. The bulk of value-added rests on downstream branding and marketing activities which are done by lead firms (e.g. H&M, Zara or Primark) (Fernandez-Stark et al. 2011).

Tourism GVCs also tend to be dominated by lead firms who are large international tour operators. They form alliances with smaller operators in destination countries, often using exploitative relationships that offer low commissions (Christian et al. 2011).

Across all sectors, lead firms not only demand compliance with standards, but also push for the lowest production costs. They do so without providing long-term contracts to Southern firms (and their labour), which has led to significant labour flexibilization and casualization.

Wages and salaries are viewed as flexible production costs that supplier firms can reduce. This drive towards low production costs has led to a move away from regular to irregular workers. The key difference here between these two groups is in the skills that they hold (and in the transferability of those skills). Workers with higher and more specialized skills are more valuable to a supplier firm and less easy to replace, so are more likely to work under a regular contract and benefit from some level of security and access to basic rights and entitlements (Barrientos 2014, Rossi 2013). In contrast, where tasks are routine and repetitive, and workers easily replaceable, having labour work under irregular contracts minimizes the costs of supply firm.

Irregular workers are hired precariously without contracts, often through third party labour brokers. They lack access to job security, work outside working rules and regulations and are given limited access to the civil society support and representation that could help them to fight for greater labour rights and standards (Standing 1999). There is a close link here with seasonal work; most labour is hired during peak seasons, be it for agriculture, apparel manufacturing or tourism. This disregard for labour conditions reduces a supplier firm's wage and non-wage (e.g. social insurance or housing costs) costs (Kaplinsky 2005), making them more profitable at the expense of labour.

It is important to note that while labour in agriculture, apparel, and tourism GVCs are dominated by lead firms, there are a number of other actors that can affect how labour are governed. For instance, government policies on labour protection and social policy; the voice of trade unions, or support by NGOs can all significantly alter how private governance works in a GVC context. The next section explores other forms of governance that may exist, and how they interact, creating varied implications on how labour is governed and on labours' own experiences of being governed when participating in GVCs.

The synergistic governance of labour in GVCs

Focusing solely on private governance within GVCs suggests a preoccupation with lead firms and fails to account for the roles of non-firm actors such as governments and civil society (Bair 2005). With GVCs spanning multiple scales, so should our understanding of their governance. Governance should be understood as synergistic and multi-layered, accounting for inputs from multiple actors, and operating at global (between Global North and Global South), regional (across the Global South or within trading blocs) and local scales (within countries and communities).

Within each form of governance, a variety of influences pushes GVCs in the direction of better labour conditions, as illustrated in Table 44.1. Thus, in addition to private governance, the remit of governance is extended to two new forms of governance. The first, public governance, is driven by supranational organizations at a global scale (such as ILO's decent work

Table 44.1 Synergistic governance in GVCs in relation to labour

Governance	Global scale	Regional/Local scale
Private	GVC lead firm governance (e.g. global buyers' voluntary codes of conduct)	Supplier firm governance; and collective efficiency (e.g. support from business associations, cooperatives)
Public	International organizations (e.g. the International Labour Organisation (ILO), Human rights	Local, regional, national government regulations (e.g. labour laws and environmental legislation)
Social	Global civil society pressure on lead firms and major suppliers (e.g. Fair Labor Association) and multi-stakeholder initiatives (e.g., Ethical Trading Initiative)	Local civil society pressure (e.g. workers, labour unions, NGOs for civil, workers, and environmental rights; gender-equity advocates)

Source: Modified from (Gereffi and Lee 2016)

Table 44.2 Strategies – to implement synergistic governance of labour in GVCs

Strategies	Synergistic governance types	Key governance instruments
Standard driven (compliance)	Private primarily, with Social and Public	Labour standards
Mentor-driven	Private primarily, sometimes Public, Social	Cooperative/ Relational partnerships
Associative (partnerships)	Private, Social Public	Cooperation/ Partnerships
Bottom-up	Public, Social	Labour agency (individual, collective and community based)

Source: Author's construction

agenda). At a regional/local level, public governance includes fulfilling the legal requirements and expectations of national or sub-national governments. Often private and public governance must work hand-in hand to ensure the smooth functioning of transactions within a GVC. The second is social governance and is exercised by civil society actors, such as NGOs and labour unions. This civil society action can take place at global and national levels, and is usually targeted at regulating workers' rights and labour conditions through various forms of activism (boycotting, petitions, and protests) and challenging private governance (Selwyn 2013, Cumbers et al. 2008).

The joint influence of GVC governance across private-public-social domains is referred to as 'synergistic.' This can be more effective in achieving sustainable improvements of working conditions in low-income countries than private, public, or social governance alone (Locke 2013). Different combinations of private, public and social governance come together to engender four strategies of synergistic governance for governing labour, which in turn facilitate social upgrading (improving labour conditions). These four strategies are: standards or compliance strategies; mentor-driven strategies, associative strategies and bottom-up strategies. A brief summary of these is presented in Table 44.2.

Standard driven or compliance strategies involve lead firms (private governance) embedding complex standards that suppliers have to comply with, often replacing local and indigenous practices (Krishnan 2018; Ponte and Ewert 2009). Knowledge (on training, best practices) is transferred down the value chain in a relatively top-down manner (De Marchi et al. 2019; Krauss and Krishnan 2021). One arena in which this has been commonly applied is in addressing issues related to working conditions in garment factories, hotels, farmers and pack houses (Alexander 2020). Lead firms may work with social governance actors – such as civil society organizations – and public governance actors, like national governments.

Compliance strategies create a hierarchy of control, with lead firms providing incentives or sanctions that force supplier firms to implement and/or meet different standards and targets (ibid). However, this strategy creates multiple challenges, such as asymmetric power relationships (Lund-Thomsen and Lindgreen 2014) or situations of collusion (De Neve 2009). For example, lead firms may be more dominant in the relationship than CSOs or governments. This creates power imbalances, which effectively reproduce the will of the lead firm rather than governing labour in a synergistic manner. Similarly, it is possible that lead firms conspire with CSOs and governments to create worse, rather than better, labour conditions. Furthermore, a compliance-driven approach requires rigorous and expensive auditing processes, which can lead suppliers to evade inspectors or to restructure systems in ways that exclude practices from inspections without alleviating labour challenges (Alexander 2020). The high costs associated with this process are pushed down the value chain, often resulting in the greater casualization of labour.

Mentor-driven strategies are also private governance-dominated. They primarily represent mutually-dependent interactions between the lead firm and suppliers in relation to acquiring specialized know-how and skills (c.f. De Marchi et al. 2019 for the case of furniture in Italy; Khattak et al. 2015 for the case of apparel in Sri Lanka). There is a greater sense of cooperation between lead and supplier firms in such strategies. For example, training and capacity building services and support materials for implementing more pro-labour processes may be provided in conjunction with social and public governance actors (Lund-Thomsen and Lindgreen 2014; Alexander 2020).

Associative strategies are primarily partnerships and collaboration across private, public and social governance actors that actively seek to address labour issues. This creates a shared value between various actors in a GVC and a cooperative atmosphere (Alexander 2020). However, cooperation can also turn to contestation if power imbalances exist between the partners (Lund-Thomsen and Lindgreen 2014).

Bottom-up strategies relate specifically to labour agency. Such strategies emerge within societally-embedded (often geographically-fixed) production contexts, and contain historical tensions between labour and private, public and civil society actors (Coe and Hess 2013). Such forms of labour agency include collective or individual exhibition of power, and have been classified as forms of resilience (coping), reworking (seeking improvements) and resistance (challenging capitalist relations) (Katz 2004).

Each of these strategies represent a combination of different forms of governance, and each is exercised under different circumstances, depending on the relationship of actors in a GVC and the country-sector context. Within GVC analysis, synergistic governance strategies directly influence the level of social upgrading that occurs. In the next sub-section, we will unpack the various form of social upgrading, which we link to governance strategies, and then study using a variety of case studies.

What is social upgrading and downgrading in GVCs?

Social upgrading is a term that encapsulates improving labour working conditions along measurable and immeasurable aspects. Within GVC analysis, social upgrading can be explicated as outcomes of the various labour governing processes and strategies. The concept of upgrading draws upon ILO's 'decent work' framework. In terms of measurable aspects the most commonly used indicators are changes in the number and permanency of jobs, improvements in the type of contracts, wage changes (including wage improvements and reductions in wage inequality between workers), and improvements in the quality of working conditions (e.g. health and safety, unemployment insurance, social protection, working hours). Immeasurable aspects relate to principles of social justice and non-discrimination (Barrientos and Smith 2007), including the empowerment of workers, greater rights and the possibilities of creating positive externalities such as improvements in health, education, and food security for workers, their households and communities (Barrientos et al. 2011, Bernhardt and Milberg 2011). In contrast the concept of social downgrading represents a decline in measurable and immeasurable aspects of working conditions. Social upgrading and downgrading are both non-linear processes and can differ significantly across worker types. For instance, home-based workers (irregular) have low wages and are often not represented, while regular workers, with medium and high-skilled labour (using technology), in knowledge intensive work, experience higher wages, and better working conditions.

It can therefore be helpful to visualize social upgrading as a relative spectrum to track changes for labour since participating in the GVC. Along this spectrum we can move from an increase to a decrease in the measurable or immeasurable aspects, and in doing so can map out the extent to which social upgrading trajectories are altered. The spectrum is as follows:

- *Increase* (improvement in measurable/immeasurable aspects of social upgrading for labour since participating in a GVC).
- *Partial change* (some improvement, but to varying degrees).
- *No change* in measurable/immeasurable aspects since participation in a GVC.
- *Decrease* or downgrading occurring in measurable/immeasurable aspects, since participation in a GVC.

In the next section, we explore how synergistic governance strategies have influenced the social upgrading of labour in the three key sectors of apparel and textiles, agriculture and agro-processing and tourism. The selection of case studies was conducted through a bibliometric analysis of journal articles and grey material (reports) between 2011 and 2020 using Web of Science and Google Scholar. Fifteen studies were selected from apparel, 15 from agriculture and 5 from the tourism sector (these can be found following the references).

What implications do synergistic governance strategies have on the social upgrading of labour?

Apparel and textiles

Across all 15 selected studies in the apparel and textiles sector, the predominant synergistic governance strategy used was the 'compliance-based' strategy, albeit to differing degrees. As Table 44.3 shows, the use of compliance strategies has differing effects across social upgrading. The table provides the average results of the dominant implications on social upgrading, in relation to each type of strategy, as elicited from the 15 selected studies. Overarchingly, compliance strategies resulted in improved social upgrading in a minority of indicators; no or partial changes, or a deterioration i.e. social downgrading occurred much more frequently when compliance strategy was used (across all the combinations).

Table 44.3 Apparel and textiles – synergistic governance strategies and social upgrading

Strategy (synergistic governance)	Social upgrading					
	No. of jobs★	Permanency of jobs★	Wages changes★	Working conditions★	Empowerment/ rights★	Secondary spillover effects★
Compliance	No change	No change	Decrease or no change	Partial changes	No change	No change
Compliance, associative	Increase	Mix between no change to decrease	Decrease	Decrease	No change to decrease	No change
Compliance, associative, bottom-up	No change	No change	Partial	Decrease	Decrease	No change
Compliance, mentoring, associative	No change	No change	Partial	Increase	Increase	Increase
Associative	Decrease	No change	No change	No change	No change	No change

Source: Author's construction (5 studies used only compliance, 5 studies used a combination of compliance and associative, 3 studies used combinations of compliance, associative and bottom-up, 1 study used combination of compliance, mentoring and associative; and 1 study used only associative) ★the table denotes the predominant changes within the social upgrading spectrum indicated by the articles. *See 'References used in systematic review'.*

Compliance strategies involved adherence to the private codes of conducts of lead firms and/ or industry standards and certifications at a global level (e.g. Fair Labour Association, ILO's decent work agenda). However, their implications on social upgrading varied. For example, in Ethiopian apparel firms, the compliance model (which involved suppliers conforming to the 'Better Work' programme of the ILO) led to support for training to move to higher value-added activities such as weaving or textile manufacturing. However, this was a contested process and was not implemented by firms or the government adequately. Ultimately, the focus continued to stay on performing business-as-usual, low-value activities like cutting and sewing. Alongside this, dis-incentivized trade union representation (indicative of a lack of empowerment of workers) resulted in low wages, high turnover rates, limited social rights and no sick leave or insurance coverage, (Staritz et al. 2017). While labour legislation exists, this was not implemented by the government (creating a public governance deficit) and allowed a private-governance dominated compliance strategy to diffuse and proliferate across the apparel industry (Staritz et al. 2017). Eventually, wildcat strikes took place due to poor working conditions, indicating some level of empowerment demonstrated through worker contestation of the private governance led-compliance model. In the same vein, inadequate public governance (including the absence of social benefit funds and the low labour standards set by the country's Wages Advisory Board) led to limited social upgrading emerging from Lesotho's compliance strategy (which involved complying to demands for a reduced product price by reducing worker wages) (Godfrey 2015).

Compliance strategies also had mixed implications on social upgrading in Morocco. Compliance to labour standards was prominent here due to the demands of European lead firms (e.g. Inditex) and the introduction of the Moroccan garment industry association-led sector-led code of conduct (Fibre Citoyenne) (Rossi 2013). However, the implications of the compliance strategy were varied, especially in relation to supplier firms working in 'fast fashion' (characterized by short-term relations with buyers and several intermediaries like wholesalers). In particular, there were big differences between the experiences of regular and casual workers. Regular workers often had permanent contracts (written or oral), were paid a minimum wage, and were offered social benefits like sick leave. In contrast, casual (irregular) workers were paid below minimum wages and faced significant discrimination because their work was considered less valuable and important than regular workers (ibid.). The use of compliance strategies escalated the prevalence of irregular workers in Morocco, rather than increasing the number (and quality of work) of permanent workers.

Table 44.3 also highlights that a considerable number of studies used associative strategies in combination with compliance strategies. Such associative forms of partnership between civil society (e.g. trade unions, NGOs and business associations) and workers created some opportunities for social upgrading. For example, in Haiti, internal pressures and organized strikes from unions in 2013 led to an increase in minimum wages from USD0.49 to USD0.59 per hour (Faucheux et al. 2014). Similarly, following the 2013 Rana Plaza collapse in Dhaka, in which 1,134 people lost their lives, pressures from international NGOs and labour movements led to the signing of the Accord on Fire and Building Safety (AFBSB) by all major international retailers (Anner 2015).

However, there were also negative consequences that emerged from associative strategies. For example, in Nicaragua, a lack of coordination between different public bodies (i.e. the Ministry of Labour and the Free Trade Zone Commission (CNZF)) on the implementation of ILO's 'Better Work' Program reduced participation in the programme. This negatively affected social upgrading opportunities for workers (Bair 2017). In Haiti, despite adhering to standards, labour conditions remain precarious. There is also an increase in short-term irregular and informal workers during peak demand season, instead of improvements of the labour conditions of the existing workforce (Faucheux et al. 2014).

There has also been social downgrading with combined compliance-associative strategies in relation to job opportunities and permanency. Despite adhering to labour standards and workers' basic rights, in Lesotho and Madagascar, Asian-owned firms serving the US market increasingly hire expatriates in nearly all 'top positions' (e.g. technical, management, supervisory, and pattern-making positions), significantly reducing opportunities for regular local labour in these countries (Staritz et al. 2014, 2017). The case is even worse in Kenya, where most of the local regular workforce is unable to gain promotion into supervisory positions which are reserved for expatriates (Staritz and Frederik 2014).

Only one case study, in Guatemala, exhibited a 'mentoring strategy.' Around 60 per cent apparel exporters here are Korean-owned and operated (Pipkin 2012). Most of these Korean firms attempted to create long-term relationships with the government and workers, thus engendering social upgrading. This occurred as Korean firms offered health services to workers, and even created spillover effects in the local economy by investing in literacy and education, nutrition, leadership and human relations (ibid.).

These 15 case studies highlight that overall, the implications of compliance-dominated strategies for social upgrading were mixed. A downgrading in working conditions and empowerment were common. This questions the efficacy of private-public governance (i.e. poorly-functioning compliance strategies) and the possible downgrading trajectories that emanate from such partnerships. This highlights the importance of local public governance to mediate and monitor lead firm-oriented compliance strategies.

Agriculture and agro-processing

Compliance strategies also dominate agriculture and agro-processing GVCs, with these privately-governed from the top by large supermarkets and wholesalers in the North (Evers et al. 2014). Similar to experiences in apparel and textiles, we see that the implications of strategy types in this sector also had mixed implications on social upgrading across the 15 case studies.

In aquaculture in Bangladesh (Ponte et al. 2017), Cocoa in Ghana (Barrientos 2014) and Cote de Ivoire (Amanor 2012), and tea in Tanzania (Loconto and Simbua 2012), small-scale farmers work under precarious contracts with European lead firms. Producers depend heavily on support from lead firms (e.g. training in pest and irrigation management, health and safety measures around agro-chemicals) to maintain high enough standards to be eligible for export markets. Frequently, the wages of farm labour and incomes earned by farmers were thin. For instance, Loconto and Simbua (2012) found the margins of Tanzanian farmers squeezed due to competition from lead firms. Lead firms could keep costs and prices low because they owned large estates and operated vertically by hiring (and paying very low wages to) contract workers. In Ghana, farmers not only had to adhere to lead firm standards (e.g. Fair Trade), but also to the intermediary codes of conduct set by processors and other large trading firms (Amanor 2012). These efforts left farmers with almost no income. However, in some cases, such as South Africa, adhering to standards was considered a major contributor to driving up minimum wages. Even these improvements, however, were not necessarily equal to a living wage (Alford 2016). Studies also revealed that labour had very low levels of trust in lead firms due to unfavourable contract terms and no enabling rights to join unions (Tallontire et al. 2014).

Some studies mentioned associative strategies alongside compliance or bottom-up strategies. These were partially successful, especially when civil society was involved. In Ghana's cocoa industry, for example, Barrientos (2014) shows how a local CSO, Kuapa Kokoo, successfully supported women by pressuring the government and lobbying lead firms to increase minimum wages and improve labour conditions (e.g. health and safety, sexual harassment at work). Similarly, the Tanzania Plantation and Agricultural Workers Union was able to form an alliance

Table 44.4 Agriculture – strategies and social upgrading

Synergistic governance strategies	Social upgrading					
	No. of jobs*	Permanency of jobs*	Wages changes*	Working conditions*	Empowerment/ rights*	Secondary spillover effects*
Compliance	No change	No change	Partial	Partial	No change	No change
Compliance, associative	No change	No change	No change	Partial	Mixed (increase / decrease)	No change
Compliance, associative, bottom-up	Increase	No change	No change	Increase	No change	No change
Compliance, bottom-up	No change	No change	Increase	Partial	Partial	Increase

Source: Author's construction (7 studies used only compliance, 5 studies used a combination of compliance and associative, 1 study used combinations of compliance, associative and bottom-up, 2 studies used combination of compliance, bottom-up)

with Fairtrade and the Tea Association of Tanzania to bargain for better rights and provide a platform for workers to voice their concerns (Loconto and Simbua 2012).

There were mixed implications on social upgrading due to compliance and associative strategies, with public governance playing an important role. For instance, in Ethiopian flori-culture, the government took several measures to benefit producers by lifting the ban on importing agro-chemicals, improving market infrastructure, and developing local labour standards (Gebreeyesus and Sonobe, 2012). In contrast, Ponte et al. (2014) find that ineffective regulatory capacity of Bangladesh's aquaculture sector manifested itself in export tax breaks and subsidies to build and operate poorly equipped processing plants and negatively impacted wages. Amanor (2012) and Barrientos (2014), found that government corruption and poor land governance policies worsened working conditions in the pineapple and cocoa industries for farmers/workers in Ghana and Cote de Ivoire respectively. In South African fruit, several policies such as Basic Conditions of Employment Act, Occupational Health and Safety Act and the Labour Relations Act (LRA) supported the improvement in minimum wages and worker tenure security and helped to enable a right to representation (Alford 2016). With support from international and local civil society and with the adoption of the Ethical Trade Initiative code, farmers and workers in Western Cape were able to gain representation. Despite these positive changes, however, their increase in minimum wage was not a sufficient living wage (Alford et al. 2017).

Only two studies suggested that bottom-up strategies were used by local cooperatives and farmers groups. For instance, in the case of Kenyan avocados and beans, farmers who were part of cooperatives were able to comply with standards (GobalGAP), and thus had more agency vis-à-vis the lead firm (as opposed to farmers and workers who were not organized into groups). They were able to improve their product quality, practices, and demand better extension support services from the government (Krishnan 2018, Krishnan and Foster 2018). This enabled them to improve their incomes and working conditions and to increase trust with their buyers. Bottom-up strategies were key in the South African fruit sector. Precarious labour on farms instigated protests with support from NGOs and trade unions. These forced the government to increase their daily wages from USD 4.5 to USD 9.9 for the lowest paid workers. However, this did not change working conditions, especially for causal and migrant labour, who lacked housing facilities or social benefits (Alford 2016).

Almost half of the studies reported that these compliance-based strategies resulted in social upgrading in terms of job creation. Yet improvements were limited beyond this, with wages kept low, contracts precarious, and labour increasingly casualized (Ponte et al. 2014). This rise in the flexibilization of the workforce was seen especially in the fruit sectors of South Africa and Cote d'Ivoire and Ghana's cocoa industry, where lead firms increasingly hired migrant and casual labour, reducing overall wage rates (Amanor 2012; Alford 2016). The trend towards cas-ualization was most prominent among women, particularly among those packaging products in factories (Barrientos 2014; Dannenberg and Nduru 2016; Gebreeyesus and Sonobe 2012). Only a few studies reported positive spillover effects in terms of skill diffusion and increased household spending. For instance, in Kenya, Krishnan (2018) suggests that good agricultural practices from GVCs were 'spilt-over' into regional markets, improving overall skill levels in the community.

Overall, the 15 case studies looked at across this sector highlight that upgrading successes in job creation and increased minimum wages were clearly overshadowed by a lack of proper working conditions, increased casualization and the increase in contestations between workers/farmers and lead firms/governments. This echoes findings from the apparel sector and casts a shadow on the efficacy of the compliance-related strategies that dominate social upgrading efforts. It also raises the question of whether private firms and the government play a hindering or facilitative role in social upgrading in LICs.

Tourism GVCs

While compliance-based strategies were the dominant form of governance in textiles and agriculture, the most common strategy within tourism GVCs are associative strategies (Table 44.5). These are frequently combined with compliance strategies that are dominated by foreign tour operators or large national private or public operators. For example, Kenyan tourism is highly coordinated by lead firm global tour operators (e.g. Thomas Cook) who work with national tour operators and hotels to create, manage, and execute tourism packages that restrict and define international tourists' experiences (Christian 2016). Lead firms use associative strategies by forming profit-sharing partnerships with national tour operators and hotels. Yet there is a cost to this international dominance of power for local tourism businesses. Several mid- and low-range hotel providers have minimal to no profits, frequent ownership turnovers, and low occupancy rates, leading to the increasing casualization of hotel labour as a way to keep costs low and flexible. Casual (irregular) workers were paid a daily wage significantly lower than mandated minimum wages There were also increased instances of sexual harassment for casualized female workers, who had limited options for reporting abuse (ibid).

The case in Uganda varied from that in Kenya. For example, almost all Ugandan hotels provided workers with on-site food while working, paid annual leave, offered appropriate working hours and social protection, including paying medical bills (Christian 2016a, Adiyia et al. 2015). However, these wages were far below living wages. To improve wages, the Uganda Hotels, Food and Tourism Allied Workers' Union (UHFTAWU) fought for and achieved collective bargaining agreements between the union and the national hotel owners' association, stressing the importance of adequate compensation for skills and increasing the minimum wage. While similar tactics were employed in Kenya by the Kenya Union of Domestic, Hotel, Educational, Institutions, Hospitals, and Allied Workers, the benefits were only limited to permanent or contract workers, and not casualized workers.

Like the agriculture and apparel sectors, public governance in a few cases perpetuated social downgrading in Uganda's tourism sector. Adiyia et al. (2015) show how the Ministry of Tourism, Wildlife and Heritage, Uganda Wildlife Authority, and the Ugandan Tourist Board failed to support the inclusion or skilling of local tourist operators. While the private sector does try to form associative relationships with government and parastatals, these are weak and uncoordinated. A further lack of training of local tour operators prevented them from fulfilling

Table 44.5 Tourism – strategies and social upgrading

Synergistic governance strategy	No. of jobs	Permanency of jobs	Wages changes	Working conditions	Empowerment/ rights	Secondary spillover effects
Associative/ bottom-up	No change	No change	Partial	No change/ decrease	No change/ Partial	No change/ Partial
Compliance, associative	No change	Increase	Increase	Increase	Partial	Partial

Source: Author's construction (3 studies used associative-bottom-up strategies, 2 studies used compliance and associative strategies)

the basic criteria (e.g. to know international languages, be certified drivers) necessary to register and gain representation under the Association of Ugandan Tour Operators. Hence, foreign tour operators dominate the space.

There was one case where the compliance-association strategy was successful. Surmeier (2020) details how certain forms of social upgrading were successful through the implementation of a South African Fair Trade Tourism (FTT) standard. The study found that wages, work contracts, and the creation of permanent positions were accelerated through the adoption of the standard. For instance, workers were paid above the minimum wage to reach a living wage, or were offered benefits (e.g. meals and transportation). But even though the standard aligned with the Broad-based Black Economic Empowerment (B-BBEE) legislation, most FTT-certified businesses were owned and controlled by white people, thus limiting social upgrading outcomes around racialization of ownership structures.

Overarchingly, case studies within the tourism sector suggests the failure of public-private governance partnerships within compliance strategies but suggest that social governance through associative strategies can create positive social upgrading for workers.

A pedagogy of labour and GVCs

The aim of the chapter was to understand how to improve social upgrading for labour in GVCs, in order to improve the working conditions and security of workers. In doing so the chapter elucidates how the expansion of GVCs has created different public, private and social governance structures which have given way to compliance, mentor-driven, associative, and bottom-up strategies for governing labour. These strategies often do not work synergistically, creating win/lose or lose/lose social upgrading implications for labour in LICs. It forces us to consider the growing dominance of the private sector in the global arena, and the power they wield over the ability of smaller firms and labour in LICs to continue to participate in GVCs. The chapter also highlights the critical role of public governance, not just in the ability of governments to create and implement regulation to support labour and facilitate better working conditions, but also in how they engage with global lead firms. The quality of such partnerships is important.

Across all three sectors we have seen that social upgrading and downgrading occur simultaneously. For instance, the results indicate that in apparel and agriculture GVCs, compliance and associative strategies have, to an extent, supported an increase in wages. In parallel, however, there exist downward trajectories in terms of poor working conditions, a lack of permanency of contracts and disempowerment. While the tourism sector consisted of associative strategies, these were clearly dominated by large public or private operators in collusion with the government, leading to a deterioration in the working conditions of local workers.

These findings open up multiple questions that must be addressed in a pedagogy of labour and GVCs. Firstly, we need to know how GVCs are governed and the related labour implications of these forms of synergistic governance. But we also need to know how GVCs impact development more broadly, and to be aware of the ways in which they can facilitate even greater inequalities and divides between high income and low-income countries. Some useful questions and thoughts to consider further are:

- Due to the expansion of GVCs (growing dominance of lead firms and private governance) is permanent labour a thing of the past, and casualization the new norm?
- How are standards designed? Are they designed only for Northern consumers and needs and fail to account for immeasurable aspects of social upgrading, leaving Southern workers with limited voice, rights, and representation?

- How should public governance be enhanced to create social upgrading opportunities for labour, while simultaneously catering to the needs of global lead firms?
- Does civil society play a critical role in GVCs? How can they be leveraged to facilitate socially equitable value chains?
- Is it possible to create inclusive and synergistic strategies to mitigate the exploitation of labour in GVCs? Who would need to be involved? How can power relations between different actors be balanced to develop shared and mutual goals for inclusion?
- Are there other strategies that exist that allow for more labour agency and voice? Think of examples from your own countries and across different sectors.
- Has COVID-19 exacerbated the worsening situation for labour employment and working conditions in GVCs and led to social downgrading across agriculture, apparel, and tourism? What needs to be done to help with recovery? Which GVC actors need to be involved?
- Are GVCs here to stay or is there a greater retreat to re-shoring and on-shoring of value chains back to nations in the aftermath of COVID-19? How will this impact the social upgrading of labour in LICs?

Key sources

1. Barrientos, S., Gereffi, G., and Rossi, A. (2011). Economic and social upgrading in global production networks: A new paradigm for a changing world. *International Labour Review, 150*(3-4), 319–340.

Explicates the links between economic and social upgrading in GVCs.

2. Gereffi, G., Humphrey, J., and Sturgeon, T. (2005). The governance of global value chains. *Review of international political economy, 12*(1), 78–104.

This is a seminal paper on explaining the different forms of private governance in GVCs and links to development.

3. Ponte, S; Gereffi, G., and. Raj-Reichert, G (Ed.). (2019). *Handbook on Global Value Chains.* Edward Elgar Publishing.

This is a compendium of literature that provides key insights into various aspects of GVCs, ranging from governance, power, to methodological approaches and different forms of upgrading.

4. Standing, G. (1999). Global feminization through flexible labor: A theme revisited. *World development, 27*(3), 583–602.

This paper highlights the implications of globalization on work and employment in an increasingly fragmented world.

5. Taglioni, D., and Winkler, D. (2016). *Making global value chains work for development.* The World Bank.

This book offers a strategic framework, analytical tools, and policy options for policymakers to explain the ways in which GVC participation can be made more economically and socially inclusive for labour.

Notes

1 Please see the second reference list for an overview of these 35 studies.
2 The related concept of Global Production Networks (GPN) suggests the need to look beyond the vertical, to include horizontal actors such as governments, Non-Governmental Organizations (NGOs), and community (Henderson et al. 2002; Coe et al. 2004). As global supply networks have become increasingly complex, scholars have noted greater synergy between the GPN and more recent GVC approaches (Bair and Palpacuer 2015; Neilson et al. 2014).

References

Alexander, R. (2020). Emerging roles of lead buyer governance for sustainability across global production networks. *Journal of Business Ethics*, *162*(2), 269–290.

Bair, J. (2005). Global capitalism and commodity chains: looking back, going forward. *Competition & Change*, *9*(2), 153–180.

Bair, J., and Palpacuer, F. (2015). CSR beyond the corporation: contested governance in global value chains. *Global Networks*, *15*(s1), S1–S19.

Barrientos, S., and Smith, S. (2007). Do workers benefit from ethical trade? Assessing codes of labour practice in global production systems. *Third World Quarterly*, *28*(4), 713–729.

Barrientos, S. (2008). Contract labour: The 'Achilles heel' of corporate codes in commercial value chains. *Development and Change*, *39*(6), 977–990.

Barrientos, S., Gereffi, G., and Rossi, A. (2011). Economic and social upgrading in global production networks: A new paradigm for a changing world. *International Labour Review*, *150*(3-4), 319–340.

Barrientos, S. (2019). *Gender and work in global value chains: Capturing the gains?* Cambridge: Cambridge University Press.

Bernhardt, T. and Milberg, W. (2011). Economic and social upgrading in global value chains: Analysis of horticulture, apparel, tourism and mobile telephones. ISBN 978-1-907247-94-1 [Capturing the Gains Working Paper 2011/06]. 113.

Christian, M., Fernandez-Stark, K., Ahmed, G., and Gereffi, G. (2011). The tourism global value chain: Economic upgrading and workforce development. In G. Gereffi, K. Fernandez-Stark, and P. Psilos, *Skills for upgrading, Workforce development and global value chains in developing countries*, 276–280, Duke University.

Coe, N. M., Hess, M., Yeung, H. W. C., Dicken, P., and Henderson, J. (2004). 'Globalizing' regional development: a global production networks perspective. *Transactions of the Institute of British geographers*, *29*(4), 468–484.

Coe, N. M., and Hess, M. (2013). Global production networks, labour and development. *Geoforum*, *44*, 4–9.

Cumbers, A., Nativel, C., and Routledge, P. (2008). Labour agency and union positionalities in global production networks. *Journal of Economic Geography*, *8*(3), 369–387.

De Marchi, V., Di Maria, E., Krishnan, A., and Ponte, S. (2019). Environmental upgrading in global value chains. In *Handbook on global value chains*. Edward Elgar Publishing.

De Neve, G. (2009). Power, inequality and corporate social responsibility: The politics of ethical compliance in the South Indian garment industry. *Economic and Political Weekly*, 63–71.

Evers, B., Opondo, M., Barrientos, S., and Krishna, A. (2014). Global and regional supermarkets: implications for producers and workers in Kenyan and Ugandan horticulture Working Paper 39. *Capturing the Gains*.

Fernandez-Stark, K., Frederick, S., and Gereffi, G. (2011). The apparel global value chain. *Duke Center on Globalization, Governance & Competitiveness*. Duke University.

Gereffi, G. (1994). The organization of buyer-driven global commodity chains: How US retailers shape overseas production networks. *Contributions in economics and economic history*, 95–95.

Gereffi, G. (1999). International trade and industrial upgrading in the apparel commodity chain. *Journal of International Economics*, *48*(1), 37–70.

Gereffi, G., Humphrey, J., and Sturgeon, T. (2005). The governance of global value chains. *Review of International Political Economy*, *12*(1), 78–104.

Gereffi, G., and Lee, J. (2016). Economic and social upgrading in global value chains and industrial clusters: Why governance matters. *Journal of Business Ethics*, *133*(1), 25–38.

Henderson, J., Dicken, P., Hess, M., Coe, N., and Yeung, H. W. C. (2002). Global production networks and the analysis of economic development. *Review of International Political Economy*, *9*(3), 436–464.

Kaplinsky, R., and Readman, J. (2005). Globalization and upgrading: what can (and cannot) be learnt from international trade statistics in the wood furniture sector? *Industrial and Corporate Change*, *14*(4), 679–703.

Katz, C. (2004). *Growing up global: Economic restructuring and children's everyday lives*. U of Minnesota Press.

Khattak, A., Stringer, C., Benson-Rea, M., and Haworth, N. (2015). Environmental upgrading of apparel firms in global value chains: Evidence from Sri Lanka. *Competition & Change*, *19*(4), 317–335.

Kowalski, P., Gonzalez, J. L., Ragoussis, A., and Ugarte, C. (2015). Participation of developing countries in global value chains: Implications for trade and trade-related policies. *OECD Trade Policy Papers*, No. 179, OECD Publishing, Paris, https://doi.org/10.1787/5js33lfw0xxn-en

Krauss, J. E., and Krishnan, A. (2021). Global decisions versus local realities: Sustainability standards, priorities and upgrading dynamics in agricultural global production networks. *Global Networks*. https://doi.org/10.1111/glob.12325

Krishnan, A., and Foster, C. (2018). A quantitative approach to innovation in agricultural value chains: evidence from Kenyan horticulture. *The European Journal of Development Research*, *30*(1), 108–135.

Locke, R. M. (2013). *The promise and limits of private power: Promoting labor standards in a global economy*. Cambridge: Cambridge University Press.

Neilson, J., Pritchard, B., and Yeung, H. W. C. (2014). Global value chains and global production networks in the changing international political economy: An introduction. *Review of International Political Economy*, *21*(1), 1–8.

Ponte, S., and Ewert, J. (2009). Which way is 'up' in upgrading? Trajectories of change in the value chain for South African wine. *World development*, *37*(10), 1637–1650.

Selwyn, B. (2013). Social upgrading and labour in global production networks: A critique and an alternative conception. *Competition & Change*, *17*(1), 75–90.

Standing, G. (1999). Global feminization through flexible labor: A theme revisited. *World Development*, *27*(3), 583–602.

Staritz, C., and Frederick, S. (2014). Sector case study: Apparel. *Making Foreign Direct Investment Work for Sub-Saharan Africa: Local Spillovers and Competitiveness in Global Value Chains. The World Bank*, 209–244.

Tallontire, A., Opondo, M., and Nelson, V. (2014). Contingent spaces for smallholder participation in GlobalGAP: insights from Kenyan horticulture value chains. *The Geographical Journal*, *180*(4), 353–364.

UNCTAD (2017). Trade and Development Report: Beyond Austerity towards a global new deal. UNCTAD: Geneva, Available at: https://unctad.org/system/files/official-document/tdr2017_en.pdf

World Bank (2019). Global Value Chains: Trading for Development. *World Development Report 2020*.

REFERENCES USED IN SYSTEMATIC REVIEW

i) Apparel and Textiles

Amankwah-Amoah, J. (2015). Explaining declining industries in developing countries: The case of textiles and apparel in Ghana. *Competition and Change*, 19(1), pp. 19–35.

Anner, M. (2015) .Worker resistance in global supply chains. *International Journal of Labour Research*, 7(1–2), pp. 17–34.

Bair, J. (2017). Contextualising compliance: hybrid governance in global value chains. *New Political Economy*, 22(2), pp. 169–185.

Curran, L. and Nadvi, K. (2015). Shifting Trade Preferences and Value Chain Impacts in the Bangladesh Textiles and Garment Industry. *Cambridge Journal of Regions, Economy and Society*, 8(3), pp. 459–474

Faucheux, B., Del Rosario, J. and Gomera Economistas Asociados (2014). Analyse des Chaînes Logistiques en Haïti Fiche Filière: Textile. *World Bank*.

Godfrey, S. (2015). Global, regional and domestic apparel value chains in Southern Africa: Social upgrading for some and downgrading for others. *Cambridge Journal of Regions, Economy and Society*, 8(3), pp. 491–504.

Goto, K. (2011). Competitiveness and decent work in Global Value Chains: Substitutionary or complementary? *Development in Practice*, 21(7), pp. 943–958.

Lund-Thomsen, P., and Lindgreen, A. (2014). Corporate social responsibility in global value chains: Where are we now and where are we going? *Journal of Business Ethics*, *123*(1), pp. 11–22.

Morris, M. and Staritz, C. (2014). Industrialization Trajectories in Madagascar's Export Apparel Industry: Ownership, Embeddedness, Markets, and Upgrading. *World Development*, 56(1), pp. 243–257.

Pipkin, S. (2011). Local means in value chain ends: dynamics of product and social upgrading in apparel manufacturing in Guatemala and Colombia. *World Development*, 39(12), pp. 2119–2131.

Rossi, A. (2013). Does economic upgrading lead to social upgrading in global production networks? Evidence from Morocco. *World Development*, 46, pp. 223–233.

Staritz, C. and Morris, M. (2017). Industrial upgrading and development in Lesotho's apparel industry: global value chains, foreign direct investment, and market diversification. *Oxford Development Studies*, 45(3), pp. 303–320.

Staritz, C., Plank, L. and Morris, M. (2016). *Global Value Chains, Industrial Policy, and Sustainable Development – Ethiopia's Apparel Export Sector*. Geneva, Switzerland: International Centre for Trade and Sustainable Development (ICTSD).Wetterberg, A. (2011) Public-Private Partnership in Labour Standards Governance: Better Factories Cambodia, *Public Administration and Development*, 31, pp. 64–73. doi: 10.1002/pad.

ii) Agriculture and Agro-processing

Alford, M. (2016). Trans-scalar embeddedness and governance deficits in global production networks: Crisis in South African fruit. *Geoforum*, 75, 52–63.

Alford, M., Barrientos, S., and Visser, M. (2017). Multi-scalar labour agency in global production networks: Contestation and crisis in the South African fruit sector. *Development and Change*, 48(4), 721–745.

Amanor, K. S. (2012). Global resource grabs, agribusiness concentration and the smallholder: Two West African case studies. *Journal of Peasant Studies*, *39*(3–4), 731–749.

Barrientos, S. (2014). Gendered global production networks: Analysis of cocoa–chocolate sourcing. *Regional Studies*, 48(5), 791–803.

Barrientos, S., Knorringa, P., Evers, B., Visser, M., and Opondo, M. (2016). Shifting regional dynamics of global value chains: Implications for economic and social upgrading in African horticulture. *Environment and Planning A: Economy and Space*, 48(7), 1266–1283.

Carter, J., and Smith, E. F. (2016). Spatialising the Melanesian Canarium industry: Understanding economic upgrading in an emerging industry among three Pacific small island states. *Geoforum*, 75, 40–51.

Dannenberg, P., and Nduru, G. M. (2013). Practices in international value chains: the case of the Kenyan fruit and vegetable chain beyond the exclusion debate. *Tijdschrift voor economische en sociale geografie*, *104*(1), 41–56.

Gebreeyesus, M., and Sonobe, T. (2012). Global value chains and market formation process in emerging export activity: Evidence from Ethiopian flower industry. *Journal of Development Studies*, 48(3), 335–348.

Krishnan, A. (2018). The origin and expansion of regional value chains: the case of Kenyan horticulture. *Global Networks*, 18(2), 238–263.

Loconto, A. M., and Simbua, E. F. (2012). Making Room for Smallholder Cooperatives in Tanzanian Tea Production: Can Fairtrade Do That? *Journal of Business Ethics*, 108(4), 451–46

Maertens, M., Minten, B., and Swinnen, J. (2012). Modern Food Supply Chains and Development: Evidence from Horticulture Export Sectors in Sub-Saharan Africa. *Development Policy Review*, 30(4), 473–497.

Ponte, S., Kelling, I., Jespersen, K. S., and Kruijssen, F. (2014). The blue revolution in Asia: Upgrading and governance in aquaculture value chains. *World Development*, 64, 52–64.

Verhofstadt, E., and Maertens, M. (2013). Processes of modernization in horticulture food value chains in Rwanda. *Outlook on Agriculture*, 42(4), 273–283.

Van den Broeck, G., Swinnen, J., and Maertens, M. (2017). Global value chains, large-scale farming, and poverty: Long-term effects in Senegal. *Food policy*, 66, 97–107.

iii) Tourism

Adiyia, B., Stoffelen, A., Jennes, B., Vanneste, D., and Ahebwa, W. M. (2015). Analysing governance in tourism value chains to reshape the tourist bubble in developing countries: The case of cultural tourism in Uganda. *Journal of Ecotourism*, *14*(2–3), 113–129.

Christian, M. (2016a). Kenya's tourist industry and global production networks: gender, race and inequality. *Global Networks*, *16*(1), 25–44.

Christian, M. (2016). Tourism global production networks and uneven social upgrading in Kenya and Uganda. *Tourism Geographies*, *18*(1), 38–58.

Mao, N., DeLacy, T., and Grunfeld, H. (2013). Local livelihoods and the tourism value chain: A case study in Siem Reap-Angkor Region, Cambodia. *International Journal of Environmental and Rural Development*, *4*(2), 120–126.

Mitchell, J. (2012). Value chain approaches to assessing the impact of tourism on low-income households in developing countries. *Journal of Sustainable Tourism*, *20*(3), 457–475.

Surmeier, A. (2020). Dynamic capability building and social upgrading in tourism-Potentials and limits of sustainability standards. *Journal of Sustainable Tourism*, 1–21.

45

INTERNATIONAL AND INTERNAL MIGRATION

Tanja Bastia and Ronald Skeldon

Introduction: patterns and trends

In 2019, some 272 million people were estimated to be living in a country or territory other than the one in which they were born, which is the standard definition for the number of international migrants (United Nations 2019; IOM 2020). This represented just 3.5 per cent of the world's population at that time and indicated an increase from 2.8 per cent in 2000. This stock figure, however, is deceptive in that it excludes large numbers of people who have ever migrated, most specifically, those who have returned to their country of origin after spending years in other countries and those who have migrated within their country of birth. One attempt at a global estimate for those who had migrated internally gave 740 million, or almost four times the number of international migrants around 2010 (UNDP 2009, 1). Again, that estimate is conservative. Return migration is even more problematic to measure and requires specialist surveys. Evidence from historical migrations shows that it can indeed be significant, with proportions of national groups that migrated across the Atlantic in the nineteenth century showing return rates of up to 40 per cent, for example (Baines 1991). Also excluded are the numbers of people who depend financially upon migrants. Hence, the significance of migration for both the movement of peoples and its contribution to development is far greater than the simple proportion of 3.5 per cent of the world's population would suggest.

Putting aside the very real difficulties of estimating global stocks, as well as flows of migrants, estimates of the volume of world migration are also not particularly informative. Migration is not randomly distributed in any population and a relatively small number of origin and destination countries dominate these global flows. The United Nations estimates for international migrants in 2019, for example, show that some two-thirds of the total number of international migrants were concentrated in just 20 countries and two-fifths had been born in two major regions: Europe (61 million) and Southern Asia (50 million). The principal country of destination for international migration was the United States, with a stock of 51 million foreign-born, followed by Germany and Saudi Arabia with 13 million each. Large absolute numbers of migrants to a large country may have a proportionally lower impact on a population than smaller numbers of migrants in a small country. For example, the 51 million stock of international migrants makes up just 15 per cent of the population of the United States, while

DOI: 10.4324/9781003017653-50

in Qatar the 2.2 million stock of migrants there made up 79 per cent of the total resident population.

At the highest level of generalization, four main foci of migration can be identified: the migration from South towards West Asia; the movement from Latin America into North America; the distress movements from West Asia, which includes the Middle East, towards Europe; and the migration among countries in Africa (Abel and Sander 2014). It is worth emphasizing that, despite the growth of those forced to move within and from countries such as Afghanistan, Syria, and Yemen, at enormous human cost in terms of lives and livelihoods, the numbers of forced migrants remain a relatively small proportion of the total number of those who move. Refugees and asylum seekers make up around 10–11 per cent of all international migrants (United Nations 2019).

Using an older version of the global stock data, estimates have shown that the movements from one country to another within the developing world (so-called 'South-South migration'), exceeded those from countries in the developing world to countries in the developed world ('South-North migration') (Parsons et al. 2007, 37). That is, most migration is not from the developing to the developed world. The vast majority of migrants from countries in both Africa and Asia move between countries in these regions but any such discussion becomes mired in easy assumptions of homogeneity in the simple categories of 'South' and 'North.' The highly urban Latin American economies are very different from the much less urbanized economies of Africa and parts of Asia, not just economically but culturally, politically, and demographically. Similarly, the contrast between settler societies, those where migration is an integral part of nation-building such as Argentina, the United States, Australia or Canada, and more exclusionary European and East Asian societies undermines any easy idea of a meaningful 'Global North.' One of the major features of global development over recent decades has been the emergence of areas of rapid economic growth within countries in the developing world itself and particularly around the metropolitan regions within those countries. Within the more developed world and in parts of the developing world, the feature has been of growing tensions between metropolitan and state governance, one that is finding expression in the realities of migration. This was the case, for example, in Argentina during the Macri administration of the City of Buenos Aires and the Cristina Fernández de Kirchner national government (2007–2015) or the current tensions between the Westminster government and the mayor of London and those between the sanctuary cities in the United States and the current (2020) Federal government.

Thus, the data on migration from one state to another, outlined at the outset of this chapter, describe but one part of a complex picture. International migration is to very specific destinations within the countries of immigration and from very specific origins in countries of emigration, in both of which urban areas play an important role. For example, in the United Kingdom in 2018, some 14 per cent of the total population was foreign-born. However, much of this population was concentrated in London, where 35 per cent of the foreign-born in the country were to be found (Vargas-Silva and Rienzo 2019). The second most important destination was the Southeast, which attracted some 13 per cent of the UK's migrant population. Almost half of all the foreign-born in the United Kingdom in 2018, therefore, were to be found in the national capital and its immediate hinterland. Similar findings have been found in other developed economies. For example, the proportion of the foreign-born in New York and Los Angeles in 2010 were respectively 36.8 and 35.6 per cent, compared with the national figure of 12.7 per cent. In France, the proportion of foreign-born in Paris in 2008 was 12.4 compared with 5.8 per cent nationally. In Argentina, the respective figures are 12.8 per cent for the capital Buenos Aires and 4.5 per cent nationally (CACS 2018; Ciudad de Buenos Aires

2017, Benencia 2012). It is hardly surprising that migrants are positively selected towards the most dynamic regions in their countries of destination.

Metropolitan areas as origins of migration are more problematic as few details on specific origins are available from the destination countries. However, given that the cities are the locations of the best schools and universities in most origin countries and that the metropolitan areas are the transport and communication hubs of their countries, it might make sense that most elite and skilled emigrants originate in those cities. It is also the case that other internal migrants might use cities as both short- and longer-term stepping-stones to further international migration. This does not mean that international migration cannot originate in specific and more isolated small-town or rural areas – as seen in cases in northern Pakistan and eastern Bangladesh – but it draws attention to important linkages between internal and international migration (King and Skeldon 2010, Skeldon 2006). Reference was already made at the outset to the significance of return migrations from international destinations. This return, after perhaps years overseas, is as likely to be to the more developed parts of origin countries – the cities – than to any isolated rural birthplace. This has implications for urbanization, but also for rural areas given a potential lack of investment of migration-derived income and resources there as a result.

While migration is often assumed to increase through time, with governments seeking to introduce limits, over the longer term, migration in fact may rise and fall as populations move through developmental, demographic, societal and even political transitions. Migration appears to be high during the redistribution of populations towards urban areas but at advanced levels of development, where urbanization levels are high, declines in internal migration have been observed to be replaced with increasing mobility (Champion et al. 2018). That is, rather than having to change their usual place of residence, people are travelling longer distances for work and leisure for shorter periods of time. They are commuting not just daily but weekly, monthly, or even for longer to both regular and variable destinations for work. Essentially, we seem to be becoming less migratory but more mobile. Whether such a pattern is sustainable in the face of growing evidence of the impact of current means of travel on climate change, and the recent diffusion of a disease such as COVID-19, remains to be seen. Patterns of migration and mobility are constantly shifting in tandem with transitions in development in its broadest sense.

The numbers and patterns of migrants also need to be set against the type of people moving. The overwhelming number of migrants into the countries of the Gulf Cooperation Council for the Arab States are labour migrants, most of whom are un- or semi-skilled from South Asian countries. Those in the United States on the other hand range from family members joining husbands, wives, parents, or other kin as immigrants, through high-skilled migrants entering by specialist work channels, to students, less-skilled workers and refugees. Virtually all countries in the more developed world operate some form of 'mixed' flows of families, workers, students and the displaced, although the settler programmes of Australia, Canada, New Zealand, and the United States – in which migration is seen as part of nation-building – are in many ways exceptional. Across the less-developed world, migration programmes tend to be specialist and temporary to provide the kinds of labour, skilled and unskilled, that are deemed to be in demand. South-South migration is also facilitated through regional economic and mobility pacts, such as the Economic Community of West African States (ECOWAS) or Mercosur, the South American trading bloc.

Having summarized the patterns and trends in global migration, we now turn to review the principal approaches to migration and development. We begin with a few general points about the linkages between the two terms, 'migration' and 'development' and then go on to examine the three major themes: remittances, skilled migration, and diaspora. We then go on to look at how policy fits in to the debate through actual practice on the ground and finish with some questions and learning goals.

Migration and development

Across all the diversity among types of migrants, one generalization appears to hold: the majority of those who move are young adults. The very young and the old do move, too, but they form a minority among those who move. Globally, the numbers of migrants are almost evenly split by gender. However, the representation of men and women in migration flows varies geographically as well as by economic sector. For example, more men than women migrate from South Asia to the Middle East, while more women migrate as nurses into Europe and the Americas, and as brides and domestic workers among countries in Asia. Nor do the poorest of the poor move internationally. Those who do migrate internationally generally need to have the physical and human capital that allows them to move. Hence, the consequences of the migration are seen to have developmental implications for both origins and destinations of the movement, implications that have been seen as both positive and negative. In fact, the way academics and policy-makers have tended to view migration has often swung like a pendulum, with positive ideas being dominant at one time only to be replaced by more negative views at another time (de Haas 2012). How this pendulum shifts in response to political change remains an area for research. Perhaps the essential issue revolves round whether migration can be managed to achieve or mitigate development as part of policy and programmes: that is, whether migration can be managed to realize desired specific development objectives, such as poverty reduction and the improved well-being of populations.

The linkages between migration and development are complex and diverse (Bastia and Skeldon 2020) but the policy debate has focused mainly on three inter-linked areas: remittances, skilled migration and diaspora. However, the results can be variable, counterintuitive, or unexpected. We explore each one of these in turn.

Remittances

Remittances, or the monies sent back by migrants to their families in areas of origin, were estimated at US$698 billion in 2018, of which some US$529 billion went to low- and middle-income economies (World Bank 2019b). The two largest recipients by far, India and China, accounted for US$146 billion or over one-quarter of that which flowed to low and middle-income economies. Nevertheless, the contribution of personal remittances to GDP in these countries is tiny, at 0.2 per cent for China and 2.9 per cent for India, with migration at the national level hardly making a significant contribution to economic development (www.theglobaleconomy.com/rankings/remittances_percent_gdp/). Much more limited absolute amounts of remittances can be significant in smaller economies, usually landlocked or small islands, where remittances contribute over 30 per cent of GDP, such as in Tonga, the Kyrgyz Republic, Tajikistan and Haiti, for example (World Bank 2019a).

The specific origins of emigration also mean that the benefits of remittances are concentrated in particular villages, towns and districts of cities and not usually among the poorest areas or among the poorest people in a country. Certainly, remittances can impact positively upon the lives of the families that receive them in terms of promoting education or improved health, but their result is likely to entrench existing wealth differences across and within communities rather than to lead to the broader alleviation of poverty that is the essence of development. This ambivalent statement must be tempered by the fact that it is reached on the basis of an examination of remittances from international sources only, given that no global database exists to measure internal remittances. Once remittances from the numerically more significant internal migrants are included, the benefits will become more widespread even if, ultimately, they may encourage more people to leave their communities of origin. Data on internal remittances, however, tend to be elusive, coming primarily from micro-surveys.

Skilled migration

Skilled migration is often associated with a loss of the skilled personnel from the developing world needed for development, for which the unfortunate term 'brain drain' has been coined. Here again, all might not be quite what it seems. Globally, most of the skilled migrants come from a relatively small number of rapidly developing economies or from the highly developed economies themselves. It is in these countries that the institutions to create the skills exist or where large numbers of students are going out to pursue further education. India and China dominate the flows but the United Kingdom and Germany are also significant countries of origin. Many of these movements are temporary as skilled migrants move through the networks of transnational corporations.

It can be argued that the exodus of relatively small numbers of skilled migrants from a small pool of skilled workers in poor economies may have a prejudicial impact on the development prospects of these countries. This is particularly difficult to demonstrate, even for countries where over two-thirds of people born in those countries with tertiary-level education, but living elsewhere, live outside their country of birth, such as Guyana, Haiti, Trinidad and Tobago, and Barbados. Many of these will have left as students and gained their higher education overseas rather than locally, where any credentials gained may not have wider recognition. More importantly, these countries may not have the positions locally to absorb adequately those with higher education, leading to a 'brain waste' and a discontented group of educated people within-country, which might cause political problems. In such a context, those who are trained to global standards will opt to migrate globally if possible, and a sound education policy will include appropriate training for local needs (Skeldon 2018).

Diaspora

The initial use of the term 'Diaspora' referred to a group that had been expelled from its homeland, virtually in its entirety, and was living in exile. More recently, it has taken on both a broader and a more limited connotation to include voluntary migrants but who may represent a minority of the origin population. Hence, diaspora has virtually come to mean migration but emphasizing continuing linkages between origins and destinations. Migrants from any country, living externally, rarely cut off their ties with their home areas but operate transnationally with family, companies and government officials. Given that these external communities include significant proportions of skilled workers from any origin, as well as those with influence and capital, the view has emerged that this group, known under the rubric of 'diaspora' can play a role in the development of their home countries. Not only can these migrants provide capital for investment but they can also return for longer or shorter periods of time in order to set up companies or otherwise support local enterprise, or provide the human capital to educate the next generation in schools and colleges. They understand the local languages and cultures but bring new ideas and ways of doing learned overseas that might help to introduce practices that promote 'development,' from management and education strategies to political views and policy approaches.

This strategy is one of insider/outsider facilitation of development, which has grown exponentially over recent decades with international organizations, developed-country aid agencies and governments of origin countries setting up institutions in an attempt to 'leverage' the diaspora (Gamlen 2019; Kuznetsov 2006). However, the reality of how effective the diaspora will be in promoting development is mixed and depends upon context. Migrants will not invest in or return to a country of origin unless the necessary environment exists. If, and when, they do return, tensions may arise between migrant and stayer groups, with the latter unwilling to be swayed by those who decided to leave and live elsewhere for what could be a number of years. Those in the diaspora can make effective contributions to their home areas but the conditions have to be right, and no automatic development outcome should be expected. The diaspora does not provide a magic development bullet but depends ultimately upon the environment created by the home government.

Lastly, is the overarching question of how development, however promoted, impacts upon migration (Clemens 2014). That is, will migration decrease if levels of development improve? The evidence suggests that developing low- and middle-income countries will increase rather than decrease migration: improved levels of education spread information on what is available 'out there' in a more comprehensive way and improved living standards give more people the resources required to travel. However, once a country has achieved levels of income towards the upper range of middle-income economies, their economies will be based more on tertiary and service activities, rather than primary or even secondary activities. The countries will be well on their way through the demographic transition to low levels of fertility, as well as mortality, and well on their way through an urban transition to a situation where the majority of the population lives in towns and cities. On reaching such a stage, internal migration appears to slow, while international migration sees a shift from emigration to immigration. These complex transitions move generally in the same direction, although not necessarily at the same speed, with periods of stasis and even reverse, and they diffuse across space and through time (Skeldon 2012). Hence, any discussion of internal and international migration needs to be located within this matrix of transition: context is all-important, and it is to an examination of specific contexts that we now turn.

Policy and practice

Policies in the area of migration tend to fall into two types: direct and indirect. Direct policies are implemented by governments specifically to impact upon the volume and direction of population flows through immigration policies in destination countries and, less commonly, through emigration policies in origin countries. These can be designed to promote or to restrict migration mainly through the promotion or restriction of particular types of migrant. Indirect policies are policies such as industrial, housing, health or education policies that are designed with other objectives, but which have a profound impact on population movement. For example, the establishment of industrial processing zones will stimulate movement towards them or the implementation of quarantine to achieve specific health objectives will restrict movements. Policy can also be developed at the multilateral level through instruments such as the Global Compact on Migration or the International Convention for the Rights of All Migrant Workers and Their Families. However, such international agreements are more normative than directive and are examples of 'soft' policy. Direct and indirect policies, whether guided by international norms or not, create regimes with which migrants have to negotiate. As it can be imagined, policy-making is not very straightforward, and policy-makers are faced with a number of challenges.

The main challenges for policy-makers lay in the definition of borders and boundaries as well as the categorization of migration and different types of migrants. Any type of migration, as we have already explained above, involves the crossing of some kind of boundary, whether from a rural area to a city or crossing an international border. By crossing these boundaries, migrants move from one legal jurisdiction to another, where they will find themselves in a different relationship with a new regime. Especially in cases of international or cross-border migration, but sometimes also in cases of internal migration, this often means that they will have a different set of rights and obligations vis-à-vis the national state or the local authority. For example, the Chinese government introduced a household registration system, called hukou, during the 1950s with the aim of controlling internal migration from rural to urban areas. Post the reforms introduced from 1979, the system did not prevent people from moving to towns and cities. However, under this system, rural migrants in cities were not able to access the basic social services such as health and education. This meant that many migrants decided to leave their children in their rural areas of origin to be looked after by grandparents (Chan 2015).

How and for what reason people migrate, places migrants in different legal categories vis-à-vis the state. Yet, some of the differences that place a migrant in one category or another can be quite subtle. For example, while the legal definition of a refugee is enshrined in the 1951 Convention relating to the Status of Refugees, in practice many migrants see themselves forced to leave their places of origin because they are unable to make a living or because their main and only source of livelihood has dried up. These migrants' experiences could be described as being 'forced' or at least, not voluntary, in that they would have preferred to stay in their places of origin. However, they would not qualify for any special treatment under refugee law. People fleeing the effects of climate change also do not qualify for international protection as refugees, yet they are clearly unable to stay in their places of origin.

Policy-makers are therefore confronted with a reality in which people move within and across international borders, for varying lengths of time and for various reasons, yet their actions and decisions are constrained by local and national boundaries, as well as quite rigid typologies of different types of migrants. These contexts then pose real difficulties in recognizing how and why people move and deciding on courses of action that would ensure that the movement is beneficial for both migrants and places of origin and destination, as well as to ensure that migrants' rights are protected.

For example, the 2000 Protocol to Prevent, Suppress and Punish Trafficking in Persons (Palermo Protocol) sets quite specific conditions for somebody to qualify as being a 'Victim of Trafficking.' The person needs to have been recruited, transported or transferred by means of the threat or use of force or other forms of coercion, of abduction, of fraud, of deception, of the abuse of power... for the purpose of exploitation (for the full definition, see UNHR 2000). It is clearly difficult to actually prove that any of this has happened to somebody who finds his or herself in a foreign country. Moreover, many migrants, especially those who move through irregular channels and for unskilled work and who seek work in countries other than the countries of their usual residence are often threatened, abused, or deceived en route (Coutin 2005) and are also often exploited at destination. Because of the way in which trafficking advocacy developed, it is generally much easier for a woman to present herself as a victim of trafficking (Macklin 2003). Men are often seen as having a higher degree of agency and therefore find it harder to fit into the 'Victim of Trafficking' image that policy-makers or lawyers have generated, despite the fact that large numbers of men are also trafficked into positions of extreme vulnerability, for example, for work in construction or agriculture. Moreover, the most common response to the problem of trafficking is the repatriation of victims of trafficking back to their countries of origin or exporting the problem outside of the boundaries of the nation state (of destination), which usually does not serve the interests of the trafficked person. It also does nothing to resolve the initial reason that led to this person to seek work abroad in the first place.

For countries of origin, the main challenges consist in retaining or encouraging the return of its brightest, most skilled and most productive citizens as well as ensuring that migrants invest their savings and remittances in productive avenues. Yet, again, they are confronted with realities that pose significant challenges to their objectives. If their educational settings train students to international standards, they are most likely going to seek work elsewhere, and will be reluctant to return to their countries of origin, especially if they are able to reap better benefits elsewhere (in terms of career progression or income). Many migrants also decide to spend their savings and remittances on improved consumption, including 'conspicuous consumption,' such as grand houses, instead of businesses or other 'productive' projects. Despite the fact that building a house will also support local businesses and help local economies, this way of spending is often frowned upon and not the policy-makers' preferences. However, these decisions make a lot of sense in contexts of weak governments, continuity and security. In such contexts, migrants decide to spend their hard-earned money on what gives them greater status as well as some sort of security for the future. For example, a 'grand house' might not be just about status, but can also be a means to earn an income through renting out rooms, which could eventually also provide an alternative to a pension in countries lacking universal pension systems (Bastia 2019, Boccagni 2020). Given that migrants' incomes are private incomes, and many have invested heavily personally in their migration projects, there are also moral questions to be asked of policy-makers' attempts to use these for boosting their countries' national development.

Pedagogy: teaching migration and development

Teaching migration and development involves debunking 'migration myths' as well as providing solid information and case studies that will help students understand the complexity of migration as well as illustrate potential avenues for harnessing the benefits of migration while ensuring that migrants' rights are protected. In this section we describe our approach to teaching migration and development and identify some key questions that we hope will be useful for others embarking on teaching this subject.

> *Q. After reading this short article, did students change their view about migration? If yes, in what way? If no, how did the article contribute to their learning, if at all? See if you can unpick the ways students think about human migration.*

We have taught a very diverse cohort of international post-graduate students at our respective universities (Sussex, Maastricht, and Manchester). Students often assume that migration is a movement from A to B. They also often think of migration as a 'discrete event.' Reading real migrants' life stories and experiences helps them understand that migration journeys are often complex and not often linear. We also emphasize that the decision-making process regarding migration often takes time, sometimes over a year, and is seldom a discrete event in which only one person, the migrant, participates.

Q: How are migration and development related? Can migration be 'controlled'?

The most common assumption that students arrive with is that migration can be controlled and that if richer countries wish to decrease the level of immigration, they should invest in the main countries of origin, in order to raise their level of development and decrease the need for people to seek work abroad. Students come from countries that have varied levels of freedom when it comes to people moving from one area to another. They also have different moral values in terms of the right that they see national governments having to control the movements of their populations.

We then move on to dispelling some of the myths that surround discussions about migration and development (de Haas 2005). The most common myth we find is that students assume that bringing development to a country or region will decrease the people's wish and need to seek work abroad. Ample evidence exists to show that the relationship between development and migration is a positive one. That is, when poorer countries become richer, their migration rates increase rather than decrease because people have greater access to knowledge, resources (skills and money) and the means to find work elsewhere. Migration rates might decrease, but only once development levels in the country of origin have started catching up with those of the country or area of destination, a process that usually takes decades, rather than years.

Close analysis and guided discussions of a range of case studies can therefore begin to unveil the complexities of the relationship between development policy and migration outcomes to students' understandings and begin to debunk pre-existing preconceptions.

Q How can policy-makers more effectively 'manage' migration for the development of their country? How should policy-makers approach the issue of skilled migration in (a) destination countries; (b) origin countries? How effective are international agreements?

We find it important to highlight the history of migration and explain that whatever assumptions we may have about government actions (whether controlling movement is acceptable; to what degree governments can go to control people's movements etc.) is a product of a place and time. That is, we explain that these assumptions are underpinned by moral values that are context specific and not universal.

Q Can we understand migration without development? Or development without migration? Is migration development? Elaborate.
(See Skeldon 1997 and Raghuram 2009 as useful starting points)

Within the context of 'migration and development,' we also aim to broaden our students' understanding of what 'development' might mean. While many come to our programmes with an interest in development as purposive action, we also aim to promote the understanding that development can also be related to a broader process of social change, not just the policies that governments might implement or NGO projects that might alleviate specific social problems. Development is related to how societies change throughout time (Hart 2010). Migration is an intrinsic part of such a change, for example, when economic depressions lead to high levels of outmigration, which in turn decreases levels of unemployment in countries of origin and relieves some of the pressure on governments for change. On the other hand, migrants have historically also built the basis of the traditional countries of immigration, such as Australia, Argentina, or the US, at the expense of the indigenous people already living there.

Conclusions

This entry has attempted to summarize the key issues in the migration and development debate. The evidence for the impact on development of remittances, the exodus of the skilled and the transnational linkages that make up the diaspora was reviewed. These impacts occurred during the last quarter century within the context of a neo-liberal economy. We can expect the nature of this development to change as has been demonstrated through the impact of the 2020 COVID-19 pandemic, which saw the spread of an infectious disease that brought global economies to a standstill. The resultant migrations were also transformed but whether the slowing in global mobility will be a permanent or a temporary trend in response to such challenges in the overall context of climate change will further illustrate the complex and often contradictory relationships between migration and development.

Key sources

1. Bastia, T. and R. Skeldon (eds.). (2020). *Routledge Handbook of Migration and Development*, London, Routledge, 2020.

This handbook contains over 50 original chapters providing state-of-the-art summaries of key areas in migration and development.

2. de Haas, H. (2012). The migration and development pendulum: a critical view on research and policy, *International Migration*, 50(3): 8–25, 2012.

This article discusses how ideas about migration and development change over time.

3. Raghuram, P. (2009). Which migration, what development? Unsettling the edifice of migration and development. *Population, Space and Place*, 15(2), 103–117. doi:10.1002/psp.536.

This article highlights the complexity of both terms as well as the relationship between both processes.

4. Skeldon, R. (2018). High-skilled migration and the limits of migration policies, in M. Czaika (ed.), *High-Skilled Migration: Drivers and Policies*, Oxford, Oxford University Press, 2018, pp. 48–64.

This chapter considers the extent to which policy on skilled migration is constrained by underlying economic and political realities in the global system.

5. World Bank. (2019b). Migration and Remittances: Recent Developments and Outlook, Migration and Development Brief 31, Washington, KNOMAD, The World Bank Group, 2019, at: www.knomad.org/sites/default/files/2019–04/Migrationanddevelopmentbrief31.pdf

This resource provides the latest data on flows of remittances among countries by level of development.

References

Abel, G. J. and Sander, N. (2014). Quantifying Global International Flows, *Science*, 343 (6178), 1520–22.

Baines, D. (1991). *Emigration from Europe 1815–1930*. London: Macmillan.

Bastia, T. (2019). *Gender, migration and social transformation: Intersectionality in Bolivian itinerant migration*. Abingdon: Routledge.

Bastia, T. and R. Skeldon (eds.) (2020). *Routledge Handbook of Migration and Development*. London: Routledge.

Benencia, R. (2012). Perfil migratorio de Argentina 2012. Retrieved from Buenos Aires: https://publications.iom.int/system/files/pdf/perfil_migratorio_de_argentina2012.pdf

Boccagni, P. (2020). So Many Houses, as Many Homes? Transnational Housing, Migration and Development. In T. Bastia and R. Skeldon (Eds.) *Routledge Handbook of Migration and Development*. London and New York: Routledge, pp. 251–260.

CACS. (2018). Informe sobre migraciones en Argentina. Cámara Argentina de Comercio y Servicios. Retrieved from Buenos Aires www.cac.com.ar/data/documentos/11_Informe%20sobre%20Migraciones.pdf

Champion, T. T., Cooke and I Shuttleworth (eds.). (2018). *Internal Migration in the Developed World: Are We Becoming Less Mobile?* Abingdon and New York: Routledge.

Chan, K. W. (2015). Five decades of the Chinese *hukou* system. In R. R. Iredale and Fei Guo (eds.), *Handbook of Chinese Migration: Identity and Wellbeing*. Cheltenham: Elgar, pp. 23–47.

Ciudad de Buenos Aires. (2017). *Migraciones. Año 2015*. Retrieved from www.estadisticaciudad.gob.ar/eyc/wp-content/uploads/2017/02/ir_2017_1118.pdf

Clemens, M. A. (2014). Does development reduce migration? In R. E, B. Lucas, (ed.), *International Handbook on Migration and Economic Development*, Cheltenham: Elgar, pp. 152–185.

Coutin, S. B. (2005). Being En Route. *American Anthropologist*, 107(2), 195–206.

de Haas, H. (2012). The migration and development pendulum: A critical view on research and policy. *International Migration*, 50(3), 8–25.

de Haas, H. (2005). International migration, remittances and development: Myths and facts. *Third World Quarterly*, 26(8), 1269–1284. doi:10.1080/01436590500336757

Gamlen, A. (2019). *Human Geopolitics: States, Emigrants and the Rise of Diaspora Institutions*. Oxford: Oxford University Press.

Hart, G. (2010). Developments after the Meltdown. *Antipode*, 41, 117–141. doi:10.1111/j.1467–8330.2009.00719.x

IOM. (2020). World Migration Report 2020, Geneva, International Organization for Migration. http://publications.iom.int/bookstore

Kuznetsov, Y. ed. (2006). *Diaspora Networks and the International Migration of Skills: How Countries Can Draw on Their Talent Abroad*. Washington: World Bank Institute.

Macklin, A. (2003). Dancing across Borders: 'Exotic Dancers,' Trafficking, and Canadian Immigration Policy. *The International Migration Review*, 37(2), 464–500.

King, R. and R. Skeldon, (2010). Mind the gap: integrating approaches to internal and international migration, *Journal of Ethnic and Migration Studies*, 36(10), 1619–1646.

Parsons, C., R. Skeldon, T. l. Warmsley and L. Alan Winters (2007). Quantifying international migration: A database of bilateral migrant stocks. In Ç. Özden and M. Schiff (eds.), *International Migration, Economic Development and Policy*. Washington: The World Bank, pp. 17–58.

Raghuram, P. (2009). Which migration, what development? Unsettling the edifice of migration and development. *Population, Space and Place*, 15(2), 103–117. doi:10.1002/psp.536.

Skeldon, R. (2006). Interlinkages between internal and international migration in the Asian region. *Population, Space and Place*, 12, 15–30.

Skeldon, R. (2012). Migration transitions revisited: Their continued relevance for the development of migration theory. *Population, Space and Place*, 18(2), 154–166.

Skeldon, R. (1997). *Migration and Development: A Global Perspective*. London: Longman.

Skeldon, R. (2018). High-skilled migration and the limits of migration policies. In M. Czaika (ed.), *High-Skilled Migration: Drivers and Policies*. Oxford: Oxford University Press, pp. 48–64.

UNDP. (2009). *Human Development Report 2009. Overcoming Barriers: Human Mobility and Development*. New York: United Nations Development Programme.

UNHR. (2000). Protocol to Prevent, Suppress and Punish Trafficking in Persons Especially Women and Children. Retrieved from www.ohchr.org/en/professionalinterest/pages/protocoltraffickinginpersons.aspx

United Nations. (2019). *International Migration Stock 2019*. New York: United Nations Population Division, www.unmigration.org

Vargas-Silva, C. and Rienzo, C. (2019). *Migrants in the UK: An Overview*. Oxford: The Migration Observatory, 2019, https://migrationobservatory.ox.ac.uk/resources/briefings/migrants-in-the-uk-an-overview/

World Bank. (2019a). Microdata on personal remittances, databank at: https://data.worldbank.org/indicator/BX.TRF.PWKR.DT.GD.ZS

World Bank. (2019b). Migration and Remittances: Recent Developments and Outlook, Migration and Development Brief 31, Washington, KNOMAD, The World Bank Group, 2019, www.knomad.org/sites/default/files/2019–04/Migrationanddevelopmentbrief31.pdf

46

FORCED MIGRATION AND ASYLUM SEEKING

Joseph Besigye Bazirake[1] and Carolina Suransky[2]

Introduction

This chapter advances a shift into a new critical development-oriented thinking on forced migration and asylum seeking. It presents the need to reframe interventions towards a broader development inclined focus, given the surge in global numbers of forced migrants and asylum seekers. The chapter emerges from Geiger and Pécoud's (2013) deliberations on a 'migration-development nexus' sustained by global development disparities, whose resultant inequalities have led to persistent migration pressures. More than a decade earlier, Castles (2003; 2006) pointed out that the perceptions of a growing migration crisis needed to be seen as a result of the gaps in economic conditions, social wellbeing and human rights between the North and South. Castles pointed towards a 'migration-asylum nexus' in recognition of the mixed motivations where, besides the traditional push factors of conflict and the fear of persecution, forced migrants are also keen on improving their families' livelihoods. Over the years, as migrancy has intensified globally, forced migration and asylum seeking have ascended as game-changers in global human population movements. This shift was recently acknowledged by Filippo Grandi, who as the United Nations High Commissioner for Refugees (UNHCR), noted that 'forced displacement is not only vastly more widespread but is simply no longer a short-term and temporary phenomenon' (UNHCR 2020, 6). By the end of 2019, at least 79.5 million people had been forced to flee their homes worldwide. This resulted in nearly 26 million refugees, around half of whom were aged under 18 (UNHCR 2020, 4). In this chapter, we interrogate the increasingly protracted dynamics of forced migration and asylum seeking by expanding the debate beyond humanitarian responses towards the inclusion of a human development approach. One focus that emerges from the chapter is how the field of education could enhance the human development capacities of both 'locals' and migrants, by creating glocal learning spaces in complex social contexts of migrancy.

The changing dynamics of forced migration and asylum seeking

Understandings of forced migration are bound to be confronted with conceptual, methodological and ethical challenges, including the separation of asylum seekers, who may ultimately be granted legal protection as refugees, from other forms of involuntary migrants (Turton

DOI: 10.4324/9781003017653-51

2003). Despite the guidance provided by the 1951 international convention and the 1967 optional protocol relating to the status of refugees, individual states continue to navigate the issue of asylum on their own terms. This has resulted in different asylum legislation grounded in states' 'different resources, national security concerns and histories with forced movements (IJRC, n.d.).' Asylum seeking remains a key prerogative for individuals and groups who may need protection outside of their countries of origin, owing to fear of persecution, conflict, violence, human rights violations, and events seriously disturbing public order. However, there have been growing concerns over the asylum system's integrity due to the mixed motivations that underline forced migration, where socio-political, economic, and environmental factors increasingly overlap as push factors. The asylum system has been slow to accommodate new and increasingly complex causes of forced migration. Thus, there is a need to note the changing global realities of shifting population dynamics, where the presumption of humanitarian relief as the primary response in situations of forced migration also needs to be re-examined.

As of 2019, 85 per cent of all displaced persons were accommodated in developing countries (UNHCR 2020). The largest hosts included: Turkey (3.6 million), Colombia (1.8 million), Pakistan (1.4 million) and Uganda (1.4 million). Despite this, there is also an increasing number of people who move from the South to the North. Of the approximately 2 million new asylum claims that were submitted globally in 2019, the United States was the world's largest recipient of new individual applications (301,000), followed by Peru (259,800), Germany (142,500), France (123,900) and Spain (118,300) (UNHCR 2020). Europe saw 'hundreds of thousands of people from Africa, the Middle East and South Asia, fleeing chronic poverty, political instability, wars, and the climate crisis in countries often laid to ruin by western-backed institutions' (Pai 2020). Similarly, Peru's dramatic increase in asylum applications resulted from its relatively low entry requirements that provided a safe haven for Venezuelans fleeing their country's precarious situation (John 2019). Nonetheless, the World Bank Group (2016) maintains that while South-South migration is more extensive than South-North migration, migration management should be considered a priority for high-income and low-income countries. This is because migration flows are expected to continue to grow due to global income gaps and demographic and environmental changes.

Of the 100+ million displaced persons in the decade preceding 2020, 79 million were internally displaced persons (IDPs). For the remaining international migrants, only 5 million received asylum – as they are entitled per the 1951 refugee convention provisions. Up to 15 million persons were recognized as refugees (outside the asylum system) and only received temporary protection. In the same period, only 3.9 million refugees and 31 million IDPs returned to their original residences, while 1.1 million refugees were resettled by other states (UNHCR 2020). These figures indicate that between 2009 and 2019, most displaced people could not return to their places of residence, and neither were they permanently resettled. Considering that the global trends of forced migration and asylum seeking continue to rise and are becoming increasingly protracted, attention needs to be directed towards expanding the efficacy of the three traditionally acclaimed durable solutions to forced migration: voluntary repatriation, resettlement and local integration. Most importantly, strategies on how local integration can be more effectively sustained across different contexts ought to be sought. This is because when displaced persons fail to be voluntarily repatriated or resettled in a third country, local integration becomes the next best alternative long-term durable solution.

A further challenge is the growing number of undocumented migrants who do not 'qualify' to benefit from the instituted forms of international protection available to asylum seekers. As the World Bank (2019) has recently noted, undocumented migrants are often the product of rejected asylum applications – rising between 2011 and 2018 from 1.4 million to 6 million in

Europe, and 1.5 million to 3.8 million in the USA. The reason for the rise in undocumented migrants can be explained by what Hansen (2014) describes as a protective strategy by nation-states to put up various institutional and legal barriers that keep asylum seekers away from their borders. States have increasingly adopted stricter border control policies linked to the growing criminalization of migration by countries that now use detention as a standard policy for dealing with asylum seekers and 'irregular migrants' (Hynie 2019; Kerwin 2019).

States increasingly use border restrictions such as visa requirements, safe country of origin and safe third country rules, carrier sanctions, interdiction at sea, and airports' declaration as international zones to restrict access (Hansen 2014). These restrictions have increased the likelihood of irregular migration, conducted outside of the laws, regulations and international agreements that govern people's entry and exit of states. As a consequence, Hansen (2014) argues that the contemporary asylum system presents a challenge to the state system on which it depends, interfering with state sovereignty by presenting states with the need to deal with long asylum clearing processes as well as the difficulty in the deportation of failed asylum applicants. The resulting standoff between the states and the asylum system emerges from a growing tendency for states to label forced migrants as 'illegal migrants' in order to delegitimize their movements (Scheel and Squire 2014).

These contemporary patterns of forced migration and asylum-seeking raise concerns regarding priorities for international protection. On the one hand, questions arise regarding the sustainability of global humanitarian commitments towards protecting the surging number of forced migrants and asylum seekers. On the other hand, there are concerns regarding whether the international protections for asylum seekers can still be safeguarded amid states' growing tendencies towards protecting their sovereignty and economic gains from 'outsiders.' These concerns notwithstanding, it is crucial to recognize that economic dynamics are increasingly playing a defining role in how the system of forced migration and asylum seeking operates. This has been demonstrated by the growing role that economic and institutional fragilities play in forced migration patterns, as the precarious economic situation of Zimbabwe (and more recently Venezuela) reveals (see Crush and Tevera 2010; John 2019). The general increase in African emigration can also be explained by the increase in young people's capabilities and aspirations to migrate, owing to contrasting development and social transformation trends in Africa and outside the continent (Flahaux and De Haas 2016).

In short, on top of the growing number of people escaping conflict or a lack of rights, an equally growing number are seeking economic betterment (Hynie 2019). It thus becomes urgent to consider stepping beyond humanitarian relief in response to forced migration, as humanitarianism's short-term focus does little to address the root causes of protracted human displacement (Harild 2016).

Transcending humanitarian approaches in forced migration and asylum seeking

When examining the growing overlaps of humanitarian and economic determinants in rising asylum seeking and forced migration, one point of significance is the 2016 incorporation of the International Organisation for Migration (IOM) as a related agency of the United Nations (UN). This reflects a strategy to integrate the IOM's work as an inter-governmental agency on general migration matters with the UNHCR's focus on refugees, both of which have operated independently of each other since their establishment in 1951. The resulting inter-agency shift provides a much-needed platform to recognize the growing number of individuals who may cross international borders, but still not qualify to receive asylum protection. This partnership is a strategic way of maintaining the asylum system's integrity since it would be expected to deliver outcomes and solutions for people who do not qualify for a refugee status (Micinski and Weiss 2016). In addition, the UNHCR adopted the Global Compact for Migration (GCM) in 2018 to synchronize humanitarian, developmental, and human rights approaches in the context of forced migration. The GCM, which seeks to advance safe, orderly, and regular migration illustrates the need for crosscutting responses to forced migration. It also aligns with the emerging UN strategy on the humanitarian-development-peace nexus linked to the 2030 Agenda for Sustainable Development (Crépeau 2018; Howe 2019; Oelgemöller and Allinson 2020).

While acknowledging these ongoing efforts to transcend humanitarian approaches in response to forced migration, this chapter also proposes the need for the integration of human capabilities as a development approach for working with migrant flows. Modelled around Sen's (1999) ground-breaking advancement of development as the expansion of capability and choice, this shift could be supported by approaches that seek to expand human freedoms and enhancements of personal and collective forms of agency. Crisp and Dessalegne (2002) argued that people often forcibly leave their countries due to a complex combination of fears, hopes, and aspirations. Beyani, Baal and Caterina (2016) similarly note that forced migration interventions need to be coordinated around humanitarian, development, and peace-building actors to adequately respond to and implement durable solutions, even for IDPs. In line with this, the capability approach to development can inspire ways to shift policy to cater to the wellbeing of all those caught within the crossroads of forced migration; asylum seekers and 'locals' alike.

According to Bazirake (2017), the need to infuse developmental aspects into humanitarian approaches while working with displaced persons would be a mutually beneficial way to address the plight of forced migrants and their hosts. Similarly, Harild (2016) emphasized that the impacts of the interactions between host communities and displaced persons have a primarily developmental character and should be handled accordingly. Harild then argues that human displacement ought to be framed as a long-term development issue with humanitarian elements. This goes against the grain of the long-held practice within the asylum system, built around humanitarian relief as the primary response to forced migration. Harild acknowledges that whereas protracted forced displacement cases may often require short-term humanitarian action, interventions also need to go ahead and address social, economic, and fiscal implications for both the displaced persons and their hosts (ibid.).

In turn, Harild (2016) and Nyce, Cohen and Cohen (2016) suggest that labour mobility improves the lives of those who are forcibly displaced. According to these authors, by upholding the rights of displaced people to move freely and work within their adoptive countries, they would be able to contribute to local growth and development while nurturing their self-reliance. This strategy would also tap into the unique skills and talents that displaced persons may already possess. Nyce et al. (2016) particularly cite the example of the recruitment pool of multinational employers, who could benefit from refugees' and forced migrants' unique skills.

Despite the level of appeal that labour mobility conjures, one persistent challenge is the potential for new forms of conflict between forced migrants and their host communities. To address this, we propose that human displacement's developmental imperatives could be engaged more peaceably through the embedding of a 'glocal' pedagogical outlook on the situation of forced migration and asylum seeking within education systems. In the following section, we explore what such a 'glocal gaze' in education could entail.

New possibilities for a 'glocal' gaze in education

As noted above, one important domain in which the predominantly humanitarian response-based framing of migrancy could be challenged is in the field of education. This could, and should, be pursued across multiple scales of learning. Day care centres, schools, colleges, and universities could all become spaces of intervention that could help to realize this proposed shift. Suransky (2020) argues that the point of departure in such a shift would be the capability enhancement, not only of those who migrated, (including the internally displaced) but also and simultaneously, the 'locals' in the 'host' environment. Since all people (not only migrants) need to learn how to live and function well in complex social contexts of migrancy, the capabilities to think about migration in complex ways need to be cultivated in both migrant and non-migrant populations. Rather than considering 'locals' and migrants to be two monolithic and antagonistic groups, there is an urgent need to acknowledge and address the 'messy' diversity that exists both within and between these dynamic communities.

Migrancy here is seen as an expression of globalization which encounters local circumstances with all its economic, social, political, and historical baggage. In the educational environments within such contexts, all students need to engage with migrancy, albeit in diverse ways. Highly educated migrants are often treated much better than those who lacked the opportunity to go to school. People fleeing conflict and violence may be traumatized, which would make it difficult to function well in the educational realm of their new environment. On the other hand, locals could learn from the resilience and agency of migrants. Migrants may bring new skills and new ways to reimagine human possibilities in life. Where the 'global' meets the 'local,' new possibilities arise to create a 'glocalized' gaze in education.

According to Mannion (2015, 29), 'the importance of place comes from the view that any given locale is always connected to many other places beyond the immediate experienced context.' Accordingly, he proposes the development of glocally-oriented pedagogies which take as a starting point the ecological, political, social and cultural dimensions of real places as a nexus of global and local flows and concerns. This approach starts from a premise of entanglement. Such entanglements may not be easy, and should not be romanticized, but are a challenge for both migrants and (local) democracy and the realization of human dignity for all in diverse communities. In a glocalized education, students would need to develop knowledge, values and skills that would enable them to creatively recognize that all local issues have global dimensions. When migration is addressed, one would explore how this is locally manifested while simultaneously exploring how local conditions are connected to global dynamics (Suransky 2020). This could mean that a school in Texas in the United States, for example, would consider different local circumstances and responses to migration than a school in Kampala in Uganda. This should include an attentiveness to the underpinning causal factors, and appropriateness, of local responses and actions. The ability to conceive of local action is an essential feature in glocal education. In both localities, however, migration can be linked to global developments and global politics which students can study.

A glocal classroom should be enriched through participation by diverse staff and students. Its pedagogy should be attentive to intersectional theory, and its assertion that people are systemically disadvantaged, or advantaged, by various sources of oppression or privilege, based on their gender, race, class, religion, or other identity markers. Indeed, intersectional theory offers an excellent conceptual framework to analyse the complexities that exist within the glocal. It provides a way of thinking that takes students away from interpreting the world through easy binary propositions or single axis perspectives, such as local versus foreigner or Christian versus Muslim (Suransky 2020).

Let us go back to the example of the school in Uganda. In the Ugandan glocal classroom, one could, for instance, study the long-lasting transnational relations between Uganda and what became South Sudan. One could analyse this like Ssentongo (2018) who states that:

> *...over time, it is becoming more and more vivid that the administrative independence gained from the colonial masters has given birth to a more sophisticated dependence network [in the context of globalization].*

Inspired by such ideas, one could study the colonial history of the region and notice that what currently is Uganda and South Sudan, have shared a long border, traversing the home areas of several ethnic groups where the:

> *... management of those people in the borderlands required coordination between the colonial governments. From the 1940s, the South Sudanese people attended schools in Uganda, and many fled across the border and sought sanctuary in 1955 following mutinies – and subsequent government repression – in Equatoria. This marked the beginning of two trends still evident today: South Sudanese searching for education in Uganda, and people in each country seeking refuge in the other.*
>
> *(Rolandsen, Sagmo and Nicholaisen 2015)*

Against the background of this (colonial) history, students could study contemporary local problems such as:

> *In the past, Ugandan cities hosted relatively affluent South Sudanese living on remittances from relatives. Now, with their reduced opportunities and increased harassment in South Sudan, Ugandans are more likely to consider refugees to be a burden than a resource. Sentiment among refugees is also shifting – from gratitude to frustration. In urban areas, local prejudices keep them unemployed – they want to be treated as equals, but instead are charged more than locals for rent or at shops. Begrudged, they recall opportunities Ugandans have enjoyed in South Sudan.*
>
> *(Rolandsen et al. 2015)*

The above example from Uganda is a specific case, and each case has its unique circumstances. However, as an example, it reveals some of the possibilities for advancing new ways of thinking about migration and asylum seeking by embedding a glocalized gaze within education. It shows us how such a gaze would allow educators to relate historical and contemporary local particularities to newcomers' ongoing and dynamic influence and thus invoke critical analyses of global-regional-local entanglements and the migration-asylum-development nexus.

Questions for developmental praxis in forced migration and asylum seeking

When considering the changing dynamics of the situation of forced migrants and asylum seekers, a number of reflective questions need to be explored. These questions stretch how one would think about these dynamics and offer further incentive to explore how the development approaches can be engaged with through the educational field as we propose in this chapter. Although humanitarian approaches continue to play a crucial role, we emphasize a need to imbue these approaches with longer-term development thinking. Therefore, the questions that follow can be used for reflection among advanced students who undertake studies in forced migration and asylum seeking.

Denoting the concepts of 'irregular migration' and 'illegal migration'

Today's migration flows result from a combination of factors. However, given the increasing tightening of borders, many would-be asylum seekers find themselves in irregular border crossing situations that are considered illegal. The question of 'irregularity' in border crossing needs to be reflected upon, particularly in as far as it pertains to the thin line that separates 'irregularity' from 'illegality' in migration. In what ways can states meet the protective requirements of would-be asylum seekers without jeopardizing their interests in regulating migration onto their territories?

State sovereignty and the right to admit [or not] individuals seeking asylum

Individuals can only seek asylum within countries, other than their own, that they have physical access to. This means that, as long as would-be asylum seekers are not within a given country's territorial confines, they cannot seek asylum there. This has been part of the motivation for high-income countries to tighten access to their borders. The critical question to ponder here is how the 'burden' that might initially come with large flows of forced migrants and asylum seekers can be equitably shared; firstly, as a humanitarian prerogative, and progressively as part of a global network for development assistance around human displacement. This fundamental question became apparent in the crisis that intensified after the refugee camp Moria on the Greek island of Lesbos went up in flames in September 2020. By and large, European nations were unwilling to accommodate devastated refugees who – once again – lost everything they had.

Protection and freedom as mutually supportive elements

The critical need addressed by asylum is human protection. However, given that asylum is increasingly a protracted condition with mixed motives, how can long-term responses be further infused within the asylum framework without vilifying the asylum system's integrity and humanitarian stance? Part of the reflection here lies in the recognition that the local communities that host asylum seekers would also need to be catered for, in a bid to address the situation of forced migration and asylum seeking more comprehensively.

Migrancy as a long-term phenomenon that changes circumstances for migrants and citizens of 'host' contexts

Migrancy, as a dimension of globalization, changes local contexts. New exchanges and interdependencies emerge, often accompanied by hardships, hostilities, and resentment. To positively address these changing circumstances, both migrants and citizens of the receiving contexts need to learn how to adjust and develop new capabilities. The reality that widespread migrancy is a 'here to stay' phenomenon across the globe demands new visions and policies on bringing migrant and local communities into contact with each other. Educational institutions could well develop learning spaces in which this could take place. How could education, as a medium for human development, play a role in rethinking local realities and address new circumstances for all involved? Acknowledging these changes as dynamic and with a long-term impact, how can students learn to imagine and articulate social change possibilities more creatively?

Conclusion

In the last few decades, migrancy has intensified across the globe. Rising global inequality is a significant factor in the continuous growth of migration pressures. Economic fragilities increasingly play an important role when people conclude that they need to leave dire circumstances at home and seek to expand their capabilities and freedoms elsewhere. Migrancy is a manifestation of globalization and a long-term phenomenon. Short-term stop-gap 'solutions' can neither adequately address the problems faced by migrants nor for those who already live in the receiving local contexts. Sheer numbers and migration patterns demand new developmental policies and interventions beyond its current humanitarian impetus. Indeed, migration changes global and local realities, which means that everyone involved needs to learn how to live together under changing circumstances and contribute to (local) wellbeing. In this chapter, we have noted that the developmental imperatives of human displacement could be further developed in the field of education. Here, a shift towards a developmental focus would be supported by a 'glocal' pedagogical outlook to engage with the migration-asylum-development nexus.

Key sources

1. Annual reports on global trends from the UNHCR. View the trends in forced displacement in 2009–2020 available at www.unhcr.org

The annual reports of the UNHCR discuss the activities related to safeguarding the rights and wellbeing of people who have been forced to flee. Together with partners and communities, the UNHCR works to ensure that everybody has the right to seek asylum and find safe refuge in another country.

2. Reports from the UNHCR, which specifically focus on education, are available at www.unhcr.org/education.html

The UNHCR operates from the principles which are captured in the New York Declaration for Refugees and Migrants. They report on initiatives which pinpoint education as a critical element of the international refugee response. They also focus on Sustainable Development Goal 4, which aims to deliver 'inclusive and quality education for all and promote lifelong learning.'

3. Migration statistics and information about global migration data provided by IOM's Global Migration Data Analysis Centre (GMDAC) available at https://migrationdataportal.org/

This Portal makes international migration data accessible and visible. It pulls together vital global data sources on migration in one place as most data are currently scattered across agencies and hidden in comprehensive reports. The Portal also reviews available migration data in various fields of migration, explains concepts and definitions and describes key strengths and weaknesses of available data sources.

4.   Annual reports from Asylum Access available at https://asylumaccess.org/

Asylum Access advocates for a world where refugees everywhere can live safely, move freely, work, feed their families, send children to school and contribute to their communities. They focus on legal empowerment, policy change, and global systems change programmes.

5.   The European Commission's information site on their policies on migrants and education available at https://ec.europa.eu/education/policies/european-policy-cooperation/education-and-migrants_en

This site informs the public about the European Commission's efforts to integrate migrants in their education and training systems. They have identified three priorities: (1) to integrate newly arrived migrants into mainstream education structures as early as possible (2) to prevent underachievement among migrants (3) to prevent social exclusion and foster intercultural dialogue. Concrete actions include tools to help assess migrants' skills and qualifications and the establishment of collaborative platforms to promote the exchange of information among education and training institutions and staff.

Notes

1 Chair for Critical Studies in Higher Education Transformation, Nelson Mandela University, South Africa.
2 University of Humanistic Studies, The Netherlands and Unit for Institutional change and Social Justice, University of the Free State, South Africa.

References

Bazirake, J. B. (2017). The Contemporary Global Refugee Crisis. *Peace Review*, vol. 29, iss. 1, pp. 61–67.

Beyani, C., Baal, N. K., and Caterina, M. (2016). Conceptual challenges and practical solutions in situations of internal displacement. *Forced Migration Review*, vol. 52, pp. 39–42.

Castles, S. (2003). The international politics of forced migration. *Development,* vol. 46, iss. 3, pp. 11–20.

Castles, S. (2006). Global perspectives on forced migration. *Asian and Pacific Migration Journal*, vol. 15, iss. 1, pp. 7–28.

Crépeau, F. (2018). Towards a mobile and diverse world: 'Facilitating mobility' as a central objective of the Global Compact on Migration. *International Journal of Refugee Law*, vol. 30, iss. 4, pp. 650–656.

Crisp, J. and Dessalegne, D. (2002). Refugee protection and migration management: The Challenge for UNHCR, Working paper, UNHCR, Switzerland.

Crush, J., and Tevera, D. S. (eds.) (2010). *Zimbabwe's exodus: Crisis, migration, survival*. Cape Town: SAMP.

Flahaux, M. L., and De Haas, H. (2016). African migration: trends, patterns, drivers. *Comparative Migration Studies*, vol. 4, iss. 1, pp. 1–25.

Geiger, M., and Pécoud, A. (2013). Migration, development and the migration and development nexus. *Population, Space and Place,* vol. 19, iss. 4, pp. 369–374. challenge for UNHCR. *UNHCR Working Paper, New Issues in Refugee Research*, vol. 64, pp. 1–20.

Hansen, R. (2014). State controls: Borders, refugees, and citizenship. In E. Fiddian-Qasmiyeh, G. Loescher, K. Long, and N. Sigona, (eds.) *The Oxford handbook of refugee and forced migration studies*. Oxford: Oxford University Press.

Harild, N. (2016). Forced displacement: A development issue with humanitarian elements. *Forced Migration Review*, vol. 52, pp. 4–7.

Howe, P. (2019). The triple nexus: A potential approach to supporting the Sustainable Development Goals. *World Development*, vol. 124, pp. 104629.

Hynie, M. (2019). Global migration trends, motivations and policies. In M.K Majmundar and S. Olson (eds.) *Forced Migration Research: From Theory to Practice in Promoting Migrant Well-Being: Proceedings of a Workshop*. National Academies Press.

International Justice Resource Center- IJRC (n.d). Asylum and the Rights of Refugees. Available at: https://ijrcenter.org/refugee-law accessed 6.12.2021

John, M. (2019). Venezuelan economic crisis: Crossing Latin American and Caribbean borders. *Migration and Development,* vol. 8, iss. 3, pp. 437–447.

Kerwin, D (2019). Global, National, and Ethical Issues in forced Migration Research. In M.K Majmundar. and S. Olson (eds.) *Forced Migration Research: From Theory to Practice in Promoting Migrant Well-Being: Proceedings of a Workshop.* National Academies Press.

Mannion, G. (2015). Towards Glocal Pedagogies: Some Risks Associated with Education for Global Citizenship and how Glocal Pedagogies might avoid them. In J. Friedman, V. Haverkate, B. Oomen, E. Park and M. Sklad (eds.) *Going glocal in higher education: the theory, teaching and measurement of global citizenship.* Middelburg: De Drvkkery.

Micinski, N. R., and Weiss, T. G. (2016). International Organization for Migration and the UN System: A Missed Opportunity. *Future United Nations Development System Briefing,* vol. 42, pp.1–4.

Nyce, S., Cohen, M. L., and Cohen, B. (2016). Labour mobility as part of the solution. *Forced Migration Review,* vol. 52, pp. 31–32.

Oelgemöller, C., and Allinson, K. (2020). The responsible migrant, reading the global compact on migration. *Law and Critique,* vol. 31, pp. 183–207.

Pai, H. H. (2020). The refugee 'crisis' showed Europe's worst side to the world'. The Guardian. Available at: www.theguardian.com/commentisfree/2020/jan/01/refugee-crisis-europe-mediterranean-racism-incarceration

Rolandsen, Ø.H., Sagmo, T.H. and Nicholaisen, F. (2015). 'South Sudan – Uganda Relations: The Cost of Peace,' *Conflict Trends,* vol. 4, pp. 33–40.

Scheel, S., and Squire, V. (2014). Forced migrants as illegal migrants. In E. Fiddian-Qasmiyeh, G. Loescher, K. Long, and N. Sigona, (eds.) *The Oxford handbook of refugee and forced migration studies,* Oxford: Oxford University Press.

Ssentongo, J.S. (ed.) (2018). *Decolonisation Pathways, Postcoloniality, Globalisation and African Development,* Kampala: Centre for African Studies, Uganda Martyrs University.

Suransky, C. (2020). Higher education in a globalizing world: The challenge of glocal education and the call to decolonize universities. In WP Wahl, and R. Pelzer, (eds.) *Leadership for Change. Developing transformational student leaders through global learning spaces,* Cape Town: AOSIS. https://doi.org/10.4102/aosis.2020.BK143.

Turton, D. (2003). *Conceptualizing forced migration,* Refugee Studies Centre Working Paper Series 12, Oxford: University of Oxford.

United Nations High Commissioner on Refugees (2020). Global Trends: Forced. Displacement in 2019. Geneva. Available: www.unhcr.org/afr/statistics/unhcrstats/5ee200e37/unhcr-global-trends-2019.html

World Bank. (2016). *Migration and Development. A Role for the World Bank Group.* Washington, DC: World Bank, Available: http://documents.worldbank.org/curated/en/690381472677671445/Migration-and-development-arole-for-the-World-Bank-Group

World Bank. (2019). *Migration and Remittances: Recent Developments and Outlooks.* Migration and Development Brief 31. www.knomad.org/publication/migration-and-development-brief-31

47

DEVELOPMENT AND CONFLICT

Jessica R. Hawkins

Introduction: making space for conflict in development studies

Globally, between 2014 and 2019, the number of fatalities from organized violence has decreased, yet the number of state-based conflicts rose to its highest level since 1945. This number reached 54 in total in 2019, of which 25 are in Africa (Pettersson and Öberg 2020, 597–598). Even if the numbers of those killed are declining, the impact of conflict on civilians and institutions is increasing and negatively effects development not only in the affected country but also in neighbouring countries (World Bank 2011). It is therefore surprising that 'theories of development and theories of conflict have largely evolved in isolation from one another' resulting in a discourse since the formation of Development Studies as a discipline where 'conflict was written out of development' (Mac Ginty and Williams, 2016, 2–3).

This chapter brings together the work of scholars who are challenging this norm to address the key debates surrounding conflict and development. The first section examines discourse on the causes of conflict as featured in my teaching on conflict and development. Five areas are discussed: fragile states, grievances, greed theory, the environment and structural violence. The second part of the chapter moves on to identify three key areas where conflict impacts development, namely displacement, gender, and finally security and peacebuilding. Specifically, I examine the cost of conflict on human development. The third section draws on my own experiences as a lecturer in Humanitarian Studies to discuss the pedagogy of teaching conflict and development within the classroom.

Conflict in the twenty-first century does not fit into the conceptual binaries of twentieth century interstate and civil wars (World Bank 2011). Instead, violence exists at multiple levels, in varying forms and intersecting with many aspects of society. State-based conflict, one-sided violence and non-state conflict all contribute to the vicious cycle of persistent poverty and underdevelopment (Mac Ginty and Williams 2016; Pettersson and Öberg 2020). Further, outbreaks of conflict are rarely isolated; often they follow a repeating cycle of latent and manifest violence that delays the attainment of social and economic development. The World Development Report (World Bank 2011) states that those living in conflict-affected areas are 'more than twice as likely to be undernourished as those in other developing countries, more than three times as likely to be unable to send their children to school, twice as likely to see their children die before age five, and more likely to lack clean water' (World Bank

DOI: 10.4324/9781003017653-52

2011, 5). Understanding the causes and nature of conflict is a priority for ensuring enduring development.

Causes of conflict

There are many theoretical approaches to the causes of conflict. These have evolved over time and space as the nature of conflict and violence has changed (Beswick and Jackson 2018; Jacoby 2008). Within my teaching on conflict and development, five theoretical approaches stand out as the most prevalent. This section elaborates on the literature on fragile states, grievances, greed theory, the environment and structural violence as causes of conflict and violence in the development context. As discussed in the second half of the chapter, when teaching case studies of conflict-affected countries, these five themes rarely stand alone as causes. Rather, they often intersect.

Fragile states

The concept of the fragile state is frequently seen as a cause of violent conflict in developing states. Often interchanged with terms such as soft states, weak states, collapsed states, crises states and failed states, fragile states describe countries where development processes cannot take place as there are minimal levels of state functioning (Turner et al. 2015). Central to this is a collapse of law and order, meaning little can be done to protect human security and ensure the basic needs of citizens are met, leaving gaps for structural or manifest violence to take place. State weakness, along with outside intervention (itself justified through a failed state rhetoric), are crucial causal factors of armed violence, often leading to regional instability (Turner et al. 2015). After 9/11 it became evident to Western countries that fragile states could no longer be considered as contained, national-scale problems, but must be viewed as challenges for global security.

The concept encapsulates some of the instigators of conflict such as insecurity, a collapse in the rule of law, corruption, a lack of state authority and legitimacy, inequalities in the distribution of basic services and a dispersal of power to non-state actors. However, the term 'fragile state' is problematic. Much early literature on failed states implies that these states must be fixed, reconstituted and developed, usually by outside intervention (Grimm et al. 2014). From 2001, fragile states were transformed into a new policy agenda by international organizations to create targeted aid, with emphasis on peacebuilding, state-building and state-capacity, legitimizing western interventions and in-turn depriving local actors of their agency (Grimm et al. 2014). The focus led to discourses of what these states cannot do to dominate, rather than investigating what is happening on a local level in spite of state collapse and violence. An alternative view by Putzel and Di John (2012) creates a more nuanced approach to state fragility and reframes responses within an agenda of creating resilience, with less neoliberal norms.

Grievances

The fragile states discourse responded to earlier debates in the literature which tried to find single causes for conflicts in the developing world. Emerging in the 1960s grievance theory stemmed from research on relative deprivation. This idea considered how people may feel deprived when they become aware of what they think they 'ought to have relative to what [they feel] others, within a referential framework, have/are getting' (Jacoby 2008, 106). The extent to which their ability to achieve goals or capabilities is interfered with, results in people choosing to act against this interference. In turn, grievance theory developed on this and argued that civil wars are caused when horizontal inequalities[1] exist within societies and governments fail to respond to demands for equality (Stewart 2008).

Processes of long-term development bring with them a host of grievances. According to Jacoby (2008), grievances may cause conflictive situations in a developing country as it experiences rapid change. For example, developmental changes can lead to some groups within a society benefitting disproportionately. Further, as change happens, some states may not be able to respond to the population's expectations due to a lack of administrative, economic and social capacity (Jacoby 2008). A clear example of grievances leading to violent conflict is South Africa where apartheid and extreme horizontal inequalities led to armed rebellions from the mid-1970s until a transfer of power in 1993 (Stewart 2008). However, despite its popularity in understanding the causes of conflict, grievances have been seen as a satisfying justification, not only for the external community but also for rebel leaders to enhance their propaganda to encourage others to join their cause, when actually greed is the rational (Collier and Hoeffler 2000).

Greed theory

Through a rational choice lens, Paul Collier and colleagues challenged grievance theory by advocating that civil wars are initiated and continue to persist through individuals' desire for greed (Collier and Hoeffler 2000). They claimed that 'opportunities for primary commodity predation cause conflict, and that the grievances which this generates induce diasporas to finance further conflict' (Collier and Hoeffler 2000, 27). Although grievances do exist in societies, especially those between fractionalized ethnic and religious groups, data demonstrated that this does not result in a bigger risk of conflict; instead, greed is the true instigator. By providing quantitative data on how access to and control of natural resources fuels conflict, greed theory appealed to international organizations as it created targeted actions for peacekeeping and helped pinpoint key perpetrators such as armed groups, local elites or organized criminals. Greed theory fit 'neatly with the neoliberal interventionist zeitgeist emanating from the United States in particular' during the 1990s and 2000s (Keen 2012, 767).

The proliferation of access to natural resources, namely diamonds and oil, are cited as motives for the civil wars in Angola, Columbia, Sierra Leone and Sudan to name a few (Jacoby 2008; Ratsimbaharison 2011). However, critiques of greed theory are plentiful, with many focusing on the fact that these perspectives ignore the roles of states and delegitimize social, economic and political inequalities and the protests that ensue in response to these (Keen 2012). Greed theory is used as a way to justify the imposition of neoliberal economic policies after some form of military intervention because regime change is easier to implement from a Western perspective than addressing entrenched inequalities and grievances (Keen 2012). Neoliberalization would ensure that the opportunities are reduced for primary commodity predation by placing tight controls over their exportation (Ratsimbaharison 2011). However, basing such interventions on statistical models created using data on conflict-affected countries is problematic, as it is difficult to account for coverage, completeness and accuracy (Beswick and Jackson 2018). A final critique to mention is from Keen (2012), that even if greed is seen to be driving violence and conflict, something drives that greed. Greed and grievances should not be seen as standalone causes of conflict as their interactions are just as important.

The environment

As discussed with reference to the greed theory, conflicts over natural resources are nothing new. Indeed, access to and control over natural resources are a common cause of conflict throughout human history. However, since the 1970s environmental issues have received increasing attention from conflict theorists, resulting in a body of scholarship on the environmental causes of conflicts.[2] Within this a number of debates have emerged on population growth and displacement, environmental degradation, resource scarcity, and climate change all situated within wider discussions of political ecology and underdevelopment (Galgano 2019). Today, environmental security (analysing the linkages between the environment, political instability and violent conflict) is now an accepted paradigm in understanding the causes of conflict and could, according to Galgano be the 'tipping point that advances violent conflict in regions that may already be on the brink of instability' (Galgano 2019, 3). In particular, states in the Global South experience rapid urbanization and rural-urban migration, which puts pressures on demands for food, water, land and energy supplies. This leads not only to further underdevelopment, but to greater security challenges within stagnant or declining economic circumstances. McMichael's (2016) analysis of land conflict in informal settlements in Juba, South-Sudan, highlights the complex nature of framing and theorizing such conflict when multiple factors are at play. For the most vulnerable in urbanizing cities, battles over resources are rarely confined to their local proximity, but are often managed and controlled by external actors (McMichael 2016).

Structural violence

When examining the relationship between conflict and development, we should not ignore the concept of structural violence, as coined by Johan Galtung (1969). This type of violence accounts for the difference between potential and actual, referring to non-behavioural or indirect violence. As Jacoby (2008) notes, 'Galtung's construction of structural violence offers a way of understanding conflict causality both freed from the constraints of behavioural evidence and as a possible instigator of overt goal incompatibility' (Jacoby 2008, 40). On this account, violence could be experienced as the difference in life expectancies between the lower and upper classes; or more manifestly, violent protests in response to a lack of employment or educational opportunities in a society. The fundamental point is that the structures in society are organized in such a way that result in unnecessary suffering or even death. As Uvin (1998, 103) has pointed out, 'the poor are denied decent and dignified lives because their basic physical and mental capacities are constrained by hunger, poverty, inequality, and exclusion.' This is exemplified by HIV and AIDS in Haiti, where there is little opportunity for patients to receive treatments for the disease thus reducing life expectancy, in stark contrast to the USA for instance. Structural violence may not always lead to manifest conflict, but it can be a precursor to small-scale violence and poor development outcomes.

Impact of conflict on development

The impact of violence and conflict on short and long-term processes of development are significant. Notwithstanding the financial and economic costs of war for the affected country, the human and social costs are high. Deaths and injuries as a consequence of violence affect the social networks, livelihoods and income generation of households and communities. Threats from unexploded ordnance also impact morbidity rates and development outcomes related to subsistence farming and school attendance (Guo 2020). Further, conflict increases the spread of disease, reduces mortality and morbidity rates and as mentioned earlier, impacts educational attainment, along with increasing flows of migration and causing the displacement of people (Chamarbagwala and Morán 2011). All of these affect any country's ability to reach the Sustainable Development Goals. Three areas tend to feature most prominently in my teaching of conflict and development, namely, displacement, gender and conflict, and security and peacebuilding. I do not imply that these are the most important or significant for processes of development, however, for my students of development and humanitarian studies, it is these three which tend to appear most in discussions pertaining to how the theories on the causes of conflict shape the impact on human and social development.

Displacement

Irrespective of whether a conflict is interstate, intrastate or more localized, an inevitable consequence is the mass movement of people. In the aftermath of World War II, it is estimated that almost eight per cent of the world's population were displaced (Gatrell 2013, 3). More recently, the number of displaced people as a percentage of the global population is lower (1 per cent by the end of 2019) but the actual numbers are still high, with UNHCR estimating that 80 million people are currently forcibly displaced of which 85 per cent are hosted in developing countries (UNHCR 2020). Displacement as a consequence of war and conflict affects people on multiple levels, including a need for food, shelter, healthcare, education and loss of livelihoods, to name a few.

While often portrayed as living in camps, refugees are rarely stationary, moving from camps to alternative spaces of refuge and back again when needs arise (Gatrell 2013). Nonetheless, development outcomes for refugees and the displaced are poor. UNHCR's data shows that refugee children are five times more likely to be out of school than their non-refugee peers (UNHCR 2020). Further, the displaced exhibit poor physical and mental health as a consequence of both trauma experienced during their forced movement and by being in a camp or temporary shelter (Khan and Amatya 2017). Displacement, whether internal or external, impacts development outcomes on host communities. Research by Maystadt and Verwimp (2014) on the consequences of Rwandan and Burundi refugees showed that there were winners and losers for the host community. Considering most refugees and internally displaced people are hosted in the developing world, pressures are placed on health systems and food distribution in particular.

Gender and conflict

When considering the impacts of conflict on gender, it is unhelpful to apply blanket statements that conflict affects men and women differently; however, it is important to interrogate the gender dynamics of different contexts and circumstances (Mac Ginty and Williams 2016). The change in scope of conflicts means that traditional armies are not always the main combatants nor victims within civil wars or new wars, instead, conflict permeates throughout the affected society. In this regard, there is a blurring between the roles which men and women play in a conflict. Yet, as has been commented on regarding the conflict in Darfur, there is very little research on the impact of conflict on women, aside from work on gender-based and sexual violence (Read 2019).

Sexual violence in conflict has always existed, but the stories have only recently started to be documented and analysed in the public domain and seen as a weapon of war since the 1990s. Prior to that, its existence in conflict has been marred by an accepted silence (Crawford 2017).

In particular, the atrocities committed during the Yugoslavian and Rwandan wars brought the topic into the scholarship on conflict (Crawford 2017). Further, the incidences of sexual violence do not evaporate once a conflict has finished, which is why it has a significant impact on processes of development. Sexual violence is a feature of the post-conflict environment, too, as gender statuses are readdressed in societies when combatants return to their civilian life (Bouta et al. 2005). The stigma for victims and any children resulting from sexual violence can lead to increased rates of poverty, although this has resulted in a plethora of initiatives aimed at victims. Hilhorst and Douma (2018) describe the case of the Democratic Republic of Congo where women who were not victims had little choice but to enrol in sexual violence programmes to receive medical, social and economic assistance, resulting in an uneven distribution of development interventions.

Women (and indeed children) however, should not only be viewed through a sexual violence lens when analysing the impacts of conflict on development. Disarmament, demobilization and reintegration (DDR) into society should also feature women. Research shows that women are involved in fighting; they may also serve as cooks, porters, administrators, medics, PR, spies, partners and sex slaves (Bouta et al. 2005). DDR programmes often fail to target women's multifaceted roles (whether voluntary or coerced), affecting the different economic, social, and psychological needs of female ex-combatants.

Security, peacebuilding, and development

Within any analysis of conflict and development, we turn to processes of reconstruction; specifically, security and peacebuilding. Since the emergence of new wars in the 1990s and more so with the War on Terror, there has been a coming together of security and aid agendas, as international development departments align with foreign policy and security interests. The most recent example of this being the British Foreign Office absorbing the Department for International Development in 2020. Within these agendas is the understanding that poverty and underdevelopment cause insecurity and vice versa (Duffield 2014). This is reflected in the remits of international organizations whereby securing development and stability mean they should be aware of 'conflict and its effects and, where possible, gear their work towards conflict resolution and helping to rebuild war-torn societies' (Duffield 2014, 1).

Agendas for peace are framed within a discourse of liberal peace, whereby a 'system of carrots and sticks' are implemented to ensure that development aid and access to systems of global governance are conditional on states willingly adopting western-directed economic, political and social processes (Duffield 2014, 34). Within this is the assumption that peace can be made, created or constructed. That this is rarely the case does not seem to affect the interventions from international peace keeping and peace building initiatives. Often, as Mac Ginty (2013, 1) points out, peacebuilding is a 'by-product or happy accident of other processes that are not directly related to peacebuilding.' In this regard, processes of development when focussed on community action, may lead to better outcomes for peace. For example, in the case of Myanmar's Kayin State, national discussions on peacebuilding had been stalled by setbacks which did little to address the original causes of the conflict. Research by Décobert (2020) demonstrated that local, community health initiatives had a much better impact on peacebuilding; through building health capacities 'structural drivers of poor health systems and outcomes were being redressed and consequently, inequalities and injustices which had contributed to the conflict were being addressed' (Décobert 2020, 9). Such examples demonstrate a move in the scholarship away from norms of liberal peace to ideas of peace formation which promotes local agency and sustainable, emancipatory and hybrid versions of peace building (Mac Ginty 2013).

Critically, this example prompts us to ask whether peacebuilding requires national or international interventions in the first place; any external involvement forges hierarchies of injustice and inequalities.

Bringing conflict and development into the classroom

As with Development Studies, teaching Conflict Studies (and specifically the theories outlined above) requires an interdisciplinary approach. In order for students to learn how development is affected by conflict, they need to understand the historical, political and economic processes which lead to violence and conflict in affected societies. Further, responding to conflict cannot just be seen through a development as intervention lens, as this negates the actions of individuals, groups, communities, and states which are working with, for and against the consequences of conflict; therefore, teaching should include lessons on public and global health, political sociology, anthropology, and economics. Multi-disciplinary approaches in conflict teaching ensures that our students, as future development and humanitarian practitioners or academics, engage with conflict-affected societies in a context-specific fashion while taking into account processes (whether historical or in the present) which affect the day-to-day outcomes of those living through conflict and violence.

In addition to providing an interdisciplinary programme, teaching violence presents a host of trauma inducing subject matter. We must therefore be mindful of those in our classroom, whether aid workers or students from conflict settings, and the mental health burden that exposing them to further trauma and violence in our teachings and materials have on their learning experience (Davey, Read and Hawkins, 2021). In this regard, we should not shy away from teaching difficult material, but we should develop a pedagogical environment where students learn in a safe manner while addressing the emotional toll this learning has. We have developed a trigger warning system for difficult material that we believe needs to remain in the curriculum. Experiences show that students who are forewarned are better able to handle such material or may then opt out of those classes if required.

Bringing conflict and development into the higher education classroom not only requires an examination of the theories and concepts pertaining to the topic, as outlined in the first section, but also a thorough analysis of what these mean in practice. In this regard, it is worth mentioning that conflict and development *in practice* usually results in some form of humanitarian intervention, which is increasingly becoming long and protracted. Conflict and violence inevitably occur on many levels and in many forms and have a multitude of consequences for development outcomes (Mac Ginty and Williams 2016). Indeed, when conflict occurs, often development as intervention does not take place. Instead, humanitarian responses are needed to ensure lives are saved and people receive food and shelter, but as conflicts increase in length, we are seeing humanitarian agencies take on increasingly more developmental roles, sometimes demonstrating explicit political engagement in doing so. That humanitarian response and development intervention have been framed as distinct enterprises has resulted in perceptions that they are different endeavours. Quite often, however, the organizations which administer both types of aid are linked or even the same (Beswick and Jackson 2018). The relationship between the two should be explored and analysed in teaching on conflict.

I have found student research visits bring pedagogical benefits to the complexities of teaching conflict, development and humanitarianism, particularly for final year undergraduates and postgraduates with little experience of development in action. However, for the purpose of introducing students to the theories, concepts and intersections of conflict, development and humanitarianism our faculty has found it beneficial to use one case study to examine these in-depth when working within the confines of a classroom. Other case studies complement the main case study throughout the module, and I do not limit students' choice of case in their assessment; but examining one case in depth allows for an exploration of the nuances of conflict and development. That there are many examples to choose from demonstrates the complex reality of protracted conflicts on development outcomes. For example, the Democratic Republic of Congo (DRC) consistently ranks repeatedly at the bottom of development indices. The country has been mired by conflict since the mid-1990s and despite containing a wealth of natural resources, in 2018 it registered 72 per cent of the population living off less than $1.90 a day (World Bank 2020). These criteria make the country an option as a consistent case study to examine throughout a course on conflict and development.

The conflicts within the DRC have multiple actors with multiple causes, intersecting the theories analysed above. Grievances, through ethnic antagonism; the proliferation of natural resources and their unequal distribution, both geographically and economically; the fragility of the national government; and structural inequalities reaching back to the colonial period have all combined to create a situation where peacebuilding requires development and vice versa, and yet both seem unachievable aims (Marriage 2013). As with the causes, the consequences of protracted conflicts in the DRC, notwithstanding the poverty rate, include massive loss of life, population displacement, degradation of government and healthcare infrastructures and human rights violations, including widespread sexual violence (Marriage 2013). The conflicts have expanded beyond the DRC's borders transforming into a global agenda of international intervention.

The case demonstrates failures of peacebuilding and humanitarian intervention for development outcomes as they centred on external intervention, disarmament, demobilization and reintegration programmes and transformation of the political and economic infrastructures along neoliberal lines (Marriage 2013). Yet, as Marriage (2013, 3) has pointed out, neither the peacebuilding processes nor humanitarian intervention had a straightforward link to the 'material condition of people in the Congo.' The variety of causes, interventions and consequences of the conflict in the DRC provide the scope for examining conflict and development in the classroom through a consistent case study which allows for intended learning outcomes such as reflecting on the shifts and changes of theories and concepts over time and space and fostering room for critical thinking of the scholarship and practice of conflict, development and humanitarian intervention.

Conclusion

This chapter has examined the often distinctly separated fields of conflict and development. The first part focussed on a thematic analysis of the theories pertaining to conflict and development which tend to feature in higher education. The two topics are often taught and studied in silos; yet many of the poorest countries in the world are or have experienced conflict on various scales. The chapter demonstrated why the two need to be taught together, outlining the linkages between conflict-affected societies and development. I discussed the different types of conflict which exist in countries with poor development outcomes and examined some of the most prominent theories on the causes of conflict. The section concluded that oftentimes in practice these theoretical causes intertwine. The chapter then moved on to discuss some of the main effects of conflict on development outcomes, namely displacement, gender inequalities and consequences of increased security and peacebuilding. Oftentimes, reconstruction efforts neglect to examine the complexities of the choices that displaced people make and the different needs and experiences of women post-conflict. Further, neoliberal understandings of how states should be reformed or rebuilt post violence have resulted in unstable democracies and the roots causes of conflicts unaddressed.

In the third part of the chapter, I discussed how I bring conflict into the Development Studies classroom. There is little research, as yet, on the effects of teaching violence and conflict in the classroom on students and faculty, yet studies on dealing with trauma in other academic fields show there is a need to develop strategies and guidance. This is an area that pedagogies of humanitarianism and development need to explore in more depth. By better understanding how to teach the subject matter related to conflict and development, academics can help to bridge the gap between the two fields of enquiry. Further, in developing this bridged agenda, educators need to understand the critical role humanitarian action plays in adjoining the two fields. Neither conflict nor development should be studied without an application of humanitarian response; in teaching, this means more collaborations between departments and institutes which may previously be seen to sit within one or the other camp. Finally, I recommended that in advocating for a single case study approach at undergraduate level, students gain the opportunity for an in-depth analysis of the thematic issues. Research on student feedback and ability to reach intended learning outcomes would be a vital addition to the literature on development pedagogy, not only for teachings of conflict but also in other areas of the field.

As development agendas move forward so should the space for understanding how conflict and its consequences negatively impact social, economic and political development outcomes. Studying the two fields together ensures that people in many of the poorest countries are included in discussions on better forms of development *and* humanitarian intervention. Studying processes needs to take into account not only the protracted nature of some conflicts but also the complex causes which underpin, and in some instances maintain, violence in these societies.

Key sources

1. Jacoby, T. (2008). *Understanding Conflict and Violence: Theoretical and Interdisciplinary Approaches*. Routledge.

In this accessible book, Jacoby provides the reader with a theoretical analysis of the causes of conflict and violence. Through nine approaches, the author not only situates the theory within their historical context but uses a variety of case-study examples to demonstrate their relevance to practice.

2. Mac Ginty, R., and Williams, A. (2016). *Conflict and Development (Second)*. Routledge.

In this second edition, the authors bring together perspectives on conflict, poverty, and development to understand the complex challenges countries *and* people experience when faced with violent conflict and its impact on underdevelopment. A really insightful text for students in both fields.

3. Marriage, Z. (2013). *Formal Peace and Informal War: Security and Development in Congo*. Routledge.

This monograph brings the debates on security and international interventions into discussions pertaining to conflict and development. Using the case of the Democratic Republic of Congo, the author argues that peace agreements established by Northern donors and political elite failed to acknowledge the security needs of the general population.

4.   Gatrell, P. (2013). *The Making of the Modern Refugee*. Oxford University Press.

This text is essential for any student trying to understand the state of contemporary displacement throughout the world. Through its historical analysis, the author demonstrates how conflict, state formation and revolution have caused mass forms of displacement in history; seen by policy makers and humanitarians as problems that have to be solved. Central to this history are the meanings and stories the refugees themselves construct of their experiences.

5.   Beswick, D., and Jackson, P. (2018). *Conflict, Security and Development: an Introduction (Third)*. Routledge.

This textbook neatly provides students with an analysis of the relationship between security and development, before, during and after conflict. Through a case-study approach, the authors account for a number of key themes within theory, practice and policy which they believe shapes the dynamics of security and development. These include an analysis of conflict, displacement, the role of international organizations, peacekeeping and peacebuilding, post-war economic development, security, and justice.

Notes

1 Of which Stewart categorizes into four areas: political participation, economic aspects, social aspects, or cultural status (2008, 13).
2 Although this body of scholarship is rather fragmented.

References

Beswick, D., and Jackson, P. (2018). *Conflict, Security and Development: An Introduction* (3rd edn.). London: Routledge.
Bouta, T., Frerks, G., and Bannon, I. (2005). *Gender, Conflict and Development*. Washington DC: The World Bank.
Chamarbagwala, R., and Morán, H. E. (2011). The Human Capital Consequences of Civil War: Evidence from Guatemala. *Journal of Development Economics*, 94, 41–61.
Collier, P., and Hoeffler, A. (2000). *Greed and Grievance in Civil War*. In Policy Research Working Paper 2355. Washington DC: World Bank.
Crawford, K. F. (2017). *Wartime sexual violence: From silence to condemnation of a weapon of war*. Washington DC: Georgetown University Press.
Davey, E., Read, R., and Hawkins, J. R. (2021.). Violence in the classroom: Navigating trauma in humanitarian studies. British Academy Journal.
Décobert, A. (2020). Health as a bridge to peace in Myanmar's Kayin State: 'Working encounters' for community development. *Third World Quarterly*, 42(1), 1–19. https://doi.org/10.1080/01436597.2020.1829970
Duffield, M. (2014). *Global Governance and the new wars: The merging of development and security* (2nd edn.). London: Zed Books.
Galgano, F. (2019). The environment–conflict nexus. In F. Galgano (Ed.), *The environment-conflict nexus: Climate change and the emergent national security landscape*. New York: Springer.
Galtung, J. (1969). Violence, peace, and peace research. *Journal of Peace Research*, 6, 167–191.
Gatrell, P. (2013). *The making of the modern refugee*. Oxford: Oxford University Press.
Grimm, S., Lemay-Hébert, N., and Nay, O. (2014). 'Fragile states': Introducing a political concept. *Third World Quarterly*, 35(2), 197–209. https://doi.org/10.1080/01436597.2013.878127
Guo, S. (2020). The legacy effect of unexploded bombs on educational attainment in Laos. *Journal of Development Economics*, 147, 102527. https://doi.org/10.1016/j.jdeveco.2020.102527

Hilhorst, D. and Douma, N. (2018). Beyond the hype? The response to sexual violence in the Democratic Republic of the Congo in 2011 and 2014. *Disasters*, 42(1), S79–S98. https://doi.org/10.1111/disa.12270

Jacoby, T. (2008). *Understanding Conflict and Violence: Theoretical and Interdisciplinary Approaches*. London: Routledge.

Keen, D. (2012). Greed and grievance in civil war. *International Affairs*, 88(4), 757–777.

Khan, F., and Amatya, B. (2017). Refugee health and rehabilitation: Challenges and response. *Journal of Rehabilitation Medicine*, 49(5), 378–384. https://doi.org/10.2340/16501977–2223

Mac Ginty, R. (2013). Introduction. In R. Mac Ginty (Ed.), *Routledge handbook of peacebuilding*. London: Routledge.

Mac Ginty, R., and Williams, A. (2016). *Conflict and Development* (2nd edn.). London: Routledge.

Marriage, Z. (2013). *Formal peace and informal war: Security and development in Congo*. London: Routledge.

Maystadt, J.-F. and Verwimp, P. (2014). Winners and losers among a refugee-hosting population. *Journal of Economic Development and Cultural Change*, 62(4), 769–809. https://doi.org/10.1515/9783110251357.1

McMichael, G. (2016). Land conflict and informal settlements in Juba, South Sudan. *Urban Studies*, 53(13), 2721–2737. https://doi.org/10.1177/0042098015612960

Pettersson, T., and Öberg, M. (2020). Organized violence, 1989–2019. *Journal of Peace Research*, 57(4), 597–613. https://doi.org/10.1177/0022343320934986

Putzel, J., and Di John, J. (2012). Meeting the challenges of crisis states. Crisis States Research Centre, London: LSE.

Ratsimbaharison, A. M. (2011). Greed and civil war in post-Cold War Africa: Revisiting the greed theory of civil war. *African Security*, 4(4), 269–282. https://doi.org/10.1080/19392206.2011.629552

Read, R. (2019). Comparing Conflict-related Sexual Violence: Expertise, Politics and Documentation. *Civil Wars*, 21(4), 468–488. https://doi.org/10.1080/13698249.2019.1642613

Stewart, F. (2008). Horizontal inequalities and conflict: an introduction and some hypotheses. In F. Stewart (Ed.), *Horizontal inequalities and conflict: Understanding group violence in multi-ethnic societies*. London: Palgrave Macmillan.

Turner, M., Hulme, D., and Mccourt, W. (2015). Governance, management and development: Making the state work (2nd edn.). London: Palgrave.

UNHCR. (2020). Figures at a Glance. www.unhcr.org/uk/figures-at-a-glance.html, date accessed 12th October 2020.

Uvin, P. (1998). *Aiding violence: The development enterprise in Rwanda*. West Hartford, Connecticut: Kumarian Press.

World Bank. (2011). Conflict, security, and development. In W. Bank (Ed.), *World Development Report*. Washington DC: The World Bank Group.

World Bank. (2020). The World Bank in DRC. www.worldbank.org/en/country/drc, date accessed 1st September 2020.

48

CHILDREN, YOUTH, AND DEVELOPMENT

Vandra Harris Agisilaou

Introduction

Children and youth are gaining increased attention in development. They are addressed directly in the Sustainable Development Goals and measured by emerging tools including the Early Childhood Development Index, Early Human Capability Index and Multiple Overlapping Deprivation Index. It would seem that they have a well-defined space in development, yet in this chapter I argue that they remain peripheral in development thinking and practice. This is largely as a result of limited understanding of the critical developmental changes and consolidation that take place throughout this formative period in a person's life, and a failure to view them as already valuable people. I argue that development practitioners need a better understanding of child development to support more effective engagement with the scale and diversity of human development, and to enable children to fully realize their capacity as active agents in the present and the future.

Children and youth

Good development action and policy are underpinned by good data and are fundamentally shaped by how the data is disaggregated; for example, disaggregating data by sex allows us to understand how men and women experience poverty differently. In national and global data produced by organizations including the World Bank, the OECD and UN agencies, the most common age disaggregation divides the global population into three age groups: 0–14, 15–64, and 65+. Measuring working age people (15–65) in opposition to younger and older populations who are viewed as dependent on them reflects an economic rationale that the ratio of workers to dependents is key to 'the economic and social functioning of societies' (Ritchie and Roser 2019). Aside from the question of counting people only in terms of their measured economic contribution, this approach characterizes 0–14 year-olds as notable only for what they need, and reflects a tendency to view children as human *becomings* – people whose value lies in what they will become not who they are now, 'redeem[ing] their promises only when they are adults' (Qvortrup 2009, 632). At the same time, counting 15–24 year-olds only as working adults ignores the 'profound importance' of their contribution now and in the future

DOI: 10.4324/9781003017653-53

(UN 1981), and overlooks the importance of the developmental and social changes in this age of transition from childhood to adulthood.

Despite these being the dominant categories for global, regional and national population counts, other specific measures use different ranges, making data comparison challenging. For example mainstream child nutrition and mortality normally count children aged 0–5 years and some measures of extreme poverty categorize a child as aged up to 18 (e.g. UNICEF and World Bank 2016, 2) consistent with the Convention on the Rights of the Child. Age disaggregated data is often simply not collected for youth, perpetuating their invisibility (see Evans and Lo Forte 2013, 2). For clarity, I will refer to 0–14 year-olds as children and 15–24 year-olds as youth. This is primarily a pragmatic choice balancing data availability, normative categories, and the significant human development that takes place across this broad age range.

With this in mind, some general figures are useful to advance this discussion. The proportion of the global population aged 0–24 peaked at 54 per cent in 1980 and has declined steadily to 41 per cent in 2020, with projections of continued decline to 35 per cent by 2050 (UN-DESA 2019). There are regional variations within this: children and youth make up around 40 per cent of population in Asia, Oceania and Latin America and the Caribbean, while Europe and North America are lower at 26 per cent and 31 per cent respectively, and Africa sits at 60 per cent. Across all regions these numbers comprise more children than youth, at a ratio of just over 3:2, except in Africa where the ratio is 2:1.While numbers are expected to decline in all regions in coming decades, they will remain above 50 per cent in Africa until after 2050 (UN-DESA 2019).

Children are over-represented in poverty figures because 'poor households in all regions typically have more children than non-poor households' (Watkins and Quattri 2016, 19). This is particularly stark in Sub-Saharan Africa, which is expected to account for 90 per cent of child poverty by 2030 (Watkins and Quattri 2016, 19). While improvements have been made, projections on the reduction of child poverty globally may prove to be overly optimistic, as the coronavirus pandemic is expected to throw a further 62–86 million children into poverty by the end of 2020 (UNICEF and Save the Children 2020, 3). The pandemic was already hitting youth hard by mid-2020, with one in eight losing access to schooling, one in six losing jobs, and two in five losing income (ILO 2020, 2).

Poverty has devastating impacts on all age groups. Importantly its many facets impact a child's development in ways that can embed disadvantage before a child turns two (Boyden 2019; Heckman 2008). Disadvantage emerges and is consolidated very early in life, and deprivation in a child's early environment will 'put the fullest realization of children's development at risk' (Degen Horowitz 2000, 6). Less attention has been given to youth poverty, despite high rates and significant impact including 'health risks as well as an inability to attain 'social adulthood,' through the markers or functions that legitimate the transition to adulthood (Ansell 2017, 55).

At a policy level there is broad acceptance that there are powerful and enduring negative impacts of childhood poverty (Boyden 2019, 49), but this is not reflected in funding of effective, child-centred programmes. A persistent view of children (and youth) as passive and powerless rather than valuable and active members of their societies leads to what Qvortrup (2009, 634) describes as 'a looming indifference' towards all aspects of childhood except those that prepare for adulthood. This erasure of young people as engaged human beings takes place when we develop programmes without a full understanding of children as complex and capable agents in the here and now, rather than confined to their trajectory towards adulthood.

Unfortunately, when children and youth are seen by policy makers, they are often viewed in negative terms. One reason is the dependency ratio described above, whereby a high proportion of children in a population is seen as inhibiting growth and development. Much of the discourse about children frames them as 'passive receptacles of provision and protection, inept and vulnerable' (Qvortrup 2009, 632). Attention to children in developing countries is similarly focused on the ways children need protection – primarily from abuse by adults in the forms of conflict, child labour, child marriage, but also protection from damage caused by poor health, poverty, forced migration, poor education, denial of rights, and loss of parental protection (UNICEF 2019, 7).

Where they are considered outside the broad 'adult workforce' category, youth inhabit a liminal space of being no longer children but not yet fully adults, and they have long been viewed in opposing stereotypes of 'problem and pathology' or 'possibility and panacea' – either the cause of or solution to all our problems (Sukarieh and Tannock 2018, 855; Ansell 2017, 47). A large proportion of youth in a population, or 'youth bulge,' is viewed as threatening due to an essentialized understanding of young men that expects negative economic outcomes because 'violence is more likely to erupt in regions with a high proportion of young men in the population' (Pruitt 2020, 711). A fuller understanding of child and youth development helps in developing approaches that truly expand opportunities and target poverty-reduction more effectively. For both children and youth, a deficit-based approach persists despite the efforts of young people, and child-focused NGOs, civil-society, and academics, who seek to promote children's desire and capacity to be agents in their own development.

Child and youth development

Just as in other fields, understanding child and youth development requires complex, specialized expertise, however even a basic foundation in this specialist knowledge can help practitioners make developmentally appropriate decisions that support child flourishing, and build foundations for future health and success. Research across disciplines shows convincingly that despite environmental variations and social and cultural diversity, there are shared developmental pathways and processes – coherent patterns of development, particularly in the earliest years of life (see Degen Horowitz 2000, 5; Spencer et al., 2006). A simple example of this is that it is extremely rare for an infant to walk before crawling, and years of close study have shown that the cognitive and physical components of crawling are important precursors to the skill of walking (Spencer et al. 2006, 1524).

The 19th and 20th-century works of European and North American men including Fröbel, Piaget and Bronfenbrenner have significantly shaped understanding of the ways learning takes place and can be fostered in childhood. Their work emphasized the importance of children's activity, particularly their self-driven learning through the critical medium of play. More recent work has been more diverse and internationalized, tending to prioritize children's agency and rights, and view childhood as expressed and experienced within diverse social, cultural, economic and political contexts. From the early 1990s, fields including the sociology of childhood, children's geographies, and childhood studies have centred on children as already valuable, competent beings (McNamee 2016, 31). Children are understood 'in their own right as socially constructed agents' (Wall 2019, 2) who are active participants 'in the construction of their own social lives, the lives of those around them and of the societies in which they live' (James and Prout 2015, 4).

There has also been enormous growth in our understanding of the critical role of brain development in young children and in youth. The first 1,000 days of life (from conception to age 2) are normally critical in establishing functioning and capacity that underpin all the learning and growth of later childhood, adolescence and adulthood. Research clearly demonstrates that 'a person's responses to internal and external stimuli depend on critical pathways and processes formed in the brain' that are created and shaped by experiences in early childhood (Young 2007, 2–3). These connections are stimulated by both repetitive and novel interactions with people and environment, and both quantity and quality of interactions matter. This early brain development impacts our cognition – our thinking and intellectual functioning – but also powerfully influences our outcomes in health, poverty, earning capacity and wellbeing (Heckman 2008, 29). Interactions between nutrition, cognition, language, movement and all aspects of development are extraordinary and mutually amplifying (Watkins and Quattri 2016, 24). Poor nutrition in this period can lead to irreversible stunting and severely limit brain development, with a lifelong impact on a child's capacity, while strong social and emotional development underpin development of healthy behaviours and strong cognitive skills (Heckman 2008, 32).

The development taking place in these earliest years, long before school or pre-school, 'can set trajectories in health (physical and mental), behaviour, and learning that last throughout the life cycle' and these pathways are hard to change later (Mustard 2007, 44). The brain development derived from these early experiences creates not only universal traits and skills such as speech and mobility, but also more specific ones. An extremely broadly conceived 'normal, highly likely range' of post-natal environments across diverse contexts provide experiences that 'shape the vast repertoire of nonuniversal behaviours important to functioning in different social, cultural, and economic societies' (Degen Horowitz 2000, 5). That is to say that brain development stimulated by conventional environments – whatever the context – also builds behaviours specific to the child's context.

What we have learned about the profound and enduring impact of development in the earliest years may apply equally to youth. In the last 50 years we have begun to understand that there is a second critical period of development in adolescence, in which there are significant changes in the structure of the brain. In particular, researchers point to emerging evidence of intense change in the areas of the brain concerned with impulse control and risk-taking (Casey et al. 2008, 63; Lee et al. 2018, 47). As in the earliest years this brain development takes place within a still growing body and in the context of an evolving environment and social demands. It therefore sees an intense 'interplay of biological, psychological and environmental changes' (Lee et al. 2018, 46; Boyden et al. 2019, 4–5).

Intervening in early child development

When we understand childhood and youth in these ways, we begin to see how important it is to actively engage with these age groups in any work to redress poverty, inequality, and rights. Where inequality and disadvantage can be firmly entrenched in a person's first 1,000 days – and nearly impossible to turn around in the ensuing 68 years – it becomes hard to think about programming that does not start here. In this section I therefore draw on this knowledge about the critical importance of a child's earliest experiences and opportunities in laying foundations for their future health, wealth, and opportunities. In particular, I focus in on how work in early childhood can address disadvantage and poverty by expanding capabilities at the earliest point. This is an example of how we can use knowledge of child and youth development to design and implement programmes that truly work to expand the opportunities they face now and in the future.

Attention to early childhood in development practice generally focuses on ages four and above, particularly in preparation for school entry. This represents a perception that this is when important development starts and can be influenced. Similarly, children's play is a critical mode of learning yet is often dismissed as unimportant because it is inconsistent with adult expressions of productive value. If we see children only in terms of their future value, we prioritize programmes that focus on childhood as preparation for productive citizenship – usually through education for future work. It has been convincingly argued, however, that waiting until school age to implement programmes for children entrenches poverty by building on, and negating, capacities already established in the early years of child development (Heckman 2000, 50).

Longitudinal studies across the North and the South demonstrate positive impacts for all children (especially disadvantaged children) of programmes engaging children and parents in the earliest years of childhood (see Mustard 2007; Boyden et. al. 2019). Indeed the impact of these early interventions for deprived and non-deprived children alike is so compelling that economists cite returns on investment that 'exceed those associated with any other investment in a country's infrastructure' (Young 2007, 3; Garcia et al. 2016, 1), significantly surpassing programmes for disadvantaged and at-risk youth (Heckman 2008, 31). The positive impacts extend beyond just economic factors, amplified by an array of positive physical and mental health impacts for the child, and further benefits for parents (Heckman 2008, 21).

Early intervention programmes have shown the best outcomes when they combine parental engagement with a form of centre-based or shared care environment. There are several integral components to impactful programmes:

> The best *[early childhood development]* interventions are comprehensive, integrated programs involving parents that combine nurturing and care, nutrition, and stimulation. They focus on the whole child and involve families and communities. Most importantly, they begin early, preferably when a mother is pregnant or soon after she gives birth.
>
> *(Mustard 2007, 78)*

Boyden et al.'s (2019) 'Young Lives' study examined such programmes in four lower and middle-income countries. Outcomes were consistent with those found for programmes in wealthier countries, namely that programme quality directly influences impact; that positive impact is enduring; and that access to the best quality programmes is often limited for those who would benefit most. When these programmes prioritize disadvantaged children and integrate both institutional and family-based components, impacts are substantial and enduring. Such programmes must retain a focus on the family as a child's primary relationship and environment, within a broader community context, because children grow in interaction with these environments.

We can enlarge the opportunity for brain development across children's crucial early years by combining good shared care provisions and support for parental development and work. In high quality shared care environments, children are engaged consistently and in a variety of ways. Engaging in shared care in addition to the family is effective for several reasons. In a home environment parents and other family members do much to enhance a child's early development, but they do so within a variety of constraints including the demands of livelihood, household management and care for other family members. Across developing countries alone, 'at least 35.5 million children under the age of 5 are being left alone, or with other young children, to look after themselves' as parents negotiate the devastating tension between caring for children and providing for them (Samman et al. 2016, 9). Particularly in poor communities, parents may be absent to seek or perform productive work, with little choice but to leave children alone or in the care of older children who are often still very young themselves.

Early childhood care programmes in developing countries can therefore serve multiple functions: keeping children safe; supporting the livelihood strategies of the poor (especially women); and extending children's brain development and supporting health, thereby expanding their future choices (Samman et al. 2016, 69). Care programmes offered for more than a few hours a day can decrease the immediate and enduring impacts of poverty by fostering child development, enabling family livelihood activities and facilitating school attendance for children relieved of caring responsibilities.

High quality programmes take culture, context and family priorities as leading drivers in design, implementation, and evaluation. They also engage children as active participants in these processes, moving away from deficit or passive recipient models even for the youngest children. Providing a strong basis for such an approach, Australia's early childhood (0–5 years) learning framework is founded on 'a view of children's lives as characterized by *belonging, being and becoming*' (DET 2000, 7). This framework is significant in its articulation of the interdependence of a child's community, present being and future realization, and how the three constantly engage in a child. An orientation like this drives a conception of children as valuable people laying down the neural, physical, and social foundations for what they can achieve and be as adults and could provide a strong starting point for development practice.

The 25 years covered in the category of children and youth is over one third of average global life expectancy and encompasses the most intensive physical, cognitive, and social change that humans experience. It is difficult to account effectively for the enormous diversity covered within this broad category. Mohanty described collective treatment of women as 'the production of the 'Third World Woman' as a singular monolithic subject' (Mohanty 1984, 333), and it is critical that we avoid replicating this with a universal child/youth who is passive, dependent, needy and potentially dangerous. Improving understanding of child and youth development throughout these years can reveal not only why these years are so important, but also young people's capacity to be actively engaged in processes in their communities and their world.

Seeing children and youth as people

It is clear therefore that 'programming across the life course needs to start early, but should also be sustained through childhood, both to avoid early gains being lost and to remediate early harms' (Boyden et al., 2019, 5). Yet child-rearing and development remain conceptualized as a parental responsibility contained within the domestic sphere, and together with stereotypes about poor communities. This can make it difficult to imagine that institutional or centre-based early childhood programmes are necessary or appropriate in developing countries even when we accept that they are beneficial in other settings. Unfortunately, even in donor countries the provision of early childhood programmes is rarely viewed in such ways. Policy and programming in much of the North remains imbued with the enduring view of children as passive recipients and a low priority for government spending. Thus, while it has been argued that development policy makers in Northern donor countries understand the importance of the earliest years and these early childhood services are 'firmly on the policy agenda', this commitment yet to be matched financially through resourcing of programmes (Boyden et al. 2019, 49, 4; Ansell 2017, 45).

As we become aware of the disproportionate number of children and youth among those poor, marginalized, and invisible, practitioners must increase their attention to young people as valued human beings in the here and now, with their own capacities, needs and dreams. The near-universal ratification of the Convention on the Rights of the Child might appear to demonstrate globally shared value of children's rights and contributions, but Ganguly Thukral argues that these rights are 'at best … seen as ornamental, and an add-on to the more 'real' problems of society' (2011, 3). Child and youth contributions and participation will necessarily take forms that differ from those of adults, and of course we need to develop appropriate ways to see, hear and be led by them – just as we have worked so hard to do with other marginalized groups.

Effective child participation begins with recognition that 'a child's personhood and her or his ability to take part in decisions that affect her or his life,' as expressed individually, socially and politically (Ganguly Thukral 2011, 22). While it may challenge our beliefs about children's capacities and roles, we must understand even very young children as 'active participants and decision makers' in their own lives, in their families and communities, and in their learning through play (DET 2000, 10; Qvortrup 2009, 632). This can expand the possibilities of what we might do in development practice grounded in communities. When we fail to think about children and youth as active, capable agents we repeatedly minimize and side-line them as human beings who matter now as well as in the future.

A pedagogy of children and youth in development

Development practice must become more attentive to the early years as a critical time of human development and an unparalleled opportunity to address lifelong disadvantage and the generational reproduction of poverty. One strategy for achieving this is by incorporating an effective understanding of child and youth development into development studies teaching programmes, just as gender has become central to these studies. There are few child-focused subjects within development studies programmes, though most universities have the opportunity to work across disciplines to develop outstanding child-focused offerings. Beyond child-focused subjects, children and youth can be central to pedagogy in a range of ways. They can be the focus of our attention when we consider important areas of development, and also a lens through which we consider those issues.

A 'childist' development pedagogy harnesses more recent child development thinking to consolidate important approaches to development. While the term childism does not explicitly include youth, this approach could have significant impact in adult-focused development practice. The focus of childist theory reaches beyond children and their relationships, and into 'the social and political foundations on which children's lives and experiences are already imagined and pre-constructed' (Wall 2019, 4). A critical aspect of childism is the notion of social interdependence, and the role it plays in reducing marginalization. Children have severely constricted opportunities to speak in ways that influence processes defined and dominated by adults, and thus childism argues that 'differences are demarginalized only interdependently: both on one's own behalf but also through responses on one's behalf from others' (Wall 2019, 10). In other words, it requires a more mutual process than a straightforward devolution of power or 'handing over the stick.'

Development is rightly concerned with pulling people out of poverty, and indeed ending it altogether as embodied in as the first Sustainable Development Goal (end poverty in all its forms everywhere). This is part of a hopeful narrative about the change that we can make in the world, but this narrative rarely addresses the physical, lifelong impact of being born into poverty. Incorporating a strong understanding of the developmental impact of poverty in the first 1,000 days of life can impact this discussion in two key ways. First, it can point to the urgency of addressing the task, for every child conceived in poverty is steeped in disadvantage that extends beyond just being hungry and poorly clothed. Rather it causes profound damage in health and cognition, which in turn reduce school success, and survivability in the dirty, degrading, and dangerous jobs available to the poor. Second, it highlights that poverty is not simply erased. Development interventions targeting adult populations may be able to increase incomes and draw people away from the perilous edge of survival, but they cannot redress stunting and other physical and cognitive impacts of early poverty. Likewise, pathways for youth to attend school that fail to address the developmental impact of poverty in the early years of brain development and deficits in their early education will achieve limited learning results. A childist development pedagogy sees the enduring impact of childhood poverty and entrenched disadvantage, and recognizes the ongoing disparities that these cement across global societies.

Children and youth are present in most development focus areas – conflict, migration, poverty, gender, sexuality, disability, employment and so on. The child-focused case studies emerging from organizations including Plan, Care, Save the Children and UNICEF offer strong opportunities for perspective-taking on contemporary issues. Employing a deliberate strategy of amplifying the voices of young people, these publications (and those that will continue to emerge) offer helpful teaching tools. They are windows into lives that are complex and multi-dimensional; lives that are not defined by any single thing, and which in particular are often characterized by hope. These are a remarkable tool for examining development concerns, drawing them from the abstract to the concrete and personal.

Within a childist development pedagogy we must understand and truly see children within a dispersed inclusive development, in which practitioners are resourced for the incredible complexity we demand of them, and are able to see beyond a 'default human' of a particular age, gender, class, colour and all the myriad nuances that cause us to overlook, disregard and exclude people. In this sense it sits within the inclusion narratives of so much of development. It also coheres with the localization agenda which is in many ways a continuation of the long struggle of development to shift the balance of control in development practice and relationships. Recognizing that children's engagement and participation require different forms of engagement and participation from adults can help us to see, for example, that localization equally requires us to transform our processes so we become interdependent development actors, not simply arms-length funding mechanisms with local implementation. We are still struggling to realize a development in which these preconceptions of people's lives and goals are relinquished and instead they truly lead and define their own development.

Finally, a childist pedagogy can also help students to engage with concepts of invisibility and exclusion. Many students are youth themselves or have been recently enough to vividly recall the experience. It is not just in developing countries that youth are treated with suspicion and seen as having little legitimate to offer to adult processes. This means that our students may be able to recall their own desire to participate in design, implementation and evaluation of processes fundamentally affecting them – and being at times unable to do so. This perspective-taking may help them to understand how people can be overlooked, seen to have little to offer in a situation that is fundamentally their own. In thinking about how they themselves might have been included, they may be better able to engage with ideas of participation and localization.

Conclusion

In this chapter I have argued that we must see children and youth in development: render them visible, present, and active. We must conceive of them not in terms of adults-in-progress, but as human beings now and well beyond our own time as adults; as people worthy of respect and dignity; and as capable actors who are subjectively engaged in their own present and future. Debates around poverty, invisibility, participation, and localization can all be expanded through an approach that accounts actively for the 3.2 billion children and youth globally, and in particular those most disadvantaged. A childist pedagogy connects with many contemporary development concerns, offering a shift in perspective that may help our students (and us) to see the issues afresh. Engaging effectively requires that we understand and value the ways that children and youth are both being and becoming, and the ways they belong as integral parts of the communities we love, reside and work in, and hope and strive for. We must see them in all points of the vast range of experience encompassed in the classifications of children and youth, and we must work with them to ensure that development is appropriate and meaningful for them now and in future.

Further reading

For further reading on this area I recommend several key sources. Nicola Ansell (2017) gives a comprehensive analysis of children and youth in development practice and contexts. Kevin Watkins and Maria Quattri (2016) discuss the distribution and impact of extreme poverty on children. Jo Boyden and colleagues (2019) present a critical study of the enduring impact of poverty on children's lives in four countries and expand this across many other publications. The work of J. Fraser Mustard (2007) constitutes a comprehensive and accessible review of early childhood brain development, and also engages with economic impacts of early childhood programs. Lee and colleagues (2018) provide a comparable discussion of the emerging understanding of brain development in adolescents.

References

Ansell, N. (2017) *Children, Youth and Development.* London and New York: Routledge.

Boyden, J., Dawes, A., Dornan, P. and Tredoux, C. (2019). *Tracing the Consequences of Child Poverty: Evidence from the Young Lives study in Ethiopia, India, Peru and Vietnam*. Bristol: Bristol University Press.

Casey, B.J., Getz, S. and Galvan, A. (2008). The adolescent brain. *Developmental Review* 28(1), 62–77.

Degen Horowitz, F. (2000) Child development and the PITS: Simple questions, complex answers, and developmental theory. *Child Development*, 71(1), 1–10.

DET (2000) Belonging, being and becoming: The early years learning framework for Australia, Australian Government Department of Education and Training for the Council of Australian Governments. www.education.gov.au/early-years-learning-framework-0

Evans, R. and Lo Forte, C. (2013). *A Global Review: UNHCR's Engagement with Displaced Youth*. Geneva: United Nations High Commissioner for Refugees – Policy Development and Evaluation Service.

Ganguly Thukral, E. (2011). Children and governance: Concept and practice. In E. Ganguly Thukral (ed.) *Every right for every child: Governance and accountability*. London: Routledge, 1–42.

Garcia. J. L., Heckman, J. J., Leaf, D. L. and Prados, M. J. (2016). The life-cycle benefits of an influential early childhood program. NBER Working Paper 22993, Cambridge: National Bureau of Economic Research. www.nber.org/papers/w22993.

Heckman, J. J. (2000). Policies to foster human capital. *Research in Economics*, 54(1), 3–56.

Heckman, J. J. (2008). Schools, Skills, and Synapses. NBER Working Paper 14064. Cambridge: National Bureau of Economic Research. www.nber.org/papers/w14064.

ILO (2020). Youth and Covid-19: Impacts on jobs, education, rights and mental well-being. www.ilo.org/wcmsp5/groups/public/--ed_emp/documents/publication/wcms_753026.pdf

James, A. and Prout, A. (2015). Introduction. In A. James and A. Prout (eds.) *Constructing and Reconstructing Childhood: Contemporary Issues in the Sociological Study of Childhood*, Oxon and New York: Taylor & Francis Group, 1–6.

Lee, N. C, Hollarek, M. and Krabbendam, L. (2018). Neurocognitive development during adolescence. In J. E. Lansford and P. Banati (eds.) *Handbook of Adolescent Development Research and Its Impact on Global Policy*, Oxford: Oxford University Press, 46–67.

McNamee, S. (2016). *The social study of childhood*. London: Palgrave.

Mohanty, C. (1984). Under Western Eyes: Feminist scholarship and colonial discourses. *Boundary*, 2(12/13), 333–358.

Mustard, J. F. (2007). Experience-based brain development: Scientific underpinnings of the importance of early child development in a global world. In M. E. Young with L. M. Richardson (eds.) *Early Child Development from Measurement to Action: A priority for growth and equity*, Washington: International Bank for Reconstruction and Development / The World Bank, 43–85.

Pruitt, L. (2020). Rethinking youth bulge theory in policy and scholarship: Incorporating critical gender analysis. *International Affairs*, 96(3), 711–728.

Qvortrup, J. (2009). Are children human beings or human becomings? A critical assessment of outcome thinking. *Rivista Internazionale Di Scienze Sociali*, 117(3/4), 631–653.

Ritchie, H. and Roser, M. (2019) Age Structure. *Published online at OurWorldInData.org.* Retrieved from: 'https://ourworldindata.org/age-structure' [Online Resource].

Samman, E., Presler-Marshall, E., Jones, N., Bhatkal, T., Melamed, C., Stavropoulou, M. and Wallace, J. (2016). *Women's Work: Mothers, children and the global childcare crisis,* London: Overseas Development Institute.

Spencer, J.P., Corbetta, D., Buchanan, P., Clearfield, M., Ulrich, B. and Schöner, G. (2006). Moving toward a grand theory of development: In memory of Esther Thelen. *Child Development*, 77(6), 1521–1538.

Sukarieh, M. and Tannock, S. (2018). The global securitisation of youth. *Third World Quarterly*, 39(5), 854–870.

UN (1981). *United Nations General Assembly Resolution 36/28. International Youth Year: Participation, Development, Peace* . New York: United Nations.

UN-DESA (2019). *World Population Prospects 2019*, United Nations, Department of Economic and Social Affairs, Population Division custom data acquired via website www.population.un.org/wpp/DataQuery/

UNICEF (2019). *The State of the World's Children 2019. Children, Food and Nutrition: Growing well in a changing world.* New York: UNICEF.

UNICEF and Save the Children (2020). Children in Monetary Poor Households and Covid-19: Technical note. Available at https://data.unicef.org/resources/children-in-monetary-poor-households-and-covid-19/

UNICEF and World Bank (2016). Ending Extreme Poverty: a Focus on Children. Briefing note October 2016, available at www.unicef.org/publications/files/Ending_Extreme_Poverty_A_Focus_on_Children_Oct_2016.pdf

Wall, J. (2019). From childhood studies to childism: Reconstructing the scholarly and social imaginations. *Children's Geographies*, 10.1080/14733285.2019.1668912.

Watkins, K. and Quattri, M. (2016). *Child Poverty, Inequality and Demography: Why sub-Saharan Africa matters for the Sustainable Development Goals.* ODI Report August 2016. Overseas Development Institute. Available at www.odi.org/publications/10520-child-poverty-inequality-and-demography

Young, M. E. (2007). The ECD agenda: Closing the Gap. In M. E. Young with L. M. Richardson (eds.) *Early Child Development from Measurement to Action: A priority for growth and equity*, International Bank for Reconstruction and Development / The World Bank: Washington, 1–16.

49

AGEING AND DEVELOPMENT

Penny Vera-Sanso

Introduction

Any study of global ageing wishing to establish its credibility will start with a selection of alarming, generic, statistics to secure attention and generate a sense of urgency. The urgency will lie in the underpinning, and unquestioned, belief that population ageing will present daunting financial and care burdens. It will point to the number of people aged 65 and over in the world today (703 million in 2019 (UNDESA 2019)) and, taking mid-level *projections* based on past trends, the number expected by 2050 (1.5 billion (ibid.)). It will point to the projected global distribution of people aged 65 and above, over two-thirds of whom will be concentrated in less-developed countries in 2050 (ibid.). This will stand as a counter to the assumption that population ageing is primarily a developed country problem, thereby establishing population ageing as a global issue for which a global policy framework must be found.

This demographic determinism, described as 'apocalyptic demography' (Gee and Gutman 2000), is rooted in an unwavering ageism that serves a number of functions. First, it positions high-income countries as the source of expertise on population ageing, despite the utterly different experience and prospects of countries ageing without the economic benefits that a history of colonial and neo-colonial power confers. Second, this manufactured discourse of old age crisis enables a 'neoliberalising of old age' (Macnicol 2015) by making changes to retirement ages and pension arrangements appear as no more than a rational, technical fix to population ageing, as can be seen in documents published by the World Bank (1994) and the UNDESA (2019). Third, it draws low and middle-income countries into a policy agenda that advances the globalizing pension market (World Bank 1994). Fourth, it pushes an ageist agenda that positions economic conflicts as the conflict between generations (Walker 2012), thereby deflecting attention from the foundational conflict in a globalizing market system, where wealth accumulation is premised on the cutting of labour costs. This chapter will argue against demographic approaches for understanding population ageing and development. It will argue for a critical, decolonial, sociology of ageing in order to understand how ageing is differently produced. Through this, we can counter the reproduction of ageism now embedded in Western knowledge production and deepen social solidarity.

DOI: 10.4324/9781003017653-54

The limits of methodological nationalism

Demography's central concern is the spatial distribution of populations by focusing on birth, deaths, migration and age structure in order to inform policy-making and planning. Demography's roots lie in the Malthusian fear that populations will outstrip their food supply, a population pessimism built into modernization and development strategies that position population control as a necessary step to raising living standards.

Demography's founding theorist, Warren Thompson, examined historical population dynamics, in an attempt to identify 'danger spots' in order to make policy recommendations on population control and land distribution (Thompson 1929) and in the context of Empire, underdevelopment and exploitation and the consequent threat to international peace (Thompson 1946). Staged, lineal models of societal progression, based on the Western experience, were fashionable at the time as a seemingly 'scientific,' hence objective, and unquestionable pathway to progress.

The failings of such methodological nationalism is most easily seen in Rostow's (1960) five stage model for countries to progress from low consumption to North American high mass consumption, which ignored the role of relations within and between countries in shaping the capacity for economic development. Inserting an evaluative tone to an 'objective' model, Thompson posited that countries would progress from a Group C stage of high birth/high death rates to Group B, of high birth/declining death rates, ending in Group A's low birth/ low death rates as a result of industrialization that generalizes the developed country experience. Subsequently the model was expanded to four stages and now constitutes the Theory of Demographic Transition (hereafter Transition Theory). The model informs how population projections are drawn up and interpreted. Demography attempts to explain the shifts from one demographic stage to another by identifying a causal mechanism, reflecting theoretical interests. The mechanisms identified include technological developments, investments in public health, the individual applying a cost/benefit analysis to fertility decisions, new norms regarding children's education and child labour and the rising status of women through education and employment. Yet these cause-effect associations are hotly disputed – do the claimed causes drive demographic transitions or are they the effects of demographic transitions (Zaidi and Morgan 2017)? Transition Theory predicts a short demographic window for a great economic leap forward created by a youth bulge moving into work, followed by the risk of decelerated economic growth when the youth bulge retires, creating the ageing population crisis.

There are six points to note here. First, that demography's central theory – that of Demographic Transition, is predicated on the idea that developed country experience is universal: that retirement is universal, universally desired, and universally feasible until a demographic tipping point is reached. We know that developed country experience is not universal.

Second, that there is an unwarranted assumption of predictability to the factors that determine population growth rates. Here Sudharsanan and Bloom's (2018) discussion of their results testing whether LMICs will grow old before growing rich is instructive. They find no evidence to support the view that LMICs will grow old before growing rich and that it could be the reverse for MICs, which could have higher levels of GDP growth and health expenditure per capita than did HICs when they reached comparable percentage shares of older people. Yet, as they state, this conclusion is only valid if average GDP and health expenditure growth rates remain the same as they have over the prior ten years.

In other words, and this is the third point to note: population growth rates are impacted by processes, events and relations operating at local, national, and global level that determine the generation and distribution of resources. Here we enter the sphere of volatile markets, geo-political events, environment and climate shocks, political regimes, and socially instituted inequalities, all of which determine the distribution of mortality *between* and *within* countries.

Fourth, that the Group B stage of high fertility/declining mortality generates a youth bulge that *may* create a demographic dividend by increasing the working population, hence the per capita productivity. But once again, it is the interlocking of local, national, and global level factors that determine whether the dividend is realized, through the creation of high productivity jobs capable of employing the youth bulge. Yet youth are twice as likely to be unemployed than adults and more than 75 per cent are stuck in informal work (UNDESA 2019) which is not noted for productivity levels, relying instead on poverty wages to generate profits.

Fifth, embedded in the heart of Transition Theory is a simplistic Malthusian pessimism and a foundational ageism that attempts to universalize a parochial perspective that young people are productive and older people are not.

Sixth, that Transition Theory's foundational ageism makes it blind to the factors that determine *whether* older people *do* become an economic or care burden. Specifically, they lose sight of the extent to which younger generations are dependent on older people for care and resources, and of the factors determining the direction of resource and care flows.

An emerging literature now critiques 'the social demographer's penchant for parsimony, at the expense of substantive plausibility (which) has led to overreliance on theories posing inevitable, irreversible and unilinear change' (Zaidi and Morgan 2017, 487). This may explain why older people have not become a central research theme despite the arrival of computers enabling demographers to shift from the use of aggregated census data to individual level survey data to produce nuanced analysis of population fractions (Caldwell 1996).

Instead, older people remained positioned as the problematic economic and care burden depicted by Demographic Transition Theory. This theory still shapes policy discourse. The model is widely used by people and institutions that find it a convenient, objective-seeming, platform from which to launch their agendas. The model's heft for policy discourse lies precisely in the paucity of data available on older people within developing country populations, making the recourse to universalized assumptions regarding later life difficult to challenge. The lack of evidence on older people is the outcome of the central reasoning in Transition Theory: limit births through contraception by building greater confidence in child survival. Consequently, national health surveys have age cut-offs that exclude older people: women from age 50 and variously for men, ranging from not included at all to age 49 for men in Afghanistan, 54 in India and Malawi, 64 in Bolivia. What is notable here is that such surveys include themes relevant to older people such as anaemia, nutrition, AIDS, domestic violence, but they are excluded from the surveys.

A key source of information on later life health are the Health and Retirement Study sister studies which provide for international, comparative analysis of the capacity of people over age 65 to undertake basic 'activities of daily living' (ADLs) such as independently bathing, dressing and eating. These have found that not only do ADL capacities vary across societies, but differentials are 'stark and do not attenuate at older ages' (Sudharsanan and Bloom 2018). Rather they found ten-to-twenty-year age gaps in ADL rates between the best and worst national averages on functional ageing as well as no association between GNP and functional age, as the United States and Spain do much worse than South Africa and Indonesia (ibid.).

Clearly, chronology is not the key factor in determining function, which directly challenges Transition Theory. The latter posits that the virtuous circle between lowered fertility and development fails when the Old Age Dependency Ratio becomes too high. This ratio, which is based on the proportion of people in the working age group, i.e. above that classed as child labour and below that classed as 'old,' sees a national level threat in the assumed burden of supporting older people, either directly through family-based care and financial support or indirectly through socialized provision (pensions, health and social care). In this formulation people aged over 60, more recently over 65 (UNDESA 2019), are identified as an inevitable burden, yet this is not the case.

Population ageing studies and ADL studies are mainly examined on the basis of national level data. There are a number of problems with this methodological nationalism. The key problem is the ageism underpinning two sets of assumptions. One set relates to people of 'working age' and to the concept of 'work' itself: the assumption that all people of working age (14–64) *do* work and are capable of work, that there is work available to them, that the work is sufficient to support that person (and possibly one or two children), but not sufficient to support parents. The second set is that people over the age of 65 do not work and that they do not contribute to family or society either financially or in an unpaid capacity but are, instead, dependent and in need of care. Yet only in certain classes around the world are children working from age 14, are older people retiring by age 65, and in which incomes are both regular and sufficient to consistently support the worker, let alone dependents.

Evidently not all working-age men work and globally a large proportion of working age women have their time taken up with unpaid care and subsistence production. All these work participation variables, which determine a family's capacity to support themselves, are exactly that, variable; and they are so in inconsistent ways, precisely because age-based employment patterns are shaped by local labour markets, not by age. For instance, in labour markets with a good deal of potential for younger people, older people who need to work will move into sectors of the economy that younger people no longer want to work in (Vera-Sanso 2013). These may be low paid, demeaning, onerous jobs. But if there are many young people seeking work in a tight labour market, such as smaller towns that have a significant proportion of migrants, or in cities where there is a sudden contraction of work, as happened with the developed country induced economic crisis of 2008 or the COVID-19 lockdowns, older people can find themselves crowded out of livelihood opportunities until the relevant sectors pick up (Harriss-White et al. 2013).

Labour market segmentation can cut the other way too: high levels of self-employment in the informal economy can help insulate older people from volatile labour markets, while younger employees and casual labourers are vulnerable to work loss or contraction (ibid.). Broadly, for people without the means to support themselves through savings or pensions – particularly people who have spent their working lives in the informal economy – they will continue to work to be self-supporting or to contribute to family-based strategies. The latter may include working as a reserve army of labour, primarily in agriculture and construction sectors. This is especially the case where child labour is prohibited. Or they may contribute to family incomes indirectly by undertaking unpaid work, thereby freeing another family member to enter the work force (Vera-Sanso 2013). What is apparent here is that it is not age that determines dependence; without effective welfare provision, it is the nature of the local economy, its relationship to national and international markets and individuals' and families' location within these markets that determine everyone's opportunities for economic independence, irrespective of age.

For older people the problem of their dependence is not generated by chronology but by age discrimination: the use of age stigmatization to reduce access to work and to lower rates of pay. Equally, older people face the dependence of their adult children, who cannot meet their own and their marital family's needs for housing, income, and care work. This discussion sets out the third problem with methodological nationalism; the focus on generating national pictures drives out the complexity needed to rise above the ageist assumption that functionality is the defining feature of later life.

A critical sociology of ageing

Far from apocalyptic narratives pushing recommendations that load later life support on to families or the individual themselves, what is needed is a critical sociology of ageing that uncovers the determinants of *whether* families *can* support older people and *whether* individuals *can* save for their own later life. Central to both is understanding the carrying capacity and the social segmentation of the economies in which people are embedded. Has there been, and does there continue to be, sufficient work available to everyone across the life course to meet their own and their family's needs? Is work consistently available across the year? Is the remuneration adequate to meet current needs and enable saving for later life? Does the distribution of work, wages, and the social wage favour certain groups over others? Does the organization of social provision actually include everyone, or is provision patchy, irregular or exclusionary in design or implementation? Does its provision conflict with peoples' goals for family care and individual saving for later life? Is social provision more gestural politics than effective safety net?

Underpinning Transition Theory is a normative model of life stages: learning, earning, retirement, death. This model is based on the male bread-winner concept underpinning standardized contracts for white collar and unionized blue collar men, which provided for work and pension continuity (Moen 2013). The model never fitted most women nor most minority experiences and does not fit the 21st century's labour market conditions of job insecurity and contracting pay (ibid.). Clearly the model never fitted the life trajectories of the vast majority of people living in developing countries either.

Under twenty-first century, pre-COVID-19 conditions, 61 per cent of the world's working people worked informally, without continuity of work or pensions (ILO 2018a). There were regional differences in informal employment: nearly 86 per cent in Africa, 67 per cent in Asia and the Pacific, 69 per cent in the Arab States, 40 per cent in the Americas and 25 per cent in Europe and Central Asia (ibid.). While 93 per cent of informal employment was in developing and emerging countries (ibid.), in the United States casual work, piece rate and gig work is on the rise: in 2017 it comprised 34 per cent of the workforce and was expected to rise to 43 per cent in 2020 (ILO, undated). The market impacts of COVID-19 are likely to cause a deeper dive in formal employment and a considerably larger leap in informal work than expected which will impact people's capacity to move into retirement, including in high-income countries. Evidently, not only was the standardized, privileged, largely white, male model of a continuous period of employment followed by a continuous period of retirement not a good fit for many in developed countries and for the overwhelming majority of people in developing countries, it will be an even poorer fit in the 21st century. Yet this remains the model that informs Demographic Transition Theory.

Underpinning normative life stage models that end in retirement and dependence is the idea that people have sufficient means to retire, via pensions, their own savings, or the support available from their adult children. But does the vast majority of the world population work in a context in which they can earn enough to save for retirement or enough to provide for dependents of any age? We have seen the high rates of informal employment globally above. On top of time-related underemployment due to a shortage of work (ILO 2018a), work in the informal economy is not just more vulnerable but can be extremely poorly paid and this is particularly the case for women (ILO 2018a, 2018b). When we turn to wage inequality, we find that the countries with the highest wage inequality are low-income countries (ILO 2018b). In South Asia, alongside the gender wage gap and concentration of women in the informal economy, there are age-based discriminations that affect younger and older people, beginning in mid-life, that impact wages, extent and conditions of work (Vera-Sanso 2013). This regional experience is likely to be widespread, as suggested by the ILO's (2018a) global data which found that over 75 per cent of young working people (to age 24) and 77 per cent of people working aged over 65 are working in the informal economy. Minority status also shapes the specificity of people's location within informal economies (Dubey 2016).

Now let us add what we know about health inequalities. Poverty disables and kills (Marmot 2005). Road traffic injuries are now the leading cause of death for people aged 5–29 years and the eighth leading cause of death for all age groups (WHO 2018). Road traffic death rates are three times higher in low and middle-income countries and affect pedestrians, cyclists and motorcyclists, the main means of travel for poorer sections of society (ibid.). It is these contextual factors, among others, that determine later life trajectories and intergenerational relations. The question is, how?

In developing countries, where income inequalities are highest, where work is insecure, inadequate and irregular, where welfare, public health, health care and pension provision, ranges from non-existent to inadequate, few people's lives resemble the work/retirement model. Nor do families match the simplistic capable prime age adult/dependent older person model, especially not over time. Instead, families are multi-generational. Multi-generational does not refer to the two or three generations modelled in standardized conceptualizations of intergenerational relations, which tend to focus on more limiting understandings of households, rather than family networks. Families potentially comprise up to five or six generations: depending on the age gaps between generations, starting at just born with a 20-year gap between each generation can produce five generations. But this is not the only failing of two or three generation models. Unhealthy and hazardous living and working conditions, inadequate nutrition and high levels of road accidents not only result in early deaths, but also punch holes into family networks. This further constrains the capacity of families to spread the risks inherent in social and economic systems that perpetuate insecurity, deprivation, and inequality. When those early deaths take out the higher earners in a family network, in the context of sharp income differentials by gender and age, it is evident that models assuming people can save for their old age or that families can support dependents are implausible. The outcome is not a capable working generation struggling to support a retired or dependent population but enormously stretched family networks pulling together as best they can in circumstances beyond their control. Younger people are dependent on older people for resources, including housing, loans, contacts, livelihood contributions, domestic work, and care inputs as much as older people are dependent on younger people to cover a similar range of needs.

Two inter-related sociological approaches help us to frame what is happening. First, is the life course approach which unseats the assumption that ageing is a physiological process with a broadly uniform trajectory of frailty, incapacity, and dependence. Instead the life course perspective locates individuals and cohorts within their contexts, thus demonstrating that ageing is socially produced; what it means to be older varies by time, place and socio-economic location, hence by connectivity with other lives/life courses, agency and the consequences of earlier contexts, opportunities and decisions (Bengston, Elder and Putney 2012).

Second, is the Cumulation of Advantage/Disadvantage approach that focuses on the systematic production of inequality across the life course, resulting in a later life that reflects the accumulation of advantage or disadvantage over the life course (Dannefer 2003). Gerontological research has become increasingly nuanced as ideas from beyond the field have been drawn into the study of social gerontology. Notable is the combining of the life course/CAD perspective with institutional theory to demonstrate how social choices are shaped by the institutional environment, creating penalties for people not conforming with institutionalized imaginings of the life course (Moen 2013). In the United States context, Moen points to a concatenation of institutional arrangements that, over the life course, determine who ends up with good pensions and the capacity to permanently retire and who does not. Further her work is suggestive of generational shifts as the opportunities for retirement contract, raising questions about the directionality of resource flows and care work between generations.

There are no standardized patterns and no stasis, demonstrating that later life is socially produced. For instance, Nigeria's women civil servants, forced out of formal government employment in mid-life due to fixed tenure employment contracts, turn to informal work to support themselves and their extended family networks in the context of inadequate and delayed contributory pensions and widespread unemployment (Beedie 2015). This experience is less often faced by male civil servants who enter the civil service with higher qualifications and retire later with larger pensions. India's impoverishing, precarious labour market that drives a wide variety of migration patterns, from seasonal bonded labour to longer term rural/urban migration is forcing 'aged women' into the rural workforce 'owing primarily to declining earnings capacity of the usual income earners' (ILO 2019, 12). Compelled to work deep into late life the women serve as an agricultural reserve army of labour (Vera-Sanso 2013). At the same time welfare measures, including the public pension, systematically exclude the poorest and most vulnerable (Sampat 2016), or trap people in locations without the physical help they need for fear of losing non-portable pensions if they move to a relative's home.

While, as we have seen, the institutional life course perspective demonstrates how class, race, gender, ethnicity, and age intersect to shape life courses to produce specific patterns of accumulated advantage or disadvantage, an intersectional life course approach would go beyond linking a structural analysis to individual experience. It powerfully links historical systems of domination between countries with contemporary processes of social differentiation to understand both diasporic later life and transnationally linked lives as happens with transnationally located family networks (Ferrer at al. 2017).

This intersectional life course approach is particularly important in three respects. First, it demonstrates the ageism embedded in a wide range of concepts, statistical measures, public discourses, and policies. Second, it demonstrates how ignoring the specificity of how later life experiences are produced not only perpetuates othering and marginalization for minority ageing (Rajan-Rankin 2018) but also ignores how the material and health consequences of later life are socially produced. The importance of an intersectional analysis is amply demonstrated by the ready, and unfounded, recourse to a genetic explanation for differential morbidity rates over social positioning explanations as happened with COVID-19 mortality in the United Kingdom (El-Khatib et al. 2020). Third, it challenges discourses that assert failing cultural norms as the source of later life vulnerability and poverty; these are hegemonic narratives that erase the conditions of people's lives.

For example, in India, gender, class, caste and religion determine women's life chances, creating divergent experiences across the life course, including radically different experiences of intergenerational relations and later life that can be explained without recourse to a 'cultural erosion' discourse. Educated, working wealthy women's life courses are marked by the means for later life self-support and significant social capital, including the greater likelihood of children surviving into adulthood who are capable of supporting themselves and their conjugal families. This contrasts markedly with the life of women oppressed by their gender, class, caste and geographical location. Their later life is much more likely to be marked by severe poverty, by a narrower family network that itself is struggling to survive, having lost multiple earners. Depending on their specific context these women may be supporting older husbands, older generations, their own generation (eg siblings), and younger family members, located near or far from them; they may be working for paid income and/or putting in the considerable labour needed to secure basic subsistence needs not, or inadequately, provided by the State or market. Neither of these common-place experiences are supported by their family because, respectively, it is either not needed or not feasible (Vera-Sanso 2004).

An intersectional, linked life course perspective gives the lie to analyses of later life and population ageing that are premised on normative notions regarding capacity, duty, frailty, dependence, culture and gendered and intergenerational relations. In their place it offers a strategy to interrogate these notions, exposing the variance between hegemonic norms, the contexts and function of their assertion as norms and how later lives and intergenerational relations are structurally, and variously, produced.

A decolonial sociology of ageing

The demand for reflexivity, including epistemological reflexivity, is longstanding. As far back as 1962 Kuhn exposed how claims to knowledge are shored up by powerful mechanisms that work against disciplinary change through publishing gateways, professional appointments, and career structures. This siege mentality constrains disciplinary relevance, the more so with the rapidity of real world change, and has underpinned real world structural violence (Chambers 2018). Training, working practices and the socioeconomic-cultural location of disciplinary leaders not only reinforce disciplinary biases, but also shape the thinking and practices of those who apply their disciplines to real world contexts. 'Thinking automatically' by using rules of thumb, 'thinking socially' by staying within the boundaries of group think and using 'mental models' rooted in a parochial world view (Chambers 2018) clearly unseat claims of objectivity and expertise. It is these conceptual walls that scholars and practitioners have been trying to break down through the criticism of working in silos, through calls for inter-disciplinarity and more recently for an unlearning and a conviviality that recognizes the incompleteness of knowledge and provides equal space and value to all forms of being, knowledge and knowledge production (Nyamnjoh 2015), creating a new geopolitics of knowledge (Bhambra 2014).

There are few attempts to understand later life beyond the frameworks driven by Western ways of knowing; a knowing that pushes a de-historicized view of ageing and later life experience as a biomedical issue with economic implications. The vast majority of the research generated in lower and middle-income countries reflects this dominant Western perspective. Why is this? First, is the politics of knowledge production: the colonial thinking that underpins careers and funding access, that defines who is classed as the 'knower' in knowledge production, and that polices what counts as valid analysis through publication practices. This requires epistemic disobedience by not allowing researchers to be dictated to by one epistemic approach, both in order to enable silenced voices to be heard and to ensure that research questions, conduct and outcomes address their concerns (Chaouni et al. 2021).

The second is the 'othering' that sees what is thought to be the minority experience as marginal (Rajan-Rankin 2018). Though as Moen (2013) points out the 'Western experience' was always a myth; it generalized the experiences of mainly white men privileged in the labour market, othering women and minority men. It is clearly even less the experience of ageing in the vast majority of the world that has neither had pensions, nor other economic resources to fit the work/retirement model that underpins Western models of ageing.

Now is the time to reconsider research approaches, methods and relations; to focus on methodologies that see ageing as primarily socially produced, that takes a linked-lives, historicized, institutionalized, intersectional, and decolonial approach. One that questions the extent to which theorizing on ageing populations, later life and older people is anything more than parochial myth making rooted in, and perpetuating, ageism. For this to happen a pedagogical shift is needed. Disciplinary boundaries must be broached, based on a recognition of the incompleteness of each discipline's assumptions, cultures, and world view. Disciplinary methods, concepts and cannon must be examined for a parochial ageism that constructs older people as dehumanized, passive objects, as readymade pawns in a governance game.

Conclusion

All research must determine how it will address real world complexity. It would be naïve to think that the choices made are in any sense neutral. They produce, or reproduce, narratives about the world, about social relations, pushing some things into the frame and pulling some out and, thereby, attempt to shape perceptions of reality and influence policy-making. Studies that focus on age-based classifications, that see ageing as a physiological process of inevitable decline and dependence, are irrefutably ageist, and play into agendas seeking to delegitimize citizen claims on their employer's balance sheets or the social wage in order to create markets by shifting economic risk and care work onto the governed. The assumption of youthful capacity and aged dependence not only constructs later life as a private, individual issue, it positions older people as a burden on family, society, and economy. The incorporation of this discourse into demographic models and factoids, such as the Dependency Ratio, which are justified on the thinnest of grounds, that of facilitating temporal and international comparisons, pulls out of view all that upends the ageist narrative. It directs our gaze from the key issues of the distribution of resources, and ecological destruction, within countries and across the globe. We need to drop the obsession with measures of population ageing and fictitious dependency ratios and refocus our attention towards social reproduction – identifying both what is necessary to produce an equitable society in which everyone's needs are met and what relations at local, national and international levels are producing societies unable to sustain themselves.

This requires two major shifts. First, dropping the assumption that developed country experience, and expertise based on those experiences, can inevitably understand and provide solutions for the 'ageing populations' issue. We need to recognize that developed countries are facing new situations, particularly increasing informalization, declining youth employment, contracting welfare provision and unaffordable housing markets: they have much to learn from developing country experiences. The second major shift needed is the recategorization of work to include all socially necessary effort and the accurate measuring of that which makes a difference: income gaps and access to social infrastructure by age, gender, class, race/ethnicity/caste, household, and family networks. This will demonstrate intergenerational *inter*dependences and will help to identify where actual risks and vulnerabilities lie across society, which can then be correlated against social, economic and environmental policy shifts in order to support an iterative process of strengthening social solidarity.

Key sources

1. Chambers, R. (2018). *Can We Know Better? Reflections for Development*, Rugby: Practical Action.

This book provides arguments and evidence for why development expertise needs to be renewed by handing the baton over to people construed as needing or benefiting from development. Readers should interrogate the merits of suggested means and mechanisms for doing so – all is open to close inspection and improvement.

2. Chaouni, S. B., Claeys, A., van den Broeke, J., and De Donder, L. (2021). Doing research on the intersection of ethnicity and old age: key insights from decolonial frameworks, *Journal of Aging Studies*, 56, pp. 1–7.

This article demonstrates how decolonial perspectives expose gerontology's inattention to the intersection of ethnicity and age and proposes strategies to produce a decolonial gerontology open to how historically-rooted hierarchies still shapes later lives.

3. Dannefer, D. (2003). Cumulative Advantage/Disadvantage and the Life Course: Cross-Fertilizing Age and Social Science Theory, *Journal of Gerontology*, 58B: 6 S327–S337.

This provides a detailed setting out of the life course and Cumulation of Advantage and Disadvantage approaches.

References

Beedie, E. (2015). Adequacy of pension income in Nigeria: The case of retired women civil servants. Unpublished PhD thesis, Birkbeck, University of London.

Bhambra, G. K. (2014). Postcolonial and decolonial dialogues. *Postcolonial Studies*, 17(2), 115–121, DOI: 10.1080/13688790.2014.966414

Bengston, V. L., Elder Jr. G. H. and Putney, N. M. (2012). The life course perspective on ageing: Linked lives, timing and history. In J. Katz, S. Peace and S. Spurr (eds.) *Adult lives: A life course perspective*, Bristol: Policy Press, pp. 9–17.

Caldwell, J. C. (1996). Demography and social science. *Population Studies*, 50(3), 305–333.

Chambers, R. (2018) *Can we know better? Reflections for development*. Rugby: Practical Action.

Chaouni, S. B., Claeys, A., van den Broeke, J. and De Donder, L. (2021). Doing research on the intersection of ethnicity and old age: Key insights from decolonial frameworks. *Journal of Aging Studies*, 56, 1–7.

Dannefer, D. (2003). Cumulative advantage/disadvantage and the life course: Cross-fertilizing age and social Science theory. *Journal of Gerontology*, 58B(6), 327–337.

Dubey, S. Y. (2016). Women at the bottom in India: Women workers in the informal economy. *Contemporary Voice of Dalit*, 30(1), 30–40.

El-Khatib, Zaid, Graeme Brendon Jacobs, George Mondinde Ikomey, Ujjwal Neogi (2020) The disproportionate effect of COVID-19 mortality on ethnic minorities: Genetics or health inequalities? *EClinical Medicine*, 23.

Ferrer, I., Grenier, A., Brotman, S. et al. (2017). Understanding the experiences of racialized older people through an intersectional life course perspective. *Journal of Aging Studies*, 41, 1–17.

Gee, E. and Gutman, G. (2000). *The overselling of population ageing: Apocalyptic demography, intergenerational challenges and social policy*. Oxford: Oxford University Press.

Harriss-White, B., Olsen, W., Vera-Sanso, P. and Suresh, V. (2013). Multiple shocks and slum household economies in South India. *Economy and Society*, 42(3), 398–429.

ILO (2018a). *Women and men in the informal economy: a statistical picture*. Geneva: International Labour Office.

ILO (2018b). *Global Wage Report 2018/19 What lies behind gender pay gaps*. Geneva: International Labour Office.

ILO (n.d.). Helping the gig economy work better for gig workers. International Labour Office for the United States.

ILO (2019). Informal employment trends in the Indian economy: Persistent informality, but growing positive development. Working Paper 254, Employment Policy Department.

Kuhn, T. S. (2012) *The Structure of Scientific Revolutions*. Chicago: University of Chicago Press, 4th Edition.

Macnicol, J. (2015). *Neoliberalising old age*. Cambridge: Cambridge University Press.

Marmot, N. (2005). Social determinants of health inequalities. *Lancet*, 1099–104.

Moen, P. (2013). Constrained choices: The shifting institutional context of aging and the life course. In L. J. Waite, T. J. Plewes (Eds.) *New Directions in the Sociology of Aging*. Washington (DC): National Academies Press, pp. 175–216.

Nyamnjoh, F. (2015). Amos Tutuola and the elusiveness of completeness. *Stichproben. Wiener Zeitschrift für kritische Afrikastudien*, 15, 1–47.

Rajan-Rankin, S. (2018). Race, embodiment and later life: Re-animating aging bodies of color. *Journal of Aging Studies*, 45, 32–38.

Rostow, W. W. (1960). *The Stages of Economic Growth: A Non-Communist Manifesto*. Cambridge: Cambridge University Press.

Sampat, K. (2016). *Public pension provisioning for older persons in India: India Exclusion Report 2016*. New Delhi: Yoda Press.

Sudharsanan, N. and Bloom, D. E. (2018). The demography of ageing in low – and middle-income countries: Chronological versus functional perspectives. In M. D. Hayward and M. K. Majmundar (Eds.) *Future directions for the demography of aging: Proceedings of a workshop*. Washington DC: The National Academies Press. www.ncbi.nlm.nih.gov/books/NBK513069/

Thompson, W. (1929). Population. *American Journal of Sociology*, 34(6), 959–75.

Thompson, W. (1946). *Population and Peace in the Pacific*. Chicago: University of Chicago Press.

United Nations Department of Economic and Social Affairs (UNDESA) Population Division (2019). *World Population Ageing 2019: Highlights* (ST/ESA/SER.A/430). New York: United Nations. www.un.org/en/development/desa/population/publications/pdf/ageing/WorldPopulationAgeing2019-Highlights.pdf

Vera-Sanso, P. (2004). They don't need it and I can't give it: Filial support in South India. In P. Kreager and E. Schröder-Butterfill (eds.) *Ageing Without Children: European and Asian Perspectives on Elderly Access to Support Networks*. Oxford: Berghahn Books, pp. 77–105.

Vera-Sanso, P. (2013). Aging, work and the demographic dividend in South Asia. In: J. Field, R. J. Burke and C. L. Cooper (eds.) *The SAGE Handbook of aging, work and society*. London: Sage, pp. 170–185.

Walker, A. (2012). The New Ageism, *The Political Quarterly*, 83(4), 812–819.

World Bank (1994). *Averting the Old Age Crisis*. Washington DC: World Bank.

WHO (2018). *Global Status Report on Road Safety 2018*. Geneva: World Health Organization.

Zaidi, B. and Morgan, P. (2017). The second demographic transition theory: A review and appraisal. *Annual Review of Sociology*, 43, 73–492.

PART 5

Reimagining futures

50

INTRODUCTION

Reimagining futures

Paul Hodge and Naohiro Nakamura

What might reimagining development futures look like and involve for development students, educators, researchers, and practitioners? In this final section of the Handbook, contributors offer a range of practices, orientations and methodologies that current and future people working in this vast and changing field might do well to consider and take on as part of re-imagining development futures beyond what we have come to know. A strong thread working through all of the chapters is the importance of attending more deeply to the peoples, knowledges, and non-human kin relations that have for far too long been relegated to development's margins. Each chapter makes a case for why development, in the diverse contexts within which the authors are writing, needs to change and what this change might encompass leading to more equitable, creative, and nourishing human/more-than-human futures (see McGregor and Alam in *this volume*). Reflecting authors' geographical location in the Asia-Pacific region, the chapters focus on specific examples of development futures from settler-colonial-Indigenous Australia, South Pacific island countries, Cambodia, Sri Lanka, and the Philippines.

Situated within recent work by Indigenous scholars on *futurities*, (ab)Original women **Michelle Bishop and Lauren Tynan** adopt Indigenous autoethnography to combine reflection, narrative and storytelling to reimagine development futures. Reframing the future as something that is created, dreamed, and *acted upon*, the authors traverse temporalities invoking more-than-human agencies as they share teachings and relational perspectives through Eagle, Ant, Grandmother and Granddaughter stories. For students of development the stories contain place-based meanings and practices enhancing their capacity to think critically and holistically while also inviting them to think about how these stories might be meaningful in their own lives. Imagining the world beyond a human perspective, the chapter teaches that human survival is deeply bound to the wellbeing of Country.

Tolu Muliaina's chapter emphasizes the holistic nature of teaching and learning informed by Pacific cultures and languages. Muliaina questions development priorities that continue to advocate and value English and French proficiency to the detriment of oral traditions of Pacific Islanders. The chapter highlights the tensions between external development educational priorities and Pacific Islander aspirations to strengthen local cultures including languages and knowledge transfer. Reflecting on teaching a third-year undergraduate course on Resource Conservation and Management at The University of the South Pacific (USP), Fiji, Muliaina

DOI: 10.4324/9781003017653-56

describes the culturally informed, innovative pedagogy and practice embedded in the course which includes aligning course content and assessment regimes to the home cultures of Pacific Islander students. Muliaina also shares how resource conservation practitioners and community elders inform the learning that takes place in the course. The chapter concludes by showing the ways in which these pedagogies are enhanced when activist platforms are created during and after the course for students to transmit their *mana* (knowledge, power, and blessing) to the next generation of Pacific leaders.

Drawing on Tuck and Yang's (2012) powerful critique of the ease with which fields and disciplines including development studies adopt the language of decolonization, **Bernard Kelly-Edwards, Gavi Duncan and Paul Hodge** share their attempts to work productively with the tensions raised by Tuck and Yang (2012) as they reflect on their teaching practice in settler-colonial-Indigenous Australia. Centring the difficult reality of teaching predominantly non-Indigenous undergraduate students a development studies course on unceded lands, the authors reflect on two custodian-led pedagogies; a yarning circle workshop and fieldtrip, to explore what these more-than-human, more-than-rational learning experiences provoke and invite for students. The chapter highlights moments of tension and discomfort, but also times of connection and attentiveness, as these future development practitioners ponder questions of colonialism, complicity, positionality, and responsibility.

Highlighting the urgent multi-disciplinary challenges facing development studies, **Yvonne Underhill-Sem** sets out ways to confront long-standing issues too easily put aside, overlooked, or ignored by the development 'canon'. Calling out injustices and profound trauma associated with racism, sexism, and intolerance as a result of imperialism and colonization, Underhill-Sem expresses a commitment to decolonial and Indigenous scholarship. She articulates alternative ways of knowing by asking; if we are not actively practicing decoloniality, are we continuing to colonize? Reflecting on personal experience, Underhill-Sem advocates the importance of positioning one's epistemic genealogy, of dismantling Western canons of knowledge production; and actively generating trans-national solidarities for decolonial gender and development. The chapter explores these commitments thus providing practical guidance for teachers, researchers, and students to do development differently.

Rebecca Bilous, Laura Hammersley and Kate Lloyd explore how community-based service-learning (CBSL) has proliferated in the education sector as high schools and universities recognize the value of work experience for key student learning outcomes. The young people enrolling in these programmes often have memorable experiences and these programmes often have considerable personal impact. Increasingly however, Australian higher education institutions are designing and promoting these experiences without questioning server-served and giver-receiver relationships underpinning CBSL. Challenging these binaries that reinforce neo-colonialism, the authors reflect on their experiences running CBSL for students offering examples of respectful and reciprocal approaches for teachers and students. Guiding teacher and student learning in CBSL, the authors foreground three interconnected themes: bringing different voices and perspectives; developing reciprocal relationships and embedding reflective practice. Significantly, the learning is also informed by perspectives from practitioners working within the development organizations that receive students adding a key viewpoint often missing in CBSL assessment and monitoring.

In their chapter on research capacity development in Vanuatu, **Krishna Kumar Kotra and Naohiro Nakamura** highlight the importance of capacity development in Small Island Developing Countries (SIDC). For universities located in SIDC countries, a key strategy for capacity development has been to build collaborations with researchers and institutions in developed countries. However, in most cases, Pacific researchers participating in collaborative projects are only seen as local 'contacts' or consultants with limited contribution to projects. This lack of parity, despite scientific research being undertaken in developing countries such as Vanuatu, has led to charges of 'neo-colonial science.' In this chapter the authors draw on collaborative research between The University of the South Pacific's Emalus Campus, Vanuatu, and institutions located in developed countries to highlight strategies and tensions of achieving capacity development. The authors outline appropriate and relevant practices for future collaborative partnerships focusing on student mentoring, training, and research participation.

Looking to adaptive approaches to development, **Aidan Craney, Lisa Denney, David Hudson and Ujjwal Krishna** show what they have to offer mainstream development discourse and practice. An underlying premise of adaptive development is that outcomes cannot be assumed or planned in advance such is the case with linear, technical approaches to development. Rather, development programmes must be responsive both to their environment, and to learning along the way in order to find successful pathways to change. Advocating adaptive development, the authors highlight the need for adaptability and reflexive practice in development work that centres a strong emphasis on cultivating a deep understanding of the local context and investing in learning. Drawing on case studies from the Philippines, Sri Lanka and Vanuatu, the authors offer examples of adaptive management, problem-driven iterative adaptation (PDIA), and thinking and working politically (TWP) as a way of providing pedagogies, strategies, and tools relevant to students and practitioners of development.

Turning to methodologies **Johanna Brugman Alvarez and Leakhana Kol** unsettle the 'worlding practices' that inform urban studies arguing that bodies of urban knowledge centred in the Global North negatively impact cities of the Global South. The influence of these 'worlding practices' on politicians, planners, architects, urban designers and citizens are increasingly apparent. Because these circulating knowledges are detached from place, history, and identity, they invariably lead to social exclusion and spatial inequalities for the most marginalized populations. Arguing for a more democratic urban studies characterized by diverse urban epistemes and imaginaries, Brugman Alvarez and Kol look to oral histories in the form of place-based storytelling as an appropriate methodology that enables 'a view from the south' in urban research. Drawing on oral histories of one informal settlement in Phnom Penh, the authors offer guidance for future researchers and practitioners as they discuss the learning that emerges through the stories of social, spatial and temporal relationships between people and place.

Jenny Cameron and Isaac Lyne's chapter reflects on a community economies approach to highlight and amplify the already existing post-capitalist economies evident in three development projects in the Asia-Pacific region. The authors trace the action research methods used by community economies scholars to work with community-based partners helping to make post-capitalist activities more visible, and then to devise ways and means to build on and strengthen these activities. Cameron and Lyne focus on the way the three projects are attentive to local conditions and to local values and aspirations. The economic development pathways that result from a community economies approach emphasize the interconnections and interdependencies between people and between people and environments. Implications for policy and practice involve recognition of the importance of attentive listening, relationship building and time. The implications for pedagogy involve deepening economic literacy, developing skills that are vital for working with others and developing a capacity for openness and to be affected by others.

Invoking Doreen Massey's (2005) work on reconceptualizing space as relations-between and stories-so-far, **Joseph Palis** reflects on five mapping workshops conducted in the Philippines to reveal the way diverse participants adopt geonarratives and countermapping to visibilize untold stories, vernacular vocabularies and lived-in experiences across place and time. Challenging the way standard maps mirror hegemonic practices and perform spatial fixities often benefitting powerful entities, Palis demonstrates that geonarrative mapping is a vigorous practice of 'unmapping' where subjectivities and cartographic stories that do not fit the development model are mobilized. Geonarrative mapping serves as a gateway to tell stories about participant's environment, particularized domestic, familial and social lives, and encounters with human, institutional, and more-than-human elements and entities. Describing the particularities of workshop design and implementation, Palis describes how this form of storytelling can be deployed by teachers, community-based groups, activists, grassroots, and peoples' organizations to generate grounded data and information as the basis to carry out specific participatory action research and development work for social justice.

In the final chapter **Sarah Wright** foregrounds the role that Indigenous literature plays in expressing Indigenous intellectual agency and political struggles for sovereignty, decolonization, and the re-establishment of Indigenous values in the Philippines. Describing the increasing militarization and violent development pressures wrought by decades of mining and logging on the ancestral lands of the B'laan people, one of 18 Indigenous Lumad tribes in Mindanao, Wright highlights the way poetry shifts the imaginary as Indigenous peoples tell their own stories in their own way as resistance. The chapter is centred around three poems written by young scholars from a Lumad school, the Community Technical College of Southeastern Mindanao (CTCSM). Lumad schools are sites of resistance, placemaking, empowerment, negotiation, and struggle and for this reason have been targeted by military forces. The poems share and hold stories, communicate pain and beauty, invoke ancestral ties to lands and continue to re-make intergenerational connections to and as place.

51

FINDING PERSPECTIVE THROUGH OUR MORE-THAN-HUMAN KIN

Michelle Bishop and Lauren Tynan

Introduction

I feel drowsy. The blinds are open but there's not enough light coming in to wake me. I hope that's just cloud outside, though the faint, familiar smell of smoke hints otherwise. Damn it. I check the air quality rating online; red warnings tell me (again) to wear a mask, run an air purifier, stay indoors and 'close your windows to avoid dirty outdoor air.' I haven't left my unit in days. The windows are closed, layers of masking tape try to stopper the gaps. The megafires seem distant, yet the smoke seeps in. I can feel the itchiness deep into my chest, a constant need to clear my throat. I miss seeing the blue sky. Is this the new 'normal'? I'm reminded of doomsday catastrophes shown in cinemas when I was a kid, the disbelieving futuristic start of the end where you shook your head and thought 'why didn't they do something sooner!?'

Do we need to reimagine the future? In a sense, the future is inevitable, it will roll around tomorrow, in ten years or ten seconds, whether we're ready or not. Some scholars are starting to think about how we can create *futurities* (cf. Goodyear-Ka'ōpua 2018; Smith, Tuck and Yang 2019; Tuck, McKenzie and McCoy 2014). Smith, Tuck and Yang (2019, 23) describe futurities as 'a word that imbibes the future with what we are doing now to bring about different futures.' Futurities are driven by action. Rather than the future being inevitable, futurities reframe the future as something that is created, dreamed, and *acted upon*. You can also think of this as agency. Futurities disrupt the linear notion of time; that the future is something *yet to come* (cf. Bawaka Country et al. 2017). From our old stories, as two (ab)Original women, we know that time is non-linear. We are all spiralling through the past, present, and future simultaneously, layered into one another through place. This chapter works from a non-linear notion of time, where we are responsive and the future can be shaped by us today, in the same way it is being shaped by our ancestors in the past. Who are your ancestors?

We are Michelle, a Gamilaroi woman, grown up on Dharawal Country, and Lauren, a trawlwulwuy[1] woman from tebrakunna country in northeast Tasmania, Australia. As granddaughters, daughters, sisters, aunties, mothers, nieces, students, and researchers, we are concerned with a blinkered and linear trajectory of 'the future.' We think decisions are being made that are not considerate of our deep future, nor respectful of our deep past – holding short-sighted visions with profit and power at the core. We believe it's important to think

DOI: 10.4324/9781003017653-57

critically about the future we want for our grandchildren's grandchildren. What kind of world do you see?

In this chapter, we utilize Indigenous autoethnography (cf. Bishop 2020; Tynan and Bishop 2019) as part of our axiology (ethics and values) and our pedagogy (way of teaching). You will notice we use italics throughout to incorporate reflection, narrative and storytelling. It is our intention to not only tell, but show you how to use the practices, pedagogies and tools we are advocating for here. We use fire as a central theme for reimagining futures and doing development differently. We do this for two main reasons. Firstly, we started writing this in January 2020, as megafires roared across most of Australia. These fires were the worst many had seen, with over 7 million hectares of land burnt, impacting many communities; human and more-than-human. At last count, more than 1 billion animals had been killed. Secondly, as Aboriginal women involved in cultural burning, we are engaged in fire practices that have been passed down to sustain this Country[2] for tens of thousands of years (cf. Steffensen 2020). We are passionate about doing/seeing fire differently and enacting a future where the survival of humans is deeply bound to the wellbeing of Country.

The chapter asks, how do we reimagine the future? There are messages in this paper for development academics and practitioners, however, we believe this paper has strong relevance for students of development, as the ones tasked with reimagining the future. We invoke pattern thinking – the ability to think critically and holistically – and look to other perspectives that span time, place, and the more-than-human as part of this reimagining. We draw from four perspectives that represent different temporalities; different ways for us to imagine the world beyond our human perspective. As two Aboriginal women, we look to entities that are familiar to us: Eagle, Ant, Grandmother, Granddaughter. You can roughly think of them as representing the macro, micro, past and future. We encourage you to find perspectives that are meaningful in your context. For example, on Darug Country, what might the perspective of Black Snapper Fish teach us about the waterways and lessons from below the surface? From our oceans, what might Whale teach us about migration and seasonality? Finally, we consider how these perspectives can inform pedagogy and practice in doing development differently.

Development as change

Development is in the business of reimagining futures. The underlying goal of much development is change and transformation – improving health outcomes, increasing family income, building infrastructure (Ife 2013). Development practitioners can be described as change-makers, maybe this is the reason you were drawn to development too! Development practices often address the inequitable living conditions many humans face, from human rights abuses to wealth distribution. Much development work is focused on improving human wellbeing (Kenny 2011).

However, in the quest to increase human wellbeing and development, have our more-than-human kin been forgotten? More-than-human refers to all elements and kin – animals, plants, weather, spirits, rocks, tides, memories, and time. McGregor (2019 n.p.) notes that 'development is measured through human indicators. When human development occurs, it is through the un-development of non-humans.' In other words, the future of development is often imagined through stories of human success and wellbeing. Can these futures come to fruition if we lose bee populations and have no fresh food? Will we feel 'well' when wildfires swallow our homes and smoke clings to our chests? Is development able to imagine a future that is not human-centric? To do development differently perhaps it is time to turn to our more-than-human kin to help us reimagine the future.

How to reimagine the future?

It's hard to reimagine a future outside of the current system. Sometimes it's so 'normal' or invisible that it's hard to see. But, from the authors' perspective, living in the place now known as Australia, the patriarchal colonial capitalist system is evident all around us. Let's try to imagine a world outside of this system. Try it. Look around you, maybe you're reading this in your office, on the beach, in a classroom. Try to imagine what this place will look like in 50- or 500-years' time. How will it feel? Often a future is imagined based on what is known, for better or worse. The current system is tweaked, adapted, mimicked. But what if the system was not sustainable, that the system itself needed to change? How might we reimagine then? For a start, we could look to our old ways; to Indigenous values and philosophies of existence that have sustained peoples and Country for thousands of generations:

> *For the greater part of human history, and in places in the world today, common resources were the rule. But some invented a different story, a social construct in which everything is a commodity to be bought and sold. The market economy story has spread like wildfire, with uneven results for human well-being and devastation for the natural world. But it is just a story we have told ourselves and we are free to tell another, to reclaim the old one.*
>
> *(Kimmerer 2013, 31)*

Let's tell a different story; one that doesn't revolve around humans. Think about the last time you saw a whale, a lizard, an insect. Perhaps there's a friendly cockatoo that visits your balcony. Now try to think about the world from their perspective. What do they see, feel, *know*? What patterns have they witnessed or been taught from their old ones?

Pattern thinking

You may have heard extraordinary accounts of Aboriginal people in Australia telling origin stories of many landscape features (such as coastlines, mountains, and craters), some that occurred over hundreds of thousands of years ago. These oral stories give witness and explanation to huge events and were told well before Western science 'discovered' how these features were made in the mid twentieth century. trawlwulwuy people hold stories of walking across land that connected the island of Tasmania from mainland Australia; a great isthmus of hunting grounds, caves and mountains (Cameron 2011). The last time the mainland of Australia was connected to Tasmania was in the most recent Ice Age, over 10,000 years ago. Aboriginal people *know* these events intimately, each generation having learnt them through story, law/lore, song, dance, and ceremony. Indigenous knowledges, then, provide insight into the deep past for the benefit of the deep future. Welcome to pattern thinking!

Pattern thinking lies at the heart of creation, and lucky for us, can also lie at the heart of our future. Through Western education systems we are taught linear thinking – how to get from *a* to *b*. Linear thinking is based on the logics of Western scientific rationality. Science often does this when it claims objectivity or isolates an object of study and its variables. Pattern thinking is about relationships, following lines of inquiry outward to other connections. Yunkaporta (2019, 14) explains,

> *Our knowledge endures because everybody carries a part of it, no matter how fragmentary. If you want to see the pattern of creation you talk to everybody and listen carefully. Authentic knowledge processes are easy to verify if you are familiar with that pattern – each part reflects the*

design of the whole system. If the pattern is present, the knowledge is true, whether the speaker is wearing a grass skirt or a business suit or a school uniform.

Pattern thinking carries important messages from the past, well into the future. Pattern thinking is a time traveller's must-have accessory! But rather than carrying it in your backpack or notebook, pattern thinking is carried in handy and portable oral stories. The details and characters of the stories may change across time, but the pattern at the heart of the story remains the same, getting passed on to new generations of listeners and storytellers.

So how can pattern thinking help us reimagine the future? Let's look at an example from our fire theme and see how pattern thinking could be used today:

Country has been torched by wildfires, everything is black and quiet. Re-growth will appear here soon, giving everyone hope that the fires weren't that bad. But what will grow? Near the coast in Eastern Australia, the seeds of the bracken fern love hot fire. They thrive under these conditions. But bracken fern doesn't belong here in such abundance, it's what we call an invasive native[3]. Bracken fern will dominate, preventing native grasses from growing and choking up Country so nothing can move through. Wallaby needs these grasses. Wallaby is our family, so if Wallaby gets sick, the people will be sick too. We need to burn this Country again, but with the right fire, a cultural burn. This will keep Country, Wallaby and humans healthy. We all rely on one another.

Pattern thinking helps us think beyond the individual human. It shows the importance of relationships, that everything is interconnected. It is impossible to reimagine the future on our own using isolated variables. For example, we can't see Wallaby if we only focus on the fire. If we can't see that Wallaby is suffering, we won't understand why the people are sick. Many researchers now use pattern thinking with our more-than-human kin to help improve social and emotional wellbeing among Aboriginal and Torres Strait Islander people (Dudgeon and Bray 2019). How can our more-than-human kin help us reimagine the future?

Perspectives: Eagle, Ant, Grandmother, Granddaughter

In this section we offer some pedagogical tools for deep, critical thinking, looking to Eagle, Ant, Grandmother, Granddaughter. We imagine the world and the future through their eyes, from their unique more-than-human multi-temporal perspectives. In this section, we try to represent how Eagle, Ant, Grandmother and Granddaughter might think about the megafires and how their perspectives reveal details that may be overlooked by dominant human thinking of fire. These perspectives are a disruption to what has become standard, linear, human-centric thought patterns. In thinking through each of the perspectives, we use the italicized text to imagine how Eagle, Ant, Grandmother and Granddaughter might experience these events, to tell their story. If you applied these same perspectives to the megafires (or any event), you would imagine a different experience. The point is, that sitting with each of these perspectives allows us to think beyond our individual human selves; to imagine a world where all entities are important and their perspectives are just as valid. Eagle, Ant, Grandmother and Granddaughter represent different *ways* of reflecting, analysing and reimagining current and future scenarios. Let's practice thinking outside ourselves, and outside time. We are not proposing to offer solutions, rather to suggest another process to critical, future thinking and pedagogies.

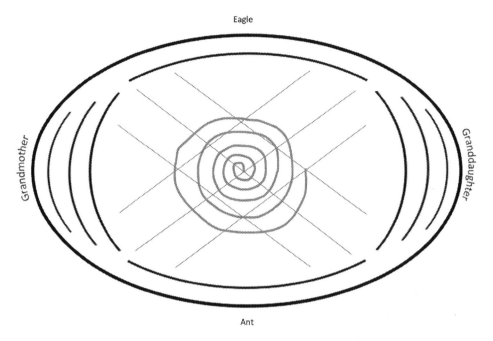

Figure 51.1   Eagle, Ant, Grandmother, Granddaughter (Bishop and Tynan 2020)

Eagle

There are many stories for Eagle. Eagle glides above, often unseen, always watching. How does Eagle view the world? What do they[4] notice?

> *All around, the land is on fire. Towering walls of smoke sweeping across Country, ocean, sky. The air is hazardous and sometimes it's hard to breathe. I have to push higher to be free from it. I find the warm air currents and let them lift me up into the sky, soaring higher with little effort. How did they let it get this bad? I've been watching this coast for decades, waiting for the signs of good fire and rarely seeing it. It's no wonder these megafires are here now. I can see them stretching up the coast, huge plumes stacked to the horizon. Too much fuel in the bush. Not enough water in the rivers. Plenty of plastic bottles though. Silly. It was only a matter of time. Lots of bright people risking their lives, helicopters bursting with water and leaving stains of red on Country. It's hard to find a safe place. I have my pick of food amongst the huge loss of life, but wish it wasn't this way. Animals torched everywhere I look, millions and millions, screeching in sorrow and pain. It's too much.*

Eagle teaches us to look around, at all of it, all at once. There are patterns to be seen, patterns to learn from. They offer a macro perspective, an ability to see across systems and beyond to other ways of knowing. From Eagle we can learn to keep an expansive but keen eye on what is happening all around us, even the bits we cannot quite see, or do not want to see.

Eagle is quite solitary but not disconnected, travelling far and sharing knowledge from different places. To travel, Eagle uses thermal soaring, relying on turbulent, warm air currents that are invisible to the human eye, showing us how to conserve energy in turbulent times. An Eagle view offers us many things. Let's look at a couple of teachings:

- How to see beyond your immediate location and context
- To embrace an openness in our thinking
- Connection between places and knowledges
- To look for routes or patterns that are sometimes hidden

Ant

There are many stories for Ant. Ant scoots below, rarely seen or cared about. Crushed often beneath the hurried feet of taller ones. How does Ant view the world? What do they notice?

> *Feeling the vibrations, a swell of blackened leaves moving towards us as the southerly catches up. Not good. The wind will feed the furious fire. We send messages ahead to our kin. But there's not much we can do; the fire is too fierce. How did it get this bad? We've been aware of the build-up of dead leaves and branches; it's made travel difficult. We look ahead to smoldering branches and plateaus of white ash, an unfamiliar and flattened horizon. The usual undulations of rich soil and tall grasses decimated. We can't even escape to the cool pockets of our nest. The heat radiates down into the soil and deep into the seed bank, incinerating what future this Country had to return to its bounty of native trees and foods. Is this hot fire happening everywhere? I know our community will get through this, it'll be tough, but we'll all come back together… whenever it ends.*

An Ant perspective to thinking about the future is to connect with the local; to truly care for/with place, people, and the more-than-human. Ant teaches us compassion and what is happening now. They offer a micro perspective, an ability to see what's in front of you, and below. Ant builds deep nests, keeping the soil aerated so water can flow down to dormant seeds. Ant perspective exposes the everyday, the microscopic details, without losing sight of what/ who is around you and its potential for the future.

Ant lives in a large community; they teach us about cooperation and interdependence. Ant uses chemical signals to create trails for other ants that lead to food and home. Much like Eagle, Ant follows paths that are invisible to the human eye. Ant teaches us that the path to safety and survival may not 'look' obvious, but if we draw on other senses, including our infinite web of relationships, we can bring our community along with us. Ant perspective is underestimated yet they see the patterns and the relationships between everything, just another way. An Ant view offers us many things. Let's look at a couple of teachings:

- Being part of a community and caring for the place where you are
- Not just living or thinking on the surface (Moreton-Robinson 2018)
- Following paths not obvious to the human eye to take others along with you
- Drawing on other senses to find your way and make decisions

Grandmother

Grandmother exists two generations before you and all around you. Did you know that when your mother was in the womb, she grew the eggs that would eventually become you? That means that your grandmother simultaneously held your mother and you in the womb; she created and held the next two generations within her body. Grandmother shares stories and knowledge from multiple generations before.[5] Don't forget, she was held in the womb of her grandmother and her grandmother was held in the womb of her grandmother…and so on.

I've been ok. A bit short of breath, but that could just be me getting old. I've never seen anything like this before … And I'm what, 82 years old, I've never even heard of anything like this. It's so sad. It's so devastating.

I was asking my Nan how she was coping with the hazardous smoke from the megafires that had permeated the sky for weeks. What Nan didn't say explicitly was that this type of crisis had not featured in any of the stories she was told by her Old People. Not only had she never 'seen' this scale of catastrophe, but she'd never 'heard' of anything like this. She carries generations of stories, sharing these with me, just as I will do with my grandchildren. They will learn from the stories and patterns of the past to shape their future.

A Grandmother perspective can jump timescales, like time travel! Grandmother can easily hold five or six generations of story and knowledge. She can see the connections, the relationships, the patterns, the bigger picture. And she is preparing the younger generations to listen carefully, to be respectful. Grandmother keeps learning, being shown new knowledge from her grandchildren to incorporate with her knowledge from the past.

Grandmother holds integrity in her teachings, having seen vast changes to the world she can discern the important messages from the frivolous. Grandmother listens. She doesn't just preach listening, she shows it. Grandmother is humble, yet strong, sticking by her values for you to learn. Grandmother perspective sometimes goes unheard, though she carries knowledge that spans generations and time. Let's look at a couple of teachings:

- Intergenerational thinking, learning, and sharing.
- Listening deeply, with respect.
- The need for balance.
- Humility and integrity.

Granddaughter

Granddaughter exists two generations after us and also within us. A pregnant person may be simultaneously granddaughter, mother, and grandmother (remember the next generation's eggs are created inside the womb). Granddaughter exists in the future and will carry forward the stories of multiple generations.

I want the world to be healthier and become pretty. It's hard to imagine but I reckon the future will be way different to how things are now. Animals will stay safe and plants will grow. People will be happy and safe and be kind to the world. Not like now. People can be mean, leaving rubbish everywhere on purpose. Don't they care that their rubbish could end up in the ocean and hurt the animals? Like when plastic makes Seagull's feet fall off! At school I'm learning about recycling and food scraps and waste, getting ready for fixing the world. Trying to make the world a better place for the earth. I feel kind of scared and kind of nervous about the big fires coming back. We had to leave our home, and our road got blocked off. It was scary and sad, and I wasn't sure what was going to happen. The fire trucks were everywhere, but it wasn't like that at the cultural burn we went to. I still have a little bit of memory of it and that didn't feel scary and there weren't any fire trucks.

This is a story from our niece, not long after her seventh birthday. A Granddaughter perspective is profound and has much to offer, yet it is often disregarded; deemed childish. Although Granddaughter has not yet lived the decades of life as Grandmother has, she carries the ancestral memory of many Grandmothers. She is aware the world needs to change, that people need to treat each other and their environment differently, better.

Granddaughter has an expansive future ahead, decades of time unrealized. She has a growing awareness that the values, priorities and actions of today inform and shape tomorrow. She tries to live consciously with this knowledge, slipping up now and then, always learning. Granddaughter sees cruelty in the world already and feels the weight of her role in fixing this. Granddaughter holds a long view towards the future, sometimes optimistic, sometimes fuzzy and unknown. Let's look at a couple of teachings:

- Desire for the world to be a 'better' place.
- To be at the centre of a world you are coming to know.
- Optimistic in her role to make change.
- That the values, priorities, and actions of today inform and shape tomorrow.

Pedagogies: Eagle, Ant, Grandmother, Granddaughter

Figure 51.1 can also be used as a pedagogical tool for learning. Students can use the tool to imagine the four perspectives on a more visual and temporal scale: Eagle above representing the macro view; Ant below representing the micro; Grandmother to the left representing the past and Granddaughter to the right looking to the future. The crossed lines in the centre of the tool represent pattern thinking, connecting all the perspectives to show how they are more meaningful together rather than separated. The spiral represents the non-linear nature of time, simultaneously past, present, and future.

Eagle, Ant, Grandmother and Granddaughter perspectives ignite processes of inquiry and ways to reimagine the future. As pedagogy in the classroom, we can encourage pattern

Figure 51.2 Authors with niece on cultural burn, 2017 (Stacy Mail, Hunter Region Landcare Network, 2017)

thinking and the perspectives of Eagle, Ant, Grandmother, Granddaughter to analyse many different situations, from global development challenges to national policies to local community initiatives. We offer examples of how this might be done below. Students can also adapt the tool to better suit their own contexts, drawing on perspectives that are more meaningful to them.

Reimagining development

Eagle, Ant, Grandmother and Granddaughter do not give us solutions to key development challenges, but they help us think about these challenges in another way. Let's try an example:

> *The Sustainable Development Goals (SDGs) were launched in 2016 to tackle global development challenges, superseding the Millennium Development Goals (MDGs). The seventeen SDGs aim to be delivered by 2030 and, currently, are not on track.*
>
> *(United Nations 2020)*

- Eagle might say, why are fourteen of the goals human-centric? Why prioritize a human and economic relationship to development?
- Ant might say, are these goals universal? What does our community need? Who benefits from these goals?
- Grandmother might say, I've seen all this before, people defining our future and relationships with Country and each other. How do our old knowledges and stories sit with these goals?
- Granddaughter might say, it's really good that someone is thinking about other people. 2030 is ages away, we've got lots of time to reach the goals. What do I need to do?

Can you extend on this analysis, using pattern thinking? Remember, pattern thinking is about interconnectedness and collapses the notion of linear time. Or try it out with your own example, what are you working on, or thinking about? Here's another example from our research:

> *In Australia, for more than a decade there's been a federal policy of 'Closing the Gap' between Indigenous and non-Indigenous people with target areas in education, employment, life expectancy and mortality. In education specifically, there are three main target areas: Attendance (close the gap); Literacy and Numeracy (halve the gap); and Year 12 Attainment (halve the gap). In isolating these variables, there has been minimal success in 'closing the gap', with only one of these targets (Year 12 Attainment) on track according to the 2020 Report.*
>
> *(Commonwealth of Australia 2020)*

- Eagle might say, why might Indigenous students not want to go to school? Are literacy and numeracy the best measurements of achievement, and how are these being assessed?
- Ant might say, which Country are you on? Do the targets need to be the same in each community? What do Indigenous people want for their children?
- Grandmother might say, schooling has been a harmful and assimilatory place for Indigenous people, has this changed? What values are our children learning in these schools?
- Granddaughter might say, am I not good enough? Who sets the measure of my success?

While we have been concentrating on using these perspectives to reimagine the future, remember that time is non-linear; there is no separation between the past, present and future. It is important to realize that drawing on these perspectives and the power of pattern thinking is also a way to become otherwise in the present. Thinking differently about/in the present can be the action we take to create futurities. Eagle, Ant, Grandmother and Granddaughter help us relate to the future in different ways, to create futurities. Goodyear-Kaʻōpua (2018, 86) notes that, 'futurity is not just another way to say "the future." Futurities are ways that groups imagine and produce knowledge about futures.' Eagle, Ant, Grandmother and Granddaughter do this because they relate to one another; they are not isolated entities but deeply connected perspectives. The same can be said for development. Lloyd, Wright, Suchet-Pearson, Burarrwanga and Bawaka Country (2012, 1076) reflect on the ethics of collaboration in development practice and conclude that development is 'inherently and always relational.' Pattern thinking sits at the heart of this relationality; a way to bring together a holistic picture and reflect deeply to impact on the future (Tynan 2021).

Conclusion

This chapter draws on Indigenous autoethnography to offer kin, more-than-human and non-linear perspectives to incorporate into your pedagogy and practice in reimagining futures and doing development differently. Pattern thinking and the perspectives of Eagle, Ant, Grandmother, Granddaughter – or other perspectives that better suit your context – enhance our ability to think critically about the world and our futurities. This thinking presents a new direction on critical development thinking, practice, and pedagogy to respond to current global and local challenges.

In this chapter we've been reimagining the future and using fire as our theme. Let's finish with a story from the future, where cultural burning continues to sustain and nurture all life:

> *The smell of smoke is back. I can see billowing white tufts finding their way into the taller leaves, giving medicine. Giving renewal. Feeling refreshed and reenergized as the flames trickle down the valley like water. We see the frogs, the spiders, the insects and lizards scatter away, scampering up the trees. They are not afraid. Moments after, they are on the cooled ground returning to their home. Head tilted down, I brush away the ash with my nose, finding small green shoots of grass, vibrant and moist. Everywhere I look there are pockets of green, the little fire meandered around our grassy patches, encouraging more to grow. As a family, we lightly bound past the humans who care for the good fire. The green shoots are bringing new life, the canopy is healing and we are with our Mother.*

Acknowledgements

We acknowledge the Old People who dreamt us into creation and guide our strong futures. We especially think of our grandparents and future grandchildren. Thank you to Paul Hodge, Kate Lloyd, Sandie Suchet-Pearson, Fiona Miller and Tiffany Jones for their feedback and yarns.

Key sources

1. Archibald Q'um Q'um Xiiem, J., Lee-Morgan, J. B. J., and De Santolo, J. (2019) *Decolonizing Research: Indigenous Storywork as Methodology*. London: Zed Books.

This book showcases the brilliance and depth of Indigenous stories and how they can be used as research methodology, to create relationships and futurities. Storywork and storytelling use pattern thinking to unlock knowledges and carry them across time. This is a great text for research methodologies and being inspired to create futurities through stories.

2. Gay'wu Group of Women. (2019). *Song Spirals: Sharing women's wisdom of Country through songlines*. Crows Nest, Australia: Allen & Unwin.

The Gay'wu Group of Women demonstrate how stories, knowledges and pattern thinking spiral throughout time and space to inspire and teach lessons in the past, present, and future. Weaving through non-linear thinking, this text shows how important Place and stories are to the creation of futurities.

3. Graham, M. (2014) 'Aboriginal Notions of Relationality and Positionalism: A Reply to Weber,' *Global Discourse* 4, 17–22.

Mary Graham takes you on a deep and accessible journey of relationality from an Aboriginal standpoint. She tackles some core assumptions of Western knowledge traditions (especially international relations and politics) and explains how Aboriginal philosophies are relational. A great read for understanding Aboriginal perspectives of place, ethics and balance.

4. Milton, V. (2018) 'Indigenous fire methods protect land before and after the Tathra bushfire,' ABC, Accessed 12 March 2020. www.abc.net.au/news/2018-09-18/indigenous-burning-before-and-after-tathra-bushfire/10258140

Dan Morgan and the crew from Bega Local Aboriginal Land Council show how cultural burning can be used to re-generate Country, even in the wake of bushfires. Filmed in Tathra, NSW, Australia, this video shows how the future is being remade through Aboriginal fire management, and the intricacies of how this is achieved.

5. Pascoe, B. (2014). *Dark Emu, Black Seeds: Agriculture or Accident?* Broome, Australia: Magabala Books Aboriginal Corporation

Bruce Pascoe revisits the journals of Australian colonists and reveals the strong evidence of Aboriginal agriculture and aquaculture practices in Australia. This book shows how the future can be re-cast by better understanding the past. Pascoe creates a futurity for Aboriginal people in Australia where our knowledges of grasses, bread-making and fish-trapping can be used to create our own economies.

Notes

1 trawlwulwuy, tebrakunna and other ancestral language words do not always use capitalization
2 The term 'Country' is purposefully capitalised to denote the sacred and specific connection to land/sky/water/place/entities that is held by many Indigenous Peoples in Australia.
3 We acknowledge the knowledge shared by Dan Morgan, Victor Steffensen and the crew of Bega Local Aboriginal Land Council who inspired this vignette. You can learn more from them in the 'Further Reading' section, 'Indigenous fire methods protect land before and after the Tathra bushfire.'
4 We use 'they' to avoid gendered binaries of she/he when referring to some of the perspectives. It is not intended to be read as plural.

5 While we draw on a biological process here, we understand that Grandmother, like many familial roles, is not determined by biology. Grandmother is a kin relationship, so can be assumed by people who haven't given birth. This kin relationship means one person can have multiple Grandmothers, sharing multiple stories and knowledges.

References

Bawaka Country including, Burarrwanga, L., Ganambarr, G., Ganambarr-Stubbs, M., Ganambarr, B., Maymuru, D., Wright, S., Suchet-Pearson, S., Lloyd, K., and Sweeney, J. (2017) 'Co-becoming time/s: time/s-as-telling-as-time/s' in J. Thorpe, S. Rutherford and L.A. Sandberg (eds.) *Methodological Challenges in Nature-Culture and Environmental History Research*, New York, NY: Routledge, 81–92.

Bishop, M. (2020) '"Don't tell me what to do": Encountering colonialism in the academy and pushing back with Indigenous autoethnography'. *International Journal of Qualitative Studies in Education* 34(5), 367–378, DOI: https://doi.org/10.1080/09518398.2020.1761475

Commonwealth of Australia. (2020) *Closing the Gap Report 2020*, Canberra: Department of the Prime Minister and Cabinet.

Cameron, P. (2011) *Grease and Ochre: The Blending of Two Cultures at the Colonial Sea Frontier*, Hobart: Fullers Bookshop.

Dudgeon, P., and A. Bray. (2019) 'Indigenous Relationality: Women, Kinship and the Law,' *Genealogy* 3, 1–11.

Goodyear-Ka'ōpua, N. (2018) 'Indigenous Oceanic Futures: Challenging Settler Colonialisms and Militarization'. In L. T. Smith, E. Tuck, and K. W. Yang (Eds.), *Indigenous and Decolonizing Studies in Education: Mapping the Long View* (pp. 82–102). New York: Routledge.

Ife, J. (2013). *Community Development in an Uncertain World: Vision, Analysis and Practice*, Port Melbourne: Cambridge University Press.

Kenny, S. (2011). *Developing Communities for the Future*, South Melbourne: Cengage.

Kimmerer RW. (2013) *Braiding Sweetgrass: Indigenous Wisdom, Scientific Knowledge, and the Teachings of Plants,* Canada: Milkweed Editions.

Lloyd, K., Wright, S., Suchet-Pearson, S, Burarrwanga, L., and Bawaka Country. (2012) 'Reframing Development through Collaboration: towards a relational ontology of connection in Bawaka, North East Arnhem Land,' *Third World Quarterly* 33, 1075–94.

McGregor, A. (2019) 'The trouble with human development: towards a more than human approach' Paper presented to Cultivating Creative Conversations and Critical Connections in Development Theory Virtual Symposium, Institute of Australian Geographers, 5 December.

Moreton-Robinson, A. (2018) 'Is Australia morally justifiable?' ABC: The Minefield 25 January, 2017, Accessed 12 March 2020. www.abc.net.au/radionational/programs/theminefield/is-%E2%80%98australia%E2%80%99-morally-justifiable/8208136

Smith, L. T., Tuck, E. and Yang, K. W. (2019) 'Introduction' in L.T. Smith E. Tuck and K.W. Yang (eds.) *Indigenous and Decolonizing Studies in Education: Mapping the Long View*. New York, NY: Routledge, 1–23.

Steffensen, V. (2020) *Fire Country*, Melbourne: Hardie Grant Travel.

Tuck, E., M. McKenzie and K. McCoy (2014). 'Land education: Indigenous, Post-Colonial, and Decolonizing Perspectives on Place and Environmental Education Research.' *Environmental Education Research* 20, 1–23.

Tynan, L. (2021) 'What is relationality? Indigenous knowledges, practices and responsibilities with kin.' *Cultural Geographies* 3, 597–610.

Tynan, L., and Bishop, M. (2019) 'Disembodied experts, accountability and refusal: An autoethnography of two (ab)Original women.' *Australian Journal of Human Rights* 25, 1–15.

United Nations (UN). (2020) 'The Sustainable Development Agenda,' United Nations, Accessed 12 March 2020. www.un.org/sustainabledevelopment/development-agenda/

Yunkaporta, T. (2019) *Sand Talk: How Indigenous Thinking Can Save The World*, Melbourne: The Text Publishing Company.

52

ACTIVISM AND DEVELOPMENT STUDIES PEDAGOGY

Tolu Muliaina

Introduction

Formal schooling has influenced the way and orientation of teaching and learning and assessment in the Pacific. Once anchored in families and communities, the introduction of formal schooling led to the building of schools to teach students different subjects (Crossley 1990) underpinned by Eurocentric, distinct disciplines and individualistic educational goals. This is far removed from the applied, holistic nature of teaching and learning informed by local cultures and languages grounded in the community where elders teach and mentor the young (Helu-Thaman 2003). The European style of classroom teaching and learning dominates at the expense of indigenous values and beliefs (Thaman 2009). Further, English or French became the language of instruction and business. Considering language is a storehouse of a culture and 'embedded within it are an ontology and epistemology of people's understanding of the world and lived experience' (Gegeo and Watson-Gegeo 2013, 138), teaching and learning in a language other than your own is a significant concern; more so when proficiency in English or French is a development priority. This is serious for Pacific Islanders whose rich oral traditions rely on many speakers of that language to transmit valuable knowledge, skills, and understandings to the next generation of users.

Box 52.1 Defining 'the Pacific'

The Pacific Islands are home to less than 5 million peoples. There are 50 independent states and self-governed territories in association with a foreign, developed nation. Except for Tonga that was not under direct foreign control, other countries added layers of complexity to an already diverse region of cultures, socio- political systems, languages and environments. The Pacific has been and is of increasing geopolitical and strategic significance to countries outside the region. Where necessary the chapter specifies issues pertaining to the region or a specific country.

Scholars in the Pacific (Thaman 2009; Tuiwhai-Smith 1999) and elsewhere (e.g. Demmert (2011), mostly in the fields of Education, Health and Pacific or Cultural Studies, have recognized

DOI: 10.4324/9781003017653-58

this serious development dilemma and suggested ways forward to address this grave challenge for peoples of the Pacific. This chapter attempts to address this dilemma between external development educational priorities and Pacific Islander aspirations to strengthen local cultures including languages and knowledge transfer. It makes a case that a culturally informed, innovative pedagogy and practice can offset the colonial and institutionalized nature of teaching and learning in Pacific classrooms. Using examples of my teaching philosophy at The University of the South Pacific (USP), I show ways to bring about educational development and outcomes that are relevant, just, and sustainable for Pacific students. I also discuss implications of these attempts for the future. The next section discusses my position and framing of this chapter.

Making connections

Traditionally, teaching and learning in all areas of life, including the conservation of the bio-physical environment, takes place in the *'aiga* (family), the *nu'u* (village) and later the *lotu* (Christianity) as Figure 52.1 shows. Every Samoan belongs to an immediate *'aiga* of parents and their children as well as the extended family of each parent. The expression *e leai se tagata noa i Samoa* speaks to this understanding that no Samoan is family-less (Meleisea 1987). By extension, the family is where Samoans initially learn lifelong skills, expected behaviours, relational conduct and values from parents and elders. Values of respect, love and humility are the most revered. When Samoans say, *O le 'aiga o le faleaoga ma faiaoga muamua*, they are saying that the family is the **first** school and the **first** teachers. They/I are talking from within that position.

Teaching and learning is largely by keen observation and performance. Elders of a family teach the young skills to farm, fish, weave and orate while also demonstrating expected values and manners. Therefore, the elders model how to live and relate well with others in the family and the wider community. Included herein is the observance of ancient protocols that guide the use and conservation of the biophysical environment. The village of Asau, for one, is forestry-rich and known for its building guild. There, the performance of a prayer chant, *fa'alanu* before cutting trees speaks of the interdependence of people and their environment (Ta'isi 2008a, 107). The ritual of burial of the *pute* (umbilical cord) in the *fanua* (land) signifies, again,

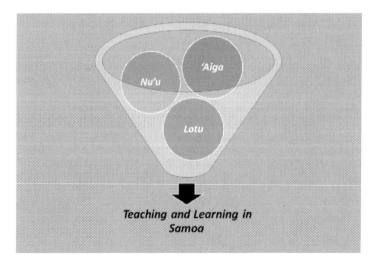

Figure 52.1 Pillars of teaching and learning (author's own)

the inseparable people-environment link. Both cases speak of spiritual and cultural continuity. Samoans have deep respect in living in harmony with their environment (see, e.g., Bhagwan, Huffer, Koya-Vakauta and Casimira 2020).

In this home/community-based education system, young Samoans receive training from elders, 'parents and immediate family members … based on the ethos of the people' (Maiai 1957, 166). Unlike formal schooling, this informal learning came in the form of 'myths and legends, in proverbial sayings, in songs and dances in reverential restraints and taboos, in parental and social comments in the action of the children' (ibid.). Myths and legends are important methods and resources of teaching and learning in Samoa. *Fagogo* (storytelling) 'is not empty theatre' (Ta'isi 2008b, 116), rather, the means through which elders transmit knowledge (some sacred) and understanding to their children, a once popular bedtime teaching and learning activity.

Extended families make up *nu'u* (villages) of varying size and history. There, family members belong to different social groups as shown in Figure 52.2. Under the guidance of the most senior members, each group has roles and responsibilities for the development of the village which includes teaching and learning responsibilities reflecting the 'ethos' described by Maiai (1957).

From their seniors, young women (*Tamaitai* and *Aualuma*) learn the crafts of weaving assorted mats, blinds, and baskets to use and sell. Also included are the skills to cultivate, use and conserve plants to make these crafts – such as *laufala* (pandanus, *Pandanus tectorius*) and

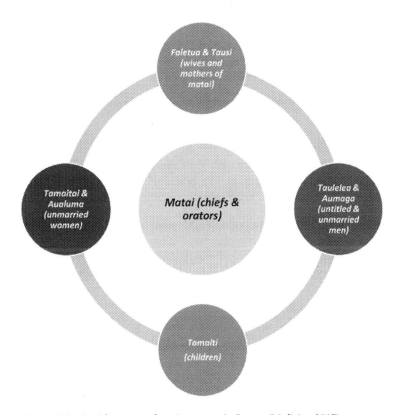

Figure 52.2 Social groups as learning spaces in Samoa (Muliaina 2017)

others of cultural, medicinal (*noni, Morinda citrifolia*) and commercial (*laga'ali, Aglaia samoensis*) importance. Samoans consider these skill sets, and associated knowledges worthy of possession and expect women to transmit them to incoming women of the village.

As important skills, women expect to pass on the skills of weaving mats especially Samoa's most treasured fine mat, *ie* Samoa. A measure of a woman's success and wealth is her ability to weave assorted mats including the *ie* Samoa to give away at funerals of elders, bestowal of *matai* (chief) title ceremonies and weddings of daughters. Women are also traditional healers and their knowledge of various plants to treat specific illnesses has long been a pillar of Samoa's health system (Whistler 1996) well before the arrival of western medicine. Not passing these skills and knowledge on is detrimental to the survival and continuity of traditional healing practices anchored in Samoan culture and ethno-biodiversity (ADAP Project 2001; Whistler 2005).

Within their domain of work and responsibilities, young men (*Taulelea* and *Aumaga*) receive their training and mentoring on skills to cultivate the land, fish and build canoes and houses through observation of the performance of adult males under the watchful eyes of a *tufuga* (master). A *tufuga* is someone the village or family considers to possess a fine skill or knowledge of boats or house making, fishing of a particular fish or the art of body tattooing. Young men also receive learning and teaching from elderly men in the house of men (*fale o matai*) such as the processes and protocols involved in an *ava* ceremony to welcome or farewell a visiting party. Of note also is the making of *afa* (sennit), a type of cordage prepared from dried coconut husk. *Afa* is used as a fastener in architecture and boat building well before nails, wood clue and related alternatives flooded the island. As providers and protectors of families and villages and their resources, Samoans expect their young men to learn appropriate skills to excel in these roles. Attaining these skills and responsibilities are central to cultural survival and continuity.

Because Samoan men conceive the spatial domain of their teaching and learning as that which stretches from the *tuasivi* (ridge) to *a'au* (reef) and everything in between, they in fact play important roles in the conservation and management of natural resources. In villages, the cultivation of land is a collaborative pursuit. Under the guidance of an adult leader, young men work cooperatively. As a group, they clear the land and plant a variety of root crops for each member's family. In so doing, young men share best farming practices and techniques to maintain soil fertility without the use of artificial fertilizers, crop rotation and poly-cultural cropping. At sea, the role of young men is critically important in the conservation, management and protection of the coastline and marine resources. While Samoa's Fisheries Management Act 2016 recognizes the precautionary approach and ecosystem approach as key principles to the conservation and management of marine resources, the success rests in the hands of village men. Their skills in traditional fishing and the use of fishing methods is highly valuable and predate these introduced legislative conventions and principles. Also contributing to conservation is the enforcement of local *tapu* that villages put in place in the harvesting of specific marine species. These examples of the types and methods of teaching and learning are 'not so different from the principle of apprenticeship today' (Maiai 1957, 167) and further exemplify the holistic nature of Samoa's place-specific sustainable practices.

Except for *Tamaiti* (Children), *Faletua* ma *Tausi* and *Matai* also conduct teaching of their members on matters related but not limited to primary health care, rearing of children, planning for development, and safeguarding village resources and people. They also oversee the work of young women and men as well as their training and protection of their environment. As leaders and decision makers, they also impose *tapu* on the use and harvest of certain resources and charge breakers of village bylaws. Further, each group conducts regular inspection of homes and village surroundings including the protection of village water sources and sites of historical significance.

Then there are, as Thomas (1993) noted, rituals and ceremonial events, funerals, weddings and bestowal of chief titles to name a few of the practices performed by Samoans. These too are teaching and learning spaces. The nature of each event requires the use of different skills and knowledge of oratory, genealogy, and the observance of expected cultural protocols and behaviours. For funerals, the exchange of cultural capital, *ie* Samoa and other material wealth, accompanies speeches between the mourners and visiting parties. However, these are not ordinary speeches. In fact, they are oral exchanges of the highest level because the speakers converse in an oratorical language, the formal language that one only acquires by years of practice and learning from elders. Such learning also means one needs to understand the *fa'alupega* (ceremonial salutation consisting of local courtesy titles) history and genealogy of each of the 340 villages and 48 political districts (SBS 2018). While all villages relate, each one is distinctly different. As in other cultural events, funerals connect generations, the past and present and collectively strengthen familial relationships. Building the *va fealoai* (relationship) is the cornerstone of the Samoan culture and governs all that Samoans do.

In the simplified illustration in Figure 52.2, Samoans understand that *Fa'aSamoa*, commonly referred to as the Samoan culture or a way of life, is more than what its English translation suggests. It is the framework which Samoans use to understand their world and others. To marginalize Samoan culture and language in development policy and practice actively devalues *Fa'aSamoa*. For teaching and learning, the latter is critical, for the socialization of Samoans takes place in the cocoon of the *Fa'aSamoa* within their *aiga* and wider community way before children start formal schooling. Overturning the colonizing nature of development priorities is a starting point to restore the place of *Fa'aSamoa* as the basis of, and for development, including teaching and learning and assessment that is culturally relevant and sits in place (Escobar 2001).

Activism and pedagogy in action: aligning course content to home cultures

Being a product of formal schooling and socialization anchored in the Samoan culture naturally shapes my views of and approaches to teaching, learning and assessment generally and in particular, the third-year course on Resource Conservation and Management, I teach at USP (Muliaina 2018). The overall aim of the course is to develop a better alignment of the course content, pedagogy and assessment regimes to the home cultures of students of USP's 12 member countries. Alignment allows for the integration of appropriate pedagogy and assessment regimes to suit diverse student interests and cultural backgrounds. Highlighted next are three examples of this experience as well as their implications for contextualized teaching and learning in the Pacific. One, negotiating selected assessment in the course; two, building bridges with resource conservation practitioners and community elders; and three, creating a platform for students to practice activism in resource conservation and management during and after the course.

Co-designed assessment: valuing student's culture

First, the fact that students have minimal say in content and assessment in formal schooling including university is unsettling for a region where principles of open participation, sense of community and consensus building are the building blocks of society. Acknowledging the value of student's cultural backgrounds, prior learning and lived experiences, led me to introduce an assessment regime where students negotiate *what* topic and *how* they would like to be assessed in one assignment worth 10 per cent of the final mark. The concept of *so'a lau pule* in the Samoan culture captures the philosophical basis of this regime. Its simplest meaning refers to the sharing of authority with one's partner or group. Sharing is, of essence, based on trust. However, its application here is more than what the translation suggests for to share authority in the Samoan culture also implies giving, conferring and delegation of ideas, opinions and power, individual rights and responsibility (Tuisuga-le-taua 2009). Sharing takes place through open negotiations before reaching a consensus. When a consensus is not forthcoming, *moe le toa*, deferment is vital. Taking time out to reflect provides insights for both parties in finding ways forward (Muliaina 2018).

So'a lau pule is an inclusive process for teachers and students to share authority and responsibility related to aspects of teaching, learning and assessment. To involve students in the design of selected aspects of their learning and their assessment encourages participation and ownership. Students negotiate ways that best reflect their culture and capabilities thus co-owning the teaching and learning process. Their involvement also encourages self-evaluation of the utility of such an assignment beyond the course, while developing a particular skill set and providing students an opportunity to further pursue a topic of interest on their terms. In addition, what students submit at the end of the semester is also negotiable. Students may use different genres of writing (essay, poetry, policy brief, letters to the editor, sermon), a combination of these or expressive arts: visual, dance/movement, drama/theater and songs as acceptable forms of assessment.

The overall aims of the Resource Conservation and Management course is consistent with regional frameworks that acknowledge the centrality of culture in educational contexts. For example, recognizing home cultures in teaching and learning aligns with Policy Area 1: Quality and Relevance of the Pacific Regional Education Framework (PacREF 2018–2030). The outcomes of Policy Area 1 state that:

i) 'curriculum programmes are embedded in the Pacific context that reflect Pacific values, cultures, traditional knowledge and skills that draw on the land that we live and exist upon and the ocean surrounds and bind us all.'
ii) 'learning is inclusive of cognitive and non-cognitive development.'
iii) 'curriculum and programmes, with appropriate pedagogy are inclusive, rights-based, promote gender equality, flexible and responsive to innovation and change and are adaptable to new learning opportunities' (PacREF 2018–2030, 8).

While most students welcome the idea to design an assessment that in their view best captures their strengths and existing skill set, a few were nervous when given the opportunity, perhaps reflecting the enduring nature of Eurocentric, formal teaching. Nevertheless, a chance to participate, in itself, represented a door to opportunities in that the popular preference among students was a combination of drama, singing and spoken poetry as forms of assessment. This indicates that teaching and learning processes need continuous refinement to account for both creative abilities and home cultural knowledges that students possess. This is in line with Taufe'ulungaki's (2002) view that Pacific students think in creative, holistic, circular and people-centred ways that reinforce cooperative learning and sharing of skills and ideas. Such skills and culturally informed learning go beyond the classroom when students find work and develop their communities. Alumni association of former USP scholars around the region is one example where the sharing of skills and ideas reinforce and strengthen Pacific-wide networks.

Building bridges: the central role of practitioners and community leaders

Second, the very nature of resource conservation and management as the basis for sustainable development requires evolving pedagogy that marries theoretical and applied knowledge. Because Pacific communities have 'lived' the conservation and management of natural resources for millennia, they, as a result, accumulate time-tested knowledge and skills. Therefore, to include Pacific-based resource conservation practitioners and champions in the community to co-teach the course and to support student learning is timely. To include them expands the information sources that students draw from as they gain hands-on experience into the complex world of resource conservation and management beyond the classroom. It also means negotiating access to knowledges that may be sacred. In the Pacific, when access to knowledge is given, recipients accept it with humility and responsibility. Given the integrated nature of life in the Pacific, negotiation is one such important process that anyone seeking knowledge cannot overlook. From a teaching and learning perspective, students need to learn and respect this process for knowledge resides with holders in the community.

Including the Fiji Spice and Nature Gardens, Shangri-la Fijian Resort, Outrigger Resort and the Sigatoka Sand Dunes as coursework partners provides links between theoretical and applied knowledge in the course. Our partnerships have grown over time and offered opportunities to exchange information and best practices. They in turn inform teaching and learning 'on-site' experiences for students, as discussed below. Different affiliation aside, working with them also shows that we share a lot in common about resource conservation and management. For Pacific peoples, it is about cultural survival and continuity.

Every year for the last five years, two cohorts of students from the course have visited these sites twice a year to observe, learn and participate in the conservation activities planned at each site such as planting (or replanting) of trees or corals. There are several highlights of these field visits. First, new students are able to see trees or corals flourishing that previous groups planted. Under the supervision of Kini Sarai, a retired Activities Manager at Outrigger Resort, students learn to build fish houses and plant them at sections of the adjacent reef in need of restoration. Students also participate in a tree-hugging activity at the Sigatoka Sand Dunes, a concept that has its origin in the Chipko Movement in 1970 India when the conservation of forests was ignored (Tewari 1995). The Park Manager, Jason Tutani explains the concept of tree hugging, a symbol of and practice to conserve forest and shows students how to make tree huggers, using dried vines, tree branches and driftwood. Finally, when practitioners point students to different species of indigenous hardwood, plants, herbs or ferns of medicinal and cultural value, they make connections to what they learn in class while also strengthening and nurturing people-environment interdependence.

Taking students to the field reaffirms the role that field labs play to scale back the confines of a Eurocentric formal education. Field labs are necessary for a course concerned with the conservation and management of natural resources in the Pacific. They are excellent outdoor teaching and learning opportunities that incorporate Pacific ways of teaching and learning where elders and champions explain, demonstrate and mentor students about resource conservation and management in the community. Therefore:

> [field labs are] important because at its heart lies a direct, active encounter with 'the other' – others who call us to attention with a jolt; who challenge us to rethink our preconceptions; who draw a sense of fellowship from us – and a new attentiveness that aids a deeper understanding.

It is this direct encounter that makes [field labs] challenging and at times difficult and it is these
features that, in my view, continue to make it a valuable mode of learning for [resource conserva-
tion and management in the Pacific.]

(Hope 2009, 182)

Nestled in the fertile foothills of Wainadoi, Fiji Spice and Nature Gardens cultivates horti-
cultural spices: vanilla, cinnamon, black pepper and nutmeg among others and an assortment
of tropical food trees/crops and medicinal and flowering plants without the use of artificial
fertilizers. Besides exports, they supply ginger, masala, turmeric, and other products locally as
well. Students learn from property owner, Veronica Hazelman and team on site and witness
for themselves the soil conservation regimes (mulching, crop rotation, terracing) they practice,
to ensure constant supply of healthy and organic species and food crops. To Veronica, working
with and caring for the soil is an inseparable part of responsible community-based development and sustainable living informed by a deep understanding and connection with the soil.
This resonates with how Pacific peoples understand their inseparable relationship with the land
noted in the symbolic burying of the umbilical cord of a new born child and continuous care
for it. That sense of connection only comes by living in harmony with resources and is the holistic ethos that Fiji Spice and Nature Gardens hopes to live by. To re-live traditional and organic
soil conservation regimes is to take care of the land and continue to support life.

Accessible by a causeway, Yanuca Island, the home of the Shangri-La Fijian Resort, takes
responsibility to treat its solid and liquid waste investing in a water treatment plant, and three
artificial wetland ponds. On the island, these ponds prevent possible discharges of additional
harmful nutrients into the sea thus controlling waste flow, which is the resort's ultimate goal.
The resort also runs a Sanctuary, Shangri-La Care for Nature Project, an environmental education facility with a designated learning space and animated displays for guests, workers, and
locals.

Students learn the steps and ways to treat wastewater from the resort's resource conservation champions – Mereoni Mataika, the Service Manager Corporate Social Responsibility and
Saimoni Komaiwaca, Sales Coordinator. Students understand that the construction of three,
60 metre by 8 metre wetlands each along a slope allows wastewater to flow from the first pond
through to the third one (Robinson 2002, 144). Students also learn about plant species (water
hyacinth, *Eichhornia crassipes*) grown in the wetlands to remove nutrients from wastewater. The
kuta grass, a plant of cultural significance in Fiji finds its way there. Women of Cuvu district,
owners of the island, make prized *kuta* mats and other handicrafts to sell at the resort. It is one
way that the resort works with and maintains links with the community. Students connect classroom learning to conservation practices in the resort mirroring much of the hands-on informal
learning widespread in the Pacific.

A healthy marine environment means protecting coral reefs from land-based harm. This
has been Outrigger Resort's preoccupation through its coral planting programme to protect
and sustain the adjacent coral reef. Over the last five years, students have contributed to the
resort's effort to plant corals while also witness abundant growth and regeneration from the
work of former students. Each time we visit, students build new coral and fish houses to add
to an existing growth pool and/or replace dying ones. More than simply a tourist destination,
the Outrigger and Shangri-la offer educational experiences. These resorts work collaboratively
with local communities to live with and conserve the natural resources while providing guests
an experience grounded in the Fijian culture and natural environment.

The Sigatoka Sand Dunes and all its beauty is also a teaching and learning laboratory for our
course. A half-day spent there offers students an opportunity to learn from the Park Manager

and energetic staff and to see for themselves the changing dynamics of a precious yet endangered resource due to natural and human activities (littering and illegal sand mining) that are not easy to control (Tutani 2021). Through planting of over four hundred native beach mahogany trees and seedlings, students help to restore vegetation of selected areas that are susceptible to illegal deforestation and bush fires. In assignments, informal discussions, and debriefing sessions in the field, students express satisfaction on the part they play in forest restoration efforts and the inspiration they receive from previous course peers and the work that Jason and his team of rangers do at the dunes. Students also learn ways to protect the fore dune by building huts along the beach made of driftwood scattered on the beachfront. The huts once built to provide shelter from the scorching sun have also turned out to be an effective defence against erosion. Controlling the movement of sand is important and students contribute to that goal by making and/or maintaining designated pathways with driftwoods and other debris. These physical structures guide users to follow a set path to avoid erosion and discourage unrestricted access on the dunes.

Creating activist spaces beyond the classroom

Finally, linking students to opportunities to advocate and expand their perspectives of resource conservation and management is critical. Teaching and learning therefore needs to guide students to practice what they learn outside of the classroom. It means connecting them to other students and youths as a power base to activate activism for positive change. This is *soalaupule* to Tuisuga-le-taua (2009). Elders share authority with the young in all areas of life including resource conservation and management. Guiding, mentoring, and sharing knowledge and skills, elders transmit their *mana* (knowledge, power and blessing) to the young and upcoming leaders to learn, grow and lead.

The Pacific Islands Students Fighting Climate Change (PISFCC), a student non-governmental organization I co-found at USP, offers students in the course a platform to advance their interests in the broader goals of resource conservation and management in the Pacific Islands. While PISFCC's core mandate is to seek an Advisory Opinion from the International Court of Justice on climate change and human rights, its broader concerns are the sustainability of natural resources, conservation and management and heritage each of which sit at the heart of the course. Realizing the potential that students have shown in the negotiated assessment as discussed earlier, there is a need for such a platform for students to practice and grow. This is a less populated teaching and learning space but a necessary one that students learn by doing if they are to contribute to the sustainable development of their countries as well as enhance professional growth. As the threats of climate change, and now COVID-19, impact resources and the livelihood of all countries including Pacific Islands, PISFCC as a teaching and learning space has an important contribution to make. With more than half of the region's 10 million people being youths (SPC 2015), there is no better way to make use of their skills and talents to engage in activism for positive change in the future.

Conclusion

Formal schooling has offered opportunities and challenges. These demand innovative pedagogy and assessment regimes that remain responsive to the evolving needs of Pacific classrooms. Including practitioners and elders in the teaching of the course builds bridges and facilitates access to community-based conservation practices and knowledge bases. Their approaches to resource conservation and management reflect the deep respect that Pacific peoples have for the land and the sea. Given their cultural backgrounds, students are familiar with them. Through soil and forest conservation regimes, the Fiji Spice and Nature Gardens and the Sigatoka Sand Dunes rebuild and restore deep respect for the soil, as provider and giver of life. The Shangri-la Fijian Resort takes onto itself to treat and control its wastes on the island reflecting a shared responsibility to protect resources. Making fish and coral houses is Outrigger Resort's attempt to restore corals as the basis for abundance and sustainable supply of fish. Sharing knowledge and skills, best practices, field demonstration and mentoring, each practitioner enriches the course and learning experiences of students. They also transmit their *mana* to the next generation of Pacific leaders.

Despite the stronghold of formal schooling, culturally informed pedagogy and practice can mediate the institutionalized character of teaching and learning in the Pacific. Several scholars in the region and elsewhere have set the foundation of this 'relevance' movement (Moorman, Evanovitch and Muliaina 2021; Nabobo-Baba 2012). However, its continuity relies on many more practitioners to advance this goal. This is an ongoing development challenge that institutions like USP needs to take a more active role in. A role that balances development educational priorities driven from outside with Pacific aspirations that continue to anchor Pacific cultures and worldviews.

Key sources

1. Freire, P (1970) *Pedagogy of the Oppressed*. 30th Anniversary edition. New York: The Continuum International Publishing Group, Inc.

Teaching liberates the oppressed individuals that they are free and in control again of their destiny.)

2. Hytten, K. (2014) 'Teaching as and for Activism: Challenges and Possibilities,' in C. Higgins, C and MS. Moses (eds) *Philosophy of Education 2014*, University of Illinois, Urbana-Champaign, 385–394.

Teaching is activist when it encourages and supports students as agents of positive change and social justice. It should also empower students with the necessary tools and direction to act.

3. Koya Vaka'uta, CF (2016) 'Straight Talk | Crooked Thinking: Reflections on Transforming Pacific Learning and Teaching, Teachers and Teacher Education for the 21st Century,' in *Weaving theory and practice in Oceania: Threads of Pacific education practitioners*. Nuku'alofa: Institute of Education, The University of the South Pacific.

Transforming Pacific learning, teaching and teacher education calls for deliberate efforts to anchor current development theory and practices on Pacific cultures as vital to achieving sustainable societies.

4. Ollis, T. (2015) 'Chapter Thirty-Three: Activism, Reflection, and Paulo Freire – an Embodied Pedagogy,' *Counterpoints* 500, 517–527.

The pedagogy of activism is about critical teaching and learning that is embodied and situated in places and practice. It is a deliberate attempt to bring about social justice and positive social change.

5. Watkins, G (1994) *Teaching to Transgress: Education as the Practice of Freedom*. New York: Routledge Taylor & Francis Group.

Teaching is to empower young people to take active role to offset the systemic injustices in schools. That teachers trusted with this role for students to find the gift of freedom to shine.

References

Agricultural Development in the American Pacific (ADAP Project). (2001) *Samoan Medicinal Plants and their Usage*. Reprint May 2001. Honolulu: University of Hawai'i.

Bhagwan, J., Huffer, E., Koya-Vakauta, F.C. and Casimira, A. (eds.) (2020) *From the Deep: Pasifiki Voices for a New Story*. Suva: Pacific Theological College.

Crossley, M. (1990) 'Collaborative research, Ethnography and Comparative and International Education in the South Pacific,' *International Journal of Educational Development* vol. 10, no. 1, pp. 37–46.

Demmert, W.G. (2011) 'What is Culture-Based Education? Understanding Pedagogy and Curriculum,' in J. Reyhner, W.S. Gilbert and L. Lockard (eds.) *Honoring Our Heritage: Culturally Appropriate Approaches for Teaching Indigenous Students*. Flagstaff: College of Education, Northern Arizona University, pp. 1–9.

Escobar, A. (2001) 'Culture sits in places: reflections on globalism and subaltern strategies of localization,' *Political Geography*, vol. 20, no. 2, pp. 139–74.

Gegeo, D.W. and Watson-Gegeo, K.A. (2013) '(Re) conceptualizing Language in Development: Towards demystifying an Epistemological Paradox,' *Journal of Pacific Studies* vol. 33, no. 2, pp. 137–55.

Helu-Thaman, K. (2003) 'Culture, Teaching and Learning in Oceania' in Helu-Thaman, K. (ed) *Educational Ideas from Oceania: Selected Readings*. Suva: Institute of Education USP/UNESCO, pp. 3–12.

Hope, M. (2009) 'The Importance of Direct Experience: A Philosophical Defence of Fieldwork in Human Geography,' *Journal of Geography in Higher Education* vol. 33, no. 2, pp. 169–82, DOI: 10.1080/03098260802276698.

Maiai, F. (1957) A study of the developing pattern of education and the factors influencing that development in New Zealand's Pacific Dependencies. Unpublished MA Thesis, Wellington: Victoria University of Wellington.

Meleisea, M. (1987) *Lagaga: A short History of Western Samoa*. Suva: The University of the South Pacific Press.

Moorman, L., Evanovitch, J. and Muliaina, T. (2021) 'Envisioning indigenized geography: a two-eyed seeing approach,' *Journal of Geography in Higher Education*, vol. 45, no. 1, pp. 1–20. https://doi.org/10.1080/03098265.2021.1872060

Muliaina, T. (2017) Grounding *Malaga* in *Aiga* Samoa: *Alofa* as manifested in Population Movement. Unpublished PhD in Development Studies, The University of the South Pacific, Suva.

Muliaina, T. (2018) 'In search of meaningful assessment in the university curriculum: the case for culturally relevant pedagogy', *Australian Geographer*, vol. 49, no. 4, pp. 517–35. DOI:10.1080/00049182.2018.1440689

Nabobo-Baba, U. (2012) 'Transformations from within: Rethinking Pacific Education Initiative. The development of a movement for social justice and equity,' *The International Education Journal: Comparative Perspectives*, vol. 11, no. 2, pp. 82–97.

Pacific Islands Forum Secretariat (PIFS) and the University of the South Pacific (USP). 2018. *Pacific Regional Education Framework (PacREF) 2018–2030: Moving Towards Education 2030*. Suva: PIFS & USP.

Robinson, F.B. (2002) Promoting a healthy environment: A Case Study of the Wai Bulabula and Coral Gardens Initiative, Cuvu District, Nadroga, Fiji. Suva: Foundation of the Peoples of the South Pacific (FSP Fiji).

Samoa Bureau of Statistics (SBS). (2018) *Statistical Abstracts 2017*. Apia: Social Statistics Division, Samoa Bureau of Statistics. www.sbs.gov.ws/digi/2017%20-%20Samoa%20Bureau%20of%20Statistics%20-%20Statistical%20Abstract.pdf

Secretariat of the Pacific Community (SPC). (2015) *The Pacific Youth Development Framework 2014–2023: A coordinated approach to youth-centred development in the Pacific*. Suva: Secretariat of the Pacific Community.

Smith, L.T. (1999) *Decolonizing methodologies: Research and Indigenous Peoples*, London: Zed Books.

Ta'isi, T.T.T.E. (2008a) 'In search of Harmony: Peace in the Samoan Indigenous Religion,' in T.S. Sauni, I Tuagalu, T.N. Kirisi-Aiai, and N. Fuamatu (eds.) *Suesue Manogi: in Search of Fragrance*, Apia: The Centre for Samoan Studies, National University of Samoa, pp. 104–14.

Ta'isi, T.T.T.E. (2008b) 'Clutter in indigenous knowledge, research and history: A Samoan Perspective,' in T.S. Sauni, I Tuagalu, T.N. Kirisi-Aiai, and N. Fuamatu (eds.) *Suesue Manogi: in Search of Fragrance*, Apia: The Centre for Samoan Studies, National University of Samoa, pp. 115–22.

Taufe'ulungaki, A.M. (2002) 'Pacific Education; Are there alternatives?' in F. Pene, A.M. Taufe'ulungaki and C. Benson (eds.) *Trees of Opportunity: Re-thinking Pacific Education*, Suva: Institute of Education, The University of the South Pacific.

Tewari, D.D. (1995) 'The Chipko Movement: The Dialectics of Economics and Environment,' *Dialectical Anthropology*, vol. 20, no. 2, pp. 133–68.

Thaman, K.H. (2003) 'Culturally Inclusive Teacher Education in Oceania,' *International Handbook of Educational Research in Asia-Pacific Region*. Dordrecht: Springer, pp. 1221–30.

Thaman, K.H. (2009) 'Towards Cultural Democracy in Teaching and Learning with specific References to Pacific Island Nations (PINs),' *International Journal for the Scholarship of Teaching and Learning* vol. 3, no. 2, pp. 1–9.

Thomas, R.M. (1993) 'Education in the South Pacific: The Context for Development,' *Comparative Education* vol. 29, no. 3, pp. 233–48.

Tuisuga-le-taua, F.A. (2009) 'O le Tofa Liliu a Samoa: A Hermeneutical critical analysis of the cultural-theological praxis of the Samoan context' Unpublished PhD thesis, Melbourne: Melbourne College of Divinity.

Tutani, J. (2021) 'Humans have been thoughtless again, says Sigatoka Sand Dunes Manager,' *Fiji Sun* Saturday, February vol. 27, pp. 26–7.

Whistler, W.A. (1996) *Samoan Herbal Medicine: 'O Lā'au ma Vai Fofō o Samoa*. Honolulu: University of Hawai'i.

Whistler, W.A. (2005) *Plants in Samoan Culture: The Ethnobotany of Samoa*. Honolulu: University of Hawai'i.

53

TENSIONS OF DECOLONIZING DEVELOPMENT PEDAGOGIES

Bernard Kelly-Edwards, Kevin Gavi Duncan, and Paul Hodge

Introduction

Decolonization brings about the repatriation of Indigenous land and life; it is not a metaphor for other things we want to do to improve our societies and schools. The easy adoption of decolonizing discourse by educational advocacy and scholarship, evidenced by the increasing number of calls to 'decolonize our schools,' or use 'decolonizing methods,' or, 'decolonize student thinking,' turns decolonization into a metaphor. (Tuck and Yang 2012, 1).

Tuck and Yang's (2012) critique of the ease with which fields, disciplines and institutions 'roll out' the language of decolonization without engaging Indigenous peoples' struggles, sovereignty or Indigenous intellectual and activist contributions, is powerful and timely. Tuck and Yang (2012) refuse the idea that existing civil and human rights-based discourses and frameworks of social justice are commensurate with decolonization. Calls for reconciliation is one such discourse that '... motivate[s] settler moves to innocence. Reconciliation is about rescuing settler normalcy, about rescuing a settler future' (2012, 35). These moves to innocence 'attempt to relieve the settler of feelings of guilt or responsibility without giving up land or power or privilege, without having to change much at all' (2012, 10). For Tuck and Yang (2012, 3), when discourses like reconciliation are used as if they are synonymous with decolonization, there is a danger that it 'domesticates decolonization.' Decolonization is not a metaphor. If it doesn't bring about the repatriation of Indigenous land and life, it is not decolonization. Tuck and Yang's (2012) important provocation is deeply unsettling to the settler. It is supposed to be.

For development studies, with moves afoot to 'decolonize' theory and practice, the message in Tuck and Yang's (2012) arguments could not be more relevant. The language of decolonization has proliferated in development studies (Langdon 2013; Patel 2020; Spiegel et al. 2017), yet the extent to which it is up to the task, as both an intellectual field and practical endeavour, is far from clear. Indeed, the kinds of material repatriation prescribed by Tuck and Yang (2012), seems hard to visualize, let alone operationalize in an area so thoroughly complicit in the colonial project and its continuation in the guise of neoliberalism. If development studies is to partake in this deeply challenging reparative work, more will need to be done to avoid the pitfalls of 'decolonization without decolonizing' (Moosavi 2020, 332). To do the hard and unsettling

DOI: 10.4324/9781003017653-59

work of decolonization, it will not be enough to add a cultural lens to definitions of 'sustainable development' or include a few Indigenous scholars to reference lists appending the latest development report or regional 'update.' One problem, as Ndlovu-Gatsheni (2012, 48) so deftly points out, is that '…development studies remain[s] deeply interpellated by its Euro-American modernist and "civilizing mission" Genealogy.'

For universities – long-time bastions of the Western canon, decolonization is also on the move. Calls to 'decolonize the academy' and curriculum, particularly in universities in the minority world, are gaining apace (Arday et al. 2021; Bhambra et al. 2020; Harvey and Russell-Mundine 2019; Howlett et al. 2013; Moosavi 2020; Patel 2020; Tamas 2019). For universities in settler-colonial and Indigenous contexts such as Australia, Tuck and Yang's (2012) important provocation demands a response. University campuses are located on stolen, unceded land. University processes and practices including curriculum, continue to marginalize Indigenous ways of knowing and being – Australia's pervasive legacies of colonial invasion and dispossession still act to silence and preserve Indigenous absence (see Kuokkanen 2007 in the Canadian context). There are exceptions emerging where Indigenous and non-Indigenous scholars and educators are centring Indigenous epistemologies and peoples in university pedagogies with promising results (Gilbert 2019; Harvey and Russell-Mundine 2019; Howlett et al. 2013; Nursey-Bray 2019). There are also sector-wide priorities in place that aim to ensure student and staff competencies and capacities 'to engage and work effectively in Indigenous contexts congruent to the expectations of Indigenous Australian peoples' (Universities Australia 2011, 3). Though, like the field of development studies, universities in settler-colonial and Indigenous contexts such as Australia, are deeply problematized by decolonization when it comes to the repatriation of Indigenous land and life. It is one thing to Indigenize the curriculum and build student and staff's cultural competencies, it is quite another to repatriate land where Indigenization and culture takes place and has done for millennia.

So where does this leave educators tasked with teaching a development studies university course to predominantly non-Indigenous undergraduate students on unceded Awabakal lands in Mulubinba (Newcastle), Australia?

In this chapter, the author educators sit with the uncomfortable tensions presented by Tuck and Yang (2012) knowing full well that while we might not fulfil the promise of repatriation through our teaching, the provocation is simply too significant pedagogically to put to one side because it is hard. For students in the course who are going to become the next development practitioners living and working here or abroad, inviting the deep questioning opened up by Tuck and Yang (2012) forces unsettling conversations about ongoing colonization and complicity, guilt, positionality, responsibility, Country and custodianship, and Indigenous ways of knowing and being. Given that universities are part of the problem and are unlikely to 'give up land or power or privilege' (Tuck and Yang 2012, 10), perhaps, as a starting point, universities, and institutions like them, should focus on truth-telling and the responsibilities that go with it?

To this end, guided by the work of Gilbert (2019), Lester (2016), others at The Wollotuka Institute at the University of Newcastle (UON), and Indigenous and non-Indigenous educators working in other university spaces (for example, Smith et al. 2019), the development studies course introduces students to a custodian-led yarning circle workshop and fieldtrip that invite embodied and emotional experiences grounded in a place-based curriculum. For Gumbaynggirr, Bundjalung and Dhunghutti custodian Bernard Kelly-Edwards and Awabakal, Gomeroi and Mandandanji custodian Gavi Duncan, these learning experiences are an invitation offered with generosity and a readiness to remind students that they are part of Country. They are Country. Yet the invitation is not offered lightly. The gift comes with difficult realities and responsibilities attached. Bernard and Gavi share personal stories of deep racism and the scarring impacts that continue to reverberate in their lives. These stories are not told to induce guilt, but they are told as truth-telling of what happened, and what continues to happen, to them and other First Nations people because of structural racism. The gift is in the invitation, the truth, and the responsibilities that students must then take on in their future lives and careers. For Paul Hodge, a settler teacher with ancestral ties to Cornwall, England and Connacht, Ireland, the gift applies in the same way as it does for the students, though with added responsibilities. The responsibility to nurture 'safe' multi-directional relationships and responsibilities means attending to the wellbeing of Bernard and Gavi as much as it is about the welfare of students. It is the custodians who are the ones willingly engaging an institution with selective and exclusionary colonial traditions of knowledge production (see Kuokkanen 2007). Attempts to hold a 'safe' space for truths to be told and new-found responsibilities to be grappled with, can never be assured. Bernard's final week twelve 'check-in' with students has become a way of following up, enabling and cultivating a reflective practice, but even then, there are no guarantees. Paul's attempts to create 'safer' spaces for difficult conversations throughout the course invariably involves risks and challenges. Yet, the gains of trying outweigh the risks of not.

In this chapter we reflect on the learning that happens in these two custodian-led practices highlighting moments of tension, discomfort, and bemusement, but also times of familiarity, connection and attentiveness. To acknowledge and situate the authorial voice, we adopt personal narratives and collective reflections throughout the chapter when highlighting particular learning experiences.

GEOG3300 Rethinking development: yarning circle workshop and fieldtrip

Rethinking Development is the capstone core course for the Program of Development Studies taught in the Discipline of Geography and Environmental Studies at UON. The course introduces students to development activities and strategies that explore and embody *in practice* what it means and involves to 'rethink' development. Embedding Indigenous-led practices in course design, content and delivery is a key feature of the pedagogical engagement with students, who predominantly come from a non-Indigenous background.[1] A formative activity in the course involves students participating in, and reflect on, a yarning circle workshop led by Bernard. Bernard takes students through a series of activities and learning experiences designed to encourage them to come into relationship with Country in deeply embodied ways. Students also undertake a one-day field trip to Darkinjung Country on the central coast of New South Wales with Gavi. Gavi takes students to sites of first colonial contact and significant creation story places as well as areas teeming with foods and medicinal plants.

Bernard's yarning circle workshop

The yarning circle workshop is in week five of the twelve-week course, so students have had several weeks to develop and nurture their relationships with each other in class.

Bernard

I arrived at class early to set up with Paul, Gagu,[2] (we arrange the chairs in a semicircle in the rectangular room we're in) and make sure everything was ready for students. They start arriving and I remember taking a big breath, saying to myself 'here is the time to help make a difference.'

The students came in and sat down, got their notebooks and laptops out and were ready like any other class. I remember looking to Gagu and asked him would it be possible for them to put them away, Paul said to me 'it's your class brother.' So, I asked all the students to put the laptops, notebooks, pens, and phones away as this class was going to be a bit different. Students followed my instructions, and all sat, relaxed, and just listened. I tell the story of how my grandmother, my father's mother, who was a midwife at Bowraville in 1938, was instructed to tell Aboriginal women they had to walk to Bellingen hospital (22 kilometres away) because they were not allowed to give birth at Bowraville because they were Aboriginal. And that when it came to my grandmother giving birth to my father, she was told the same thing. It was okay to help birth the babies of white women but not our own. My mother's mother had to walk from Nambucca to Bellingen too (24 kilometres) for the same reason. I tell them the story that not all the women made it. I tell them other difficult stories too. I finished the yarn by talking about connections to the land bringing in new awareness tools for them to take away from the class.

Having opened up the students to this awareness, we moved outside the four walls of the classroom into the ball of yarn activity (Figure 53.1). Students, Gagu and myself moved into a circle and I started by throwing the ball of yarn to a random person in the circle. The main practical process for this activity is that the first thrower, me in this case, holds the end of the yarn. As each person catches and throws (to the next person), they hold onto the yarn eventually producing a physical web of connection between everyone. At the same time, on catching the ball of yarn, students introduce themselves, share two things they want to share about themselves, and finally, if they can, recount one thing the person who threw the yarn to them shared. This final step invites a way of connecting their story to the previous student's story. It also ensures everyone is listening deeply!

Figure 53.1 Bernard explains what we'll be doing as he holds the yarn for the ball of yarn activity. Photographed by Paul Hodge

Once everyone shared a bit about themselves, I invited students on the way back to the class-room to connect with the land in a very simple way. I said to them if there is something that they feel drawn to while they're walking back, pick it up and bring it back to the class. Once we returned to the semicircle, I invited students to voluntarily share what they connected with. One time a student held a branch in his hand. His background was Sudanese, and he shared his story that the branch reminded him of his grandfather back in Sudan teaching him about the medicine plants when he was younger. Other students have leaves and rocks. Another student raised his hand and with no object in his hands simply shared with the class that he became more aware of the wind when he was walking back to the class, and it reminded him of the simple things in life.

Paul

On yarning day Bernard, Gagu, immediately elicits a response. His strong presence and big smile ignite something in the students as they enter the space and take a seat (what exactly, I can't quite put my finger on yet). Due to the elongated room, the chairs are set up in a crest shape, just enough of an angle to allow everyone to see each other. Having put their laptops and note pads away, Bernard invites students to take their shoes off. There is a palpable shift in the room. For some students this bodily invitation to connect physically with the floor offers a sense of ease. For others, there is a hint of trepidation, an unknowing that brings with it dis-comfort, especially once familiar technologies were removed.

Moving outside we began the ball of yarn activity. As the yarn is thrown from one person to the next, I try to stay focused on the task at hand – listening closely to each introduction and story shared…in case I'm next! Occasionally, the ball of yarn is dropped, or students forget to hang on to the yarn as they're throwing it to the next person. There are also moments when the yarn is thrown too far or too short, mingling with the forming web. Sometimes students hesitate, perhaps lost in a thought, or distracted by (or connecting to) the sounds of Awabakal Country around them.

Having completed the ball of yarn activity, Gagu invited students on their walk back to the classroom to bring something from Mother Earth into the classroom circle. Students picked up leaves, sticks, bark, rocks, food wrappings – things that in some way resonated with them. When back in class, I often start the sharing if nobody volunteers immediately, trying to model the practice for others. I am always struck by what students share as they recount memories of moments and that what they hold provides a passage of sorts to these recollections. Students allude to the shape of a twig, the texture of the bark or the smell of a broken leaf to tap into a thought and memory. Other times it is the feeling of the wind!

Guided by Bernard, what became apparent was that students were drawing on their many senses to come into relationship with Mother Earth and that this was somehow comforting. The more people shared their stories, the more people chose to participate and reflect on their connection to Country. Even the discarded food wrapper, someone's carelessness, enabled a student to share a story of human impacts on the earth. It was in the sharing of the stories in the circle, connecting things from the earth with memories and senses, that a realization rippled through the room – everything is connected, everything comes from Mother Earth.

The practice is the learning

The practice is the learning. The trepidation and discomfort in the classroom as we started the yarning circle workshop disrupts students well-versed educational routines and expectations.

Whether decentring 'expertise' through a particular use of space in the ball of yarn activity or inviting more-than-rational ways of knowing and being as part of the story sharing of connection with Awabakal Country, students are confronted with an unfamiliar pedagogy.

Yet, while the form of the learning is new for many, the intuitive and sensory way that students come into relationship with Country and become comfortable with it, drawing on memories and feelings, reveal a kind of innate capacity. For some students, there is an ease in their attunement to the connections and relationships that are being formed during the two hours of the workshop between people and between people and place. For others, any kind of comfort or connection is fleeting. Some find difficulty in the prospect of tapping into emotional registers so rarely invoked and supported in university learning.

Whatever way the students engage in the activities the learning is felt and deeply embodied and somehow profound, or at the very least, deeply reflective. On forming the web of connection with the yarn, Bernard poses questions about what we have just created together reminding students that the learning is in the process and the product. What do you notice when one of us pulls the yarn? What does it mean to you to be physically connected through the web of the yarn? What did it feel like hearing about each other and making connections across stories? Having invited this discussion, which invariably leads to a general sense of oneness or connection, Bernard ends with a challenge to see if we can unravel the yarn to reform the ball. Of course, we can't. And therein lies another lesson. Once we embody and come to understand our interconnectedness and reliance on one another a bond has been struck. Everything comes from Mother Earth, and everything (everybody) is connected.

Gavi's fieldtrip on/with Darkinjung Country

The fieldtrip to Darkinjung Country is undertaken in week six following Bernard's yarning circle workshop. Building on the connections established the previous week, students are taken to three sites of deep significance on Darkinjung Country in the one-day fieldtrip. We begin at Pearl Beach, one of the first contact sites between Europeans and First Nations people only weeks after the First Fleet sailed into the Gadigal lands to the south in 1788. Gavi retells the stories that have been passed down to him about these first interactions. He also shares his own experiences of growing up in a rural town under racist government policies. Gavi finishes by taking students through the meanings and practices of the kinship system. We then go to Mount Ettalong lookout, where Gavi talks about the many medicinal plants, foods and stories related to that place. Students are given an opportunity to physically connect with plants such as native wattle as he introduces cleansing and medicinal practices. Our final destination is the highly significant and sacred Bulgandry (Bungary) Aboriginal Art and engraving site. Here Gavi shares stories of creation beings and deep spiritual connections with Country, Sea and Sky.

Gavi

It is about connection. Australia is a strange land to the students. They are from a European culture, so they are stuck to that completely. On the fieldtrip at the different sites, I walk them through our eyes and feelings; we are children of Mother Earth. I guide them into that connection that we are born into since creation. I teach them to tread softly, to listen to it. Tenderly introducing them to those things, as they wouldn't usually see the bush in that way. I want them to find out more, physically, and spiritually. Walk with us, listen to us, take in our cultural spiritual practices and respect that, and listen to it.

For millennia, we were born into having responsibilities to our totems and to look after the land – never walk with heavy feet on Country and disturb things in a big way. Same for students, we hope they can take these learnings from us and build that responsibility. I want the students to know, it's not just my story it's their story too. As a human being, it's yours too. Stories relate to us as human beings. When we acknowledge this, we are spiritually one in that moment.

I introduce students to the stories of the places we go to on the fieldtrip. The coolaman bowls (for water, food or carry babies) and stories gifted by the angophora and the medicinal and health properties of the sweet wattle at Mount Ettalong.

At Bulgandry, I tell the story of Baiyami, the creator. The engravings contain lessons on how to survive when the earth changes (www.youtube.com/watch?v=l6X6oxKUiXA&t=4s). When at Pearl Beach I share stories of the kindness of the Darkinjung peoples, the European fascination with local women's swimming skills, and the brutality of Joseph Banks[3] and how he pickled Aboriginal people's heads for European audiences. Students need to take that in too; stories of truth about what happened.

I also want the students to know about the kinship system that was here for millennia. I take them through what that means and extend that out. Kinship means all of us. We are all part of one kinship system here. What effects the earth affects us all. I want students to feel and experience this wholeness; that they are family too. They are not separate. If students could take this wholeness out of their studies and take this into their own families it would change the mindset.

Whatever student's study for their career, knowing these stories and their responsibilities mean they can be better placed to influence in the wider world. They can influence others, their kids, their parents as they give them the responsibility. When students feel whole within themselves, they can understand this story is their story. Australians are trying to find what their identity is – what their responsibilities are now in this new land. For student teachers, for example, they can influence curriculum the way they teach Aboriginal culture and history. Aboriginal history needs to be compulsory learning and should not be taught separately. It's Australian history. Aboriginal history is part of who we all are.

Paul

Before making our way to Pearl Beach, the students, Gavi and myself stand in a circle in the carpark at the Darkinjung Local Aboriginal Land Council. Gavi starts singing the welcome song in language. Clearly, this was not something students anticipated. There was a slight sense of bemusement among the group, how often does the 'teacher' sing a song, for them, at the start of the day? The words sung have a purpose. The song begins the relationship between the students, Gavi and Country. They don't know it yet, but this is Gavi's way of centring Country's language and culture, marking a moment where a relationship of responsibility had begun.

There is a gentleness to Gavi, a humility that invites comfort for students. I notice it first at Pearl Beach as we huddle in around the plaque marking the spot of first contact (Figure 53.2). Gavi generously shares personal stories too from when he was young growing up in a racist rural town. These are horrific stories. Damning. He talks about the lines painted on the playing field at school to keep him and other Aboriginal children from mixing with the white kids. The cruel and violent suppression of language and culture on the missions. The students are left speechless as they stand in that place, where it all began. There are looks of bewilderment and disgust. These are poignant moments of self-reflection. Several students move away from the group needing time alone. These stories linger for some well beyond the day, retold in their written reflections weeks later.

Figure 53.2   Gavi sharing stories at Pearl Beach with the plaque behind him denoting contact on 2 March 1788. Photograph by Paul Hodge

By the time we reach the Bulgandry site, students have relaxed into the rhythm of Gavi's storytelling anticipating stops and starts as he describes features of the landscape and their significance as we wander down the slope from the carpark to the sacred engraving site. As we sit on the walkway surrounding the engravings, Gavi shares how his own and other Indigenous kids aren't going into the bush anymore. They're not sitting on rocks and listening, learning the language of the land, the songs, dances, and art. He uses this example to tell a story of bringing his grandson into this place to teach him about the spirituality of the bush and how to celebrate that and have responsibility. That we are born to have this responsibility, to keep strong and alive. This story of intergenerational learning lifts the students, and as Gavi finishes with the story of Baiyami, the creator, they are hit with a sense of awe and a feeling of being part of something much bigger than themselves. When Gavi invites question time, few, if any, take up the opportunity. It is hard to know what they're thinking and feeling though some of what has just transpired for students is shared on the drive home; snippets of the impact of Gavi's words and Darkinjung Country.

Learning on/with Country

The stories are the learning. There is something about hearing stories about a place, on and in that place, that deepens the learning. And when spoken and shared by a custodian who has had stories passed down to them from decedents of the First Nations people of that place, the resonances and connections reverberate beyond time and space. The knowledge shared is sacred. The gifting of the stories to students on and with Country, is the first step of their new-found responsibility. Starting with the invitation in the welcome song, stories shared throughout the day finishing with the story of Baiyami creates a relationship with Country.

The experiential learning on and with Country invites an embodied and felt connection between people and between people and place. At Mount Ettalong, as Gavi shows students how to form a lather as the water mixes with the sweet wattle seeds, their senses are aroused as the students smell the strong tangy fragrance and feel the smooth texture of the lather on their hands. The joy and appreciation reflected in the way students react to the practice and the knowledge it holds is infectious. Other students move forward to be part of the experience invited to engage sensorially with Darkinjung Country. Building on the more-than-rational ways of knowing and being that begun the previous week with Bernard, students start to know what it means and feels like to come into relationship with Country.

The Kinship story of connection to the earth and to that sense of wholeness is a story of accountability as much as it is a story of interconnectedness. The invitation to come into a relationship of responsibility is not to be taken lightly. It is a gift, yes, but it is also a calling to reciprocity. Wanting students to feel and experience wholeness comes with the knowledge that what happens next is what matters most.

Concluding reflections: staying in the tensions of decolonization

Australian universities are beginning to recognize their complicity in the disruptive and violent impacts of colonizing educational structures by introducing new processes, protocols and curriculum that centre Indigenous ways of knowing and being (Gilbert 2019; Harvey and Russell-Mundine 2019; Nursey-Bray 2019). For students who undertake the Rethinking Development course at UON, a settler-colonial and Indigenous context situates and guides the learning.[4] Central to this learning lies the inherent tension of being nurtured by the very lands that have been stolen. Attempts to grapple with how to live and learn respectfully and appropriately on unceded land, particularly given the continued absence of these harder conversations in university settings, makes the task exceedingly difficult. For the small number of Indigenous students in the course, it is no less challenging as they are often coming to terms themselves with ongoing intergenerational impacts of systemic institutional racism and other forms of social exclusion. For these students an invitation is made to meet custodians prior to, and after, week five and six to create a space to connect outside class time though these opportunities are rarely taken up. The impacts of colonizing educational and societal structures run deep and are experienced differently.

For our part, the difficulties of teaching in a system that has, and continues to, marginalize Indigenous ways of knowing and being presents a major challenge, one made more difficult by the fact that all this learning is playing out on unceded land. The attempts we are making in this course to embed activities and practices that actively confront conventional educational pedagogies is one way to respond. The Indigenous ways of knowing and being that students come into relationship with during the yarning circle workshop and fieldtrip, invite students into a different relationship with Country, themselves and each other. But this alone is not enough. More needs to be done. We close the chapter by reflecting further on ways of coming to terms with Australia's settler-colonial-Indigenous pasts, presents and futures.

Bernard

We need a refresh button to shift from rethinking to re-being. To do this we need to teach deep listening skills. Everybody needs to be a deep listener. It is that surface listening that has got us to where we are now. This leads to teaching how to 'be it.' You've heard that term 'you are what you eat.' 'I'm being the shade, that's making me cool.' Students could magnify that and 'be it' using their senses – seeing, feeling, tasting. We are now what we see. See the ocean; hear the sound of the tide; feel the water, taste the salt. We're that ocean out there. Being differently is the refresh and acknowledging where we are 'being,' the place itself, is an important part of this. If we are on Awabakal Country, we need to bring in the Awabakal name and definition of dolphin and we breath with that. We ask the Awabakal custodians what is a dolphin? The undercurrent of all this is the deep listening.

Gavi

Non-Aboriginal students need to learn the language and stories of where they are from and live. Young Australian's 'rites of passage' is to go off on their treks around the world. When they are off doing this, wouldn't it be great if they introduced themselves as 'I'm from Gadigal land, the spiritual home of the whale' (instead of 'I'm from Sydney'). This would be the result of students being immersed spiritually. They identify where they come from in this way. 'I'm from Gomeroi lands and we have amazing stories. I'm from the land of the sand goanna people.' Students don't know this but if they do start to learn the language and stories of these areas, they will respect that culturally. Stories create how we live and our identity. Our Aboriginal names mean something, a connection to the earth. It is the same for non-Aboriginal people's names. All names have a meaning and connection to the earth. This is the common link. 'By the rivers of Babylon' (the song). 'How do we sing the same song in a strange land.' It is good to think about the words in this song. Realizing these same connections is to sing the same song.

Paul

While we are not fulfilling the promise of repatriation of Indigenous land, are we fulfilling another promise? Is the discomfort that student's feel when listening to Bernard and Gavi's truth-telling; their uneasiness when tapping into bodily emotions, their realization that the invitation comes with responsibilities, not a promise of a different kind? Perhaps the deeply felt encounters that students have with Country is laying the groundwork and creating the conditions of possibility for land justice? Perhaps by teaching the deep listening to 'be it,' and the message that the language and stories of place are 'our' stories (the good and the bad), we are going some way to honouring the practices that enable and legitimize the kinds of Indigenous futures that Bernard and Gavi's children and grandchildren can flourish in.

Acknowledgements

Bernard attributes his knowledge to his parents, grandparents, and Ancestors. Bernard also acknowledges Michael Williams from the University of Queensland for introducing him to the ball of yarn activity. Gavi acknowledges his Gomeroi/Awaba family Ancestors and Darkinjung peoples and Paul Hodge for his special acknowledgment for having Aboriginal cultural knowledge as part of learning institutions and studies. Paul acknowledges and pays his respects to custodians and Elders of the Awabakal, Worimi, Gumbaynggirr and Darkinjung Nations where he lives and works. Paul also acknowledges the careful work and support of Lauren Tynan before and during the inaugural Development Studies Association of Australia (DSAA) Conference, February 2020 when ideas explored in this chapter were first discussed. The authors would like to thank the students who made these reflections possible. The authors would also like to thank Fee Mozeley, Sarah Wright, Lara Daley and Naohiro Nakamura for taking the time to read and comment on earlier versions of this chapter.

Key sources

1. Gilbert, S. (2019) 'Embedding Indigenous epistemologies and ontologies whilst interacting with academic norms,' *The Australian Journal of Indigenous Education*, First View, 1–7.

This article offers an insight into the challenges of incorporating Indigenous ontologies and epistemologies into university curriculum and how the Wollotuka Institute met these challenges through a culturally and pedagogically rigorous process.

2. Kuokkanen, R. (2007) *Reshaping the University: Responsibility, Indigenous Epistemes, and the Logic of the Gift*, Toronto: UBC Press.

This powerful book offers both a scathing critique of the academy and a way of thinking and practicing that opens up possibilities that await if only the gift of Indigenous Epistemes is taken up according to principles of responsibility and reciprocity.

3. Moosavi, L. (2020) 'The decolonial bandwagon and the dangers of intellectual decolonization,' *International Review of Sociology*, 30(2), 332–354.

This article outlines common limitations of intellectual decolonization some of which re-inscribe coloniality. The paper focuses on neglected theorists of the Global South and what they bring to decolonial theory.

4. Smith, L.J., Tuck, E. and K.W. Yang (eds) (2019) *Indigenous and decolonizing studies in education: mapping the long view*. Routledge: New York.

This impressive book brings together leading Indigenous scholars as they sketch out the futurities of Indigenous and decolonizing studies in education. Drawing on examples of First Nations struggles, activisms and agency, the chapters in the book offer a plethora of creative engagements with decolonization.

5. Tuck, E. and Yang, K.W. (2012) 'Decolonization is not a metaphor,' *Decolonization: Indigeneity Education and Society*, 1(1), 1–40.

This article is a tour de force in debates relating to decolonization. Refusing anything less than the repatriation of Indigenous land and life, the article is a must read for students, academics and practitioners working in the development field particularly those situated in the minority world.

Notes

1 According to the 2016 Census, people identifying as Aboriginal or Torres Strait Islander in Australia made up 3.3% of the total population. There has been a 17.4% increase since the 2011 Census (Diddle and Markham 2017).
2 Gagu in Gumbaynggirr language means brother. Over the years we have come to refer to each other in this way.
3 The British naturalist and botanist continues to be lauded for his accomplishments and exploits yet there is very little mention of the horrific actions he was responsible for in the early days.
4 In this chapter we have only reflected on week's five and six of the twelve-week course. The work of preparing students for these experiences starts in week one.

References

Arday, J., Belluigi D.Z. and Thomas, D. (2021) 'Attempting to break the chain: reimaging inclusive pedagogy and decolonising the curriculum within the academy,' *Educational Philosophy and Theory*, 53(3), 298–313.

Bhambra, G.K., Nişancıoğlu, K. and Gebrial, D. (2020) 'Decolonising the university in 2020,' *Identities*, 27(4), 509–516.

Diddle, N. and Markham, F. (2017) 'Census 2016: what's changed for Indigenous Australians?' *The Conversation*, https://theconversation.com/census-2016-whats-changed-for-indigenous-australians-79836 (accessed 19 April 21).

Gilbert, S. (2019) 'Embedding Indigenous epistemologies and ontologies whilst interacting with academic norms,' *The Australian Journal of Indigenous Education*, First View, 1–7.

Harvey, A. and Russell-Mundine, G. (2019) 'Decolonising the curriculum: Using graduate qualities to embed Indigenous knowledges at the academic cultural interface,' *Teaching in Higher Education*, 24(6), 789–808.

Howlett, C., Ferreira, J., Seini, M. and Matthews, C. (2013) 'Indigenising the Griffith School of Environment Curriculum: Where to from here?' *The Australian Journal of Indigenous Education*, 42(1), 68–74.

Kuokkanen, R. (2007) *Reshaping the University: Responsibility, Indigenous Epistemes, and the Logic of the Gift*, Toronto: UBC Press.

Langdon, J. (2013) 'Decolonising development studies: reflections on critical pedagogies in action', *Canadian Journal of Development Studies/Revue canadienne d'études du développement*, 34(3), 384–399.

Lester, J. (2016) *Why do Aboriginal kids switch off school?* Unpublished PhD Thesis, University of Newcastle, Australia.

Ndlovu-Gatsheni, S.J. (2012) 'Coloniality of Power in Development Studies and the Impact of Global Imperial Designs on Africa,' *ARAS*, 33(2), 48–73.

Nursey-Bray, M. (2019) 'Uncoupling binaries, unsettling narratives and enriching pedagogical practice: lessons from a trial to Indigenize geography curricula at the University of Adelaide, Australia,' *Journal of Geography in Higher Education*, 43(3), 323–342.

Moosavi, L. (2020) 'The decolonial bandwagon and the dangers of intellectual decolonization,' *International Review of Sociology*, 30(2), 332–354.

Patel, K. (2020) 'Race and a decolonial turn in development studies,' *Third World Quarterly*, 41(9), 1463–1475.

Smith, L.J., Tuck, E. and K.W. Yang (eds) (2019) *Indigenous and decolonising studies in education: mapping the long view*. Routledge: New York.

Spiegel, S., Gray, H., Bompani, B., Bardosh, K. and Smith, J. (2017) 'Decolonising online development studies? Emancipatory aspirations and critical reflections–a case study,' *Third World Quarterly*, 38(2), 270–290.

Tamas, S. (2019) 'Tricky Stories: Settler–Academic Reflections on Anti-Colonial Teaching,' *GeoHumanities*, 5(2), 376–385.

Tuck, E. & Yang, K.W. (2012) 'Decolonisation is not a metaphor,' *Decolonisation: Indigeneity Education and Society*, 1(1), 1–40.

Universities Australia. (2011). *Guiding principles for the development of Indigenous cultural competency in Australian universities*. Canberra: ACT.

54

DECOLONIAL GENDER AND DEVELOPMENT

Yvonne Underhill-Sem

Introduction

Development studies is a rapidly growing multi-disciplinary field addressing urgent multi-disciplinary challenges. But, it needs to be done differently. The urgency and the demand for different approaches can result in some issues being considered less pressing and purposely placed aside, unintentionally ignored, knowingly overlooked, or assumed 'too hard.' Long-standing forms of injustice and profound trauma associated with racism, sexism and intolerance as a result of imperialism and colonization, routinely fall into this last category. Increasingly though, decolonizing research and pedagogies provide intentional epistemic practice (Tuhiwai-Smith 2013) and agile intellectual approaches (Underhill-Sem 2016) that keep sight of injustice and trauma in all their intersectional complexity. This is challenging work, so in this chapter, I focus on how feminist decolonial research and pedagogies work with this intersectional complexity and how this generates ways of doing development differently.

The first decolonial challenge in 'doing development differently' is a two-fold movement: to resist the erasure of diversity within colonial differences while simultaneously finding epistemic strength in one's partial position. Increasingly, positioning one's epistemic genealogy is the preferred approach for critical scholars, but it has been a long- standing oral practice of Indigenous researchers (Tuhiwai-Smith 2013) and, increasingly, published Indigenous scholars (Ngurra 2020; Thomsen 2020; Tynan and Bishop 2018). This practice works two ways: it allows for new relationships, and it signals the presence of new knowledge. In the spoken version of this paper to the inaugural Development Studies Association of Australia (DSAA) Conference in February 2020, it was also possible to have an embodied engagement with the audience, adding another dimension to knowledge-making in all its unknown partialities. Face to face introductions are often done purposely through the tone and pace of presentation and facial, hand and other bodily gestures. Such encounters also swivel around intersecting embodiments such as race, gender, class, age, sexuality, and religious beliefs, and influence the engagement, often in intangible ways. The decolonial challenge is to take up practices that explicitly render visible these partialities in ways that recognize a diversity of worldviews, permit expressions of historical and contemporary trauma and dis-ease, and in the process generate new knowledge.

As part of the process of reconstituting colonial relations, the practice of positioning oneself does not just inform; it also must unsettle. As a multi-hyphenated Pacific feminist decolonial

DOI: 10.4324/9781003017653-60

development geographer, I am cognizant of the distraction of personal identity journeys that sometimes accompany the practice of positionality. These can be problematic as scholarly histories with extended positioning practice can function as self-aggrandizement. However, the imperatives of decolonizing knowledge mean we need to encourage diverse practices for authors to acknowledge various ways of positioning themselves in written work and recognize the powers that such positioning can enliven for Indigenous scholars. Learning the ever-changing political art of positioning for challenging and reconstituting is critical for doing development differently (Underhill-Sem 2017). We can learn to be more attuned to the intricacies of non-referential scholarship, in contrast to the practice of, for instance, some Australian or British anthropologists working in Papua New Guinea, Vanuatu or Solomon Islands who comfortably called themselves 'Melanesianists.' Instead, radical reflexivity demands we consider, and sit with, the discomfort of our complicities. Doing this requires well-established critical interrogation and deep dives into academic material and creative techniques of wayfinding through social media.

Another associated critical practice is identifying the author's ideological position, which is rarely explicit. We must think carefully through one's ideological formation and transformation. Decolonial radical reflexivity comprises the pursuit of knowledge as an enduring political act – something that if one is not knowingly attending to, means they are unknowingly perpetuating the status quo, which is still predominantly extractive and colonizing of development knowledge.

This chapter expands on the pursuit of knowledge as an enduring political act which make pedagogical practices possible within global development studies by: (i) curating positionality; (ii) critical interrogation of published material; and, (iii) recognizing ideological, as distinct from epistemological, standpoints – the latter invites further discussion, the former closes it down. Together, these practices offer ways to do development differently. I begin by drawing attention to the political positioning I took at the Development Studies Association of Australia (DSAA) conference. I then speak to the process of launching a systematic literature review. Finally, I amplify my epistemological position as an Indigenous Pacific feminist with reference to how transnational solidarities can be generated by decolonial gender and development. These all underpin long-standing practices that can be mobilized as teachers, researchers, and practitioners to decolonize gender and development. It also contributes to the broader work of dismantling long-standing and hegemonic patriarchal practices of uninformed ignorance, self-confident arrogance, and discriminatory ideology.

Positioning

I am a cis-gendered woman with heritage and close kin from the Cook Islands, Niue, Aotearoa, England, and Papua New Guinea. At the DSAA conference, I acknowledged and extended thanks to the people of the Wurundjeri – past, present, and future – on whose unceded lands we gathered. These acknowledgements have become a 'customary' refrain at conferences in Australia – significantly missing in 2002 when I worked at an Australian university. I continued by recognizing the careful nurturing of the Wurundjeri, who had provided longevity of life, even as it suffered the worst effects of Australian racist colonization. In more recent times, the ravages of 'wildfires' have been exacerbated by extractive economic practices and climate change. I wanted to engage with the Wurundjeri people of the Melbourne area and the Dja Dja Wurrung people of central Victoria, but this was almost impossible in the multi-storied central city building we were meeting. I was uncomfortable in the lead up to my keynote, especially when I realized that I was one of only two keynotes at this inaugural conference. Although we were both non-white women, neither of us had a connection to the heritage of those on whose lands we were meeting. Fortunately, from the tinted windows of the conference rooms, I could see the lands and waters of the Wurundjeri people. I also learnt the meaning of the local welcome – 'Wominjeka.' According to Indigenous scholar Professor Liz Cameron, this means 'I see you …' I responded by offering a way of 'seeing' me. By introducing myself in an Indigenous Pacific language, I offered an arc of connection to the Wurundjeri people and simultaneously added another Indigenous epistemic positionality to the connective tissues constituting diverse worldviews in development.

Ko Moana Nui a Kiwa te moana.	The Pacific Ocean is my ocean/water.
Nō Ngā Pū Toru au. Nō Alofi au.	I am Nga Pu Toru (The three islands of Mauke, Mitiaro and Atiu in the Southern Cook Islands); Alofi in Niue.
Kō Ngati Akatauira te ivi.	Ngati Akatauira is my tribe in Ngā Pū Toru.
Kō Tepari te enua	Tepari is my land.
Tōku tūpuna ko Koumu, ko Pamatatau, ko Underhill.	Koumu is one of my ancestors from Ngati Akatauira, Pamatatau is one of my ancestors from Niue, Underhill is one of my ancestors from England.
Kō Te Ruki Rangi o Tangaroa tōku ingoa.	Te Rūki Rangi o Tangaroa is my name.

Positioning ones indigenious self is an intentional act of exposure, and this is a delicate act. The intent behind positioning myself as a keynote speaker at that particular international conference was simply to signal the existence of yet another Indigenous worldview which was open to engagement. But the particularities that fashion diverse worldviews which we share at any introduction are sometimes foregrounded by the embodied expression of strong feelings and emotions, while other times they are part of a disembodied context. Strong emotions may emerge over the untimely deaths of family members to preventable health conditions like malnutrition, domestic violence, police violence and trauma; or fear of more frequent cyclones with screaming high winds, battering vertical rain, and devastated homes and livelihoods. Sometimes, diverse worldviews are less obvious such as in discussions about Indigenous philanthrocapitalism or inequitable multilateral trade agreements. However, in discussions with Indigenous scholars about these contemporary issues, analytical concepts like exchange, gifting and reciprocity emerge quite differently to the published canon. The critical pedagogical point is that space needs to be created in decolonial development teaching to allow for new and varied worldviews, the dimensions of which may be as unfamiliar as the languages in which they are expressed. Unsettling mainstream entry-points into the development dialogue by decentring the mainstream with the diverse, is a critical decolonial practice for educators because they invite practices of positionality both by example, and by encouraging students to also position themselves, however partially (Underhill-Sem 2020).

As an educator, I have learnt to willingly open myself up to these partialities and to leave open other ways of coming into relation with students. This is one way of creating space for knowledge-making. Given that the English language will likely retain its hegemony in teaching development studies, we can purposely invite other languages into our discussions – simply via greetings and the invitation to consider 'translations' of key concepts into student's preferred languages in multi-lingual classrooms. It may also involve the use of visual material in languages other than English; the development of relationships with communities in our cities who have lived experiences of 'development'; and favouring academic literature authored by scholars from the Global South whose experience of development is different from the outset. It is the careful curation of the positionality of educators and knowledge-makers that enables a precise and intentional decolonial turn. It also opens up the possibilities and imperatives for students as knowledge-makers and future practitioners of gender and development to begin to position themselves in their own histories. So, while educators may be tempted to include their published work in reading lists for graduate teaching, a more exciting gauge of the extent to which scholarly work has relevance is to see if it is discoverable, and by which students.

Dismantling

To engage in decolonial practices in development studies means to assemble, articulate and embed decolonized forms of knowledge production into teaching, and research. It also means to practice good citizenship in the academy, as an educator and as a student. Guided by values emergent from one's situated experience, in my case the gendering processes of Pacific peoples, my ambition is to reimagine worldviews, knowledges, structures and processes so that they are recognized as valid parts of ways of thinking and working in the Pacific, in Aotearoa and beyond. This is a bold ambition, but it is not mine alone, especially not for the researchers and educators of gender and development. One way to achieve this ambition is to use mainstream knowledge-making practices but dismantle their hegemonic reach by starting in places that are meaningful for marginalized people. The following is an example of undertaking a systematic literature review, which began a process of dismantling the 'canons' of knowing.

Gender equality research exists in one way or another in the Pacific, but there are many unknowns about the scope, nature, and quality of this research. In 2015, a group of gender specialists based in Fiji met to express their concern about the need to strengthen gender-responsive policy development in the Pacific and to continue to build research capacity in the region. Close attention to gender inequality in the Pacific began in 1994, associated with the 1995 Fourth World Conference on Women in Beijing. The Pacific Platform for Action on the Advancement of Women and Gender Equality (PPA) was the first attempt to state a regional position and initiate new interventions. Pacific leaders adopted the PPA, with the support of metropolitan governments of New Zealand and Australia. It took almost two decades before Pacific leaders made another serious step towards addressing gender inequality, again with some serious nudging from metropolitan interests (see Molyneux et al. 2020 and Laó-Montes 2016 for discussions on the Beijing conference). In 2012, the Pacific Leaders Gender Inequality Declaration was adopted. Simultaneously, the Australian Government launched Pacific Women Shaping Pacific Development (Pacific Women), a ten-year, $320 million programme. The aim was to improve Pacific women's political, economic and social opportunities and support countries to meet their commitments to the Pacific Leaders' Gender Equality Declaration 2012. This was a significant investment over a substantial time period and, despite being informally known with intended irony as 'Australian Women Shaping Pacific Development,' extra resourcing has been welcomed in many places in the Pacific.

However, more was needed to understand the nature of the entrenched gender inequalities in the Pacific, especially given social, economic, political, cultural, and environmental changes. It seemed apparent to Pacific gender researchers that the programme also needed a research component. As a result of the 2015 meeting of gender specialists in Fiji, Pacific Women commissioned a study with three aims: to provide a 'gap analysis' of existing gender research in eight thematic areas related to the Pacific Platform for Action; to assess quality, reliability and usefulness and; to provide recommendations relating to capacity building on gender research in the Pacific (Underhill-Sem et al. 2016). All members of the study team, selected as a group through the procurement processes of Cardno, a global infrastructure, environmental and social development company, were Pacific women with doctoral degrees (two were resident in a Pacific country). All had been active as researchers, teachers, and consultants for a combined period of over four decades. The study methodology involved three complementary methods: an internet search and literature review, an online survey, and in-depth interviews. As a result, the project generated a bibliography of over 400 citations, an annotated bibliography of 135 pieces of research and a literature review of eight areas of 'gender concern.' For many, including those in the study team, this was a welcome result. For others, it was deeply flawed and exclusionary.

The key reason for claiming the work was flawed and exclusionary was that the starting point for the literature review and the key questions driving the search were, ironically, considered to be 'stacked' against non-Pacific gender researchers. This was partly because the initial search platform was the personal bibliographic databases of two senior members of the team, but also the definition of a 'Pacific gender' researcher, which focused on Pacific heritage and enduring personal connections to the Pacific. Both of these decisions came from the study team and represented our knowledge base and group discussions. There were other 'databases' and scholarly works that could have been the starting point, but by working from databases built by the Pacific women involved in the project, we shifted the centre of knowledge. The study team's senior members had both taught gender and development courses at universities in the Pacific region as well as in Australia and New Zealand. They had a long commitment to sourcing research by Pacific scholars for their students as part of the process of Pacific knowledge-making and/or co-production. While bibliographic databases and research from scholars who were not from the Pacific were also considered valuable, the decolonial practice was to explicitly begin the systematic search for research in geographies closest-in – that is in the bodies, minds, and words of diverse Pacific women scholars. In addition to building new knowledge, this practice addresses the tensions that decolonial sensibilities draw attention to: that is, the easily-discoverable corpus of existing scholarship on gender inequality in the Pacific, overlooks much that is authored by Pacific gender researchers, yet, starting with the latter, is considered exclusionary. Long-standing scholarly regimes of knowledge in the Pacific remain a powerful impediment for the emergence of Indigenous feminist decolonial knowledge on gender inequalities. While not all Pacific gender researchers identify as Indigenous feminist decolonial scholars, contrary to mainstream arguments, those who do, remain inclusive, intentional, and unapologetic in their pedagogies, research, and practice.

The following steps in the literature review were more routine with citations shared and searched for via library search engines of the University of Auckland, the University of the South Pacific and Google Scholar. The team's senior researchers had excellent knowledge of gender research in the Pacific, which was supplemented by robust searching, key informant interviews, questions asked in a survey and input from a Research Peer Group (comprising members of the initial group of gender experts in Fiji). Further steps in this systematic literature review involved the standard practice of searching databases via the University of Auckland library using keywords 'Pacific Islands and Gender Research'; 'Gender Research in the Pacific Islands'; 'Pacific Gender Research.' For some, the number of citations exceeded their expectations. For others, it remained incomplete. This example demonstrates how a decolonial sensibility provided the vital initial impulse which shaped the otherwise routine research process of undertaking a systematic literature review. The 'deep dive' into literature was a critical interrogation of published material, but it was shaped in a way that still allowed research undertaken by non-Pacific researchers and institutions to be more widely recognized in mainstream gender policy spaces. This means that research that is referenced regularly can have an impact beyond its intrinsic value (Mott and Cockayne 2017). Decolonial gender and development requires that we are intentional about the places we begin our learnings so that in the crafting of our projects we can dismantle canons of knowledge, but still be comprehensible to broader audiences. So, researchers, educators, practitioners and students must continually ask questions about authors, their practice of research and their approach to questioning long-standing canons of knowledge

Generating

Another critical axis of analysis in decolonial gender and development is how to deal with ideological viewpoints. In contrast to feminist standpoint theories, which offer a new epistemic lens and subsequently new forms of knowledge, ideological viewpoints offer an epistemic lens that is hostile to forms of knowledge that challenge long-standing racist and sexist hegemonic canons. They are often marked by their desire to close down discussions. I explain this further by moving away from the Pacific and swivelling towards Latin America which is home to many key feminist decolonial scholars (Lugones 2010, Belausteguigoitia 2016)

The public display of a painting of the revolutionary hero Emiliano Zapata on horseback is not the familiar macho image usually associated with Zapata – yet it was part of a 2019 exhibition at the Palace of Fine Arts in Mexico City called 'Emiliano. Zapata Después Zapata' ('Emiliano. Zapata After Zapata'). It commemorated the centenary of Zapata's death, a leading figure in the Mexican revolution in 1910, assassinated when he was only 39 years old. His perceived martyrdom has made him one of the most revered figures in Mexican history. The exhibition was one of the numerous events marking 2019 as the year of Zapata, as designated by Mexican president Andrés Manuel López Obrador. The Mexican painter Fabian Cháirez wanted the work, which was featured prominently on a poster promoting the exhibition, to stand in contrast to patriarchal depictions of the Mexican farmer-turned-revolutionary, in which 'Zapata's masculinity is glorified' (Chairez, as cited in Cascone 2019). The silence of indigenous women in the Zapitisa movement has been documented without controversy, and further complicated in the construction of Indigenous subjects (Belausteguigoitia 2000, 2016). Indigenous women played critical roles in the liberation and emancipation movement which elevated Zapata to hero status, but despite demanding attention, their lives remained burdened by the struggles of maintaining livelihoods and caring for the home. There was no dis-ease over their realities. However, Cháirez's representation generated considerable unease.

There were passionate demonstrations and strong calls to remove the offending picture from the exhibition because close supporters and family members said it denigrated Zapata by depicting him as gay. However, the museum stood by its decision to exhibit the work and kept it on display despite protesters' threats. A joint statement from Mexico's culture ministry and the National Institute of Fine Arts and Literature said, 'Welcome to the world of political and aesthetic debate and discrepancy' (*Telegraph 2019*). The painting 'raised the level of debate about what constitutes femininity and masculinity.' Cháirez, the artist, added on Twitter, 'If you use the feminine, race, or social class as an insult, you are part of the problem.' This is a familiar refrain: if you are not part of the solution by engaging in decolonial practices, then you are part of the problem, in perpetuating colonization. But the situation is getting more complicated because of the effect of right-wing gender ideology and rhetoric (Corredor 2019). Development projects rarely pay attention to what can be learnt from the creative sector. Yet it has become a space that has the power to both challenge ideas, concepts and characters whose development impacts seem to have been taken-for-granted messages, as well as a space to amplify new ideas, concepts and characters, often with more troubling impacts for emancipatory development projects.

I saw the exhibition while on a feminist intergenerational dialogue in Mexico City organized by DAWN – Development Alternatives with Women for a New Era. At our event, there were 21 of us spanning four different moments in DAWN's history; we were from ten countries in four regions of the globe. We were leaders of NGOs, academics, community educators, journalists, and development bureaucrats. We looked for collective actions and transformative strategies that different generations and organizations could carry into the future. It was the first time that we were explicitly building collective memory and intergenerational dialogues. Analytical insights from our differently positioned experiences were shared: including the troubling governance of digital platforms (Gurumurthy et al. 2019); the frightening violence erupting around new exclusionary citizenship laws in India (Chapparban 2020), the violent growth of militarization and securitization in the Global South (Samuel 2019), and the new age of the Anthropocene with corporate extractivism and associated models of regional governance pushing the planet to its limits of survival (Slatter 2020, Penjueli 2016). Our challenge was to articulate a comprehensive and meaningful feminist framework that accommodated structural and representational concepts in women's local and transnational struggles. We were also committed to understanding the transnational convergence of local and global racial hierarchies (Thomsen 2020). In our analysis of various spaces, dis-ease over gendered silences was apparent, at times despite considerable engagement as allies of the associated movements, or as colleagues of scholars, academics and policy-makers working on related issues. Our exasperation rose and eventually emerged as a renewed commitment to new rhythms of feminist action and movement building, and informed by critical feminist analysis from the Global South.

During the last days of our meeting – just before the 25 November, the International Day for the Elimination of Violence Against Women – dozens of women gathered as feminist flash mobs in public squares throughout major cities in Latin America. In Santiago, Chile – a country beset by popular uprisings against inequality – a local Chilean feminist collective *Las Tesis* gathered outside the supreme court building. It delivered a powerful pulsating performance called 'Un violador en tu camino' (A rapist in your path). Similar versions were performed throughout Latin America, including in the Plaza del Zócalo – the main square in Mexico City. The flash mob song and accompanying dance explicitly charge patriarchy with violating and

Figure 54.1   Video link via YouTube to Performance colectivo Las Tesis 'Un violador en tu camino.' Uploaded by the Colectivo Registro Cellejero (2019, 3.44)

victim shaming women. '*Y la culpa no era mía, ni dónde estaba, ni cómo vestía,*' they sang ('and the fault wasn't mine, not where I was, not how I dressed'). These performances went viral.

In the Latin world, a new pulse of feminist actions and movement building had begun and was travelling (Méndez 2020). Most notable was that young women were generating this pulse – both from the sense of outrage at being denied justice in the light of the growing incidence of rape, sexual harassment, femicide and illegal abortion. This expression of solidarity carries a palpable sense from young women of 'we are calling you out ... we have had enough.' These are brave actions in places where public displays of displeasure with state agencies are violently suppressed, but they are likely to continue to travel virtually especially in the solidaristic global context of #metoo and #BLM. While this expression of solidarity did not take root in the same performative manner in the Anglophone world and the palpability of street activism like this with its particular Latina flavour, key activists acknowledged that they draw inspiration from widely acknowledged gender theories. Feminist theorist Judith Butler looms large but is not alone (Méndez 2020; Wasser and Lins Franca 2020) in challenging well-established modes of knowledge production about gendered bodies. In one way, these progressive ideologies allowed subaltern voices to emerge, disrupt patriarchal spaces and challenge critical institutions. In another, right-wing gender ideologies were also present as a powerful counter-movement (Antić and Radačić 2020). Ideologies of whichever persuasion matter because they operate as radical statements of truth based on epistemic hegemonies, but they are not invincible. I argue that allowing for epistemic freedom, rather than demanding conformity, generates arcs of connection, not without some friction, but with the ability to energize new ways of knowing. So, for researchers, educators and students, actively identifying ideological, distinct from epistemological, standpoints is an essential part of feminist decolonial development pedagogies. Together with positioning oneself when one introduces themselves and critically interrogating all published works; the identification of the ideological underpinnings of contemporary political and cultural practices are key pedagogical practices.

Indigeneity, relationality, decoloniality

There is another emergent comparative and global studies movement that pivots on the kinds of critical self-reflexivity I am advocating in this chapter – identified as 'critical Indigenous studies' or 'Indigenous critical studies' (Moreton-Robson 2009). This movement refuses to take indigeneity – the claims and conditions of aboriginal belonging to specific places – as self-evident or given. Instead, it problematizes key categories (such as 'Native,' women, Indigenous people, the poor, white folk) by requiring that multiple, intersecting and interlocking forces of power, inequalities and relations are understood historically and politically (see Thomas 2020). However, this does not assume a unified or coherent, even self-evident, sense of, for instance, Pacific indigeneity when we know 'it' is not self-evident. Instead, we have the task of working with indigeneity's radical differences, specificities, and particularities.

Pacific indigeneity and knowledge production pivot on many different formations and logics of power (e.g. gender, neoliberal state formation, race, modernity, colonialism, to name some). What this means is that we have to work differently – in radical ways. Indigeneities matter for their relational racialized radicality – especially in the area of languaging – and in their temporal and spatial dimensioning of the world. We recognize languages not just for communication but also because competency in a language provides insights into relationships. All development educators, practitioners and students need to keep working to improve our competence in the languages of the peoples we work with. Otherwise, we perpetuate the many colonial and imperial practices that created the space for development studies to take root and flourish. The challenge is to express a commitment to decolonial and Indigenous scholarship, which articulates alternative ways of knowing by asking; if we are not actively practising decoloniality, are we continuing to colonize?

In addition to transforming knowledge production, structural issues in the academy and other development spaces also need transformation. Aotearoa New Zealand is moving closer to gender equity in terms of equal numbers of men and women in the academic workforce, although labour market segmentation means compared to men, women are proportionately more likely to be in lower-paid positions. The lifetime economic effect on women is that they are paid about NZD$400K less than men (Brower and James 2020).

The situation for Māori and Pacific academics reveals even more significant challenges. It has been estimated that 'Māori may account for a share of the research workforce corresponding to their share of the national population around the year 2096, and Pacific around the year 2150.' (Tertiary Education Review Commission of New Zealand 2020). That is almost two generations for Māori and over four generations for Pacific peoples (based on a generation span of 30 years). That is too long and a very dismal thought. Committing to decolonial and Indigenous scholarship, which articulates alternative ways of knowing, means we need more Indigenous scholars in the academy.

Positioning oneself in a large and complex academic infrastructure needs to be done intentionally especially when not everyone has the same access to resources, decision-making or even questioning. So, researchers and teachers must take advantage of the access that comes our way. Active membership of hiring committees, research committees, ethics committees, course development committees may seem a chore, but these are places where we can push open the boundaries of possibility by practicing new forms of relating to other colleagues and students. The often-thankless tasks of organizing public seminars to discuss 'sensitive issues' are also critical spaces to dismantle long-standing knowledge-making practices and generate new ways of looking at development. Encouraging panels of diverse speakers rather than sole expert voices are tasks of those who take their role as 'critic' and 'conscience' of society seriously. I would add that there is an additional role as 'decolonizer' of society. In this chapter, I have tried to demonstrate that positioning oneself, dismantling canons of knowledge and generating new knowledge constitute the practice of decoloniality. This is critical because in development studies, if we are not actively doing this, we are colonizing development studies. Similarly, if we are not actively addressing gender equality, we are perpetuating inequality; and if we are not intentionally engaging with indigeneity, we are not doing development differently. This is not easy work on top of teaching, writing research grants and being a thoughtful academic citizen – and student – but we ignore it at our peril.

So, while context matters, it matters a lot; we cannot lose sight of macro and meso level insurgencies – and in particular, gender ideologies. We need to resist actively. We need to carve out new spaces of diverse knowledge generation and actively hire, promote, and publish scholars whose work pivots on different worldviews. In my work, the temporal and spatial coordinates of indigeneity that I used above to introduce myself signal relational positionalities from which my intellectual trajectories originate. Indigeneity systematically animates the conceptual and material characteristics of my scholarship and teaching as a feminist decolonial development geographer. The decolonial project is multi-faceted and always contested, so we need to be open to different and contested notions of indigeneity; recognize shifting positionalities; enact generous indigeneities; nurture careful relationalities, and train agile intellects (see also Teasley and Butler 2020).

> Some guidance for practising decoloniality, working with intersectional complexity and doing
>   development differently, concerning non-human others....
> Prune, sometimes gently, sometimes more harshly.
> Provide a solid base – perhaps not complete, but reliable.
> Decide if it's a weed or a new shoot, or if someone else can tell you?
> Nourish regularly.

Acknowledgements

Warm gratitude to the Development Studies Association of Australia for inviting me to deliver a keynote address at their inaugural conference in 2020. Scholarly solidarity to the editors of this book, especially Dr Paul Hodge, for careful editing and a judicious ability to ensure the critical scholarly arguments are made clear. Warm thanks to Professors Linda Tuhiwai-Smith, Robyn Longhurst and Lynda Johnston for their critical scholarly collegiality over the years. Warmly to Professor Liz Cameron for her timely insight. Finally, to my treasured feminist colleagues in DAWN, I remain ceaselessly grateful for their unyielding feminist solidarity and astute, timely, and critical analysis.

Key sources

1. Lugones, M. (2010). Toward a decolonial feminism. *Hypatia*, *25*(4), 742–759.

This is the classic text in English that outlines the argument for decolonial feminist theorizing

2. Teasley, C., and Butler, A. (2020). Intersecting Critical Pedagogies to Counter Coloniality. *The SAGE Handbook of Critical Pedagogies*, 186.

Welcome to a comprehensive collection of critical pedagogies that counter coloniality

3. Thomas, K. B. (2020). Intersectionality and Epistemic Erasure: A Caution to Decolonial Feminism. *Hypatia*, *35*(3), 509–523.

This paper makes a nuanced argument about the key concepts of decoloniality and intersectionality are intertwined.

4. Tuhiwai-Smith, L. T. (2013). *Decolonizing methodologies: Research and Indigenous peoples*. Zed Books Ltd.

This is a classic text that reminds all scholars of the imperatives of decolonizing research, especially for Indigenous peoples, but also for people marginalized by processes of colonization, racism, sexism, and other forms of intolerance

5. Underhill-Sem, Yvonne, (2020), The audacity of the ocean: Gendered politics of positionality in the Pacific. *Singapore Journal of Tropical Geography*, 41(3), 314–328.

This paper offers a challenge to think and do development differently.

References

Antić, M., and Radačić, I. (2020) The evolving understanding of gender in international law and 'gender ideology' pushback 25 years since the Beijing conference on women. *Women's Studies International Forum*, 83, 102421.

Belausteguigoitia, M. (2000) The Right to Rest: Women's struggle to be heard in the Zapatistas' movement. *Development*, 43, 81–87. https://doiorg.ezproxy.auckland.ac.nz/10.1057/palgrave.development.1110176

Belausteguigoitia, M. (2016). From Indigenismo to Zapatismo: Scenarios of construction of the Indigenous subject. In *Critical terms in Caribbean and Latin American thought* (pp. 23–36). Palgrave Macmillan, New York.

Brower, A., and James, A. (2020). Research performance and age explain less than half of the gender pay gap in New Zealand universities. *Plos One*, 15(1), e0226392.

Cascone, S. (2019). Protestors storm Mexico museum over a painting that depicts revolutionary Emiliano Zapata nude (and wearing a pink sombrero). *Artnews News* Dec 11. https://news.artnet.com/art-world/nude-zapata-painting-sparks-protests-1729050

Chapparban, S. N. (2020). Religious Identity and Politics of Citizenship in South Asia: A Reflection on Refugees and Migrants in India. *Development*, 63(1), 52–59.

Corredor, E. S. (2019). Unpacking 'gender ideology' and the global right's antigender countermovement. *Signs: Journal of Women in Culture and Society*, 44(3), 613–638.

Gurumurthy, A., C. Nandini, and C Alemany (2019) *Gender equality in the digital economy: a new social contract for women's rights in the data economy,* IT for Change and DAWN. https://itforchange.net/sites/default/files/1785/Feminsit%20Digital%20Justice%20Issue%20Paper%201_%20updated%20name%20and%20logo.pdf

Laó-Montes. A. (2016). Afro-Latin American Feminisms at the Cutting Edge of Emerging Political-Epistemic Movements. *Meridians, 14*(2), 1–24. doi:10.2979/meridians.14.2.02

Méndez, M. (2020). Operación Araña: reflections on how a performative intervention in Buenos Aires's subway system can help rethink feminist activism. *Estudos Históricos (Rio de Janeiro), 33*(70), 280–297. Epub June 08, 2020. https://doi.org/10.1590/s2178–14942020000200004

Molyneux, M., Dey, A., Gatto, M. A., and Rowden, H. (2020). Feminist activism 25 years after Beijing. *Gender & Development, 28*(2), 315–336.

Moreton-Robinson, A. (2009). Introduction: critical Indigenous theory. *Cultural Studies Review, 15*(2), 11.

Mott, C., and Cockayne, D. (2017). Citation matters: mobilizing the politics of citation toward a practice of 'conscientious engagement'. *Gender, Place & Culture, 24*(7), 954–973. DOI: 10.1080/0966369X.2017.1339022

Ngurra, D., Dadd, U. L., Glass, P., Norman-Dadd, A. C., Hodge, P., Suchet-Pearson, S., … and Lemire, J. (2020). Yanama Budyari Gumada, Walk with Good Spirit as Method: Co-creating Local Environmental Stewards on/with/as Darug Ngurra. In *Located Research* (pp. 15–37). Palgrave Macmillan, Singapore.

Penjueli, M. (2016). Civil society and the political legitimacy of regional institutions: An NGO perspective. In *The New Pacific Diplomacy*. Edited by Fry, G., and Tarte, S. ANU Press, pp. 65–78.

Tertiary Education Commission of New Zealand, (2020), *E koekoe te tūī, e ketekete te kākā, e kūkū te kererū* Toward the Tertiary Research Excellence Evaluation (TREE): Report of the PBRF Review Panel, www.education.govt.nz/assets/Documents/Further-education/PBRF-Review/The-Report-of-the-PBRF-Review-panel-E-koekoe-te-tuie-ketekete-te-kaka..-.pdf

Thomsen, P. S. (2020) Transnational interest convergence and global Korea at the edge of race and queer experiences: A Talanoa with Gay Men in Seoul. *Du Bois Review: Social Science Research on Race*, 1–18.

Tynan, L., and Bishop, M. (2019). Disembodied experts, accountability and refusal: an autoethnography of two (ab) Original women. *Australian Journal of Human Rights*, 25(2), 217–231.

Samuel, K., Slatter, C., and Gunasekara, V. (Eds.). (2019). *The Political Economy of Conflict and Violence against Women: Cases from the South*. Zed Books Ltd.

Slatter, C., (2020), Intersecting interests in deep-sea mining: Pacific SIDS, venture capital companies and institutional actors, DAWN INFORMS, https://dawnnet.org/wp-content/uploads/2020/07/DAWN-Informs-on-Blue-Economy_2020.pdf

Underhill-Sem, Y., J., (2016). Critical gender studies and international development studies: interdisciplinarity, intellectual agility and inclusion. *Palgrave Communications, 2.*

Underhill-Sem, Y., J., (2017). Academic work as radical practice: getting in, creating a space, not giving up. *Geographical Review, 55*(3), 332–337.

Underhill-Sem, Y. J., A. Chan Tung, E. Marsters and S. Eftonga Pene, (2016). *Gender Research in the Pacific 1994–2014: Beginnings.* Suva: Pacific Women Support Unit. Retrieved from www.pacificwomen.org/

Wasser, N., and Lins França, I. (2020). In the Line of Fire: Sex (uality) and Gender Ideology in Brazil. *Femina Politica–Zeitschrift für feministische Politikwissenschaft, 29*(1), 27–28.

55

COMMUNITY BASED SERVICE LEARNING FOR DEVELOPMENT

Rebecca Bilous, Laura Hammersley, and Kate Lloyd

Questioning Community Based Service learning

On my way to delivering a workshop on Child Rights I am stopped by students. They are handing out pamphlets supporting the work of a youth-run global volunteer programme. I don't know much about the organization but the flyer shows a young Caucasian student with her arms around three smiling Vietnamese children. Among other things, the flyer is inviting young people to 'Empower Vietnam.' Of course! How can I encourage my students to critically question these organizations and the roles they play in majority countries? (Rebecca)

As a student I spent three months living, learning, and working in a remote village in Northern Vanuatu. It was an exciting but increasingly disillusioning experience. I found myself purporting to teach life skills to local youth with whose culture and language I had no prior knowledge or experience. To my horror, I realized I was an ignorant and active participant in a form of neo-colonialism being enacted around the world. The volunteer tourism industry is now worth an estimated $173 billion and continues to grow. Was it appropriate for me, as a student, to take on this role given my lack of experience? (Laura)

My daughter comes home with a flyer; 'Laos Service Learning Adventure.' Years 9 and 10 students have been given the opportunity to travel to Laos and among other things, work alongside the community to construct dormitory residences for school students. My daughter would like to go. How do I encourage my daughter to contribute positively to the world but to do this in an ethical way? Is doing nothing better than doing something if that something is recreating neo-colonial attitudes and structures? (Rebecca)

Working with students going on CBSL activities for many years I regularly receive e-mails from third party providers selling experiences to students that promise to 'solve the world's social issues through the creation and implementation of tailored products and services for communities who need it the most.' How closely do these providers work with communities to understand the issues or with universities to understand the capabilities of undergraduate students? (Kate)

DOI: 10.4324/9781003017653-61

Introduction

Community based service-learning (CBSL) is everywhere. These programmes are available to everyone, from all walks of life, but they are particularly prevalent in universities and high schools. In high schools the programmes are neatly tailored to meet a range of curriculum outcomes across multiple key learning areas and promise to develop resilience, independence, confidence, collaboration, a civic responsibility.... the list goes on. In Australian higher education institutions these often very similar programmes are promoted as opportunities to gain valuable work experience. The young people enrolling in these programmes often have memorable experiences and these programmes often have considerable personal impact. However often this type of western led expertise can lead to the beneficiaries of development programmes becoming products of a development industry that profits from their 'disadvantage' (Tynan and Bishop 2019). Do these target audiences or those providing the opportunities stop to consider the impact of these programmes on the community 'beneficiaries' or how this phenomenon may be just another wave of deep colonizing that constitutes a global challenge?

Building on volunteerism and social activism, CBSL was initially described as the practice of joining formal learning and teaching with the experience of volunteer community service and was, and still is, seen as a way of responding to issues of social justice and effective pedagogy (Hammersley 2015). A continuing driver of the service-learning movement is a perceived disconnect between universities, the complex communities in which they are situated (globally and locally), and the interconnected social, economic, environmental and political realities confronting these communities. In this context, higher education institutions, the instructors who lead the programmes and the students who participate are making a commitment to civic engagement and to larger public purposes (Kliewer 2013).

Like volunteer tourism, service-learning has a strong historical association with one-way service and action, often spurred on by an institutional and individual moral obligation to help solve community problems (Simpson 2004). The server-served and giver-receiver relationship becomes particularly problematic when sites of service take place in complex postcolonial contexts, including international service-learning with communities in the 'developing South' (Mellom and Herrera 2014) and Indigenous communities in settler states such as Australia. In the context of dominant, usually Western discourses of development and aid, service-based activities can reinforce structures of power and privilege (Hartman and Kiely 2014) and exacerbate ethnocentric attitudes about culture, language, and social norms (Mellom and Herrera 2014).

Positioning universities and students within this charity/helping model, even if the focus of engagement is around co-learning, tends to reinforce neo-colonial paternalistic binaries that sees the 'expert/beneficiary' relationship and experiences between student and host as fixed and static. Critics of volunteer tourism have argued for the need to engage with the complexities around practices, outcomes and effects of tourists on local communities (Sin et al. 2015). How can we in minority world settings engage with CBSL in reciprocal and ethical ways?

Through our engagement with CBSL we raise many questions that students, teachers and institutions need to critically engage with. We do this by drawing on approaches that shed light on the importance of the 'more-than-rational' (Wright 2012), relationality and the process of co-becoming where everything is constantly made and defined by and through relationships (Bawaka Country et al. 2016). In her work Everingham (2016: 521) concludes that 'the ambiguity of emotional connections and embodied encounters with local people opens up spaces for hopeful possibilities.' These languages of hope and possibilities have been used in a number of disciplines in order to affect social change (Anderson 2006; Ateljevic et al. 2007; Dinerstein and Deneulin 2012; Everingham 2016; McGregor 2009).

As female settler academics our research and teaching has intersected over many years as we engaged in exploring respectful and reciprocal approaches to CBSL. While in this chapter we pose questions for teachers and students to ask themselves before, during and after taking part in CBSL we are also contributors in this space as a result of ongoing reflections on our own thoughts, behaviours, biases, and assumptions. We have engaged as teachers – designing, leading, and assessing learning programmes for CBSL students. As learners – building and sustaining relationships with our CBSL partner organizations, constantly renegotiating as institutional policies change and shape our formal roles. We ask questions of our work and of each other, challenging the deep colonizing that so often accompanies well-intentioned student engagement with communities. Responding to these questions has changed the way we see, think, do and be methodologically, conceptually, ethically, and epistemologically. We are still asking questions.

This chapter shares some of the teaching and learning practices we have used with students undertaking CBSL and with the development organizations that receive them. Using a cycle of reflection our practice has informed and been informed by a range of values, frameworks and principles including: co-creation principles (Bilous et al. 2018); intercultural communication principles (Bawaka Collective 2019); orientations of reciprocity (Dostilio et al. 2012); and benefits to stakeholders (Lloyd et al. 2017). Reflecting on our CBSL practice we find that three interconnected themes emerge: bringing different voices and perspectives; developing reciprocal relationships and embedding reflective practice.

Questions to consider:

- Is the CBSL programme designed with communities or for communities?
- How do we challenge Eurocentrism and neo-colonial thinking when we prepare for CBSL projects?
- Have you considered the issues that will inform the context in which your CBSL activity will take place and how these might impact your work?

Bringing different voices and perspectives to CBSL practice

The conceptualization that 'development' is about assisting communities that are in some way deficient in resources or expertise has been challenged in recent years with post-development practitioners and scholars calling for an approach that focuses on creating 'new networks and spaces of opportunity for people and communities' couched in 'languages of hope and possibility' (McGregor 2009, 1688).

This is not, however, the same discourse used by many of the organizations providing CBSL opportunities for students. Instead, students are being 'sold' opportunities to 'Empower Vietnam' (AISECC), 'solve the world's social issues through the creation and implementation of tailored products and services for communities who need it the most' (Project Everest) or 'fight poverty, improve health, educate and empower change' (IVI). Many organizations position the student volunteer and the sending organization as an expert, rich in resources or education, and the beneficiary partner organization as being somehow deficient. Without thoughtful preparation and ongoing critical analysis and reflection, programmes can easily reinforce a dichotomy of 'us and them' and become 'small theatres that recreate historic cultural misunderstandings and simplistic stereotypes' (Grusky 2000, 858).

CBSL programmes are scaffolded and embedded in academic programmes and usually include pre-departure preparation and re-entry as a way of supporting students' learning, often using reflective practice to facilitate their experiences into learning outcomes. Embedding post-development theory into pre-departure training and preparation is also an effective way to develop student's awareness of the broader issues surrounding development, inequality, and poverty. It also provides students with an analytical framework to critically reflect on their own positionality and involvement in these spaces. Encouraging students to situate their experience within broader historical, geographical, political and ideological contexts also highlights how global/local forces affect the everyday lives of individuals and communities they engage with. Bishop and Tynan (this volume) encourage students to think outside their individual perspectives by offering pedagogical tools that consider more-than-human and temporal perspectives. They advocate for students to use pattern thinking to disrupt linear notions of time and rethink global development challenges by considering the perspectives of Eagle, Ant, Grandmother and Granddaughter (the macro, micro, past and future).

When we began teaching in CBSL many of the resources we used to prepare students were developed within the context of western patriarchal ways of knowing and teaching about development (Oldfield 2008; Tynan and Bishop 2019; Hammersley et al. 2018). Feedback from partners was that this current curriculum lacked the 'partner voice' in preparing students to undertake international activities. This was despite the fact that the partners were in the best place to contribute and were acutely aware of student misconceptions and failures to prepare adequately. We felt that by ignoring the 'partner voice,' we were not only dismissing the expertise in community development organizations but as a result, students had a simplified and singular understanding of key themes.

We invited long-term partner organizations, who were familiar with our programme and the needs of the students to co-create a preparatory and re-entry curriculum. In this way the hosting community development partners had a role to play in determining what and how students were taught (for more on this process see Hammersley et al. 2018; Bilous et al. 2018). Positioning community partners as co-educators/ co-teachers can be difficult as it requires vulnerability, humility and a willingness to not only relinquish expertise, but to sit with the discomfort of not knowing (Bilous et al. 2018).

The result was the co-creation of an open access resource that includes 35 activities divided into six core thematic modules: Developing Reciprocal Relationships, Team Building and Group Reflection, Challenging Perspectives, Children's Wellbeing and Empowerment, Workplace Cultures, and Creating Videos for Community Advocacy. These modules were supported by 53 video clips that shared community development partner insights. Within each curriculum teaching module, specific learning activities comprise lesson plans, each following a consistent structure to provide clear guidance to the instructor (the resources can be accessed through classroomofmanycultures.com). As an example, one of the co-created activities focused on how poverty is conceptualized. Rather than referring to the many definitions developed by academics or Western development organizations, our community development partner organizations were asked on video to define how poverty is experienced for their community. In the activity students are first asked to define poverty for themselves and then asked to view the videos and reflect on poverty as it is experienced differently by our partner representatives. Another activity asks representatives to provide a definition of a 'child,' allowing students to critically engage with the Convention of the Rights of the Child and through discussion of other contextual factors are encouraged to become aware of other approaches to the concepts of Child rights as experienced by partner organizations.

Including different voices in these activities enables students to better understand the lived experience of communities and multiple ways of seeing and being in the world. For example, through the CoMC resources partner organizations shared vernacular stories and understandings about education and the understanding that this is not equivalent with schooling. This challenges students to think more deeply about alternative educational practices and the needs of particular groups of young people and their lived realities (Hammersley et al. 2018).

Questions to consider:

- What does it mean to be reciprocal in CBSL projects?
- How do we ensure greater reciprocal, collaborative, and mutually enriching relationships between the community, students, and the academy?
- What are your motivations, expectations, and assumptions for the CBSL placement?

Developing reciprocal relationships in practice

In CBSL projects, students and partners enter into relationships of exchange. Students often leave with the best of intentions. In our experience, students indicate a desire to 'help others that are less fortunate,' 'make a difference' and 'gain international work experience and skills' (CoMC 2020). However, we need 'to abandon a naive assumption that universities 'service' communities' (Oldfield 2008, 284) as no simple scheme proposing reciprocal relationships as linear, transactional and fixed is sufficient to capture what is complex, fluid and multi-directional (Hammersley 2015). Indeed, campus-community partnerships 'can be dynamic, joint creations in which all the people involved create knowledge, transact power, mix personal and institutional interests, and make meaning' (Enos and Morton 2003, 24–25).

Putting learning at the centre of reciprocity acts as a point of common ground. Placing emphasis on a 'here to learn' rather than a 'here to help' or 'here to make a difference' mentality builds a CBSL model based on 'empathy learning' rather than 'sympathetic service' (Papi 2013). Community partners are often offering an educational service, and recognizing this perspective nurtures reciprocal relationships as they challenge more simplistic notions of charity (Hammersley 2015). We asked our partner organizations what reciprocity meant to them. Here are two examples from one particular partner, PACOS Trust in Sabah, Malaysia.

> …*the main thing that I really value is the interaction that the programme provides. It's the interaction that is most important. What the students bring provides motivation and empowerment to staff.*
>
> (Yoggie Lasimbang, PACOS Trust)

> *The community often feels that the students are somebody that they can rely on to help them convey their issue and message back to the student's countries or help the community issue be heard by the world.*
>
> (Atama, PACOS Trust)

During a workshop with many of our community development partners we explored some of the common processes that have enabled reciprocity to take place. We found that most important was the simple yet often undervalued process of making space to be together (Bilous et al. 2018). Rather than valuing the tangible outputs of the CBSL project, which might include a report, video or research project, they valued the intangible. Much more important to our partner organizations were the opportunities to improve their conversational English, to expose students to a different perspective and for students to become advocates on their return home. Most important of all were the opportunities to build relationships, to have informal conversations, to teach a student to play a gong, to swim or dance together, or to share in the preparation of a meal.

This is so important for students to understand prior to undertaking a CBSL project. For students it means developing caring interpersonal relationships with their hosts by privileging acts of relating and moments of personal connection, not just the project outputs or object-ives. This can often mean blurring the boundaries between work and personal, de-centring academic priorities and engaging in activities that actively promote relationship building. It also means valuing the spontaneous, serendipitous and in-the-moment aspects of relationship building, the unplanned occurrences and informal interactions that provide life, learning and reciprocity (Hammersley et al. 2014):

> *100% of the success that we find when we have people coming in from outside… comes through relationships and those relationships come from when people feel like they are being respected. We've had volunteers that come in, come in to the office to do the work and leave and they will go home with a fairly mediocre experience because they haven't been able to engage, they haven't been embraced by local people…regardless of any of the technical official output that you achieve, I think that's where the connections are made and you will go leaps and bounds ahead if people feel you are making an effort.*
>
> *(Cath Scerri, Bahay Tuluyan)*

Of course, in the context of research, learning and teaching, building relationships of care and reciprocity is also an attempt to reframe unequal power dynamics that dominate researcher/researched and academy/community relationships while being attentive to gender, class and ethnicity and how they shape interactions (Kirsch 1999). While the power imbalances ingrained within university bureaucracy remain problematic, some of these are disrupted in the process of developing personal relationships, as these relationships, which take time to develop and require and deserve an ongoing and long-term commitment, extend beyond the scope and life of spe-cific projects. This can be challenging as there are so many external factors over which you have little control. Relationships are built between people, rarely between organizations and relationships are not static. We have all seen partnerships between organizations and institutions fall apart as people have changed roles and moved on. We have also all established relationships with individuals that continue to grow despite changes to employment.

Students embarking on a CBSL project need to be taught to pay attention to the lives of people that exist outside of the immediate concerns of the project, and the interactions they have, so that the community partners can benefit from the relationship in the ways that they desire. For some students this is about learning how to develop relationships in cultur-ally appropriate ways. We asked representatives from our partner organizations to tell us how students might work to build deeper relationships and what damages relationships. We share these answers with our students. Students are then able to really focus on the more intangible aspects of the CBSL project. Students too focused on creating an output run the risk of missing those moments when community participants might be interested in communicating a range of different things important to them but perceived by students as irrelevant to completing the research task at hand. Of course, most important of all, and arguably difficult to explicitly teach, is that students remain open to the diverse and unexpected ways reciprocity can occur, unfold, and evolve at different points in time. So much is out of the predetermined control of the participants and instead can occur in-the-moment or arises in the thick of things. (Hammersley et al. 2014).

> *So the second day, [during the] interview with elders, I thought the students would ask more, not just plants they saw at the jungle but I'm very interested if they asked more on traditional knowledge but my expectation was wrong because this is very important for us as community here, very useful for young generation to come in the future, so I thought that they would ask more and more questions. …but no questions and I don't want to speak and answer because they don't want to know more on culture. We wasted a lot of time, so [it's] not good for me, because I always work and work. I see the students are bored when done interview, so I ask them to go to [the] swimming pool, and that is why it was a lot of fun.*
>
> (Hilda Pius, Kipouvo Community Homestay)

To bring value to the intangible and interpersonal process and outcomes that exist, students must be prepared for a role that is just as much about co-creating material outputs as it is about learning from, and relating with, community participants.

Questions to consider:

- Do you perceive yourself as the expert or provider of solutions to the organization or community?
- Are you there to learn and gain a better understanding of the complex and relational processes of poverty, globalization, and inequality?
- Do you understand the ways in which your education, culture, gender, sexuality, age, and belief system might impact the way you see the world? How will these impact the ways in which you are perceived?
- Have opportunities to debrief on your CBSL placement been arranged and how will you reflect on your experiences going forward?
- Have you thought about how your experiences will contribute to your lifelong learning and impact on the world?

Embedding reflective practice in CBSL

Reflective practice is well established in the academic literature around CBSL because of its role in promoting deep learning and potentially transformative learning experiences (Harvey et al. 2016). Emerging research on CBSL shows that the learning value of projects, for both self-efficacy and global awareness, increases when students are given the tools to discuss and think critically about those experiences.

We have found that the best learning outcomes are achieved when reflection is embedded and scaffolded in curriculum for students to learn from their CBSL pre, during and post experience. Before embarking on a CBSL project, students are asked to challenge their own assumptions, to ask questions of themselves; Who am I? Where am I coming from? What are my expectations? By questioning their own thoughts, behaviours, biases and assumptions, students can then understand how they might be perceived by their host community. They are asked to question the ways in which their education, culture, gender, sexuality, age and belief system might impact the way they see the world. If students can do this, they are better equipped to embark on a CBSL project with an open mind and to learn from these experiences.

In our experience, even in university-led programmes that emphasize reciprocal and ethical partnerships, and acknowledge the expertise within communities, many students assume the role of 'expert' or 'saviour' when preparing for and contributing to development projects. They assume that their university education is superior, and this is sometimes reflected in the way they are treated by their hosts. Simpson (2004, 690) reflects that such projects often result in 'simplistic, consumable and ultimately do-able notions of development.' While some students will return home self-congratulating themselves for their 'good work' many more will experience feelings of desperation or even failure – failure to have solved global poverty / hunger / issues of child labour. In other cases, stereotypes previously held by the students persevered through their experiences. Experience alone was not sufficient to bring about transformational learning, especially when the prejudices that obstructed insight were so durable and deeply ingrained. We share a poem from a student who undertook a CBSL project in Peru and reflected on these issues.

> Buzzwords of development, aid, and need,
> Language of power, of money, of greed.
> Decisions to change people's lives;
> How can I make them?
> Responsibility to an organisation;
> How can I help them?
> 21, standing on the edge of the next chapter.
> How can I save the world?
>
> *(CBSL student)*

We also encourage students to reflect throughout their CBSL project. We draw on Kolb's philosophy which acknowledges learning as 'the process whereby knowledge is created through the transformation of experience' (Kolb 1984, 38). Essential to this process is the learner being able to reflect on the experience, conceptualize the experience and use the experience to make decisions and solve problems. Prior to their departure we expose students to a number of established strategies to use while they are away. These include Bolton's (2010) 'Through the Mirror Writing' and Gibbs' (1988) 'Reflective Cycle,' all very accessible methodologies. We also expose students to Patel et al.'s (2013) 'Vyaktitva Explorer,' a process used by a youth-based organization in India to explore their vyaktitva, or character. For Patel et al. (2013, 100) 'Refl-action' is a learning process that 'must precede and succeed action.' Students are given the freedom to select the method or process that works for them but are required to continually question what they see, think and experience while away in order to make sense of what they are learning; to look inward instead of outward and to continually question their 'good intentions.'

Finally, on their return to Australia we use creative approaches to reflection, which encourage students to integrate the learning from both affective and cognitive perspectives (Harvey et al. 2016). Creative reflection can include visual (drawing, photography, vlogs, sculpture and films), performative (dance, plays), auditory (songs, poems, storytelling) and written (blogs, websites) approaches. We find this facilitates students' ability to better understand themselves, their privilege and their place in the world.

There are a number of excellent resources (see e.g., Comlanh https://comhlamh.org and Raidiaid) which provide a range of activities that also assist students to process their experiences on their return home. For many students, this process will take much longer than the length of their university course or degree programme. It may be many years later and therefore they need to be provided with tools they can use again and again, so that they continue to reflect, to ask questions of themselves and the world they live in.

Reflections by students on their learnings below reveal one learning.

> *A lot needs to change in order to achieve equality, both in Peru and in Australia, and the first step is making those in comfortable, powerful positions deeply uncomfortable with the state of inequality. As distressing as my experience in the Amazon was, I am grateful that it happened, as it has taught me to see my tourist identity as one fraught with the potential to harm, and therefore equally laden with the responsibility to actively seek the path of least harm. This path may go against the ready-made tourist experience so easily available, but it's the only one I want to take in future.*
>
> *(CBSL student)*

Conclusion

This chapter has posed a range of questions which can be used to help institutions, teachers and students from the minority world to think about the type of CBSL activities that they engage in and the impacts their engagement may have. As teachers, researchers, and practitioners we have a significant role to play in reimaging and reconceptualizing CBSL away from paternalistic neo-colonial models. We have shared various strategies developed and trialled through our research, teaching and personal lives to try and do CBSL differently in ways that prioritize an ethics of reciprocity and focus on respect and co-constructed ways of knowing. In doing so we seek to provide ethical and reciprocal CBSL engagement and contribute to a growing community of practice that aims to decolonize CBSL.

While our impact is limited mainly to those in the minority world our co-creation methodology with many community partners means that these strategies, resources and questions have the potential to be used beyond our institutions. We continue to engage and ask questions of our work and of each other by embedding reflective practice in our life/work and challenging the deep colonizing that so often accompanies well-intentioned student engagement with communities. There is space and indeed a need for further research and practice that brings different voices and perspectives, reciprocal relationships, and reflective practice to CBSL practice. We are still asking questions.

Key sources

1. Ateljevic, I., Morgan, N., and Pritchard, A. (2007) 'Introduction: Creating an academy of hope: An enquiry learning action nexus.' in I. Ateljevic, N. Morgan and A. Pritchard (eds.) *The critical turn in tourism studies: Creating an academy of hope*, Oxford: Elsevier, 1–10.

This paper evaluates the scope and potential for a revitalized radical critique of tourism that engages with issues of power, inequality, and development processes in tourism while acknowledging the significance of cultural diversities. In doing so, this paper puts forward the case that tourism research needs to further engage with some of the major themes and theoretical debates related to processes of globalization, capitalism, and structural power if it is to engage with issues of substantive import related to critical scholarship and social justice.

2. Bawaka Country, Wright, S., Suchet-Pearson, S., Lloyd, K., Burarrwanga, L., Ganambarr, R., and Sweeney, J. (2016) 'Co-becoming Bawaka: Towards a relational understanding of place/space,' *Progress in Human Geography* 40(4), 455–475.

The article illustrates the limits of western ontologies and opens up possibilities for other ways of thinking and theorizing. With Bawaka Country as lead author, it makes a significant contribution through its reconceptualization of place and space as more-than-human, relational and emergent.

3. Sin, H., Oakes, T., and Mostafanezhad, M. (2015) 'Traveling for a cause: Critical examinations of volunteer tourism and social justice,' *Tourist studies* 15(2), 119–131.

This paper provides a theoretically rich analysis of emerging critical research agendas at the intersection of volunteer tourism and social justice. It considers these agendas – focusing on the theoretical themes of neoliberal development, governmentality, geographies of care and responsibility, and the dilemmas found at the frequently encountered intersection of ethics and aesthetics.

4. Tynan, L., and M. Bishop (2019) Disembodied experts, accountability and refusal: an autoethnography of two (ab)Original women, *Australian Journal of Human Rights*, 25(2), 217–231.

Winner of the 2019 Andrea Durbach Prize for Human Rights this paper is a powerfully written call for re-examining how 'expertise' is constructed and for the inclusion of and respect for voices that are too often only present in human rights scholarship as the 'objects' – rather than the agents and authors – of study. The article demonstrates that when Indigenous peoples' perspectives on the violations they experience shape and lead conversations, they can more effectively destabilize racial hierarchies and enliven the transformative promises of human rights.

5. Wright, S. (2012) 'Emotional geographies of development,' *Third World Quarterly* 33(6), 1113–1127.

This article considers the importance of emotions in a development context. It argues that emotions are central to theory and practice in development and is a must
read paper on the removal of feeling from development studies.

References

Anderson, B. (2006) '"Transcending without transcendence": Utopianism and an ethos of hope,' *Antipode* vol. 38, no. 4, pp. 691–710.
Ateljevic, I., Morgan, N., and Pritchard, A. (2007) 'Introduction: Creating an academy of hope: An enquiry learning action nexus.' in I. Ateljevic, N. Morgan and A. Pritchard (eds.) *The critical turn in tourism studies: Creating an academy of hope*, Oxford: Elsevier, vol. 1–10.

Bawaka Country, Wright, S., Suchet-Pearson, S., Lloyd, K., Burarrwanga, L., Ganambarr, R., and Sweeney, J. (2016) 'Co-becoming Bawaka: Towards a relational understanding of place/space,' *Progress in Human Geography* vol. 40, no. 4, pp. 455–475.

Bilous, R., Hammersley, L. A., Lloyd, K., Rawlings-Sanaei, F., Downey, G., Amigo, M., Gilchrist, S. and Baker, M. (2018) "All of us together in a blurred space': Principles for co-creating curriculum with international partners,' *International Journal for Academic Development*, vol. 23, no. 3, pp. 165–178.

Bishop, M., and Tynan, L. (2022) 'Reimagining futures: Finding perspective through our more-than-human kin', in Sims, K., Banks, N., Engel, S., Hodge, P., Makuwira, J., Nakamura, N., Rigg, J., Salamanca, A. and Yeophantong, P. (eds) *The Handbook of Global Development*. London: Routledge.

Bolton, G. (2010) *Reflective Practice: writing and professional development*. Sage: London.

CoMC (2020). Retrieved from http://classroomofmanycultures.net

Dostilio, L.D., Brackmann, S.M., Edwards, K.E., Harrison, B., Kliewer, B.W., and Clayton, P.H. (2012) 'Reciprocity: saying what we mean and meaning what we say,' *Michigan Journal of Community Service Learning* vol. 19, no. 1, pp. 17–32.

Dinerstein, A. C., and Deneulin, S. (2012) 'Hope movements: Naming mobilisation in a post-development world,' *Development and Change* vol. 43, no. 2, pp. 585–602.

Everingham, P., (2016) 'Hopeful possibilities in spaces of 'the-not-yet-become': Relational encounters in volunteer tourism,' *Tourism Geographies* vol. 18, no. 5, pp. 520–538.

Enos, S., and Morton, K. (2003) 'Developing a theory and practice of campus–community partnerships,' in. B. Jacoby and Associates (eds.), *Building partnerships for service-learning*. San Francisco: Jossey-Bass, pp. 20–41.

Gibbs G (1988) *Learning by Doing: A guide to teaching and learning methods*, Further Education Unit. Oxford Polytechnic: Oxford.

Grusky, S. (2000) 'International service learning: a critical guide from an impassioned Advocate,' *American Behavioral Scientist* vol. 43, no. 5, pp. 858–867.

Hammersley, L.A. (2014) 'Volunteer tourism: building effective relationships of understanding,' *Journal of Sustainable Tourism* vol. 22, no. 6, pp. 855–873.

Hammersley, L. A., Bilous, R. H., James, S. W., Trau, A. M., & Suchet-Pearson, S. (2014) 'Challenging ideals of reciprocity in undergraduate teaching: The unexpected benefits of unpredictable cross-cultural fieldwork,' *Journal of Geography in Higher Education* vol. 38, no. 2, pp. 208–218.

Hammersley, L. A. (2015) '*It's about dignity not dependency': Reciprocal relationships in undergraduate community-based service learning'*, Doctoral Dissertation Macquarie University. Sydney, Australia.

Hammersley, L., Lloyd, K., and Bilous, R. (2018) 'Rethinking the expert: Co-creating curriculum to support international work integrated learning with community development organisations,' *Asia Pacific Viewpoint* vol. 59, no. 2, pp. 201–211.

Hartman, E., and Kiely, R. (2014) 'Pushing boundaries: introduction to the global service-learning,' *Michigan Journal of Community Service Learning* vol. 21, no. 1, pp. 55.

Harvey, M., Walkerden, G., Semple, A., McLachlan, K., Lloyd, K. and Baker, M. (2016) 'A song and a dance: Being inclusive and creative in practicing and documenting reflection for learning,' *Journal of University Teaching & Learning Practice* vol. 13, no. 2, pp. 1–17.

Kliewer, B.W., (2013) 'Why the Civic Engagement Movement Cannot Achieve Democratic and Justice Aims,' *Michigan Journal of Community Service Learning* vol. 19, no. 2, pp. 72–79.

Kolb, D., (1984) *Experiential Learning: Experience as the Source of Learning and Development*. Prentice-Hall, Inc., Englewood Cliffs, NJ.

Kirsch, G. (1999) *Ethical dilemmas in feminist research: The politics of location, interpretation, and publication*. Albany, NY: State University of New York Press.

Lloyd K., Bilous, R., Clark, L., Hammersley, L. A., Baker, M., Coffey, E. and Rawlings-Sanaei, F. (2017) 'Exploring the reciprocal benefits of community-university engagement through PACE,' in J. Sachs and L. Clark L. (eds) *Learning through community engagement*. Singapore: Springer, pp. 245–261.

McGregor, A. (2009) 'New possibilities? Shifts in post-development theory and practice,' *Geography Compass* vol. 3, no. 5, pp. 1688–1702.

Mellom, P., and Herrera, S. (2014) 'Power relations, north and south: Service in the context of imperial history,' in P. M. Green, and M. Johnson (eds.), *Crossing boundaries: Tension and transformation in international service-learning*. Sterling, VA: Stylus Publishing, LLC. 1–30.

Oldfield, S. (2008) 'Who is serving whom? Partners, process, and products in service-learning projects in South African urban geography,' *Journal of Geography in Higher Education*, vol. 32, no. 2, pp. 269–285.

Papi, D. (2013) 'Learning service – what's wrong with volunteer travel?,' ASHOKA. Retrieved from: http://startempathy.org/blog/2013/01/learning-service-what%E2%80%99s-wrong-volunteer-

Patel, A., Venkateswaran, M., Prakash, K. and Shekhar, A. (2013) *The Ocean in a Drop: Inside-Out Youth Leadership*, New Delhi: Sage.

Simpson, K. (2004) 'Doing development: The gap year, volunteer-tourists and a popular practice of development,' *Journal of International Development* vol. 16, no. 5, pp. 681–692.

Sin, H., Oakes, T., and Mostafanezhad, M. (2015) 'Traveling for a cause: Critical examinations of volunteer tourism and social justice,' *Tourist studies* vol. 15, no. 2, pp. 119–131.

Tynan, L., and M. Bishop (2019) Disembodied experts, accountability and refusal: an autoethnography of two (ab)Original women, *Australian Journal of Human Rights*, vol. 25, no. 2, pp. 217–231.

Wright, S. (2012) 'Emotional geographies of development,' *Third World Quarterly* vol. 33, no. 6, pp. 1113–1127.

56

CAPACITY DEVELOPMENT AND HIGHER EDUCATION

Krishna Kumar Kotra and Naohiro Nakamura

Capacity development in small island developing countries

Capacity development is set as a related topic to United Nation's Sustainable Development Goal 17: 'Strengthen the means of implementation and revitalize the global partnership for sustainable development,' under the 2030 Agenda for Sustainable Development launched in 2015 (UN n.d.). The term capacity development is broad, including development of human resources, improved access to physical infrastructure, development of research policies in support of sustainable development objectives, and so on (cf. Miloslavich et al. 2019). As Miloslavich et al. (2019, s147) state, 'When planning a capacity development programme or activity, it is important to first define what the objectives are and whom the programme or activity is targeting.' For some small island developing countries (SIDCs), capacity development 'often refers to the provision of training to scientific staff and students,' and 'implies a long term strategy to ensure that the skills and knowledge acquired are applied to the development of [science] in those countries' (Miloslavich et al. 2019, s147; see also Boshoff 2009). This chapter looks at how an SIDC can achieve capacity development in science and what strategies can be adopted, based on the case study of Vanuatu and The University of the South Pacific's Emalus Campus (EC), particularly focusing on student mentoring, training, and research participation.

SIDC's challenges in capacity development in science and collaboration

The role of science in economic development has been widely recognized. Scientific research can be a driving force for positive evolution in developing countries, because technological innovation produced from new information and scientific knowledge invariably produces quality goods and services (Nguyen et al. 2017, 1036). Indeed, some developing countries have substantially invested in the development of scientific research and achieved economic growth. A good example of this is India's IT industry. In 2019, India's IT industry contributed to 7.9 per cent of the country's gross domestic product and employed 4.1 million professionals. In 2019/2020, it was expected to grow at 8.4 per cent (Sharma 2020). Recently Sub-Saharan African countries have implemented new science policies and initiatives to demonstrate their

DOI: 10.4324/9781003017653-62

commitment to science and technology (Boshoff 2009). Nevertheless, unlike developed countries, developing countries (in particular least economically developed countries) still face many challenges in capacity development in science.

Based on interviews with students and teaching staff at five Southern African universities, Zdravkovic et al. (2016) found challenges when conducting research in Mathematics, Physics, and Chemistry. First, universities in those countries have poor access to funding and facilities. This is because university's primary mission is teaching, not research, and the government does not dedicate funding for research. The second challenge is a very high student–teacher ratio. For instance, a professor at the University of Zambia stated that they teach a first-year class with about 1190 students, in addition to postgraduate students and supervising research students, subsequently compromising the professor's research. Third, these universities have poor access to scientific journals, because of the high subscription fees charged by the publishers. Schneider et al.'s (2016) research lists additional challenges in capacity development when relating to mental health research in developing countries. These challenges included limited culture of research and few skilled researchers who can train and mentor students. Although Zdravkovic et al' and Schneider et al' studies are discipline-based, researchers in developing countries share similar challenges across disciplines. Hence, despite the majority of the world's population living in developing countries, researchers from developing countries are significantly underrepresented in academic publications (Bai 2018).

In order to solve these challenges, researchers and institutions in developing countries have actively sought collaboration with those in developed countries. As Zhang et al. (2020, 214) have identified, scientific collaborations have various positive outputs, including higher publication productivity, higher citation rates, relatively higher quality of publications, and international visibility of publications for national researchers and institutions. However, for researchers and institutions in developing countries, one of the largest benefits is that collaboration provides access to research funds, facilities, and scientific journals (Zdravkovic et al. 2016; Nguyen et al. 2017). Establishing a formal institutional partnership between developing and developed countries have also enabled researcher/student exchanges for training and mentoring.

Despite the increasing cases of scientific collaboration, developing countries, especially SIDCs are still facing considerable challenges and disadvantages. Namely, SIDCs do not have their own tertiary education institutions (in particular, research-focused) and often satisfactory research capacities do not exist. Collaborative possibilities are sometimes sought from researchers and institutions in developed countries; however, in most cases, individuals participating in a collaborate project are only seen as a local contact or consultant, *not* as a researcher who equally contributes to the project (Bai 2018; Dahdouh-Guebas et al. 2003; Zdravkovic et al. 2016). Because of the lack of research-focused institutions, there are no suitably qualified SIDC students who could be part of collaborative projects; hence, collaboration does not result in capacity development in the country (Schneider et al. 2016). These examples of unequal relations have led to some scholars, such as Bai (2018), stating that Global South countries have become a playground for scholars from the Global North. Making a similar observation, Dahdouh-Guebas et al. (2003, 341) revealed the discrepancy between developed countries' research priorities, such as global equality, and the reality that developing country contacts, while 'partners,' 'are totally neglected when it comes to peer-reviewed publishing.' Dahdouh-Guebas et al. (2003) have labelled the lack of authorship in prestigious academic publications, despite the scientific research being undertaken in developing countries, as 'neo-colonial science.'

Science education in Vanuatu

Vanuatu, a small South Pacific Island country, is an archipelago of 83 islands surrounded by narrow fringing reefs. Tourism and blue economy are the main income sources supporting and driving the population, which is close to three hundred thousand. Geologically, Vanuatu is also the home for active volcanoes in some islands. Given its location, the country is exposed to high intensity tropical cyclones on a yearly basis, which substantially affects the country's economy and development. Category 5 tropical cyclones hit the country in 2015 and 2020, which substantially ruined many islands. According to the UN World Risk Index, Vanuatu is the most disaster-prone country in the world for natural hazards (UN World Risk Index).

The University of the South Pacific's (USP) Emalus Campus (EC) is located in the capital city of Port Vila and since its establishment in 1989, it has been predominantly functioning as USP's Bachelor of Laws programme campus. Concurrently, EC had long been the country's only tertiary education institution until February 2020, when Vanuatu National University opened. It is still the country's sole tertiary education institution that offers science courses (Biology, Chemistry, Computer Science, Engineering, Earth and Environmental Science, Marine Science, Mathematics and Physics). However, course delivery is limited within introductory level, with a few exceptions in Mathematics and Physics, and the majority of courses do not involve face-to-face lecture delivery. Instead, students are expected to watch recorded lectures and complete assignments online. Since most advanced level science courses are not offered at EC due to lack of laboratory facilities and teaching staff, to complete all the required courses in the final year, Vanuatu students need to travel to USP's Laucala campus in Fiji, where laboratory facilities are equipped. Since 2015, Krishna (Lead author) has been the sole lecturer-level teaching staff in science at EC. As the Science Programme Coordinator, his role includes student consultation, communication with Laucala-based teaching staff (ensuring smooth delivery of science programmes), and science programme promotion in the country.

Vanuatu, as an SIDC located in the Pacific ring surrounded by six active volcanoes and increasingly impacted by global climate change, has attracted many scientists (e.g. Firth et al. 2014; Fitzgerald et al. 2020; Le Dé et al. 2018; Simons et al. 2020a; 2020b; Webb 2020; Westoby et al. 2019). The level of scientific attention and lack of scholarly recognition certainly classifies Vanuatu as a neo-colonial scientific playground for Global North scholars (Dahdouh-Guebas et al. 2003). The kinds of research studies range from 2–3 day single visits to 2–3 year long with periodic visits carried out by groups of researchers or at times being conducted by a single visiting researcher. Many of these studies are conducted in a relatively short period as field visits by masters and doctoral students, or as feasibility studies by Global North researchers. Research projects sometimes involve local stakeholders and/or communities when requiring logistics support on remote islands or when needing relevant permissions to undertake research. Some consultancy studies for various agencies and donor countries are also conducted with existing collaborations; however, researchers and students, primarily coming from Global North institutions, rarely revisit to give feedback to communities or stakeholders who made the research possible in the first instance once data is collected. Indeed, research results in these cases are often only presented in theses or academic journals; documents local people hardly access and read due to expensive paywalls and inaccessible, technical language. Moreover, local students are rarely part of these research studies at any level, despite the existence of a tertiary educational institute in the country. The upshot of all this is that these existing scientific collaborations have had negligible impact on Vanuatu's capacity development, reflecting a persistent 'neo-colonial science' agenda by Global North institutions even when their core priorities run contrary to such extractive research relations. Future impacts of this discrepancy are felt most in terms of student mentoring and training despite global declarations to 'implement' and 'revitalize' global partnerships ala the UN's Sustainable Development Goals.

Concurrently, in Vanuatu, there has been a widely shared view among locals that postsecondary students have not been passionate to study science, nor strongly been encouraged to pursue a career in science. There are some reasons for this. Firstly, as stated, EC, as the country's only tertiary education institution, lacks a full science programme delivery and laboratory facilities. Secondly, advocacy in science-focused communication at schools has been lacking. This is partly because of inadequate or under-qualified science teachers in the country. These two factors are linked. Institutions without a full programme and facilities cannot produce the kinds of quality teachers who can convey the importance and attractiveness of science to students. The size of Vanuatu's population was another factor. The population is scattered across the country, with many remote islands with a population of a few hundred. The national government had long assumed the lack of demand for science at tertiary level and investing into science education facilities may not be feasible and sustainable. The lack of facilities has made it difficult for EC's staff to conduct science research and accommodate visiting researchers from overseas. In this sense, experiences of 'neo-colonial science' in the country has been exacerbated by the country's structural problems and undervaluing of the benefits of supporting science. It was in this context that EC started to look for creative ways to change the status quo and find ways to initiative collaborative capacity development.

Collaboration for capacity development – a creative way forward

In 2015, EC started an intensive agenda to promote science programmes in the country, believing that demand for science programmes could be created and that increasing student enrolments would eventually result in academic excellence and capacity development. Because of limited staff and facilities, a creative and advocacy-focused tactic was necessary. In terms of student enrolment, EC sought a means for slow and steady increases along with the establishment of research facilities. Using the help from USP alumni, EC also sought the possibility of collaboration with the ministries and departments of Vanuatu Government to acquire first-hand information on their research activities, as well as various stakeholders and NGOs for information on activities being carried out on remote islands. Additionally, invitations for possible collaboration for research studies in the country and for the staff and student exchange programmes were sent to universities and national research centres in the Oceania region.

After year-long communications with educational institutions and stakeholders, it became apparent that discrete science research activities and capacity development at EC would be very challenging due to the lack of full science course delivery, staff, and facilities. Despite these setbacks, EC never lost hope to become a research and education hub for science in the country. What EC learned from the communications with the government and other institutions is that various research activities are conducted in the country by researchers from overseas. However, as stated, many of these projects were short-term field visits by student researchers, and local partners were used merely as a 'data source.' Overseas researchers did not visit EC, and EC science students were never part of these research projects hence perpetuating neo-colonial science (Dahdouh-Guebas et al., 2003). EC therefore decided to strategically utilize those researchers' visits.

A key strategy was to request information from time to time from all ministries and departments to inform EC of any research permit applications being lodged. EC also continuously requested local collaborators to inform any incoming research visit. With continued advocacy, the authorities eventually confirmed USP's EC to be the best institution to host visiting researchers. At this point, EC finally succeeded establishing collaboration networks with visiting researchers. The plan was that visiting scholars are offered use of the campus as a platform to carry out their research. They are given access to facilities and services at EC and encouraged to involve EC students and staff in research activities such as data collection, participation in research skills workshops, or presenting findings. In this way, visiting researchers contribute to student mentoring and training as part of capacity development and science promotion in Vanuatu.

Nevertheless, initially, no researchers or institutions approached EC for possible collaborations. Given this poor response, an additional strategy formulated by EC was to offer logistic support on research grant applications as a local collaborator. With this approach, the number of inquiries on collaboration from overseas, including potential research areas, available resources, cultural protocols, funding agencies, and local consultancy, gradually increased. In this process, EC established a world-wide network on science research. EC also decided not to decline any collaboration request, even if the project area was not guided or led by staff expertise. Notwithstanding these strategies, to begin to decentre 'neo-colonial science' practice in Vanuatu, EC staff members ideally participate in all projects as specialists and are always included in project design, development and publications as co-authors. However, as Bukvova (2010) states, not all collaborations lead to publications. For EC, the priority was student mentoring, training and participation in research projects and science promotion, which in Vanuatu's case, is the cornerstone of capacity development.

Collaborative research impact at Emalus campus: participation in research activities

The successes of EC in terms of developing research collaborations is reflected in the number of Memorandums of Understandings (MoUs) signed with other institutions These include: University of Oxford (www.usp.ac.fj/news/usp-collaborates-with-oxford-university-in-research/), Griffith University (www.usp.ac.fj/news/usp-griffith-university-signs-mou-for-collaborative-research/) (Figure 56.1), Centre for Environment, Fisheries and Aquaculture Science (Cefas) in the UK (www.usp.ac.fj/news/usp-signs-mou-with-international-research-institute/), Swiss Federal Institute of Technology (www.usp.ac.fj/news/usp-and-swiss-federal-institute-of-technology-starts-collaborative-research/), and Vanuatu Cultural Centre (www.usp.ac.fj/news/usp-vanuatu-cultural-centre-signs-mou/). EC is also a co-investigator in a number of projects, including with DFAT (Australia), Water for Women, National Science Foundation, and the Pacific Islands Universities Research Network. Over the past few years, EC has been occupied with various collaborations and activities. Of note is the substantial support (approximately USD 160,000) that has flowed in through donations from various collaborators for laboratory equipment thus strengthening capacity development. Some collaborative projects also increased local authorship adding scientific credibility to EC as a research institution (Devlin et al. 2020; Faivre et al. 2021; Foster et al. 2019; Foster et al. 2021a; Foster et al. 2021b; Kotra et al. 2017; Smith et al. 2021).

EC students too have had a number of opportunities to participate in research activities led by overseas researchers. One such examples is research on groundwater resources in Vanuatu, a collaboration with University of New South Wales (UNSW), Australia (www.usp.ac.fj/news/usp-unsw-collaborate-on-groundwater-studies-in-vanuatu/). In this project, EC students and staff from the Department of Water Resources participated in water sampling and preliminary observations, together with the research team from UNSW. Other research projects involving

Figure 56.1 (a) Student exchange visit from Griffith University to Emalus Campus and (b) students and staff at ocean acidification monitoring deployment site at Erakor Lagoon, Port Vila

EC students include: collection of sediment cores with USA's Woods Hole Oceanographic Institution (www.usp.ac.fj/news/usp-collaborates-with-usa-oceanographic-institution/); ocean acidification monitoring with University of Otago, New Zealand (www.usp.ac.fj/news/usps-emalus-campus-introduces-ocean-acidification-monitoring-in-vanuatu/; see also Figure 56.1); and workshops as part of the Commonwealth Marine Economies Programme with Cefas, London, UK (www.usp.ac.fj/news/usps-emalus-campus-hosts-international-workshop/). Recently staff and students of EC were part of consultancy studies: a DFAT funded study on Water, Sanitation and Hygiene (WASH) status in market houses in Vanuatu under the Covid-19 pandemic (Love et al. 2021); and developing Vanuatu National Chemical Profile for the Department of Environmental Protection (vanuatuwok.vu/jobs/invitation-for-expression-of-interest-2/). EC believes that student interactions with internationally renowned researchers have led to high-quality research possibilities and outcomes, while establishing a science 'community of practice' on campus. These developments have contributed to Vanuatu's capacity development by raising the profile of science in the country.

Science education communication among students

These examples of participation of students and staff in collaborative research projects has had positive outcomes for Vanuatu's capacity development. The most important outcome has been student academic performance in science units and communication. Over the past few years, students who took science units participated in multiple projects as assistants, enumerators, translators, or research volunteers. Students have accumulated academic experiences with international researchers and substantially benefitted from science courses with field components. Students are now enrolling in science units with expectations to undertake similar field components. While honorarium/remunerations may to some extent explain higher student uptake of courses, it has been proven that students who were part of the research studies performed well and achieved better grades. Exposure to high-quality research projects and interaction with international researchers has also changed students' attitude: students are now more motivated and seriously engaging in course content. In addition to those academic research skills, students have gained experiences in peer learning, leadership skills, and community engagement.

The senior students' experiences from science units have also been shared with junior students or secondary school students, inspiring enrolments in science courses and developing peer-to-peer capacity development. Constant press coverage on EC science courses has also

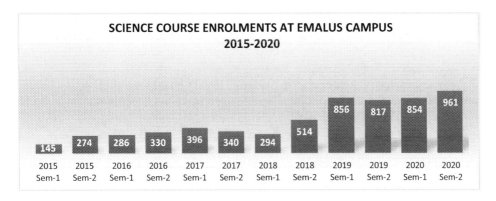

Figure 56.2 Science course enrolments at Emalus Campus 2015–2020

reached remote island populations transforming long-held community views on the relevance of pursuits in science. As a result, as shown in Figure 56.2, EC has observed an excellent progression rate on student enrolment in science courses over the past six years.

Challenges and way forward in collaborative partnerships

Of course, EC has not been free from challenges and constraints in developing research collaborations despite these promising examples of more equitable partnerships (Brown 1994; Jones et al. 1998). Ironically, reflecting the success of the strategies EC has put in place, one challenge has been dealing with increasing collaboration requests given the limited number of EC staff available to ensure the needs of visiting researchers are met. Other challenges have included requests to change field sites at the last minute, community leaders refusing to host research visits, and the difficulty in negotiating the diversity of needs and requirements within the same research group. In addition, there have been a few cases where research projects had to be aborted in the middle of study due to culturally insensitive approaches made by the research team. And in one instance, data required by EC for a particular project was not made available for reasons unknown to local staff leading to the rescheduling of the project. In this case, despite these concerns, EC was urged to work on the project to meet the deadlines of the collaborators. In other examples, EC staff and students were requested to work voluntarily on several projects because of sudden budget reductions for site visits. While we cannot be certain, we wonder whether such 'unreasonable' requests are remnants of colonial attitudes of researchers of the Global North counting on the 'accommodating' nature of South Pacific Islanders. Would Global North researchers make the same requests of their northern colleagues? One way to ameliorate these challenges and constraints is to incorporate local staff into the early drafting stage of research project design and discussions on implementation. By including EC staff in these formative processes, cultural sensitivities to certain sites, for example, can be anticipated in advance and community leader's needs met and concerns addressed.

Despite these ongoing challenges and colonial remnants, EC has been able to navigate these collaborative partnerships and accommodate research visitors. Increasing inclusion of EC staff in research projects has meant that appropriate processes and accumulated knowledge on cultural protocols has ensured more relevant and sensitive research being conducted in the country. Our philosophy is that the challenges and constraints are temporary and worth the pain for capacity development. If any research project is terminated, it will be our loss. Together with all of these experiences, EC is currently reviewing the progress and elaborating plans to establish a research centre with a more effective and logistical friendly administrative procedures for collaborators.

Another aspect we still need to consider is what students can do with scientific skills they learned from collaborations (cf. Miloslavich et al. 2019, s147). Currently secondary science teachers are not well trained, as stated; hence, with science capacity development, skills and knowledge of science, teachers' overall capabilities will be enhanced. As trainees become trainers, science capacity development will further advance. The collaboration with United Nations Institute for Training and Research (UNITAR) since 2018 has now made the participants act as facilitators in the Geographic Information System workshops and are well equipped with the latest software for cyclone prediction and flood mapping (www.usp.ac.fj/news/usp-emalus-workshop-to-progress-git-in-vanuatu/). In addition, in Vanuatu, a sustainable research trainee workforce has been slowly developed where they will be capable of handling in-country studies/consultancies in the near future. This collaborative approach is greatly helping the current 'Vanuatu 2030 The Peoples Plan' for building a stable, sustainable, and prosperous nation by supporting quality education and research skill development. However, despite these promising shifts demand for science graduates may not be that large in the country. Hence EC still needs to explore various possibilities of career path for science graduates both within Pacific Island countries and beyond.

Conclusion

We have discussed how The University of the South Pacific's Emalus Campus has attempted to enhance science capacity through international collaborations. We understand that the targeted readership of this volume is those who are teaching and/or studying social science and developing practitioners, not necessarily physical/natural scientists. However, our attempt to decentre neo-colonial science practice and develop capacity is certainly of interest to many of the readers of this volume. In settler/colonial countries, it is now nearly impossible and certainly inappropriate to develop a research project related to Indigenous peoples without a partnership and prospective benefits flowing to Indigenous collaborators and communities. Our hope is that such practices will also be extended to research projects in SIDCs. Regarding research projects in Vanuatu, many projects are still conducted by overseas researchers and local authorship is limited. How do local people benefit from these research projects? Do such projects contribute to capacity development in the country? Our attempt at EC provides an example that responds to these questions while setting out more appropriate and relevant practices for future collaborative partnerships.

Although our case study is based on a small island country, we propose two lessons. First, as stated, the targeted readership of this volume is social scientists. If you are based on a small developing country and your physical/natural science colleagues are studying/teaching with limited staff and research facilities, you can inform them of our practices as a potential strategy for science capacity development. Of course, you need to consider the country's political environment and administrative structures but working with the immigration office may be effective to bring visiting researchers into your institution. Second, if you are a researcher/student who is developing a research project in a developing country, you should seriously think how you can contribute to capacity development of the country. Without local involvement and consent, your research may be perpetuating 'neo-colonial science.' According to Miloslavich et al. (2018, s139), capacity development is 'identified as a major need by at least half of the 24 international conventions that relate to biological and/or ecological aspects of the ocean.' Several other reports such as UN Sustainable Development Goals, the Convention on Biological Diversity (CBD) Aichi Targets also refer to the need for capacity development to support SIDCs (Miloslavich et al. 2019, s139). Depending on the type of research, if involving non-professional community members becomes difficult, why not involve students who are studying science instead? Contributing to capacity development is a great way to move away from the practice of 'neo-colonial science' (Dahdouh-Guebas et al. 2003). Although our hope is that the EC will eventually become South Pacific's research hub and equally contribute to academic publishing in science, this will take time. In the meantime, we hope that EC's attempt to develop collaboration for capacity development in Vanuatu, especially in terms of student mentoring, training and research participation, becomes a mainstream practice for Global North researchers. No longer a 'playground' for these collaborators, countries in the South Pacific and elsewhere need to be valued for their contribution to knowledge creation and for the existing expertise that invariably comes from there.

Key sources

1. Bai, Y. (2018). Has the Global South become a playground for Western scholars in information and communication technologies for development? Evidence from a three-journal analysis. *Scientometrics*, 116(3), 2139–2153. https://doi.org/10.1007/s11192–018–2839-y

This article analyses the role of the Global South countries and the representation of scholars from the Global South in information and communication technology for development. The author finds that although many studies on the countries of the Global South, scholars from the Global South are generally underrepresented.

2. Miloslavich, P., Seeyave, S., Muller-Karger, F., Bax, N., Ali, E., Delgado, C., ... and Urban, E. (2019). Challenges for global ocean observation: the need for increased human capacity. *Journal of Operational Oceanography*, 12(sup2), S137–S156. https://doi.org/10.1080/1755876X.2018.1526463.

This article argues the importance of capacity development in less developed countries for global ocean observation. To achieve capacity development, the authors stress the need of the political will and the institutional engagement. The article includes examples of capacity development training and practices.

3. Schneider, M., Van de Water, T., Araya, R., Bonini, B., Pilowsky, D., Pratt, C. and Susser, E. (2016). Monitoring and evaluating capacity building activities in low and middle income countries: Challenges and opportunities. *Global Mental Health*, 3, E29. https://doi.org/10.1017/gmh.2016.24.

This article describes the range of capacity building activities, the process of monitoring, and the early outcomes of capacity building in mental health studies. Although capacity building activities conducted in five regional hubs established in less developed countries have been successful in terms of more publications and successful grant applications, the remaining challenge is sustainability of such activities.

4. Zdravkovic, M., Chiwona-Karltun, L. and Zink, E. (2016). Experiences and perceptions of South–South and North–South scientific collaboration of mathematicians, physicists and chemists from five southern African universities. *Scientometrics*, 108(2), 717–743. https://doi.org/10.1007/s11192–016–1989-z.

Based on the interviews with scientists based at five southern African Universities, this article discusses the challenges faced by scientists in the Global South and how collaborations, both south–south and north–south, have helped southern scientists in terms of acquiring research funds and facilities, increasing authorships, and enhancing scientific productivity.

5. Zhang, Q., Abraham, J. and Fu, H.Z. (2020). Collaboration and its influence on retraction based on retracted publications during 1978–2017. *Scientometrics*, 125(1), 213–232. https://doi.org/10.1007/s11192–020–03636-w.

This article discusses negative outcomes of collaboration in scientific research, focusing on retracted publications. The authors find that international collaborations tend to increase retraction rates and the occurrence of plagiarism tends to be low when a paper is contributed by more authors; however, falsification or manipulation, errors or unreliable results appear to be frequent in a paper contributed by more authors.

References

Bai, Y. (2018). Has the Global South become a playground for Western scholars in information and communication technologies for development? Evidence from a three-journal analysis. *Scientometrics*, vol. 116, no. 3, pp. 2139–2153. https://doi.org/10.1007/s11192–018–2839-y

Boshoff, N. (2009). Neo-colonialism and research collaboration in Central Africa. *Scientometrics*, vol. 81, no. 2, pp. 413. https://doi.org/10.1007/s11192–008–2211–8

Brown, G.T. (1994). Collaborative research between clinicians and academics: Necessary conditions, advantages and potential difficulties. *Australian Occupational Therapy Journal*, vol. 41, no. 1, pp. 19–26. https://doi.org/10.1111/j.1440-1630.1994.tb01807.x

Bukvova, H. (2010). Studying research collaboration: A literature review. *Sprouts: Working Papers on Information Systems,* vol. 10, no. 3. http://sprouts.aisnet.org/10–3

Dahdouh-Guebas, F., Ahimbisibwe, J., Van Moll, R. and Koedam, N. (2003). Neo-colonial science by the most industrialised upon the least developed countries in peer-reviewed publishing. *Scientometrics*, vol. 56, no. 3, pp. 329–343. https://doi.org/10.1023/A:1022374703178.

Devlin, M., Smith, A., Graves, C.A., Petus, C., Tracey, D., Maniel, M., Hooper, E., Kotra, K., Samie, E., Loubser, D. and Lyons, B.P. (2020). Baseline assessment of coastal water quality, in Vanuatu, South Pacific: Insights gained from in-situ sampling. *Marine Pollution Bulletin*, vol. 160, pp. 111651. https://doi.org/10.1016/j.marpolbul.2020.111651

Faivre, G., Sami, E., Mackey, B., Tomlinson, R., Zhang, H., Kotra, K.K., Aimbie, J., Maniel, M., da Silva, G.V. and Rand, E. (2021). Water circulation and impact on water quality in the southwest of Efate Island, Vanuatu. *Marine Pollution Bulletin*, vol. 173, pp. 112938. https://doi.org/10.1016/j.marpolbul.2021.112938.

Firth, C.W., Handley, H.K., Cronin, S.J. and Turner, S.P. (2014). The eruptive history and chemical stratigraphy of a post-caldera, steady-state volcano: Yasur, Vanuatu. *Bulletin of Volcanology*, vol. 76, no. 7, pp. 837. https://doi.org/10.1007/s00445–014–0837–3

Fitzgerald, R.H., Kennedy, B.M., Gomez, C., Wilson, T.M., Simons, B., Leonard, G.S., … and Garaebiti, E. (2020). Volcanic ballistic projectile deposition from a continuously erupting volcano: Yasur Volcano, Vanuatu. *Volcanica*, vol. 3, no. 2, pp. 183–204. https://doi.org/10.30909/vol.03.02.183204

Foster, T., Willetts, J. and Kotra, K.K. (2019). Faecal contamination of groundwater in rural Vanuatu: prevalence and predictors. *Journal of Water and Health*, vol. 17, no. 5, pp. 737–748. https://doi.org/10.2166/wh.2019.016

Foster, T., Priadi, C., Kotra, K.K., Odagiri, M., Rand, E.C. and Willetts, J. (2021a). Self-supplied drinking water in low-and middle-income countries in the Asia-Pacific. *npj Clean Water*, vol. 4, no. 1, pp. 1–10. https://doi.org/10.1038/s41545-021-00121-6

Foster, T., Rand, E.C., Kotra, K.K., Sami, E. and Willetts, J. (2021b). Contending with water shortages in the Pacific: Performance of private rainwater tanks versus communal rainwater tanks in rural Vanuatu. *Water Resources Research*, vol. 57, no. 11, e2021WR030350

Jones, S.L., Myers, S.L., Biordi, D.L. and Shepherd, J.B. (1998). Advantages and disadvantages of collaborative research: a university and behavioral health care provider's experience. *Archives of Psychiatric Nursing*, vol. 12, no. 5, pp. 241–246. https://doi.org/10.1016/S0883-9417(98)80033-2.

Kotra, K.K., Samanta, S. and Prasad, S. (2017). Rainwater harvesting for drinking: a physiochemical assessment in Port Vila, Vanuatu. *The South Pacific Journal of Natural and Applied Sciences*, vol. 35, no. 2, pp. 33–44. https://doi.org/10.1071/SP17004

Le Dé, L., Rey, T., Leone, F. and Gilbert, D. (2018). Sustainable livelihoods and effectiveness of disaster responses: a case study of tropical cyclone Pam in Vanuatu. *Natural hazards*, vol. 91, no. 3, pp. 1203–1221. https://doi.org/10.1007/s11069–018–3174–6

Love, M., Kotra, K.K. and Souter, R. (2021). *WASH in the market house-a situation analysis of water, sanitation and hygiene services in market places in Vanuatu.* International Water Centre, Griffith University, Nathan, Queensland, Australia / The University of the South Pacific, Emalus Campus, Port Vila,

Vanuatu. www.watercentre.org/wp-content/uploads/2021/07/WASH-in-the-market-house_FINAL_July_2021-1.pdf

Miloslavich, P., Bax, N.J., Simmons, S.E., Klein, E., Appeltans, W., Aburto-Oropeza, O., et al. (2018). Essential ocean variables for global sustained observations of biodiversity and ecosystem changes. *Global Change Biology*. vol. 24, pp. 2416–2433. https://doi.org/10.1111/gcb.14108.

Miloslavich, P., Seeyave, S., Muller-Karger, F., Bax, N., Ali, E., Delgado, C., ... and Urban, E. (2019). Challenges for global ocean observation: the need for increased human capacity. *Journal of Operational Oceanography*, vol. 12(sup2), S137–S156. https://doi.org/10.1080/1755876X.2018.1526463

Nguyen, T.V., Ho-Le, T.P. and Le, U.V. (2017). International collaboration in scientific research in Vietnam: an analysis of patterns and impact. *Scientometrics*, vol. 110, no. 2, pp. 1035–1051. https://doi.org/10.1007/s11192-016-2201-1

Schneider, M., Van de Water, T., Araya, R., Bonini, B., Pilowsky, D., Pratt, C. and Susser, E. (2016). Monitoring and evaluating capacity building activities in low and middle income countries: Challenges and opportunities. *Global Mental Health*, vol. 3, E29. https://doi.org/10.1017/gmh.2016.24

Sharma, S. (2020) IT industry may become lighthouse for India's growth; here's how many IT firms operate in India. Financial Express. February 17. www.financialexpress.com/industry/it-industry-may-become-lighthouse-for-indias-growth-heres-how-many-it-firms-operate-in-india/1870795/.

Simons, B.C., Cronin, S.J., Eccles, J.D., et al. (2020a). Spatiotemporal variations in eruption style, magnitude and vent morphology at Yasur volcano, Vanuatu: insights into the conduit system. *Bulletin of Volcanology* vol. 82, no. 59. https://doi.org/10.1007/s00445-020-01394-4.

Simons, B.C., Cronin, S.J., Eccles, J.D., Jolly, A.D., Garaebiti, E. and Cevuard, S. (2020b). Spatiotemporal variations in eruption style and magnitude at Yasur volcano, Vanuatu: part 2 – extending Strombolian eruption classifications. *Bulletin of Volcanology*, vol. 82, no. 11, pp. 1–18. https://doi.org/10.1007/s00445-020-01404-5

Sharma, S. (2020) IT industry may become lighthouse for India's growth; here's how many IT firms operate in India. Financial Express. February 17. www.financialexpress.com/industry/it-industry-may-become-lighthouse-for-indias-growth-heres-how-many-it-firms-operate-in-india/1870795/

Smith, A.J., Barber, J., Davis, S., Jones, C., Kotra, K.K., Losada, S., Lyons, B.P., Mataki, M., Potter, K. D. and Devlin, M.J. (2021). Aquatic contaminants in Solomon Islands and Vanuatu: Evidence from passive samplers and Microtox toxicity assessment. *Marine Pollution Bulletin*, vol. 165, 112118. https://doi.org/10.1016/j.marpolbul.2021.112118

United Nations (UN), (n.d.). Capacity development. *Department of Economic and Social Affairs: Sustainable Development*. https://sdgs.un.org/topics/capacity-development.

Webb, J. (2020). What difference does disaster risk reduction make? Insights from Vanuatu and tropical cyclone Pam. *Regional Environmental Change*, vol. 20, no. 1, pp. 1–13. https://doi.org/10.1007/s10113-020-01584-y

Westoby, R., McNamara, K.E., Kumar, R. and Nunn, P.D. (2020). From community-based to locally led adaptation: Evidence from Vanuatu. *Ambio*, vol. 49, no. 9, pp. 1466–1473. https://doi.org/10.1007/s13280-019-01294-8

Zdravkovic, M., Chiwona-Karltun, L. and Zink, E. (2016). Experiences and perceptions of South–South and North–South scientific collaboration of mathematicians, physicists and chemists from five southern African universities. *Scientometrics*, 1 vol. 08, no. 2, pp. 717–743. https://doi.org/10.1007/s11192-016-1989-z

Zhang, Q., Abraham, J. and Fu, H.Z. (2020). Collaboration and its influence on retraction based on retracted publications during 1978–2017. *Scientometrics*, vol. 125, no. 1, pp. 213–232. https://doi.org/10.1007/s11192-020-03636-w

57

ADAPTIVE PROGRAMMING, POLITICS AND LEARNING IN DEVELOPMENT

Aidan Craney, Lisa Denney, David Hudson, and Ujjwal Krishna

Introduction

Since 1980, women have won fewer than 1 per cent of all contests for a seat in Vanuatu's national parliament. At the most visible levels of leadership, ni-Vanuatu women do not have the same levels of power, money, and influence as their male counterparts. This, despite the fact that public confidence in elected officials is low and that 14 national parliamentarians (all men) were found guilty of bribery in 2015 (Rousseau and Kenneth-Watson 2018). However, since 2013, women's representation in politics at the municipal level has been enshrined in law through Temporary Special Measures (TSMs). For a minimum of four electoral cycles, 30–34 per cent of seats have been reserved for women in the municipal district of the two largest cities, Port Vila and Luganville.

The campaign for securing seats for women was driven primarily by senior bureaucrat Dorosday Kenneth-Watson (Rousseau and Kenneth-Watson 2018, 7). As the then Director of Women's Affairs, she assembled a coalition of allies – of all genders – committed to greater female political representation under the banner Women in Shared Decision Making (WISDM). WISDM determined that TSMs would be their focus. Learning from previous legislative reforms that failed due to perceptions they were women's-only issues, WISDM determined not to attach the campaign for reserved seats to other feminist movements in Vanuatu (Rousseau and Kenneth-Watson 2018, 5). The group used contacts in development agencies, government departments, and within the national government to foster political support – mostly working with key male politicians, recognizing that it was these men who would ultimately vote on the TSMs and had to be won over.

These experiences from Vanuatu carry multiple lessons for social change movements that embrace adaptive development measures. The movement utilized the political capital of Kenneth-Watson and other members with influence, built a grassroots coalition that worked with the grain of the Vanuatu cultural ecosystem known as *kastom*, and worked to a shared objective to address a clearly identified problem (Rousseau and Kenneth-Watson 2018).

There is growing recognition within development studies and practices that many of the challenges we face have no clear, predetermined pathways to change. To provide just two

DOI: 10.4324/9781003017653-63

examples, how can we change social norms that perpetuate inequality? Or how might corruption be reduced? The pathways for these processes of change are uncertain and therefore require interventions to adapt as they go, in response to context and learning. This has led to the rise of adaptive approaches to development.

In this chapter, we discuss adaptive development. We summarize three adaptive development approaches – adaptive management, problem-driven iterative adaptation (PDIA), and thinking and working politically (TWP) – and provide short case studies of each approach. We tease out the subtle differences between approaches, but also emphasize that there are overlaps in their logics and methodologies. These three approaches should not be read as the only flavours of adaptive development. Two other prominent approaches that are beyond the scope of this chapter, for example, include doing development differently (ODI 2016) and navigation by judgement (Honig 2018). For a more complete picture of adaptive management, we strongly recommend engaging with the texts referenced in this chapter – many of which are working papers accessible to audiences beyond academic paywalls. Finally, the chapter concludes with some ideas for teaching adaptive approaches in the classroom.

What is adaptive development?

Adaptive development is best understood as a school of complementary approaches, rather than a single, cohesive method. As the name suggests, these approaches all have a focus on adaptability at their heart. They begin from the premise that for some development challenges, it is not possible to plan how change can be achieved at the outset. Too little is known about the context, the likely pathway to change, and the impacts of interim steps along the way, to rely upon the simple cause-and-effect logic of traditional development projects. Development policymakers and practitioners cannot control the conditions in which development occurs as a scientist can control conditions in the lab. As a result, adaptive approaches to development argue that programmes must be responsive both to their environment, and to learning along the way to find successful pathways to change.

As a result of this focus on adapting to context and pathways to change, adaptive development approaches all share a strong emphasis on cultivating a deep understanding of the local context and investing in learning. Understanding the local context means getting to grips with how society really operates – not just according to formal, written laws and policies but in the day-to-day reality. This means getting to grips with power and politics and understanding that any development intervention (if at all successful) will have political implications (see Box 1 for what we mean by 'politics').

The importance of learning means not becoming too heavily invested in particular assumptions about how change happens. In the WISDM coalition at the beginning of this chapter, for instance, conventional wisdom suggests that improving women's political participation requires working collaboratively with women's organizations to advocate for gender equality. Yet in the case of Vanuatu, previous reform efforts that pursued this approach failed. Success came from a different approach, led primarily by one woman leveraging her political and customary relationships, particularly with male politicians, to build support. Adaptive approaches emphasize remaining critically aware of our assumptions about how change happens and testing whether they are indeed the best pathways to change.

Box 57.1 What do we mean by politics?

The word 'politics' often brings to mind Presidents and Prime Ministers, Parliaments and Congresses, political parties and policy debates. All of this is certainly 'politics,' but can be termed 'big P' Politics. It relates to the formal bodies and individuals invested with State power. But politics, by definition, refers to how power is distributed and managed within a society (Leftwich 2004). This of course includes 'big P' Politics. But it also includes what is sometimes called 'small p' politics. 'Small p' politics includes how gender relations in the household constrain women's access to household resources; it includes how the dominance of particular groups in the economy (be they ethnic, religious, gendered or class-based) structure wealth inequality; and it includes a whole host of other negotiated relationships that are laden with power exchanges between different actors across a whole host of contexts and scales. In short, 'small p' politics attends to the politics of the everyday. When we speak of aid being political, we mean this in that it has implications for how power and resources are distributed and managed.

Adaptive development approaches acknowledge that formulaic and linear approaches to development can work in settings where inputs can be reliably predicted to result in certain outputs, such as mosquito nets to reduce malaria or distribution of food aid in response to crisis events. However, they also recognize that driving social change is much more complex. As Andrews et al. (2015, 123) write: 'the global community has been far less proficient at addressing non-simple, non-technical problems that are implementation intensive – such as reducing corruption in procurement, providing dispute resolution, ensuring student learning, and administering land and natural resources.'

This begs the question, when is conventional development programming acceptable and when is adaptive development needed? It depends on the problem encountered. More is known about the pathways to achieve change in relation to some problems than others. For instance, persistence of disease may be a significant problem for much of the world's population. But despite presenting a real challenge, the pathway to address this problem is reasonably well known – vaccination, handwashing and sanitation are known solutions that will work with relative certainty, even if they can sometimes be difficult to implement. Other problems are more complex, because less is known about the pathways to change.

Why the adaptive turn?

The emergence of adaptive programming approaches in development can be located in critiques of development practice, and the turn to 'governance' that took place in many Western donor agencies in the 1990s. First, mainstream development practice has long been criticized for being apolitical[1] and static. For example, James Ferguson (1990) famously described development as an 'anti-politics machine' in his analysis of livestock management programmes in Lesotho that treated deeply political questions about how power and resources were divided in society as technical problems, requiring solutions delivered by development 'experts.' In fact, these solutions were themselves infused with politics that were simply ignored or swept aside, with detrimental impacts on 'aid recipient' communities as a result.

This apolitical nature of development endeavours has been further entrenched by the new public management of the 1980s, which sought to rationalize the way in which development was delivered

(Cooke and Dar 2008). Greater attention to systematized planning and accounting was intended to deliver more tangible results. This meant trying to measure the relationship between what development agencies delivered (workshops, campaigns, infrastructure), what resulted (numbers of people trained and buildings built) and their impacts (better education or health outcomes).

The idea that discrete deliverables lead to results and impacts in a linear causal relationship is encapsulated in programming tools like logical frameworks ('logframes') that are now standard practice. They have made development programming increasingly rigid – unable to respond to often fluid or rapidly changing contexts, or indeed to respond to learning about what does or does not work in achieving social change. Such tools are increasingly seen as a straitjacket, prioritizing the accounting and compliance needs of development agencies rather than on-the-ground realities of the countries where development takes place (see for instance, Natsios 2010).

Second, partly in recognition of the above shortcomings, from the mid-1990s a much stronger focus on governance became apparent within the development industry (Hout and Robison 2008). The rise of governance enabled greater focus on the institutional and regulatory aspects of 'technical' areas of development, such as health, education, water and sanitation, and financial management. This, in turn, opened up discussion on the politics inherent within these technical areas – for instance, around what constitutes quality services, who benefits from them, and how they are best delivered. These issues have long been recognized as deeply political within donor countries (think about debates surrounding the privatization of healthcare) but were treated as technical issues for donors in other countries.

Increasingly it is recognized that what has prevented progress towards improved development outcomes for many people is not only – or even mostly – an absence of technical knowledge or infrastructure, but rather the persistence of political incentives that serve other, non-developmental interests (Leftwich 2000). To do development more effectively, therefore, programmes must navigate or influence this politics to achieve real change. For example, achieving better education outcomes will probably not come about by building more schools; it will likely come about through influencing political leaders to invest in education so that schools are equipped, teachers properly trained and remunerated, curriculums developed, and the teaching profession more broadly valued. As Carothers and de Gramont note, this embrace of politics has resulted in 'efforts by development aid actors intentionally and openly to think and act politically for the purpose of making aid more effective in fostering development' (2013: 10).

This confluence of critiques of development practice as apolitical and rigid, as well as the growing recognition of the need to engage with the politics of change, has paved the way for adaptive programming approaches. While such approaches currently occupy somewhat of a darling status among the development industry, it is important to note that the ways of working embodied in adaptive programming are not entirely new (Derbyshire et al. 2018). Some movements and organizations have already been working in ways that share many similarities to adaptive approaches. In particular, feminist movements have long had to work in politically astute ways to carve out space for progressive reform that goes against the grain (O'Neil 2016). This has implicitly involved many of the principles now more clearly articulated as being part and parcel of adaptive programming. In this sense, adaptive approaches are not entirely new – although the hold that they have within mainstream development discourse and (to a lesser extent) practice certainly is.

Adaptive management

Adaptive management advances the idea that responses to complex development challenges cannot be accurately designed in advance of applying interventions. Interventions seeking to promote positive broad-scale developmental change must acknowledge that the conditions (economic, political, environmental, and so on) that an intervention operates within will change during the course of the intervention. These changes will be driven both by factors beyond the intervention, as well as by facets of the intervention.

Adaptive management approaches to development emerged as part of the adaptive turn, mentioned above. Adaptive management specifically arose in response to an overemphasis on the need for development interventions to be able to clearly demonstrate efficacy and causality at all levels. The results-based management (RBM) approach to development helped to entrench accountability as a proxy for success in driving developmental change (Cooke and Dar 2008; Honig and Pritchett 2019; Natsios 2010). As Honig (2020, 4) writes: 'some of the projects that succeed less look very successful, by focusing on numbers that are unrelated to success.' Adaptive management does not necessarily conflict with results-based management, though; rather its focus on results is measured over longer time scales and by assessing broader social change instead of discrete input-outcome relationships. In this way, it is similar to PDIA (below) in that it deliberately avoids creating success indicators that may provide the superficial appearance of positive results.

How can adaptive management practices be implemented? Bond (2016, 1) offers that organizations 'that embrace learning, flexibility and have humility' internally and with donors who support such characteristics are best placed to implement adaptive management principles in practice. They add, however, that examples of adaptive management are more likely to emerge immanently, writing, 'many of the best early examples of adaptive management come from projects in which committed individuals and teams find enough 'wiggle room' even within a constraining organizational and funding environment to work in adaptive ways' (2016, 1).

To implement effectively, adaptive management also requires ongoing monitoring and evaluation of interventions that allows for real-time learning and course correction. This includes embedding 'double feedback' loops that not only assess the effectiveness of inputs of an intervention but also critically reflect on the suitability of interventions at a holistic level – questioning the accuracy of problem identification, project design, human and material resources involved and beyond. In this way, adaptive management connects to recent movements in the development sector to incorporating research, learning and adaptation into monitoring and evaluation frameworks (Pett 2000).

Case study: Coalitions for Change

The Coalitions for Change (CfC) program is a partnership between the Australian Embassy and The Asia Foundation (TAF) in Manila. It has supported policy reform initiatives in the Philippines, in sectors as diverse as excise tax, elections, education, land governance, local roads management, and disaster risk reduction, as well as the formal peace process and conflict resolution efforts in the country's south. CfC describes itself as encouraging civil society, the private sector, academia, and government to 'work together and bring about public policies that contribute to development reform priorities for the Philippines' (TAF 2017, 1).

Sidel and Faustino (2019, 28–29) note that CfC's coalition-building efforts comprise recruiting and activating tight-knit teams combining experienced TAF staff with 'policy wonks,' political analysts, and networkers, as well as 'insiders' within diverse realms of the Philippine government. They find that CfC has 'proven most innovative and effective and achieved the greatest impact when and where it has operated through reform advocacy work focused on policy arenas, rather than through more established and open-ended forms of promoting engagement, consultation, and cooperation between governments and civil society.'

Comparing CfC with other adaptive management programmes in Tanzania, Nigeria, and Myanmar, Green (2019, 1) finds that a key factor that differentiates CfC is the history and culture of TAF. Given its 65-year-long experience of working in the Philippines, TAF has 'developed deep roots of understanding and partnership' and is seemingly more embedded than 'fly-in-fly-out' management consultants, as well as 'more entrepreneurial and agile than many international NGOs.' More importantly, the long presence and commitment of Australia's foreign ministry in the Philippines has engendered a 'high level of trust' and 'strategic patience,' which give CfC sufficient leeway to 'dance with the system.' This essentially means that CfC can adjust its pace to the various possibilities and pathways for reform, as opposed to being strictly bound to 'promises in the project document,' consistent with adaptive management principles. There are also significant elements of PDIA and TWP to CfC, showing how these approaches can overlap and be blended.

Problem-Driven Iterative Adaptation

Problem-Driven Iterative Adaptation (PDIA) is an approach to development that focuses on addressing a tangible problem identified by those who experience it, and iteratively testing potential solutions that are continually adapted in response to ongoing feedback loops. Initial approaches to addressing the identified problem are brainstormed and quickly trialled, with an emphasis on learning and (re)designing by doing. In effect, this collapses the 'design' and 'implementation' stages of conventional development programmes, preferring to quickly mobilize and learn from implementation, rather than invest significant time in perfecting plans during a distinct 'design' phase. As approaches are implemented, PDIA involves continued assessment of the intervention and preparedness to change established approaches if they are not meeting the intended end goal (Andrews et al. 2015, 125).

PDIA was designed in response to what was identified as repeated failures to adapt development thinking and practice to the political, economic, and environmental realities that development challenges exist within. Instead, development interventions tended to result in 'isomorphic mimicry' (Andrews et al. 2012) where the *form* of successful institutions was established – giving the impression of change – but the intended *function* of those institutions wasn't delivered.

Case study: learning to target for economic diversification in Sri Lanka

In 2016–17, academics at Harvard University's Center for International Development (CID) collaborated with officials in the Sri Lankan government on a program aimed at building capabilities for economic diversification. Three linked initiatives were pursued, adopting a PDIA approach: learning to target (Andrews et al. 2017d); improving the investment climate (Andrews et al. 2017c); and engaging new investors (Andrews et al. 2017b).

The challenge of supporting economic diversification had been the focus of many economists for over a decade – most of whom worked in isolation from one another and did not engage with Sri Lankan officials. This isolation led to multiple lists being developed that identified target sectors to improve diversification. As a result – and despite working to the same identified problem – different donor bodies supported different actors to engage different sectors. More importantly, the absence of involving Sri Lankan officials meant that state capacity to develop targeting lists domestically remained constrained. Spotting the limitations resulting from reliance on external targeting advice, Sri Lankan bureaucrats began engaging with CID to address the problem through a rapid, learning-by-doing approach using PDIA processes. This effort comprised local teams working 'consistently for a six or seven-month period, stopping every two weeks to assess progress and determine next steps' (Andrews et al. 2017d, 5).

At the end of this phase of PDIA-driven engagement, Andrews et al. (2017d, 37–38) explored whether the PDIA process promoted 'a different type and pace of progress,' and if so, how. They compared their results with the end product of a similar project developed by a donor agency for the government almost simultaneously, noting that it was delivered completely outside of government, and no state capabilities were built as a result. The CID-led product was, on the other hand, 'developed in-house, with expanded in-house capability and engagement and ownership,' their experience comparing extremely positively to the 'counterfactual.'

Thinking and working politically

Thinking and Working Politically (TWP) is an approach to development that emphasizes that interventions are more likely to be effective if they have the support of people who carry political influence within that environment. It is 'an approach which emphasizes that developmental change requires a knowledge of how to navigate power, vested interests and who or what is deemed locally legitimate at critical junctures' (Craney and Hudson 2020, 1653–1654). Key to good TWP practice is being aware of the political economic environment within which an intervention is proposed. Indeed, development and social change cannot occur *without* navigating the political environment in which those changes are sought (Hudson et al. 2018).

Thinking politically involves analysis of how change is possible in light of the local context. This involves understanding relevant formal and informal institutions (essentially, how power works *de jure* through formally agreed laws, policies, and rules; and how power works *de facto* through often unwritten social norms), as well as the relevant players on the ground and their relative power, interests, ideas, and incentives around a given issue. Thinking politically in this way is often undertaken through political economy analysis and requires deep local knowledge of how societies work (Hudson, Marquette and Waldock 2016).

Working politically involves applying politically astute thinking into practice. This can take many forms but boils down to acting in ways that recognize how power operates in a given society. For instance, it may be that changing the way health services are delivered in a given country depends less on the support of the Health Minister (who, formally, appears to be powerful) and more on addressing the interests of private sector businesses who benefit from contracts from the Health Ministry (who might hold the real power and political influence). Working politically can mean engaging with a wider set of actors that go beyond those who – at first glance – appear to be relevant, recognizing that there may be others who in fact influence the prospects of change underneath the surface.

It can be particularly challenging for expatriate development workers to 'think and work politically,' given the need for a deep and nuanced understanding of the local context, as well as extensive networks and relationships. As David Booth and Sue Unsworth (2014, 70) write, 'locals are more likely than outsiders to have the motivation, credibility, knowledge and networks to mobilize support, leverage relationships and seize opportunities in ways that qualify as "politically astute".' Indeed, while TWP might be a new concept for the development industry, local reformers and activists have long had to navigate power and politics to achieve change.

A key feature of TWP is working with coalitions of influential people, rather than individuals, to both increase the scope of influence and minimize the risk of providing outsized power to any one person (DLP 2012). TWP thus faces the challenge of being selective about which persons of power are supported in ways that contribute to transformation and positive social change (Craney and Hudson 2020; Denney and McLaren 2016). There is a real risk that focusing on engaging people who already wield influence and power may act to further entrench power differentials within communities (Craney 2020; Roche et al. 2020) and that working politically can carry risks for individuals (Sims 2021). This is a fine line as demonstrated by the CfC partnership with a multinational cigarette manufacturer to reform tobacco and alcohol taxation in the Philippines, directly leading to significantly increased funding for public accessible healthcare (Sidel and Faustino 2019).

Case study: The Pacific Leadership Program

The Pacific Leadership Program (PLP) was an Australian Aid funded programme that supported development initiatives in the Oceania region between 2008 and 2017. It was designed specifically to support emerging and innovative approaches to development that were being driven by locals. A study of PLP's approach to development notes that the 'criteria that PLP staff came to see as most important in deciding whether to support a coalition was whether a partner wanted the change more than PLP' (Denney and McLaren 2016, 4).

PLP took a 'portfolio' approach to the suite of programmes and projects for which it provided funding, technical assistance and/or material support. In total, PLP worked with 77 organizations and supported over 30 initiatives that it identified as significant partnerships. Supporting so many initiatives allowed PLP to diversify its exposure to risk by increasing the scope for positive outcomes. Further, it allowed for comparative monitoring and evaluation of interventions, such that PLP could support its partners to adapt ways of working but also so that staff could consider how they could improve their own practice. Such consideration was embedded through six-monthly reflection and refocus workshops (Denney and McLaren 2016).

Although not all supported initiatives achieved positive social change outcomes during the tenure of partnership, some significant achievements were realized. For example, PLP supported the WISDM initiative in Vanuatu that led to the creation of reserved seats for women politicians at municipal level (Rousseau and Kenneth-Watson 2018), the Simbo for Change project in Solomon Islands that created livelihood opportunities and enhanced social cohesion on a small island (Suti et al. 2021), and the Green Growth Leaders Coalition, a regional initiative recognized as shifting development norms and influencing government development plans across the region (Craney and Hudson 2020). Even after its closure, the influence of PLP continues, with some former staff being actively recruited by other development programmes in the region to share their skills in partnership building and navigating politics (Roche et al. 2020).

Teaching and learning adaptive approaches

There are lots of great pedagogical opportunities that adaptive approaches make available, whether in the classroom or as part of a training course. In our experience the more hands-on the better. But equally, for many of the examples below, the learning experience can work just as well remotely in an online environment or virtual classroom.

One of the key exercises that the PDIA materials get participants to do is the so-called '1804 challenge.' The challenge asks participants to imagine themselves in St Louis, in the United States in 1804, when there was no settled knowledge of the route to the West Coast, and to figure out how to find their way to it. The exercise, admittedly US- centric, is used to illustrate how to respond to a situation where the roadmap is not known, which is of course the situation that many face day-to-day and is the whole point of adaptive approaches. Ultimately, whatever specific solution participants come up, the key lesson is 'Action cannot be predefined but must rather emerge through experimental iterations where teams take a step, learn, adapt, and take another step' (Andrews et al. 2015, 130).

A second PDIA exercise is the use of *Ishikawa* or fishbone diagrams (Ishikawa, 1985). This is a deeply practical exercise. A fishbone diagram has the problem identified as the head of the fish and then all possible causes and sub-causes of the problem as the fish bones. For example, if the problem is money lost in service delivery, then the bones could be improper disbursement, inflated procurement costs, corruption, etc. And then off each of these bones are sub-causes/bones, for example, insufficient skills, poor systems, loopholes (Andrews et al. 2017a, 152). The idea is to break down problems from large and intractable ones to manageable, bite-size ones. Deconstructed problems are manageable problems. As recounted above, the purpose is to empower people to define and understand their own problems rather than waiting for an expert to come along! Get small groups to work on their own fishbone diagrams and to compare notes.

When it comes to teaching TWP, one of the main challenges is that it either sounds abstract and general or obvious ('Of course politics matters!'). A good way of making it more compelling is through concrete case studies. One of the best known and well-documented examples of TWP is the Coalitions for Change program in the Philippines (Booth 2014; Sidel 2014; Sidel and Faustino 2020). This makes it particularly well-suited for converting into a classroom setting and allows the instructor to tweak the materials to suit their learning outcomes. The detail provided by such cases is most illuminating. For example, the lesson that building coalitions of unusual suspects can be pretty meaningless, but when they learn how and why British American Tobacco was key to the successful introduction of new taxes on, then the lesson tends to land and get remembered. A Harvard Medical School teaching note provides a succinct and effective account of the reform process providing a perfect set reading (Madore et al. 2015). Another useful teaching aide is the Everyday Political Analysis paper (Hudson et al. 2016). It gives people a really clear and straightforward set of questions to analyse a case with.

For those looking to mix things up in their classroom yet further, there is a particularly exciting example of an escape room being used to convey the key elements of TWP – but which could be used to teach any of the approaches (Roche 2020). The escape room described by Roche (2020) required participants to solve a series of challenges such as using clues to identify the other people in the classroom that they needed to form a coalition with, completing a crossword puzzle based on documents to proxy for doing political analysis, and so forth. The pedagogical key is that the puzzles and exercises should effectively convey the political process. The key pedagogical intervention is the reflection session and a debrief of what actually happened in real life and reinforcing the experiences of the participants in line with the conceptual learning outcomes.

Beyond these exercises, we would recommend the suite of free materials and existing courses provided by the Building State Capability team at Harvard University. These include the excellent Building State Capability book (Andrews et al. 2017a) and the flagship online 10 week course (https://bsc.cid.harvard.edu/online-course).

Conclusion

Adaptive approaches to development, although not new, now enjoy a place within mainstream development discourse and practice that is novel. As this chapter has set out, adaptive development is more of an umbrella of approaches that enjoy a family resemblance but have slightly different emphases, tools, and histories. Adaptive approaches to development share a rejection of linear, technical approaches to development, instead emphasizing the need to learn, adapt, and take politics seriously.

We have surveyed three of the most well known here: adaptive development, problem-driven iterative adaptation (PDIA), and thinking and working politically (TWP). Adaptive management approaches tend to emphasize the importance of structured feedback and learning to guide an informed process of iteration and adaptation. PDIA starts with a focus on identifying the problem from the lived experience of those it affects, not by outsiders, and breaking the problem down into manageable parts before addressing them. TWP recognizes that change happens – or doesn't – because of political interests and influence more than inadequate design or resources.

Of course, in the real world, many initiatives – including the cases introduced here – mix and match elements from the different approaches. In addition, there are many other similar approaches that we haven't been able to cover that also put learning and politics at their heart. But, as one overly long piece concludes, 'If you wish to destroy poverty, prepare to address the power and the politics that keeps people poor. The rest is detail' (Hudson and Leftwich 2014, 108).

Key sources

1. Andrews, M., Pritchett, L. and Woolcock, M. (2012) *'Escaping Capability Traps Through Problem-Driven Iterative Adaptation,'* Center for Global Development, Washington DC.

Link: www.cgdev.org/publication/escaping-capability-traps-through-problem-driven-iterative-adaptation-pdia-working-paper

This paper argues that many developmental reform initiatives fail as a result of emphasizing institutional forms, rather than focusing on the functions they perform in practice. By investing resources in reforming institutional forms, human and organizational capital is diverted from more meaningful reform efforts. The authors propose that complex reform efforts are more likely to be effective if they are centred on specific problems determined by locals and take an experimental approach to finding solutions, and evaluation and feedback from each experimental strategy is rapid and ongoing, enabling adaptation in real time.

2. Booth, D. and Unsworth, S. (2014) *'Politically Smart, Locally Led Development,'* Discussion Paper, Overseas Development Institute, London.

Link: https://odi.org/en/publications/politically-smart-locally-led-development/

This paper examines seven successful development initiatives across multiple sectors and countries that were supported by external donors but fundamentally locally led and politically smart in design and implementation. Booth and Unsworth argue that when locals lead development initiatives, they are more likely to achieve success due to deeper contextual knowledge, networks, credibility, and motivation. This, in turn, enables them to deliver more politically astute programmes that mobilize support, seize opportunities, leverage relationships, and manage opposition.

3. Carothers, R. and de Gramont, D. (2013) *Development Aid Confronts Politics: The Almost Revolution*, Carnegie, Washington, DC.

Link: https://carnegieendowment.org/2013/04/16/development-aid-confronts-politics-almost-revolution-pub-51407

This book places politics at the centre of development. Carothers and de Gramont suggest that long-term, sustainable developmental change that addresses issues from service provision through to anticorruption and beyond is most likely to occur when local civil society and political capacity is strong. They argue that development donors should prioritize supporting interventions that explicitly address politics and advance democratic governance. They chart the growing influence of politically informed approaches to development but also the obstacles that remain.

4. Hudson, D., Mcloughlin, C., Roche, C. and Marquette, H. (2018) '*Inside the black box of political will: 10 years of findings from the Development Leadership Program,'* Birmingham: Developmental Leadership Program.

Link: www.dlprog.org/publications/research-papers/inside-the-black-box-of-political-will-10-years-of-findings-from-the-developmental-leadership-program

This report summarizes the findings from the Developmental Leadership Program, a research program focused on how leadership, power and political processes drive or block successful development. It argues that social change requires developmental leadership – that is, the strategic, collective, and political process of building political will to make change happen. The report concludes that the process of developmental leadership can be carefully supported by external actors if they think and work politically, facilitate effective coalitions, and navigate the politics of legitimacy.

5. Sidel, J.T. and Faustino, J. (2019) *Thinking and Working Politically in Development: Coalitions for Change in the Philippines*, The Asia Foundation, Pasig City

Link: https://asiafoundation.org/publication/thinking-and-working-politically-in-development-coalitions-for-change-in-the-philippines/

This open access book offers a detailed case study of Thinking and Working Politically in practice. Sidel and Faustino document the history, processes, and lessons of the Coalitions for Change program in the Philippines, which has a strong track record of successfully supporting diverse developmental reform efforts. The book is an excellent example of how adaptive development approaches that are politically astute can achieve significant change.

Note

1 Refer to Box 1: apolitical in this context refers to being unattentive to, or inconsiderate of, how aid impacts local distribution of power and resources.

References

Andrews, M., Pritchett, L., Samji, S. and Woolcock, M. (2015) 'Building Capability by Delivering Results: Putting Problem-Driven Iterative Adaptation (PDIA) Principles into Practice', in A Whaites, et al. (eds), *A Governance Practitioner's Notebook: Alternative Ideas and Approaches*, OECD, Paris, pp. 123–133.
Andrews, M., Pritchett, L. and Woolcock, M. (2012) '*Escaping Capability Traps Through Problem-Driven Iterative Adaptation,'* Center for Global Development, Washington DC.

Andrews, M., Pritchett, L. and Woolcock, M. (2017a) *Building state capability: Evidence, analysis, action*, Oxford: Oxford University Press.

Andrews, M., Ariyasinghe, D., Britto, K., Harrington, P., Kumaratunga, N., Lawrance, M.K.D., McNaught, T., Naotunna, H., Palaketiya, G., Poobalan, A., Samarasinghe, D. and Wijayathilake, P. (2017b) '*Learning to Engage New Investors for Economic Diversification: PDIA in action in Sri Lanka*,' Cambridge, MA: Center for International Development, Harvard University.

Andrews, M., Ariyasinghe, D., Beling, A.S., Harrington, P., McNaught, T., Niyas, F.N., Poobalan, A., Ramanayake, M., Senavirathne, H., Sirigampala, U., Weerakone, R.M. and Wijesooriya, W.A.F.J. (2017c) '*Learning to Improve the Investment Climate for Economic Diversification: PDIA in action in Sri Lanka*,' Cambridge, MA: Center for International Development, Harvard University.

Andrews, M., Ariyasinghe, D., Batuwanthudawa, T., Darmasiri, S., de Silva, N., Harrington, P., Jayasinghe, P., Jayasinghe, U., Jayathilake, G., Karunaratne, J., Katugampala, L., Liyanapathirane, J., Malalgoda, C., McNaught, T., Poobalan, A., Ratnasekera, S., Samaraweera, P., Saumya, E., Stock, D., Senerath, U., Sibera, R., Walpita, I. and Wijesinghe, S. (2017d) '*Learning to Target for Economic Diversification: PDIA in Sri Lanka*,' Cambridge, MA: Center for International Development, Harvard University.

Bond (2016) '*Adaptive management: What it means for CSOs*', London: Bond.

Booth, D. (2014) '*Aiding Institutional Reform in Developing Countries: Lessons from the Philippines on What Works, What Doesn't and Why*,' London: The Asia Foundation, San Francisco; Overseas Development Institute.

Booth, D. and Unsworth, S. (2014) '*Politically Smart, Locally Led Development*,' Discussion Paper, London: Overseas Development Institute.

Carothers, R. and de Gramont, D. (2013) *Development Aid Confronts Politics: The Almost Revolution*, Washington, DC: Carnegie.

Cooke, B. and Dar, S. (2008) 'Introduction: The New Development Management,' in B. Cooke and S. Dar (eds) *The New Development Management*, London: Zed Books, 1–17.

Craney, A. (2020) 'Local participation or elite capture in sheep's clothing? A conundrum of locally led development,' *Politics and Governance*, 8(4), 191–200.

Craney, A. and Hudson, D. (2020) 'Navigating the dilemmas of politically smart, locally led development: the Pacific-based Green Growth Leaders' Coalition', *Third World Quarterly*, 41(10), 1653–1669.

Denney, L. and McLaren, R. (2016) 'Thinking and Working Politically to Support Developmental Leadership and Coalitions: The Pacific Leadership Program', DLP Research Paper 41, Birmingham: Developmental Leadership Program.

Derbyshire, H., Siow, O., Gibson, S., Hudson, D. and Roche, C. (2018) '*From Siloes to Synergy: Learning from politically informed, gender aware programs*,' DLP Gender and Politics in Practice Research Paper, Birmingham: Developmental Leadership Program.

DLP (2012) 'Coalitions in the politics of development: Findings, insights and guidance from the DLP Coalitions Workshop, Sydney, 15–16 February 2012', Birmingham: Developmental Leadership Program.

Ferguson, J. (1990). *The Anti-Politics Machine: 'Development,' Depoliticization and Bureaucratic Power in Lesotho*. Cambridge: Cambridge University Press.

Green, D. (2019) '*How does Coalitions for Change in the Philippines Compare with other Adaptive Management Programmes?*', From Poverty to Power, Oxfam Great Britain.

Honig, D. (2018) *Navigation by Judgment: Why and when top down management of foreign aid doesn't work*, Oxford: Oxford University Press.

Honig, D. (2020) 'Actually Navigating by Judgment: Towards a New Paradigm of Donor Accountability Where the Current System Doesn't Work', CGD Policy Paper 169, Washington DC: Center for Global Development.

Honig, D. and Pritchett, L. (2019) 'The Limits of Accounting-Based Accountability in Education (and Far Beyond): Why More Accounting Will Rarely Solve Accountability Problems,' CGD Working Paper 510, Washington DC: Center for Global Development.

Hout, W. and Robison, R. (2008) 'Development and the Politics of Governance: Framework for analysis,' in W. Hout and R. Robison (eds) *Governance and the Depoliticisation of Development*, London: Routledge, 1–12.

Hudson, D. and Leftwich, A. (2014) 'From Political Economy to Political Analysis,' Research Paper 25, Birmingham: Developmental Leadership Program.

Hudson, D., Marquette, H. and Waldock, S. (2016) 'Everyday political analysis,' Birmingham: Developmental Leadership Program.

Hudson, D., Mcloughlin, C., Roche, C. and Marquette, H. (2018). 'Inside the black box of political will: 10 years of findings from the Development Leadership Program,' Birmingham: Developmental Leadership Program.

Ishikawa, K. (1985) *What Is Total Quality Control? The Japanese Way*. Translated by Lu, D.J., Englewood Cliffs, NJ: Prentice-Hall.

Leftwich, A. (2000) *States of Development: On the primacy of politics in development,* London: Polity Press.

Leftwich, A. (2004) *What is Politics? The activity and its study*, London: Polity Press.

Madore, A., Rosenberg, J. and Weintraub, R. (2015) 'Sin Taxes and Health Financing in the Philippines,' Module Note for Cases in Global Health. Boston, MA: The President and Fellows of Harvard College, Online at https://hbsp.harvard.edu/product/GHD030-PDF-ENG

Natsios, A. (2010) 'The Clash of the Counter-bureaucracy and Development,' Center for Global Development Essay, July 2010, Online at https://files.ethz.ch/isn/118483/file_Natsios_Counter bureaucracy.pdf

ODI (2016), 'Doing Development Differently: two years on, what have we done?' Overseas Development Institute, Online at https://odi.org/en/publications/doing-development-differently-two-years-on-what-have-we-done/

O'Neil, T. (2016) 'Using adaptive development to support feminist action,' London: Overseas Development Institute.

Pett, J. (2020) 'Navigating adaptive approaches for development programmes: A guide for the uncertain,' ODI Working Paper 589, London: Overseas Development Institute.

Roche, C. (2020) 'Escape the room: Take the facts with you!,' Developmental Leadership Program, Online at https://dlprog.org/opinions/escape-the-room-take-the-facts-with-you

Roche, C., Cox, J., Rokotuibau, M., Tawake, P. and Smith, Y. (2020) 'The characteristics of locally led development in the Pacific,' *Politics and Governance*, 8(4), 136–146.

Rousseau, B. and Kenneth-Watson, D. (2018) 'Supporting coalition-based reform in Vanuatu,' Research Paper 51, Birmingham: Developmental Leadership Program.

Sidel, J.T. (2014) 'Achieving Reforms in Oligarchical Democracies: The Role of Leadership and Coalitions in the Philippines,' Birmingham: Developmental Leadership Program.

Sidel, J.T. and Faustino, J. (2019) *Thinking and Working Politically in Development: Coalitions for Change in the Philippines*, Pasig City: The Asia Foundation.

Sims, K. (2021) 'Risk navigation for Thinking and Working Politically: The work and disappearance of Sombath Somphone,' *Development Policy Review*, 39(4), 604–620.

Suti, E., Hoatson, L., Tafunai, A. and Cox, J. (2021) 'Livelihoods, leadership, linkages and locality', *Asia Pacific Viewpoint*, 62(1), 15–26.

TAF (2017) 'Coalitions for Change,' Manila: The Asia Foundation.

58

SOUTHERN RESEARCH METHODOLOGIES FOR DEVELOPMENT

Johanna Brugman Alvarez and Leakhana Kol

Introduction

'Seeing from the South' (Watson 2009) in urban studies remains a crucial task because bodies of urban knowledge in this part of the world have been detached from place, history, and identity. It is increasingly common that southern cities borrow research and planning expertise to address local problems, usually embedded within the dominant development paradigm that the north replicates (Patel, Greyling, Parnell and Pirie 2015). Also, 'worlding practices' (Roy and Ong 2011) are circulated globally and have become increasingly attractive to politicians, planners, architects, urban designers and citizens alike. These mobile planning knowledges have had negative effects on places where they have been promoted and implemented. Watson (2014) argues that in the poor, weakly governed, rapidly growing, largely informal, and often culturally different cities of the Global South, such planning ideas have promoted social exclusion and spatial inequalities and have been insensitive to sustainability issues.

Despite the validity of these arguments, there are divergent views on whether a claim for southern urban knowledge is beneficial for urban studies, as well as the meaning of southern urban knowledge itself. Roy (2014) explains that the intent to craft new geographies of theory that can draw upon the urban experience of the Global South is not simply to study Southern cities because of being interesting or different, but to re-calibrate urban theory. This is needed because our conceptions of 'the city' are often premised on the experiences and theoretical work based upon cities in Western Europe and North America (McFarlane 2010). However, the idea of 'southern urban knowledge' is problematic because it perpetuates a North-South binary which not only reproduces colonial power relationships, but also fails to acknowledge other types of urban contexts such as settler colonial cities (Blatman-Thomas and Porter 2018). In this light, claims have been made for comparative urban research across the North-South divide to allow any city in the world to be a place from which urban life can be theorized (Robinson 2006, 2016; McFarlane 2010).

Despite the diverse views on the topic, there is a strong call from urban theorists in the South to reposition the point of departure when generating urban knowledge in which the detail and experience of cities of the South offers the basis for learning, instead of abstract rules, frameworks or theories derived from Northern contexts (Oldfield and Parnell 2014; Pieterse 2010; Watson 2009). Furthermore, there are moves to consider how to bring urban experiences

DOI: 10.4324/9781003017653-64

and knowledges *beyond academia* into urban theory to produce more democratic urban studies characterized by diverse urban epistemes and imaginaries (Allen and Lambert 2015; McFarlane 2010, 2017). This repositioning foregrounds thinking critically and creatively about appropriate methodologies that allow the construction of 'a view from the South' in urban research.

Oral history in the form of place-based storytelling has been identified as a productive methodology for this task. This is because of stories' capacity to capture and convey detailed and contextualized knowledge about a particular place over time and provide intimate narratives about individual and collective experiences attached to place across generations. Furthermore, stories are effective as a strategy to include non-specialist knowledges in urban research, not just from research participants but also local practitioners. Thus stories facilitate an empirical engagement with the ordinary and everyday life that constitute 'cityness' in the Global South (Pieterse 2010) and are able to generate solidarity and empathy with the different struggles encountered by people and professionals living in these places. Overall, oral history helps make explicit the lived experiences and embodiment of place-based knowledge in Southern contexts and can be used to challenge the universality of worlding practices in urban planning. This facilitates the construction of bodies of urban knowledge based on careful empirical analysis of the realities of place and a critical reflection on how it has come to be that way (Duminy, Watson and Odendaal 2014).

In this chapter we discuss oral history in the form of storytelling. This research methodology unlocks 'a view from the South' in ways that other approaches do not and contributes to planning and urban knowledge based on a research project on urban informality conducted in Phnom Penh, Cambodia in 2016. We discuss the strengths of this method in making explicit the practices that emerge from social, spatial, and temporal relationships between people and place in the consolidation of informal settlements. We also explain the capacity of this methodology to recognize the marginal voices of the urban poor and their meaning for the city, and support the political struggle of urban poor groups in Phnom Penh by enabling a learning process between various actors in the city. Ultimately, we reflect on the potential of oral history to allow researchers to experience the research process differently and reimagine the academic project towards developing more socially just approaches for urban learning embedded in the realities of Southern cities. While the chapter seeks to provide understandings on how oral history can provide a 'view from the South,' the intention is not to perpetuate the division between 'Global North' and 'Global South.' Instead the chapter highlights the importance of taking a place-based approach to the generation of knowledge in urban studies; one that is embedded in the realities of place and history.

Seeing from the South through oral history in urban studies

Using oral history as a research methodology is particularly relevant in many contexts of the Global South, where storytelling is a cultural practice and a fundamental form of meaning making and 'passing on' of knowledge and experience (Duminy, Odendaal, Watson 2014; Flyvbjerg 2011). Oral history allows for the embodiment of place-based knowledge over time by capturing the lived experiences and memories of research participants, as well as the various connections between people and place across generations. For instance, stories told across generations in a particular place can make visible what is often invisible and allow for understanding of how communities and individuals are shaped and the issues that affect them. This type of understanding is important in rapidly developing cities where the individual and collective memory of people and place are being lost. Here the use of oral history to recover and maintain this memory is fundamental to rescue identity and sense of belonging of citizens to the territory and to rescue the emerging city from becoming an urban space defined by economic growth and exchange value (Golda-Pongratz 2016).

Furthermore, oral history is appropriate for researchers to engage ethically with marginalized groups as placed-based stories foreground participant's knowledges, skills and experiences as they become active subjects rather than passive objects in the research (Scheyvens, Scheyvens and Murray 2011). Stories invite research participants to express their feelings and life trajectory, the decisions they make and what unites and distance themselves from one another. Thus, the use of this research method is highly appropriate; adding non-specialist knowledges thus challenging existing historical narratives that actively exclude particular knowledges from the research process and outcomes (Gosseye 2019). This opens the transformative potential of oral storytelling to account for those voices that are typically unheard, repressed and erased in cities and make their knowledge visible and recognized.

Oral history allows researchers to focus on learning from real-life experience rather than relying on hypotheses or abstract theories. Also, the methodological flexibility of oral history can support researchers in navigating difficult practical, ethical, and personal challenges when working in Southern contexts, especially if the research is designed based on the rules of Northern universities and institutions. Thus, oral history can support researchers to communicate complex information from the research to broader audiences and transform research results in forms of activism ensuring the research process is not only a self-serving exercise (Greenop 2019). Furthermore, using oral history with the intention to learn from the depth of experience of Southern contexts and people can lead to research processes where relationships between participants and researchers, as well as local and international co-researchers, are developed and nurtured. These relationships enable social learning and an empathetic production of knowledge, different from an objective, instrumental and extractive research process.

In our research, we used oral history to understand formal and informal relationships in informal settlement upgrading practices and their contribution to planning knowledge and social and spatial justice. We undertook fieldwork in one informal settlement excluded from systematic land registration to uncover systemic problems with the management of public land, including lake in-filling practices and the exclusion of informal settlements from the land registry. Our research evidenced the importance of collective action as a source of power and support for the urban poor, and the limitations of political recognition under market-led formalization programmes in contexts of rapid urbanization and power inequalities. In the next section we focus on our experience using oral history in the form of collecting the life stories of residents of one informal settlement. We explain the contribution of these research methods to generating 'a view from the South' in Phnom Penh and enabling more socially just approaches for urban learning.

This research shows the benefits of establishing relationships and collaboration between a local researcher (Leakhana) and an international researcher (Johanna). Our collaboration was formed out of work experiences and social networks established while Johanna lived and worked in Phnom Penh in 2013. From this past experience it became clear that in order to conduct responsible and sensitive research in this Global South context, there was a need to establish a collaboration with a local researcher and the research needs to be informed and guided by local knowledge. The research was designed in consultation with Leakhana and local NGOs who helped to establish the necessary relationships to gain access to the location, build trust and confidence among the participants of the study, and properly inform on the historical, cultural, social, political, and economic context of Phnom Penh. With these collaborative processes underway, the research was guided by local knowledges and contexts as it unfolded at its own pace based on local conditions and circumstances which at times meant the modification of research practices. This required flexibility and adaptability to uncertainty, sensitive communication, respect for local culture and persistence. This collaboration continues and is enriched by sharing knowledge and friendship as well as our passion for producing quality research outcomes that can benefit the situation of the urban poor in Phnom Penh. The research process and outcomes explained in the following section show the great value of establishing partnerships and sharing knowledge between international and local researchers and the need for these partnerships to gain 'a view from the south' in urban research.

Learning from people and place in one informal settlement in Phnom Penh

The settlement we studied, Phka (as we researchers name it to ensure confidentiality) is located within ten kilometres of Phnom Penh's city centre and close to public and private services. After the fall of the Khmer Rouge, the Ministry of Defence settled soldiers and their families in Phka. During the late 1980s the area experienced an influx of people returning from rural and refugee camps after the civil war. In 2016, the settlement was home to 48 households. Because of its attractive location, newcomers had arrived in the area, such as people living in central Phnom Penh and a number of rural migrants. Thus, residents of Phka had diverse socio-economic backgrounds. The level of poverty that people experienced in Phka had decreased over time due to the benefits that some families had been able to gain from urban development. The research was particularly focused on collecting stories from the original settlers in Phka who had experienced these changes.

The process of collecting stories involved visiting each household and having an informal conversation with the residents about their life. Through this initial conversation we explained the purpose of our research and why it was important for us to learn from each person's experience. We then started asking questions about people's life trajectory and this motivated residents to tell us stories about their past and present experiences. This involved talking about the changes they had experienced over time as a result of various investments in their houses, livelihoods, and community as well as their feelings, fears, and future life expectations. The nature of this process was highly informal and flexible, allowing us to prioritize building a relationship with our participants rather than simply collecting information for our research.

The stories were not recorded; instead, notes were taken and typed into a computer. We decided not to record the audio of the stories following advice from community leaders. This allowed residents of Phka to feel more comfortable to share information with us, as well as building trust and relationships. This decision prioritized the participant's feelings, rather than our own interests as researchers. We also ensured confidentiality of all participants and provided open information about the research. For this, consent and information forms were developed and translated into Khmer. However, in most cases, rather than asking participants to sign the form we read or explained it to them. This happened as most people felt uncomfortable signing the form. We also found that asking residents to sign a paper at the first encounter damaged opportunities to build trust and made them feel insecure in case information shared was disclosed to government authorities or other parties. This shows some of the complexities discussed in the literature of translating research guidelines originated in a Northern University to the Global South, and the need for researchers to be flexible, make compromises and take action to make the research process culturally appropriate and sensitive to the characteristics of people and place (Scheyvens, Scheyvens and Murray 2011).

Figure 58.1 Phka in the 1980s. Photograph provided by community leaders of Phka

The stories that we collected from residents uncovered the memories from the early days of settlement and provided understanding of the deep relationships that residents had with their place of residence through an account of their incremental individual and collective development practices. In the past, Phka was a wetland surrounding a lake as seen in Figure 58.1. Residents arrived in Phka to find a place to live and rebuild their lives after years of suffering under the Khmer Rouge regime and having to live in refugee and rural camps. The stories we collected showed that in Phka, 'place' was consolidated out of the struggle of many of its residents to rebuild their lives after the civil war and overcome poverty and precarious living conditions. Residents expressed how they found peace and security in Phka after the Pol Pot regime, and how they worked by themselves, with their families and neighbours, to rebuild their lives.

Residents particularly expressed a deep relationship with their house, as a place which had grown together with their needs and families. Houses started as a basic wooden structure but over time had been upgraded to a two-storey house which maintained the original wooden structure in the top as seen in Figure 58.2. Housing was an important source of economic livelihood. Many residents, especially women had home-based businesses such as shops and sewing enterprises and had developed extra rooms to rent out to migrants coming from other Cambodian provinces. Also, most young people lived and relied on their parents' land and housing to have a place to live and make a livelihood. Residents' stories recognized collective development practices among neighbours as the principal mechanism to negotiate power relationships with government officials and wealthy residents monopolizing the access to basic services to address their basic needs in the early days of settlement. These practices included developing a community fund, negotiating connections to water and electricity, and building common infrastructure such as an access road as seen in Figure 58.3. The access road was particularly important as it was used as public space where children played, residents gathered to

Figure 58.2   An example of a house developed incrementally by residents in Phka. Photograph by Johanna Brugman

chat in the afternoon, and occasional celebrations were organized. This piece of infrastructure enhanced social ties between residents and nurtured feelings of belonging and sense of place in Phka.

Stories uncovered the way individual and collective development practices born out of residents' struggle transformed this settlement into a place where a sense of belonging was constructed. As other studies of informal settlements in the Global South have shown, these feelings allowed for the construction of plural meanings between people and the different elements constituting place in informal settlements including land, housing, infrastructure and community (Lombard 2014). These meanings transcend the commercial value of land and housing which characterizes Phnom Penh today, making explicit the social relationships that are attached to the urban space and the embodiment of place in Phka. Furthermore, the stories were able to capture how these practices evolved over time supporting an incremental development process. The incremental nature of development in informal settlements in the Global South such as Phka is a response to exclusions experienced from formal planning processes and is recognized as a central process of urban life in contexts of extreme inequality (McFarlane

Figure 58.3   Access road built by residents of Phka (Source: Community leaders of Phka)

2017). Stories allowed for the deep understanding of the tangible and intangible outcomes of incremental development practices in Phka. In other words, storytelling created knowledge with the potential to enrich the responses driven by built environment professionals to address the issues experienced by informal settlements in Phnom Penh, and rather than relying on 'best practice' recipes added contextual, place-based knowledges of Southern realities.

Residents of Phka did not have formal land titles but possession rights over their land. Possession rights gave Cambodian citizens the opportunity to claim land ownership if they had been in possession of their land prior to the passing of the law in August 2001. For residents of Phka possession rights gave them a higher level of tenure security compared to other informal settlements in Phnom Penh where residents did not have the legal basis to claim land ownership. However, possession rights positioned Phka residents in a 'grey space' between the whiteness of legality and the blackness of eviction (Yiftachel 2009). This happened because the claim for ownership was subject to the state's power and decision-making under the Land Management and Administration Programme (LMAP). The LMAP followed the worldwide pattern of formalizing land through the large-scale distribution of land titles. This was done through Systematic Land Registration (SLR). Access to legal title as a form of tenure security had been a struggle for residents of Phka since the settlement was excluded from SLR in 2007. The exclusion was caused by the interest of the government in two sites of public state land adjoining the settlement, and the capacity of the Cambodian state to manipulate land legal frameworks to satisfy development interests.

Residents told us that they built on their collective spirit developed throughout the consolidation of the settlement to scale-up their collective responses to the threats of eviction they experienced when the settlement was excluded from SLR by government authorities. This follows the experiences of other communities in Phnom Penh and the Global South who use collective action to build internal power and secure their rights to housing, land and citizenship in cities (Herrle, Ley and Fokdal 2015). Residents of Phka built power by making alliances with other informal communities facing exclusion from SLR, non-governmental organizations and international donors in Phnom Penh funding land rights programmes. Residents also produced legal, demographic, and spatial information to build a case to contest their exclusion. These strategies were used to make their case visible and gain recognition from the state by attending public forums and other events organized by civil society organizations. After seven years of advocating their exclusion case residents of Phka obtained SLR and land title thus demonstrating the power of collective action.

We found that the stories about residents' struggles during the consolidation of the settlement and their collective practices, were used by community leaders as a principal strategy to present their case to government authorities and make arguments to obtain land title. In this light, place-based stories acted as a political strategy to gain visibility and recognition from the state. Also, stories were the main means used by community leaders of Phka to share their knowledge with other informal communities in Phnom Penh. This included the methods used to produce demographic and spatial information, and the facilitation techniques to motivate community members to be involved in collective activities. As researchers attentive to these place-based stories and the power of storytelling, we contributed to the legitimation of these individual and collective stories of residents in documents form that can be used by community leaders, residents, and NGOs to continue to advocate for urban justice in informal settlements in Phnom Penh.

This is important because studies conducted in Phnom Penh show that informal residents have been forcibly evicted and displaced over recent years by the state and the real estate market (STT 2018). With forced evictions a loss of individual and collective memory and places has occurred, threatening the belonging and citizenship rights of people living in informal settlements. Forced evictions of informal residents are exacerbated by Phnom Penh experiencing one of the highest urbanization rates in Southeast Asia, and carrying within it the scars left by civil war and the legacies of the Khmer Rouge regime, such as a form of 'disjointed governance' (Paling 2012) in which state-private informal alliances and relationships bypass urban legislation threatening the rights of urban poor communities. This highlights the importance of using research for advocacy purposes in this context and the political agency within case study research and oral history.

Also, through the stories we found that older people who had lived in Phka for a long time and had gone through the struggle to obtain services, built a place and community and advocate for land title, and valued their land and community differently to younger generations. Older people that had experienced the trauma of being a refugee in the Cambodia-Thai border camps during the war, had a stronger sense of belonging and attachment to place in Phka. This led to an understanding of the value of keeping the land for the future and ensuring their family wellbeing over time. In contrast, most young people in Phka saw land as a profitable asset, a view that presented risks to the security of tenure of families, especially within Phnom Penh's urban environment experiencing gentrification and fast transformations. Older women, particularly, were aware and concerned about this pattern and thus reluctant to give control of their land to their children. These accounts revealed the importance of stories and place-based knowledges in understand the connections to place that residents developed through their involvement in the consolidation of the settlement, particularly important in the face of the changes in generational values in society. Recovering the memory of struggle and connection to place through the stories raised awareness of these connections within the younger generations and opened possibilities for a different perception of the city that respects its past and struggles to urban justice.

Conclusion

This chapter reflected on how oral history can generate 'a view from the South' in urban studies. The chapter highlights the importance of constructing knowledge embedded in the specificities of Southern cities with the aim not to perpetuate a North–South division; but instead emphasize the importance of taking a locally specific, place-based approach to the generation of knowledge in urban studies embedded in the realities of oral histories. This form of knowledge production is particularly important for the Global South in order to move beyond colonial power dynamics leading urban development practices based on abstract theories, 'best practice' and/or external frameworks. Despite this, this form of knowledge production is also relevant for a variety of urban contexts such as the Global North and settler colonial cities.

The accounts presented in the chapter are useful for researchers and practitioners alike as these show how oral history can be used as a tool to understand more deeply the realities of Southern cities and their inhabitants. We found storytelling a useful methodology at making explicit the embodiment of place-based knowledge over time and showing the intimate relationships between residents of informal settlements and their built environment. In the context of urban planning, this detailed and locally informed understanding is important to generate appropriate planning, legal and design responses to the complex realities of Southern cities such as urban informality while moving away from following 'best practice' models (see Devi and Nakamura, Chapter 43 in *this volume*). Lastly oral history can allow researchers and practitioners to experience the knowledge production process differently. Stories permitted us to have a close connection to our participants and learn about their lives, hopes, sadness, and joy. These emotional attachments created a relationship among us and the participants which deepened mutual respect throughout the research process. In this way stories can be transformative for urban research and practice by changing the production of knowledge from being extractive to prioritizing social relationships, and respect with all those involved. We also prioritized establishing long-lasting relationships and collaboration between us as researchers, reflecting on the value of sharing knowledge between international and local researchers and the importance of this relationship to contribute towards shaping a 'view from the South' in research processes. Overall, our experience shows how place-based oral history is a useful method for urban research, with the potential to generate more socially just approaches to urban learning embedded in Southern contexts.

Key sources

1. Allen, A, Lambert, R. and Yap, C. (2017). Co-learning the city: Towards a Pedagogy of poly-learning and planning praxis. In Bhan, G, Srinivas, S and Watson, V (eds). *The Routledge Companion to Planning in the Global South*, pp. 355–367, Great Britain: Routledge.

This chapter offers critical reflections on methodological and epistemic conditions underpinning a pedagogical approach in Lima, Peru able to identify strategic opportunities for transformative change through the strategic inclusion of a plurality of knowledges in the city.

2. Chambers, R. (1997) Professional Realities. In Chambers, R. *Whose reality counts? Putting the first last*. London: ITDG Publishing/ Practical Action Publishing.

This chapter offers critical reflections on the role of professionals in the process of knowledge production in development studies.

3. Gosseye, J., Stead, N. and Van der Plaat, D. (Eds). (2019) *Speaking of Buildings: Oral history in Architectural Research*. New York: Princeton University Press.

The chapters in this book discuss the role of oral history in including marginalized knowledges and stories in architecture research.

4. Luansang, C., Boonmahathanakorn, S. and Domingo-Price, M. (2012). The role of community architects in upgrading; reflecting on the experience in Asia. *Environment and Urbanization*, vol. 24, no. 2, pp. 497–512.

This paper discusses participatory methodologies and approaches used by community architects for including the knowledges of urban poor communities in informal settlements upgrading in Asia.

5. Osuteye, E., Ortiz, C., Lipietz, B., Castan Broto, V., Johnson, C. and Kombe, W. (2019) Knowledge co-production for urban equality. *KNOW Working Paper Series*, No 1, May 2019.

This paper discusses the theoretical basis for the co-production of knowledge in urban research in the Global South.

References

Allen, A. and Lambert, R. (2015) Learning through mapping. In Alfaro, P, Campkin, B, Gupte, R, Mkhabela, S, Novy, J. and Savela, M. (eds) *Urban Pamphleteer*, pp. 40–42, London: UCL Urban Laboratory.

Blatman-Thomas, N. and Porter, L. (2018) Placing Property: Theorizing the Urban from Settler Colonial Cities. *International Journal of Urban and Regional Research*, vol. 43, no. 1, pp. 30–44.

Duminy, J., Odendaal, N. and Watson, V. (2014) Case study research in Africa: methodological dimensions. Duminy, J., Andreasen, J., Odensaal, N., and Watson, V. (Eds) *Planning and the case study method in Africa: the planner in dirty shoes*. Hampshire: Palgrave Macmillan.

Flyvbjerg, B. (2011) Case Study. Denzin, N.K. and Lincoln, Y.S. (eds) *The Sage Handbook of Qualitative Research*, 4th edn, Thousand Oaks, CA: Sage.

Golda-Pongratz, K. (2016) Urban Memory – Palimpsests, traces and demarcations in Metropolitan Lima. *Trialog*, vol. 118/119, pp. 4–18.

Gosseye, J. (2019) A Short History of Silence: The Epistemological Politics of Architectural Historiography. Gosseye, J., Stead, N. and Van der Plaat, D. (Eds). *Speaking of Buildings: Oral history in Architectural Research*. New York: Princeton University Press.

Greenop, K. (2019) Taking my place/talking your place: race, research and Indigenous architectural history. Gosseye, J., Stead, N. and Van der Plaat, D. (Eds) *Speaking of Buildings: Oral history in Architectural Research*. New York: Princeton University Press.

Herrle, P., Ley, A. and Fokdal, J. (2015) *From Local Action to Global Networks: Housing the Urban Poor*. London, New York: Routledge.

Lombard, M. (2014) Constructing ordinary places: place-making in urban informal settlements in Mexico. *Progress in Planning*, vol. 94, pp. 1–54.

McFarlane, C. (2010) The comparative city: knowledge, learning and urbanism. *International Journal of Urban and Regional Research*, vol. 34, no. 4, pp. 725–742.

McFarlane, C. (2017) Learning from the city: a politics of urban learning in planning. Bhan, G., Srinivas, S. and Watson, V. (Eds) *The Routledge Companion to Planning in the Global South*, pp. 323–333, London: Routledge.

Oldfield, S. and Parnell, S. (2014) From the South. Parnell, S and Oldfield, S. (Eds) *The Routledge Handbook on Cities of the Global South*, pp. 1–4, London: Routledge.

Paling, W. (2012) Planning a future for Phnom Penh: Mega projects, aid dependence and disjointed governance. *Urban Studies*, vol. 49, no. 13, pp. 2889–2912.

Patel, Z., Greyling, S., Parnell, S. and Pirie, G. (2015) Co-producing urban knowledge: Experimenting with alternatives to 'best practice' for Cape Town, South Africa. *International Development Planning Review,* vol. 37, no. 2, pp. 187–203.

Pieterse, E. (2010). Cityness and African Urban Development. *Urban Forum*, vol. 21, pp. 2015–2019.

Robinson, J. (2006) *Ordinary cities: between modernity and development*. London: Routledge.

Robinson, J. (2016) Thinking cities through elsewhere: Comparative tactics for a more global urban studies. *Progress in Human Geography*, vol. 40, no. 1, pp. 3–29.

Roy, A. (2014) Worlding the South: Toward a post-colonial urban theory. Parnell, S and Oldfield, S. (Eds) *The Routledge Handbook on Cities of the Global South*, pp. 9–20, London: Routledge.

Roy, A and Ong, A. (2011) *Worlding Cities: Asian Experiments and the Art of being Global*. Malden, MA: Blackwell.

Scheyvens, R., Scheyvens, H. and Murray, W. E. (2011) Working with marginalized, vulnerable or privileged groups. *Development Fieldwork*. London: Sage.

Sahmakum Teang Tnaut (STT). (2018) *The Phnom Penh Survey: A study on urban poor settlement in Phnom Penh*. STT: Phnom Penh.

Watson, V. (2009) Seeing from the South: Refocusing urban planning on the globe's central urban issues. *Urban Studies,* vol. 46, no. 11, pp. 2259–2275.

Watson, V. (2014) Learning planning from the South: Ideas from the new urban frontiers. Parnell, S and Oldfield, S. (Eds). *The Routledge Handbook on Cities of the Global South*, pp. 98–108, London: Routledge.

Yiftachel, O. (2009) Theoretical Notes On 'Gray Cities': The Coming of Urban Apartheid? *Planning Theory*, vol. 8, pp. 88–100.

59

COMMUNITY ECONOMIES

Jenny Cameron and Isaac Lyne

Introduction

A community economies approach is concerned with creating post-capitalist worlds.[1] In this approach, these worlds are understood as already existing. In place of the refrain 'another world is possible,' a community economies approach says 'another world is already here' – but these worlds are hard to detect because of the dominant Western way of understanding economies and societies. Thus, a community economies approach aims to seek out and strengthen already existing post-capitalist worlds. This involves community economies scholars using action research methods to work with community-based partners to help make post-capitalist activities more visible, and then to devise ways and means to build on and strengthen these activities. In this chapter, we demonstrate this community economies approach by discussing three development projects in the Asia-Pacific region. These projects are characterized by attentiveness to local conditions and to local values and aspirations. Thus, a community economies approach to doing development differently starts by acknowledging the local context and valuing the diverse economic activities and possibilities that are already present.

The economic development pathways that result from a community economies approach emphasize the interconnections and interdependencies between people and between people and environments. This differs from mainstream development pathways which generally are based on the Western assumption of individuated and economically rational actors disconnected not just from other people but from the environments around. As we discuss in this chapter, sometimes this means acknowledging, valuing and tracking those things that are already important to local communities, sometimes this means taking steps to actively foster community-based economic initiatives, and sometimes what is important is to 'leave the villagers alone' (Somsak 2005 cited in McKinnon 2017, 344). Such an approach has crucial implications for policy and practice, and for pedagogy. Before discussing the three projects and their implications, it is important to clarify what is meant by post-capitalist worlds in a community economies approach.

Post-capitalist worlds

For some, post-capitalism is understood in temporal terms as an economic phase that follows capitalism (e.g. Mason 2016). This is not the understanding of a community economies approach.

DOI: 10.4324/9781003017653-65

In a community economies approach, post-capitalism refers to a way of thinking about economies and societies (with implications for policy and practice, and pedagogy). Rather than understanding the world as dominated by one form of economy (capitalism) which might be replaced by another (post-capitalism), a community economies approach understands the world as being comprised of multiple and diverse economic activities (including capitalist activities, but also non-capitalist and even alternative capitalist activities).[2] One analogy used in a community economies approach is that of the iceberg. Visible above the waterline are a relatively small set of economic activities associated with capitalism (capitalist firms employing paid workers producing goods and services that are transacted through markets); but hidden below the waterline are an immense array of economic activities. Post-capitalism refers to a shift in thinking – from thinking that the economy is only those set of activities above the waterline to thinking that the economy is vast and diverse.

Thinking in this way is crucial to a community economies approach as it helps to shed light on the economic possibilities that already exist in the world (but are hidden by blinkered economic thinking). In particular, a community economies approach is interested in those diverse economic activities that contribute to wellbeing of both people and the planet. These are the economic activities that help people to survive well; to produce and distribute surplus in ways that take into account the wellbeing of other people and the planet; to transact goods and services more fairly; to care for shared resources; and to invest in ways that will support a better future (Gibson-Graham et al. 2013). Thus, a community economies approach involves identifying and acknowledging the economic activities that contribute to wellbeing and considering ways that these activities might be strengthened and multiplied. In what follows, we discuss three projects that illustrate various aspects of a community economies approach. In these sections, we include discussion of the implications for policy and practice. In the penultimate section we consider the overall implications for pedagogy.

Economic development pathway 1: acknowledging and valuing community economic priorities

The first project was based in Fiji and the Solomon Islands and here community economies scholars worked with the NGO International Women's Development Agency (IWDA) and their local NGO partners. These NGO partners were looking to develop community-based indicators to be able to track the gendered impacts of economic social and environmental changes, especially in settings where subsistence and semi-subsistence livelihoods are being impacted by growth of export-oriented agriculture, spread of a cash economy and rapid urbanization (McKinnon et al. 2016). These current changes are compounding the gendered impacts of colonization in the late nineteenth century on Solomon Islanders and iTaukei (Indigenous) Fijians. For example, through colonization, women's traditional work was devalued and women's input into community-level decisions eroded; meanwhile, men 'gained status, power and recognition in the cash economy' (McKinnon et al. 2016, 1378).

This project sought to develop indicators of the gendered impacts of change that incorporated community-level perspectives. This meant tapping into what mattered for community members, including what their experiences had been to date and what their aspirations were. It also meant putting to one side Western assumptions which generally frame human rights as individual rights and development as resulting from the actions of economically rational individuals. These assumptions are reflected in the way that programmes for women's economic empowerment tend to prioritize developing the entrepreneurial skills of individual women so they can run micro-enterprises and participate in local markets. McKinnon et al. (2016,

1379–1380) highlight that the assumption of individualism is at odds with collective ways of being that are important for Solomon Islanders and iTaukei Fijians. As well, such agendas ignore (and potentially undermine) the diverse economic activities on which the wellbeing of families and community members relies, including non-market transactions such as reciprocity and gifting, and unpaid work such as caring labour, volunteering and shared labour.

To inquire into what mattered for community members, the research team engaged in a series of community workshops conducted in local languages. Workshops were held in two rural settlements in the Solomon Islands (in Western Province), two urban settlements around Honiara, and two urban settlements in Suva (with iTaukei Fijians). The workshops were facilitated by the local NGO research partners in Solomon Islands Pidgin and Bauan Fijian. Between ten to fifteen community members participated in each workshop, with roughly equal numbers of women and men, and older (40 years+) and younger (18–40 years) participants (Carnegie et al. 2019, 255). Participants included those from nuclear and extended families, and female-headed households; those who were single or married; and those with or without children. At the workshops the researchers ran four group exercises: an inventory of diverse economies focusing on individual daily work activities; a mapping of distributions of labour, goods and cash within the networks of family and community members; a timeline of events that produced significant change; and role plays on desired changes in relations between women and men. Some exercises were conducted in gender- and age-segregated groups with participants coming together for shared discussion (Carnegie et al. 2019, 255).

These exercises were an opportunity for community members to talk more about their everyday experiences of gendered and economic relationships and their aspirations for how these relationships might be. The exercises were also an opportunity for the team of researchers to listen attentively to the understandings and aspirations of community members:

> *we listened for and acknowledged the diversity of economic activities in which women and men were involved … We also listened for the culturally important and different ways that women and men's contributions are valued in the community … [We listened] to how community members foresaw possibilities of cooperation, negotiation and change for the better.*
> *(Carnegie et al. 2019, 255–256)*

Based on this exercise in listening, the research team (involving community economies scholars, and staff from IWDA and their local NGO partners) devised an initial series of indicators based on what participants said mattered to them. The indicators were then tested with groups of local volunteers from villages around Honiara and, based on the feedback, further refined. These indicators were organized in terms of four domains related to gendered and economic relationships. The domains were named by incorporating participants' phrases from the workshops: Women 'Come Up' (related to women's individual agency and access to opportunities); Household Togetherness (related to relations between women and men in households); Women's Collective Action (related to mutual support and collaboration between women); and Leadership, Say and Role Modelling (related to women's opportunities for participation and leadership, and the quality of men's leadership) (Carnegie et al. 2019, 256). Of these four, Women 'Come Up' was most aligned with mainstream development and human rights policy agendas. But in this project all four domains were given equal importance. This was reflected in the representation of the four domains as being tributaries of a River of Change. The metaphor of a river also helped to communicate the role of indicators with comparisons being made to how water samples along the course of a river can provide vital information about water quality and interventions that might be needed to improve water quality. With these indicators (which encapsulate both existing economic activities as well as the way that community members would like things to be) local NGOs and community members can track what is happening and intervene when activities are being weakened or are under threat.

The indicators that were developed and the process used to generate the indicators are documented in resource materials that other communities can apply (Carnegie et al. 2012). In turn, those who have used these materials have documented their processes and the outcomes. For example, Live and Learn International's Western Pacific Sanitation, Marketing and Innovation Project produced a short 6-minute documentary on how their staff have used the floating coconut tool (which had been developed as part of the original project to translate the image of the diverse economy iceberg to the context of the Pacific Island countries) (see www.youtube.com/watch?v=t0zQjGtg2d0). Staff from the Solomon Islands and Vanuatu report how the tool provides an effective means of making women's economic contribution more evident to women and men, and how this has especially helped men to better understand the crucial contribution of women's work to the wellbeing of families and communities (see also Thomas 2015).

In terms of the implications for policy and practice, this project highlights the importance of attentive listening. Based on the understanding that post-capitalist possibilities are already present, the community economies researchers asked questions, observed and listened for these existing possibilities. This led to an appreciation of how wellbeing is sustained by a raft of individual and collective practices. The process of attentive listening includes putting aside pre-existing assumptions. For example, mainstream development and human rights policy agendas tend to rely on imported assumptions about economically rational individuals and gender equity. This is reflected in programmes for women's economic empowerment that are based on the development of individual entrepreneurial skills and micro-enterprises. This research found that such programmes are of interest to women but there was also a strong desire for economic empowerment that incorporated greater acknowledgement of women's household contribution, opportunities for women to work collectively and avenues for women to contribute to decision-making. Prioritizing only economic activities that have an individual focus overlooks and potentially undermines group, household and community economic activities and relationships which are equally important for wellbeing. Listening attentively and putting to one side familiar assumptions are crucial if the range of activities that sustain wellbeing are to be heard.

Economic development pathway 2: building community-based initiatives

The Jagna Community Partnering Project was based in the Philippines (in Jagna on the island of Bohol, in the country's Central Visayas region). The project involved a collaboration between community economies researchers, the NGO Unlad Kabayan Migrant Services Foundation Inc. and the Municipal Planning and Development Unit from the Jagna Municipal Government. Unlad Kabayan is largely funded through the remittances of overseas migrant workers from the Philippines and it has pioneered a strategy of Migrant Savings for Alternative Investment. Here the goal is to fund alternative investment to help foster economic opportunities within the Philippines, so the next generation of family members do not have to support their families by moving overseas as migrant workers and sending remittances back home. Unlad Kabayan is therefore interested in generating endogenous economic development pathways, and especially the potential for economic initiatives based on the collective effort of groups of people (Gibson et al. 2010).

The Jagna Community Partnering Project employed seven researchers: four community members recruited from economically marginalized groups in Jagna, and staff from the university partner (Australian National University), Unlad Kabayan and the Municipal Planning and Development Unit (Cahill 2008). The action research project started with an assets-based approach of working with community members to document the diversity of assets in Jagna, especially the diversity of economic activities.[3] This exercise revealed the extent of available resources, knowledge and skills, and diverse economies (including small-scale rice and fruit farming, and customary exchange practices based on bartering, gleaning, labour sharing and gifting) (Gibson et al. 2010). Community members were then supported to develop economic initiatives that built on these assets. For example, a group of older women were interested in putting together the abundance of ginger in the region with their skills, gained over generations, in processing ginger into tea powder (known locally as *salabat*). Initially, the women felt that an economic initiative was beyond their reach as it would require the input of significant monetary start-up capital, funding they were unlikely to be able to access because of their age and background. But with a tiny grant from the Community Partnering Project they were supported to conduct their own market research and mobilize goodwill from local traders who agreed to sell their product. They learned, 'we can start with what we have' (Community Economies Collective and Gibson 2009, 130).[4] Other initiatives that were started through the project included a coconut confectionary enterprise that allowed a group of farmers to value-add to their coconut harvest, and a sewing enterprise that filled the demand for specialized clothing including gowns for school graduations. These enterprises provided households with a cash income (for things such as medicines, schooling and transport) which supplemented their largely subsistence-based livelihoods.

The steps and outcomes of the Community Partnering Project have been made readily available through resources that include a Community Partnering website (which includes short videos of all the economic enterprises that were initiated) (www.communitypartnering.info/) and a 50-minute documentary (Gibson et al. 2008). Here the intention is to expand economic development pathways in the Philippines (and in other countries that are heavily reliant on exporting their citizens as overseas migrant workers or importing exploitative labor and resource stripping practices from the so-called developed world). Unlad Kabayan continues this work through its mission of 'mobilizing migrant workers, the marginalized in the community and their resources to build a sustainable local economy' (www.unladkabayan.org/vision-and-mission/), and through the partnerships it continues to build with municipal authorities and others.

Similar to the research project in the Solomon Islands and Fiji, the Jagna Community Partnering Project involved attentive listening to learn more about the post-capitalist possibilities that were already present. Responding to the concerns of local NGOs, the research in the Solomon Islands and Fiji focused on designing a set of indicators that captured those things that community members said they valued and the direction of change they said they wanted (and as outlined above, these concerns are resonating in other contexts in the Pacific as the indicators and associated tools are being applied and adapted). In the context of the Philippines, the research focused on developing a process for stimulating endogenous economic development. One of the lessons from the Community Partnering Project is the value of genuine partnerships. The project serves as an illustration of what can be achieved by partnerships between academic researchers, practitioners and communities, especially when all parties are involved at every step from the initial conceptualization of the project to the ongoing work of supporting the outcomes of the project (including promoting the resource material). In the case of this project, the partners already had relationships with each other, and had spent time developing trust with each other. They were in no rush to enter into the formal partnership that became the Community Partnering Project but waited until they were ready to proceed on their own terms. This approach speaks to the importance of allowing time for relationships to develop and for genuine partnerships to take root.

Economic development pathway 3: 'leave the villagers alone'

The third project was based in two villages in eastern Cambodia near the town of Kampong Cham, the country's sixth largest urban settlement. As across other parts of Cambodia, daily life in these villages occurs in the shadow cast by the violence and terror of the Khmer Rouge's regime of the late 1970s. Even after the Khmer Rouge was ousted by Vietnamese forces at the end of 1978, local despots stepped into the political vacuum and continued a reign of violence and terror in this part of the country until the 1990s (Hinton 2005). Any development project in Cambodia has to be alert to the ongoing effects of trauma; nevertheless, the effects run deep and even the most attentive of projects can overlook and potentially destabilize the ways and means that people have developed to survive and live together post-conflict. In this setting, Somsak's (2005) admonishment to 'leave the villagers alone' (cited in McKinnon 2017, 344) is a credible development pathway.

The project in these two villages was modelled on the action research approach used in the Jagna Community Partnering Project. Local community members were invited to participate as community researchers; and through workshops, field trips and walking tours an inventory of local economic activities was assembled. The next phase focused on social enterprise development based on some of the activities identified in the inventory. The inventory phase identified a diverse range of local economic activities, including subsistence agriculture and livestock raising; informal trading in and around the local markets; informal construction work; harvesting, cutting, and selling bamboo as barbeque skewers; and in a few cases formal employment in the nearby Medtecs International Corporation which employs around 5,000 workers to manufacture medical clothing.

Participants expressed an interest in developing additional activities based on the bamboo resources in the area. Some requested workshops so they could learn additional skills in bamboo furniture making. A trainer was secured from the Rattan Association of Cambodia (which supplies products to the multinational furniture retailer, IKEA). Despite initial enthusiasm, attendance was poor, and no project participants came to the final workshop. Even those who had prepared bamboo for this workshop did not attend. Indeed, on the day before the final workshop, on being asked by phone about preparations, one participant proclaimed to the researcher that 'tomorrow I will have diarrhoea' (Lyne 2017, 178). The drop-off in interest and attendance can certainly be explained by practical concerns. The participants have to spend most of their time securing what is a precarious existence; they do not have time to participate in an unpaid workshop that is only likely to bring a return on their investment of time at some distant point in the future (if there is any return on this investment).

However, it also became clear that there was a highly nuanced and finely balanced bamboo skewer community economy that could potentially be destabilized by the additional use of bamboo resources in the area. The villagers (mostly women) make bamboo skewers that are sold to one middleman who then sells the skewers further afield to barbeque businesses and street food stalls. The work of making skewers is conducted at home, and it is easy to fit in and around women's other home-based activities. Even though villagers receive around US$1.50 for six hours work making skewers, it is an income that can be relied on, especially during the dry season which can be a time of high outward migration with some temporarily moving to urban areas to pick up informal work. Making bamboo skewers also serves as an economic safety net (in the absence of a national welfare system). This became evident when one villager was fired from the Medtecs garment factory because he participated in a strike. The family was dependent on his wage; with that lost, the family immediately turned to skewer making.

There is a tacit agreement that all villagers are entitled to this economic safety net. The bamboo is on public land, but each household has a claim to a specific plot. Those who deplete their plot can borrow from other households and return it when their bamboo regrows. It is unknown for any household to reject the request to borrow. Indeed, there is no theft of bamboo, perhaps because these requests are not refused. When one household borrows the bamboo from another's plot the bamboo is carefully cut so that the bamboo is thinned to let in light and promote growth. The idea of using the bamboo for furniture had the potential to destabilize this finely-tuned community economy. As one village elder later commented about the workshops, 'They were worried about losing the bamboo.' Indeed, the villages had this experience when the nearby lake was designated private and dredged for sand. Their fisheries were lost and the soil quality for agriculture was compromised. As well as an economic safety net, the bamboo skewer community economy served an additional role – it balanced household and community commitments in a way that enabled villagers to build relationships of trust with each other in the post-conflict era.

In terms of implications for policy and practice, this project illustrates how important it is to listen attentively but how difficult this can be. In this instance, the participants were initially keen to develop an enterprise based on an additional use for bamboo, but away from the eyes and ears of the researcher, the villagers reconsidered their initial decision and found a powerful way to relay their revised decision (by not turning up). Certainly, the decision to not attend reflects practical concerns (about the immediate priorities for villagers, including their use of time). These concerns are important to consider in any development project. But in the post-conflict context of Cambodia a deeper concern is being relayed about the pivotal role that resources play in households and communities, and how sometimes it is best to leave matters as they are. Even though the furniture making did not proceed, the lessons learned about this resource meant that when changes were being made to a nearby bamboo bridge the researcher was able to hear more clearly. In this case, a hand-built bamboo bridge that is reconstructed each dry season to provide access between an island on the Mekong River and the nearby town of Kampong Cham was being replaced by a much larger concrete bridge that would permanently connect the island to the mainland. The bridge builders and the villagers on the island had mixed feelings about this symbol of 'progress.' In response, the researcher initiated, with others, a 60-minute documentary film that captures the final making of the bamboo bridge and the crucial role that bamboo has played in the lives of the bridge builders and the villagers for centuries (Gibson and Salazar 2019). This film reflects back to people their everyday lives, and the resources and relationships that have been important to making their lives what they are.

Implications for pedagogy

The community economies approach has three main implications for pedagogy. First, it is important that the language of economics is expanded so the crucial role of current and potential diverse economic activities is more readily apparent. This would both expand the economic understanding of students who are unfamiliar with economic diversity, and value what is already known by students whose lived experience of economic diversity is not reflected in mainstream economic language. Tools such as the iceberg and the floating coconut provide a good starting point to help make diversity more visible and for valuing diversity (and these tools are readily available online, for example, www.communityeconomies.org/resources/diverse-economies-iceberg). There are also associated tools such as Asset-Based Community Development that can be augmented by including economic diversity as a crucial community asset (e.g. Cameron and Gibson 2001). Such tools that document economic diversity provide the basis for students to appreciate the contribution that diversity makes to people's lives. Here a community economies approach has sought to develop questions that can be asked to help bring to the fore the ways diverse economic activities contribute to people and planetary well-being. This includes questions about how activities contribute to surviving well, to just and sustainable production and distribution of surplus, to the fair transaction of good and services, to care for shared resources, and to forms of investment that are oriented towards a better future (e.g. Gibson-Graham et al. 2013). From this, students might then be able to recognize the multitude of post-capitalist possibilities that already exist and how these possibilities can provide the basis for more equitable and sustainable economic development pathways.

Second, it is important that students develop skills in working with others, including the capacity to listen to and hear others. As community economies researchers point out, the 'community' in community economies is not about pre-existing communities (such as those based on a shared identity or location). Rather, community is a process of 'being-with' others (Gibson-Graham 2006, 81–82), including the world around. In other words, fundamental to our existence is our interdependence with others and how we ethically negotiate ways of living together on a finite planet (captured in different contexts in terms such as living well and *buen viver*). In the classroom setting, doing group work is one way to help develop relevant skills, and here there are opportunities for students with a range of life experiences to work together to try out the various tools that community economies researchers have developed for working with communities. For example, the Community Partnering for Local Development website (www.communitypartnering.info/), which is based on the Jagna Community Partnering Project, includes exercises in needs and assets mapping and a 'portrait of gifts' that can be adapted for the classroom. The caution with group work (especially assessable group work) is that it has to be scaffolded so that students have the opportunity to learn both cognitively and experientially about group processes, forms of negotiation, and ways of 'being-with' others.

Third, it is important that students have the opportunity to cultivate an open and enquiring mind especially by putting aside pre-existing assumptions. In community economies schol-arship, this is associated with 'refusing to know too much' (Gibson-Graham 2006, 8). This disposition fits well with the affective and embodied forms of learning that can be prompted through exercises such as field trips and exposure to cultural artefact such as films. One of the benefits of these types of exercises is that they create opportunities for students to 'stand in the shoes' of others and to see things from others' perspectives. Field trips have been a mainstay of the community economies approach and the types of outcomes that can be provoked through this method have been reflected on in Cameron (2011), Cameron and Gibson (2005) and Lyne (2017). Films have also been produced by community economies scholars as provocations to encounter the world differently.[5] For example, the Bamboo Bridge documentary sheds light on little-known things that communities have done well for generations and that intuitively balance social and environmental sustainability and resilience. It also gives room for the world-at-large to speak through evocative images and sounds of bamboo and water. This documen-tary might help to challenge generalized preconceptions or caricatures of a country such as Cambodia and help to show how sometimes our best inclinations to 'help' or 'develop' com-munities represents a liability.

Conclusion

In this chapter, we have discussed three projects that exemplify the ways in which a community economies approach to doing development differently starts by acknowledging the local context and valuing the diverse economic activities and possibilities that are already present. This gives rise to economic development pathways that are appropriate to the context. In the indicators project in Fiji and the Solomon Islands this meant acknowledging the current range of activities that contribute to individual and community wellbeing as well as acknowledging community members' aspirations for how these activities might be changed for the better. In the Jagna Community Partnering Project (in the Philippines) this meant working with groups of community members to build on existing individual and community assets. In the work in Cambodia, the appropriate development pathway was to 'leave the villagers alone.' The implications for policy and practice involve recognition of the importance of attentive listening, relationship building and time. The implications for pedagogy involve deepening economic literacy, developing skills that are vital for working with others and developing a capacity for openness and to be affected by others.

Key sources

1. Cameron, J. and Gibson, K. (2020) 'Action Research for Diverse Economies,' in J.K. Gibson-Graham and K. Dombroski (eds) *Handbook of Diverse Economies*, Cheltenham: Edward Elgar, 511–519.

A discussion of the research steps in three action research projects, including the Jagna Community Partnering Project in the Philippines. The chapter highlights the role of a politics of language, a politics of the self and a politics of collective action as post-structural research tools that can play a role in policy formulation.

2. Gibson, K., Astuti, R., Carnegie, M., Chalernphon, A., Dombroski, K., Haryani, A.R., Hill, A., Kehi, B., Law, L., Lyne, I., McGregor, A., McKinnon, K., McWilliam, A., Miller, F., Ngin, C., Occeña-Gutierrez, D., Palmer, L., Placino, P., Rampengan, M., Lei Lei Than, W., Isiyana Wianti, N. and Wright, S. (2018), Community Economies in Monsoon Asia: Keywords and Key Reflections, *Asia Pacific Viewpoint* 59(1), 3–16.

This collaborative project documents the extent of diverse economic practices across Monsoon Asia that the modernist imaginary assumes have been superseded by the growth of the cash economy. By highlighting the persistence, and even flourishing, of these economic practices, the paper contributes to a post-development agenda.

3. Lyne, I. (2017) 'Social Enterprise and the Everydayness of Precarious Indigenous Cambodian Villagers: Challenging Ethnocentric Epistemologies,' in C. Essers, P. Dey, D. Tedmanson and K. Verduyn (eds) *Critical Perspectives on Entrepreneurship; Challenging Dominant Discourses in Entrepreneurship*, London: Routledge, 36–50.

This paper uses a diverse economies perspective to analyse a development intervention which aimed to increase resin tappers' incomes by developing opportunities along the value chain while instigating or supporting resin producer associations. The paper finds that an independently founded producer association was more sustainable than one marshalled together by consultants who were focused on the value chain primarily. It shows how wellbeing is promoted by a closely integrated subsistence, barter, and gift economy intermeshed with spiritual practices.

4. Lyne, I. and Madden, A. (2020) 'Enterprising New Worlds: Social Enterprise and the Value of Repair,' in J.K. Gibson-Graham and K. Dombroski (eds) *Handbook of Diverse Economies*, Cheltenham: Edward Elgar, 74–81.

This chapter examines social enterprise through the lens of decolonial thinking and loving kindness, and highlights the role that everyday activities can play in repairing or maintaining communities in damaged and traumatized societies. The analysis is an alternative to the notion that social enterprises are subject to the logic of capital and should act in ways that are competitive and business-like.

5. Mathie, A., Cameron, J. and Gibson, K. (2017) 'Asset-based and Citizen-led Development: Using a Diffracted Power Lens to Analyze the Possibilities and Challenges,' *Progress in Development Studies* 17(1), 54–66.

This paper draws on projects from Africa and Asia to show how an asset-based development approach can be used to reverse internalized powerlessness, strengthen opportunities for collective endeavours and help to build local capacity for action. The paper is a counter to the modernist assumption that power is a resource held by some and not others, and that strategies such as asset-based and citizen-led development are an extension of the neoliberal project.

Notes

1 The community economies approach discussed in this chapter is associated with the work of J.K. Gibson-Graham (e.g. Gibson-Graham 1996; 2006) and their collaborations with other members of the Community Economies Research Network (e.g. Gibson-Graham et al. 2013; Gibson-Graham and Dombroski 2020; Roelvink et al. 2015).
2 More recently, some community economies scholars have used the term more-than-capitalist rather than post-capitalist (e.g. Gibson-Graham and Dombroski 2020).
3 This was an adaptation of an earlier process developed in Australia (Cameron and Gibson 2005), which in turn was based on an adaptation of the Asset-Based Community Development work of Kretzmann and McKnight (1993).
4 For this publication, the Community Economies Collective writers included May-An Villalba and Benilda Flores-Rom from Unlad Kabayan Migrant Services Foundation Inc; and Maureen Balaba and Joy Miralles-Apag from Bohol Initiatives for Migration and Community Development.
5 Along with the films mentioned in this chapter, other films include It's in Our Hands (Parts 1 and 2) (see www.youtube.com/watch?v=I1r_h9GeV2Q and www.youtube.com/watch?v=ptLjXNECrFs) and Social Enterprise and Asset Based Community Development in Cambodia (see www.youtube.com/watch?v=ux2DZ5fGPw0&t=2949s).

References

Cahill, A. (2008) 'Power Over, Power To, Power With: Shifting Perceptions of Power for Local Economic Development in the Philippines,' *Asia Pacific Viewpoint* 49(3), 294–304.
Cameron, J. (with Manhood C. and Pomfrett J.) (2011) 'Bodily Learning for a (Climate) Changing World,' *Local Environment* 16(6), 493–508.
Cameron, J., and Gibson, K. (2001) *Shifting Focus: Alternative Pathways for Communities and Economies*, Victoria: Latrobe City Council and Monash University.

Cameron, J., and Gibson, K. (2005) 'Participatory Action Research in a Poststructuralist Vein,' *Geoforum* 36(3), 315–31.

Carnegie, M., McKinnon, K. and Gibson, K. (2019) 'Creating Community-based Indicators of Gender Equity: A Methodology,' *Asia Pacific Viewpoint* 60(3), 252–266.

Carnegie, M., Rowland, C., Gibson, K., McKinnon, K., Crawford, J. and Slatter, C. (2012) *Gender and Economy in Melanesian Communities: A Manual of Indicators and Tools to Track Change*, Sydney: University of Western Sydney, Macquarie University and International Women's Development Agency.

Community Economies Collective and Gibson, K. (2009) 'Building Community-based Social Enterprises in the Philippines,' in A. Amin (ed.) *The Social Economy: International Perspectives on Economic Solidarity*, London: Zed Books, 116–39.

Gibson-Graham, J.K. (1996) *The End of Capitalism (as we knew it): A Feminist Critique of Political Economy*, Oxford: Blackwell.

Gibson-Graham, J.K. (2006) *A Postcapitalist Politics*, Minneapolis: University of Minnesota Press.

Gibson-Graham, J.K., Cameron, J. and Healy, S. (2013) *Take Back the Economy: An Ethical Guide for Transforming our Communities*, Minneapolis: University of Minnesota Press.

Gibson-Graham, J.K. and Dombroski, K. (eds) (2020) *The Handbook of Diverse Economies*, Cheltenham: Edward Elgar.

Gibson-Graham, J.K. and Dombroski, K. (2020) 'Introduction', in J.K. Gibson-Graham and K. Dombroski (eds) *The Handbook of Diverse Economies*, Cheltenham: Edward Elgar, 1–24.

Gibson K. (Executive Producer) and Salazar JF. (Director), (2019) *The Bamboo Bridge [Motion picture]*, Australia: Matadora Films.

Gibson, K., Cahill, A. and McKay, D. (2010) 'Rethinking the Dynamics of Rural Transformation: Performing Different Development Pathways in a Philippine Municipality,' *Transactions of the Institute of British Geographers* 35(2), 237–55.

Gibson, K., Hill, A. and McLay, P. (2008) *Building Social Enterprises in the Philippines, (DVD)*, Canberra: The Australian National University. Online at www.youtube.com/playlist?list= PLKN9NChecdBKp-TicDgbVWp3hm3ifLLbr

Hinton, A. (2005) *Why Did They Kill? Cambodia in the Shadow of Genocide*, Berkeley CA: University of California Press.

Kretzmann, J. and McKnight, J. (1993) *Building Communities from the Inside Out*, Chicago: ACTA.

Lyne, I. (2017) *Social Enterprise and Community Development: Theory into Practice in Two Cambodian Villages*, PhD thesis, Australia: Western Sydney University.

Mason, P. (2016) *Postcapitalism: A Guide to Our Future*, Basingstoke: Macmillan.

McKinnon, K. (2017) 'Naked Scholarship: Prefiguring a New World Through Uncertain Development Geographies,' *Geographical Research* 55(3), 344–349.

McKinnon, K., Carnegie, M., Gibson, K. and Rowland, C. (2016) 'Gender Equality and Economic Empowerment in the Solomon Islands and Fiji: A Place-based Approach,' *Gender, Place and Culture* 23(10), 1376–1391.

Roelvink, G., St. Martin, K. and Gibson-Graham, J.K. (eds) (2015) *Making Other Worlds Possible: Performing Diverse Economies*, Minneapolis: University of Minnesota Press.

Somsak, J. (2005) *Developing One's Brother: Good Intentions, Unnatural Practices*, Bloomington, IN: Author House.

Thomas, A. (2015) Floating Coconuts in Vanuatu. Stories from the Field, Online at https://thewashbusiness. wordpress.com/2015/03/07/floating-coconuts-in-vanuatu/

60

GEONARRATIVES AND COUNTERMAPPED STORYTELLING

Joseph Palis

Introduction

[T]o tell other stories than the official sequential or ideological ones produced by institutions of power.

(Said 2000, 154)

How do stories come about and why are there so many of these stories that do not get told? In arguing for new conceptions of space, Doreen Massey advocated for multiplicities, of relations-between and of stories-so-far (2005, 9). This chapter re-evaluates the use and functional possibilities of geonarratives as a method and approach used in a subaltern setting that allows participants to tell stories on their own terms and in a manner they deem best captures their place-based often-untold narratives. By enabling and encouraging the use of available materials at their immediate disposal, the participants produced drawings, collages, technology-aided illustrations, and other forms of visualizations to tell various spatial stories. The outcome is then used as visual basis and prompt to discuss, tell and perform their stories-so-far in a manner that allows stories to flow, meander, and circle back consistent to the chosen style and modality of the participant.

In the five mapping workshops, geonarrative mapping was used, approached and practiced with the aim of making invisible stories visible, and enabling the storytelling of untold stories. The urban spaces of Metro Manila and a flood-prone rural area in Bulacan – a province adjacent to Metro Manila – were the settings of these workshops. These were conducted from March 2019 to October 2020 both in a face-to-face mode, and remotely in the case of two workshops during the COVID-19 pandemic where a lockdown was in effect.

Although the use and deployment of story maps and geonarratives have been in use in various iterations for pedagogical (GIS, field geography, primary education), methodological and didactic purposes (Carroll 2020, Kwan and Dong 2008, Lee 2017, Marta and Osso 2015, Mukherjee 2019, Reutzer 1985), my use and practice of geonarratives in these mapping workshops owe their inspiration to Massey's stories-so-far that promote a 'dimension of multiplicity' (Massey in Featherstone et al. interview 2013, 265), unhampered by time constraints and limitations of place.

I define geonarratives as place-writing – subjective stories that define, portray, delineate, emphasize, expand, rewrite, and imagine a place. It is world making that is grounded on

DOI: 10.4324/9781003017653-66

realities and emotions specific to the individual in relation to class and other identity politics and markers. Geonarratives employ the act and process of storytelling to depict, portray, represent, and un-think spaces of habitation (Bell et al. 2017, Hawkins 2015, Massey 2005). In the five geonarrative undertakings involving a group of indigenous K–12 students, female participants, differently-abled individuals, urban and rural dwellers in disaster-prone areas, and children, maps created by these participants served as a gateway to tell stories about their environment, their particularized domestic, familial and social lives, and their own encounters with human, institutional, and more-than-human elements and entities. The completed maps are a combination of hand-drawn, collaged and a pastiche of meaningful fragments of images derived from available resources such as magazines, newspapers, old calendars, and other picture books. In one geonarrative workshop among differently-abled participants, maps are submitted as snapshot realities that can change when these map images are revisited to reflect newer experiences and engagements. As these maps remain incomplete and ongoing, the act of mapping is subverted as the previous one can be unmapped or countermapped. In other cases, specific markers in an image are enlarged and given particular attention as these markers represent ongoing negotiations. Geonarratives are embodied and practice-based geographical projects that entangle the individual with the created image through the act of storytelling. These geonarrative maps are active forms of countermapping that do not conform to the standard mapmaking practices which are produced by state institutions that give premium to technical precision. Geonarrative mapping is a vigorous practice of 'unmapping' where subjectivities and cartographic stories that do not fit the development model encouraged by institutions of power are emphasized and accentuated. What geonarrative mapping does is to foreground an alternative modality of storytelling that brings the emergent, emotional, and performative dimensions to the centre. Geonarrative mappings, storytelling and storying are about encouraging and celebrating different and differentially calibrated versions of development that consider the specific and particular in placemaking and worlding.

The geonarrative storying that utilizes countermapping and unmapping practices can be deployed by community-based groups, activists, grassroots, and peoples' organizations to generate grounded data and information as basis to carry out specific participatory action research and development work for social justice. This approach gives authorial agency to individual participants to shape their geonarratives using non-traditional and unconventional resources and modalities. Action researchers gain nuanced lessons and insights from the stories generated from maps that deal with the participant's subjective representation of facts and realities. This methodology has the potential to disrupt the propensity to link gathered data from stories to general and universalized knowledge. Ground-truthed data from geonarratives mapping are derived from vernacular vocabularies and lived-in experiences of the participants. Likewise, map stories acknowledge the polyvocalities, relational pluralities and multiplicities of individual narration stemming from their engagements with their diverse environments.

The following geonarrative mapping workshops were undertaken between 2019 and 2020. Permission was obtained from the participants and institutions to use the images for publication.

The Lumad's maps of meanings

In March 2019, a geonarrative mapping workshop was undertaken involving K–12 students belonging to the Lumad of southern Mindanao. Lumad comes from 'Katawhang Lumad' (literally 'indigenous people') which was adopted in 1986 by the delegates of the founding assembly of the Lumad Mindanao Peoples Federation (Alamon 2017, Lee 2017, Raymundo 2019). The Lumad youth who participated in the mapping workshop that I conducted with a faculty colleague are part of the Bakwit School (*bakwit* is a colloquialized term for 'evacuate') who travelled from Mindanao to Metro Manila, specifically, the University of the Philippines campus in Diliman, Quezon City. The purpose of the education caravan was to exchange and share knowledge with college instructors who in turn team-taught classes for the Lumad K-12 students. Save Our Schools (SOS) – a collective of academics and Lumad activists – organized and facilitated the formation of Bakwit School. The Lumad community also faces harassment and military intimidation in Mindanao because of their activism regarding extractive mining operations and state-sanctioned human rights violations (Raymundo 2019).

The information shared about mapping with students involved how a map is seen as a visualization of the world, of a particular environment, of a region and territory, as well as an instrument of power where information is contained that assess the extent and value of the resources endemic to a region. Examples were given that illustrate how maps mirror the hegemonic practices that were undertaken to forcibly obtain lands, extract resources, and organize landscapes that benefit the more powerful entities. Finally, examples were shown how maps can be dynamic and not rooted in any spatial fixity, and that a progressive place and landscape can be imagined and realized through collective action. The students were divided into five groups having chosen their own members.

The primary aim and intention of the geonarrative mapping workshop was to introduce the concept of a map and of mapmaking practices by highlighting that maps are also means to allow for the visualization of the milieu where the students live as well as to narrate stories that animate their lifeworlds. To make these visual narratives come to life, pencils, paper, crayons, markers, and colour pens were provided. The instruction was to create a map of their community. It can be their domestic dwelling place, or their street, or a cross-section of the public areas they frequent, or areas marked as important by individuals or of a collective. The Lumad youth created maps that tell their individual and collective stories. The process and act of creating a map among the Lumad was far from harmonious. Because the Lumad participants simultaneously work singularly and as part of another map project of another group, there were arguments, verbal bickering in relation to the 'accuracy' of the place being drawn, and discussions that aim to recollect a common experience and to accurately portray that in the group map. Some participants circulate throughout the mapping workshop and comment on other map projects without being shackled by the directive to work and focus only on one map. Being part of a collective creates a map that incorporates many stories from various participants.

The geonarrative mapping entails that the map that is produced should be able to communicate to an audience. When it was time to talk about their maps that contain various stories, the students were excited to have created an ongoing image that is a dynamic geovisualization not only of the general and typical, but also of the quotidian, the particular and the specific. This geonarrative mapping activity offers a snapshot of the tiny geographies that were drawn and co-created by the Lumad youth in narrativizing their home, community, and multiple entities that simultaneously animate, terrorize and defamiliarize their worlds.

The maps very interestingly show performative activities of particularly memorable moments. The scale of the place being depicted is adjusted to accommodate a series of tableaus. Each of the five maps that were created in the mapping workshop was intended as pedagogical illustrations in cartography in higher education to uncover and recover geo-stories emphasized in the map drawings.[1]

All of the maps show the spatial extent of the students' community in Mindanao as drawn and imagined in their hand-drawn illustrations. While their homes are represented as being lined up in the street, trees and vegetation figure prominently in all of the depictions of their

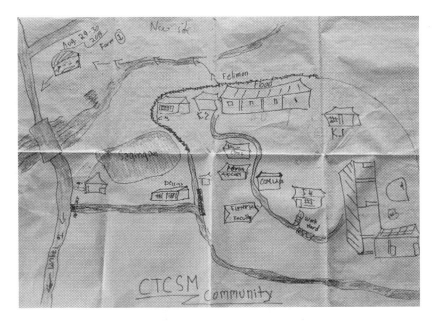

Figure 60.1 A map showing Felimon's odyssey. Lumad map, 2019

chosen landscape. The school in particular – Community Technical College of Southeastern Mindanao (CTCSM) – looms large in their cartographic illustration.

Figure 60.1 tells the story of Felimon who, according to the group's storytellers, is a free spirit who would cut classes, avoid school work, and leave the school premises to go to a nearby community to dance with similarly-aged females. The map becomes Felimon's odyssey from school premises to a nearby community dance event. After the event, Felimon re-entered the school premises where he and the rest of the CTCSM students stay. The cartographic portrayal of Felimon's journey ended with the school's punitive measure towards truant students. A section of the map was marked 'work hard' conveying the disciplinary action against Felimon's non-compliance to school policies. Although the map did not indicate this, Felimon later rejoined his peers to the classroom after completing his hard work. What is evident from this map is the recognition of Felimon's fearless and outrageous behaviour that became a legendary cautionary tale told with a mixture of derision and admiration. The CTCSM school's premises are delineated by border fences that students cannot cross as it was deemed illegal and unlawful.

The other maps are fanciful in their portrayal of their community and general lifeworlds. When asked if there are ghosts or ghoulish creatures in their community, the students looked perplexed. Among the Lumad, the concept of ghosts and ghouls does not match the popular notion held in urbanized and urbanizing areas, until the word *busaw* (roughly translated as monster) is mentioned as being the closest description. One of the storytellers in the group pointed to the four soldiers outside the school premises as *busaw*. Although not shown in the map, a military jeep is constantly patrolling and surveilling the school. There are obvious similarities between CTCSM which is being run by the Lumad community, with the Zapatista schools in Chiapas, Mexico, as well as the community-designated peace zones in Colombia. CTCSM was not given an accreditation by the Department of Education therefore outlawing the community-led programme being used for Lumad Youth education (Palicte 2020).[2] A CTCSM volunteer

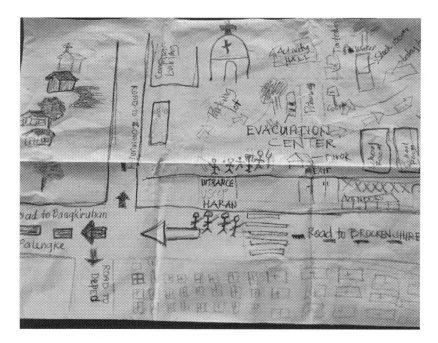

Figure 60.2 Linear street with barracks and soldiers. Lumad map, 2019

teacher informed me that harassment of the school and students happen frequently before the closure and that the pupils learn to duck automatically when they hear the loud rattling sound of gunfire. The identified *busaw* as drawn in the map of the Lumad youth is placed outside of the school premises as the proverbial monster in Philippine folklore that continually harass, intimidate, and terrorize children and adults.

Another map (Figure 60.2) emphasizes the linearity of the streets in a particular neighbourhood – clearly delineating the direction of public transportation to specific destinations. Big square blocks with no marked texts or details are drawn which was later revealed as the barracks of the military. The hand-drawn cartographic illustration provided no information or additional details about the barracks indicating their unfamiliarity or non-engagement with these spaces. This active form of unmapping negates the space occupied by non-allies of the school and community. Stories reveal that the school children rarely venture into these areas to avoid untoward violent incidents. These blank spaces are in stark contrast to the lush greeneries of trees within the community premises. The trees have their own stories that elicit hilarity and feigned terror from the participants.

Landscape of barricade and disrupted home

In March 2020, a geonarrative mapping workshop was undertaken with female participants from the University of the Philippines (UP).[3] This workshop aimed to gather stories within UP campus from various sectoral representatives 'whose map stories were not told before.' Aside from soliciting spatial narratives from those in attendance who answered our call, the workshop was described as a safe space where participants can freely and without intimidation create their maps to tell their stories as individuals occupying spaces in campus. The choice of a female constituency for this workshop was to hear a multiplicity of stories from a female group whose lives intersect with the university, but mostly through their experiences with unequal access to opportunities, the often-underappreciated and silenced emotional labour undertaken, and the paths taken in their daily negotiations. Ten female participants were represented by both senior and junior faculty from the social sciences and fine arts, alumni, and a currently enrolled student.

Materials provided were colour pens, crayons, pencils, blank papers, an unmarked map of the campus as well as donated old magazines, newspapers, and old calendars with images. The participants cut up these materials to produce a collage while others used crayons and colour pencils to colour their maps. Still others used the blank map of the university campus to tell their stories.

Faculty members who made up the majority of the participants prominently featured their homes in relation to their work offices, while participants who are alumni represented and highlighted spaces of gaiety and conviviality as a fond remembrance of the university as space when they were still students on campus. These are spaces of food, leisure, and buildings that they consider personal favourites. One participant showed her broken family life through vividly coloured houses that resulted in marital separation. Another showed a glimpse of the historical landscape of campus during the tumultuous year of protest in 1971 when students and faculty walked out of their classes to join mass rallies to protest then President Marcos' dictatorial leadership that resulted in the imposition of Martial Law.[4] Of the maps that were produced from the workshop, I will focus on two geonarrative maps: one that serves as an historical map of the university campus under government lockdown in 1971, and the other, a map that recorded the disintegration of marital life in 2017.

Figure 60.3 is a remembered snapshot of what the university campus looked like when state government attempted to quell incipient activism and discipline the academic landscape. Much as the map depicts an image of trees and the academic oval, the spoken narrative that accompanied the map provided details of progressive movements leading to mass actions of academe-based activists. The map shows how the campus became a fortress against the military that were deployed by the government to disperse a congregation of students and faculty decrying human rights violations and upholding social justice. As the participant tells the story from her recollection using the map as a visualization of remembered activism, additional illustrations were added to emphasize the point and also underline important and relevant markers in re-creating the story of activism from her perspective. Barricades made up of classroom armchairs represented by x's in the map were put in place by students and professors to prevent the military from entering campus grounds. Specific stories arising from detailed recollections provided a lived-in familiarity with these subversive spaces. As the participant recalled:

> *Back in the day, the academic oval was not easily traversable as it is now, so I had to walk quite a distance from my dormitory to the barricade and get through an area with thick foliage so I can help put more fortifications to ensure that the police and the military cannot enter our campus.*

The stories told were also about sensorial details that accompany the fear and exhilaration of being present in that particular historical moment of the university.

The map also shows the configuration of the university before the physical landscape was domesticated. The academic oval cannot be traversed easily because of untamed vegetation and foliage. Mobility was restricted to the traversal of official paths and roadways and the building of additional infrastructure would not happen after another decade.

The other maps were created by mostly junior faculty members and colourfully brought to life various transitions indicating how embedded professional careers are in the personal lives of

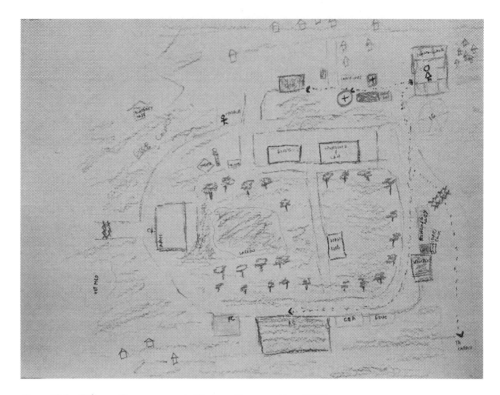

Figure 60.3 Diliman Commune map. Geonarratives mapping, 2020

academics. Detailed and highly descriptive elements were vividly captured to showcase lived-in work and personal spaces.

Figure 60.4 shows a blank map of the university and collaged with irregularly-cut fragments taken from old calendars bearing images of houses. These images of houses were pasted around the university map. Above an image of a woman sitting alone in a desk, bears cut-up letters that spell out the word 'remember' in capital letters. Two other cut-ups read '1997' and '2017.' The participant (tearfully) recounted her career as an academic and the dissolution of her nuclear family resulting from spousal separation. Both years indicated in the map represent the union and dissolution of her marriage. What the houses represent are various domiciles that she inhabited with her children during and after the end of her marital life. The houses were intimately described corresponding to the struggles that come from the physical and emotional fragmentation of a family. The participant offers a clear-eyed plan how life is going to progress for an individual living in a society that does not legally recognize marital divorce and separation. The geonarrative map offers a visual testament to the challenges of single motherhood and a resolve to keep on moving.

Quarantined landscape

A workshop I conducted in October 2020 was done remotely with the assistance of two faculty colleagues in geography, and an artist collective for a joint project that narrates disaster from two communities: two flood-prone rural sitios in a province adjacent to Metro Manila (San

Figure 60.4 Disruption of home. Geonarratives mapping, 2020

Miguel, Bulacan), and an urban community located inside the university (UP campus). The two workshops were held on successive Saturdays.

For the Barangay UP campus participants, all of them self-declared as PWD (person with disability). The aim of the geonarrative workshop was to amplify the stories of differently-abled individuals as they relate, interact, navigate, and negotiate with a landscape on a state-ordered lockdown due to the pandemic. The PWD participants self-declare as having psychosocial disabilities and having Down Syndrome condition. The geonarrative maps that were created were a combination of hand-drawn maps and a map that was generated through the use of online apps. In the flood-prone sitio of San Miguel in Bulacan, an intimate knowledge of specific areas became apparent where residents experience and encounter rising waters from torrential rains. In these workshops, the focus of discussion is the geonarrative map produced in San Miguel to get an insight how spatial stories centring on floods were visualized. Figure 60.5 situates flood (labelled 'bahâ') in relation to the residence (labelled 'bahay namin' or 'our house') of the participant, the intersecting roads and the prominence of vegetation ('manggahan' or mango trees; 'palayan' or rice fields) and river ('ilog').

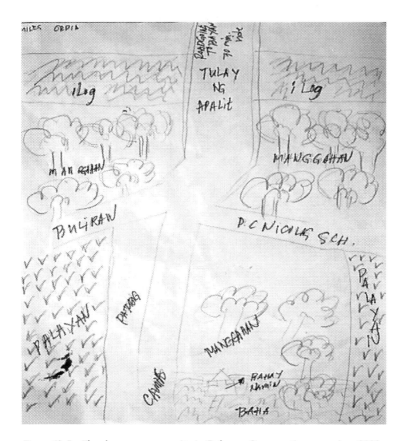

Figure 60.5 Flood-prone community in Bulacan. Geonarratives mapping, 2020

Geonarrative maps are interesting visual texts but the more animated and colourful stories accompanying these maps are full of details and information some of which do not get portrayed in the map. In Figure 60.5, stories of evacuation from the flood were given more heft than the drawing suggests.

The participant remembered the long duration for the flooded waters to subside and how this necessitated her family to take refuge in a residence from a relatively higher elevation. The sensory details that accompanied hunger, thirst and exhaustion were described more fully in a deadpan vocal delivery that belied the intensity of the situation. As the participant recalled:

> The rising waters initially did not bother us because we measure the gravity of the situation by the amount of rain that comes to our village. As rain continues to pour, we started to pack our important belongings to seek higher grounds. By this time, the rising waters made mobility very difficult, but we managed slowly to reach the higher elevations for safety. We were exhausted from our long and arduous journey by wading through the waters and find a place we can stop and rest. We experienced the first pangs of hunger and extreme thirst but we were thankful that at least if we die from this calamity, it won't be from drowning. But we were without food and water for hours and hours and we wondered how long we can last before someone comes along to give us something to eat and drink.

Likewise, the reconstruction of the house damaged by the flood was narrated with an attention to tiny details and information that highlight the emotional as well as financial toll on the participant and her family. The participant said:

> *We initially wanted a specific type of building material for our home's reconstruction that will outlast succeeding calamities, but we cannot afford those building materials because having food on the table is far more important. But we know another calamity will happen at some point and even with the new construction that we plan to undertake, we will be back to square one because not investing in good building materials will not solve our problems.*

While admitting that the experience made her resilient to face the vicissitudes of living in an environmentally precarious place, she does not see relocation as an option. The place matters because of memories and emotional attachments to the land. The illustration of lush vegetation in the hand-drawn map indicates the negotiation a household confronts in the habitation of a place, at once hazardous and uncertain and also binding and familiar.

Children's perception of place

Finally, a geonarrative mapping workshop intended for children aged 9 to 11 was undertaken in April 2019 to elicit insights as to how they perceive and expand on the notion of home and community. The children were accompanied by a parent/guardian during the session. The maps provided stories that reveal narratives undergirding class and domestic and familial relations. In one instance, a child drew his home in relation to school by mapping the route from origin to destination (Figure 60.6). Owing to his family's changed economic status, he would take public transport locally by tricycle which is a repurposed motorcycle with an additional compartment to accommodate a passenger. Tricycles are ubiquitous in urban and rural places in the country with an added layer of class-related identity marker: people who regularly patronize tricycles belong to the working class. The child would get off at a convenient stop and walk to school to hide his working-class family life background. He recounted that when his father lost his job, belt-tightening was enforced in the household and not having a private car to go to school, meant he had to ride the tricycle. The narratives that spring from the hand-drawn map allowed the child to express his own disappointments and a keen awareness of his class and social status while attending a privately-run school with affluent classmates.

Despite the children's unabashed retelling of their feelings that veer towards personal details, the parent/guardian who was present did not interrupt the child's oral narrative adding a complicit solidarity with the stories being told. One child's version of home offers a spacious one which was brought to life by very detailed recollections of how each room plays a specific part in the daily routine. It is worth noting that when a child participant presents their version of hand-drawn maps using crayons and pencils, the attentive audience comprises of a parent/guardian, a colleague who acts as a facilitator, a museum administrator who organized the workshop and other children who are also participants, and myself. The storytelling session becomes a monologue about one's self and performance to the members of the audience. The child participant's map is similar to a cartographic story-so-far as they keep adding details in the map when a recollected fact is included. Massey (2005) argued that stories-so-far are ongoing narratives that do not have formal endings as new stories are added and modified.

Figure 60.6   My route from school going home. Mapmaking for kids, 2019

Mapped lifeworlds: concluding reflections

The maps produced by participants in the five mapping workshops serve as illustration guides acting as visual cues for storytelling that intersect class, gender, age, and other markers of identity. Permission was solicited to ensure the privacy of participants whose maps and stories serve as pedagogical tools to teach the importance of maps, mapmaking practices and the role of the mappers in representing subjective realities. These maps that function as snapshots serve to emphasize stories-so-far which are ongoing negotiations with the environment and encounters with people, institutions, and situations. The stories and the performativities of storytelling that accompany these maps oftentimes take on a vibrancy that are not present in maps, even if the maps themselves stimulate storytelling.

In the classroom setting where geonarrative map exercises are undertaken in undergraduate and graduate human geography courses, most students actively participate in the storytelling aspect. While the creative aspect of mapmaking excites a majority of students, others are less willing to reveal details of their lifeworlds. Still others are emboldened to share when the class environment encourages a safe space among equals. The technical quality of the maps is also a recurring issue: students preface their storytelling with a caveat that the technical execution of their maps is not professional. There is a propensity to assume that the word 'map' is synonymous with the professional cartographic images produced by private and state institutions.

Doreen Massey argued that 'if you think of space and landscape as stories-so-far, then the stories are ongoing' (Massey interview by Featherstone et al. 2013, 264). Geonarrative maps and storytelling provide an avenue for people to create versions of their environment and their engagements to it. It is a snapshot of a story captured creatively and cartographically. In several workshops conducted to elicit spatial stories, participant narratives oftentimes include untold and invisible fragments of their lives. Whether these are therapeutic exercises or as avenues for various performativities, place-writing in the form of geonarratives allows the fleeting to be captured textually and cartographically, elaborating lives as ongoing stories.

This intervention endeavours to create and encourage the formation of a productive and creative space for an alternative representation and imagination of a life through countermapped storying. It acknowledges multiple and plural realities and experiences of geonarrative participants towards their environments, encounters with hazards and disasters, and state institutions. These geonarratives do not profess to be authoritative or official but an awareness and acknowledgement of the relational and multi-layered realities as embodied by participants through storytelling and mapping of their everyday geographies of emotion, natures, and bodies.

Acknowledgement

J.M. Adams, D.S. Amorsolo, C. Bautista, M.C. Conaco, E. Garcia, M.C. Larobis, A.L. Pacaña, M.A.J. Pernia, N. Swanson, Museum of Contemporary Art and Design (MCAD), Prod Jx Artist Community, University of the Philippines Department of Geography, and UP-OVCRD's Extension grant for the Geonarratives Mapping Project.

Key sources

1. Hawkins, H. (2015) 'Creative geographic methods: knowing, representing, intervening. On composing place and page,' *Cultural Geographies*, vol. 22, pp. 247–268.

An insightful piece that establishes the relationship between geography, the creative arts and humanities. It focuses not only on the finished products but also in the creative process.

2. Kwan, M.P. and Ding, G. (2008) 'Geo-Narrative: Extending Geographic Information Systems for Narrative Analysis in Qualitative and Mixed-Method Research,' *The Professional Geographer*, vol. 60, no. 4, pp. 443–465.

Geo-narrative as defined here is GIS-based narrative analysis to interpret qualitative data. This study is considered as the forerunner of most geonarrative studies undertaken by other scholars.

3. Lee, D.M. (2019) 'Cultivating preservice geography teachers' awareness of geography using Story Maps,' *Journal of Geography in Higher Education*, vol. 44, pp. 387–405.

This is a survey of preservice geography teachers (PGTs) and their awareness of geography (geographic literacy) using story maps. A phenomenographic analysis was deployed to evaluate the PGTs' geographic awareness. The study concludes that immersion to story maps vastly improves awareness of geographic issues helpful in teaching introductory geography classes.

4. Marta, M. and Osso, P. (2015) 'Story Maps at school: teaching and learning stories with maps,' *Journal of Research and Didactics in Geography*, vol. 2, no. 4, pp. 61–68.

This is an elaboration of Map-based Storytelling that encourages students to use stories to understand and interpret geographic phenomena. The authors propose that the use of modern technology spark the students' imaginative geographies.

5. Mukherjee, F. (2019) 'Exploring cultural geography field course using story maps,' *Journal of Geography in Higher Education*, vol. 43, pp. 201–223.

The use and application of ESRI's Story Map to visualize community stories with the use of videos, photographs, and student journals. This place-based approach shows that qualitative data provide more nuance and depth.

Notes

1 Permission was asked and granted to use their maps for pedagogical and publication purposes.
2 CTCSM ceased to exist as of 2020. It was issued a closure order by the Department of Education (Palicte 2020).
3 A written permission for each participant was granted to use their maps for pedagogical and publication purposes. Names are withheld to protect privacy of these individuals.
4 The 1971 uprising is called the Diliman Commune and is being hailed as the refusal of the university to be silenced as the country's social critic.

References

Alamon, A. (2017) *Wars of Extinction,* Iligan City: RMP-NMR.
Bell, S., Wheeler, B.W. and Phoenix, C. (2017) 'Using Geonarratives to Explore the Diverse Temporalities of Therapeutic Landscapes: Perspectives from "Green" and "Blue" Settings', *Annals of the Association of American Geographers*, vol. 107, no. 1, pp. 93–108.
Carroll, A (2020) 'Story maps' in D. Richardson et al. (Eds) *The International Encyclopedia of Geography*, John Wiley, pp. 1–6.
Featherstone, D., Bond, S. and Painter, J. (2013) "'Stories so far": a conversation with Doreen Massey'. In: Featherstone, D. and Painter, J. (eds.) *Spatial Politics: Essays for Doreen Massey*. Oxford: Wiley-Blackwell, pp. 253–266.
Lee, P.T. (2017) 'Stop the Lumad Killings', *Indigenous Environmental Network*, December 24, 2017, www.ienearth.org/stop-the-lumad-killings/
Massey, D. (2005) *For space,* London: Sage.
Palicte, C. (2020) 'DepEd Closes Davao Oro School for "Deficiencies"' Philippine News Agency, June 1, 2020, www.pna.gov.ph/articles/1104552
Raymundo, S.J. (2019) 'Elsewhere schooling: The Lumad bakwit school at the national university', Bulatlat, July 11, 2019, www.bulatlat.com/2019/07/11/elsewhere-schooling-the-lumad-bakwit-school-in-the-national-university/
Reutzer, D.R. (1985) 'Story Maps Improve Comprehension', *The Reading Teacher*, vol. 38, pp. 400–404.
Said, E. (2000) *Reflections on Exile and Other Essays*, London: Granta Publications.

61
POETRY AS DECOLONIAL PRAXIS

Sarah Wright

If the Land can Speak

Manama [a Manobo god] blessed us
with our land
for us to live in peace.

But now, only few people
benefit from you.
No land to till
for the farmers.

They stole the lands
of the Lumad,
along with our culture.
Now, children are drowned
with their tears.

Do we allow this to happen?
Deteriorated land
and destroyed forest.
If the land can speak,
You will hear, 'Enough!'

But the busaw [monsters] are here,
ready to empty the land.
We become homeless
with empty stomach
while other's
with their full stomach
cannot hear.

DOI: 10.4324/9781003017653-67

We exist because of you,
our land.
To defend you,
We learn how to fight.
                    Lino[1]

The size of the open pit that stretched before us seemed of a different order of magnitude to the small wooden huts that teetered at its lip. Beside it, our group and what was left of the village above, were like toys. Children played on the dirt track that stretched along the rim; the only thing, apart from a small fence, that separated the village from the pit. The mine, as it expanded, was literally eating the village. Later, or earlier, a B'laan priest, active in resisting exploration for a new mine, tells us he prays with his eyes open, for when he closes them, they steal the land. At night, I watch performances by the light of a kerosene lamp. There are dances and poems, there is song. My heart cracks open.

This was during my first visit to Mindanao in the Philippines. It was in the mid-1990s and I visited as part of a fact-finding mission looking at impacts of an Australian mining company, Western Mining Corporation, that had been granted rights over 99,387 hectares of land for 50 years as they explored for copper-gold[2] (Ambay 2016). The exploration, as well as the mine we visited, were on the ancestral lands of the B'laan people, one of 18 Indigenous Lumad tribes in Mindanao.[3] The B'laan communities, especially those that strongly resisted the mine, were facing increasing militarization; militarization that would later lead to killings of leaders, a massacre of a tribal leader's family. Several of the people we met spoke about how their land was taken after they were misled into adding their thumbprints as a signature to documents they could not read. Since that time, intense and often violent development pressures, particularly around mining and logging, have continued in Indigenous lands in Mindanao including in the Pantaron mountain range, the 'heart of Mindanao,' on the lands of the Manobo peoples.

In the poem, *If the Land can Speak*, which opens this chapter, these are monsters: those mining trucks, the pit itself, the military and paramilitary groups that guard the mine. They have come to empty the land of Lumad people, of culture, connection, and song.

There are many monsters, with many aspects, that have appeared to try and empty Lumad lands for centuries, and many complex responses on the part of diverse Lumad peoples. The development aggressions faced by Lino and many other Lumad people, lands and communities sit within a longer history of colonization, exploitation and marginalization that began under Spanish colonization (1565–1898) and intensified with US colonization (1898–1946) when large areas of Mindanao were expropriated from Lumad and Moro (the Muslim people of Mindanao) peoples, and granted to an emerging oligarchy for logging, plantations and extractive industries (Alamon 2017). Along with vast in-migration of settlers from elsewhere in the Philippine archipelago driven by class-inequalities and official colonial policies, such moves rendered Lumads and Moros, vastly outnumbered and pushed off ancestral lands (Bengan 2015; Chanco 2017; Tiu 2008). Since independence, the Philippine state has used military force and a wide raft of policies and pressures to defend the interests of settlers, plantations, and extractive industries, including through promoting the formation of paramilitary groups and fomenting a culture of impunity around violence (Alamon 2017).

And there are perhaps other monsters, more subtle but no less pernicious, those who, with their full stomach, cannot or will not, hear? Those, like myself, whose national wealth including the public education systems that it supports, is built in part on the activities of mining companies such as WMC and their practices overseas.[4] Those with shares and stocks in these companies, their cosy retirements funded, even through compulsory superannuation, by violent extraction. There is also the monster of romanticizing stereotypes that might deny Lumad agency, complexity and vitality, requiring Lumads to 'perform to certain stereotypes in order to claim the particular rights and benefits that, by law, are supposed to be their due' (Paredes 2019, 87). There are the monsters of ongoing racial discrimination and marginalization, the daily challenges of survival, the judgements, and conditions of mainstream development. It is all these monsters, and more, that might listen and learn from Lino and heed the call of his poem: settlers in Mindanao from other parts of the Philippines, potential allies, those from other countries, those that might be brought to listen to the land as it cries, 'Enough!'

In *If the land can Speak*, the student author Lino ends the poem with a call for resistance, 'We exist because of you, / our land//' he declares. 'To defend you, / We learn how to fight.' And, indeed, in the midst of this violence and suffering, Lumad people have spoken back, they have fought, do fight, continue to resist. The use of the term Lumad itself, a Cebuano word embraced and re-fashioned through social movements of Lumads in the mid-1980s, is illustrative of some of the ways people from many diverse tribes, as diverse historical/contemporary agents, have asserted an identity, political consciousness and distinctive Indigenous collectivity (Bengan 2015; Paredes 2013; Vidal 2004). As part of such movements, Lumads have rallied, created schools to support their young people, performed cultural events, mobilized on social media, written poetry, stories, songs, newspaper articles. They have been forced into bakwit, evacuation, creating centres and schools. They have sustained their Elders and young people, their culture and ancestral lands, and been sustained by them, as well as they could. And, of course, the ways they have done this have been many and complex, sometimes riven with internal conflict and always informed by multifaceted, intricate negotiations around what it means to be Lumad, to assert Lumad identity and claims around Lumad land, in an often-fraught, contemporary reality.

In this chapter, I look to one important expression of this resistance, poetry, written by young scholars from a Lumad school in Mindanao. The poems are by students from the Community Technical College of Southeastern Mindanao (CTCSM) and were shared through a project working with several Lumad groups in Mindanao that aimed to centre Lumad connections with place. Written by students as part of their creative writing programme, the poems were highlighted by students and teachers for the ways they communicated some of the complex beauty and pain of the students' experiences.[5] The poems now form part of a book, *Scent of Sun, Rain and Soil* documenting the experiences of the students and the school (Wright 2020).[6]

Lumad schools such as CTCSM are sites of resistance, placemaking, empowerment, negotiation, and struggle (Yambao et al. 2021). As such, they have been relentlessly targeted by military and paramilitary forces, by the state and by conservative forces more broadly. I aim to use this chapter both to highlight the situation faced by Lumad people, particularly as attacks against them and specifically against Lumad schools, have so intensified under the administration of the current president Rodrigo Duterte, and to highlight the ongoing resistances of Lumad people, and particularly Lumad youth. More broadly, my aim is to attend to the way poetry can be used as a point of resistance, and to consider what poems and poetry may do as decolonial[7] praxis.

Poetry and other forms of literature communicate on a visceral level. They distil vast experience and bring reader and writer into affective contact (Coleman 2016; Paiva 2020). Many Indigenous scholars and others working towards decolonization, point to poetry and literature for their potential in resisting colonization, calling attention to the ongoing violences of colonial and neocolonial processes, and asserting sovereignty including by imagining and so bringing about different kinds of pasts, presents and futures (Hargreaves 2017; Ortiz 1997;

Sium and Ritskes 2013; Womack 2009). As Colorado-born Canadian academic and member of the Cherokee Nation/ ᏣᎳᎩ ᎠᏰᎵ Daniel Heath Justice observes, Indigenous literature is both an expression of Indigenous intellectual agency, an aesthetic accomplishment, and a political intervention with a 'role to play in the struggle for sovereignty, decolonization, and the re-establishment of Indigenous values' (Justice 2018, 336–7). An important part of the decolonizing potential of literature lies in the power of Indigenous peoples telling their own stories in their own way. This is not only a matter of fiction, and shifting imaginations, but also of analysis and, as Justice says, intellectual agency. As I look to poems, then, in this academic space, I aim to support this decolonizing potential both by sharing poems from Lumad students and also prioritizing analysis and academic literature from and by Indigenous scholars (for example, Bawaka Country 2019; Episkenew 2002; Hunt 2014; Justice 2018). In sharing the poems, I am responding to a specific task that was given to me by the students and teachers of the school, one re-iterated by Lumad Elders and organisers: to share the words widely and in as many forms as possible.

I structure the piece around three poems, *If the Land could Speak, Witness* and *Land to us is Life*. As I do so, I engage directly with the poems, their content, meaning, and the ways they communicate and create resonances between authors and readers. I also highlight the political context, the real-life situations, from which these poems emerge. This is not to subsume the works into politics, something Justice cautions against (2018, 335–6), but to situate these poems in Lumad struggles in the way they were offered. Cree-Métis literary scholar Jo-Ann Episkenew (2002, 65) calls this approach 'socially responsible criticism' that examines 'the text of works of Aboriginal [Indigenous] literature [within] the *context* from which it is written.'

The poems shared by the students in this chapter are a way they have chosen to make sense of some of their pain, communicate ties to ancestral lands, and continue to re-make intergenerational connections to and as place. They are a resistance in the face of militarization and contribution to re-membering and a re-making of different kinds of worlds (Smith et al. 2020). The poems share and hold stories, they seek to keep Lumad places, people and culture alive and Lumad connections strong.

And they reach out to the reader, invite connection, potentially creating significant affective resonances across time and space. They challenge the reader to *witness*: to hear and be brought into relationship, despite their full stomach, and, more, to act, to begin to both understand their complicity and to find ways to support change. In this chapter, then, I consider the ways these poems are points of decolonial praxis, with the potential to support intra-active, relational, and more-than-human witnessing between distant others.

Witness

I am the land that can see,
loved by farmers
and cared for by them.
I am happy because they are happy.

But a typhoon came
along with a monster machine.
Funeral for the crops came after,
fear drowned the smiles.

Bai [friend], look around,
on your travel you'll see
crops gasping for air
it's not the typhoon's fault alone.

The monster machine is smiling
after vomiting some smoke.
Endless destruction,
how will their tribe survive?
I am the land and I am their witness.
                                        Doris.

In the poem *Witness* Doris centres the reader's attention on the idea of witnessing. In doing so, she opens up the need for a multi-layered and relational conception of what witnessing might mean, and what it might do. In putting forward the notion of witnessing, she joins other Indigenous scholars who speak to the potential power of witnessing in diverse situations and traditions (Hunt 2014; Iseke 2011; Scofield 2016). For Kwagiulth legal geographer Sarah Hunt, witnessing is a Kwagiulth methodology that draws upon the principles of Kwagiulth potlatch and stems from an intimate and specific set of relations; a process that evokes connection and responsibility. For Sarah Hunt, witnessing is an inherent part of her taking up her life-long responsibilities as a Kwagiulth woman:

> *As I have come to understand witnessing methodology, I view these experiences as shared with me in the context of reciprocal relationships with an Indigenous cultural framework, and in witnessing the stories, I am obligated to ensure they are not denied, ignored or silenced. Further, if I see them being denied, it is my responsibility to recall both the truths of what I have witnessed and the ways in which their erasure is being accomplished…*
>
> *… witnesses are called upon to collectively recall what happened, in order to reinforce, rather than replace, the individual whose ceremonial act was in question. (Hunt 2014, 37–38)*

Witnessing then, involves a responsibility to support and co-create spaces within which the voices that might otherwise be silenced, ignored or even unhearable, become heard. In *Witness*, Doris directly works to create such a space, acknowledging a wide range of agencies and relationships with whom/which this space of witnessing is co-created. Indeed, for Doris, witnessing is deeply relational, affective and more-than-human; a way to speak back, voice her experiences, on her own terms.

The ways the poems do this is deeply affective. Poems communicate emotions and powerfully evoke specific relationships and places. Indeed in bringing together the different emotional states, from the ways that the 'land is happy because they are happy' in the opening stanza of the poem, to the fears sparked by the monster machine and typhoon, to the desperation of 'crops grasping for air,' the poem invites the reader on an emotional journey, encouraging affective attunement (de Leeuw 2019; de Leeuw and Hawkins 2017; Paiva 2020). In *Witness*, Doris engages intimate, affective, and pluralistic understandings of the world in ways that speak directly to some of the violences perpetrated against Lumad people, to Lumad resistance, and to decolonizing ideas and relationships with land.

Doris begins her poem with the assertion, 'I am the land that can see…' With this assertion, she at once articulates her connection with land and acknowledges the more-than-human agency of place, of land that can and does see. For Lino, the land calls 'Enough!', for Doris, she *is* the land that witnesses. In this way, her poem is a powerful assertion of co-becoming, centring Lumad ontologies that deny separation of people from place.

With this articulation, both Doris and Lino join Indigenous and decolonial scholars who centre the need to attend to, and respect, Indigenous understandings of place, space and land (Larsen and Johnson 2017). This means understanding that, rather than being a backdrop to human agency, or an inert thing to be possessed, exploited, bought and sold, land has intelligence and law (Bawaka Country 2019; Daigle 2016; Watts 2013). The land that sees is part of a nourishing complex of relationships.

These are sacred, vibrant and vital connections that are nurtured in many ways including through poems, stories, songs and songspirals (Gay'wu Group of Women 2019). As Laguna Pueblo Indian woman and acclaimed poet Leslie Marmon Silko (1996, 58) says, Pueblo stories 'cannot be separated from their geographic locations, from actual physical places on the land.' The Lumad students' poems like *Witness* and *If the Land can Speak*, acknowledge, and continue to create and share, storied and sung space. This is particularly important, in the case of bakwit, where the students are often forced to flee from their ancestral domains due to violence. The poems acknowledge the land's agency and help students draw upon its strength to learn and resist, and, in doing so, actively re-make the storied lands themselves and Lumad peoples' connections with them. As such, the witnessing that is called for by Doris is more-than-human, asserting both more-than-human agency and relationality. The land, as a witness, is brought into relationships of witnessing. As a witness, the land is a source of strength, it works against the silence. It refuses to be quiet.

Attention to connections and to the knowledge and agency of land is not only needed to share and re-member Lumad connections but also to bring focus to the ways that non-Indigenous relationships with land may lead to violence, and exploitation. Doris calls on the reader to attend: 'Bai [friend], look around,' she requests. 'On your travel you'll see/ crops gasping for air/ it's not the typhoon's fault alone.' In calling on the reader, the friend, to look around, she draws the reader in, creating at once an invitation and an affective tie that begins to situate the reader within a web of relationships, and gestures, gently but insistently, towards emerging responsibilities. For it is not, as she points out, the typhoon's fault alone.

As Lumads have continued to struggle to counter violence against themselves, their lands, languages and cultures, and as they have continued to assert and sustain their identities and connections, they have sought many forms of resistance, connection and placemaking. An important aspect of this has been the movement to create, sustain and protect Lumad schools. Doris, as with the other authors of poems in this piece, was a student at a Lumad school when she wrote this poem. The school, CTCSM, was established in 2013 and was the largest single-campus Lumad school in Mindanao. Children from 13 Indigenous communities came to study at the school: the school has hosted Bagobo, Manobo, B'laan, Teduray, T'boli, Mandaya, Mansaka, Matigsalog, Kulamanon, Tagkaulo, Higaonon, Tinonanon and Kalagan students.

The school is one of over 200 Lumad schools that started on the island of Mindanao. The schools were a response to the ongoing marginalization of young Lumad people from mainstream education; there are often simply no schools in Lumad areas, and many families cannot afford the costs associated with reaching or attending other schools. The schools are also an important response to militarization and development aggression in Lumad lands. Many leaders feel that if their children are educated within a Lumad-led school, they will be better equipped to defend their lands and cultures. In addition to educating schools, some of the buildings are also used for adult education and as centres for evacuees.

At a meeting of women Lumad leaders at an evacuation centre in Davao, we gather around a cool bamboo hut, where the women talk of the schools. One explains:

> *We have an experience with our parents, we are gathered in a meeting and we place thumb marks. We cannot read and write. After a few months it turned out to be an approval for a mining and logging company. It was in 1994. We are women, we want the schools. We want the schools back in our communities. We will not stop in calling, we will continue to advocate in radio, on TV, to share our stories…We want to express our feelings, not to be kept silent.*

The women's discussion gives a sense of the ways the schools go beyond providing a basic education for youth. The schools become sites of intergenerational placemaking that nurture young generations of Lumad youth like Lino, Doris and Welgen; sites of cultural transmission, and Lumad resistance (Yambao et al. 2021). Of course, the schools are not perfect and should not be romanticized (Paredes 2019). They navigate a complex pathway between mainstream education with mainstream development goals of literacy and numeracy on the one hand and culturally embedded learning practices on the other, while also seeking to support the needs of diverse Lumad peoples, lands, languages and cultures against a deeply adversarial government, widespread development aggression and orchestrated right-wing trolling campaigns. While often fraught and always working within a complex political terrain, the Lumad schools are deeply important. As another datu (chief) powerfully observed, 'Our schools are being attacked because it is our weapon, our strength to stand for our ancestral lands against plantations and mining corporations, to defend the Pantaron [mountain] range.'

The importance of the schools to the Lumad cause is also clear to those who seek to exploit their lands and in recent years, under the Duterte government, militarization and violence against Lumad people, their culture and land, and particularly against the schools, has intensified. Indeed, President Duterte is on record saying, 'Get out of there, I'm telling the Lumads now. I'll have those bombed, including your structures' and 'I will use the armed forces, the Philippine air force. I'll really have those bombed!'

He, the president of the Philippines, was talking about bombing Indigenous schools.

His stance promoted brutal harassment and military suppression in Lumad communities. There have been extrajudicial killings of teachers and students, military, and paramilitary harassment, extensive anti-Lumad propaganda where schools are vilified as communist army training encampments, and the weaponized use of the legal system with arrests of teachers and parents on trumped up charges. Some schools are taken over by the military and used as military encampments, some are suddenly stripped of their licence to operate, nominally due to small infractions with paperwork etc, some forcible shut down by military and paramilitary forces (Alamon 2017).

As Doris calls for witnessing of these violences and losses, her relational conception includes the reader-as-witness, the friend who looks around. The reader is invited to witness something closer to the lived reality of the students, the multiple concerns and stories that the students share, so that the students and the schools are not there as some kind of political football or victim but as real people, inspiring young people, intellectual agents, living their lives in the most arduous of circumstances.

Here, as witnesses, readers become present; we are invited to open ourselves up, in all our diverse 'we-ness' because readers are different, from different places and with different relationships to Doris, to CTCSM and Indigenous struggles. Through these different but real and placed and important connections we are invited to open ourselves to some of the realities, connections, care, complicities, and responsibilities that are part of our witnessing.

Land to us is Life

> *Lupa sa amin ay buhay*
>> *Land to us is life*
> *Yaman na nagbibigay kulay sa amin*
>> *Wealth that gives us color*
> *Nakaugat sa buhay ng sariwang dugo*
>> *Rooted in the life of fresh blood*
> *Ng mapanuring henerasyon.*
>> *Of a critical, analytical generation.*
>>> *Welgen.*

The poems shared in this chapter were written at a time that pressure on CTCSM was building. In writing and sharing these poems the young authors and the teachers bear witness to their own lives, that which they want to share, their own struggles, and hopes and dreams.

In May 2020, CTCSM was ordered closed. Throughout 2020 and 2021 several of the teachers and volunteers were hit with trumped-up charges. The students had to leave, in the midst of a pandemic. Many were forced to other sites of evacuation and internal displacement, including other bakwit schools. Others were able to return home, often to traumatized families and communities, many with their dreams of an education in tatters.

In an online event in February which included a poetry reading, testimonials and presentations, one of the senior teachers began crying.

'Why are they attacking these children?' she asked.

Many of those listening were moved to tears both by her testimony and the heartbreaking and beautiful reading of the poem *Witness* by one of the ex-students. The student read the poem from a bakwit school on another island in the predawn light. She read far from home, courage radiating from her voice and in her face, despite the sometimes-poor internet connection, the bustle around her.

The students are clear on the links between their stories and their struggles: 'We are happy to let you know our stories and experiences of agroecology but we hope you could also immerse with us, fight with us,' one CTCSM Lumad student said.

As Syilx poet and novelist Jeanette Armstrong (2005, 244) observes, the 'dominating culture's reality is that it seeks to affirm itself continuously.' It continually pursues ways to silence, deny, even annihilate those that do not conform, and to ignore, sometimes wilfully, sometimes violently, any potential for transformation. Part of this will to affirm dominant reality means that most of those who benefit from the colonizing and neocolonial order, who are supported by the status quo, will neither 'desire or choose to hear these truths unless they are voiced clearly' (Armstrong 2005, 245). In the poems of the Lumad students, such realities, such potential for transformation, are placed powerfully at the centre; spoken clearly, resonantly. The students refuse the will of dominant culture to endlessly affirm itself at their expense.

For students, scholars and development practitioners, the call to witness comes with implications. One important aspect is the way witnessing suggests a way to listen, to attend to some of the stories, the pain, hope, resistances big-and-small, of Indigenous people. Witnessing means seeking ways to reinforce Lumad voices, not appropriating them, not saturating them with colonial desire, but, through the affective ties that may be established, to participate in a collective recollection of what has happened, what is happening. Witnessing through these poems means not turning away from complexities or from pain, not denying experiences, but being open to the ways that, through these poems, the stories, and experiences they tell of may become more deeply understood, more richly felt and more visible (Hunt 2014). Because through this witnessing, as the land witnesses, comes the call to responsibility, not responsibility *for* someone separate from ourselves, not responsibility *for* someone who lacks agency, but responsibilities that stem from, and are part of, the relationships within which all are enrolled (Bawaka Country 2019).

And, in this, there are implications too for a decolonizing pedagogy. Poetry is one way to communicate affectively and effectively, for knowledge shared to resonate. It can centre power and agency of Indigenous people on the terms on which authors choose to present themselves. In doing so, poetry not only may amplify Indigenous voices in established spaces but can be a way to create new affective-political spaces, to reach across difference and generate new dialogues, new ways of dialoguing. In this, there can be deep learning for those that read, that listen, that witness; learning that might shift understandings.

For poetry to be part of such a decolonizing pedagogy, the work must be also read within its context, in the spirit in which it is offered, recognizing the entangled nature of art, land, sovereignty, politics and practice. This means, as I have tried to do in this Chapter, working with the constructive tensions between text and context, not seeing poems and poetry as divorced from life and politics (Episkenew 2002). Rather, witnessing through the poetry becomes an opportunity of taking up, what Muskogee Creek literary critic Craig Womack (2009, 96) calls, 'the challenge of relating literature to the real world in the hopes of seeing social change.' A chance, as one CTCSM student says above, to 'immerse with us, fight with us.'

This is the power of the poems as decolonial praxis. The poems of the Lumad students work against colonizing relationships. They speak to the intensity of violence that is experienced in their communities, to the pain and suffering that is felt by communities and the land. They speak too, to the strength of resistance to this violence, from the land, from the communities, from the young people. They point to the bittersweet pain of the present and the possibilities of an entangled future. And they, through their more-than-human, affective and resonant witnessing, invite others into relationships of responsibility. They invite readers to listen and act, to attend to and help bring about the possibility of a future and a present where land, care and connection are wealth that gives meaning to life, where Indigenous Lumad youth are respected and heard, leading with loud voices as a critical, reflective, and powerful generation.

Key sources

1. Alamon, A. (2017) *Wars of Extinction: The Lumad Killings in Mindanao, Philippines.* Quezon City: RMP-NMR.

This work analyses the experiences of Lumad people and the many violences they have faced. It discusses the history of aggressions against Lumad people, culture, and land, and provides an analysis that links their experience to colonialism and capitalism, resource extraction and development aggression.

2. Hunt, S. (2014) *Witnessing the Colonialscape: lighting the intimate fires of Indigenous legal pluralism.* PhD Thesis, Simon Fraser University. Accessed http://summit.sfu.ca/item/14145 accessed 18.2.21.

This thesis, able to be accessed online, is a grounded analysis of law, violence and place through the perspectives of Indigenous people working against violence in BC, Canada. Kwagiulth author Sarah Hunt uses a Kwagiulth witnessing methodology to centre the agency of Indigenous people while acknowledging the violences that they face.

3. Justice, D. (2018) *Why Indigenous Literatures Matter*. Waterloo: Wilfrid Laurier University Press.

This book powerfully speaks to the importance of Indigenous literature particularly as it relates to sovereignty, creativity, and intellectual agency. It discusses the way that stories – who tells them and why – matter and have the potential to re-imagine and re-create worlds.

4. Ortiz, J. S. (1997) *Speaking for the Generations. Native Writers on Writing,* Tucson: University of Arizona Press.

This anthology brings together experiences from Indigenous writers from North America. Poets, novelists, and playwrights speak from their personal experiences about writing and what it means to them, touching on themes of sovereignty, relationships with land, history, place and politics.

5. Wright, S (2020) *Scent of Rain, Sun and Soil: Stories of Agroecology by Lumad Youth in the Philippines.* Davao: Southern Voices Press and CTCSM.

This work is a collection of poems, stories, and experiences of agroecology from students and staff at a Lumad school in Mindanao. It is a short, creative work that aims to work within a decolonizing methodology to share some of the students' and teachers' stories in ways both inspiring and heartbreaking.

Notes

1 The authors of these poems were students at the Community Technical College of Southeastern Mindanao at the time of writing. They have used pseudonyms (or nicknames) to protect their anonymity.
2 The extremely controversial Philippine Mining Act 1995 had been recently passed. Western Mining Company transferred its lease to Sagittarius Mining Incorporated in 2001. The operation was subsequently owned and co-owned with other companies including Glencore Xtrata and the local Alcantra group. The operation is currently in abeyance.
3 Lumad is a self-ascribed term that was adopted by a gathering of members and affiliates of Lumad-Mindanao in 1986 as part of a strengthening political movement of Lumad tribes during the time of martial law under Ferdinand Marcos. Lumad-Mindanao called for self-determination for member-tribes: the ability to self-govern in accordance with cultural practice and customary law (Paredes 2013; Rodil 1994; Vidal 2014).
4 Though not it must be said as much as mining companies might have the Australian public believe.
5 The poems were originally written in Filipino and translated into English in the hope more people could read the work. The students read and commented on the draft translations.
6 The project is part of my ongoing collaborative engagement and research with the organization MASIPAG which began in 2002. The stories for this article were mostly gathered during a three-week field visit in 2019 which included a visit to the school and other Lumad sites of bakwit, as well as follow-up web-based connections in 2020 and 2021.
7 In speaking of decolonial work and praxis, my intention is not to diminish the many complex contemporary power structures associated with extractive development, class, and capitalism in the

Philippines but rather to name and disrupt the enduring colonial and neocolonial power structures and logics that underpin them.

References

Ambay, M. (2016) *Blood and Gold: Tampakan and the B'laan resistance*. Manilatoday.net, accessed on 18.2.21, at https://manilatoday.net/blood-and-gold-tampakan-and-the-blaan-resistance/

Armstrong, J (2005) 'The disempowerment of First North American Native Peoples and Empowerment through their writing.' In *An Anthology of Canadian Native Literature in English*, edited by Daniel David Moses and Terry Goldie, Don Mills, ON: Oxford University Press.

Bawaka Country including Suchet-Pearson S, Wright S, Lloyd K, Tofa M, Sweeney J, Burarrwanga L, Ganambarr R, Ganambarr-Stubbs M, Ganambarr B, Maymuru D (2019) Goŋ Gurtha: Enacting response-abilities as situated co-becoming, *Social and Cultural Geography*, vol, 37, no. 4, pp. 682–702.

Bengan, J. B. (2015). Hunger, Desire and Migratory Souls: Interethnic Relations in Three Short Stories by Satur P. Apoyon. *Asiatic: IIUM Journal of English Language and Literature*, vol, 9, no. 2, pp. 122–138.

Chanco, C. J. (2017). Frontier polities and imaginaries: the reproduction of settler colonial space in the Southern Philippines. *Settler Colonial Studies*, vol. 7, no. 1, pp. 111–133.

Coleman, D. (2016) Indigenous place and diaspora space: of literalism and abstraction, *Settler Colonial Studies*, vol. 6, no. 1, pp. 61–76.

Daigle, M. (2016) Awawanenitakik: The spatial politics of recognition and relational geographies of Indigenous self-determination, *The Canadian Geographer*, vol. 60, no. 2, pp. 259–269.

de Leeuw, S. (2019) Novel poetic protest: what Your Heart is a Muscle the Size of a Fist offers geographers writing to attune to and express rather than convey. *Emotion Space Society*, vol. 30, pp. 58–61.

de Leeuw, S. and Hawkins, H. (2017) Critical geographies and geography's creative re/turn: poetics and practices for new disciplinary spaces. *Gender, Place and Culture* vol. 24, pp. 303–324.

Episkenew, J. (2002) 'Socially Responsible Criticism: Aboriginal Literature, Ideology, and the Literary Canon.' *Creating Community: A Roundtable on Canadian Aboriginal Literature*. Eds. Eigenbrod, Renate, and Jo-Ann Episkenew. Penticton, BC: Theytus Books.

Gay'wu Group of Women (2019). *Songspirals*. Allen and Unwin: Melbourne.

Hargreaves, A (2017). *Violence Against Indigenous Women: Literature, Activism, Resistance*. Waterloo: Wilfrid Laurier University Press.

Iseke, J.M. (2011) Indigenous Digital Storytelling in Video: Witnessing with Alma Desjarlais. *Equity & Excellence in Education*, vol. 44, no. 3, pp. 311–329.

Larsen S and Johnson J (2017) *Being Together In Place: Indigenous coexistance in a more than human world*. Minneapolis: University of Minnesota Press.

Paiva, D. (2020) Poetry as a resonant method for multi-sensory research. *Emotion, Space and Society,* vol. 34, 100655.

Paredes, O. (2013) *A mountain of difference: the Lumad in early colonial Mindanao*. Ithaca, New York: Southeast Asia Program Publications, Southeast Asia Program, Cornell University

Paredes, O. (2019) Preserving 'tradition': The business of indigeneity in the modern Philippine context, *Journal of Southeast Asian Studies*, vol. 50, no. 1, pp 86–106

Rodil, B. R. and Alternate Forum for Research in Mindanao. (1994). *The minoritization of the indigenous communities of Mindanao and the Sulu Archipelago*. Davao City, Philippines: Alternate Forum for Research in Mindanao

Scofield, G. (2016) *Witness, I am*. Gibsons, B.C: Nightwood Editions.

Silko, L. M. (1996) Language and Literature from a Pueblo Indian Perspective. *Yellow Woman and a Beauty of the Spirit: Essays on Native American Life Today*. New York: Simon, pp 48–59.

Sium, A. and Ritskes, E. (2013). Speaking truth to power: Indigenous storytelling as an act of living resistance. *Decolonization: indigeneity, education & Society*, vol. 2, no. 1.

Smith, A.S., Smith, N., Wright, S., Hodge, P. and Daley, L (2020) Yandaarra is living protocol, *Social & Cultural Geography*, vol. 21, no. 7, pp. 940–961.

Tiu, M. D. (2008) *Davao: Reconstructing History from Text and Memory*. Davao: Ateneo de Davao University Research and Publication Office.

Vidal, A. T. (2004) *Conflicting Laws, Overlapping Claims: The Politics of Indigenous Peoples' Land Rights in Mindanao*. Davao: Alternate Forum for Research in Mindanao, pp. 17–61.

Watts, V. (2013). Indigenous place-thought & agency amongst humans and non-humans (First Woman and Sky Woman go on a European world tour!). *Decolonization: Indigeneity & Society*, vol. 2, no. 1, pp. 20–34.

Womack, C. (2009) *Art as Performance, Story as Criticism: Reflections on Native Literary Aesthetics*. Norman: University of Oklahoma Press.

Yambao, C., Wright, S., Theriault, N. and Castillo, R. (2021). Placemaking amid displacement among Lumads in the Philippines. *Submitted for publication*.

INDEX

Note: Page numbers in *italics* and **bold** denote figures and tables, respectively. Page numbers with an 'n' denote notes.

double tax treaties and 330–331; tax heavens blacklisting and 327–330
politics: in aid delivery 77; 'big P' and 'small p' 667
Pollin, R. 25
Ponte, S 525
Poole, L. 110
Popular Anthropocene 161
Population and Sustainability Network website 186
population policies 186–187
Portfolio Earth 224
post-agrarian land alienation 368
post-capitalism 689–690
postcolonial and subaltern studies, in critical SSC research 84
post-colonial countries, and debt 61
post-development and critical development 292
post-industrial globalized modernization 37
poststructuralism 185–186
Potts, D. 505, 510
poverty 311–312, 314, 487; under colonialism and capitalism 316–317; data to identify 314–315; distribution, between low-income and middle-income countries 318; extreme poverty 311, 312, 313–314, 320; global, at different thresholds 313; and health 472; identification of 317–319; justice and 320; in Laos 321; measurement of 315–316; past and 316; questioning of 313–314; reduction 494; regional distribution of 319; as singular body image 319–320; in United Kingdom 321
poverty reduction: and development priorities 76–77; in retroliberalism 34, 35, 39, 40; and World Bank and MDBs orientation 94
power inequalities, in aid chain 115
'practice is the learning' concept 621–622
practitioners, role in activism and pedagogy development 611–613
precautionary principle 150–151
Prevention of Violence against Women and Children programme (India) 409
primary health care 458–459; comprehensive 456–458, 464; selective 457, 460; *see also* health
Primary School Leaving Examination (PLSE) (Singapore) 397
Principled Aid Index 76
private capital: facilitation of 33, 39, 43; markets funding 95–96
private governance, in global value chains 517, 519, 522, 527
private philanthropy 18, 124
private sector, in post-GFC policies 34–35
privatization 21, 48, 63, 96, 98, 99; of education 400, 419; of health services 458, 668; of housing 505; of land 364

problem-driven iterative adaptation (PDIA) 591, 666, 669, 670–671, 674; 1804 challenge 672; use of Ishikawa/fishbone diagrams 673
pro-globalization agenda 21
Programme for International Student Assessment (PISA) 308, 392, 393
project quality and corruption, and new multilateral development banks (MDBs) 99
pro-poor growth 317, 354
Proshika 108
Protocol to Prevent, Suppress and Punish Trafficking in Persons (Palermo Protocol) 539
public debt 59; defaulting of 64; servicing, in Latin American countries 66
public governance, in global value chains 518–519, 522, 525, 526, 527
public health 475; and COVID-19 463; definition of 462; *see also* health
Pueblo stories 718
Puerto Rico 330
Putrajaya, Malaysia 352
Putzel, J. 556

Queersxclimate.org website 189
Qvortrup, J. 567

racism, as anti-globalization agenda 21, 30
Rademeyer, J. 237
radical philanthropy 126
Randall, C. 189
Randers, J. 152
rating agencies and private investors, role in multilateral development banks (MDBs) 95–96
Rattan Association of Cambodia 694
Rauniyar, G. 131
Ravallion, M. 317, 319–320
reciprocity, and learning 643, 645–647
Recommendations on Anti-Money Laundering and Combating the Financing of Terrorism (Financial Action Task Force) 335
Redefining Progress 191
reflection, as learning process 647, 648–649
refugees 534, 539; and displacement 559; and health inequalities 471–472; health of 474
regional development 16, 48–52; bottom-up approaches in 52; comparative advantage promotion in 49, 51, 52, 56; competitive advantage in 47, 50, 53; endogenous and exogenous interventions in 49, 51, 54; new regionalism and localism 49–50; place-based development 50–51; place-neutral interventions in 53; policy and practice in 48; political-economy discourse, shifts in 48–49; regional resilience and evolutionary economic geography 51–52; state-centric interventions in 47; uneven geographical development 49
regional disadvantages 47–48, 52, 53

Printed in the United States
by Baker & Taylor Publisher Services